D0206276

Mechanics of Solids and Fluids

Franz Ziegler

Mechanics of Solids and Fluids

With 352 Figures

 Springer-Verlag   New York   Vienna

Franz Ziegler
Technische Universität Wien
Institut für Allgemeine Mechanik (E201)
Wiedner Hauptstrasse 8–10
A-1040 Wien
Austria

Library of Congress Cataloging-in-Publication Data
Ziegler, Franz, 1937–
    [Technische Mechanik der festen und flüssigen Körper. English]
    Mechanics of solids and fluids / Franz Ziegler.
        p.   cm.
    Translation of: Technische Mechanik der festen und flüssigen
Körper.
    Includes bibliographical references and index.
    ISBN 0-387-97529-2 (alk. paper)
    1. Mechanics, Applied.   I. Title.
TA350.Z5413   1991
620.1—dc20   91-9323

Printed on acid-free paper.

©1991 Springer-Verlag New York, Inc.
All rights reserved. This work may not be translated or copied in whole or in part without the written permission of the publisher (Springer-Verlag New York, Inc., 175 Fifth Avenue, New York, NY 10010, USA), except for brief excerpts in connection with reviews or scholarly analysis. Use in connection with any form of information storage and retrieval, electronic adaptation, computer software, or by similar or dissimilar methodology now known or hereafter developed is forbidden.
The use of general descriptive names, trade names, trademarks, etc., in this publication, even if the former are not especially identified, is not to be taken as a sign that such names, as understood by the Trade Marks and Merchandise Marks Act, may accordingly be used freely by anyone.

Camera-ready copy prepared by the author.
Printed and bound by Edwards Brothers, Inc., Ann Arbor, MI.
Printed in the United States of America.

9 8 7 6 5 4 3 2 1

ISBN 0-387-97529-2 Springer-Verlag New York Vienna
ISBN 3-211-97529-2 Springer-Verlag Vienna New York

WIDENER UNIVERSITY
WOLFGRAM
LIBRARY
CHESTER, PA.

To the Students Who Accept the Challenge of Science
and to My Children Robert and Eva C.

Preface

This book offers a unified presentation of the concepts and most of the practicable principles common to all branches of solid and fluid mechanics. Its design should be appealing to advanced undergraduate students in engineering science and should also enhance the insight of both graduate students and practitioners. A profound knowledge of applied mechanics as understood in this book may help to cultivate the versatility that the engineering community must possess in this modern world of high-technology.

This book is, in fact, a reviewed and extensively improved second edition, but it can also be regarded as the first edition in English, translated by the author himself from the original German version, "Technische Mechanik der festen und flüssigen Körper," published by Springer-Verlag, Wien, in 1985.

Although this book grew out of lecture notes for a three-semester course for advanced undergraduate students taught by the author and several colleagues during the past 20 years, it contains sufficient material for a subsequent two-semester graduate course. The only prerequisites are basic algebra and analysis as usually taught in the first year of an undergraduate engineering curriculum. Advanced mathematics as it is required in the progress of mechanics teaching may be taught in parallel classes, but also an introduction into the art of design should be offered at that stage. The book is divided into 13 chapters that are arranged in such a way as to preserve a natural sequence of thought and reflections. Within a single chapter, however, the presentation, in general, proceeds from the undergraduate to intermediate level and eventually to the graduate level.

The first three chapters are devoted to the basic components of the mechanical modeling of systems at rest or in motion. They are followed by a chapter on constitutive relations ranging from Hooke's law and Newtonian fluids to that of visco-plastic materials. Vector and cartesian tensor notation is applied consequently, but one-dimensional relations of standard material testing are always given priority. Students familiar with the programming techniques of field variables will have no difficulty following this text.

Kinematics of material points and particle fields including both pathlines and streamlines, as well as the conservation of mass; statics, providing experience with forces and stresses including hydrostatic pressure fields; mechanical work, and force potentials, are the prerequisites that, in Chap. 5, allow early exposure to the powerful principle of virtual work. Since force is recognized to be flux in the potential field, even an extension to nonmechanical systems with other driving agents becomes evident.

In the lengthy Chap. 6, material on linearized elastostatics is assembled but selected to serve the needs of structural mechanics. Thermal effects in beams, plates, and shells are given special attention in relation to load stresses in order to prepare the student for design considerations of the high-temperature environment of modern structures. The elastic visco-elastic correspondence principle serves as the vehicle for applying elastic solutions in order to determine stationary creep and solve the associated lifetime problems. At this stage, the Laplace transformation enters like the previous considerations of Chap. 4 in the form of operational calculus. Straight and curved beams, simple frames, plates, and shells of revolution are considered; torsion and the Hertz theory of contact are given special attention.

While disks and rings in stationary rotation are already considered in elastostatics, the Euler-Cauchy equations of motion are the starting point of Chap. 7 and highlight the transition from the statics of a material point at rest to the dynamics of a moving point. The conservation of both momentum and angular momentum is formulated for moving bodies enclosed in a material volume, and for the flow through a control volume that is fixed in space, or for the case in which the control surface is in prescribed motion. Prerequisites are the Lagrangean and Eulerian kinematics of Chap. 1. The control volume concept is immediately applied to determine the guiding forces of stationary flow and to explain the thrust of rockets and jet engines. Euler's turbine equation is derived, and the drag and propelling forces in a parallel viscous main stream are determined. The vector of angular momentum is further defined for rotating rigid bodies, and the generally valid formula for taking the time derivative of a vector with respect to an intermediate (rotating) reference frame is quite naturally derived. Euler's equations of gyros are discussed as well. In connection with the material on rigid-body kinematics of Chap. 1, the important field of (nonlinear) multibody dynamics (eg of vehicles and satellites) is addressed. Sections on linear and nonlinear vibrations present not only useful integration techniques in both the time and frequency domain, but also illustrate pure dynamic phenomena, like resonance and phase shift. Blake's logarithmic diagram is introduced.

The coupled equations of motion of an unbranched chain of spring-mass systems are derived by means of Newton's law, and the modal properties of its natural vibrations are determined by means of the Holzer-Tolle procedure. Stodola's matrix iteration scheme is sketched in Exercise A 11.11. The practically important design of vibrational absorbers is considered in general in Chap. 7 and for torsional vibrations in Chap. 10.

Using the free-body diagram of a high beam element, the partial differential equations of a vibrating Timoshenko beam are derived. Plane body waves and the associated linear eigenvalue problems and the Rayleigh surface wave are included in this chapter to enhance the understanding, eg of ultrasonic techniques. They are applied in material testing and for medical diagnosis. Seismic waves, namely, the loadings in earthquake engineering, should be mentioned here, as well as the water hammer in hydraulic engineering. Some illustrations are given in Chap. 11 and 12.

A first integral of the equations of motion is useful for both solids and fluids. A first example, the planar pendulum, is already analyzed along these lines of general validity in Chap. 7. Thus, in Chap. 8 not only the work theorem of dynamics and its special form of conservation of mechanical energy is derived by proper integration over the material volume, but also integration performed along a given streamline, keeping the time constant, renders the generalized Bernoulli equation of fluid dynamics. The latter is specialized to the stationary flow of ideal fluid and recognized to be the law of conservation of the specific mechanical energy of a particle moving along the streamline and pathline. Interpretations of hydraulic measurements of stationary flow are given by means of that original Bernoulli equation. Generalizations to include the power supply of a stationary stream and, thus, the loss of pressure head of a guided ideal flow through a turbine or the pressure gain of a flow through a pump and the loss of energy head in a viscous stream are discussed in some detail. Relative streamlines with respect to a stationary rotating reference frame are considered, and a proper form of a Bernoulli-type equation is derived to further ease the application of fluid dynamics to rotating machines. The extension to the first law of thermodynamics (of material and control volumes) concludes that introductory chapter on energy conservation, and it is hoped that the gap between the mechanics course and a parallel course on thermodynamics is thereby somewhat narrowed. The Clausius-Duhem inequality is only mentioned. Fourier´s law of heat conduction, however, as an outcome of an irreversible process is stated and applied in Chap. 6.

Chapter 9 on stability starts out with the derivation of the Dirichlet criterion in an energy norm by means of small perturbations applied to a conservative mechanical system at rest. Thus, the dynamic nature of instability is stressed from the very beginning. The conservation of mechanical energy of the perturbing motion and the assumption of minimal potential energy of the equilibrium configuration render the proper inequalities that bound the kinetic and potential energy in the nearfield of the equilibrium configuration. Bifurcation, snap-through, and imperfection sensitivity are discussed and well illustrated in the load factor deformation diagram. The Euler buckling of slender columns and the buckling of plates are generally discussed; further examples are given in Chap. 11. In addition, the method of small perturbations is applied to consider the stability of a principal motion.

The limits of stability of ductile structures are discussed and the ultimate loads of simple beams and frames determined. Safety analysis within structural mechanics like a cross-sectional and plastic system reserve follows quite naturally. Melan-Koiter´s shake-down theorems are formulated and applied to a plastic thick-walled spherical pressure vessel under lifepressure loading. Consideration of the stability of an open-channel flow and of instability as a result of flutter round off this first overview based on phenomena, rather than mathematics.

The knowledge of dynamics of MDOF-systems is further expanded in Chap. 10, where the Lagrange equations of motion (of the second kind), the outcome of the more general D´Alembert´s principle are presented. The latter is derived in quite the same fashion as the principle of virtual work in statics. Only dynamic systems under holonomic constraints are considered and a few applications to vibrational systems given. A spring-mounted foundation in coupled translational-rotational motion is treated in some detail to illustrate the dangerous beat phenomenon. Parametric excitation is shown to occur in a pendulum with a periodically moving support. Matrix structural dynamics is encountered when considering a simple beam with lumped masses.

The principle of virtual work as presented in Chap. 5 and 10 is the basis for the important approximation techniques and discretization procedures associated with the names of Rayleigh, Ritz, and Galerkin. Chapter 11 offers a rather complete account from a purely mechanical standpoint and also gives a short introduction to the finite-element method (FEM). Convergence in the mean square of the outcome of the Galerkin procedure is conserved if the equations of equilibrium or motion containing the forces are subject to approximations. A generalized form is discussed that makes application as convenient as the original Ritz approximation. Any practical application of some complexity, however, requires further consultation of the specialized literature and a gradual buildup of experience. Priority is given in this context to examples in which additional mechanical insight can be gained: For example, the buckling of a slender rod under the influence of a Winkler foundation exhibits mode jumping, and flexural vibrations under moving load excitation illustrate another type of effective structural bending stiffness in addition to critical speeds, to name just two illustrations of considerable engineering importance. The reduction in time of a nonlinear ordinary differential equation of motion is shown by the Ritz-Galerkin approximation of the Duffing oscillator, by harmonic balance, and by means of the Krylow-Bogoljubow approximation.

In Chap. 12, which deals with impact dynamics, most of the material is related to the simplest possible modeling. The exchange of momentum is assumed to be a sudden process, ie the velocity fields of the colliding masses suffer a jumplike redistribution. Only the two extreme physical cases of idealized elastic impact with conservation of total mechanical energy and the inelastic impact of extreme dissipation are elaborated within that context. To illustrate and justify some of these assumptions, a thin elastic rod of finite length is considered, taking into account the back and forth running waves following a short and hard impact. Thereby, the sound speed of rods is introduced, in addition to the wave speeds derived in Chap. 7. Similarly, water hammer in a suddenly shut down lifeline is reconsidered, examining the compressibility of the fluid and the elasticity of the walls of a cylindrical pipe. The crude approximation derived in Chap. 7 for a draining pipe may be seen as limited to the quasistatic closure of the gate. The sound speed in that case depends on the stiffness of springs in a series connection.

The last, Chap. 13, completes to some extent the discussion of fluid dynamics. More important, the lift exerted on a body in an ideal flow is related to circulation, in addition to being just the result of the surface tractions. The Navier-Stokes equations of viscous flow and their nondimensional form are explored, providing further motivation for the introduction of the Reynolds´ and Froude´s numbers. Similarity solutions with respect to the drag coefficient, the viscous flow through a pipe, and the boundary layer of a flow along a semiinfinite plate are the few applications possible due to the limited length of this book. Since the singular nature of perturbation of the ideal flow through viscosity becomes evident in many cases to be limited to the boundary layer, the importance of the outer flow that may be assumed the ideal is recognized. Therefore, potential flows are discussed and some fields of streamlines derived. The singularity method is a major tool of analysis and its basic idea is sketched; the formulas of Blasius are then derived. The force exerted on a body by a von Karman street of vortices is calculated and the Strouhal number introduced. The boundary value problem of kinematic waves excited in a fluid strip of finite depth by a moving rigid and linear elastic wall is solved to illustrate one of the important interaction problems and to show mode coupling. The Mach number and supersonic flow of gas dynamics are discussed by considering the isentropic outflow of gas through a nozzle from a pressure vessel.

Each chapter not only contains more or less concise derivations, but also exclusively shows practical applications. A book on mechanics cannot be read like a novel. The reader is expected to work out examples with paper and pencil or a personal computer and to have a good command of a suitable programming language, eg FORTRAN. Exercises are presented in an appendix to each chapter. Some test general problem-solving skills, some contain new material. However, hints when possibly needed are always given, together with complete solutions. Other collections of examples should be consulted.

This book was influenced by the standards of teaching mechanics at Northwestern University, Stanford University, and Cornell University, schools the author visited several times and in the order cited. It is a textbook designed for classroom teaching or self-study, not a treatise reporting new scientific results. The author is obviously indebted to many investigators over a period of more than two centuries, as well as earlier books on mechanics. Extensive bibliographies may be found in the *Encyclopedia of Physics,* published by Springer-Verlag, Berlin, and various handbooks of engineering mechanics and fluid mechanics. Also, the regular volumes of *Applied Mechanics Reviews* , published by ASME, New York, should be consulted. The classic textbooks written by the late Professor Timoshenko must be mentioned here. The first edition of this book published in German by Springer-Verlag Wien, had its roots in H. Parkus´s *Mechanics of Solids* (in German). Most of the figures of the first edition have been used to illustrate this book. The author is indebted to several colleagues at the Technical University of Vienna for promoting this book in their classes and to a number of former graduate students, including Dr. P. Fotiu (UC-San Diego), Dr. N. Hampl (now with Getzner Chemie), Dr. R. Hasenzagl (now with Control Data), Dr. H. Hasslinger (now with AMAG), Dr. H. Hayek, Dr. R. Heuer, Dr. F. Höllinger (now with the Danube-Power), Dr. H. Irschik (Professor of Mechanics, U-Linz), Dr. F. Rammerstorfer (Professor of Light-Weight Structures and Aircraft Design), Dr. W. Scheidl (chief engineer of Elin). Their encouragement and helpful criticism and the numerous other contributions by students of civil and mechanical engineering made this book possible.

The author is indebted to the reviewers who enthusiastically suggested the publication of this text by Springer-Verlag, New York and, last but not least, to the engineering editor and his staff for copyediting and valuable technical advice during the preparation of the camera-ready manuscript. The latter the author himself completed using an Apple Macintosh II, with Microsoft Word and Expressionist. My thanks go to my wife, Dr. Waltraud Ziegler, for her patience and linguistic advice.

Please forward any suggestions that might lead to an improvement of the text to the author or Springer-Verlag.

Franz Ziegler

Vienna, Austria

Contents

8 First Integrals of the Equations of Motion, Kinetic Energy 475

1
Kinematics

Kinematics is that division of Mechanics which describes the geometry of motion (deformation) of a body, regardless of the forces and stresses, the sources of that motion. Either the position or displacement vector, both the velocity and acceleration vector are key to analyzing the motion of a point. The fields of those vectors determine the kinematics of simple continua (point-continua). The deformation gradients, the spatial derivatives of the displacements, determine the local deformations and, hence, define the strains. Thus, the elongation of a fiber, the extension of a volume element, and the angular change of a configuration of two perpendicular fibers can be calculated. The kinematic model of a rigid (undeformable) body is characterized by a constant distance between any pair of points in motion. It is the principal reference system (eg for the deformations) and the velocity field is represented by means of the angular velocity vector. Polar cones of a spatial pendulum and polar curves (centrodes) associated with a rigid body in plane motion illustrate the velocity field of such an idealized model and introduce the notion of pure rolling contact.

Streamlines, on the other hand, reflect the motion of a fluid in those cases where the path of an individual fluid particle is unimportant. A study of the conservation of the mass of solids and fluids concludes this introduction.

1.1. Point Kinematics

The position of an individual (material) point P of a body in motion measured against a reference point 0 (eg fixed in an inertial system) at any instant of time t is given by the radius vector $r_P(t)$. The components of that position vector are just the coordinates of the point P in the reference system [eg the cartesian coordinates $x_P(t)$, $y_P(t)$, $z_P(t)$ in the directions of the mutual orthogonal unit vectors e_x , e_y , $e_z = e_x \times e_y$ and, hence, $r_P(t) = x_P(t)\, e_x + y_P(t)\, e_y + z_P(t)\, e_z$]. With subscripts, the unit vectors are renamed e_i, $i = 1, 2, 3,$ and the components of r_P are denoted $x_i(t)$, $i = 1, 2, 3.$ In matrix notation,

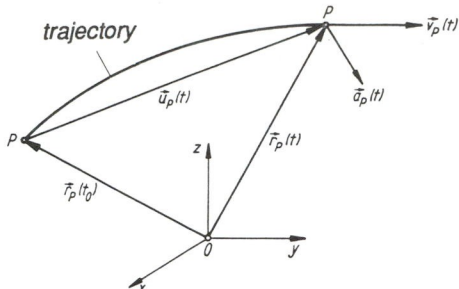

Fig. 1.1. Position vector, displacement vector, velocity and acceleration vector of a point P and its pathline

the radius vector is a column matrix $\{x_i(t)\}$, or transposed, it becomes a row vector $\{x_i(t)\}^T$. The tip of the position vector describes the pathline of P as a function of time (Fig. 1.1). The commonly smooth curve spans from the initial position to the terminal point. Every instant position may be considered a point mapping from the initial location at time t_0 . Hence, the displacement vector $u_P(t) = r_P(t) - r_P(t_0)$ also describes the pathline. It has three components, eg in cartesian coordinates, $u_P = u\, e_x + v\, e_y + w\, e_z$ or $u_P = \sum_i u_i\, e_i$, $u_i = x_i(t) - x_i(t_0)$, $i = 1, 2, 3$. In the case of a free pathline (without any kinematic constraints of the motion of P), the coordinates x_i or the displacements u_i , $i = 1, 2, 3$ are independent, and point P has three degrees of freedom. If the motion of P is constrained to a surface $F(x, y, z; t) = 0$ (a guiding system in a prescribed motion), the number of degrees of freedom is reduced to two. For example, the control condition for z_P may be solved to become a function of x_P and y_P . Curved surface coordinates form the natural pair of independent generalized coordinates. A further restriction on the motion - namely, that it follows a prescribed trajectory - reduces the number of degrees of freedom to one. The natural generalized single coordinate is the arc-length of the guiding pathline.

Considering a second point Q on the pathline $r_Q(t)$ in a motion without any kinematic constraints, ie with three degrees of freedom, we see that the set of two points P and Q in free motion has altogether six degrees of freedom. If a control keeps the distance $PQ = |r_Q - r_P|$ constant, the joint motion obviously has just five degrees of freedom left. The kinematic model of a rigid body is achieved by considering a third noncollinear point R in constant distance to P and Q , $PR = const$ and $QR = const$. The whole set of infinite material points in a rigid continuum has thus six degrees of freedom. The six independent coordinates of the rigid body may be chosen to be the cartesian coordinates of P, and the motion of Q is then constrained to the surface of a sphere centered at P. Thus, the latitude and degree of longitude form an additional pair of angular coordinates, and finally, with these two positions determined, R has only one degree of freedom left over. It can only rotate about the axis through P and Q.

In general, the number of degrees of freedom of a system is the number of independent generalized scalar coordinates that are necessary and sufficient to determine the configuration at any instant of time. A deformable body has infinitely many material points in motion without kinematic constraints and thus possesses an infinite number of degrees of freedom. The instant configuration relative to the initial configuration is determined through the field of displacement vectors $\boldsymbol{u}_P(t)$ for all material points P.

The displacement vector between two instant positions on the pathline of a single point P, $\Delta\boldsymbol{r}_P = \boldsymbol{r}_P(t + \Delta t) - \boldsymbol{r}_P(t)$, when stretched by the reciprocal of the time interval $(1/\Delta t)$, is called the vector of the mean velocity. For a smooth and differentiable trajectory (with a tangent), the limit $\Delta t \to 0$ exists and defines the instantaneous velocity vector of the point P at time t ,

$$\lim_{\Delta t \to 0} \frac{1}{\Delta t} \Delta\boldsymbol{r}_P = \frac{d\boldsymbol{r}_P}{dt} = \dot{\boldsymbol{r}}_P = \boldsymbol{v}_P \ .$$

This vector is parallel to the tangent unit vector $\boldsymbol{e}_t$ of the pathline. With arc-length $s(t)$, measured, eg in meter m, and, by noting $d\boldsymbol{r}_P/ds = \boldsymbol{e}_t$, $|\boldsymbol{e}_t| = 1$, the chain rule of differentiation yields,

$$\boldsymbol{v}_P = d\boldsymbol{r}_P/dt = (d\boldsymbol{r}_P/ds)\,(ds/dt) = v_P \, \boldsymbol{e}_t \ . \tag{1.1}$$

The component $v_P(t) = ds/dt$ is the scalar speed with the dimension of length over time, eg meters per second m/s . The reciprocal is called slowness. The velocity vector is also defined by the time derivative of the displacement vector

$$\boldsymbol{v}_P(t) = d\boldsymbol{u}_P/dt \ . \tag{1.2}$$

The difference between the velocity vectors of P measured at two distinct instants $\Delta\boldsymbol{v}_P = \boldsymbol{v}_P(t + \Delta t) - \boldsymbol{v}_P(t)$, when stretched by the factor $(1/\Delta t)$, is the vector of the mean acceleration. Excluding any jump in velocity, the limit $\Delta t \to 0$ renders the acceleration vector $\boldsymbol{a}_P(t)$

$$\lim_{\Delta t \to 0} \frac{1}{\Delta t} \Delta\boldsymbol{v}_P = \frac{d\boldsymbol{v}_P}{dt} = \boldsymbol{a}_P \ .$$

A further interpretation of the rate of change of the velocity vector is easily recognized in the velocity space. From the origin, the velocity vectors $v_P(t)$ are drawn likewise to the radius vector $r_P(t)$ in the configurational space. The tip of the vectors traces the hodograph and the acceleration vector is the "velocity" of the image to the moving point P and, thus, is parallel to the tangent of this orbit (see Sec. 1.1.1). By generalizing the usual meaning of the hodograph, the derivative of any vector is the "velocity" of its end point along that line.

From the definition of velocity follows

$$\mathbf{a}_P(t) = d\mathbf{v}_P/dt = d^2\mathbf{r}_P/dt^2 = d^2\mathbf{u}_P/dt^2 \quad . \tag{1.3a}$$

Acceleration is collinear to the direction of motion in the case of a straight path or instant if the speed vanishes. The dimension of a_P is speed over time and, hence, length over squared time, eg meters per squared second m/s^2 .

The motion of a continuous body may be described by considering the displacement, velocity and acceleration vector of each individual particle. Instead of assigning a name like P, a material point may be identified by its initial or material coordinates, eg in a cartesian reference frame by X_i, $i = 1, 2, 3$. Upper case letters are used to distinguish the former from the position coordinates $x_i(t)$, $i = 1, 2, 3$ at time t . Hence, the vector fields in the *lagrangean* or material representation, mainly applied in solid mechanics, become

$$\mathbf{u}_P(t) = \mathbf{u}(t; X_1, X_2, X_3) = \sum_{i=1}^{3} u_i(t; X_1, X_2, X_3)\, \mathbf{e}_i \quad ,$$

$$\mathbf{v}_P(t) = \mathbf{v}(t; X_1, X_2, X_3) = \sum_{i=1}^{3} v_i(t; X_1, X_2, X_3)\, \mathbf{e}_i \quad ,$$

$$\mathbf{a}_P(t) = \mathbf{a}(t; X_1, X_2, X_3) = \sum_{i=1}^{3} a_i(t; X_1, X_2, X_3)\, \mathbf{e}_i \quad .$$

Since the material coordinates X_i, $i = 1, 2, 3$ are independent of time, the total time derivative and the partial time derivative that keeps the initial location of a point unchanged are equal, $v = du/dt = \partial u/\partial t$, $a = dv/dt = \partial v/\partial t$. Symbolically, the time-derivative is denoted by

$$\mathbf{v} = \dot{\mathbf{u}} \quad , \qquad \mathbf{a} = \dot{\mathbf{v}} \quad . \tag{1.3b}$$

The vector field may be projected into any convenient coordinate system. For cylindrical and natural coordinates, see Secs. 1.4.4 and 1.1.3; for the spherical frame, see Sec. 1.2 .

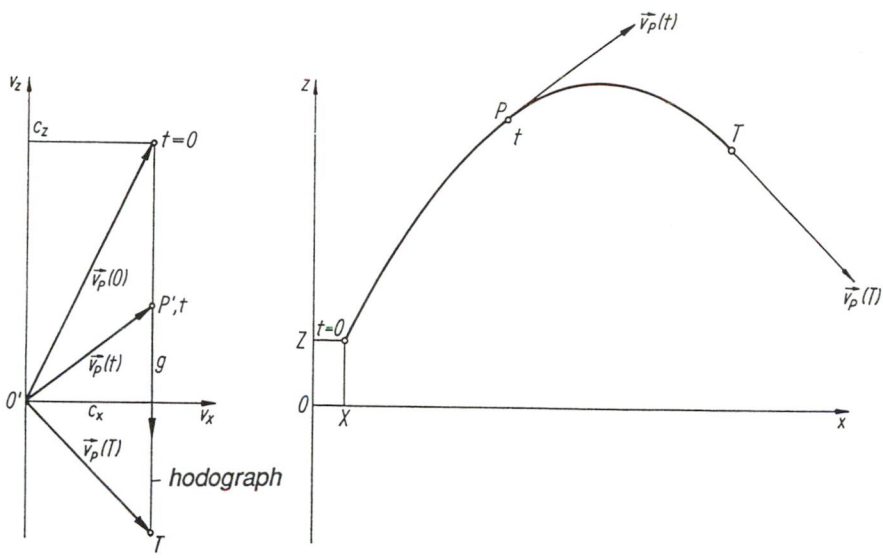

Fig. 1.2. Trajectory and hodograph, acceleration g = *const*

1.1.1. Example: The Trajectory in a Homogeneous Gravity Field Above a "Flat Planet"

A point P moves with constant vertical acceleration $\boldsymbol{a}_P = -g\boldsymbol{e}_z$ in an (x, z) plane (thus in a vacuum). Its velocity is found by integration over time, keeping g = *const* (= 9.81 m/s^2 on Earth) to be $\boldsymbol{v}_P = c_x \boldsymbol{e}_x + (c_z - gt) \boldsymbol{e}_z$; c_x and c_z are integration constants, namely, the velocity components at time $t = 0$. The motion of the image of P in velocity space, a straight-line hodograph, has constant "velocity" since the acceleration vector $\boldsymbol{a}_P = \boldsymbol{const}$. A second time integration renders the pathline, a parabola, $\boldsymbol{r}_P = (X + c_x t) \boldsymbol{e}_x + (Z + c_z t - gt^2/2) \boldsymbol{e}_z$; the initial values of the position vector components are denoted by X and Z (see Fig. 1.2). The elimination of time gives the relation of height over distance, $z_P - Z = (c_z/c_x) (x_P - X) - (g/2c_x^2) (x_P - X)^2$.

1.1.2. Example: Guided Motion of a Point

Using the arc-length s of the prescribed pathline as the natural coordinate yields the speed $v = ds/dt$ and the tangential acceleration component $a_t = d^2s/dt^2$, which is the component of principal kinematic interest. In practical applications, the latter is a prescribed function of time, or speed, or arc-length s, respectively. A second, orthogonal component of acceleration is discussed in Sec. 1.1.3.

The speed and arc-length are found by proper integration as functions of time; the initial conditions at $t = 0$ are prescribed to be $s = s_0$ and $v = v_0$.

(a) $a_t = dv/dt = f(t)$

Integration over time gives the speed

$$v(t) = \int_0^t f(\tau)\, d\tau + v_0 \ .$$

A second integration by parts renders the convolution integral

$$s(t) = \int_0^t v(\tau)\, d\tau + s_0 = \int_0^t \left(\int_0^\tau f(\mu)\, d\mu \right) d\tau + v_0 t + s_0 =$$

$$= s_0 + v_0 t + \left[\tau \int_0^\tau f(\mu)\, d\mu \right]_0^t - \int_0^t \tau f(\tau)\, d\tau = s_0 + v_0 t + \int_0^t (t - \tau) f(\tau)\, d\tau \ .$$

The average speed, derived from the path time function, in general, is unequal to the absolute value of the mean velocity

$$v_m = \frac{s(t) - s_0}{t} = \frac{1}{t} \int_0^t v(\tau)\, d\tau = v_0 + \frac{1}{t} \int_0^t (t - \tau) f(\tau)\, d\tau \ .$$

(b) $a_t = dv/dt = f(v)$

The separation of variables of this first-order differential equation yields $dt = dv/f(v)$. Integrating both sides yields the "inverse relation"

$$t = \int_{v_0}^{v(t)} \frac{du}{f(u)} = F[v(t), v_0] \ ,$$

and $v(t)$ is found by solving the nonlinear equation. The path time function is determined by inserting $d\tau = du/f(u)$ into the defining time-integral

$$s(t) = s_0 + \int_0^t v(\tau)\, d\tau = s_0 + \int_{v_0}^{v(t)} \frac{u}{f(u)}\, du \ .$$

The average speed is

$$v_m = \frac{s(t) - s_0}{t} = \frac{1}{t} \int_0^t v(\tau)\, d\tau = \frac{1}{F[v(t), v_0]} \int_{v_0}^{v(t)} \frac{u}{f(u)}\, du \quad .$$

(c) $a_t = dv/dt = (dv/ds)(ds/dt) = v\,(dv/ds) = f(s)$

The separation of variables gives $vdv = f(s)ds$, and integrating both sides renders

$$\frac{1}{2}(v^2 - v_0^2) = \int_{s_0}^s f(\xi)\, d\xi \quad ,$$

and, hence, $v(s)$. Since $ds/dt = v$, the inverse of the path time function is found by integrating $dt = ds/v(s)$

$$t = \int_{s_0}^s \frac{d\xi}{v(\xi)} \quad .$$

The reciprocal of the average speed, the average slowness, is given by

$$\frac{1}{v_m} = \frac{t}{s(t) - s_0} = \frac{1}{s - s_0} \int_{s_0}^s \frac{d\xi}{v(\xi)} \quad .$$

1.1.3. The Natural Coordinates of the Trajectory

In the case of a given (apriori prescribed) pathline of a point P, velocity and acceleration are conveniently projected onto the three orthogonal unit vectors of the natural coordinate system: e_t, tangent; e_n, principal normal; $e_m = e_t \times e_n$, binormal (note the vector or cross product). If we use the arc-length $s(t)$ in the position vector $r(t)$ to build the box function $r = r\,[s(t)]$, the velocity becomes by applying the chain rule

$$v = \frac{dr}{ds}\frac{ds}{dt} = v\, e_t \quad .$$

Time-differentiation renders two components of the acceleration

$$a = \frac{dv}{dt} = \frac{dv}{dt}\, e_t + v\, \frac{de_t}{dt} \quad .$$

During the time differential dt , the point moves a distance ds and a change of the slope of the trajectory renders $de_t/ds = [1/r(s)]\, e_n$, one of *Frenet´s* formulae. The curvature of the pathline *1/r* is the length of this vector that is orthogonal to e_t. Hence, the chain rule yields $de_t/dt = (de_t/ds)\ (ds/dt) = (v/r)\ e_n$, and, finally, the tangential and normal component of acceleration are recognized

$$a = \frac{dv}{dt}\ e_t + \frac{v^2}{r}\ e_n \quad.$$

The acceleration vector lies in the plane of principal curvature determined by e_t and e_n. There is no component pointing in the binormal direction. By means of a formula for the change of a material vector fixed to a rigid body that rotates with the angular velocity $\omega = (v/\tau)\ e_t + (v/r)\ e_m$, the time derivative of e_t is found directly and to be independent of the second curvature $1/\tau$, see Sec. 1.2,

$$\frac{de_t}{dt} = \omega \times e_t = \frac{v}{r}\ (e_m \times e_t) = \frac{v}{r}\ e_n \quad.$$

1.2.   Kinematics of Rigid Bodies

After selecting two individual points, a reference point A, and a general point P in a distance according to the material vector r_{PA}, the position vector of P with respect to origin 0 is given by summing the vectors: $r_P = r_A + r_{PA}$. Differentiation with respect to time renders, in general, the combination of velocity vectors: $v_P = v_A + v_{PA}$, $v_{PA} = dr_{PA}/dt$. In the case of the undeformable body under consideration, the material vector r_{PA} is kinematically constrained to constant length during the motion $r_{PA} \cdot r_{PA} = r_{PA}^2 = const$, and the time derivative yields $v_{PA} \cdot r_{PA} = 0$, ie v_{PA} is orthogonal to r_{PA} by definition of the scalar (dot) product. Orthogonality in mathematical terms can be expressed by the vector product $v_{PA} = \omega \times r_{PA}$; the dimension of ω must be *1/time* (eg s^{-1}). Since $v_{PA} = 0$ for all the material points P on the action line of the vector ω, that straight line through point A must be the instantaneous axis of the rotation of the rigid body. Hence, ω is called the angular velocity vector. The velocity field of the rigid body is determined by two vectors, v_A and ω, and, it is a linear function of the material radius vector r_{PA}

$$v_P = v_A + \omega \times r_{PA} \quad. \tag{1.4}$$

According to the six degrees of freedom of the motion of the rigid body, the velocity v_P is determined by the six components of v_A and ω. The angular velocity is expressed in radiants per second *rad/s*.

(§) Show ω to be a Free Vector. For ω to be a free vector, it must be independent of the choice of the reference point A. For that sake we select two pairs of vectors (v_A , ω) and $(v_A{}' , \omega')$ and apply the above given basic formula for the velocity field of a rigid body twice

$$v_P = v_A + \omega \times r_{PA} = v_{A'} + \omega' \times r_{PA'} \quad , \quad v_{A'} = v_A + \omega \times r_{A'A} \quad .$$

Inserting $r_{PA} = r_{A'A} + r_{PA'}$ and, eliminating $v_{A'}$ render $\omega \times r_{PA'} = \omega' \times r_{PA'}$, for all the material points P. Hence, $\omega = \omega'$. <<The instantaneous angular velocity of a rigid body is independent of the choice of the material reference point A and the rotational axes through A and A′ are parallel.>> Such a rotation about one instantaneous axis can be substituted with a rotation with the same ω about a parallel axis superposed by a translation with velocity $\omega \times p$ in the direction orthogonal to the plane through both axes (p is the distance between the axes).

(§) Reduction of Angular Velocity Vectors. At time t , a rigid body is supposed to rotate instantaneously about various moving axes passing through the reference points A_1, A_2 , ..., A_n with the angular velocities ω_1 , ω_2 , ..., ω_n . After the selection of a common reference point A , positive and negative rotations ω_1 , ω_2 , ..., ω_n , $-\omega_1$, $-\omega_2$, ..., $-\omega_n$, are attached to this point without changing the kinematic state of the rigid body. The resulting angular velocity (at A) is the sum $\omega = \Sigma_i \, \omega_i$. The actual angular velocities ω_i at their points of action A_i and the negative counterparts $-\omega_i$, attached to the single point A, form pairs of vectors. Each pair renders a translation of the rigid body with a velocity ($\omega_i \times r_{AAi}$). Summation gives the resulting translational velocity $-\Sigma_i \, (\omega_i \times r_{AiA})$.

Taking cartesian components of the basic kinematic formula involving the vector product $\omega \times r$ and considering the quadratic and skew-symmetric angular velocity matrix

$$\boldsymbol{\Omega} = \begin{pmatrix} 0 & -\omega_z & \omega_y \\ \omega_z & 0 & -\omega_x \\ -\omega_y & \omega_x & 0 \end{pmatrix} \quad ,$$

which has tensor properties, renders in matrix notation

$$v_P = v_A + v_{PA} \quad , \quad v_{PA} = \boldsymbol{\Omega} \, r_{PA} \quad .$$

With respect to subsequent applications to time-independent and small displacements δr and small rotations $\delta \alpha$, $|\delta \alpha| << 1$, of a rigid body, Eq. (1.4) is multiplied by the time differential dt to render

$$\delta r_P = \delta r_A + \delta\alpha \times r_{PA} \quad . \tag{1.5}$$

<<The small displacement δr_P of a point P of a rigid body is the sum of the small displacement δr_A of a material reference point A and that from a rotation through a small angle $\delta\alpha$. Small angles have vector properties. However, large rotations about different axes cannot, in general, be superposed by adding vectors. It is easy to prove that the commutative law does not apply when changing the sequence of large rotations. For the proper matrix notation of superposition, see Sec. A 1.4.>>

The selection of a reference point A in a rigid body still keeps v_A and ω varying with time during a general motion. However, at every instant, it is possible to find a special reference point A′ such that its velocity is collinear to ω. The instantaneous velocity field of a rigid body is a combination of a rotation about the "central" axis passing through such a special reference point and a translation in the direction of this axis. The latter is called the *velocity winder* of the rigid body. The material coordinates of A′ of the winder follow from the mathematical condition of parallelism

$$v_{A'} \times \omega = 0 \quad , \quad v_{A'} \parallel \omega \quad ,$$

with $v_{A'} = v_A + \omega \times r_{A'A}$, when searching perpendicular to ω,

$$r_{A'A} = \left(\frac{\omega}{\omega} \times v_A\right)\omega^{-1} \quad .$$

The application of Eq. (1.4) simplifies the time derivative of any vector of constant length. For example, a unit vector when fixed to a rigid body that has the angular velocity ω assigned changes direction by the rate

$$\frac{de}{dt} = \omega \times e \quad . \tag{1.6}$$

If we consider a frame of cylindrical coordinates (r, φ, z), which rotates instantly with the angular velocity $\omega = d\varphi/dt$ about the z-axis, φ is the polar angle, the radial and circumferential unit vectors change during the motion according to their rates

$$\frac{de_r}{dt} = \omega \times e_r = \dot{\varphi}\, e_\varphi \quad , \quad \frac{de_\varphi}{dt} = \omega \times e_\varphi = -\dot{\varphi}\, e_r \quad . \tag{1.7}$$

e_r, e_φ, $e_z = e_r \times e_\varphi$, are mutually perpendicular.

In a frame of spherical coordinates, however, φ denotes the longitude and $\vartheta - \pi/2$ the northern latitude. Hence, the angular velocity vector

$$\boldsymbol{\omega} = \dot\varphi \, \mathbf{e}_z - \dot\vartheta \, \mathbf{e}_\varphi \ , \quad \mathbf{e}_z = -\, \mathbf{e}_r \cos\vartheta + \mathbf{e}_\vartheta \sin\vartheta \ ,$$

renders the rates, with $\mathbf{e}_r$ as the unit radius vector,

$$\frac{d\mathbf{e}_r}{dt} = \boldsymbol{\omega} \times \mathbf{e}_r = \dot\varphi \sin\vartheta \, \mathbf{e}_\varphi + \dot\vartheta \, \mathbf{e}_\vartheta \ , \quad \frac{d\mathbf{e}_\varphi}{dt} = \boldsymbol{\omega} \times \mathbf{e}_\varphi = -\, \dot\varphi \sin\vartheta \, \mathbf{e}_r - \dot\varphi \cos\vartheta \, \mathbf{e}_\vartheta \ ,$$

$$\frac{d\mathbf{e}_\vartheta}{dt} = \boldsymbol{\omega} \times \mathbf{e}_\vartheta = -\, \dot\vartheta \, \mathbf{e}_r + \dot\varphi \cos\vartheta \, \mathbf{e}_\varphi \ .$$

$$(1.8)$$

The field of *acceleration vectors* in a rigid body follows by the termwise differentiation of Eq. (1.4),

$$\mathbf{a}_P = \frac{d\mathbf{v}_P}{dt} = \mathbf{a}_A + \mathbf{a}_{PA} \quad , \quad \mathbf{a}_A = \frac{d\mathbf{v}_A}{dt}$$

$$(1.9)$$

and

$$\mathbf{a}_{PA} = \frac{d\mathbf{v}_{PA}}{dt} = \frac{d\boldsymbol{\omega}}{dt} \times \mathbf{r}_{PA} + \boldsymbol{\omega} \times (\boldsymbol{\omega} \times \mathbf{r}_{PA}) = \dot{\boldsymbol{\omega}} \times \mathbf{r}_{PA} + \omega^2 \mathbf{n}_P \ ,$$

$$(1.10)$$

where, by expanding the vector triple product, the vector

$$\mathbf{n}_P = \left(\frac{1}{\omega} \, \boldsymbol{\omega} \cdot \mathbf{r}_{PA}\right) \frac{\boldsymbol{\omega}}{\omega} - \mathbf{r}_{PA} \ ,$$

$$(1.11)$$

is recognized to be the normal distance of P pointing toward the line of action of $\boldsymbol{\omega}$ through the reference material point A ; $\omega^2 \mathbf{n}_P$ is the essential normal component of the acceleration $\mathbf{a}_{PA}$. If the axis through A is fixed in space, it represents the centripetal acceleration.

When considering the kinematics of relative motion, a rigid body serves as an intermediate reference system and Eqs. (1.9) and (1.10) determine the guiding acceleration, see Sec. 8.5.10. Applications are encountered in multi-body dynamics like robotics, vehicle and flight dynamics, etc.

1.2.1. Special Motions of a Rigid Body

(§) Pure Translation. In this case, $\boldsymbol{\omega} = d\boldsymbol{\omega}/dt = 0$ and the kinematic fields are determined by a single point's velocity $\mathbf{v}_P = \mathbf{v}_A$ and acceleration $\mathbf{a}_P = \mathbf{a}_A$. The trajectories are parallel straight lines.
(§) Rotation About a Fixed Point. If a material point is also fixed in space and, hence, considered the reference point A = 0 with $\mathbf{v}_A = 0$

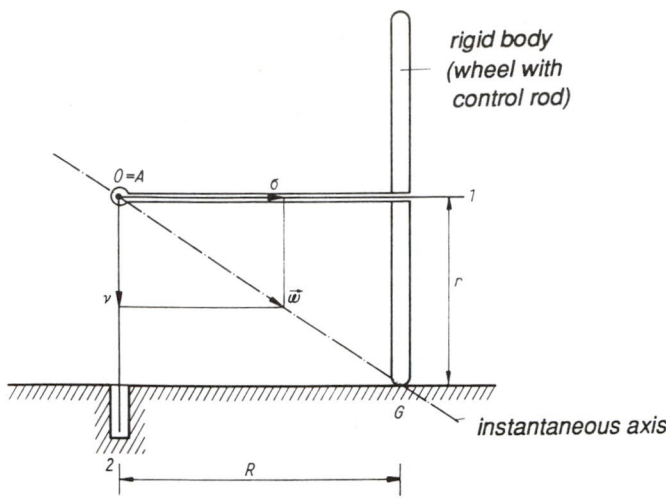

Fig. 1.3. Kinematics of a crushing roller

and $a_A = 0$, the instantaneous axis ω passes through the origin 0. In the course of time, the action line of ω generates two cones with a common tip at 0: One at rest in space; the other, described in material coordinates, moving with the rigid body and in rolling contact along the instantaneous axis. These cones are called the fixed and moving centrodes, respectively.

Good examples include the rigid wheel on a circular path, similar to crushing rollers, and the wheels of a car in a circular turn (see Figs. 1.3 and 1.4).

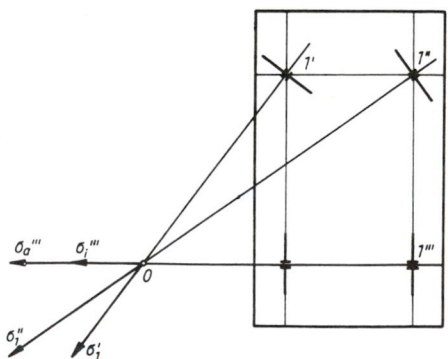

Fig. 1.4. Car in a circular turn

One point of the action line of ω is $0 = A$; another one is the point G of contact with the rigid and plane surface since $v_G = 0$ (the condition of pure rolling contact, no slip). Figure 1.3 shows the ratio $\sigma/v = R/r$ of the angular velocity components. The circular cones have a common tip at 0 and a base radius R (fixed) and r (moving), respectively. Keeping the axis 01 fixed renders the bevel pinions with gear ratio R/r, and $v_G = r\sigma = R v$ is the common circumferential speed.

In the case of the car of Fig. 1.4, v points upward and the four wheels shown must have a different spin σ to preserve rolling contact.

(§) Plane Motion of Rigid Bodies. If the distance of the material points in a plane parallel to an (x, y) plane, which is assumed fixed in space, remains constant, the rigid body is in plane motion. The angular velocity has a constant direction parallel to the normal e_z

$$\omega = \omega\, e_z \quad , \quad e_z = e_x \times e_y \ .$$

In matrix notation, the cartesian components of v_{PA} are

$$v_{PA} = \Omega\, r_{PA} \quad ,$$

where the 2 x 2 angular velocity matrix is simply

$$\Omega = \begin{pmatrix} 0 & -\omega \\ \omega & 0 \end{pmatrix} .$$

The introduction of the rotated vector that is orthogonal to the original one yields a short-hand notation of the vector product with e_z

$$v_{PA} = \omega\, \hat{r}_{PA} \quad , \quad \hat{r}_{PA} = e_z \times r_{PA} \quad ,$$

and the velocity of a material point in vector notation is simply

$$v_P = v_A + \omega\, \hat{r}_{PA} \ . \tag{1.12}$$

In the case of $\omega \neq 0$, a point $P = G$ with vanishing velocity exists that is considered the instantaneous center of plane rotation of the rigid body with respect to the velocity field. In general, the acceleration of G is nonzero. Its material position with respect to a reference point A is easily calculated from the condition

$$v_G = 0 = v_A + \Omega\, r_{GA} \quad ,$$

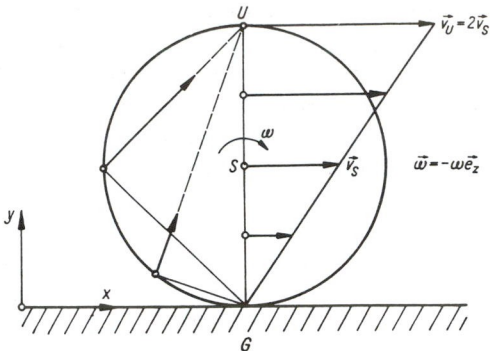

Fig. 1.5. Velocity field of a rolling wheel

by inverting the matrix Ω

$$r_{GA} = -\Omega^{-1} v_A \ , \qquad \Omega^{-1} = \begin{pmatrix} 0 & \omega^{-1} \\ -\omega^{-1} & 0 \end{pmatrix} . \tag{1.13}$$

Selecting the instant locus of G as the reference point A renders the velocity field in the form of an instant rotation

$$v_P = \omega \, \hat{r}_{PG} \ . \tag{1.14}$$

The successive loci of the velocity-center G in the moving body and in the plane (x, y) are the moving and fixed centrodes, respectively, in rolling contact. The cones of the spatial rotation about a point become cylinders, since, in the limit, the common tip moves to infinity in the case of plane motion.

(§) Example: The Wheel in a Straight Rolling Motion. The girth of the wheel and the straight track coincide with the moving and fixed centrodes, respectively. The point of contact has speed zero, no slip is assumed, and, hence, it is the instantaneous center (see Fig. 1.5).

(§) Acceleration. The time derivative of Eq. (1.12) yields

$$a_P = \frac{dv_P}{dt} = a_A + \dot{\omega} \, \hat{r}_{PA} + \omega \frac{d\hat{r}_{PA}}{dt} = a_A + \dot{\omega} \, \hat{r}_{PA} - \omega^2 r_{PA} \ . \tag{1.15}$$

The plane field of acceleration vectors

$$a_{PA} = \dot{\omega} \, \hat{r}_{PA} - \omega^2 r_{PA} \ , \tag{1.16}$$

has a *tangential component* $\dot{\omega} \, \hat{r}_{PA}$ and a *normal component* $-\omega^2 r_{PA}$.

In every instant configuration, there is a point P = B with vanishing acceleration. The material coordinates of this acceleration-center follow at once from the condition $a_B = 0$

$$x_B - x_A = (\omega^2 \ddot{x}_A - \dot{\omega}\,\ddot{y}_A) / (\dot{\omega}^2 + \omega^4)$$

$$y_B - y_A = (\omega^2 \ddot{y}_A + \dot{\omega}\,\ddot{x}_A) / (\dot{\omega}^2 + \omega^4) \ . \tag{1.17}$$

If the reference point selected is A = B, the acceleration field has two special components

$$a_P = a_{PB} = \dot{\omega}\,\hat{r}_{PB} - \omega^2 r_{PB} \ .$$

<<The instant plane state of acceleration of a rigid body corresponds to a rotation about the center B.>> The velocity of B, in general, does not vanish.

1.3.   Kinematics of Deformable Bodies

The displacement vector field u maps the initial (undeformed) configuration into the instant one, point by point, and thus determines the configuration of the deformed body in a reference space. The deformation state of the body is defined by considering the change of the length of an arc-element between two neighboring material points $P(X, Y, Z)$ and $Q(X+dX, Y+dY, Z+dZ)$ in the initial state, where $(dl_0)^2 = (dX)^2 + (dY)^2 + (dZ)^2$, and $P(x, y, z)$ and $Q(x+dx, y+dy, z+dz)$ in the instant configuration at time t , where $(dl)^2 = (dx)^2 + (dy)^2 + (dz)^2$. The assumption that P and Q remain in close distance during deformation is understood; this constraint holds especially for solid bodies under (admissible) structural deformations. Reference is made to one and the same cartesian frame (e_x , e_y , e_z) . If we consider the displacement vector $u(t; X, Y, Z)$ of P and $u(t; X+dX, Y+dY, Z+dZ) = u(t; X, Y, Z) + du$ of Q, the components of the small vector difference are given simply by

$$du = dx - dX \ , \quad dv = dy - dY \ , \quad dw = dz - dZ \ .$$

The metric of the configurational space, ie the measure of the length and angle of the deformations, is defined by the squared length element. The increment with the historical factor $1/2$ is then

$$\left(dl^2 - dl_0^2\right)/2 = du\,dX + dv\,dY + dw\,dZ + (1/2)\left(du^2 + dv^2 + dw^2\right)$$

$$= \varepsilon_{xx}\,(dX)^2 + \varepsilon_{yy}\,(dY)^2 + \varepsilon_{zz}\,(dZ)^2 + 2\left(\varepsilon_{xy}\,dX\,dY + \varepsilon_{yz}\,dY\,dZ + \varepsilon_{zx}\,dZ\,dX\right). \tag{1.18}$$

The six dimensionless coefficients ε_{ij} $(i, j = x, y, z)$ are the strains of the body at point P. Substituting the differentials of u, v, w,

$$du_i = \frac{\partial u_i}{\partial X}\, dX + \frac{\partial u_i}{\partial Y}\, dY + \frac{\partial u_i}{\partial Z}\, dZ \quad , \quad u_x = u \ , \ \ u_y = v \ , \ \ u_z = w \ ,$$

and comparing coefficients in Eq. (1.18) render the set of six nonlinear geometric relations

$$\varepsilon_{xx} = \frac{\partial u}{\partial X} + (1/2)\left[\left(\frac{\partial u}{\partial X}\right)^2 + \left(\frac{\partial v}{\partial X}\right)^2 + \left(\frac{\partial w}{\partial X}\right)^2\right] \ ,$$

$$\varepsilon_{yy} = \frac{\partial v}{\partial Y} + (1/2)\left[\left(\frac{\partial u}{\partial Y}\right)^2 + \left(\frac{\partial v}{\partial Y}\right)^2 + \left(\frac{\partial w}{\partial Y}\right)^2\right] \ ,$$

$$\varepsilon_{zz} = \frac{\partial w}{\partial Z} + (1/2)\left[\left(\frac{\partial u}{\partial Z}\right)^2 + \left(\frac{\partial v}{\partial Z}\right)^2 + \left(\frac{\partial w}{\partial Z}\right)^2\right] \ ,$$

$$\varepsilon_{xy} = (1/2)\,(\frac{\partial u}{\partial Y} + \frac{\partial v}{\partial X}) + (1/2)\left[\left(\frac{\partial u}{\partial X}\frac{\partial u}{\partial Y}\right) + \left(\frac{\partial v}{\partial X}\frac{\partial v}{\partial Y}\right) + \left(\frac{\partial w}{\partial X}\frac{\partial w}{\partial Y}\right)\right] \ ,$$

$$\varepsilon_{yz} = (1/2)\,(\frac{\partial v}{\partial Z} + \frac{\partial w}{\partial Y}) + (1/2)\left[\left(\frac{\partial u}{\partial Y}\frac{\partial u}{\partial Z}\right) + \left(\frac{\partial v}{\partial Y}\frac{\partial v}{\partial Z}\right) + \left(\frac{\partial w}{\partial Y}\frac{\partial w}{\partial Z}\right)\right] \ ,$$

$$\varepsilon_{zx} = (1/2)\,(\frac{\partial w}{\partial X} + \frac{\partial u}{\partial Z}) + (1/2)\left[\left(\frac{\partial u}{\partial Z}\frac{\partial u}{\partial X}\right) + \left(\frac{\partial v}{\partial Z}\frac{\partial v}{\partial X}\right) + \left(\frac{\partial w}{\partial Z}\frac{\partial w}{\partial X}\right)\right] \ .$$

(1.19)

In compact notation

$$2\varepsilon_{ij} = \left(\frac{\partial u_i}{\partial X_j} + \frac{\partial u_j}{\partial X_i}\right) + \sum_{k=1}^{3}\frac{\partial u_k}{\partial X_i}\frac{\partial u_k}{\partial X_j} \quad , \quad (i, j = 1, 2, 3) \ ,$$

(1.20)

and the partial derivatives $\partial u_i / \partial X_j$ are related to the *deformation gradients* $F_{ij} = \partial x_i / \partial X_j$. The strain components ε_{ij} are functions of time t and in the *lagrangean* representation depend on the material coordinates X, Y, Z . They vanish for the rigid-body motion. In the case of the deformable body, the instant configuration is determined by the field of displacement vectors, thus, for point P by three components. Since the six differential equations (1.19) when integrated must render the displacements superposed by any arbitrary rigid-body motion, the six strain components in point P are not independent. The integrability conditions of Eq. (1.19) are called the compatibility conditions of the strains.

For the practically important case of *linear* , more precisely, *linearized geometric relations* in which $|\partial u_i / \partial X_j| \ll 1$ and, hence, $|(\partial u_i / \partial X_j)(\partial u_k / \partial X_l)| \ll |\partial u_m / \partial X_n|$, Eq. (1.20) reduces to

$$\varepsilon_{ij} = (1/2)\,(\partial u_i / \partial X_j + \partial u_j / \partial X_i) \quad , \quad (i, j = 1, 2, 3) \ .$$

(1.21)

St.Venant derived 81 compatibility conditions from the second spatial derivatives of Eq. (1.21) under the condition of independence of the mixed differentials from the order of differentiation. Six necessary conditions on the linearized strain components are, $i \neq j$ and with no summation over the common double index,

$$2 \, \partial^2\varepsilon_{ij}/\partial X_i \, \partial X_j = \partial^2\varepsilon_{ii}/\partial X_j^2 + \partial^2\varepsilon_{jj}/\partial X_i^2 \quad , \tag{1.22}$$

and

$$\partial^2\varepsilon_{kk}/\partial X_i \, \partial X_j = (\partial/\partial X_k)(- \partial\varepsilon_{ij}/\partial X_k + \partial\varepsilon_{jk}/\partial X_i + \partial\varepsilon_{ki}/\partial X_j) \quad , \quad i \neq j \neq k \quad . \tag{1.23}$$

Also, Eqs. (1.22) and (1.23) are dependent. Three independent compatibility relations are derived from still higher-order derivatives. The conditions are also sufficient for the strains in a simply connected body (without holes).

Considering the displacement vector **u** to be a function of the instant coordinates of P*(x, y, z)* and eliminating the difference in the material coordinates of the neighboring points P and Q in *(dl₀)²* $(dl_0)^2$ in Eq. (1.18) render formally the strains by changing the plus to a minus sign before the sum of the nonlinear terms in Eq. (1.20) and replacing the material coordinates X_i , by the spatial coordinates x_i . This *eulerian* representation of the strains takes the instant coordinates of P - namely, the spatial coordinates as the independent variables - and thus considers the inverse mapping into the initial configuration **u** = **u***(x, y, z; t)* . A large class of problems in Fluid Mechanics is formulated without displacement vector fields and without considering individual material points in these spatial coordinates. The *eulerian* representation is also convenient for the conservation laws and for the constitutive equations. The linearized geometric relations (1.21) do not anymore reflect the differences in the *lagrangean* and *eulerian* representations.

The strain components are sampled in a quadratic and symmetric 3 x 3 matrix. Due to the special properties of the strains during transformation by a rigid rotation of the reference frame *(**e**$_x$,* **e**$_y$, **e**$_z$)* , the matrix in the *lagrangean* representation is called *Green's strain tensor* ; it was introduced independently by *Green* and *St.Venant* . The matrix in the *eulerian* representation carries the name *Almansi's strain tensor* (*Almansi* and *Hamel* being the inventors). Some important properties of (symmetric) tensors are discussed in Secs. 1.5.3. and 2.1, [see also Eq. (1.50)].

Several authors circumvent the introduction of displacement vectors and define the strains directly through the *deformation gradients* F_{ij} ,

$$F_{ij} = \partial x_i \, / \, \partial X_j \quad , \tag{1.24a}$$

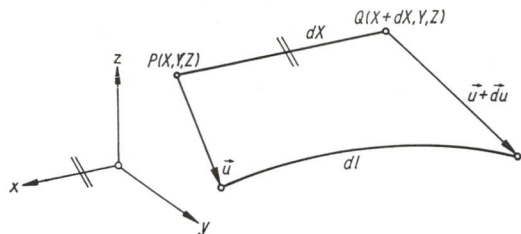

Fig. 1.6. Deformation of the arc-element $dl_0 = dX$

equivalent to Eq. (1.20) ; δ_{pq} is the *Kronecker* symbol,

$$\varepsilon_{pq} = \frac{1}{2}\left(\sum_{i=1}^{3} F_{ip}F_{iq} - \delta_{pq}\right) \quad , \quad p, q = 1, 2, 3 \quad , \quad \delta_{pq} = \begin{cases} 0...p \neq q \\ 1...p = q \end{cases} \quad . \quad (1.24b)$$

The nonlinear geometric relations are mainly applied within stability analysis, in which the *lagrangean* representation is superior.

1.3.1. Elongation and Shear

Selecting the initially straight arc-element $dl_0 = dX$ to be aligned with the $\boldsymbol{e}_x$-direction (Fig. 1.6), renders the length after deformation $dl = (1 + 2\,\varepsilon_{xx})^{1/2}\,dX$ by properly specializing Eq. (1.18). The dimensionless elongation is the length increment divided by the initial arc-length

$$\varepsilon_x = (dl - dX)/dX = (1 + 2\,\varepsilon_{xx})^{1/2} - 1 \quad . \tag{1.25}$$

A nonlinear relation between the strain ε_{xx} and the specific elongation ε_x results

$$\varepsilon_{xx} = \varepsilon_x\,(1 + \varepsilon_x/2) \quad . \tag{1.26}$$

Only for sufficiently small elongations $|\varepsilon_x| \ll 1$, the strain $\varepsilon_{xx} = \varepsilon_x$. According to Eq. (1.26)

$$\varepsilon_{xx} = \varepsilon_x + o(\varepsilon_x)^2 \quad . $$

The elongations ε_y and ε_z of the arc-elements $dl_0 = dY$ and $dl_0 = dZ$ are analogously determined by the strains ε_{yy} and ε_{zz} , respectively.

In the case of a special deformation, eg uniaxial deformation, or the uniform elongation of a tensile rod, which keeps the line elements before $dX\,\boldsymbol{e}_x$ and after deformation $dl\,\boldsymbol{e}_x$ parallel, Eq. (1.25) may be integrated over a finite initial length l_0 to render:

$$\int_0^{l_0} \varepsilon_x\, dX = \int_0^{l_0} dl - l_0 = l - l_0 \ .$$

In the case of a uniform elongation of all elements within l_0, the integral renders $\varepsilon_x\, l_0$ and the constant elongation can be determined by means of the increment of finite length

$$\varepsilon_x = (l - l_0)/l_0 \ . \tag{1.27}$$

Large displacements are likely to be encountered in uniaxial creep or plastic deformations. They are considered by integrating the specific elongation increment, which in the *eulerian* representation is $d\varepsilon = \partial(du)/\partial x$, to define the *logarithmic strain* measure. Change to the *lagrangean* derivative first to find

$$\frac{\partial}{\partial x}(du) = \frac{\partial}{\partial X}(du)\,\frac{\partial X}{\partial x} = \frac{d(\frac{\partial u}{\partial X})}{\frac{\partial x}{\partial X}} = \frac{d(\frac{\partial u}{\partial X})}{1 + \frac{\partial u}{\partial X}} \ ,$$

and integrate

$$\varepsilon = \int d\varepsilon = \int \frac{d\left(\frac{\partial u}{\partial X}\right)}{1 + \frac{\partial u}{\partial X}} = \ln\left(1 + \frac{\partial u}{\partial X}\right) = \ln\left(\frac{\partial x}{\partial X}\right) \ . \tag{1.28}$$

For uniform specific elongation ε_x, the "effective," or "natural," or, as above, logarithmic strain becomes

$$\varepsilon = \ln(1 + \varepsilon_x) = \ln(l/l_0) \ . \tag{1.29}$$

For a generalization to three-dimensional principal strains, see R. Hill: *The Mathematical Theory of Plasticity* . Oxford: University Press, 1967, p.31.

Configurational changes are described by considering the angular decrement of two initially orthogonal arc-elements with a common point of intersection, eg of the elements dX and dY (see Fig 1.7). The lengths after deformation are $(1 + \varepsilon_x)\,dX$ and $(1 + \varepsilon_y)\,dY$,

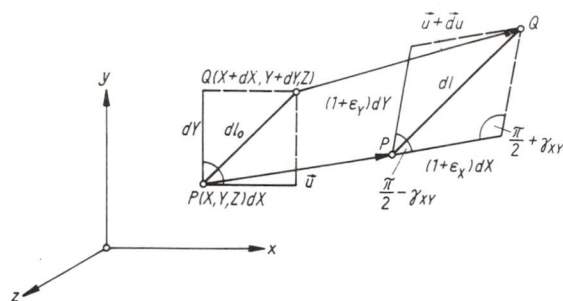

Fig. 1.7. The shear angle γ_{xy}

respectively. The decrement of the right angle is denoted as shear γ_{xy} and the base-length is dl after and dl_0 prior to deformation. The cosine theorem renders

$$dl^2 = (1 + \varepsilon_x)^2 \, dX^2 + (1 + \varepsilon_y)^2 \, dY^2 - 2 \, (1 + \varepsilon_x) \, (1 + \varepsilon_y) \, dX \, dY \cos(\gamma_{xy} + \pi/2) \quad ,$$

and

$$(dl_0)^2 = dX^2 + dY^2 \quad .$$

Putting $dZ = 0$ in Eq. (1.18) yields

$$dl^2 = dX^2 + dY^2 + 2 \, \varepsilon_{xx} \, dX^2 + 2 \, \varepsilon_{yy} \, dY^2 + 4 \, \varepsilon_{xy} \, dX \, dY$$
$$= (1 + \varepsilon_x)^2 \, dX^2 + (1 + \varepsilon_y)^2 \, dY^2 + 4 \, \varepsilon_{xy} \, dX \, dY \quad .$$

Equating dl^2 gives the final result

$$\sin \gamma_{xy} = \frac{2 \, \varepsilon_{xy}}{(1 + \varepsilon_x) \, (1 + \varepsilon_y)} \quad , \tag{1.30}$$

which is a nonlinear function of the strains ε_{xx}, ε_{yy}, ε_{xy} by the relations of Eq. (1.25). The shear strains ε_{ij}, $i{\neq}j$, hence, determine the angular deformations. In the case of small specific elongations $|\varepsilon_x| \ll 1$ and $|\varepsilon_y| \ll 1$, and if also $|\varepsilon_{xy}| \ll 1$, Eq. (1.30) when linearized gives the shear angle to be proportional to ε_{xy}

$$\gamma_{xy} = 2\varepsilon_{xy} \quad . \tag{1.31}$$

Analogous expressions are derived by the cyclic permutation of X, Y, Z. For polar coordinates, see Eq. (6.204).

1.3.2. Dilatation and Strain Deviations

An important measure of deformation is the dilatation (specific volume extension) that is determined by considering the extension of the initial volume element $dV_0 = dX\,dY\,dZ$

$$\frac{dV - dV_0}{dV_0} = \det\{F_{ij}\} - 1 \ ,$$

(1.32)

"det" stands for the determinant; for F_{ij} , see Eq. (1.24). For small strains, the *linearized dilatation e* becomes

$$\frac{dV - dV_0}{dV_0} = e = \varepsilon_{xx} + \varepsilon_{yy} + \varepsilon_{zz} \ ,$$

(1.33)

the so-called *first invariant* (under the rotation of the reference frame) of the strain tensor. It controls the deformation of viscoplastic bodies since the plastic deformation, in general, is isochoric (no change of volume).

The average strain,

$$\varepsilon = e/3 \ ,$$

(1.34)

is the measure of isotropically constant straining and defines the strain deviations, the *deviatoric strain* $\varepsilon_{ij}{}'$ by taking the difference

$$\varepsilon'_{ij} = \varepsilon_{ij} - \varepsilon\,\delta_{ij} \quad , \quad i, j = 1, 2, 3 \ ,$$

(1.35)

where δ_{ij} is the *Kronecker* symbol. Hence, $\varepsilon_{ij}{}' = \varepsilon_{ij}$ for the shear strains, where $i \neq j$, since then, $\delta_{ij} = 0$.

1.3.3. Streamlines and Streamtubes: Local and Convective Acceleration

Above, the displacement vectors of individual material points and their pathlines from the initial to terminal configuration are considered. For some problems, however, it suffices to consider the velocity field in an instant configuration of the body without regard to displacements and their spatial derivatives. Every material point has a definite location in the reference space, the spatial coordinates *x, y, z,* coincide with the instantaneous coordinates of a material point P, and the velocity field is a vector function of these spatial coordinates at the time instant *t*

$$\mathbf{v} = d\mathbf{r}/dt = \mathbf{v}(x, y, z; t) \ .$$

(1.36)

v is the velocity vector of that material point P whose instant coordinates x_P *(t)*, y_P *(t)*, z_P *(t)* take on the values of the selected spatial coordinates *x, y, z* . The velocity can change at a fixed point in space *nonstationary* with time, since the continuous flux of particles through that point may take on different velocity vectors. The *eulerian* representation (1.36) of the velocity vector in all spatial points does not take into account individual material points (eg by their material coordinates in the initial configuration) that pass through the space.

At constant time, the vector lines of the velocity field of Eq. (1.36), the *streamlines*, visualize the motion: These are spatial curves with the tangent pointing in the direction of the velocity that, in general, change their shape with time. The parametric representation of such a streamline at constant *t* may be *r(θ)* ; hence, *dr/dθ* parallels the tangent and by definition so does the velocity vector *v[r(θ); t]* . Therefore,

$$\frac{d\mathbf{r}}{d\theta} \times \mathbf{v}\big[\mathbf{r}(\theta); t\big] = 0 \quad , \tag{1.37}$$

renders in cartesian coordinates three equations for the components of the radius vector $\mathbf{r}(\theta) = x(\theta)\, \mathbf{e}_x + y(\theta)\, \mathbf{e}_y + z(\theta)\, \mathbf{e}_z$:

$$\frac{dx}{d\theta} v_y\,(x, y, z; t) - \frac{dy}{d\theta} v_x\,(x, y, z; t) = 0 \quad ,$$

$$\frac{dy}{d\theta} v_z\,(x, y, z; t) - \frac{dz}{d\theta} v_y\,(x, y, z; t) = 0 \quad ,$$

$$\frac{dz}{d\theta} v_x\,(x, y, z; t) - \frac{dx}{d\theta} v_z\,(x, y, z; t) = 0 \quad .$$

This set of first-order differential equations is integrated (numerically in small steps $\Delta\theta$ in the directions of the velocity vectors) by considering the boundary conditions at $\theta = \theta_0 : x = x_0$, $y = y_0$, $z = z_0$. Cancelling the differential $d\theta$ renders the canonical representation of the streamline

$$\frac{dx}{v_x} = \frac{dy}{v_y} = \frac{dz}{v_z} \quad , \qquad t = const \quad . \tag{1.38}$$

The tangent in every point of the streamline is also tangent to the pathline of the material point with an instantly coinciding location. In a stationary flow ("stream" in the *eulerian* representation), the particles move with constant velocity through the spatial point *(x, y, z)*, $\mathbf{v} = \mathbf{v}(x, y, z)$, and the streamline passing through that spatial point is the pathline of all the particles sitting on that streamline. The parameter θ can be identified with time *t* .

The introduction of a *streamtube*, whose surface consists of a densely packed set of streamlines, is very convenient. The design is rather simple. After selecting a closed curve C_1, the control surface of the streamtube is given by the streamlines passing through the points of C_1. The instant motion of the body contained in the streamtube follows those lines, since in the nonstationary case only instantly, in the stationary case permanently, there is no flux through the surface. The streamtube is kept to a finite length by selecting a proper terminal closing curve C_2. In the case of a stationary stream, the surface could be imagined to be rigidly materialized and those parts of a three-dimensional flow are visualized according to the one-dimensional flow through a pipe with a varying cross-section. The first application of this concept follows subsequently when considering the conservation of mass in a general flow (Sec. 1.6).

The vector field **a** of acceleration in the *eulerian* representation is the vector function **a**$(x, y, z; t)$ and, by definition, it is given by the total time derivative

$$\mathbf{a} = d\mathbf{v}(x, y, z; t)/dt \quad .\tag{1.39}$$

Differentiation is shown for the x-component, where v_x is considered to be a function of four variables

$$a_x = dv_x(x, y, z; t)/dt = \partial v_x/\partial t + (\partial v_x/\partial x)(dx/dt) + (\partial v_x/\partial y)(dy/dt) + (\partial v_x/\partial z)(dz/dt) \quad .$$

Since, by definition, in cartesian coordinates, $v_x = dx/dt$, $v_y = dy/dt$, $v_z = dz/dt$, the scalar spatial differential operator can be factored out

$$\left(v_x \frac{\partial}{\partial x} + v_y \frac{\partial}{\partial y} + v_z \frac{\partial}{\partial z} \right) ,$$

which renders the spatial change of v_x in the direction of the velocity vector **v** when applied to the function $v_x (x, y, z; t)$. The same differential operator appears in the components a_y and a_z. By means of the vector *Hamilton* differential operator ∇ (del), in cartesian coordinates,

$$\nabla = \mathbf{e}_x \frac{\partial}{\partial x} + \mathbf{e}_y \frac{\partial}{\partial y} + \mathbf{e}_z \frac{\partial}{\partial z} \quad ,\tag{1.40}$$

it becomes the formal scalar product, independent of the special choice of the coordinate system

$$\left(\mathbf{v} \cdot \nabla \right) \quad .\tag{1.41}$$

The acceleration vector in the *eulerian* representation, thus, takes on the form

$$\mathbf{a} = \frac{\partial \mathbf{v}}{\partial t} + (\mathbf{v} \cdot \nabla) \mathbf{v} \quad , \tag{1.42}$$

and so it is quite naturally split into the vector of the local acceleration $\partial \mathbf{v}/\partial t$ when keeping the spatial coordinates fixed and into the vector of the convective acceleration $(\mathbf{v} \cdot \nabla)\mathbf{v}$, which is a spatial derivative. The first is the measure of the nonstationary acceleration that is experienced by a particle passing through the spatial location and which is due to the nonstationary change in the time of the velocity vector in that point in space. The second vector gives the necessary acceleration of the material point when it moves in the direction of $\mathbf{v}$ to a neighboring spatial location and, hence, there it takes on a different velocity. The local acceleration vanishes in a stationary motion and the acceleration field is purely convective and time is formally eliminated

$$\mathbf{a}_S = (\mathbf{v} \cdot \nabla) \mathbf{v} \quad . \tag{1.43}$$

Using cartesian components of acceleration and substituting identities, eg those in a_x,

$$v_x \frac{\partial v_x}{\partial x} = \frac{1}{2} \frac{\partial (v_x{}^2)}{\partial x} \quad ,$$

and adding $(1/2) \, \partial(v_y{}^2 + v_z{}^2)/\partial x$ and subtracting the same expression $v_y \, (\partial v_y/\partial x) + v_z \, (\partial v_z/\partial x)$, analogous for the two other components, render by means of the del-operator, Eq. (1.43) changed to *Weber´s* form,

$$(\mathbf{v} \cdot \nabla) \mathbf{v} \equiv \nabla \left(\frac{v^2}{2}\right) - \mathbf{v} \times (\nabla \times \mathbf{v}) \quad , \tag{1.44}$$

where $v^2 = (\mathbf{v} \cdot \mathbf{v})$.

The application of the del-operator ∇ on the scalar function above on $v^2(x, y, z; t)$ yields the vector that is known as the *gradient*, symbolically

$$\nabla \left(\frac{v^2}{2}\right) \equiv \mathrm{grad} \left(\frac{v^2}{2}\right) \quad . \tag{1.45}$$

The formal vector multiplication of the del-operator ∇ with a vector function symbolically renders an orthogonal vector that is called the *curl*

$$\nabla \times \mathbf{v} \equiv \operatorname{curl} \mathbf{v} \quad . \tag{1.46}$$

The cartesian components are derived by the formal rules of vector multiplication

$$\operatorname{curl} \mathbf{v} = \left(\frac{\partial v_z}{\partial y} - \frac{\partial v_y}{\partial z}\right) \mathbf{e}_x + \left(\frac{\partial v_x}{\partial z} - \frac{\partial v_z}{\partial x}\right) \mathbf{e}_y + \left(\frac{\partial v_y}{\partial x} - \frac{\partial v_x}{\partial y}\right) \mathbf{e}_z \quad . \tag{1.47}$$

A physical definition is given by Eq. (3.16) with respect to the force-vector field. The advantage of *Weber's* form of $\mathbf{a}_S$ is seen for the class of irrotational flow problems in which the velocity field $\mathbf{v}$ has the property of *curl* $\mathbf{v} = \mathbf{0}$, and, hence,

$$(\mathbf{v} \cdot \nabla) \mathbf{v} = \operatorname{grad}\left(\frac{v^2}{2}\right) \quad , \quad \operatorname{curl} \mathbf{v} = 0 \quad , \tag{1.48}$$

reduces the convective acceleration to a simple gradient vector field. The notion "irrotational" is deducted from the kinematic interpretation of *curl* $\mathbf{v}$. When we consider the z-component of that vector and the velocity components at four corners of a rectangular element in the (x, y) plane with sides dx and dy, the first-order expansion of the velocity components suffices, local rotations can be deducted,

At (x, y) : v_x , v_y ;
at $(x + dx, y)$: $v_x(x + dx, y) = v_x + (\partial v_x /\partial x) \, dx$,
                         $v_y(x + dx, y) = v_y + (\partial v_y /\partial x) \, dx$,
at $(x, y + dy)$: $v_x(x, y + dy) = v_x + (\partial v_x /\partial y) \, dy$,
                         $v_y(x, y + dy) = v_y + (\partial v_y /\partial y) \, dy$.

The lower side which parallels x, thus, rotates instantly like a rigid edge with the angular velocity $[v_y (x + dx, y) - v_y]/dx = \partial v_y /\partial x$ about the z-axis, ie counterclockwise. The left edge, however, rotates clockwise with $[v_x (x, y + dy) - v_x]/dy = \partial v_x /\partial y$ about the same axis. Hence, the mean angular velocity of a rigid rotation of the volume element about the z-axis is given by the algebraic average,

$$(1/2) \, (\partial v_y /\partial x - \partial v_x /\partial y) = \omega_z \quad , \tag{1.49}$$

which is the z-component of the so-called vorticity vector

$$\boldsymbol{\omega} = (1/2) \operatorname{curl} \mathbf{v} \quad . \tag{1.50}$$

It takes that vector into account as the first-order approximation of the average angular velocity of rotation of a small mass element of a point continuum in the spatial point *(x, y, z)*. So-called irrotational flows are free of vorticity under the condition $\boldsymbol{\omega} = \boldsymbol{0}$ in all material points (for singular points, see Sec. 13.4).

It is noted here that irrotational flows are potential flows, since due to the condition $\boldsymbol{\omega} = (1/2)$ *curl* $\boldsymbol{v} = \boldsymbol{0}$, a velocity potential $\Phi(x, y, z, t)$ exists, such that

$$\boldsymbol{v} = \text{grad } \Phi \ , \tag{1.51}$$

renders a gradient vector field that is irrotational (when free of any jumps)

$$\text{curl grad } \Phi \equiv \boldsymbol{0} \ .$$

1.3.4. Kinematic (Geometric) Boundary Conditions

Most of the technically important motions of bodies are constrained. A subclass of the constraining conditions are the so-called kinematic or geometric boundary conditions to be prescribed in singular points or at some parts of the surface of the body under consideration. Boundary conditions on the displacement vector **u** or on the gradient $\partial \boldsymbol{u}/\partial X_j$, or alternatively on the velocity **v** or on the spatial derivatives $\partial \boldsymbol{v}/\partial x_i$, are called kinematic or geometric. Examples of these boundary conditions are (idealized) structural supports where in spatially fixed points the displacement vector is

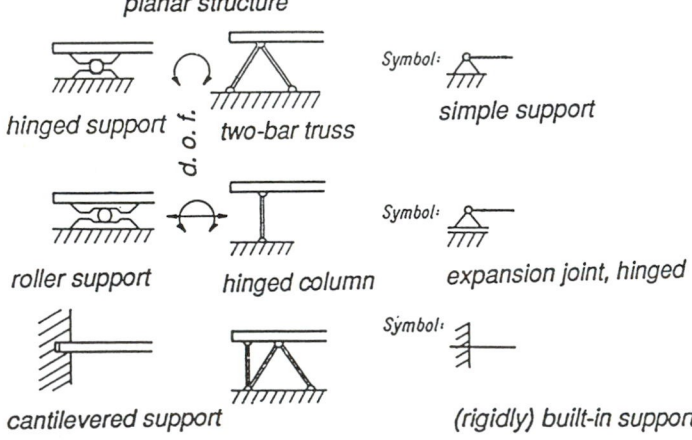

Fig. 1.8. Some symbolic drawings of supports of "plane structures" (beams)

prescribed to vanish, or where in partially constrained supports some displacement components still are supposed to be zero. In the case of a rigidly clamped support, the boundary conditions contain the first derivatives of the displacements; see Fig. 1.8. A simple example of a boundary condition in velocity is the interface of fluid flow and a rigid wall at rest, where the condition of nonintrusion is expressed by a vanishing normal velocity component. In a viscous flow the fluid particles are at rest at the wall (no-slip condition), and since the tangential component is also supposed to vanish, the boundary condition is expressed by $v = 0$ at all points of the interface. Conclusions about these boundary conditions and their influence on body motion and deformation are discussed later on for structures like deformable rods, beams, plates and shells, but also for systems of rigid bodies connected in joints, where they take on the form of contact and pure rolling conditions.

1.4. Supplements to and Applications of Point and Rigid-Body Kinematics

1.4.1. The Velocity Diagram of Plane Motion

The velocity and acceleration field of a rigid body in planar motion is easily determined by graphic methods. The velocity state is subsequently considered. The instant scaled plane configuration is drawn first; see, eg Fig. 1.9 . Assuming that v_1 of the point P_1 and the action line of a second velocity, say, of point P_2 , are given renders the velocity field at time t . The velocity-center G in the configurational plane is determined by the intersection of the normals to the velocities of P_1 and P_2 . The image of G is the origin G´ of the velocity plane. In addition to the length scale of the configurational plane, a scale for the speed in the velocity plane is selected. Velocity vectors are attached on G´ like the common radius vectors; hence, v_1 can be drawn. Since v_{21} (it was named above v_{PA}) is normal to P_1P_2 , the velocity triangle is determined according to the basic formula

$$v_2 = v_1 + v_{21} \quad ,$$

by intersecting the action line of v_{21} through the image point $P_1´$ and the action line of v_2 through the origin G´. Thus, the scaled speed v_2 is determined, and as a by-product also $v_{21} = - v_{12}$. The velocity vector of a third point P_3 of the rigid body is analogously found: The action line through G´ is orthogonal to P_3G. Two subsequent applications of the basic formula

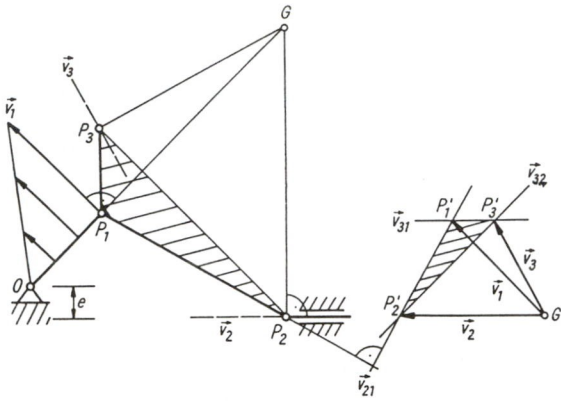

Fig. 1.9. Eccentrical crank mechanism. Rigid body is the connecting rod $P_1 P_2 P_3$

$$\mathbf{v}_3 = \mathbf{v}_1 + \mathbf{v}_{31} = \mathbf{v}_2 + \mathbf{v}_{32} \quad,$$

not only yield the speed v_3 by intersecting the action line of $\mathbf{v}_{31}$ that is normal to P_3P_1 and runs through the tip $P_1{'}$ of $\mathbf{v}_1$, with that of $\mathbf{v}_3$, but also give the location $P_3{'}$ independently of the previously determined direction of $\mathbf{v}_3$. This proves the accuracy of the design, but even more important, it makes the velocity diagram independent of the prior determination of the center G in the configurational space. Quite often, the determination of G is unsafe or impossible due to lack of space. Note the similarity of the triangle P_1, P_2, P_3, which in general is a polygon in the configurational plane, to the rotated one (by $\pi/2$ in the sense of ω) in the velocity diagram, $P_1{'}$, $P_2{'}$, $P_3{'}$ (see again Fig. 1.9).

1.4.2. Kinematics of the Planetary Gear Train

An assembly of meshed gears consisting of a central gear (sun), a coaxial ring gear, and one or more intermediate pinions (planets) supported on a revolving carrier has many technical possibilities; see Fig. 1.10. Three coaxial shafts rotate with three different angular velocities. The contacting radii r_1, r_3 are given, the transmission ratios are to be found. The constraint of rigid-body, slip-free motion is expressed by a common velocity at the point of contact. The basic formula is applied to the planets

$$v_1 = r_1\,\omega_1 \quad , \quad v_2 = r_2\,\omega_2 = v_1 + r_P\,\omega_P \quad , \quad v_3 = v_2 + r_P\,\omega_P = r_3\,\omega_3 \quad ,$$

where

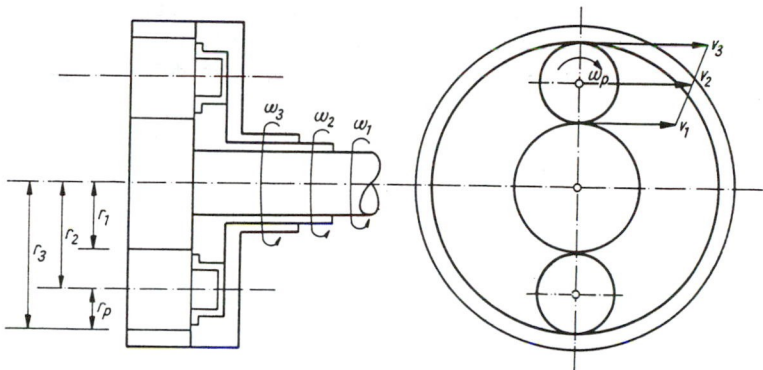

Fig. 1.10. An epicyclic planetary gear train

$$r_2 = (r_1 + r_3)/2 \quad , \quad r_P = (r_3 - r_1)/2 \quad .$$

Thus, evaluating the two resulting equations

$$r_2\,\omega_2 = r_1\,\omega_1 + r_P\,\omega_P \quad ,$$
$$r_3\,\omega_3 = r_1\,\omega_1 + 2\,r_P\,\omega_P \quad ,$$

and keeping ω_2 as a control parameter render the gear ratios

$$\omega_3/\omega_1 = (r_1/r_3)\,(2r_2\,\omega_2/r_1\,\omega_1 - 1) \quad ,$$

$$\omega_P/\omega_1 = (r_1/r_P)\,(r_2\,\omega_2/r_1\,\omega_1 - 1) \quad .$$

Some special cases for an input-output power relation are

$$\omega_1 = 0 \quad , \quad \omega_3 = 2\,r_2\,\omega_2/r_3 \quad , \quad \omega_P = r_2\,\omega_2/r_P \quad ,$$
$$\omega_2 = 0 \quad , \quad \omega_3 = -\,r_1\,\omega_1/r_3 \quad , \quad \omega_P = -\,r_1\,\omega_1/r_P \quad ,$$
$$\omega_3 = 0 \quad , \quad \omega_2 = r_1\,\omega_1/2\,r_2 \quad , \quad \omega_P = -\,r_1\,\omega_1/2\,r_P \quad .$$

The revolutions per minute n are related to the angular velocity ω, rad/s , by

$$\omega = \pi n/30 \quad , \quad [\omega] = rad/s \quad , \quad [n] = rpm \quad .$$

1.4.3. The Universal Joint (after Kardan)

Two rotating nonaligned shafts are connected without slip by a universal joint (after *Kardan*); see Fig. 1.11. The shaft ends are forked and joined through a rigid cross. The gear ratio is to be found when the angle of misalignment α is given. Since three rigid bodies are connected in such a way that point A of the axes' intersection is fixed in space and is also a material point fixed in all three bodies, respectively, the velocities of the end-points of the forks are expressed by

$$v_P = \omega \times r_P = \omega_2 \times r_P \quad , \quad v_Q = \omega_1 \times r_Q = \omega_2 \times r_Q \quad ,$$

where ω_2 denotes the angular velocity of the cross. The resulting vector equations

$$(\omega - \omega_2) \times r_P = 0 \quad , \quad (\omega_1 - \omega_2) \times r_Q = 0 \quad ,$$

have a nontrivial solution if $(\omega - \omega_2) \,//\, r_P$ and $(\omega_1 - \omega_2) \,//\, r_Q \perp r_P$. Scalar multiplication by r_Q and r_P, respectively, yields the intermediate result when a theorem for the scalar triple product is applied,

$$[(\omega - \omega_2) \times r_P] \cdot r_Q = (\omega - \omega_2) \cdot (r_P \times r_Q) = 0 \quad ,$$
$$[(\omega_1 - \omega_2) \times r_Q] \cdot r_P = (\omega_1 - \omega_2) \cdot (r_Q \times r_P) = 0 \quad .$$

Addition eliminates ω_2 and renders the relation,

$$(\omega - \omega_1) \cdot (r_P \times r_Q) = 0 \quad ,$$

which can be reformulated by means of the normal e_n of the midplane of the cross

$$(\omega - \omega_1) \cdot e_n = 0 \quad .$$

If we put $(\omega \cdot e_n) = \omega \cos \varphi$ and $(\omega_1 \cdot e_n) = \omega_1 \cos \varphi_1$, the ratio becomes

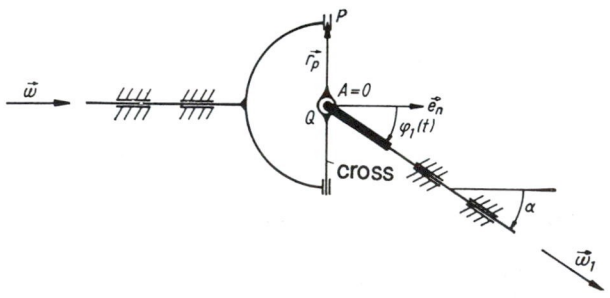

Fig. 1.11. Geometry of the universal joint. Cross positioned at $\varphi = 0$

$$\omega/\omega_1 = \cos \varphi_1/\cos \varphi \quad .$$

A relation between α , the constant angle of ω and ω_1 , and the angles $\varphi(t)$ and φ_1 (t) results, under the condition of orthogonal components of the angular velocities in the cross midplane, from

$$\boldsymbol{\omega} \cdot \boldsymbol{\omega}_1 = \omega \omega_1 \cos \alpha = \omega \cos \varphi \, \omega_1 \cos \varphi_1 \, (\mathbf{e}_n \cdot \mathbf{e}_n) \quad ,$$

as *cos φ_1 = cos α/cos φ* . Thus, φ_1 can be eliminated and the time-dependent ratio of the rotational speeds is finally given by

$$\omega/\omega_1 = \cos \alpha \, / \cos^2\varphi(t) \quad ,$$

in the limits

$$\cos \alpha \le \omega/\omega_1 \le (\cos \alpha)^{-1} \quad .$$

1.4.4. *Central Motion (The Kepler Problem): Polar Coordinates*

The plane motion of a point is considered whose acceleration $\boldsymbol{a}$ is permanently directed to an origin O fixed in space. In polar coordinates, the position vector is $\boldsymbol{r(t) = r(t) \, e_r(t)}$.
      The velocity is the time derivative,

$$\mathbf{v}(t) = \dot{r} \, \mathbf{e}_r + r \, \dot{\mathbf{e}}_r = \dot{r} \, \mathbf{e}_r + r\dot{\varphi} \, \mathbf{e}_\varphi \quad ,$$

where the formula for the differentiation of the material vector $\boldsymbol{e_r}$ has been applied with the proper angular velocity *(dφ/dt) $\boldsymbol{e_z}$* . Thus, the radial speed is $v_r = dr/dt$, and the angular component $v_\varphi = rd\varphi/dt$. Acceleration in general motion is expressed by

$$\mathbf{a}(t) = \ddot{r} \, \mathbf{e}_r + \dot{r} \, \dot{\mathbf{e}}_r + \dot{r}\dot{\varphi} \, \mathbf{e}_\varphi + r\ddot{\varphi} \, \mathbf{e}_\varphi + r\dot{\varphi} \, \dot{\mathbf{e}}_\varphi = (\ddot{r} - r\dot{\varphi}^2) \, \mathbf{e}_r + (r\ddot{\varphi} + 2 \, \dot{r}\dot{\varphi}) \, \mathbf{e}_\varphi \quad .$$

In central motion, the angular acceleration component should vanish

$$(r\ddot{\varphi} + 2 \, \dot{r}\dot{\varphi}) = 0 \quad .$$

Putting $\omega = d\varphi/dt$ renders a first-order differential equation, which after the separation of variables $d\omega/\omega = - \, 2dr/r$ is integrated, if we assume the initial values r_0 and ω_0 are given

$$\ln \omega = - \, 2 \ln r + \ln C \quad \text{or} \quad \omega r^2 = C = \omega_0 r_0^2 \quad .$$

Thus, in central motion, the area-velocity [the area per unit of time of a rectangular triangle with base $r(t)$ and height $v_\varphi = r\,\omega$] $(rv_\varphi/2)$ = $\omega r^2/2$ = $C/2$ is conserved. This general theorem was formulated by *Kepler* around *1610* in his *second law of the planetary motion*. The radial acceleration takes on the general form

$$a_r(t) = \ddot{r} - r\omega^2 = \ddot{r} - \frac{C^2}{r^3} \ .$$

Substituting the inverse square law, ie *Newton´s* law of gravitation,

$$a_r(t) = -k/r^2 \ ,$$

yields the nonlinear second-order differential equation for the radius $r(t)$ of the pathline

$$\ddot{r} - \frac{C^2}{r^3} + \frac{k}{r^2} = 0 \ .$$

Time can be eliminated by considering $r(t) = r[\varphi(t)]$, where $dr/dt = \omega\, dr/d\varphi$, and

$$\ddot{r} = r''\omega^2 + r'\dot{\omega} = C^2\frac{r''}{r^4} - 2C^2\frac{r'^2}{r^5} = \frac{C^2}{r^6}(r''r^2 - 2\,r\,r'^2) = -\frac{C^2}{r^2}\left(\frac{1}{r}\right)'' , \quad r' = \frac{dr}{d\varphi} \ .$$

Substitution yields a linear differential equation for the reciprocal $1/r(\varphi)$, which is readily solved by the superposition of a particular solution to the integral of the homogeneous equation

$$(1/r)'' + (1/r) = k/C^2 \ , \quad (1/r)_p = k/C^2 \ , \quad (1/r)_h = A\cos\varphi + B\sin\varphi \ , \rightarrow$$

$$(1/r) = A\cos\varphi + B\sin\varphi + (k/C^2) \ .$$

The pathline $r(\varphi)$ is easily recognized to be a plane conical section line with one focal point at the origin O. Putting one principal axis in the direction $\varphi = 0$ renders $B = 0$, and the equation of the pathline becomes, with ε denoting "numerical eccentricity",

$$r(\varphi) = C^2/k(1 + \varepsilon\cos\varphi) \ , \quad \varepsilon = (r_0 v_0^2)/k - 1 \ , \quad r_0, v_0 = r_0\omega_0 \ \text{at} \ \varphi = 0 \ .$$

In a first-order approximation, this solution gives the planetary orbits with the Sun in point 0 and renders also the orbits of satellites, eg with the Earth in the focal point 0

$$0 \le \varepsilon < 1 \quad \text{closed orbits (ellipses);}$$
$$\varepsilon \ge 1 \quad \text{open trajectories (hyperbolas or } \varepsilon = 1, \text{ a parabola).}$$

The escape velocity in the perihelion (at $\varphi = 0$) in the cases where $\varepsilon \geq 1$ is

$$v_0 \geq \sqrt{(2k/r_0)} \quad .$$

Since the value of k for the Earth at 0 equals approximately 4×10^{14} m^3/s^2 and the Earth's radius r_0 is $6.37 \times 10^6\ m$, the escape velocity is $11.2 \times 10^3\ m/s$. Rough values for the Sun in 0 are $k \approx 133 \times 10^{18}$ m^3/s^2, with the Sun's radius $r_0 \approx 6.96 \times 10^8\ m$.

1.5. Supplements to and Applications of Deformation Kinematics

1.5.1. The Uniaxial Homogeneous Deformation

One end of a rod of undeformed length l_0 moves with constant speed v_e; the other supported end remains fixed in space. The condition of homogeneous deformation renders the displacement of a cross-section that is located at X in the undeformed configuration as

$$u(t; X) = x(t) - X \quad ,$$

where

$$x(t)/X = l(t)/l_0 = 1 + t\, v_e/l_0 \quad .$$

Eliminating the instant coordinate $x(t)$ yields the *lagrangean* representation

$u = u(t; X) = t\, X\, v_e/l_0$, and, hence, $v = v(t; X) = \partial u/\partial t = X\, v_e/l_0 = \text{const}$, $a = 0$, $0 \leq X \leq l_0$.

Eliminating the material coordinate X renders the *eulerian* representation

$$u = u(x; t) = x\,(1 - l_0/l) = x\, v_e\, t/(l_0 + t\, v_e) \quad , \quad 0 \leq x \leq l(t) \quad .$$

The linear equation is solved for the velocity v

$$v(x; t) = du/dt = \partial u/\partial t + v\,\partial u/\partial x = [v_e\,/(l_0 + t\, v_e) - v_e^2\, t/(l_0 + t\, v_e)^2]\, x + t\, v\, v_e\,/(l_0 + t\, v_e)$$

to render

$$v = v(x; t) = x\, v_e\,/(l_0 + t\, v_e) \quad ,$$

where, asymptotically, $\lim_{t \to \infty} v(x; t) = 0$, for all values of x .

A second differentiation with respect to time determines the acceleration, where the positive convective acceleration cancels with the local acceleration

$$a = a(x; t) = dv/dt = \partial v/\partial t + v \, \partial v/\partial x = x \, [-v_e^2/(l_0 + t \, v_e)^2 + v_e^2/(l_0 + t \, v_e)^2] = 0 \ .$$

The *eulerian* representation clearly indicates the nonstationary motion of the cross-sections through the spatial point x .

1.5.2. The Natural Coordinates of the Streamline

The *eulerian* velocity and acceleration of a material point in a cartesian frame of reference *(s, n, m)* with the base vectors e_t, e_n , $e_m = e_t \times e_n$ given by (see Fig. 1.12)

$$v = v(t; s, n, m) = v_t \, e_t + v_n \, e_n + v_m \, e_m \ ,$$

$$a = \frac{dv}{dt} = \frac{dv_t}{dt} \, e_t + \frac{dv_n}{dt} \, e_n + \frac{dv_m}{dt} \, e_m \ , \qquad (1.52)$$

where the components of acceleration are further expanded according to the total time derivatives

$$dv_t /dt = \partial v_t /\partial t + v_t \, \partial v_t /\partial s + v_n \, \partial v_t /\partial n + v_m \, \partial v_t /\partial m \ ,$$

$$dv_n /dt = \partial v_n /\partial t + v_t \, \partial v_n /\partial s + v_n \, \partial v_n /\partial n + v_m \, \partial v_n /\partial m \ ,$$

$$dv_m /dt = \partial v_m /\partial t + v_t \, \partial v_m /\partial s + v_n \, \partial v_m /\partial n + v_m \, \partial v_m /\partial m \ .$$

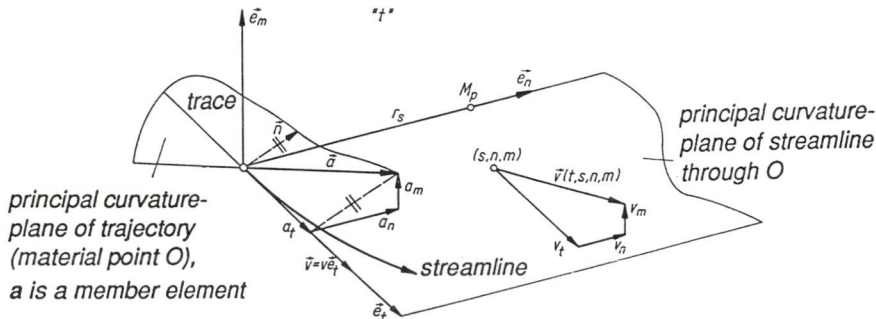

Fig. 1.12. Acceleration vector *a* with components shown in the natural coordinates of the streamline of a nonstationary flow at a constant time

At some time instant t , the reference frame may coincide with the natural coordinates at a point of the streamline under consideration; the tangent vector $\boldsymbol{e}_t$ points in the direction of the velocity of the material point sitting at the origin 0; $\boldsymbol{e}_n$ is assumed to be the principal normal of the streamline. Hence, at that origin $v_t = v$, $v_n = 0$, $v_m = 0$

$$dv/dt = \partial v/\partial t + v \, \partial v/\partial s = \partial v/\partial t + (1/2) \, \partial(v^2)/\partial s \ ,$$

$$dv_n/dt = \partial v_n/\partial t + v \, \partial v_n/\partial s \ ,$$

$$dv_m/dt = \partial v_m/\partial t + v \, \partial v_m/\partial s \ .$$

The convective acceleration component in the $\boldsymbol{e}_n$-direction, $v\partial v_n/\partial s$, corresponds to the changing velocity in the arc distance ds measured along the streamline since $t = const$. It results from a rotation of the vector $\boldsymbol{v}$ about the axis parallel to the binormal $\boldsymbol{e}_m$ through the center of curvature of the streamline, through the infinitesimal angle ds/r_S , which is also expressed by $ds/r_S = dv_n/v$. Hence, the familiar expression results

$$v \, \partial v_n/\partial s = v^2/r_S \ .$$

$r_S{}^{-1}$ is the principal curvature of the streamline. The *convective acceleration* vector lies in the plane of curvature of the streamline; hence, $\partial v_m/\partial s = 0$. Thus, in general, the three components of acceleration in the natural coordinates of the streamline are determined as follows

$a_t = dv/dt = \partial v/\partial t + (1/2) \, \partial(v^2)/\partial s$ (tangential acceleration)   (1.53)
$a_n = dv_n/dt = \partial v_n/\partial t + v^2/r_S$ (principal normal acceleration)   (1.54)
$a_m = dv_m/dt = \partial v_m/\partial t$ (nonstationary binormal acceleration)   (1.55)

If we compare a_n and a_m with the two acceleration components in the natural coordinates of the pathline through the origin 0 (see Sec. 1.1.3), a difference in the nonstationary flow is noted, whereas the tangential component a_t is the same. In a stationary flow, all components become identical, since the streamline and pathline coincide and $r_S = r$ of the pathline.

In the nonstationary case

$$(\partial v_n/\partial t + v^2/r_S) \, \boldsymbol{e}_n + (\partial v_m/\partial t) \, \boldsymbol{e}_m = (v^2/r) \, \boldsymbol{n} \ ,$$

must hold, where $\boldsymbol{n}$ is the principal normal and $1/r$ is the principal curvature of the pathline. $\boldsymbol{n}$ is also a (general) normal of the streamline.

1.5.3. The Strain Tensor. The Plane Strain State

(§)   The Tensor Property of the Strain Matrix. Considering two rotated cartesian frames, e_x , e_y , e_z *(X, Y, Z)* and n, m, k *(X´, Y´, Z´)* , for reference of one and the same deformation and noting the invariance of the material arc-element during the rotation of the reference frame yield

$$(1/2) (dl^2 - dl_0{}^2) = \varepsilon_{xx} (dX)^2 + \varepsilon_{yy} (dY)^2 + \varepsilon_{zz} (dZ)^2 + 2 (\varepsilon_{xy} \, dX \, dY + \varepsilon_{yz} \, dY \, dZ + \varepsilon_{zx}$$
$$dZ \, dX) = \varepsilon´_{xx} (dX´)^2 + \varepsilon´_{yy} (dY´)^2 + \varepsilon´_{zz} (dZ´)^2 + 2 (\varepsilon´_{xy} \, dX´ \, dY´ + \varepsilon´_{yz} \, dY´ \, dZ´ + \varepsilon´_{zx}$$
$$dZ´ \, dX´) \ .$$

Scalar multiplication of the equivalent representation of the position vector increment

$$dr = dX \, e_x + dY \, e_y + dZ \, e_z = dX´ \, n + dY´ \, m + dZ´ \, k \ ,$$

by the unit vector e_i , $i = x, y, z$, renders the proper linear vector transformation between the rotated increments

$$\begin{pmatrix} dX \\ dY \\ dZ \end{pmatrix} = \begin{pmatrix} n_x & m_x & k_x \\ n_y & m_y & k_y \\ n_z & m_z & k_z \end{pmatrix} \begin{pmatrix} dX´ \\ dY´ \\ dZ´ \end{pmatrix} \ .$$

Thus, the differentials *dX, dY, dZ* can be eliminated, and comparing the coefficients yields the full set of transformations of the nine strain components. For example for the rotated normal strain $\varepsilon´_{xx}$,

$$\varepsilon´_{xx} = \varepsilon_{xx} \, n_x{}^2 + \varepsilon_{yy} \, n_y{}^2 + \varepsilon_{zz} \, n_z{}^2 + 2 (\varepsilon_{xy} \, n_x \, n_y + \varepsilon_{yz} \, n_y \, n_z + \varepsilon_{zx} \, n_z \, n_x) \ , \quad (1.56)$$

and for the rotated shear strain $\varepsilon´_{xy}$,

$$\varepsilon´_{xy} = \varepsilon_{xx} \, n_x \, m_x + \varepsilon_{yy} \, n_y \, m_y + \varepsilon_{zz} \, n_z \, m_z + \varepsilon_{xy} (n_x \, m_y + n_y \, m_x) +$$
$$\varepsilon_{yz} (n_y \, m_z + n_z \, m_y) + \varepsilon_{zx} (n_z \, m_x + n_x \, m_z) \ . \quad (1.57)$$

The other strains are determined analogously by cyclic permutations of the subscripts. A square matrix, whose elements are transformed according to those rules during rotation of the reference frame, is denoted a *tensor* (of second order). For the stress tensor, see Eqs. (2.22) and (2.23) and for the inertia tensor, see Eq. (2.123) and Eq. (7.54). The symmetric strain tensor may be transformed to diagonal form by a proper rotation of the reference frame to become the orthogonal set of principal axes, ie under the extreme condition *max* $[\varepsilon´_{xx} (n_x , n_y , n_z)]$, with the constraint $n_x{}^2 + n_y{}^2 + n_z{}^2 - 1 = 0$, etc. The non-vanishing elements in the main diagonal are called the principal normal strains, the corresponding shear strains ε_{ij} $(i \neq j)$ are zero. According to Eq. (1.30), three orthogonal directions exist in every material point that remain orthogonal after deformation; they are called the principal axes of strain.

(§)   The Principal Axes Transformation, *Mohr's* Circle. The state of plane strain is considered that is defined by $\varepsilon_{zi} = 0$ $(i = x, y, z)$ and $\partial \varepsilon_{ij} / \partial z = 0$. Illustrative problems are, eg the half-space and long cylinders, like long dams etc. The symmetric plane strain tensor has three independent elements

$$\begin{pmatrix} \varepsilon_{xx} & \varepsilon_{xy} \\ \varepsilon_{yx} & \varepsilon_{yy} \end{pmatrix} , \quad \varepsilon_{xy} = \varepsilon_{yx} \ . \tag{1.58}$$

The coordinates *(x, y)* are rotated about the *z* -axis through the angle α , and the direction cosine becomes $n_x = \cos \alpha$, $n_y = \sin \alpha$, $m_x = - \sin \alpha$, $m_y = \cos \alpha$ to render (addition theorems for double angles are applied)

$$\varepsilon_{xx \atop yy}' = \frac{\varepsilon_{xx} + \varepsilon_{yy}}{2} \pm \frac{\varepsilon_{xx} - \varepsilon_{yy}}{2} \cos 2\alpha \pm \varepsilon_{xy} \sin 2\alpha \ , \tag{1.59}$$

$$\varepsilon_{xy}' = - \frac{\varepsilon_{xx} - \varepsilon_{yy}}{2} \sin 2\alpha + \varepsilon_{xy} \cos 2\alpha \ . \tag{1.60}$$

The plus sign refers to ε'_{xx} , the minus sign to ε'_{yy} . The necessary condition that $\varepsilon'_{xx} (\alpha)$ takes on an extreme value is

$$\partial \varepsilon'_{xx} / \partial \alpha = 0 = - (\varepsilon_{xx} - \varepsilon_{yy}) \sin 2\alpha + 2 \varepsilon_{xy} \cos 2\alpha = 2 \varepsilon'_{xy} \ .$$

That is, the vanishing shear component renders the principal strain axis in the direction α_1 ,

$$\tan 2\alpha_1 = 2 \varepsilon_{xy} / (\varepsilon_{xx} - \varepsilon_{yy}) \ , \tag{1.61}$$

and the second principal axis points in the orthogonal direction $\alpha_2 = \alpha_1 + \pi/2$. The corresponding shear $\varepsilon_{12} = \varepsilon'_{xy} = 0$, and the principal normal strains become

$$\varepsilon_{1, 2} = (\varepsilon_{xx} + \varepsilon_{yy})/2 \pm (1/2) \left[(\varepsilon_{xx} - \varepsilon_{yy})^2 + 4 \varepsilon_{xy}^2 \right]^{1/2} \ . \tag{1.62}$$

Using the principal axes for reference and denoting the coordinates rotated through α by *(x, y)* reduce Eq. (1.59) accordingly

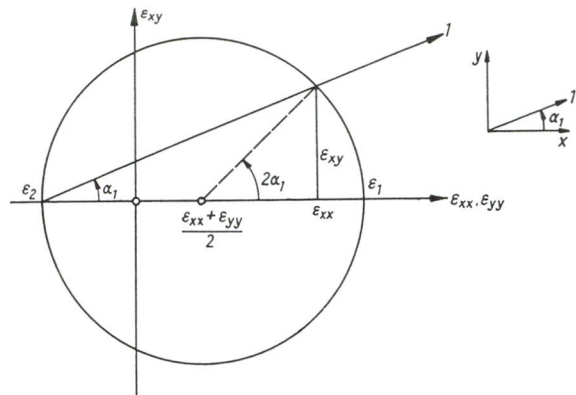

Fig. 1.13. *Mohr´s* circle of the plane strain tensor. Principal strain axis 1. Given: ε_{xx}, ε_{yy}, ε_{xy}. Wanted: ε_1, ε_2, α_1. See also Fig. 2.7 .

$$\varepsilon^{xx}_{yy} = \frac{\varepsilon_1 + \varepsilon_2}{2} \pm \frac{\varepsilon_1 - \varepsilon_2}{2} \cos 2\alpha \quad , \quad \varepsilon_{xy} = -\frac{\varepsilon_1 - \varepsilon_2}{2} \sin 2\alpha \quad . \tag{1.63}$$

This is the equation of a circle expressed in polar coordinates in the $[(\varepsilon_{xx}, \varepsilon_{yy}), \varepsilon_{xy}]$ plane. The circle associated with a plane tensor is called *Mohr´s circle*, see also Fig. 2.7.

     If we assume that the strain tensor is known in general form, *Mohr´s* circle can be drawn with the center located on the abscissa at $(\varepsilon_{xx} + \varepsilon_{yy})/2 = (\varepsilon_1 + \varepsilon_2)/2$ and passing through the points $(\varepsilon_{xx}, \varepsilon_{xy})$, $(\varepsilon_{yy}, \varepsilon_{xy})$. Besides the double angle $2\alpha_1$ at the center, the direction α_1 of the principal strain axis, measured against the x-axis, is given directly by the peripheral angle; see Fig. 1.13.

     The sum of the normal strain components remains constant under the rotation of the reference frame (x, y) (the center of *Mohr´s* circle does not move); it is called the first invariant of the plane strain tensor.

1.6. Conservation of Mass: The Continuity Equation

The kinematics of a body consisting of material points is considered above. Each body has a mass assigned that is a fundamental physical property. It is measured in multiples of the unit "one kilogram", *1 kg* and, in the absence of any external sources, is assumed to be constant during motion. Effects described in the *Relativistic Mechanics* of high-speed motion with respect to the velocity of light *(3 x 10⁸ m/s)* are out of the scope of this textbook. The body is considered to be a continuum, with a mass Δm assigned to every partial material volume ΔV . A *mass density*, the mass per unit of volume, is assumed to exist with the (mathematical) definition

$$\rho(x, y, z; t) = \lim_{\Delta V \to 0} \frac{\Delta m}{\Delta V} \quad , \quad [\rho] = kg/m^3 \quad , \tag{1.64}$$

in every material point of the body. The initial density distribution at $t = 0$ is denoted $\rho_0(X, Y, Z)$. The conservation of mass during the motion of the body from the initial to instant configuration is expressed by keeping the sum of small mass elements, $dm = \rho dV$, ie the volume integral, constant

$$m = \int_{V(t)} \rho(x, y, z; t) \, dV = \int_{V_0} \rho_0(X, Y, Z) \, dV_0 \quad , \quad [m] = kg \quad , \tag{1.65}$$

where $dV = dx \, dy \, dz$ and $dV_0 = dX \, dY \, dZ$ are related by Eq. (1.32). Since Eq. (1.65) must hold for every partial volume of the body, the integrands must be the same and the local condition

$$\rho_0(X, Y, Z) = \rho(x, y, z; t) \det \{F_{ij}\} \quad ,$$

when inverted renders the *lagrangean* continuity equation

$$\rho(x, y, z; t) = \rho_0(X, Y, Z) \det \{F_{ij}^{-1}\} \quad . \tag{1.66}$$

The point mapping of V_0 to $V(t)$ is complex and Eq. (1.66) contains the determinant of the deformation gradients. It is much simpler to select a proper space volume V , a *control volume,* with a permeable surface ∂V , the *control surface.* At some time instant t , the control volume contains the mass $m(t)$

$$m(t) = \int_V \rho(x, y, z; t) \, dV \quad , \tag{1.67}$$

where (x, y, z) denotes a spatial point in V . During the flow through the control volume V , the mass $m(t)$ may change according to the sum of the nonstationary density variations in the spatial points (x, y, z) ; hence,

$$\frac{dm(t)}{dt} = \int_V \frac{\partial \rho(x, y, z; t)}{\partial t} \, dV \quad . \tag{1.68}$$

In the absence of any external sources in V , such an increase in mass corresponds to a net influx of mass through the control surface ∂V . When we denote the external normal unit vector of ∂V by e_n , the velocity component of a particle normal to the control surface is

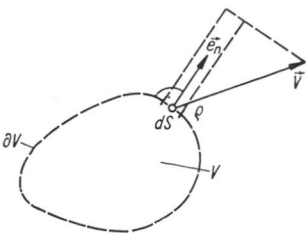

Fig. 1.14. Control volume V fixed in space. Flow through the element dS of the control surface ∂V

the scalar product $(\mathbf{v} \cdot \mathbf{e}_n)$, and the mass flow through the surface element dS , per unit of time, is given by (see Fig. 1.14)

$$\rho \, \mathbf{v} \cdot \mathbf{e}_n \, dS \ . \tag{1.69}$$

The specific mass flow rate (per unit of the control surface) is denoted by

$$\mu = \rho \, \mathbf{v} \cdot \mathbf{e}_n \ , \quad [\,\mu\,] = kg/s \, m^2 \ . \tag{1.70}$$

The influx of mass per unit of time is the sum of the partial mass flow rates, ie the surface integral over the closed control surface ∂V, and must equal Eq. (1.68)

$$-\oint_{\partial V} \mu \, dS \ = \ \frac{dm(t)}{dt} \ . \tag{1.71}$$

Thus, the balance of mass flow rate with respect to the control volume V requires

$$\int_V \frac{\partial \rho}{\partial t} \, dV \ + \oint_{\partial V} \mu \, dS \ = \ 0 \ . \tag{1.72}$$

Instantly, the control volume V contains a mass of a body with the same material volume $V(t)$, and the rate form of Eq. (1.65), $dm/dt = 0$, m of Eq. (1.65), when expressed in *eulerian* form exactly renders Eq. (1.72), where ρ, dV , and dS are expressed in spatial coordinates. This interpretation of Eq. (1.72) is called *Reynolds´ transport theorem* , a generalization of *Leibniz´* rule of differentiation of an integral (see also Sec. 7.1).

By means of the *Gauss* integral theorem that has the general form

$$\int_V \{\nabla g\} \, dV = \oint_{\partial V} \{\mathbf{e}_n g\} \, dS \ ,$$

the surface integral in Eq. (1.72) is changed to a volume integral (by the divergence theorem); the control volume must be simply connected and fixed in space

$$\oint_{\partial V} \rho \mathbf{v} \cdot \mathbf{e}_n \, dS = \int_V \nabla(\rho \mathbf{v}) \, dV \ .$$

Hence, Eq. (1.72) becomes a volume integral

$$\int_V \left[\frac{\partial \rho}{\partial t} + \nabla(\rho \mathbf{v}) \right] dV = 0 \ ,$$

(1.73)

which vanishes for any control volume fixed in space. Thus, the integrand must vanish and the local condition, the *eulerian* continuity equation, holds in every spatial point

$$\frac{\partial \rho}{\partial t} + \text{div} \, (\rho \mathbf{v}) = 0 \ ,$$

(1.74)

$div(\rho \mathbf{v}) \equiv \nabla(\rho \mathbf{v})$ is the divergence of the vector specific mass flow rate $\rho \mathbf{v}$. By combining with the total time derivative of the density $\rho(x, y, z; t)$ [compare with Eq. (1.42)]

$$d\rho/dt = \partial \rho/\partial t + (\mathbf{v}.\nabla)\rho \ ,$$

the continuity equation takes on the form

$$\frac{d\rho}{dt} + \rho \, \text{div} \, \mathbf{v} = 0 \ ,$$

(1.75)

or, identically, when we note the differential of the logarithmic function,

$$\text{div} \, \mathbf{v} = -\frac{d}{dt} (\ln \rho) \ .$$

(1.76)

The latter form not only expresses the kinematic constraint on the velocity field of an incompressible flow (deformation) of a homogeneous continuum, $\rho = const$, namely,

$$\text{div } \mathbf{v} = \nabla \cdot \mathbf{v} = 0 \quad , \tag{1.77}$$

but also indicates that Eq. (1.77) is a good approximation, even in the case of compressible flow with light to moderate fluctuations of density, since the logarithmic function is slowly varying.

In the case of an incompressible and irrotational flow [see Eq. (1.50)] where the additional kinematic constraint *curl* $\mathbf{v} = \mathbf{0}$ applies, the velocity is a potential vector field, $\mathbf{v} = grad\ \Phi = \nabla\Phi$, [see Eq. (1.51)], and renders upon substitution into Eq. (1.77)

$$\nabla \cdot \left(\nabla\Phi \right) = \Delta\Phi = 0 \quad , \tag{1.78}$$

the *Laplace* differential equation for the velocity potential $\Phi(x, y, z; t)$. Since $\Delta = \nabla^2 = (\partial^2/\partial x^2 + \partial^2/\partial y^2 + \partial^2/\partial z^2)$ is a spatial operator, time is only a parameter of the motion. The flow becomes nonstationary only in the case of time-varying external actions: For example, *kinematic waves* are solutions of the *Laplace* equation. The velocity potential is a harmonic function of the spatial coordinates.

In some applications, the introduction of a moving control surface considerably simplifies the problem. A surface point has a prescribed velocity $\mathbf{w}$, and the mass flow rate changes accordingly against that of Eq. (1.69) to become (see Fig. 1.15)

$$\rho(\mathbf{v} - \mathbf{w}) \cdot \mathbf{e}_n\, dS \quad . \tag{1.79}$$

The time rate of the mass $m(t)$ [see Eq. (1.67)] that is enclosed in the moving control volume V^* must equal the net influx of mass through the moving control surface ∂V^*

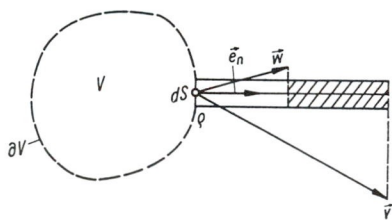

Fig. 1.15. Moving control surface ∂V. Flux through the element dS with prescribed self-velocity $\mathbf{w}$

$$-\oint_{\partial V^*} \mu \, dS = \frac{dm(t)}{dt} \quad .$$

$$(1.80)$$

The specific mass flow rate at the moving control surface,

$$\mu = \rho(\mathbf{v} - \mathbf{w}) \cdot \mathbf{e}_n \quad ,$$

$$(1.81)$$

is to be substituted. The mass balance takes on the most general form (no external sources are considered)

$$\frac{\partial}{\partial t} \int_{V^*} \rho \, dV + \oint_{\partial V^*} \mu \, dS = 0 \quad .$$

$$(1.82)$$

It is useful if, eg a streamtube is selected to be the control surface in a nonstationary flow; see Sec. 1.3.3.

In many applications, the assigned motion of the control volume is considered to be rigid-body motion; Eq. (1.4) applies in this special case to the field $\mathbf{w}$, and the volume V remains constant. A cartesian reference frame is attached to the rigidly moving control volume, (x', y', z') , and since ρ is a scalar function, differentiation and integration may be interchanged to render

$$\frac{\partial}{\partial t} \int_V \rho \, dV = \int_V \frac{\partial \rho(x', y', z'; t)}{\partial t} \, dV' \quad .$$

$$(1.83)$$

Such a moving control volume is most suitable if the relative flow $(\mathbf{v} - \mathbf{w})$ becomes stationary. In that case, the volume integral (1.83) vanishes and

$$\oint_{\partial V} \mu \, dS = 0 \quad , \quad \mu = \rho(\mathbf{v} - \mathbf{w}) \cdot \mathbf{e}_n \quad .$$

$$(1.84)$$

In stationary flow, the streamlines are also pathlines and the control surface according to a streamtube remains fixed in space, $\mathbf{w} = 0$. Since $\partial \rho / \partial t = 0$, Eq. (1.72) reduces to

$$\oint_{\partial V_{St}} \mu \, dS = 0 \quad , \quad \mu = \rho \mathbf{v} \cdot \mathbf{e}_n \quad .$$

$$(1.85)$$

The remaining surface integral is further reduced and simplified if plane cross-sections A_1 and A_2, which are orthogonal to a mean streamline, are selected instead of end caps of a more general form

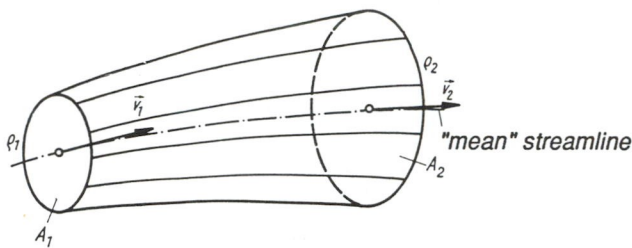

Fig. 1.16. Streamtube. Area of the cross-section of inlet A_1, of the outlet A_2. Control surface fixed in space according to a stationary flow, $\dot{m}$ = const

for closing the streamtube properly. The mean value theorem applied to the cross-sectional integral renders (see Fig. 1.16)

$$\oint_{\partial V_{st}} \mu \, dS = \mu_1 A_1 + \mu_2 A_2 = 0 \ , \ \ \mu_1 = -\rho_1 v_1 \ , \ \ \mu_2 = \rho_2 v_2 \ ,$$

or in the engineering form of the mass flow rate equation,

$$\dot{m} = \rho_1 v_1 A_1 = \rho_2 v_2 A_2 \ . \tag{1.86}$$

<<The intake of mass per unit of time through the cross-section A_1 with average speed v_1 must equal the outflow of mass through the cross-section A_2 with average velocity v_2, in order that the mass flow through the surface of the stream tube vanishes.>> This equation is applied during the stage of design of the net of streamlines. The analogy to the flow through a pipe with a varying cross-section is quite instructive and Eq. (1.86) gives the mean velocity distribution along the pipe´s arc-length.

1.6.1. Stationary Flow Through a Conical Pipe: Eulerian and Lagrangean Representations

The flow is assumed to be incompressible over the interval $0 \le x \le l$, where the cross-section varies linearly according to $A(x) = A_1 - (A_1 - A_2) \, x/l$. The *eulerian* continuity equation renders the average speed, $v(x = 0) = v_1$ [see Eq. (1.86)],

$$\dot{m} = \rho v_1 A_1 = \rho v(x) A(x) = \text{const} \ , \tag{1.87}$$

and, hence,

$$v(x) = v_1 \frac{A_1}{A(x)} = \frac{v_1}{1 - (1 - \frac{A_2}{A_1}) x/l} \quad .$$

Acceleration in the *eulerian* representation due to the stationary flow, $\partial v(x)/\partial t = 0$, with $v = dx/dt$, becomes

$$a(x) = \frac{dv}{dt} = \frac{\partial v}{\partial x} \frac{dx}{dt} = \frac{v_1^2}{l} \frac{(1 - \frac{A_2}{A_1})}{\left[1 - (1 - \frac{A_2}{A_1}) \frac{x}{l}\right]^3} \quad ,$$

independent of time.

Alternatively, if we consider an individual particle moving through the pipe, the speed $v(t; X)$ changes with time and depends on the initial position, say, $X = 0$ at time $t = 0$. At time t, the speed should be $v(t; X = 0) = v(x)$ at the position x. The integration of $dt = dx/v(x)$ renders

$$[t]_0^X = \int_0^X \frac{d\xi}{v(\xi)} = \frac{1}{v_1} \left[x - (1 - \frac{A_2}{A_1}) \frac{x^2}{2l}\right] \quad ,$$

which constitutes a quadratic equation for the *lagrangean* path time relation $x(t; 0)$. The proper solution, with the upper sign, is

$$x(t; 0) = \frac{l}{1 - \frac{A_2}{A_1}} \left[1 \overset{-}{\underset{(+)}{}} \sqrt{1 - 2 \frac{v_1}{l} (1 - \frac{A_2}{A_1}) t}\right] \quad .$$

The velocity is the time derivative

$$v(t; 0) = \frac{\partial x}{\partial t} = \frac{v_1}{\sqrt{1 - 2 \frac{v_1}{l} (1 - \frac{A_2}{A_1}) t}} \quad .$$

Finally, the acceleration in the *lagrangean* representation is given by the second time derivative

$$a(t; 0) = \frac{\partial v}{\partial t} = \frac{v_1^2}{l} \frac{1 - A_2/A_1}{\left[1 - 2 \frac{v_1}{l} (1 - \frac{A_2}{A_1}) t\right]^{3/2}} \quad ,$$

as a function of time in the course when the particle passes through the pipe.

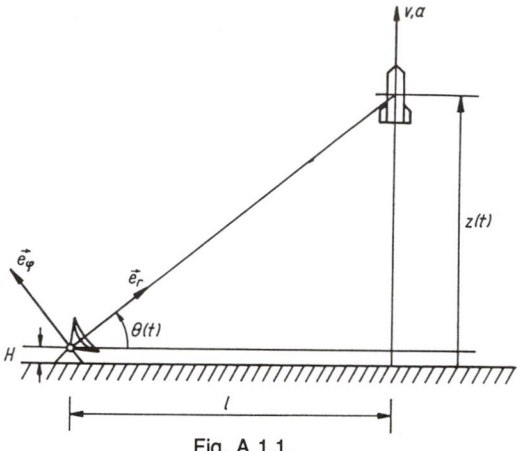

Fig. A 1.1.

1.7. Exercises A 1.1 to A 1.8 and Solutions

A 1.1: A rocket in a vertical ascending flight is watched by a radar monitor that is positioned at a horizontal distance l .

The angle θ is measured, and, furthermore, from a real-time calculation, the angular speed and its acceleration, $\dot{\theta}, \ddot{\theta}$, both are assumed to be known. Determine the height $z(t)$, velocity $v(t)$, and acceleration $a(t)$. See Fig. A 1.1.

Solution: Inspection of the triangle gives $z(t) = H + l \tan \theta$, and differentiation renders $v = dz/dt = l (1 + \tan^2\theta) d\theta/dt$, $a = d^2z/dt^2 = l (1 + \tan^2\theta) [d^2\theta/dt^2 + 2(d\theta/dt)^2 \tan \theta]$. With $\boldsymbol{e}_z = \boldsymbol{e}_r \sin \theta + \boldsymbol{e}_\varphi \cos \theta$, the representation in polar coordinates yields the same result. Since $r = l/\cos\theta$ and $\varphi \equiv \theta$, the radial component of the velocity becomes $v_r = dr/dt = l (d\theta/dt) \sin \theta/\cos^2\theta$, and $v_\varphi = r d\varphi/dt = l (d\theta/dt)/\cos \theta$,

$$a_r = \ddot{r} - r \dot{\varphi}^2 = l \frac{\sin \theta}{\cos^3\theta} (\ddot{\theta} \cos \theta + 2 \dot{\theta}^2 \sin \theta) \ ,$$

$$a_\varphi = r \ddot{\varphi} + 2 \dot{r} \dot{\varphi} = \frac{l}{\cos^2\theta} (\ddot{\theta} \cos \theta + 2 \dot{\theta}^2 \sin \theta) \ .$$

A 1.2: Substitute the cam mechanism in the instant configuration shown in Fig. A 1.2 by an equivalent three-bar-system such that the velocity and acceleration remain unchanged.

Solution: In a second-order geometric approximation, the centers of curvature of the cam profiles M_1, M_2, at the point of contact P, determine the connecting rod. Hence, the radius of the crank shaft is

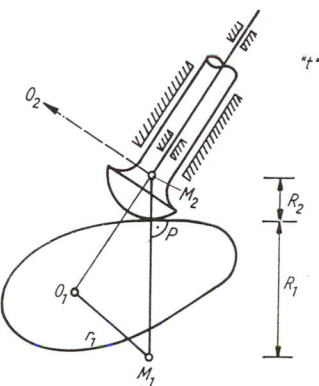

Fig. A 1.2. Cam mechanism with equivalent three-bar-system $O_1M_1M_2$ at time instant t (relative motion of P is an instant rotation about M_1 and M_2, respectively)

found to be O_1M_1. The center point of the more general four-bar-system O_2 moves to infinity: The third bar is in translational motion.

A 1.3: A propeller attached to an airplane in a cornering flight is considered. The center A, according to Fig. A 1.3, is assumed to move on a circular path of radius R with constant speed v_A. Determine the velocity and acceleration of the propeller-tip when $|AP| = l$ and when the spin σ of the propeller about an axis fixed to the airplane is assumed constant.

Solution: The angular velocity ω of the rigidly rotating propeller is the vector sum of the spin $\sigma \, \boldsymbol{e}_\varphi$ and the rotational speed of the

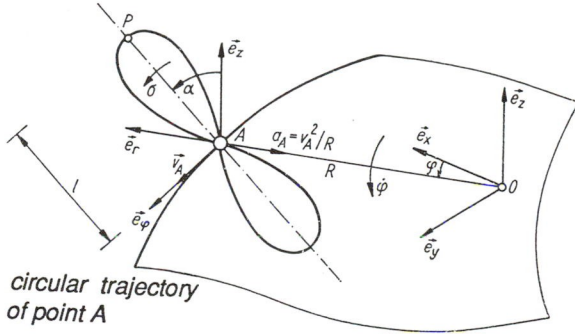

circular trajectory
of point A

Fig. A 1.3.

curving airplane $e_z \, d\varphi/dt$: $\omega = \sigma \, e_\varphi + (v_A/R) \, e_z$, since $d\varphi/dt = v_A/R$. The application of Eq. (1.4) renders with $r_{PA} = I \, (sin \, \alpha \, e_r + cos \, \alpha \, e_z)$

$$v_P = \sigma \, I \, cos \, \alpha \, e_r + v_A \, [1 + (I/R) \, sin \, \alpha] \, e_\varphi - \sigma \, I \, sin \, \alpha \, e_z \quad .$$

Extreme values of the tip-speed v_P occur when $\alpha = \pm \, \pi/2$. Acceleration follows by taking the time derivative and by considering the relations

$$\dot{e}_r = \dot{\varphi} \, e_z \times e_r = \frac{v_A}{R} \, e_\varphi \, , \quad \dot{e}_\varphi = \dot{\varphi} \, e_z \times e_\varphi = - \frac{v_A}{R} \, e_r \quad \text{to be}$$

$$a_P = - \left[I\sigma^2 \, sin\alpha + \frac{v_A^2}{R} \left(1 + \frac{I}{R} \, sin\alpha \right) \right] e_r + 2 \frac{v_A}{R} \, \sigma I \, cos\alpha \, e_\varphi - I\sigma^2 \, cos\alpha \, e_z \quad .$$

Note the e_φ -component, the *Coriolis acceleration* , which vanishes only in the positions $\alpha = \pm \, \pi/2$, when the propeller is in the aligned configuration. See Eq. (8.62) and Exercise A 7.1.

A 1.4: Large Rotations . When considering large rotations of a rigid body, three independent angular coordinates determine the final configuration according to the three rotational degrees of freedom. Classically, *Euler* angles are selected with the following intermediate steps: A rotation ψ about the $3 = 3^*$ -axis of a reference system *(1, 2, 3)* with mutually orthogonal axes, followed by a θ -rotation about the $1^* = 1^0$ -axis and finally, by a rotation through the angle φ about the $3^0 = 3'$ -axis into the terminal configuration (1´, 2´, 3´). Linearization for small angles of rotation is difficult. Therefore, the superposition of large rotations is shown for the three *Kardan* angles α, β, γ. In that case, the first rotation through α is about the $1 = 1^*$ -axis, followed by a rotation through β about the already rotated $2^* = 2^0$ -axis. The third rotation is about the $3^0 = 3'$ -axis through the angle γ and finally gives the, generally rotated, terminal configuration which is represented by the primed reference system *(1´, 2´, 3´)* . By means of the rotational matrices $D \, (\alpha), \, C \, (\beta), \, B \, (\gamma)$, a given vector is successively transformed according to the matrix products.
For the α -rotation, $x^* = D \, x$ [in components $x_i^* = \Sigma_l \, a_{il}(\alpha) \, x_l$]

$$D \, (\alpha) = \begin{pmatrix} 1 & 0 & 0 \\ 0 & cos \, \alpha & sin \, \alpha \\ 0 & - sin \, \alpha & cos \, \alpha \end{pmatrix} \, ,$$

β -rotation, $x_0 = C \, x^* = C \, D \, x$ $[x_j = \Sigma_i \, a_{ji}(\beta) \, x_i^* = \Sigma_i \Sigma_l \, a_{ji}(\beta) \, a_{il}(\alpha) \, x_l]$

$$C\,(\beta) = \begin{pmatrix} \cos\beta & 0 & -\sin\beta \\ 0 & 1 & 0 \\ \sin\beta & 0 & \cos\beta \end{pmatrix} \,,$$

γ-rotation, $x' = B\,x_0 = B\,C\,D\,x = A\,x$
[$x_k' = \Sigma_l\,a_{kl}\,x_l$, where obviously the elements of the matrix A are expressed by the double summation of the products $a_{kl} = \Sigma_j\Sigma_i\,a_{kj}(\gamma)\,a_{ji}(\beta)\,a_{il}(\alpha)$]

$$B\,(\gamma) = \begin{pmatrix} \cos\gamma & \sin\gamma & 0 \\ -\sin\gamma & \cos\gamma & 0 \\ 0 & 0 & 1 \end{pmatrix} \,,$$

and,

$$A\,(\alpha,\beta,\gamma) =$$

$$\begin{pmatrix} \left(\cos\beta\cos\gamma\right) & \left(\cos\alpha\sin\gamma + \sin\alpha\sin\beta\cos\gamma\right) & \left(\sin\alpha\sin\gamma - \cos\alpha\sin\beta\cos\gamma\right) \\ \left(-\cos\beta\sin\gamma\right) & \left(\cos\alpha\cos\gamma - \sin\alpha\sin\beta\sin\gamma\right) & \left(\sin\alpha\cos\gamma + \cos\alpha\sin\beta\sin\gamma\right) \\ \left(\sin\beta\right) & \left(-\sin\alpha\cos\beta\right) & \left(\cos\alpha\cos\beta\right) \end{pmatrix} \,.$$

In the case of rotations through sufficiently small angles, the matrix A is linearized to become

$$A \approx \begin{pmatrix} 1 & \gamma & -\beta \\ -\gamma & 1 & \alpha \\ \beta & -\alpha & 1 \end{pmatrix} \,, \quad \begin{array}{l} |\alpha| \ll 1 \\ |\beta| \ll 1 \\ |\gamma| \ll 1 \end{array} \,.$$

This result may also be derived by adding the small vectors; see Eq. (1.5).

The inverse transformation $x = A^{-1}\,x'$ is simply given by the transposed matrix A^T , since for so-called orthogonal vector transformations $A^{-1} = A^T$, the inverse equals the transposed matrix ($x_k = \Sigma_l\,a_{lk}\,x_l'$), the length of the vector remains unchanged. <<Hence, the superposition of successive rotations through large angles is expressed by the product of the associated rotational matrices.>>

Find the transformation formula for the angular velocity vector $\omega = \omega_\alpha + \omega_\beta + \omega_\gamma$, where

$$\omega_\alpha = \dot{\alpha}\,e_1\,, \quad \omega_\beta = \dot{\beta}\,e_2^0\,, \quad \omega_\gamma = \dot{\gamma}\,e_3' \,.$$

Solution: The sequence of rotations as explained above gives at once the partial transformations,

$$\omega'_\alpha = A\,\omega_\alpha\,,\ \ \omega'_\beta = B\,\omega_\beta\,,\ \ \omega'_\gamma = \omega_\gamma\ \ \text{and}\ \ \omega' = \omega'_\alpha + \omega'_\beta + \omega'_\gamma\ \ ,$$

all vectors expressed in the body-fixed frame.

Note: If the given vectors x and y are related by a linear vector transformation, $y = K\,x$, the relation after rotating the coordinate system by the matrix A is to be determined, $y' = K'\,x'$. Inserting the transformation $y' = A\,y$ and $x' = A\,x$ and further using the identity $x = (A^{-1}\,A)\,x$ render $y' = A\,y = A\,K\,(A^{-1}A)\,x = (A\,K\,A^{-1})\,A\,x = K'\,x'$. Hence, the *similarity transformation* between the matrix K and K' is easily recognized to provide the proper relation,

$$K' = A\,K\,A^{-1}\ .$$

The determinant of K remains unchanged.

A 1.5: Determine the large deformation gradients and strains of an initially straight and thin leaf spring when deformed to a semicircle of radius R without the stretching of the central fiber, see Fig. A 1.5. Thickness is approximately constant.

Solution: When we consider point P with the material coordinates (in the initially straight configuration) $X = R\,\varphi$, $Y = R + P$, P is the distance from the axis measured in the cross-section, in the final position $x = Y\,\sin(X/R)$, $y = Y\,\cos(X/R)$. The deformation is constrained by the geometric relations (see Fig. A 1.5) $x^2 + y^2 = (R + P)^2 = Y^2$ and $x/y = \tan\varphi = \tan(X/R)$. Deformation gradients are derived by partial differentiation: $\partial x/\partial X = (1 + P/R)\cos\varphi$, $\partial x/\partial Y = \sin\varphi$, $\partial y/\partial X = -(1 + P/R)\sin\varphi$, $\partial y/\partial Y = \cos\varphi$, $0 \le \varphi \le \pi$. The gradients of the displacements, $u = x - X$, $v = y - Y$, are therefore not small with respect to one; they are of the order of two. The strains $\varepsilon_{xx} = (1 + P/2R)P/R$, $\varepsilon_{xy} = \varepsilon_{yy} = 0$, on the other hand, are small, since $|P|/R \ll 1$.

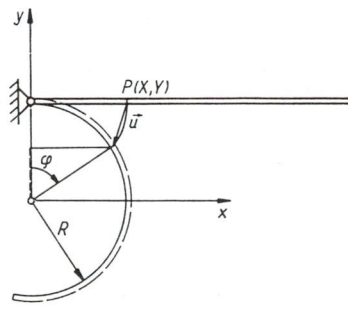

Fig. A 1.5.

A 1.6: Find the geometric relations, Eq. (1.20), and their linearized form, Eq. (1.21), in cylindrical and spherical coordinates, respectively.

Solution: The material coordinates of point P are (R, Φ, Z) and (R, Φ, Θ) before and (r, φ, z), (r, φ, θ) after the deformation, respectively [see also Eqs. (1.7) and (1.8)]. The displacement in the radial direction is commonly denoted by $u = r - R$; the other coordinate increments are χ, w and χ, ψ (angular increments), respectively. The squared arc-elements before and after deformation, due to the orthogonal coordinates, are given by

$$(dl_0)^2 = dR^2 + (R'd\Phi)^2 + dZ'^2 \quad, \quad (dl)^2 = dr^2 + (r'd\varphi)^2 + dz'^2 \quad,$$

where

$$R' = \left\langle \begin{matrix} R \\ R\sin\Theta \end{matrix} \right. \quad, \quad dZ' = \left\langle \begin{matrix} dZ \\ R\,d\Theta \end{matrix} \right. \quad, \quad r' = \left\langle \begin{matrix} r \\ r\sin\theta \end{matrix} \right. \quad,$$

$$dz' = \left\langle \begin{matrix} dz \ldots \text{ cylindrical} \\ r\,d\theta \ldots \text{ spherical} \end{matrix} \right\rangle \text{coordinates} \quad.$$

Considering material coordinates renders the strain components by comparing coefficients in the relation,

$$\frac{1}{2}(dl^2 - dl_0{}^2) = \varepsilon_{rr}\,dR^2 + \varepsilon_{\varphi\varphi}(R'd\Phi)^2 + \varepsilon_{z'z'}\,dZ'^2$$
$$+ 2(\varepsilon_{r\varphi}\,R'\,dR\,d\Phi + \varepsilon_{\varphi z'}\,R'd\Phi\,dZ' + \varepsilon_{rz'}\,dZ'\,dR) \quad,$$

where subscript z' equals z for cylindrical coordinates and is to be substituted by θ for spherical coordinates. Only the linearized expressions are given below, where $\chi = v/r'$ and $\psi = w/r$

$$\varepsilon_{rr} = \frac{\partial u}{\partial r} \quad, \quad \varepsilon_{\varphi\varphi} = \frac{u}{r} + \frac{1}{r'}\frac{\partial v}{\partial\varphi} + \alpha\frac{w}{r}\cot\theta \quad, \quad \varepsilon_{z'z'} = \alpha\frac{u}{r} + \frac{\partial w}{\partial z'} \quad,$$

$$2\,\varepsilon_{r\varphi} = \frac{1}{r'}\frac{\partial u}{\partial\varphi} + r'\frac{\partial(v/r')}{\partial r} \quad, \quad 2\,\varepsilon_{\varphi z'} = r'\frac{\partial(v/r')}{\partial z'} + \frac{1}{r'}\frac{\partial w}{\partial\varphi} \quad,$$

$$2\,\varepsilon_{rz'} = \frac{\partial u}{\partial z'} + \frac{\partial w}{\partial r} - \alpha\frac{w}{r} \quad.$$

and $\alpha = 0$ and subscript $z' = z$ apply in the case of cylindrical coordinates, $\alpha = 1$ and subscript $z' = \theta$ are proper for the spherical reference system. Point or axial symmetry considerably simplifies the expressions to

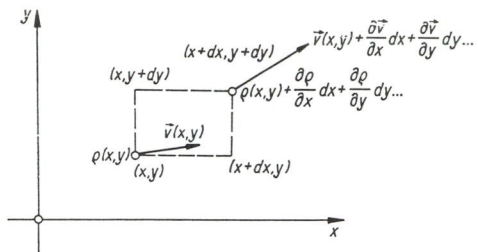

Fig. A 1.7.

$$\varepsilon_{rr} = \partial u/\partial r \quad , \quad \varepsilon_{\varphi\varphi} = (\varepsilon_{\theta\theta} =) \, u/r \quad , \quad (\varepsilon_{zz} = \partial w/\partial z) \quad .$$

A 1.7: Determine the equation of continuity (1.75) for a two-dimensional compressible flow by considering the balance of the mass flow rate in a plane, differential control volume *(dx, dy)* . See Fig. A 1.7.

Hint: The first-order expansion in the spatial coordinates suffices to derive the complete equations when keeping in mind the subsequent limiting process of contraction to a point. Note the nonstationary changing mass in the differential control volume first.

Solution: The net rate of outflux of mass (per unit of length of the *z* coordinate) according to Fig. A 1.7 is, the terms which cancel each other are not shown,

$$\frac{\partial(\rho v_x)}{\partial x} \, dx \, dy \; + \; \frac{\partial(\rho v_y)}{\partial y} \, dy \, dx \; + \ldots$$

The absence of any external source of mass supply requires an equally nonstationary decreasing rate of mass inside the control volume $- \, (\partial\rho/\partial t) \, dx \, dy$. Division through $dx \, dy$ and taking the limit $dx \to 0, \, dy \to 0$ yield the local *eulerian* continuity equation

$$\frac{\partial\rho}{\partial t} + \frac{\partial(\rho v_x)}{\partial x} + \frac{\partial(\rho v_y)}{\partial y} = 0 \quad .$$

A 1.8: Determine the strain ε_x of a fiber which is oriented parallel to the axis of a tensile rod of constant cross-section, $dl_0 = dX$, which remains parallel to the axis after deformation, $dl = dx$. Calculate the average strain from the measured elongation Δl of control length l_0 . Finally, find the relation between the normal strain component ε_{xx} and the specific elongation ε_x from $(1/2) \, (dl^2 - dl_0^2) = \varepsilon_{xx} \, dX^2$.

Solution: Two neighboring cross-sections in distance dX are differently displaced during deformation by $u(X)$ and $u(X + dX) = u(X) + du$, respectively. The fiber between the cross-sections of initial length $dl_0 = dX$ is elongated, $dl = dX + du$. The specific elongation, the stretch of the fiber, is given by

$$\varepsilon_x(X) = \frac{dl - dl_0}{dl_0} = \frac{dX + du - dX}{dX} = \frac{du}{dX} \ .$$

The elongation of a finite control length l_0 is determined by integration,

$$\Delta l = \int_0^{l_0} \varepsilon_x \, dX = \overline{\varepsilon_x} \, l_0 \quad ,$$

and the average specific elongation becomes

$$\overline{\varepsilon_x} = \frac{\Delta l}{l_0} \ .$$

By comparing coefficients in the expression

$$(1/2)\,(dl^2 - dl_0^2) = (1/2)\,(2\,du\,dX + du^2) = [(du/dX) + (1/2)\,(du/dX)^2]\,dX^2 = \varepsilon_{xx}\,dX^2 \ ,$$

the nonlinear relation of the strain results,

$$\varepsilon_{xx} = \varepsilon_x + (1/2)\,\varepsilon_x^2 \ .$$

In the case of $|\varepsilon_x| \ll 1$, linearization renders $\varepsilon_{xx} = \varepsilon_x$.

A 1.9: By means of three *strain gauges* (a commercially available rosette), the small normal strains ε_x , ε_y , ε_ξ (the ξ-axis is inclined at 45°) are measured at a point of a traction-free surface of a body (with the local normal z). Determine *Mohr´s* strain circle in that plane and, hence, the principal strains $\varepsilon_{1,2}$ as well as the directions of the principal strain axes in the *(x, y)* plane under the assumption $\varepsilon_{xz} = \varepsilon_{yz} = 0$, $\varepsilon_3 = \varepsilon_{zz}$.

Solution: The center of the circle is located at $(\varepsilon_x + \varepsilon_y)/2$. Equation (1.59) applies since the ξ-axis is in the *(x, y)* plane. Putting $2\,\alpha = \pi/2$ renders the small angle of the shear deformation

$$\gamma_{xy} = 2\,\varepsilon_{xy} = 2\,\varepsilon_\xi - (\varepsilon_x + \varepsilon_y) \ .$$

The three-dimensional problem is reduced to the plane (x, y) since ε_{zz} is assumed to be a principal strain. Equation (1.62) yields the in-plane principal strains

$$\varepsilon_{1,2} = (1/2) (\varepsilon_x + \varepsilon_y) + (1/2) [(\varepsilon_x - \varepsilon_y)^2 + \gamma_{xy}^2]^{1/2} \quad ,$$

and the radius of *Mohr's* circle, $|\varepsilon_1 - \varepsilon_2|/2$. The directions of the principal strain axes are determined by Eq. (1.61). The graphic solution in the $(\varepsilon_x , \varepsilon_{xy})$ plane provides a quick check of the numerical results. The extreme of the shear strain is given by $|\varepsilon_1 - \varepsilon_2|/2$ in the (x, y) plane, or, according to the solution of the fully three-dimensional problem by $|\varepsilon_1 - \varepsilon_3|/2$.

2
Statics, Systems of Forces, Hydrostatics

In Kinematics, the geometry of motion is studied without consideration of the sources of the deformation or acceleration. From observations, it is concluded that the driving agent of any change of the state of motion is a force. Thus, the notion of force is at first heuristically introduced, and later physical definitions of force are given by *Newton´s law* or through the notion of power or work (and energy). Another definition of force is given by the flux of a potential field, the gradient, which is used also in nonmechanical fields like thermodynamics, chemistry, and electrical sciences for the driving agent. In this section, the geometry of forces is studied, eg by considering the reduction of a system of forces and the equilibrium state; thereby, the occurrence of a double force with moment is detected. Theory and applications presented in this chapter are basic to the field of Statics and, hence, for the equilibrium of balanced forces acting on bodies at rest.

2.1. Forces, Body-Forces, Tractions, Stresses, Equilibrium

At the beginning, some experience with gravitation, ie with the mutual attraction between all masses and particles of matter in the universe, is helpful. For two masses and *Newton´s* law of gravitation, see Eq. (3.23). For our sake it suffices, however, to consider the *gravity* of a mass in the gravitational field of the Earth, the force called *weight*, $G = mg$ ($g = 9.80665$ m/s^2 is the precise average of acceleration of free fall at the "surface" of the Earth, the technical mean is $g = 9.81$ m/s^2 ; the rough approximation of $g = 10$ m/s^2 is sometimes used). Since g varies between a maximum of 9.832 m/s^2 at the Poles and a minimum of 9.780 m/s^2 at the Equator, the weight of a mass changes with location. *Mass m* is an intrinsic property of a body (within the limits of newtonian mechanics), weight G is not. The force of gravity is the source of downward motion in free fall and visibly deforms slender structures, eg under self-weight loading conditions. Its value is

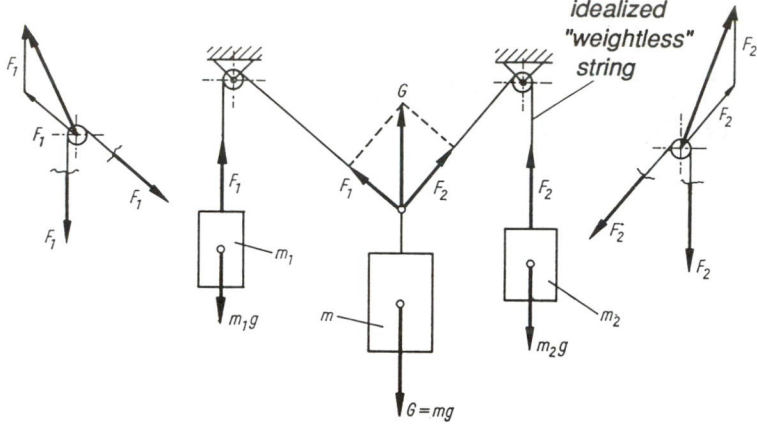

Fig. 2.1. Determination of nonvertical forces through the force of gravity

measured in derived SI units in *newton, 1N = 1kg x 1m/s²* (one kilogram mass corresponds to *2.2046 lb* , one meter is equal to *39.3701 in.*); the direction is vertically downward.

Forces in other directions are simply derived by the cable support of a mass. Free-fall motion is avoided by the equilibrium of the three forces acting on the mass: the self-weight *G* and the tensile forces F_1 and F_2 acting in the suspension cables to the left and to the right in Fig. 2.1, respectively. The value of the forces F_k , *k = 1, 2* , can be checked by running the cable over reel supports and by attaching proper masses at the ends of the vertically oriented straight parts. According to the simple experiment of Fig. 2.1, equilibrium is achieved when the tensile forces in the cable are added according to the triangle or parallelogram rule like vectors to render the resulting force −*G* that corresponds to an equivalent support of the mass *m* by a single vertically stretched rope.

Hence, a simple definition can be given. <<A force is a vector that can be combined with the force of gravity according to the rules of adding two vectors.>> The vector of a single force has three properties: its absolute value measured in *newton*, direction in space, and point of application at a body.

In the example of Fig. 2.1, those single forces were detected through the resulting actions of distributed forces, eg *G* denotes the total weight of the mass *m* that is the sum of the small gravity forces acting on all the mass elements *dm = ρ dV* . If we assume all the forces *gdm* are parallel (ie a locally "flat" Earth), integration over the volume renders

$$G = \int_m g \, dm = \int_V \rho g \, dV \ .$$

(2.1)

ρ is the mass density, $[\rho] = kg/m^3$. Hence, the gravity forces are distributed over the volume V of the body with a force density of ρg $= \gamma$ (the weight per unit of volume or the specific weight) and are acting at all material points in the vertical direction downward. Such *body forces* are spatially distributed and are defined in general by the force density per unit of volume $k(x, y, z; t)$, a vector field. The dimension is $[k] = N/m^3$. The points of application are the material points.

When we consider the cross-section of the cable stretched by the force F in the axial direction and the parallel system of small forces σdA acting at every cross-sectional element dA in the same direction, the result is the sum

$$F = \int_A \sigma \, dA \quad , \tag{2.2}$$

where σ is the force density per unit of area, the (normal) *stress*, with the point of application coinciding with the material points in the cross-section, $[\sigma] = N/m^2$. Due to its importance, the derived SI unit is called *pascal* ($1 \, Pa = 1 \, N/m^2$). Forces per unit of area distributed over a surface, the *tractions* , are in general given by stress vectors that are not normal to the area element. Thus, a *stress vector* has four characteristic properties: its value, direction, point of application (like a force) and in addition, the spatial orientation of the area element. The traction t_n or the *Cauchy* stress vector σ_n carry the subscript n of the normal e_n of the area element and are mathematically defined by the limit

$$\sigma_n = \lim_{\Delta S_n \to 0} \frac{\Delta F}{\Delta S_n} = \frac{dF}{dS_n} \quad , \tag{2.3}$$

ΔF denotes the resulting force on the generally curved area element ΔS with the normal e_n after taking the limit.

2.1.1. Stresses in a Tensile Rod: Mohr's Circle

A tensile rod with a constant cross-section of area A is stretched by a load F in the axial direction x. With $F = F \, e_x$ and the assumption of uniformly distributed stresses over the cross-section, the *uniaxial stress state* is given by

$$\sigma_x = \frac{F}{A} \, e_x = \sigma_{xx} \, e_x \quad . \tag{2.4}$$

A is the deformed plane cross-sectional area with the normal (after deformation) $e_n = e_x$; $\sigma_{xx} = F/A$ is the normal stress (normal to the area element). Considering a plane section with normal $e_n = e_x \cos \alpha + e_y \sin \alpha$ and area $A' = A/\cos \alpha$ renders the stress vector

$$\sigma_n = \frac{F}{A'} \, e_x = \frac{F \cos \alpha}{A} \, e_x = \sigma_{xx} \cos \alpha \, e_x \quad .$$

(2.5)

It has the same direction as σ_x , and, hence, also that of the normal force $F = F \, e_x$, but the value has changed to $\sigma_{xx} \cos \alpha$ and the orientation is inclined to A' . The stress vector or traction σ_n is decomposed into the normal-stress component orthogonal to A' and into the shear-stress component pointing in the direction e_m in A' ,

$$e_n = \cos \alpha \, e_n + \sin \alpha \, e_m \quad ,$$

$$\sigma_n = \sigma_{xx} \cos^2 \alpha \, e_n + \sigma_{xx} \cos \alpha \sin \alpha \, e_m = \sigma_{nn} \, e_n + \sigma_{nm} \, e_m \quad .$$

(2.6)

Hence, the normal stress is still uniformly distributed over A' but with intensity $\sigma_{nn} = (\sigma_{xx}/2)(1 + \cos 2\alpha)$, and the shear stress becomes $\sigma_{nm} = (\sigma_{xx}/2) \sin 2\alpha$. In the stress plane $(\sigma_{nn}, \sigma_{nm})$, *Mohr's* stress circle is derived, centered on the abscissa at $\sigma_{xx}/2$ and passing through the origin (see Fig. 2.2). For a more general situation, see the section 2.1.2 on the plane stress state.

Thus, the maximal normal stress $\sigma_{xx} = F/A$ acts in a cross-section that is free of any shear. Any other skew section carries normal and shear stresses. The largest shear stress equals $\sigma_{xx}/2$ in the section with $\alpha = 45°$. The results can be applied (with care, see Sec. 9.1.4) to a column with compressive loading $F < 0$. The stress circle moves to the left of the origin. In materials that are sensitive

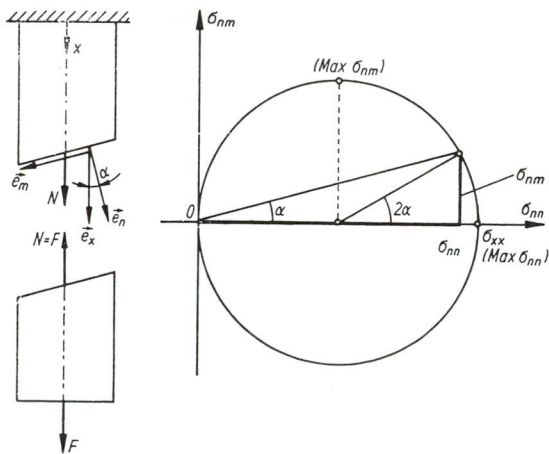

Fig. 2.2. Tensile rod and *Mohr's* circle of the uniaxial stress state

to tension, fracture may occur in the cross-section. In shear-sensitive materials, however, fracture occurs in sections inclined to the axis.

Due to the dependence of the stress vector σ_n on the normal e_n, a superposition of stresses by the addition of their vectors is possible only in the case of a common point of application. Thus, stress components in a point can be added if the first of their subscripts is the same. Forces with a common point of application can be added according to vector rules.

The *principle of virtual sections* has been quietly applied during the above definition of forces and during the discussion of stresses. Thus, it is necessary during an investigation of internal forces to virtually cut a body in motion into sections, keeping the time constant. Since a continuum is considered, the pieces can be made infinitesimally small; it is assumed that the limits $\Delta V \rightarrow 0$ and $\Delta S \rightarrow 0$ exist and that the virtual parts are separate. To keep the motion and deformation unchanged, tractions have to be applied to the material surfaces with the same distribution as the internal forces acting at the corresponding points of application, or with proper correspondence, the tractions are the stress vectors with respect to the surface elements. It is easily recognized that the stress vectors in corresponding surface elements must be equal in value but opposite in direction (see Fig. 2.3). This *law of equal action and reaction* in opposite points of corresponding material surfaces is named after *Euler* and *Cauchy* . It can be generalized at large to certain force fields and especially it holds (also in the far-field) for the forces of attraction between two masses with separate centers, the gravity forces. Figure 2.3(b) illustrates equal

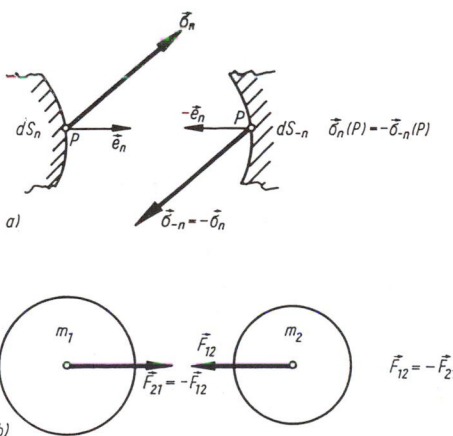

a)

b)

Fig. 2.3. (a) Law of equal action and reaction for the stress vector. (b) *Newton´s third law* for the gravity force

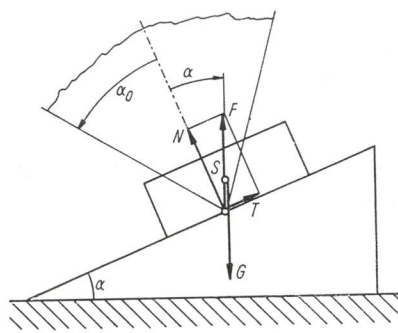

Fig. 2.4. A rigid body at rest in contact with an inclined plane. The cone of static friction.

attraction according to *Newton´s* (third) law of universal gravitation; see Eq. (3.23).

In analogy to the cable support of Fig. 2.1, the equilibrium of a body resting on a dipping plane is considered. Figure 2.4 shows the weight G in equilibrium with the resulting force F of the complex and unknown stress distribution exerted from the underlying wedge upon the material surface in the contact zone. Thus, $F - G = 0$ and the forces act along the same line. The projection of the force F onto the surface normal is $N = G \cos \alpha$ and the tangential component is $T = G \sin \alpha$. Experience shows that a limitation of the shear component exists when α is increasing. A maximum of α, the *angle of static friction* α_0 , limits the range of equilibrium. The condition of static friction may be expressed by the inequality $\alpha < \alpha_0$ or $T < T_0 = G \sin \alpha_0 = N \tan \alpha_0$. The *coefficient of static friction* $\mu_0 = \tan \alpha_0$ depends on the smoothness and cleanness of the contacting surfaces and on the materials in contact. The numerical values for absolutely dry contact are determined experimentally often in connection with abrasive testing (a field called tribology) and are available in tabulated form. For reasons of safety, μ_0 is set equal to the generally smaller value of the coefficient of friction μ in sliding contact [in that case *Coulomb´s* law of dry friction $T = T_R$, $T_R = \mu N$ holds with constant coefficient for a limited range of "slow" relative velocities; see Sec. 7.4.8 (§)]. If we assume that μ is known, the condition of static friction is easily proved by drawing a cone of half aperture α_0 , with its axis in the N -direction and tip at the point of application of F in the zone of contact. Equilibrium is save as long as F points into the interior of this *cone of static friction*. Some average values of μ are the following: for steel, *0.1* ; for steel contacting bronze, *0.16* ; for metal or wood in contact with wood, *0.5* and so on.

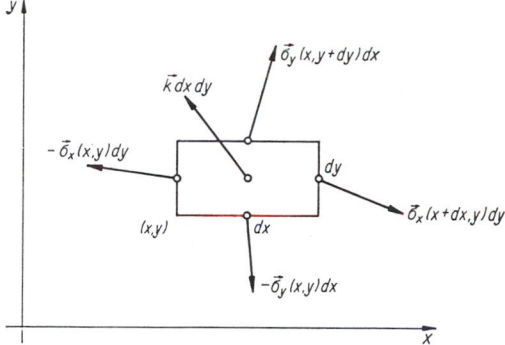

Fig. 2.5. Plane stress state. In-plane forces per unit of length acting on a plate element

2.1.2. Plane State of Stress: Mohr's Circle

More experience with stresses and the principle of virtual sections is gained when the plane stress state is considered. Such a stress state is encountered in the points of a traction-free surface, and approximately in thin plates with in-plane loading. In Fig. 2.5, the free-body diagram of an infinitesimal plate element with in-plane forces per unit of length is shown, $\sigma_z = \boldsymbol{0}$. It is important to notice the subscripts of the stress vectors that are attached to the surface elements when the normal points in the coordinate direction. For example, the element at $(x + dx)$ with outward normal $\boldsymbol{e}_x$ carries the positive traction σ_x; that element at (x) with normal $-\boldsymbol{e}_x$ is loaded by $-\sigma_x$. At the limit $dx \to 0$, the law of action and reaction must hold. All forces are reduced to the common point of application (x, y) (see Fig. 2.10) and their vectors are added to gain the resultant $\boldsymbol{dR}$

$$[\sigma_x(x + dx, y) - \sigma_x(x, y)] \, dy + [\sigma_y(x, y + dy) - \sigma_y(x, y)] \, dx + \mathbf{k} \, dx \, dy = \mathbf{dR} \quad .$$

It is sufficient to substitute the linear approximation of the stress increments to render with $dA = dx \, dy$

$$\left[\frac{\partial \sigma_x}{\partial x} + \frac{\partial \sigma_y}{\partial y} + \mathbf{k} \right] dA = \mathbf{dR} \quad .$$

Dividing through the area element dA and subsequently taking the limit $dx, \, dy \to 0$ give the exact expression of the force density $\boldsymbol{f}$

$$\frac{\partial \boldsymbol{\sigma}_x}{\partial x} + \frac{\partial \boldsymbol{\sigma}_y}{\partial y} + \mathbf{k} = \mathbf{f} \; ,$$

$$(2.7)$$

since the higher-order terms of the *Taylor* expansion of the stress increments go to zero. A necessary condition of equilibrium in that point is $\mathbf{f} = \mathbf{0}$. If the material point (x, y) remains at rest, the vector condition must hold

$$\frac{\partial \boldsymbol{\sigma}_x}{\partial x} + \frac{\partial \boldsymbol{\sigma}_y}{\partial y} + \mathbf{k} = 0 \; .$$

$$(2.8)$$

It is equivalent to two local differential conditions of equilibrium. Substituting the stress vector

$$\boldsymbol{\sigma}_i = \sigma_{ix}\, \mathbf{e}_x + \sigma_{iy}\, \mathbf{e}_y \quad , \quad i = x, y \; ,$$

$$(2.9)$$

yields,

$$\frac{\partial \sigma_{xx}}{\partial x} + \frac{\partial \sigma_{yx}}{\partial y} + k_x = 0 \; , \quad \frac{\partial \sigma_{xy}}{\partial x} + \frac{\partial \sigma_{yy}}{\partial y} + k_y = 0 \; .$$

$$(2.10)$$

$\mathbf{k}$ is the externally applied body force. It will be shown below that $\sigma_{yx} = \sigma_{xy}$ follows from an additional consideration that becomes necessary when moving forces to another point of application. If the body force vanishes, or if it is equal to a constant $\mathbf{k} = \mathbf{c}$, the partial differential equations (2.10) are solved identically by means of the *Airy stress function* $F(x, y)$ and by putting

$$\sigma_{xx} = \frac{\partial^2 F}{\partial y^2} \quad , \quad \sigma_{yy} = \frac{\partial^2 F}{\partial x^2} \quad , \quad \sigma_{xy} = \sigma_{yx} = -\frac{\partial^2 F}{\partial x \partial y} - y\, c_x - x\, c_y \quad .$$

$$(2.11)$$

Instead of the determination of the plane stress state by the three stress components $\sigma_{xx}, \sigma_{yy}, \sigma_{xy}$, it suffices to know the scalar function $F(x, y)$ analytically and to apply Eq. (2.11) subsequently. The plane stress state is statically determinate if the stress function is derived from the boundary conditions of the plate. For in-plane loaded elastic plates, see Sec. 6.5.

The plane stress state in a point (x, y) is given by the 2 x 2 stress tensor or by the two stress vectors $\boldsymbol{\sigma}_x, \boldsymbol{\sigma}_y$. Thus, it must be possible to calculate the stress vector acting upon a rotated surface element with normal $\mathbf{e}_n$ rotated through the angle α against the x - axis. According to Fig. 2.6, a rectangular triangle of infinitesimal extensions is considered, dx, dy, ds, loaded by surface and body forces. Equilibrium requires

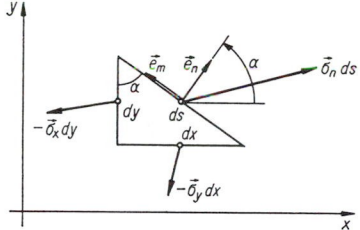

Fig. 2.6. Plane stress state. Tractions per unit of length on the surface of a triangular plate element

$$\mathbf{dR} = (1/2)\,\mathbf{k}\,dx\,dy - \vec{\sigma}_x(x, y)\,dy - \vec{\sigma}_y(x, y)\,dx + \vec{\sigma}_n\,ds = \mathbf{0}\quad.$$

Dividing through ds and subsequently taking the limit $dx, dy \to 0$, eliminate the external body force $\mathbf{k}$ and render the general *Cauchy-*traction, with $dx/ds = \sin \alpha,\ dy/ds = \cos \alpha$,

$$\vec{\sigma}_n = \vec{\sigma}_x \cos \alpha + \vec{\sigma}_y \sin \alpha = n_x\,\vec{\sigma}_x + n_y\,\vec{\sigma}_y\quad. \tag{2.12}$$

The normal stress points in the direction $\mathbf{e}_n = n_x\,\mathbf{e}_x + n_y\,\mathbf{e}_y$, $\vec{\sigma}_n \cdot \mathbf{e}_n =$

$$\sigma_{nn} = \sigma_{xx}\,n_x{}^2 + \sigma_{yy}\,n_y{}^2 + 2\,\sigma_{xy}\,n_x\,n_y =$$

$$\frac{\sigma_{xx} + \sigma_{yy}}{2} + \frac{\sigma_{xx} - \sigma_{yy}}{2}\cos 2\alpha + \sigma_{xy}\sin 2\alpha\,. \tag{2.13}$$

The shear-stress component in the perpendicular direction, $\mathbf{e}_m = m_x\,\mathbf{e}_x + m_y\,\mathbf{e}_y$ is given by

$$\sigma_{nm} = \sigma_{xx}\,n_x\,m_x + \sigma_{yy}\,n_y\,m_y + \sigma_{xy}\,(n_x\,m_y + n_y\,m_x) =$$

$$-\frac{\sigma_{xx} - \sigma_{yy}}{2}\sin 2\alpha + \sigma_{xy}\cos 2\alpha\,. \tag{2.14}$$

The coefficients are $n_x = \cos \alpha,\ n_y = \sin \alpha,\ m_x = -\sin \alpha,\ m_y = \cos \alpha$.
By arranging the stress components in a 2 x 2 matrix

$$\Sigma = \begin{pmatrix} \sigma_{xx} & \sigma_{xy} \\ \sigma_{yx} & \sigma_{yy} \end{pmatrix}, \tag{2.15}$$

and considering the transformation rules of the elements under coordinate rotation, the tensor property of the *plane stress tensor* becomes evident. In analogy to the plane strain tensor, the principal

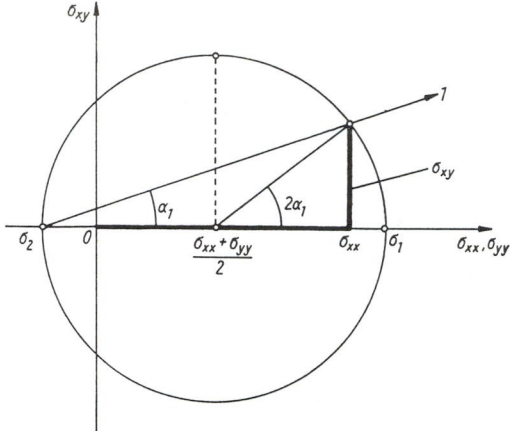

Fig. 2.7. *Mohr's* stress circle of the plane stress state

stress axes are determined by Eq. (1.61) and the principal normal stresses become [see Eq. (1.62)]

$$\sigma_{1,2} = \frac{\sigma_{xx} + \sigma_{yy}}{2} \pm \frac{1}{2} \left[(\sigma_{xx} - \sigma_{yy})^2 + 4\,\sigma_{xy}^2 \right]^{1/2} , \qquad (2.16)$$

acting on the surface elements free of any shear, $\sigma_{12} = 0$, with orientations

$$\tan 2\alpha_1 = \frac{2\sigma_{xy}}{\sigma_{xx} - \sigma_{yy}} \quad , \quad \alpha_2 = \alpha_1 + \pi/2 \quad , \qquad (2.17)$$

Mohr's stress circle is drawn in the $[(\sigma_{xx}, \sigma_{yy}), \sigma_{xy}]$ -plane with its center at $(\sigma_{xx} + \sigma_{yy})/2$ and points of the circle at $(\sigma_{xx}, \sigma_{xy})$ or $(\sigma_{yy}, \sigma_{xy})$ according to Fig. 2.7 (see also Fig. 1.13). The direction of the principal stress axis 1 against the x -axis is shown. An extreme value of the shear stress is found in the element rotated 45° as $(\sigma_1 - \sigma_2)/2$. If both principal normal stresses are tensile, $\sigma_1 > \sigma_2 > 0$, the largest shear stress is $\sigma_1/2$ in the out-of-plane element; see the three-dimensional stress state, Eq. (2.31), $k = 2$, $\sigma_3 = 0$.

2.1.3. General State of Stress

Generalization to three dimensions remains to be done. The force density is derived by considering a small volume element dx, dy, dz, in equilibrium. Positive tractions σ_x $(x + dx, y, z)$, σ_y $(x, y + dy, z)$, and σ_z $(x, y, z + dz)$ are attached to the (positive) surface elements with normal vectors e_x , e_y , e_z , respectively. The opposite sections carry the negative stress vectors according to the common point *(x, y, z)* . Vector addition (*k* is the given body force per unit of volume) renders the resultant force

dF = **k** dx dy dz + [σ_x(x + dx, y, z) − σ_x(x, y, z)] dy dz + [σ_y(x, y + dy, z) − σ_y(x, y, z)] dz dx + [σ_z(x, y, z + dz) − σ_z(x, y, z)] dx dy .

The difference in neighboring tractions is sufficiently well, ie linearly approximated, and after division through the volume element *dV = dx dy dz* , the limit *dx* , *dy* , *dz* → *0* is performed to render the exact expression of the force density in the point *(x, y, z)*

$$f = \frac{dF}{dV} = k + \frac{\partial \sigma_x}{\partial x} + \frac{\partial \sigma_y}{\partial y} + \frac{\partial \sigma_z}{\partial z} \quad .$$

$$(2.18)$$

The necessary condition of equilibrium in vector form is *f = 0* , and the three conditions of equilibrium of the three-dimensional stress state in subscript notation become accordingly,

$$\sum_{j=1}^{3} \frac{\partial \sigma_{ji}}{\partial x_j} + k_i = 0 \quad , \quad i = 1, 2, 3 \quad .$$

$$(2.19)$$

In Sec. 2.2.2, the symmetry of the stress tensor is shown, $\sigma_{ji} = \sigma_{ij}$, to hold in general in a point-continuum.

The three-dimensional stress state at a material point *(x, y, z)* is determined by the three stress vectors σ_x , σ_y , σ_z . That is, the stress vector σ_n on a sectional element *dA* with orientation e_n is determined by these three stress vectors by considering the equilibrium of a tetrahedral element according to Fig. 2.8. The

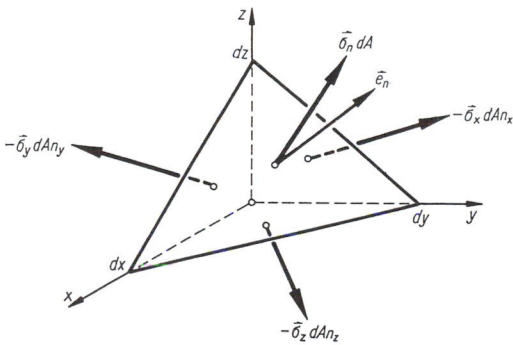

Fig. 2.8. Three-dimensional stress state. Free-body diagram of a tetrahedral element

vector sum of the surface tractions and the body force $\mathbf{k} \, dA \, dh/3$ must vanish

$$\mathbf{dF} = \mathbf{k} \, dA \, dh/3 - \sigma_x \, dA \, n_x - \sigma_y \, dA \, n_y - \sigma_z \, dA \, n_z + \sigma_n \, dA = 0 \quad.$$

Dividing through dA , subsequently followed by the limits $dA \to 0$ and height $dh \to 0$, yields the important formula for the stress vector (the *Cauchy* -traction) referred to a sectional element of arbitrary orientation, the normal is $\mathbf{e}_n$,

$$\sigma_n = n_x \, \sigma_x + n_y \, \sigma_y + n_z \, \sigma_z \quad. \tag{2.20}$$

In tensor notation

$$\sigma_{ni} = \sum_{j=1}^{3} \sigma_{ji} \, n_j \quad, \quad i = 1, 2, 3 \quad, \tag{2.21}$$

n_j denotes the direction cosine of the unit normal vector $\mathbf{e}_n$ in the cartesian frame (x, y, z) . Hence, the *normal stress component* becomes, we take the scalar product $(\sigma_n \cdot \mathbf{e}_n) = \sigma_{nn}$, $\sigma_{ij} = \sigma_{ji}$,

$$\sigma_{nn} = \sigma_{xx} \, n_x^2 + \sigma_{yy} \, n_y^2 + \sigma_{zz} \, n_z^2 + 2 \, (\sigma_{xy} \, n_x \, n_y + \sigma_{yz} \, n_y \, n_z + \sigma_{zx} \, n_z \, n_x) \quad. \tag{2.22}$$

The *shear stress component* in the tangential direction $\mathbf{e}_m \perp \mathbf{e}_n$, where $\mathbf{e}_m \cdot (\mathbf{e}_n \times \sigma_n) = 0$, is the projection given by $(\sigma_n \cdot \mathbf{e}_m) = \sigma_{nm}$,

$$\sigma_{nm} = \sigma_{xx} \, n_x \, m_x + \sigma_{yy} \, n_y \, m_y + \sigma_{zz} \, n_z \, m_z + \sigma_{xy} \, (n_x \, m_y + n_y \, m_x) + \\ \sigma_{yz} \, (n_y \, m_z + n_z \, m_y) + \sigma_{zx} \, (n_z \, m_x + n_x \, m_z) \quad. \tag{2.23a}$$

Its vector is also conveniently determined by the vector triple product:

$$\sigma_{nm} \, \mathbf{e}_m = \sigma_n - \sigma_{nn} \, \mathbf{e}_n = \mathbf{e}_n \times (\sigma_n \times \mathbf{e}_n) \quad. \tag{2.23b}$$

Above, the transformation formulas of tensor elements under coordinate rotations are recognized. The *Cauchy* stress tensor is symmetric, see Sec. 2.2.2,

$$\Sigma = \begin{pmatrix} \sigma_{xx} & \sigma_{xy} & \sigma_{xz} \\ \sigma_{yx} & \sigma_{yy} & \sigma_{yz} \\ \sigma_{zx} & \sigma_{zy} & \sigma_{zz} \end{pmatrix} \quad, \tag{2.24}$$

and determines the stress state at a point (x, y, z) [the traction is given by the linear vector transformation $\sigma_n = \Sigma \cdot \mathbf{e}_n$ (see Eq. 2.20)]. Associated are three mutually orthogonal, principal stress axes.

Hence, the mathematical notion of a tensor is interpreted, since the properties were found historically by the consideration of a stress state. The corresponding sectional elements of the principal normal stresses are free of any shear stress

$$\Sigma = \begin{pmatrix} \sigma_1 & 0 & 0 \\ 0 & \sigma_2 & 0 \\ 0 & 0 & \sigma_3 \end{pmatrix} . \tag{2.25}$$

Those principal normal stresses (matrix eigen-values) are the roots of the characteristic cubic equation

$$-\sigma^3 + I_1 \sigma^2 - I_2 \sigma + I_3 = 0 \ , \tag{2.26}$$

where

$$I_1 = \sigma_{xx} + \sigma_{yy} + \sigma_{zz} = \sigma_1 + \sigma_2 + \sigma_3 \ , \tag{2.27}$$

$$I_2 = \begin{vmatrix} \sigma_{yy} & \sigma_{yz} \\ \sigma_{zy} & \sigma_{zz} \end{vmatrix} + \begin{vmatrix} \sigma_{xx} & \sigma_{xz} \\ \sigma_{zx} & \sigma_{zz} \end{vmatrix} + \begin{vmatrix} \sigma_{xx} & \sigma_{xy} \\ \sigma_{yx} & \sigma_{yy} \end{vmatrix} =$$

$$\sigma_1 \sigma_2 + \sigma_2 \sigma_3 + \sigma_3 \sigma_1 \ , \tag{2.28}$$

$$I_3 = \det\{\sigma_{ij}\} = \sigma_1 \sigma_2 \sigma_3 \ , \tag{2.29}$$

are the coefficients and, thus, determine these eigenvalues of the stress tensor: They remain constant during any rotation of the coordinate system and the coefficients, therefore, are called the *invariants of the stress tensor* . The first and second invariants are of major importance when considering isotropic constitutive laws.

The directions of the principal stress axes $n_j^{(k)}$ are the solutions of the linear system of homogeneous equations when substituting the principal normal stresses σ_k successively

$$\sum_{j=1}^{3} (\sigma_{ij} - \sigma_k \delta_{ij}) \, n_j^{(k)} = 0 \ , \quad \begin{array}{l} i = 1, 2, (3) \\ k = 1, 2, (3) \end{array} \tag{2.30}$$

with the auxiliary condition $\Sigma_j \, n_j^2 = 1$; δ_{ij} is the *Kronecker* symbol.

The sectional elements where the shear stresses take on the extreme values τ_1, τ_2, τ_3 have normal vectors rotated 45° against the principal stress axes

$$\tau_k = (1/2)|\sigma_i - \sigma_j| \ , \quad i, j, k = 1, 2, 3, \ k \neq i \neq j \ . \tag{2.31}$$

The normal stress components σ_k in those elements are given in terms of the principal normal stresses by

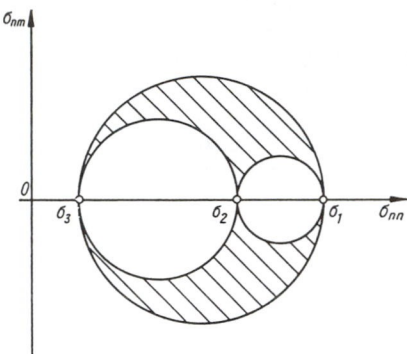

Fig. 2.9. *Mohr´s* circles of the three-dimensional stress state

$$\sigma_k = (1/2)(\sigma_i + \sigma_j) \quad , \quad i, j, k = 1, 2, 3, \quad k \neq i, k \neq j, i \neq j \quad . \tag{2.32}$$

Altogether, three of *Mohr´s* circles are associated with the three-dimensional stress tensor. The points of all possible pairs of stress components $(\sigma_{nn}, \sigma_{nm})$ in a given stress state fall within the hatched area in Fig. 2.9.

2.1.4. Mean Normal Stress and Stress Deviations

The stress tensor $\Sigma = (\sigma_{ij})$ is decomposed into a spherical tensor and the stress deviation. The deviatoric stress components, ie the deviations from the hydrostatic stress state that is given by the mean of the normal stresses at a material point *(x, y, z)*, denoted p, are,

$$p = I_1/3 = (1/3)(\sigma_{xx} + \sigma_{yy} + \sigma_{zz}) \quad , \tag{2.33}$$

$$s_{ij} = \sigma_{ij} - p\,\delta_{ij} \quad , \quad i, j = 1, 2, 3 \quad , \quad \delta_{ij} = \begin{cases} 1, & i = j \\ 0, & i \neq j \end{cases} . \tag{2.34}$$

Hence, the sum of the deviatoric normal stresses vanishes

$$\sum_{i=1}^{3} s_{ii} = \sum_{i=1}^{3} \sigma_{ii} - 3p \equiv 0 \quad .$$

The principal values of the deviatoric normal stresses are the roots of the reduced cubic equation

$$s^3 - J_2 s - J_3 = 0 \quad , \tag{2.35}$$

with coefficients equal to the remaining invariants under coordinate rotation J_2 and J_3 of the deviatoric tensor,

$$J_2 = 3p^2 - I_2 = (1/2) \sum_{i=1}^{3} \sum_{j=1}^{3} s_{ij}^2 = (3/2) \tau_0^2 ,$$

(2.36)

$$J_3 = 2p^3 - p\, I_2 + I_3 = (1/3) \sum_{i=1}^{3} \sum_{j=1}^{3} \sum_{k=1}^{3} s_{ij}\, s_{jk}\, s_{ki} .$$

(2.37)

τ_0 , the *octahedral shear stress* , has been originally introduced by *R. von Mises*. It is the resultant shear stress in sectional elements, with orientations under equal angles to the principal stress axes

$$n_1 = n_2 = n_3 = \frac{1}{\sqrt{3}} .$$

$$9\,\tau_0^2 = (\sigma_1 - \sigma_2)^2 + (\sigma_2 - \sigma_3)^2 + (\sigma_3 - \sigma_1)^2 = 2\,I_1^2 - 6\,I_2 ,$$

is proportional to the sum of the areas of *Mohr´s* circles. Applications of τ_0 are to be found in the theory of the plastic flow of ductile materials.

The state of *plane stress* is recognized to be a special case of the three-dimensional state of stress when $\sigma_{zi} = 0$, $i = x, y, z$. The deviatoric stress components in that case become by means of the mean normal stress

$$p = (\sigma_{xx} + \sigma_{yy})/3 ,$$

(2.38)

$$s_{xx} = \sigma_{xx} - p , \quad s_{yy} = \sigma_{yy} - p , \quad s_{xy} = \sigma_{xy} ,$$

(2.39)

and,

$$J_2 = 3p^2 - (\sigma_{xx}\, \sigma_{yy} - \sigma_{xy}^2) .$$

(2.40)

Since $\sigma_3 = 0$, the extreme values of the shear stress are

$$\tau_3 = (1/2)\,|\sigma_1 - \sigma_2| , \quad \tau_2 = (1/2)\,|\sigma_1| , \quad \tau_1 = (1/2)\,|\sigma_2| .$$

(2.41)

The associated sectional elements are oriented under 45° to the principal stress axes.

In the *uniaxial stress state*, the nonvanishing principal normal stress is $\sigma_{xx} = \sigma_1$; $\sigma_{zi} = 0$ and $\sigma_{yi} = 0$, $i = x, y, z$. The deviatoric stress component becomes with

$$p = \sigma_{xx}/3 ,$$

(2.42)

$$s_{xx} = (2/3)\ \sigma_{xx}\ ,\qquad (2.43)$$

and the second invariant takes on the value

$$J_2 = 3\ p^2 = \sigma_{xx}^2/3\ .\qquad (2.44)$$

2.2. Systems of Forces

The superposition of forces is a crucial static problem. In the case of the superposition of internal forces, the stresses with a common point of application, but with different sectional elements, must be multiplied with the proper area elements of reference before the vector summation renders the infinitesimal resultant force. In other considerations, such a system of external or internal forces with a common point of application, a *central force system*, is fundamental. The resultant force

$$R = \sum_{i=1}^{n} F_i\ ,$$
$$(2.45)$$

is attached to the common point and there it is "statically equivalent" to the central force system. Consequently, a central force system with $R = 0$ is self-equilibrating; the forces F_i, $i = 1,...n$, are in a state of equilibrium. A single joint of a truss or the head of a guyed tower (where the cables are attached) are examples of such a common point of application of the truss member forces or cable tensions, respectively.

The resultant force is calculated by considering the components in a cartesian reference frame. By means of the decomposition

$$F_i = X_i\ e_x + Y_i\ e_y + Z_i\ e_z\ ,\qquad (2.46)$$

the components of the resultant force are expressed by summing the individual components

$$R_x = \sum_{i=1}^{n} X_i\ ,\quad R_y = \sum_{i=1}^{n} Y_i\ ,\quad R_z = \sum_{i=1}^{n} Z_i\ ,$$
$$(2.47)$$

and the resultant force vector becomes

$$R = R_x\ e_x + R_y\ e_y + R_z\ e_z\ .\qquad (2.48)$$

Thus, a central force system has three conditions of equilibrium

$$R_x = \sum_{i=1}^{n} X_i = 0, \quad R_y = \sum_{i=1}^{n} Y_i = 0, \quad R_z = \sum_{i=1}^{n} Z_i = 0 \ .$$

$$(2.49)$$

The more general problem of the superposition of forces with noncommon points of applications is frequently encountered. In such a case, a parallel translation of each force to a common point is found to be necessary before the vector summation can be performed. In Fig. 2.10, this basic static problem is shown for a single force F attached to point A of a body. A simple parallel translation to another point A´ changes the static configuration. Thus, a special self-equilibrating central force system of F and $-F$ is added at the reference point A´ that (statically) does not alter the force F in point A. When considering the translated force F in A´, the pair of parallel forces, F in the actual point of application A and $-F$ in the reference point A´, is left over. Such a *double force with moment*, a *couple*, is a new static element of fundamental importance. Thus, the static action of the force F in A with respect to a reference point A´ is statically equivalent to the translated force F in A´ superposed by a couple, as described above. The static action of the couple is determined by the *moment Fa* , ie by the product of $|F|$ with the normal distance between F and $-F$, together with the orientation of the plane of the couple in space, ie the positive oriented (in the sense of a right-handed screw motion) normal vector e_n of the plane defined by the parallel lines through F and $-F$. Hence, the *moment vector*

$$\mathbf{M} = \mathbf{r} \times \mathbf{F} \ , \qquad\qquad (2.50)$$

determined by the vector product of a position vector r , pointing from a point on the line of action of $-F$ to a point on F , eg from A´ to A, with the given force vector F , has the desired properties: The hatched area of Fig. 2.10 equals

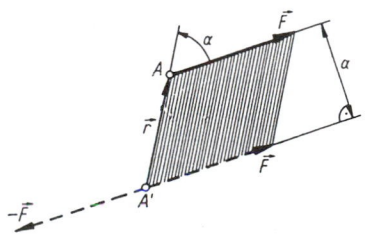

Fig. 2.10. Statically equivalent reduction of a single force F to a point of reference A´. Moment of a single force. Double force with moment, a couple

$$|\mathbf{M}| = |\mathbf{r}|\,|\mathbf{F}|\,\sin\alpha = a\,|\mathbf{F}| \quad , \quad [|\mathbf{M}|] = Nm \quad , \tag{2.51}$$

since $|\mathbf{r}|\,\sin\alpha = a$, and the unit vector

$$\mathbf{e}_n = \mathbf{M}/|\mathbf{M}| \quad , \tag{2.52}$$

has the proper orientation. The moment vector is a short-hand notation of a couple. It is a free vector and the point of application in the plane is not fixed; all parallel vectors of the same length are statically equivalent <<By definition, ($\mathbf{F}$ at A) and ($\mathbf{F}$ at A´ with $\mathbf{M}$) are statically equivalent.>>

The components of the moment vector in a cartesian reference frame with A(x, y, z), A´(x´,y´,z´) and

$$\mathbf{r} = (x - x')\,\mathbf{e}_x + (y - y')\,\mathbf{e}_y + (z - z')\,\mathbf{e}_z \quad ,$$

are given by expanding the determinant

$$\mathbf{M} = M_x\,\mathbf{e}_x + M_y\,\mathbf{e}_y + M_z\,\mathbf{e}_z = \begin{vmatrix} \mathbf{e}_x & \mathbf{e}_y & \mathbf{e}_z \\ x-x' & y-y' & z-z' \\ X & Y & Z \end{vmatrix}$$

$$= \begin{array}{l} [(y-y')\,Z - (z-z')\,Y]\,\mathbf{e}_x + \\ [(z-z')\,X - (x-x')\,Z]\,\mathbf{e}_y + \\ [(x-x')\,Y - (y-y')\,X]\,\mathbf{e}_z \,. \end{array} \tag{2.53}$$

and by inspection,

$$\begin{array}{l} M_x = [(y-y')\,Z - (z-z')\,Y] \quad , \\ M_y = [(z-z')\,X - (x-x')\,Z] \quad , \\ M_z = [(x-x')\,Y - (y-y')\,X] \quad . \end{array} \tag{2.54}$$

They represent the *axial moments* of the force $\mathbf{F}$ with respect to the axes through point A´ that are parallel to the *x, y, z* coordinates. Thus, the force component that parallels an axis does not contribute to the axial moment.

By means of the skew-symmetric distance matrix (in cartesian coordinates)

$$r = \begin{pmatrix} 0 & -(z-z') & (y-y') \\ (z-z') & 0 & -(x-x') \\ -(y-y') & (x-x') & 0 \end{pmatrix} \quad ,$$

the moment vector is defined by matrix multiplication with $\mathbf{F}$, that is a column matrix

$$\mathbf{M} = r \cdot \mathbf{F} \quad ,$$

(see the linear vector transformation $\mathbf{v}_{PA} = \mathbf{\Omega} \cdot \mathbf{r}_{PA}$, an expression discussed in Sec. 1.2) .

Consider a force system of n single forces $\mathbf{F}_i$ with points of application A_i, $i= 1, 2, ..., n$, at a space structure, with respect to static reduction to a reference point A' that, for convenience, is selected to be the origin of a cartesian coordinate system 0. Addition of a self-equilibrating central group of forces consisting of the pairs $\mathbf{F}_i + (-\mathbf{F}_i)$ at point 0 does not change the statics of the given force system. Hence, the statically equivalent reduction renders the central force system at 0 with the resultant

$$\mathbf{R} = \sum_{i=1}^{n} \mathbf{F}_i = R_x \, \mathbf{e}_x + R_y \, \mathbf{e}_y + R_z \, \mathbf{e}_z \text{ , at } A' \text{ ,}$$

$$R_x = \sum_{i=1}^{n} X_i \text{ , } R_y = \sum_{i=1}^{n} Y_i \text{ , } R_z = \sum_{i=1}^{n} Z_i \text{ ,}$$

(2.55)

superposed by n couples, $\mathbf{F}_i$ at A_i, $(-\mathbf{F}_i)$ at A', with the individual moment vectors

$$\mathbf{M}_i = \mathbf{r}_i \times \mathbf{F}_i \text{ , } (i = 1, 2, ...n) \text{ .}$$

$\mathbf{r}_i$ is, eg the position vector of the point of application A_i. It can be easily verified that the resulting moment is given by the sum of the individual moment vectors

$$\mathbf{M} = \sum_{i=1}^{n} \mathbf{M}_i = \sum_{i=1}^{n} \mathbf{r}_i \times \mathbf{F}_i = M_x \, \mathbf{e}_x + M_y \, \mathbf{e}_y + M_z \, \mathbf{e}_z \text{ .}$$

(2.56)

(Superposition of two couples with nonparallel planes of action is left as an exercise. A hint may be to shift two opposing forces into the common line of intersection of the two planes after properly manipulating and rotating the couples in their planes).

Also, the resulting axial moments are given by summing the individual axial moments of the forces $\mathbf{F}_i$

$$M_x = \sum_{i=1}^{n} y_i Z_i - z_i Y_i \text{ , } M_y = \sum_{i=1}^{n} z_i X_i - x_i Z_i \text{ , } M_z = \sum_{i=1}^{n} x_i Y_i - y_i X_i \text{ .}$$

(2.57)

<<The resultant force $\mathbf{R}$ attached to the reference point $A' = 0$ and the resulting moment $\mathbf{M}$ are statically equivalent to the space force system $\mathbf{F}_i$ at A_i, $i = 1, 2, ...n$.>>

The spatially distributed forces $\mathbf{F}_i$ are in a state of equilibrium if

$$\mathbf{R} = 0 \text{ and } \mathbf{M} = 0 \text{ .}$$

(2.58)

Thus, the *conditions of equilibrium* are the *six linear equations*

$$R_x = \sum_{i=1}^{n} X_i = 0 \,,\; R_y = \sum_{i=1}^{n} Y_i = 0 \,,\; R_z = \sum_{i=1}^{n} Z_i = 0 \,,$$

$$M_x = \sum_{i=1}^{n} y_i Z_i - z_i Y_i = 0 \,,\; M_y = \sum_{i=1}^{n} z_i X_i - x_i Z_i = 0 \,,$$

$$M_z = \sum_{i=1}^{n} x_i Y_i - y_i X_i = 0 \,.$$

$$(2.59)$$

They are independent of the special choice of position of the reference point $A' = 0$.

If $R \neq 0$, the moment vector changes by moving the reference point to $A''(a, b, c)$, $\mathbf{a} = a\,\mathbf{e}_x + b\,\mathbf{e}_y + c\,\mathbf{e}_z$. The new position vectors of the points of application of the forces $\mathbf{F}_i$ are $\mathbf{r}_i''$ and are related to $\mathbf{r}_i = \mathbf{a} + \mathbf{r}_i''$. Thus, the resulting moment with respect to A'' is

$$\mathbf{M}'' = \sum_{i=1}^{n} \mathbf{M}_i'' = \sum_{i=1}^{n} \mathbf{r}_i'' \times \mathbf{F}_i = \sum_{i=1}^{n} (\mathbf{r}_i - \mathbf{a}) \times \mathbf{F}_i$$

$$= \sum_{i=1}^{n} \mathbf{r}_i \times \mathbf{F}_i - \mathbf{a} \times \sum_{i=1}^{n} \mathbf{F}_i = \mathbf{M} - \mathbf{a} \times \mathbf{R} \,.$$

$$(2.60)$$

If $\mathbf{R} = 0$ and $\mathbf{M} = 0$, all moments $\mathbf{M}''$ vanish with respect to reference points at any $\mathbf{a}$, respectively. Prior to the formulation of the equilibrium conditions, such a point of reference may be selected at random.

The final version of Eq. (2.60) can be derived directly by the reduction of the single resulting force $\mathbf{R}$ at $A' = 0$ to the point A''; the moment is $(-\mathbf{a}) \times \mathbf{R}$, which must be added to the (free) moment vector $\mathbf{M}$.

The solution of Eqs. (2.59), which are linear in the force components X_i , Y_i , Z_i , follows according to the rules for a nonhomogeneous system of linear equations. Every spatial force system may be changed to an equilibrating one by applying additional forces.

The six conditions (2.59) are necessary for the equilibrium of the force system $\mathbf{F}_i$ and, hence, for the body loaded by these forces. For a rigid body with six degrees of freedom, the conditions are necessary and sufficient for equilibrium; see also Chap. 5. A deformable body, on the other hand, may continuously creep or yield when loaded by such an equilibrating force system. If the unknown force components can be determined by just solving the equilibrium conditions, the group of forces, or, alternatively, the material system loaded by the forces, is called *statically determinate*, otherwise *statically indeterminate*.

A general, spatially distributed system of forces can be reduced to the resultant **R** with a point of application in the reference point A´ that is also the center for the moments with the resulting moment vector **M**. Selecting a different point of reference A in a distance **a** from A´ renders the new moment vector

$$M_A = M - a \times R \quad .$$

At a special point A, the moment vector M_A becomes parallel to the resultant **R**. The mathematical condition

$$R \times M_A = 0 \quad ,$$

renders the position vector **a** that, without loss of generality, is assumed to be orthogonal to **R**. Expansion of the vector triple product renders in that case

$$R \times (M - a \times R) = R \times M - R \times (a \times R) = R \times M - a\,R^2 = 0 \quad , \quad a \cdot R = 0 \quad ,$$

and, explicitly,

$$a = (1/|R|)\,[(R/|R|) \times M] \quad .$$

The moment vector M_A parallel **R** that remains after the reduction to this special point of reference

$$M_A = M - [(R/|R|) \times M] \times R/|R| = [(R/|R|) \cdot M]\,R/|R|$$

is the component of **M** in the direction of **R**. The latter even does not change during a general translation which keeps the resulting force parallel. The pair of parallel vectors **R**, M_A is called a *force-winder* and is statically equivalent to the original force system.

Another, sometimes more convenient reduction of a given spatial system of forces to three mutually perpendicular single forces that are equal to the cartesian components of the resulting force $R = \Sigma_i\,F_i : X_r = R_x$, $Y_r = R_y$, $Z_r = R_z$, with crossing but in general not intersecting lines of action, renders, under the condition of static equivalence, the three points of application located on the coordinate axes in the distances (solution is not unique)

$$a_x = M_z/R_y \text{ for } Y_r, \quad a_y = M_x/R_z \text{ for } Z_r \text{ and } a_z = M_y/R_x \text{ for } X_r \quad ,$$

to be measured from the point of reference A´ = 0, the origin, respectively.

2.2.1. The Plane Force System: Computational and Graphic Reduction, Conditions of Equilibrium

If all lines of action of the forces F_i, $i = 1, ..., n$, are bound to a single plane, they form a planar system of forces. After selecting a common point of application A in that plane and attaching the two-dimensional, self-equilibrating central forces $F_i + (-F_i)$, the statically equivalent reduction renders the in-plane resulting force $R = \Sigma_i F_i$ in A and the resulting moment vector orthogonal to the (x, y) plane and, hence, orthogonal to R

$$M = \sum_{i=1}^{n} M_i = \sum_{i=1}^{n} r_i \times F_i = \sum_{i=1}^{n} (x_i Y_i - y_i X_i)\, e_z \ .$$

$$(2.61)$$

After reduction, only three components remain

$$R_x = \sum_{i=1}^{n} X_i \ , \quad R_y = \sum_{i=1}^{n} Y_i \ , \quad M_z = \sum_{i=1}^{n} (x_i Y_i - y_i X_i) \ .$$

$$(2.62)$$

The forces in the plane are in equilibrium if the three (necessary) linear conditions

$$R_x = \sum_{i=1}^{n} X_i = 0 \ , \quad R_y = \sum_{i=1}^{n} Y_i = 0 \ , \quad M_z = \sum_{i=1}^{n} (x_i Y_i - y_i X_i) = 0 \ ,$$

$$(2.63)$$

hold. It is suspected that the three conditions of equilibrium are also sufficient, if the forces F_i are the in-plane loadings of a *rigid* plate with three degrees of freedom of the in-plane motion (see also Chap. 5). The reduction of the in-plane forces is conveniently performed graphically by means of the *force* and *funicular polygon* . A scaled configurational plane of the force system is drawn; Fig. 2.11(a) illustrates four forces F_i and, separately, after a proper

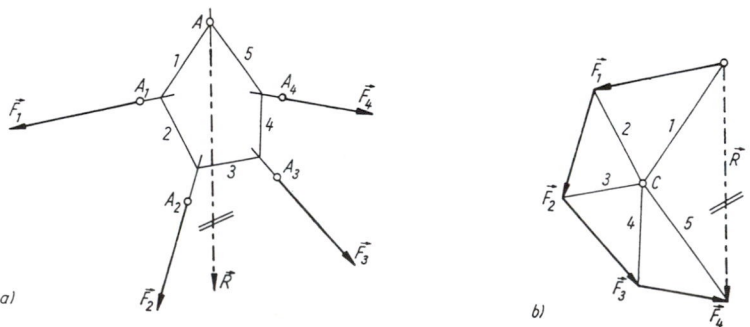

Fig. 2.11. Resultant R of a plane system of forces. Force and funicular polygon

force scale is chosen, the vector summation of the given forces is performed graphically and renders the resulting force $R = \Sigma_i F_i$, $i = 1$ to 4 [see Fig. 2.11(b)] in the force plane. In the case of nonvanishing R, its point of application A in the configurational plane is determined such that the resulting moment is zero. Since M is orthogonal to R, Eq. (2.60) can be solved for a. For the plane problem,

$$M = M_z - a R = 0 \ , \tag{2.64}$$

gives a under the assumption $R \neq 0$. Graphically, this special point of reference A is determined by the following procedure: In the force plane, a point, the *Culman* pole C, is chosen and the rays are drawn according to Fig. 2.11(b). Mapping into the configurational plane by parallel translation is done such that the pair of rays which encloses a given force forms a pair of funicular rays intersecting on the line of action of that force. The point of intersection of the first and last funicular ray [1 and 5 in Fig. 2.11(a)] is A. In the static interpretation of this design, the rays at C are considered dummy forces equilibrating pairwise the given forces: The triangles with common pole C are closed, their partial resultant vanishes. When proceeding successively, eg starting at C running along ray *1* to F_1 and back to C along ray *2*, forward to F_2, thereby passing the dummy force *2* twice in opposite directions and, hence, in a self-equilibrating manner, it is easily recognized that R is in equilibrium with the first and last of the dummy forces. Three forces in the plane with its partial resultant zero are equilibrating iff they form a central force system. Thus, they must have a common point of intersection A in the configurational plane.

The notion of *funicular rays* and the *funicular polygon* of Fig. 2.11(a) is very practical, since a ring of an ideal string (or cord) deforms to the polygon when loaded by the single forces F_i, $i = 1,...n$, and, in addition, by $F_{n+1} = -R$, $n = 4$ above. Hence, the dummy forces are the sectional cable tensions. If the pole C is outside of the area of the closed force polygon, some of the dummy forces are compressive and need a jointed column support in any realization of Fig. 2.11(a). Two more cases of reduction of a plane force system are possible:

$R = 0$. The force polygon of the sum of F_i is closed, but occasionally it happens that the funicular polygon is open, the first and last ray are parallel. A couple $M_z \neq 0$ that which corresponds to the moment of these dummy forces. Consider, eg $F_5 = -R$ passing not through the point A in Fig. 2.11.

$R = 0$ and $M_z = 0$. The force and funicular polygon are closed, the system of plane forces is in equilibrium. Consider, eg $F_5 = -R$ through point A in Fig. 2.11.

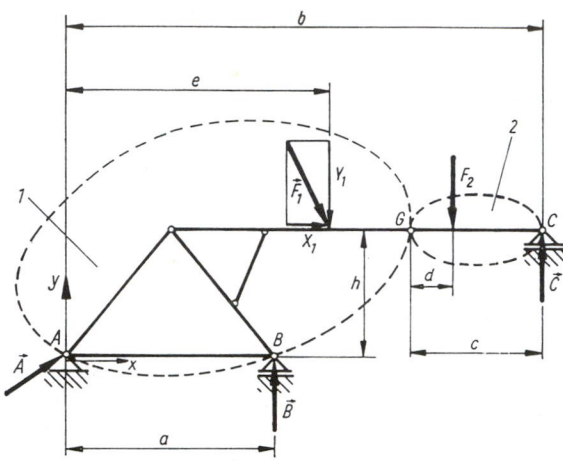

Fig. 2.12. Plane structure consisting of two substructures 1 and 2

The force and funicular polygons are mainly applied to laterally loaded beams; see, eg Fig. 2.40.

(§)   Example: Support Reactions of an In-Plane Loaded Structure
The structure of Fig. 2.12 is loaded by the given external forces F_1 and F_2. In reaction to that loading, supporting forces act at the points A, B, and C on the structure to render a plane system of forces equilibrating the given ones. Supports B and C allow free displacement in the horizontal direction; thus, the direction of the supporting forces is known apriori to be vertical. The hinged support at A is a joint fixed in space and may transmit a reaction force of general direction; see Sec. 1.3.4. Thus, four unknown force components should be determined by the three equilibrium conditions (2.63). The fourth, linearly independent equation necessary for the solution is given by the special joint condition at C, where no moment is transmitted between the substructures denoted by 1 and by 2 in the figure. Hence, with this reference point C, the partial sum of the moments of the external forces acting on the substructure 2 or 1 vanishes

$$A_x + X_1 = 0 , \tag{a1}$$
$$A_y + B + C - Y_1 - F_2 = 0 , \tag{a2}$$
$$Ba + Cb - X_1 h - Y_1 e - F_2(b - c + d) = 0 , \quad \text{reference point A,} \tag{a3}$$
$$\text{hinged joint condition: } Cc - F_2 d = 0 , \quad \text{substructure 2 .} \tag{a4}$$

One of the conditions of equilibrium [(a1) to (a3)] may be replaced by the second hinged-joint condition for the substructure 1 and thereby takes on the role of an independent control equation. Since all unknown forces are determined by the solution, the structure is considered to be externally supported in a statically determinate manner.

2.2.2. *Symmetry of the Stress Tensor*

The free-body diagram of an infinitesimal element renders a three-dimensional force system of body force and surface tractions with six conditions of equilibrium. Three conditions based on the vanishing resultant force have been discussed in Sec. 2.1. In addition, the axial moments dM_x, dM_y, dM_z must be zero. With reference point (x, y, z), which has been selected as the point of reduction for the force system, the axial moment dM_z becomes by inspection

$$dM_z = -[\sigma_{xx}(x + dx, y, z) - \sigma_{xx}(x, y, z)]\, dy\, dz\, dy/2 + \sigma_{xy}(x + dx, y, z)\, dy\, dz\, dx$$
$$+ [\sigma_{yy}(x, y + dy, z) - \sigma_{yy}(x, y, z)]\, dx\, dz\, dx/2 - \sigma_{yx}(x, y + dy, z)\, dx\, dz\, dy$$
$$- [\sigma_{zx}(x, y, z + dz) - \sigma_{zx}(x, y, z)]\, dx\, dy\, dy/2$$
$$+ [\sigma_{zy}(x, y, z + dz) - \sigma_{zy}(x, y, z)]\, dx\, dy\, dx/2 - k_x\, dx\, dy\, dz\, dy/2 +$$
$$k_y\, dx\, dy\, dz\, dx/2 = 0 \ .$$

Linear approximation with subsequent division through $dV = dx\, dy\, dz$ renders in the limit $dx, dy \to 0$, exactly [note Eq. (2.19)]

$$\sigma_{xy}(x, y, z) - \sigma_{yx}(x, y, z) = 0 \ .$$

The other two conditions $dM_x = 0$ and $dM_y = 0$, yield, analogously to $dM_z = 0$,

$$\sigma_{xy} = \sigma_{yx} \ , \quad \sigma_{yz} = \sigma_{zy} \ , \quad \sigma_{zx} = \sigma_{xz} \ . \tag{2.65}$$

Hence, in a point-continuum, the symmetry of the stress tensor follows from equilibrium considerations. Consequently, Eq. (2.65) renders the following lemma on *adjoint shear stresses* : <<At two mutually orthogonal surface elements, the shear-stress components that are normal to the line of intersection have equal values and are both pointing toward the edge of the elements or away from it.>> See Fig. 2.13.

In a continuum of higher order, like the *Cosserat* -continuum, rotational degrees of freedom are assigned to the material point and so-called *couple stresses* are introduced. In a further generalization physical properties are assigned to these (deformable) cells and a theory of micromechanics results. Problems are encountered with respect to the boundary conditions.

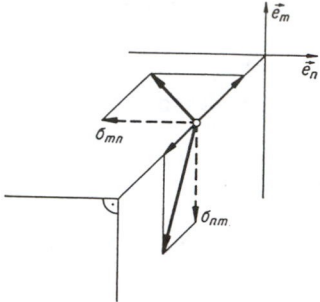

Fig. 2.13. Result of the lemma on adjoint shear stress: $\sigma_{nm} = \sigma_{mn}$

2.2.3. The Parallel Force System: Center of Forces, Center of Gravity (Centroids), Static Moments

In the special case of a parallel force system, all the lines of action of the contributing forces are parallel. Statically equivalent reduction to the single resultant force $R \neq 0$, in a special point of application A, where $M_A = 0$, is possible also in three dimensions, since for any arbitrary A′, the moment $\boldsymbol{M}$ is orthogonal to $\boldsymbol{R}$. Equation (2.60) may be solved for the vector $\boldsymbol{a}$ pointing from A′ to A:

$$\boldsymbol{M_A} = \boldsymbol{M} - \boldsymbol{a} \times \boldsymbol{R} = 0 \quad , \quad (\boldsymbol{M} \cdot \boldsymbol{R}) = 0 \quad . \tag{2.66}$$

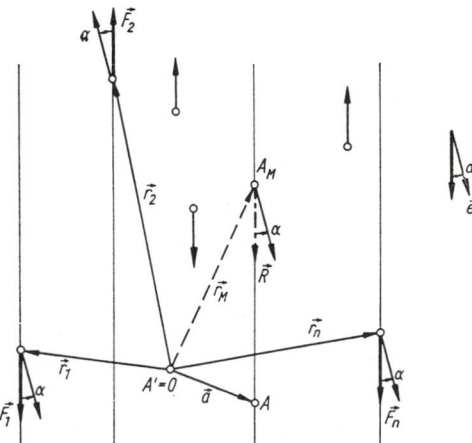

Fig. 2.14. Resultant $\boldsymbol{R}$ of the spatially distributed parallel force system and the true point of application, the force center A_M, after a common axial rotation through the angle α

Selecting the origin 0 for A' and taking the position vectors to the given points of application A_i of the parallel forces $F_i = F_i e$, static equivalence when expressed in the resulting couples and in the moment of $R = \Sigma_{i=1..n} F_i e$, render

$$\mathbf{a} \times \mathbf{e} \sum_{i=1}^{n} F_i = \sum_{i=1}^{n} F_i \mathbf{r}_i \times \mathbf{e} \ ,$$

or equivalently,

$$\left[\sum_{i=1}^{n} (\mathbf{a} - \mathbf{r}_i) F_i \right] \times \mathbf{e} = 0 \ .$$

In this equation, the line of action of $\mathbf{a}$ can be arbitrarily selected, eg orthogonal to $\mathbf{e}$, and the solution renders a point A of application of R with $M_A = 0$. By stiffening the condition at $\mathbf{a}$, such that A becomes the true point of application of the resultant R , the force center, which remains fixed in space, when the given parallel forces F_i are rotated through the common angle α about parallel axes through their points of application A_i (see Fig. 2.14), yields, since the unit vector $\mathbf{e}$ is directionally arbitrary,

$$\sum_{i=1}^{n} (\mathbf{r}_M - \mathbf{r}_i) F_i = 0 \ . \tag{2.67}$$

The unique solution $\mathbf{a} = \mathbf{r}_M$ determines the position vector of the force center, pointing from the origin $0 = A'$ to A_M

$$\mathbf{r}_M = \frac{\displaystyle\sum_{i=1}^{n} F_i \mathbf{r}_i}{\displaystyle\sum_{i=1}^{n} F_i} \ . \tag{2.68}$$

The force center of two parallel forces with the nonvanishing resultant $R = F_1 + F_2 \neq 0$, is thus located on the straight line passing through their points of application.

The *central forces of gravity* acting on a body are commonly approximated by a parallel force system in those cases where the angle of inclination between the actual forces remains small (there are exemptions where this approximation fails, eg there have been long-time effects observed in Space-Dynamics). The force per unit of volume attached to every material point becomes vertical

$$\mathbf{k} = \rho g \, \mathbf{e}_z \ ,$$

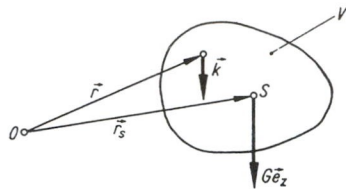

Fig. 2.15. Center of gravity C = S is the force center of the parallel body forces denoted by $\mathbf{k} = \rho g\,\mathbf{e}_z$, g = const

with $\mathbf{e}_z$ normal to a horizontal plane (flat Earth model), pointing downward (Fig. 2.15). The summation of the parallel forces over the volume renders the weight of the body

$$G\,\mathbf{e}_z = \mathbf{e}_z \int_V \rho g\, dV \ .$$

(2.69)

The center of the parallel forces of gravity, the *center of gravity* C = S of the body, is the true point of application of the resulting weight, with the additional approximation of constant gravitational acceleration g (for a body in a homogeneous field of gravity). Equation (2.67) takes on the proper form of distributed forces,

$$\int_V (\mathbf{r}_S - \mathbf{r})\,\rho g\, dV = 0 \ .$$

(2.70)

Its solution is the radius vector pointing from the origin to the center of gravity

$$\mathbf{r}_S = \frac{1}{G} \int_V \mathbf{r}\,\rho g\, dV \ .$$

(2.71)

With those idealizations, the center remains fixed to the body when it is rigidly rotated. When we use $G = mg$, the constant acceleration g cancels in Eq. (2.71), and the *center of mass* M is determined by

$$\mathbf{r}_M = \frac{1}{m} \int_V \mathbf{r}\,\rho\, dV = \frac{1}{m} \int_m \mathbf{r}\, dm \ .$$

(2.72)

The latter is independent of any assumptions on gravity.

In the case of a homogeneous body, mass density ρ = const cancels since $m = \rho V$, and the geometric center of volume V is located at

$$r_V = \frac{1}{V} \int_V r \, dV \ .$$

$$(2.73)$$

The radius vector r_V is the vector mean of the point vectors measured from the origin to all the points in the volume V (cf. with the statistical mean value). Geometrical centers are analogously determined for surfaces S of arbitrary curvature

$$r_S = \frac{1}{S} \int_S r \, dS \quad (dS \text{ is the two-dimensional surface element}) \ ,$$

$$(2.74)$$

and for (spatially) curved lines L ,

$$r_L = \frac{1}{L} \int_L r \, ds \quad (ds \text{ is the arc-length element}) \ .$$

$$(2.75)$$

The domain integrals are linear operations that allow superposition, eg in the case of composite simple domains with known partial integrals just a summation remains.

The stretched vectors are of mathematical importance and do appear in many technical applications

$$G \, r_S = \int_V r \, \rho g \, dV \ , \quad m \, r_M = \int_V r \, \rho \, dV \ , \quad V \, r_V = \int_V r \, dV \ ,$$

$$S \, r_S = \int_S r \, dS \ , \quad L \, r_L = \int_L r \, ds \ ,$$

$$(2.76)$$

as the *static moments (moments of first order)* of weight, mass, volume, surface area, and arc-length, respectively, taken about the origin A′ = 0. Especially, the static moment of a plane area A with respect to a point 0 in that plane is given by the *(x, y)* components

$$A \, x_S = \int_A x \, dA \ , \quad A \, y_S = \int_A y \, dA \ ,$$

$$(2.77)$$

which are the axial static moments about the y and x axis, respectively.

In the case of axial symmetry about the axis through a point 0′ in the direction *e* with *r = a + r′* and *r_M = a + r_M′*, the first moment of the mass distribution becomes, *r′= r′e_r + z′e_z , (e_r . e) = 0 ,*

$$m \, \vec{r}_M = \int_V r' \, e_r \rho \, dV + e \int_V z' \rho \, dV = e \int_V z' \rho \, dV \quad ,$$

r' is measured from $0'$, ie the center of mass distribution is located on the axis of symmetry, $r_M' = z_M' \, e$.

2.3. Hydrostatics

Problems related to the equilibrium of fluids are rather simple and easy to solve. An intrinsic property of a fluid body at rest is the vanishing of all the shear stress components, $\sigma_{ij} = 0, i \neq j$. Consequently, the remaining three normal stress components at each material point become equal to the isotropic hydrostatic pressure p $= - \sigma_{xx} = - \sigma_{yy} = - \sigma_{zz}$. Since fluids mainly carry compressional stresses (liquids under tension start to evaporate) pressure p is considered positive.

Recalling Eq. (2.20), which is material-independent,

$$\sigma_n = \sigma_x \, n_x + \sigma_y \, n_y + \sigma_z \, n_z \quad , \tag{2.78}$$

and inserting the isotropic normal stresses $\sigma_x = \sigma_{xx} \, e_x$, $\sigma_n = \sigma_{nn} \, e_n$, renders by scalar multiplication with the unit vectors e_x , e_y , e_z ,

$$\sigma_{nn} \, e_n \cdot e_x = \sigma_{xx} \, n_x , \quad \sigma_{nn} \, e_n \cdot e_y = \sigma_{yy} \, n_y , \quad \sigma_{nn} \, e_n \cdot e_z = \sigma_{zz} \, n_z ,$$
$$n_x = e_n \cdot e_x , \text{ etc.}$$

Since the direction of e_n is arbitrary, the state of hydrostatic stress is derived to exist in a fluid at rest. Substituting

$$p = - \sigma_{xx} = - \sigma_{yy} = - \sigma_{zz} = - \sigma_{nn} \quad , \tag{2.79}$$

into Eq. (2.18) yields

$$f = k - \left(\frac{\partial p}{\partial x} \, e_x + \frac{\partial p}{\partial y} \, e_y + \frac{\partial p}{\partial z} \, e_z \right) = k - \nabla p = k - \text{grad } p = 0 \quad .$$

The vector condition of local equilibrium of a fluid at rest

$$k = \text{grad } p \quad , \tag{2.80}$$

holds only, iff the external body force k is a gradient vector field; $p(x, y, z)$ is a scalar function. Such a force is derived from a potential density $W_P'(x, y, z)$,

$$k = - \text{grad } W_P' \quad . \tag{2.81}$$

Hence, surfaces of equal pressure, *p(x, y, z)* = *const*, have assigned orthogonal trajectories that are the vector lines of such a given force field **k** . In every point, the normal to the surface is the tangent to the vector line and parallel to **k** . Eq. (2.80) is generalized to include the inertia force of an ideal flow, see Eq. (8.30).

Pressure is a (normal) stress and as such has the same dimension *[p]* = *N/m²* . Its unit of measure is named *pascal* , 1 *N/m²* = 1 *Pa* . A technical unit is *1 bar* = *10⁵ N/m²* , which is approximately the differential pressure of a vertical column of water with a height of *10 m* . Atmospheric pressure should be measured in *hectopascal* , 1 *hPa* = *10² Pa* . Pressure enters the constitutive relation of compressibility, see, eg Eqs. (2.87) and (4.60).

2.3.1. *Fluid under Gravity*

A fluid at rest in a homogeneous gravity field having such limited horizontal dimensions that the body force is approximately a field of parallel forces is considered. The vertical unit vector e_z is pointing upward. Equation (2.80) reduces to one component of pressure variation in the vertical z direction

$$\mathbf{k} = - \rho g \, \mathbf{e}_z = \text{grad } p = \frac{\partial p}{\partial z} \, \mathbf{e}_z \ .$$

$$(2.82)$$

The surface of the *isobars* becomes a horizontal plane.

(§) Incompressible Fluid. Integration of the differential equation (2.82) yields (the density of an inhomogeneous fluid may depend on z but not on the pressure p , ie the fluid is assumed incompressible)

$$p = -g \int_0^z \rho \, dz + C \ ,$$

$$(2.83)$$

C is the reference pressure in a plane $z = 0$. An incompressible and homogeneous fluid, where ρ = *const* , is a good approximation of a liquid layer of small to moderate thickness

$$p = -\rho g z + C \ ,$$

with the pressure decreasing linearly with height. It is convenient to prescribe the pressure at some height $z = H$, eg at the free surface of a water body, the interface to air at rest, by p_0 , and to measure the depth h from this plane and against the direction of the coordinate z . Hence,

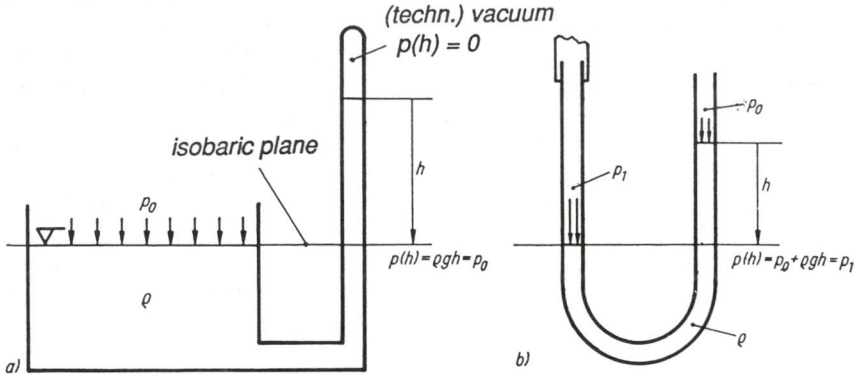

Fig. 2.16. (a) Mechanical principle of a manometer (mercury, $\rho = 13600 \ kg/m^3$).
(b) U-gaged manometer (water, $\rho = 1000 \ kg/m^3$)

$$p(h) = p_0 + \rho g h \ . \tag{2.84}$$

<<Pressure increases linearly with depth h.>>

$$[p(h) - p_0] / \rho g = h \tag{2.85}$$

is called the (differential) *pressure head* . Barometers and manometers, based on liquid-column gages, use mercury or water with proper scales of h to measure the absolute or differential pressure, respectively (see Fig. 2.16). In the case of $p_1 = p_0$ in the U-shaped gage of Fig. 2.16(b), the head h is zero, which is an important observation in any *communicating system of pipes*, where the interface of the liquid and air is a horizontal plane at the highest level of the isobars.

The effects of *surface tension* caused by cohesion and adhesion, like the *meniscus* and *capillary* action, are of importance in some special problems, but are assumed negligible in many technical applications and are not considered in this book. (However, it is quite an experience to float a steel razor blade at the interface of air and water).

Layered fluids in a parallel gravity field have horizontal interfaces of constant pressure. A stable equilibrium configuration exists if the density increases with depth. In homogeneous layers, the differential pressure increases linearly with depth, being similar to the gage pressure from the air-fluid interface (see Fig. 2.17).

A *compressible, homogeneous barotropic fluid* has a constitutive relation of density and pressure, $\rho = \rho(p)$. Temperature, possibly, may have been eliminated by means of the thermodynamic

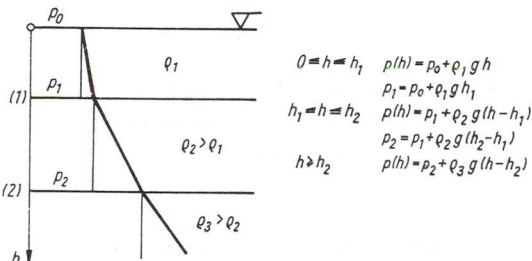

Fig. 2.17. Absolute, gage and differential pressure distribution in a layered fluid at rest

state equation, and Eq. (2.82) in that case can be integrated by a separation of variables

$$\frac{dp}{\rho(p)} = -g\,dz \;, \quad \int_{p_0}^{p}\frac{dp}{\rho(p)} = -g(z-z_0) \;, \quad p_0 \text{ at } z_0 \;.$$

$$(2.86)$$

Simple numerical integration reconsiders the layered structure of the fluid at rest (see again Fig. 2.17)
(§) A Linear Compressive Spring. In the case of linear (volume) compressibility, the constitutive relation becomes

$$dp = K_F\,d\rho/\rho_0 \quad \text{or} \quad p - p_0 = K_F\,[(\rho/\rho_0)-1] \;, \quad p,\rho > 0 \;, \qquad (2.87)$$

where K_F denotes the bulk modulus. For liquids, the modulus is practically independent of temperature and, for fresh (distilled) water, takes on the value of 20.9×10^8 N/m². On the contrary, for an ideal gas under isothermal conditions, like dry air, $K_F = p_0 = \rho_0\,RT$, where $R = 287$ Nm/kgK, is the gas-constant and T is the absolute temperature measured in degrees *kelvin*, $[T] = K$. Equation (2.86) yields upon integration of (2.87) and after inversion of the logarithmic function

$$p = p_0 + K_F\,\{[\exp\,(g\rho_0\,h/K_F)]-1\} \;, \quad h = z - z_0 \geq H \;. \qquad (2.88)$$

<<Pressure increases exponentially with depth h .>> In a plane of reference, $h = 0$, $p = p_0$. Since $K_F \geq p_0$, the pressure vanishes in a height $(-H)$ above that plane, where $H = (K_F/g\rho_0)\,\ln(1 - p_0/K_F)$.
    A linear approximation of the exponential function in Eq. (2.88) renders the linear pressure distribution of Eq. (2.84) and, thus, holds for small arguments $(g\rho_0\,h/K_F)$.
    The inversion of Eq. (2.88) gives for an isothermal layer of dry air ($K_F = p_0$ is the reference pressure at z_0 and $H \to -\infty$)

$$z - z_0 = (R\,T/g)\,\ln\,(p_0\,/p) \quad . \tag{2.89}$$

The barometric formula for the altitude is quite useful but needs updating. It is valid within layers of thickness $max\,|z - z_0| \approx 100\ m$. Thus, measuring pressure renders the altitude by a simple calculation.

(§)   A Nonlinear Spring. A homogeneous layer of a gas at rest in an adiabatic state of equilibrium, in a parallel gravity field, is considered to illustrate nonlinear compressibility. The barotropic material law is

$$p/p_0 = (\rho/\rho_0)^{\kappa} \quad , \quad \rho/\rho_0 = (p/p_0)^{1/\kappa} \quad , \tag{2.90a}$$

where $\kappa = 1.4$ denotes the *adiabatic coefficient* valid for dry air and two-atomic gases. The incremental version of that constitutive relation can be cast in a form using the notion of an adiabatic tangent modulus K_a [see Eq. (2.87)]

$$dp = K_a\,d\rho/\rho \quad , \quad K_a = \kappa\,p \quad . \tag{2.90b}$$

Substitution into Eq. (2.86) renders in that case

$$\int_{p_0}^{p} \left(\frac{p_0}{p}\right)^{1/\kappa} dp = -\,g\rho_0\,(z - z_0) \quad . \tag{2.91}$$

Performing the integration yields the pressure, decreasing with height $(z - z_0)$ above niveau $z = z_0$ according to the potential law; p_0 is an assigned pressure

$$p = p_0\,\{1 - [(\kappa - 1)/\kappa]\,g\rho_0\,(z - z_0)/p_0\}^{\kappa/(\kappa - 1)} \quad . \tag{2.92}$$

Temperature is proportional to the pressure, $T = p/R\rho$. Replacing κ in Eq. (2.92) by the *polytropic exponent* $n < \kappa$ yields the quite general pressure distribution in a layer of air at rest under the conditions of a polytropic process.

2.3.2. Pressurized Fluids

Fluid in closed piping systems or in containers is often exposed to high pressures. Consider, eg a large vessel with an attached displacement pump where a piston with a small cross-sectional area A under axial force F is in direct contact with the (incompressible) fluid (see Fig. 2.18). A mean pressure $p_1 = F/A$ is

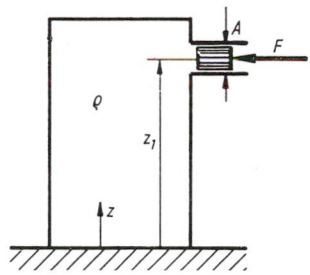

Fig. 2.18. Pressurized tank

initiated at the level z_1 of the pump, which is independent of the shape of the piston surface. This is easily verified by projecting the infinitesimal force $p_1 \, dS \, \mathbf{e}_n$ ($\mathbf{e}_n$ is the normal of dS) in the axial direction $\mathbf{e}$ of the driving force F

$$\overline{p_1} \, dS \, \mathbf{e}_n \cdot \mathbf{e} = \overline{p_1} \, dA \ .$$

dA is the cross-sectional element. Summing the axial components to the resulting force parallel $\mathbf{e}$ yields

$$F = \int_A \overline{p_1} \, dA \ = p_1 \, A \ ,$$

with p_1 as the mean pressure. Due to gravity, the pressure varies linearly

$$p(z) = p_1 + g\rho \, (z_1 - z) \ ,$$

where $p = p_1$ at $z = z_1$. In the case of sufficiently high pressure $p_1 \gg$ max $[g\rho|z_1 - z|]$ (in Fig. 2.18, $p_1 \gg g\rho z_1$), the variation due to gravity is negligible, and the pressure in the whole fluid body is assumed to be constant

$$p \approx p_1 = \text{const} \ . \tag{2.93}$$

This approximation holds also for compressible fluids.

(§)   Principle of the Hydraulic Pump. Two cylinders of strongly differing cross-sections are attached to a rigid container (see Fig. 2.19). The force F_1 produces the high pressure $p = F_1 / A_1$ in the transmitting fluid. Thus, neglecting the influence of gravity, the pressure p acts on the working piston with a large cross-sectional

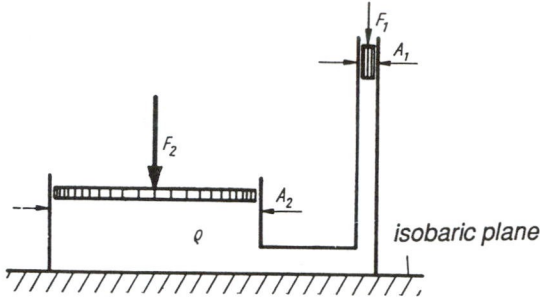

Fig. 2.19. The mechanical principle of a hydraulic pump

area A_2 and yields a large "lifting" force $F_2 = p \, A_2 = F_1 \, A_2 \, /A_1$ according to the ratio $A_2 \, /A_1 \gg 1$. Note that the large force F_2 is also transmitted to the ground and special support is required in field applications.

(§)   Vessels and Pipes. Pressurized fluids are often stressing thin-walled structures of circular cylindrical shape. Spherical containers are used for the storage of fluids and, eg of highly pressurized natural gas. The hydrostatic pressure p acts normally on each exposed surface element and, thus, is considered a loading of the container wall (see Fig. 2.20).

The axial force F in a closed cylindrical shell of volume $l \times A$ (Fig. 2.20 shows the uniformly distributed parallel forces $p \, dA$ in the case of a flat plate bottom) is in general given by

$$F = \int_A p \, dA = p \, A \, , \quad A = \pi \, a^2 \, .$$

$$(2.94)$$

Hence, like a tensile rod, the axial normal stress in the circular wall is uniformly distributed and given by

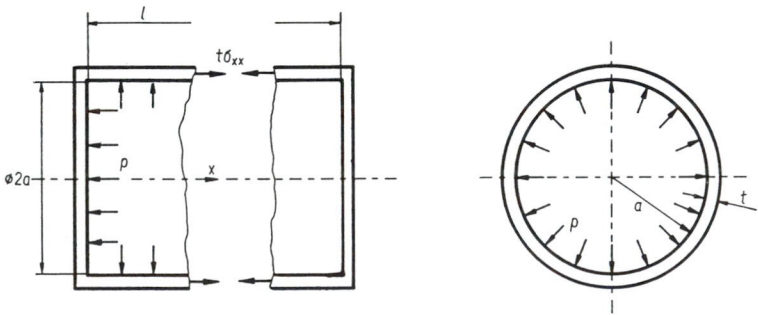

Fig. 2.20. The circular cylindrical and spherical pressurized vessel, $t \ll a$

$$\sigma_{xx} = \frac{F}{2\pi a t} = \frac{p\,a}{2\,t} \quad . \tag{2.95}$$

A central force field acts on the cylindrical and spherical wall and renders a circumferential stress field. Because of its symmetry, the upper half of the shell is considered in Fig. 2.21, where the equilibrium of forces in the vertical z direction requires (the cylindrical shell is assumed to be very long in terms of its diameter)

$$\int_S p\,dS\,n_z - \sigma_{\varphi\varphi}\,t\,L = 0 \quad . \tag{2.96}$$

Since by projection $dS\,n_z = dA_z$ and $p = const$, the integral is simply $p\,A_z$, and the solution of the equation

$$p\,A_z - \sigma_{\varphi\varphi}\,t\,L = 0 \quad , \tag{2.97}$$

yields the celebrated "pressure-vessel formula" of hoop-stresses

$$\sigma_{\varphi\varphi} = p\,\frac{A_z}{t\,L} \quad , \tag{2.98}$$

where A_z is the projection of the loaded shell surface on the plane $z = const$.

The *membrane stresses* in a closed cylindrical shell of length l become upon the substitution of $A_z = 2a\,l$ and $L = 2l$ (two edges) into Eqs. (2.95) and (2.98)

$$\sigma_{\varphi\varphi} = p\,a/t = 2\,\sigma_{xx} \quad . \tag{2.98a}$$

They approximate the principal normal stresses in the thin wall,

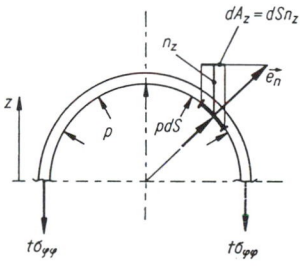

Fig. 2.21. Upper half of a symmetric shell. Circumferential force $t\,\sigma_{\varphi\varphi}$ per unit of length

sufficiently distant from the endplates. The bending state at the boundaries diminishes rapidly with axial distance (see Sec. 6.7.1). Note that the circumferential stress is twice the axial stress, ie the failure mode of a bursting cylindrical shell commonly shows an axially oriented crack.

The spherical shell has a projected area $A_z = \pi a^2$ and a perimeter of $L = 2\pi a$, and Eq. (2.98) yields upon the substitution the principal normal stresses (the pressurized shell is in a membrane state of stress)

$$\sigma_{\varphi\varphi} = \sigma_{\vartheta\vartheta} = p\, a/2\, t \; .$$

φ and ϑ are the spherical angles.

The radial stress at $r = a$ equals the pressure p and diminishes linearly to zero at the outer traction-free surface. For thin shells, $a/t \gg 1$, the radial stress component is negligibly small when compared to the membrane stresses. The latter have been determined by equilibrium considerations and, thus, define a *statically determinate state of stresses* (see Sec. 6.7 for other shells of revolution and distributed loads).

2.3.3. The Gravitational Hydrostatic Pressure in Open Containers

In general, the linearly increasing pressure $p = p_0 + g\rho h$ in an incompressible fluid at rest acts normal to retaining walls or to selected portions of the fluid-solid interface in the form of spatially distributed forces $p\, dS\, \mathbf{e}_n$. In the case of a plane section, a nonuniform parallel force field results.

(§) A Flat Horizontal Base of Area A. This base is loaded by uniformly distributed parallel forces and a vertical force F results,

$$F = p(H)\, A = (p_0 + g\rho H)\, A \; , \tag{2.99}$$

which depends on the depth H, but not on the shape of the container or the weight of the fluid body (see Fig. 2.22). Historically, this fact is called a hydrostatic paradoxon.

(§) A Plane Retaining Wall of Area A. This wall is also loaded by parallel forces that are statically equivalent to a single resulting force attached to the force-center or, specifically, to the center of pressure. The uniform reference pressure field p_0 has a partial resultant $p_0 A$, with its line of action normal to A passing through the (geometric) centroid of A, analogous to the plane horizontal bottom case. Since superposition applies, the action of the linearly varying differential pressure $p(h) = g\rho h$, which is the gage pressure

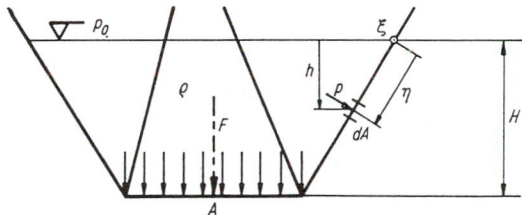

Fig. 2.22. Pressure on plane rigid surfaces

in case of a free surface under atmospheric conditions, is considered subsequently. The resulting force on an inclined retaining wall is

$$R = \int_A p \, dA = g\rho \int_A h \, dA \ ,$$

or, upon substitution of $h = \eta \cos \alpha$,

$$R = g\rho \cos \alpha \int_A \eta \, dA = g\rho \cos \alpha \, \eta_S A \ .$$

According to Fig. 2.22, the static moment of the loaded area A about the ξ -axis, which is the line of intersection of the free surface and the wall, is expressed by $\eta_S A$, where η_S is the coordinate of the centroid of the area A . Hence, denoting the depth at the centroid of A by $h_S = \eta_S \cos \alpha$,

$$R = g\rho h_S A \ . \tag{2.100}$$

<<The value of the resultant force is given by the product of the difference pressure at the centroid of the area A , which is also the mean pressure of the linear distribution, times the area A .>> The point of application of $\boldsymbol{R}$ (without moment) is determined by considering static equivalence, eg through the common axial moment about the ξ -axis

$$R \eta_M = \int_A (p \, dA \, \eta) = g\rho \cos \alpha \int_A \eta^2 \, dA = g\rho \cos \alpha \, J_\xi \ , \tag{2.101}$$

and, in addition through the axial moment about the η -axis

$$R \, \xi_M = \int_A (p \, dA \, \xi) = g\rho \cos \alpha \int_A \eta \xi \, dA = g\rho \cos \alpha \, J_{\xi\eta} \; .$$

(2.102)

(ξ_M, η_M) are the desired coordinates of the center of pressure, and the integral J_ξ denotes the (axial) *moment of inertia* of the area A about the ξ-axis; the integral $J_{\xi\eta}$ is the *deviatoric (or centrifugal) moment* or the *product of inertia* of the area A with respect to the orthogonal planes $\xi = 0$ and $\eta = 0$. Mathematically, both are area moments of second order, since the area-element is multiplied by the squared normal distance to the axis (cf. to statistical moments of second order). Substituting the value for R renders purely geometric expressions

$$\eta_M = \frac{J_\xi}{A \, \eta_S} \; , \quad \xi_M = \frac{J_{\xi\eta}}{A \, \eta_S} \; ,$$

(2.103)

where

$$J_\xi = \int_A \eta^2 \, dA \quad \text{and} \quad J_{\xi\eta} = \int_A \xi\eta \, dA \; , \quad [J] = m^4 \; .$$

(2.104)

It is easy to show that $\eta_M > \eta_S$ for all shapes of the area A. Considering a (x, y)-coordinate system parallel to (ξ, η), but shifted to the centroid S, and substituting $\eta = \eta_S + y$ into Eq. (2.104) yield

$$J_\xi = \int_A (\eta_S + y)^2 \, dA = \eta_S^2 A + 2\eta_S \int_A y \, dA + \int_A y^2 \, dA = \eta_S^2 A + J_x \; ,$$

(2.105)

since the static moment about the central axis x vanishes by definition. Equation (2.103) renders the inequality

$$\eta_M = \eta_S + \frac{J_x}{A\eta_S} > \eta_S \; , \quad J_x = \int_A y^2 \, dA > 0 \; .$$

(2.106)

The sign of $J_{\xi\eta}$ depends on the shape of the area A. The deviatoric moment vanishes if one of the axes (the η-axis above) is an axis of symmetry. For further considerations, see Sec. 2.4 .

In the case of a pressure loaded *rectangular area* $A = b \times t$, of width b, extending downward from the free surface, the axial moments to be inserted are

$$J_\xi = \int_0^t \eta^2 b \, d\eta = \frac{b \, t^3}{3} \quad \text{and} \quad J_x = \int_{-\frac{t}{2}}^{\frac{t}{2}} y^2 b \, dy = \frac{b \, t^3}{12} \; ,$$

and, hence, $\eta_M = 2t / 3$. If we choose η to be the axis of symmetry yields $\xi_M = 0$. It is left as an exercise to determine the center of pressure when the area $A = b \times t$ is moved downwards to a distance c from the free surface.

The pressure loading of a *spherical* or circular *cylindrical* portion of a container wall gives a spatially distributed or plane central force field, respectively. In both cases, a reduction to a single force is possible with a line of action passing through the common point of application.

(§)   Circular Cylindrical Surface (Fig. 2.23). The force per unit of length $pa \, d\varphi$ is projected onto the x and z axes and the resulting parallel components are summed to render

$$R_x = \int_{-\alpha_2}^{\alpha_1} (pa \, d\varphi \cos \varphi) \quad , \quad R_z = -\int_{-\alpha_2}^{\alpha_1} (pa \, d\varphi \sin\varphi) \; .$$

Substitution of the differential pressure, expressed as a function of the polar angle φ , yields

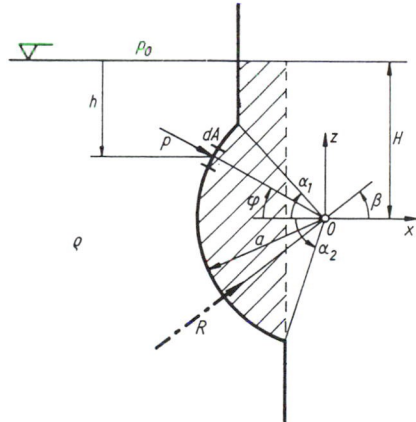

Fig. 2.23. Hydrostatic pressure acting on a circular cylindrical wall

$$R_x = g\rho aH \int_{-\alpha_2}^{\alpha_1} \left(\cos\varphi - \frac{a}{2H}\sin 2\varphi\right) d\varphi = g\rho h_{S_x} a\,(\sin\alpha_1 + \sin\alpha_2)\ ,$$

$$R_z = -g\rho aH \int_{-\alpha_2}^{\alpha_1} \left[\sin\varphi - \frac{a}{2H}(1-\cos 2\varphi)\right] d\varphi = g\rho V ,$$

where

$$V = Ha\ \{\cos\alpha_1 - \cos\alpha_2 + (a/2H)\ [\alpha_1 + \alpha_2 - (\sin 2\alpha_1 + \sin 2\alpha_2)/2]\}$$

is the hatched volume per unit of length shown in Fig. 2.23. The horizontal component R_x can be directly calculated, such that the pressure distribution acts normally on the area A_x in a vertical plane by projecting the cylindrical surface in the x direction. Hence,

$$h_{S_x} = H\ [1 - (a/2H)\ (\sin\alpha_1 - \sin\alpha_2)]\ \ .$$

The resultant force $R = (R_x^2 + R_z^2)^{1/2}$ passes under the angle β against the x-axis statically equivalent through the common point of application, the center of the circle, $\tan\beta = R_z\,/\,R_x$.

(§)   Hydrostatic Pressure on a Spherical Surface (Fig. 2.24)
Consider the vertical plane of symmetry (x, z) at longitude $\varphi = 0$. The force component in the horizontal x direction in the depth h becomes

$$dR_x = g\rho h\ dA\cos\vartheta\cos\varphi = g\rho h\ dA_x\ ,$$

with the angle of latitude ϑ. Integration over the elliptical area $A_x = \pi a^2\ \sin^2((\alpha_1 + \alpha_2)/2)\ \cos\ ((\alpha_1 - \alpha_2)/2)$ of the x-projection of the spherical surface yields

$$R_x = g\rho \int_{A_x} h\ dA_x = g\rho h_{S_x} A_x\ ,$$

where $h_{S_x} = H\ [1 - (a/2H)\ (\sin\alpha_1 - \sin\alpha_2)]$ is the depth at the centroid of A_x . Integration of the z-components over the projected area A_z , however,

$$dR_z = g\rho h\ dA\sin\vartheta = g\rho h\ dA_z\ ,$$

renders the difference of V_1 , the horizontally hatched volume in Fig. 2.24 , and V_2 , which is vertically hatched

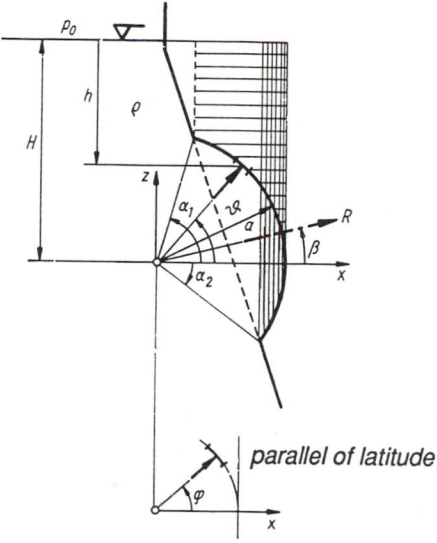

Fig. 2.24. Hydrostatic pressure acting on a spherical portion of the wall

$$R_z = g\rho \int_{A_z} h \, dA_z =$$

$$- g\rho \frac{\pi a^3}{3} \left\{ \frac{3H}{a} (\alpha_2 - \alpha_1) + \frac{1}{2} \left[(1 - \cos \alpha_1)^2 (2 + \cos \alpha_1) + \right.\right.$$
$$\left.\left. (1 - \cos \alpha_2)^2 (2 + \cos \alpha_2) \right] \right\} = g\rho \, (V_1 - V_2) \ .$$

(§) Hydrostatic Loading of a Doubly Curved Surface. Implicitly, such a wetted surface in space, eg of an arch dam, is given by the equation $f(x, y, z) = 0$, which is assumed differentiable. The normal vector $\boldsymbol{e}_n$ gives the direction of the forces of the gage pressure in depth h

$$p(h) \, dS \, \boldsymbol{e}_n \ , \quad p = g\rho h \ , \quad n_x / (\partial f / \partial x) = n_y / (\partial f / \partial y) = n_z / (\partial f / \partial z) \ . \quad (2.107)$$

The horizontal components of the loading are $dX = p \, dS \, n_x$ and $dY = p \, dS \, n_y$, where the direction cosine are $n_{x,y} = (\boldsymbol{e}_n \cdot \boldsymbol{e}_{x,y})$. The vertical component, positive upward, is $dZ = p \, dS \, n_z$, $n_z = (\boldsymbol{e}_n \cdot \boldsymbol{e}_z)$. The surface element, when multiplied with the direction cosine of its normal, becomes the projected plane element. Hence, a summation of the parallel force systems, $dX = p \, dA_x$, $dY = p \, dA_y$, and $dZ = p \, dA_z$, has to be performed over the projected plane areas A_x [in the (y, z) - plane], A_y [in the (x, z) -plane], and A_z [in the horizontal (x, y) -

plane], respectively, to render three resulting single forces. Two horizontal forces have orthogonally crossing lines of action,

$$X_r = g\rho \int_{A_x} h \, dA_x = g\rho h_{S_x} A_x \ , \quad Y_r = g\rho \int_{A_y} h \, dA_y = g\rho h_{S_y} A_y \ ,$$

(2.108)

and the vertical force is proportional to the volume V "above" the wetted surface and bounded by the horizontal surface $h = 0$,

$$Z_r = g\rho \int_{A_z} h \, dA_z = g\rho \, V \ .$$

(2.109)

In addition to the depths $h_{Sx,y}$ at the centroids of A_x and A_y, the coordinates of the centers of pressure of these plane areas also have to be determined by means of Eq. (2.103), where η_M vertically becomes $h_{Mx,y}$, respectively. The line of action of Z_r passes through the geometric centroid of the volume V. The three mutually perpendicular forces at these positions are statically equivalent to the spatially distributed hydrostatic loading (see Sec. 2.2).

In the case of arch dams, scaled models are usually built and, hence, the plane areas A_x, A_y may be determined by planimetric measurements of the properly projected surface. Furthermore, the volume V is measured by properly embracing the model and by filling it with water (or mercury of the scaled static testing). Hanging twice rotated homogeneous plates of midplane area A_x and A_y renders h_{Sx} and h_{Sy}, respectively. Also, a free-hanging solid model of V, supported by a string, may be used to determine the geometric center experimentally. The inertia moments of the projected areas must be calculated by numerical integration (ie approximating the integrals by finite sums).

(§)   Illustrative Example: Uplift. A rather "unexpected" effect of hydrostatic pressure loading is observed by filling an open and free-standing vessel in the form of a parabolic thin-walled shell of revolution of density ρ_K; the fluid density is ρ. The depth H is quasistatically, ie slowly increased. With base radius a, height H_0, and shell thickness $t \ll a$, the material volume is approximated by the product of the surface area times wall thickness t, and the weight becomes (see Fig. 2.25, the opening in the shell is negligibly small)

$$G = g\rho_K t \,(2/3)\, 2\pi a H_0 \{[1 + (a/2H_0)^2]^{3/2} - (a/2H_0)^3\} \ .$$

The vertically, upwardly oriented resultant of the pressure loading acts along the shell axis and is given by

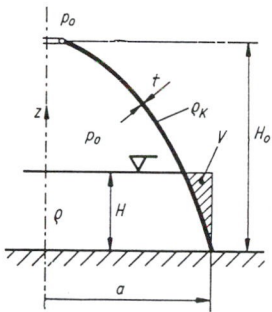

Fig. 2.25. Open, free-standing parabolic shell of revolution. "Uplifting"

$$R_z = g\rho V \quad , \quad V = \pi a^2 \, (H^2/2H_0) \quad .$$

A positive contact pressure at the base of the shell requires

$$G - R_z > 0 \quad ,$$

and, hence, the first inequality may be more severe,

$$(H/H_0)^2 < 2G/g\rho \, \pi a^2 \, H_0 \quad , \quad (H/H_0) < 1 \quad .$$

The filling height is to be constrained even in the case of concrete walls to avoid uplift.

2.3.4. The Hydrostatic Buoyancy

A rigid body is considered to be submerged in a homogeneous and incompressible fluid under the action of gravity. The surface elements of the body are loaded normally by the hydrostatic pressure $p = p_0 + g\rho h$. Virtually, the rigid body may be replaced by the same volume of the surrounding fluid that is at rest and in equilibrium under the same field of surface tractions. Thus, the vertical resultant of the pressure distribution at the interface equilibrates the weight of the fluid displacement V_D , which is $G_D = g\rho V_D$. The resultant $R_z = A_S$, positive upward, is the hydrostatic *buoyancy force* and follows from the condition of equilibrium of the virtual fluid displacement

$$A_S - G_D = 0 \quad \text{to be} \quad A_S = g\rho V_D \quad . \tag{2.110}$$

The point of application of A_S is the geometric centroid S_D of the displacement volume V_D . The weight G_K of the body is attached to the center of gravity S_K , which differs in general from S_D. The necessary condition of equilibrium of the fully submerged body is

$$G_K - A_S = 0 \; , \tag{2.111}$$

which happens to be independent of the average depth in the case of the rigid body having a constant volume of fluid-displacement. Since the moments must vanish, two equilibrium configurations are possible: S_K above or below S_D ($S_K = S_D$ is just a mathematical case). It is easy to verify that the configuration S_K below S_D is not sensitive to small perturbing moments and, thus, is the stable equilibrium configuration; see Sec. 9.1. The rigid body is insensitive to a translation. A small rotational disturbance inclines the line $S_K S_D$ against the vertical direction and renders the double-force G_K and $A_S = G_K$ with moment. In the stable configuration, the restoring moment counteracts the disturbance and (quasistatically) drives the body back into equilibrium. In the other, unstable position, the moment tends to enlarge the perturbation and acts as an overturning moment. The static situation is analogous to that of a rigid pendulum when supported at S_D.

The buoyancy of a partly submerged, floating body is determined by the above given formulas when considering the surrounding fluid inhomogeneous with a horizontal interface. The displacement is separated by this plane in two volume portions, and the resulting force of the surface tractions is the sum $A_S = g\rho V_D + g\rho_A V_A$, where V_D is the displacement in the lower and denser liquid and V_A is the volume portion in the upper and less dense fluid, usually air. For the latter $\rho_A << \rho$, and, in that case,

$$A_S = g\rho V_D \; , \tag{2.112}$$

with sufficient accuracy. V_D is the displacement volume in the liquid only. Since the effects of the surface tension are not considered here, all the related special problems like the "floating of a razor blade" on water, etc., are to be excluded. Equilibrium in the floating configuration requires

$$G_K = g\rho V_D \; . \tag{2.113}$$

Stability is given for the strongly inhomogeneous floating body when S_K is situated below S_D, analogous to the submerged body. A small general disturbance renders restoring forces and moments. A different situation is encountered for the homogeneous floating body or, generally, in the case, where S_K is above S_D . Restoring forces are still present for a translational motion in the vertical direction but, contrary to the submerged body, a rotational disturbance may

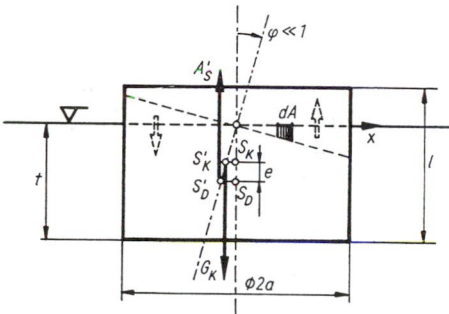

Fig. 2.26. Cylinder, floating in the upright position. Perturbation of the equilibrium configuration

produce a restoring moment (eg the cross-section of a ship is optimized for stabilizing moments). Subsequently, the floating of a homogeneous circular cylinder of density ρ_S is considered in the upright and in the lying horizontal position, respectively.

(§) The Upright Floating Cylinder: Stability (Fig. 2.26). The dipping depth t is found from the equilibrium condition

$$G_K - A_S = 0 \quad \text{to be} \quad t/l = \rho_S/\rho < 1 \ .$$

A rotational disturbance $\varphi \ll 1$ produces an axial moment that is decomposed according to Fig. 2.26

$$\delta M_y = g\rho \int_A (dA \times \varphi \ x) - A_S' e \ \varphi > 0 \ , \quad A_S' = A_S \ . \tag{2.114}$$

The inequality gives a restoring moment and, thus, expresses stability. Integration is performed over the horizontal cross-section $A = \pi a^2$ in the (x, y) -plane, at the surface level, and gives the moment of inertia

$$J_y = \int_A x^2 \, dA = \frac{\pi}{4} a^4 \ ,$$

which is substituted together with $A_S = g\rho \ \pi a^2 t$ to render the geometric stability condition

$$(2a)^2 / l^2 > 8(\rho_S /\rho) \ (1 - \rho_S /\rho) \ .$$

By considering the $max\{(\rho_S /\rho) \ (1 - \rho_S /\rho)\} = 1/4$, a sufficient condition follows:

$$(2a / l) > \sqrt{2} \ .$$

<<Only short, overquadratic cylinders are floating in an upright position.>> The restoring moment may be expressed by the actual couple, namely, G_K in S_K' , and the buoyancy force A_S attached to the geometric centroid S_D'' of the actual, perturbed displacement volume,

$$\delta M_y = A_S H_M \ \varphi > 0 \ , \tag{2.115}$$

where φH_M is determined by Eq. (2.114) to be the normal distance between the couple forces in the perturbed configuration. The angle φ cancels, and the inequality is reduced to $H_M > 0$, where H_M is denoted the *metacentric* height. The intersection of the inclined cylinder axis and the vertical line of actual action of $\boldsymbol{A}_S$ is a point above S_K for stable equilibrium. In this example,

$$H_M / l = (a / 2l)^2 \rho / \rho_S - (1 - \rho_S /\rho) / 2 > 0 \ .$$

(§) Floating Cylinder with Horizontal Axis: Stability. In that case, the arc-length $a \ \alpha$ of the wetted surface is used instead of the depth t , and the displacement becomes $V_D = (1/2) \ a^2 \ (\alpha - sin \ \alpha) \ l$. Equilibrium requires $A_S = G_K$ and a nonlinear equation results for α

$$\alpha - sin \ \alpha = 2\pi \ \rho_S /\rho \ \ , \ \ \alpha < 2\pi \ .$$

Since

$$\overline{S_K S_D} = \frac{2}{3\pi} \frac{\rho}{\rho_S} a \ sin^3 \frac{\alpha}{2} \quad \text{and} \quad J_y = \frac{l^3 \ 2a \ sin \dfrac{\alpha}{2}}{12} \ ,$$

the metacentric height becomes

$$H_M / a = \frac{2\rho}{3\pi\rho_S} \left(\frac{l^2}{(2a)^2} - sin^2 \frac{\alpha}{2} \right) sin \frac{\alpha}{2} > 0 \ .$$

The geometric condition for stable floating becomes by inspection *(l/2a) > sin α/2* . A sufficient condition is simply *(l/2a) > 1* ; ie underquadratic cylinders float horizontally. Between upright and horizontal floating, stable configurations exist with an inclined cylinder axis. Due to the more involved geometry, those intermediate cases are not considered here.

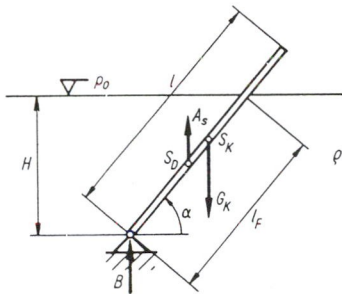

Fig. 2.27. "Buoyancy" of a rigid rod hinged at a point of the wetted surface

(§)   Buoyancy. The concept of hydrostatic lift is preserved also for submerged bodies with supports that are, in a mathematical approximation, attached at points or along lines of the interface to the surrounding fluid. The floating of a hinged, homogeneous, and rigid rod of constant cross-section A illustrates the approximating application of Eq. (2.110) (Fig. 2.27). By neglecting also the inclined intersection of the free surface with the rod, the displacement $V_D = A\, l_F$ is used to approximate the resultant of the hydrostatic pressure by the buoyancy force $A_S = g\rho V_D$ in the moment equilibrium condition

$$(1/2)(G_K\, l - A_S\, l_F)\cos\alpha = 0 \quad , \quad G_K = g\rho_K\, A\, l \ .$$

For $\alpha \neq \pi/2$, the wetted length becomes

$$l_F / l = H / l \sin\alpha = \sqrt{(\rho_K /\rho)} \leq 1 \ .$$

A vertical reaction force B is transmitted at the hinged support that follows from the equilibrium condition $B + A_S - G_K = 0$ to be

$$B = G_K\,[1 - \sqrt{(\rho / \rho_K)}] < 0 \ .$$

The second solution of the factorized moment equation gives $\alpha = \pi/2$, and, for the rod in the upright position, $A_S = g\rho AH$, $l_F = H < l$, the reaction force is

$$B = G_K\,(1 - \rho H / \rho_K l) \ .$$

The upright configuration is stable under the condition

$$\delta M_y = A_S\,\frac{H}{2}\,\varphi - G_K\,\frac{l}{2}\,\varphi > 0 \quad , \quad 0 < \varphi \ll 1 \ ,$$

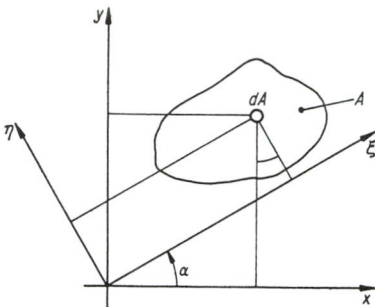

Fig. 2.28. Area A and two rotated coordinate systems

which becomes finally, φ cancels,

$$\rho_K / \rho < \left(\frac{H}{I}\right)^2 < 1 .$$

2.4. Moments of Inertia of a Plane Area A and Their Rules of Transformation

After selection of the cartesian coordinates (x, y) in the plane of the given area A (see Fig. 2.28), the axial moments of inertia are defined by the double integrals over the given domain

$$J_x = \int_A y^2 \, dA , \quad J_y = \int_A x^2 \, dA ,$$

$$(2.116)$$

and the moment of deviation, the deviatoric or centrifugal moment which is also named the product of inertia, is

$$J_{xy} = \int_A xy \, dA .$$

$$(2.117)$$

These second-order moments depend on the shape and the size of the area A and its location and orientation with respect to the coordinates (x, y). Their dimension is length to the power of four, eg $[m^4]$. For many shapes and cross-sections of (milled) metal profiles, these moments are tabulated in handbooks of mechanical and civil engineering. See, eg
Dubbels Taschenbuch für den Maschinenbau (W. Beitz and K. H. Küttner, Eds.), New York: Springer-Verlag, 14th ed. 1981, in German.

The deviatoric moment may be positive or negative or zero. It is a measure of the deviation of the area distribution from symmetry. If either one of the axes x or y is an axis of symmetry, $J_{xy} = 0$. If Eqs. (2.116) and (2.117) are (analytically or numerically) integrated for a selected pair of axes, the second-order moments with respect to parallel translated coordinates or a coordinate system rotated through the angle α can be determined without further and possibly cumbersome integrations.

(§)   Moments of Inertia About Parallel Axes. For the ξ-axis parallel x in normal distance b, the relation $\eta = y + b$ is substituted in the definition of the axial moment of inertia to render

$$J_\xi = \int_A \eta^2 \, dA = \int_A y^2 \, dA + 2b \int_A y \, dA + b^2 A = J_x + b^2 A + 2b \int_A y \, dA \quad .$$

In general, the static moment about the x-axis is needed. If the x-axis passes through the geometric centroid of the area, the static moment vanishes by definition with respect to that central axis, and the transformation reduces to *Steiner's lemma* :

$$J_\xi = J_x + b^2 A \quad . \tag{2.118}$$

x is the central axis, and ξ parallel to the central axis in normal distance b. Analogously, for the axis η parallel to y in the normal distance a, where $\xi = x + a$ holds,

$$J_\eta = \int_A \xi^2 \, dA = \int_A x^2 \, dA + 2a \int_A x \, dA + a^2 A = J_y + a^2 A + 2a \int_A x \, dA \quad .$$

For the central axis y, the static moment vanishes, and

$$J_\eta = J_y + a^2 A \quad . \tag{2.119}$$

η is parallel to the central axis y in a distance a.

The deviatoric moment $J_{\xi\eta}$ becomes with $\xi = x + a$ and $\eta = y + b$ by definition

$$J_{\xi\eta} = \int_A \xi\eta \, dA = \int_A xy \, dA + a \int_A y \, dA + b \int_A x \, dA + ab \, A$$

$$= J_{xy} + ab \, A + a \int_A y \, dA + b \int_A x \, dA \quad .$$

With both x and y orthogonal central axes, all static moments vanish, and the transformation is reduced to

$$J_{\xi\eta} = J_{xy} + ab\, A \ , \tag{2.120}$$

ξ and η are considered axes in the directional distances a and b from the central axes x and y, respectively.

(§) Moments of Inertia About Rotated Axes (*Mohr's* Circle). A rotation of coordinates about the z-axis through the angle α renders according to Fig. 2.28 the linear relations $\xi = x\, \cos\alpha + y\, \sin\alpha$, $\eta = - x\, \sin\alpha + y\, \cos\alpha$. The orthogonal vector transformation

$$\left\{ \begin{matrix} \xi \\ \eta \end{matrix} \right\} = D \left\{ \begin{matrix} x \\ y \end{matrix} \right\} \ , \quad D = \left(\begin{matrix} \cos\alpha & \sin\alpha \\ -\sin\alpha & \cos\alpha \end{matrix} \right) , \tag{2.121}$$

combines the above given relations and contains the well-known, skew symmetric rotational matrix D . Substitution into the definitions of the second-order moments gives, using proper formulas for the product of trigonometric functions,

$$J_\xi = \int_A \eta^2\, dA = J_x \cos^2\alpha + J_y \sin^2\alpha - J_{xy} \sin 2\alpha$$

$$= \frac{J_x + J_y}{2} + \frac{J_x - J_y}{2} \cos 2\alpha - J_{xy} \sin 2\alpha \ ,$$

$$J_\eta = \int_A \xi^2\, dA = J_x \sin^2\alpha + J_y \cos^2\alpha + J_{xy} \sin 2\alpha$$

$$= \frac{J_x + J_y}{2} - \frac{J_x - J_y}{2} \cos 2\alpha + J_{xy} \sin 2\alpha \ ,$$

$$J_{\xi\eta} = \int_A \xi\eta\, dA = \frac{J_x - J_y}{2} \sin 2\alpha + J_{xy} \cos 2\alpha \ . \tag{2.122}$$

Equation (2.122) represents the transformation formulas of the elements of a plane tensor; see Eqs. (2.13) and (2.14). Thus, the inertia tensor with respect to the rotated coordinates is given by the product of the rotational matrix times the given moment tensor times the transposed rotational matrix

$$\left(\begin{matrix} J_\xi & -J_{\xi\eta} \\ -J_{\xi\eta} & J_\eta \end{matrix} \right) = D \left(\begin{matrix} J_x & -J_{xy} \\ -J_{xy} & J_y \end{matrix} \right) D^T \ , \tag{2.123}$$

which is a *similarity transformation* . Equation (2.122) determines *Mohr´s* circle (compare with *Mohr´s* plane stress circle) in the right half of the $[(J_\xi , J_\eta), - J_{\xi\eta}]$ -plane with the center on the abscissa at $(J_x + J_y)/2 = (J_\xi + J_\eta)/2$ and the points $(J_\xi , - J_{\xi\eta})$, $(J_\eta , J_{\xi\eta})$ at the arc. The principal moments of inertia are

$$J_{1, 2} = (J_x + J_y) / 2 \pm (1/2)[(J_x - J_y)^2 + 4J_{xy}^2]^{1/2} \quad , \qquad (2.124)$$

referred to the orthogonal pair of axes 1 and 2 , where the angle α_1 measured against the x -axis,

$$\tan 2\alpha_1 = - 2J_{xy} /(J_x - J_y) \quad , \qquad (2.125)$$

gives the direction of the principal axis 1, which follows also directly from Fig. 2.29. The deviatoric moment $J_{12} = 0$. The *polar moment of inertia* , which is the moment of inertia about the out-of-plane axis z , is considered in some applications

$$J_p = \int_A r^2 \, dA = \int_A (x^2 + y^2) \, dA = J_x + J_y \quad .$$
$$(2.126)$$

Since J_p is independent of the special orientation of the coordinates x, y , Eq. (2.126) expresses the first invariant of the inertia tensor, $J_x + J_y = J_1 + J_2$, under coordinate rotation about the z -axis.

(§)   The Ellipse of Inertia. By means of the *radius of gyration* $i_x > 0$, defined by

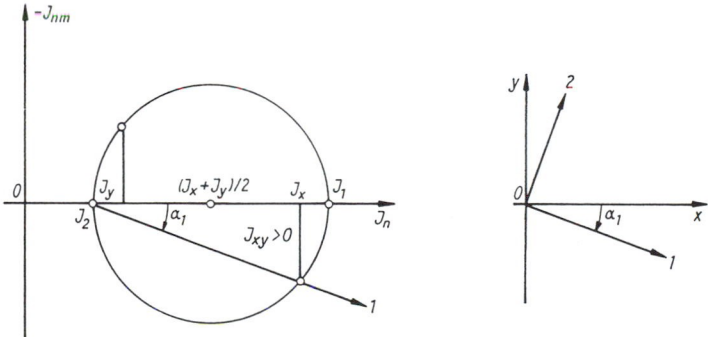

Fig. 2.29. *Mohr´s* circle of the inertia tensor. $J_x , J_y , J_{xy} > 0$ is given.
Wanted: J_1 , J_2 , α_1

$$J_x = A \, i_x^2 \quad , \quad i_x = (J_x / A)^{1/2} \quad , \tag{2.127}$$

with the principal radii $i_1 = (J_1 /A)^{1/2}$ and $i_2 = (J_2 /A)^{1/2}$, a radial coordinate that is inversely proportional to the above radius

$$r = i_1 \, i_2 / i_x \quad , \tag{2.128}$$

is defined. When we refer to the principal axes, the first of Eqs. (2.122) gives $J_x = J_1 \cos^2 \alpha + J_2 \sin^2 \alpha$, where α is now the angle measured from the principal axis 1 to the rotated axis x. Substituting the above definitions, yields the ellipse of inertia (invented by *Culman*.)

$$1 = [(r / i_2) \cos \alpha]^2 + [(r / i_1) \sin \alpha]^2 \tag{2.129}$$

in principal-axes representation, where $x = r \cos \alpha$ and $y = r \sin \alpha$ are the cartesian coordinates of the points of the ellipse. Since $i_1 > i_2$ is understood, the larger semiaxis i_1 is orthogonal to the principal axis 1. The *central ellipse of inertia* is derived when considering the geometric centroid of the area A to be the origin. The inertia tensor is accompanied by an ellipse since $J_1 , J_2 > 0$. A conic section is associated with a general plane symmetric tensor (like the plane strain and the plane stress tensor), eg a hyperbola or parabola is also possible in addition to the ellipse. The three-dimensional mass inertia tensor of Eq. (7.54) is represented by an ellipsoid with all special cases of symmetry possible.

(§)   Example: The Central Ellipse of a Rectangle, $A = B \times H$. The pair of central axes parallel to the edges are axes of symmetry and, hence, are the central pair of principal axes. With $J_1 = BH^3/12 > J_2 = HB^3/12$, the ratio $i_1 /i_2 = H/B$ and the principal radius of gyration is $i_1 = H \sqrt{3}/6 \cong 0.289 \, H$.

2.5.   Statics of Simple Structures

Slender, one-dimensional structures like rods, beams, frames, and cables are considered together with trusses under the action of prescribed external loadings. The stress state in these structures is commonly simplified, and furthermore, statically equivalent internal forces replace the distributed stresses. Equilibrium considerations for a beam element are performed, in general, in a "frozen" state of deformations. For an idealized cable element, the equilibrium configuration is subsequently determined under the assumption of inextensibility.

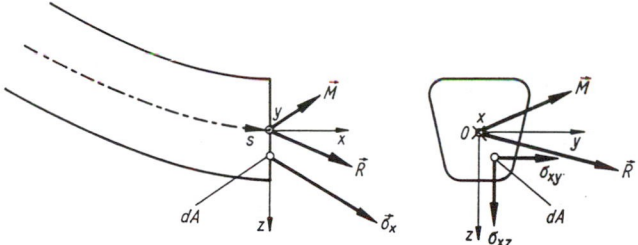

Fig. 2.30. Cross-sectional resultants of the internal forces $\sigma_x \, dA$

2.5.1. Beams and Frames

A slender and curved beam under general loading is cut along a cross-section at a position given by the arc-length s, and a local coordinate frame is considered (see Fig. 2.30). The stress vector σ_x is decomposed into the normal stress σ_{xx} in the tangential direction of the beam axis (the latter is quasidefined by the line which connects the geometric centroids of the cross-sections and has the arc-length s) and into two shear components, σ_{xy} and σ_{xz}, lying in the cross-section. A rod or beam is a slender body whose characteristic dimensions in the cross-sections are small when compared to the overall arc-length (span). Consequently, the stress tensor is approximated by considering the stress components σ_{xx}, $\sigma_{xy} = \sigma_{yx}$, $\sigma_{xz} = \sigma_{zx}$ and neglecting the other four components, σ_{yy}, σ_{zz}, $\sigma_{yz} = \sigma_{zy}$. There are cases in which the latter stress components are not small, eg in some composites, and *generalized beam theories* must be applied. Subsequently, the simplified stress state is assumed to give a good approximation and the internal forces $\sigma_x \, dA$ distributed over the plane cross-section A, are reduced to a point 0, the origin of the in-plane coordinates y, z, in a statically equivalent manner (see Fig. 2.30)

$$\mathbf{R} = \int_A \sigma_x \, dA \quad , \quad \mathbf{M} = \int_A (\mathbf{r} \times \sigma_x) \, dA \quad , \quad \mathbf{r} = y \, \mathbf{e}_y + z \, \mathbf{e}_z \ .$$
(2.130)

The directions of the coordinates with respect to the beam axis indicate the technical characteristics of the components of the internal resultant force $\mathbf{R} = N \, \mathbf{e}_x + Q_y \, \mathbf{e}_y + Q_z \, \mathbf{e}_z$:

Axial or *normal force*

$$N(s) = \int_A \sigma_{xx} \, dA \ ,$$
(2.131)

where

$$N/A = (1/A) \int_A \sigma_{xx} \, dA \quad ,$$

gives the mean of the normal stress distribution at $s = const$, which becomes the uniformly distributed normal stress in the case of pure tension or compression; see Eq. (2.4) with $F = N$.

Shear forces

$$Q_y (s) = \int_A \sigma_{xy} \, dA \quad , \qquad Q_z (s) = \int_A \sigma_{xz} \, dA \quad , \tag{2.132}$$

where

$$Q_i / A = (1/A) \int_A \sigma_{xi} \, dA \quad , \quad (i = y, z)$$

determines the mathematical mean of the shear-stress components at $s = const$. Engineering applications of such a constant shear stress are found in approximating formulas of lightweight structural design.

In the case of a straight beam, $s = x$, a central loading in the axial direction renders $Q_y = Q_z = 0$ and also $\boldsymbol{M} = \boldsymbol{0}$ and, hence, reconsiders the stress state of a tensile or compressive rod under the action of the axial force N with mean normal stress N/A .

The axial moments of the internal forces are found by expanding the formal determinant of the vector product

$$\boldsymbol{M} = \int_A \begin{vmatrix} \boldsymbol{e}_x & \boldsymbol{e}_y & \boldsymbol{e}_z \\ 0 & y & z \\ \sigma_{xx} \, dA & \sigma_{xy} \, dA & \sigma_{xz} \, dA \end{vmatrix} = M_x \boldsymbol{e}_x + M_y \boldsymbol{e}_y + M_z \boldsymbol{e}_z \quad ,$$

$$\tag{2.133}$$

$$M_x = \int_A (y\sigma_{xz} - z\sigma_{xy}) \, dA \quad , \qquad M_y = \int_A z\sigma_{xx} \, dA \quad , \qquad M_z = -\int_A y\sigma_{xx} \, dA \quad .$$

Changing the point of reference from 0 to $A'(y = b, z = c)$ in the cross-section yields by the specialization of Eq. (2.60) the new moment vector

$$\boldsymbol{M'} = \boldsymbol{M} - \boldsymbol{a} \times \boldsymbol{R} = M'_x \boldsymbol{e}_x + M'_y \boldsymbol{e}_y + M'_z \boldsymbol{e}_z \quad , \quad \boldsymbol{a} = b \boldsymbol{e}_y + c \boldsymbol{e}_z \quad ,$$

and the new axial moments

$$M'_x = M_x - (b\,Q_z - c\,Q_y) \quad , \quad M'_y = M_y - c\,N \quad , \quad M'_z = M_z + b\,N \quad . \quad (2.134)$$

The appearance of the components of R within these expressions suggests the consideration of the two groups of internal forces separately: $(Q_y,\,Q_z,\,M_x)$ and $(N,\,M_y,\,M_z)$. The latter may be reduced in a statically equivalent manner to the normal force $N \neq 0$, attached eccentrically at the special point $A' = A_N$ with the coordinates

$$y_N = b = -M_z/N \quad , \quad z_N = c = M_y/N \quad , \quad M'_y = M'_z = 0 \quad , \qquad (2.135)$$

with respect to the origin 0. Another special point of reduction $A' = S$ may be chosen such that the axial moments M'_y and M'_z become independent of the axial force N or, equivalently, independent of the normal stress distribution $\sigma'_{xx}\,(y,\,z;\,s)$ due to N (that is, eg independent of the uniform mean stress $\sigma'_{xx} = N/A$). If we put $\sigma_{xx} = \sigma'_{xx} + \sigma''_{xx} + \sigma'''_{xx}$, where

$$\int_A \sigma'_{xx}\,dA = N \quad , \quad \int_A \sigma''_{xx}\,dA = 0 \quad ,$$

holds, and where σ'''_{xx} denotes a self-equilibrating stress distribution that renders neither a normal force nor a moment,

$$\int_A \sigma'''_{xx}\,dA = 0 \quad , \quad \int_A z\sigma'''_{xx}\,dA = \int_A y\sigma'''_{xx}\,dA = 0 \quad ,$$

Eq. (2.134) yields upon substitution

$$M'_y = \int_A z(\sigma'_{xx} + \sigma''_{xx})\,dA - cN \quad , \quad M'_z = -\int_A y(\sigma'_{xx} + \sigma''_{xx})\,dA + bN \quad .$$
$$(2.136)$$

The assumption of independence of the moments of the forces $\sigma''_{xx}\,dA$ of the location of the point of reference, ie

$$M'_y = \int_A z\,\sigma''_{xx}\,dA \quad , \quad M'_z = -\int_A y\,\sigma''_{xx}\,dA \quad ,$$

determines the coordinates, by comparing the remaining moments in Eq. (2.136),

$$b = (1/N) \int_A y \, \sigma'_{xx} \, dA \quad , \quad c = (1/N) \int_A z \, \sigma'_{xx} \, dA \quad .$$

$$(2.137)$$

If there is any axis of symmetry through the geometric centroid of the cross-section, eg the z-axis, and if the normal stresses σ'_{xx} are at least evenly distributed with respect to this axis, the first of the above equations yields, if we put $y = y_S + y'$ (the integrand $y' \sigma'_{xx}$ is skew-symmetric)

$$b = (1/N) \int_A y_S \, \sigma'_{xx} \, dA + (1/N) \int_A y' \, \sigma'_{xx} \, dA = y_S \quad .$$

$$(2.138)$$

The point of application of the axial force is located on the axis of symmetry at c. For a doubly symmetric cross-section, and if σ'_{xx} is evenly distributed also with respect to the y-axis,

$$c = (1/N) \int_A z_S \, \sigma'_{xx} \, dA + (1/N) \int_A z' \, \sigma'_{xx} \, dA = z_S \quad .$$

$$(2.139)$$

The reference point approaches the centroid S of the cross-section. Usually, in this doubly symmetric case of a homogeneous cross-section, the stresses $\sigma'_{xx} = N/A$ are the constant mean stresses. Hence, the cross-sectional integrals

$$N = \int_A z \, \sigma_{xx} \, dA \quad , \quad M_y = \int_A z \sigma_{xx} \, dA \quad , \quad M_z = -\int_A y \sigma_{xx} \, dA \quad ,$$

(y and z are the central axes) are the axial force N and the *bending moments* M_y, M_z, respectively. The latter are independent of the axial force N attached at the centroid S of the cross-section.

The remaining group of forces (Q_y, Q_z, M_x) at 0 may be reduced, in a statically equivalent manner to a point where $M'_x = 0$. Equation (2.134) under this condition gives a straight line ($y = b, z = c$),

$$b \, Q_z - c \, Q_y = M_x \quad .$$

$$(2.140)$$

At any of these points *(b, c)*, the internal group of forces (Q_y, Q_z, M_x) is statically equivalent to an eccentrically applied resulting shear force $Q = (Q_y^2 + Q_z^2)^{1/2}$.

The search for a special point A' = A_N, where the axial moment M'_x becomes independent of Q, is to be performed analogously.

Putting the shear stresses $\sigma_{xi} = \sigma'_{xi} + \sigma''_{xi} + \sigma'''_{xi}$, $i = y, z$, where σ'_{xi} denotes the *torsional shear* with a vanishing mean,

$$\int_A \sigma'_{xy}\, dA = \int_A \sigma'_{xz}\, dA = 0 \ ,$$

and furthermore, where the shear forces are the resultants of σ''_{xi} ,

$$Q_i = \int_A \sigma''_{xi}\, dA \ , \quad i = y, z \ ,$$

(whereas σ'''_{xi} denotes a self-equilibrating shear-stress distribution) renders upon substitution in the first of Eq. (2.134)

$$M'_x = \int_A [y\,(\sigma'_{xz} + \sigma''_{xz}) - z\,(\sigma'_{xy} + \sigma''_{xy})]\, dA \ - (bQ_z - cQ_y) \ .$$

Hence, the *torsional moment* , the *torque* , $M_T = M'_x$, by definition

$$M_T = \int_A [y\sigma'_{xz} - z\sigma'_{xy}]\, dA \ , \tag{2.141}$$

follows, if the point of reference is selected to be located on the straight line which parallels **Q** ,

$$\int_A [y\sigma''_{xz} - z\sigma''_{xy}]\, dA = bQ_z - cQ_y \ . \tag{2.142}$$

Since $\sigma''_{xi} \neq const$, in general, $b \neq y_S$ and $c \neq z_S$. The reference point $A' = A_M$ is called the *center of shear* . In the case where an axis of symmetry parallels z , and where the shear-stress components σ''_{xz} are evenly distributed, the center of shear is located on that symmetry axis. Putting $y = y_S + y'$ and comparing terms in the equation render that result,

$$\int_A [(y_S + y')\sigma''_{xz} - z\sigma''_{xy}]\, dA = y_S\, Q_z - \int_A z\sigma''_{xy}\, dA = bQ_z - cQ_y \ ,$$

$$b = y_S \ , \quad c = (1/Q_y)\int_A z\sigma''_{xy}\, dA \ . \tag{2.143}$$

Analogously, it follows for a doubly symmetric cross-section, in the case of an evenly distributed shear stress σ''_{xy} with $z = z_S + z'$,

$$\int_A z\sigma''_{xy}\, dA = \int_A (z_S + z')\sigma''_{xy}\, dA = z_S Q_y = cQ_y \quad , \quad c = z_S \quad .$$

(2.144)

Thus, the geometric centroid of the cross-section becomes the shear center and, hence, is the special point of reference for both systems of internal forces only for doubly symmetric "homogeneous" cross-sections, ie the axial moment $M_x = M_T$ is the torque (independent of the shear forces), and M_y, M_z are the bending moments (independent of any axial force N). In all the other, more general cases, the determination of the torsional moment for any loading in torsion and shear requires knowledge of the location of the center of shear A_M [see Eq. (6.118)] in addition to the geometric centroid S. The latter point S looses its static importance, eg for composite beams; for thermal stresses, see Eq. (6.93).

2.5.1.1. Local Equilibrium of a Plane Arch and Plane Beam Element (Fig. 2.31)

The curved beam is assumed free of torsion, and the equilibrium of an element (the free-body diagram is shown in Fig. 2.31) gives the coupled relation between the axial and shear forces and the bending moment with the given external loading per unit of length, q_r and q_φ. The latter components are determined by the body forces (weight) and surface tractions through their reduction to the beam axis. No externally distributed moment loadings are considered subsequently. In addition to the local polar coordinates r, φ, a cartesian frame y, z in the cross-section is used. Resultants of the internal forces in positive coordinate directions are applied to the positive cross-section, with its normal pointing in the direction of increasing arc-length, $(s + ds)$; the opposite forces act on the section at s .

The planar field of forces N , $Q_z = Q$, $M_y = M$, at s and $s + ds$, respectively, and $q_r\, ds$, $q_\varphi\, ds$, is subject to the following three conditions of equilibrium.

In the radial direction (r or z), with a linear approximation of $Q_z (s + ds)$ shown explicitly in the vanishing sum of forces, $o(ds^2)$ means terms of the second order

$$Q(s) + (dQ/ds)\, ds + q_r\, ds - Q(s) - N(s)\, d\varphi + o(ds^2) = 0 \quad .$$

Substitution of the arc-element $ds = r(s)d\varphi$, where r^{-1} is the curvature of the beam axis, and taking the limit $d\varphi \to 0$ render the exact differential condition

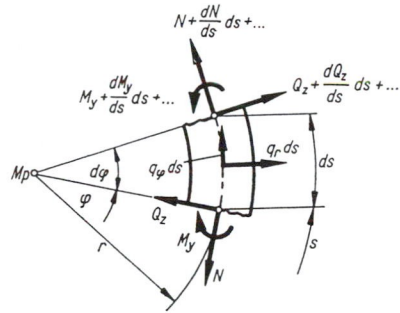

Fig. 2.31. Free-body diagram of a plane arch element in shear bending

$$dQ/ds - N/r = -q_r \quad . \tag{2.145}$$

In the tangential direction (ds or x), the vanishing sum of forces yields [a linear expansion of $N(s + ds)$ is shown explicitly]

$$N(s) + (dN/ds)\, ds + q_\varphi\, ds - N(s) + Q(s)\, d\varphi + o(ds^2) = 0 \quad ,$$

and, in the limit $d\varphi \to 0$,

$$dN/ds + Q/r = -q_\varphi \quad . \tag{2.146}$$

The sum of the axial moments about the y-axis vanishes; $M(s + ds)$ is linearly approximated in the worked out part of the equation

$$M(s) + (dM/ds)\, ds - M(s) - Q(s)\, ds + o(ds^2) = 0 \quad ,$$

which becomes in the limit $d\varphi \to 0$ the exact relation between the shear force and the bending moment

$$Q_z = dM_y/ds \quad . \tag{2.147}$$

These three differential conditions of local equilibrium decouple at once when the beam axis is straight (approximately or exactly after the deformation of a properly predeformed beam), since $r^{-1} \to 0$, $d\varphi \to 0$. But $ds = r\, d\varphi = dx$, and in that case, they are easily integrated. Hence,

$$\frac{dN}{dx} = -q_x \quad , \qquad N(x) = N_0 - \int_0^x q_x(\xi)\, d\xi \quad , \qquad N(0) = N_0 \quad . \tag{2.148}$$

$$\frac{dQ_z}{dx} = -q_z \quad , \quad Q_z(x) = Q_0 - \int_0^x q_z(\xi)\,d\xi \quad , \quad Q_z(0) = Q_0 \ .$$

$$(2.149)$$

Equation (2.147) now reads $dM_y/dx = Q_z$ and becomes, after differentiation, the linear second-order differential equation for the bending moment

$$d^2M_y/dx^2 = -q_z \ . \tag{2.150}$$

Integrating twice, or alternatively, a single integration of Q_z in Eq. (2.149), yields

$$M_y(x) = Q_0 x + M_0 - \int_0^x \left(d\eta \int_0^\eta q_z(\xi)\,d\xi \right) \quad , \quad M_y(0) = M_0 \ ,$$

with the two integration constants Q_0 and M_0. The box integral is changed to a convolution integral by partial integration

$$\int_0^x (d\eta\, Q_z(\eta)) = [\eta\, Q_z(\eta)]\Big|_0^x - \int_0^x \eta\frac{dQ_z}{d\eta}\,d\eta = x\, Q_z(x) + \int_0^x \xi q_z(\xi)\,d\xi$$

$$= -\int_0^x (x-\xi)\, q_z(\xi)\,d\xi \ ,$$

and the bending moment can be expressed by a single weighted integration of the transverse loading

$$M_y(x) = Q_0 x + M_0 - \int_0^x (x-\xi)\, q_z(\xi)\,d\xi \ .$$

$$(2.151)$$

The above representation of the cross-sectional resultants can be verified by considering the equilibrium of a finite element of the straight beam of length x . According to the free-body diagram shown in Fig. 2.32 that also applies in a problem-oriented manner to any given special beam, the planar system of forces is determined by $q_x(\xi)\,d\xi$, N , $q_z(\xi)\,d\xi$, Q_z , M_y . Thus,

$$N(x) + \int_0^x q_x(\xi)\,d\xi - N_0 = 0 \quad , \quad Q_z(x) + \int_0^x q_z(\xi)\,d\xi - Q_0 = 0 \quad ,$$

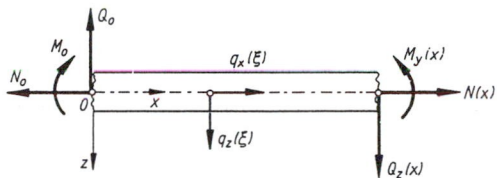

Fig. 2.32. Free-body diagram of a finite and straight beam element, loaded in tension and shear-bending

is the vanishing sum of forces in the horizontal and vertical direction, respectively, and reproduces the formal solution of Eqs. (2.148) and (2.149). The equilibrium of moments about the point S at x requires

$$M_y(x) + \int_0^x (x - \xi)\, q_z(\xi)\, d\xi - Q_0 x - M_0 = 0 \quad .$$

Also, Eq. (2.151) is reproduced, and the integration constants N_0, Q_0, M_0 have, according to Fig. 2.32, the static meaning of axial and shear force and bending moment, respectively, at the "negative" cross-section at $x = 0$.

   If the beam is supported at $x = 0$, all three integration constants may be non-zero for a rigidly *clamped* edge and equal the reaction forces acting on the beam: $N_0 = A_h$, $Q_0 = A_v$, $M_0 = M_e$. In the case of a *simple* support, $M_0 = 0$, at a *free* end of the beam, $Q_0 = M_0 = 0$. The latter are homogeneous *dynamic boundary conditions.*

   In general, the point 0 at $x = 0$ is considered in the close neighborhood to the right of any single force loadings $F_z = P_1$, P_2, ..., $F_x = F_1$, F_2,... and of any concentrated external couples $M = M_1$, M_2,..., and N_0, Q_0, M_0 are the resultant internal forces in the adjacent negative cross-section.

   In the case of a *biaxial bending* of a straight beam (a result of oblique bending, see Exercises A 6.1 and A 11.6) a lateral loading $q_y(x)$ is given in addition to $q_z(x)$, and Eqs. (2.149) and (2.150) are to be supplemented by the separate relations in the (x, y)-plane

$$dQ_y/dx = -q_y(x) \quad , \quad dM_z/dx = -Q_y \quad ,\text{or} \quad d^2M_z/dx^2 = +q_y(x) \quad . \quad (2.152)$$

The central pair of axes y, z, must be selected as the principal axes 1, 2 of the cross-section; see, eg Eq. (6.39) and Exercise A 6.1.

   Subsequently, the statics of a *circular arch* of radius $r = a = const$ is developed as a special case of a curved beam. By means of the generalized force $N_H(s) = N + M_y/a$, with the derivative dN_H/ds

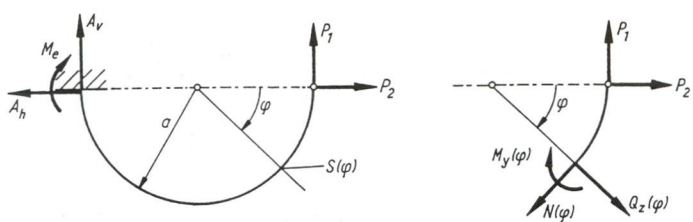

Fig. 2.33. Semicircular clamped beam and the free-body diagram of a finite element

$= dN/ds + Q_z /a$, Eq. (2.146) is simplified and can be integrated at once

$$\frac{dN_H}{d\varphi} = - aq_\varphi \ , \quad N_H(\varphi) = N(\varphi = 0) - a \int_0^\varphi q_\varphi \, d\varphi \ .$$

Eliminating the shear force in Eq. (2.145) by means of Eq. (2.147) and substituting for N give a linear second-order differential equation for the bending moment

$$d^2 M_y/d\varphi^2 + M_y = aN_H - a^2 q_r \ . \qquad (2.153)$$

Since $N_H (\varphi)$ is calculated in advance, the forcing function is known, and the solution of the linear equation (2.153) is given by the superposition of a particular integral to the homogeneous solution

$$M_y(\varphi) = M_0 \cos \varphi + M_1 \sin \varphi \ . \qquad (2.154)$$

The special *radial loading* , $q_r = q = const$ and $q_\varphi = 0$, renders $N_H = N_H$ $(\varphi = 0) = const$, and the moment becomes explicitly $M_y(\varphi) = M_0 \cos \varphi$ $+ M_1 \sin \varphi + aN_H - a^2 q$. The constants of integration, $N_H = N_0 + M_y (\varphi = 0)/a$, M_0 and M_1 , are to be determined by consideration of the boundary conditions and in the statically indeterminate case by additional conditions on the deformations. See Sec. 6.4.1.

Analogously to the straight beam, the cross-sectional resultants $N(\varphi)$, $Q_z(\varphi)$, $M_y(\varphi)$ are expressed by the given loading through the equilibrium consideration of a finite element of the circular arch. This is shown in the simple illustrative example of Fig. 2.33. The in-plane system of forces [P_1 , P_2 are given and $N(\varphi)$, $Q_z = Q(\varphi)$, $M_y = M(\varphi)$ wanted] is in the state of equilibrium, if in $0 \le \varphi \le \pi$,

$$N \cos \varphi + Q \sin \varphi - P_1 = 0 \ ,$$
$$N \sin \varphi - Q \cos \varphi - P_2 = 0 \ ,$$

and with the point of reference $S(\varphi)$,

$$M - aP_1(1 - \cos \varphi) + aP_2 \sin \varphi = 0 \ .$$

The solution of the system of linear equations is

$$N = P_1 \cos \varphi + P_2 \sin \varphi = dQ/d\varphi \ , \qquad (q_r = 0) \ ,$$
$$Q = P_1 \sin \varphi - P_2 \cos \varphi = dM/(ad\varphi) = - dN/d\varphi \ , \qquad (q_\varphi = 0) \ ,$$
$$M = aP_1(1 - \cos \varphi) - aP_2 \sin \varphi \ .$$

The support reactions are found by putting $\varphi = 0$, or quite independently, in a poblem oriented manner, by considering the equilibrium of the gross system consisting of the forces and the clamping moment $(P_1 , P_2 , A_v , A_h , M_e)$. The latter exercise is left to the reader.

2.5.1.2. Straight Beams, Force and Funicular Polygon

A cantilever beam and a single-span hinged-hinged beam are both considered in plane bending by the application of problem-oriented analytical and graphic methods for illustrative reasons, and for the sake of checking formal solutions derived by computer routines. When we work in the (x, z) -plane, a given torsion-free loading $q_z (x) = q(x)$ renders the internal resultants $Q_z = Q(x)$ and $M_y = M(x)$, the shear force and bending moment, respectively.

(§)   The Cantilever Beam. This beam in Fig. 2.34 is loaded by a linearly distributed lateral force per unit of length, $q = q_0 (1 - x/l)$, and in addition, by a single load $F = P$, at the tip, $x = l$. The given loading and reaction forces at the clamped edge, $x = 0$, are in-plane forces $(P, q \ dx, A_v = A, M_e)$. Since the sum of forces in the axial direction is identically zero, two conditions of equilibrium of the

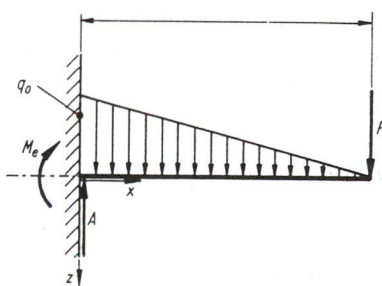

Fig. 2.34. Cantilever loaded by a linearly varying distributed load and a single force, free of torsion

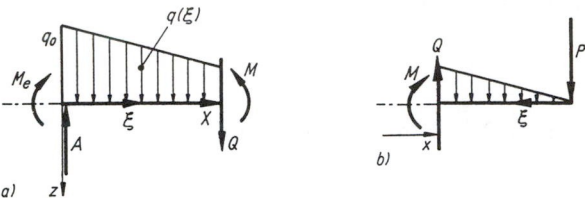

Fig. 2.35. Finite elements of the cantilever and free-body diagrams

gross system remain for the determination of the unknowns A and M_e

$$A - P - R = 0 \; , \quad R = \int_0^l q \, dx = q_0 \frac{l}{2} \; ,$$

(at $x = l/3$, statically equivalent).

The selection of the centroid S of the clamped cross-section at $x = 0$ renders

$$M_e + P\,l + R\frac{l}{3} = 0 \; , \quad R\frac{l}{3} = \int_0^l xq \, dx \; .$$

Analogously, $Q(x)$ and $M(x)$ are determined by solving the linear conditions of equilibrium of the left or right finite element shown in Fig. 2.35. The loading of the element in Fig. 2.35(a) is more complicated, and moreover contains the previously determined reaction forces at the support, which may be erroneous. Thus, the second system of Fig. 2.35(b) is preferred.

$$\text{Forces } (qdx, A, M_e, Q, M): \; A - \int_0^x q \, d\xi - Q = 0,$$

$$S(x): \; M + \int_0^x (x - \xi)\, q(\xi) \, d\xi - M_e - A\,x = 0.$$

$$\text{Forces } (qdx, P, Q, M): \; Q - \int_x^l q \, d\xi - P = 0,$$

$$S(x): \; M + \int_0^{l-x} (l - x - \xi)\, q(\xi) \, d\xi + P\,(l - x) = 0.$$

The analytical solution in $0 \le x < l$ is

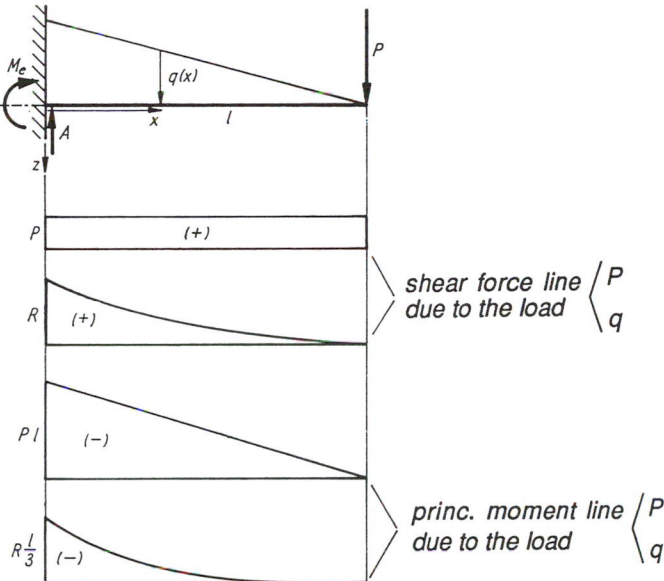

Fig. 2.36. Distribution of shear force and bending moment of the cantilever due to the individual loadings by P and $q(x)$, respectively

$$Q(x) = P + q_0(1 - x/l)(l - x)/2 \quad ,$$
$$M(x) = -P(l - x) - q_0(1 - x/l)(l - x)^2/6 \quad . \tag{2.155}$$

Its proof is given by checking $dM/dx = Q$ and by inspection of $M(x = 0) = M_e$, $Q(x=0) = A$.

Due to the linearity of the equilibrium conditions, superposition applies to the loading [see Eq. (2.155)], where P and $q(x)$ appear in separated terms. In engineering practice, such individual loadings are considered successively (see Fig. 2.36), and the results of such analysis are superposed. The boundary conditions at the clamped, $x = 0$, and free edge, $x = l$, of the cantilever are considered in problem-oriented analysis by inspection of Fig. 2.34 .

(§)   The Hinged-Hinged Beam. This beam of span l has an overhang a (see Fig. 2.37) and is loaded uniformly by $q_z = q = const$ and by a single force $F = P$ at the outmost point of the overhang. The uniform loading may be due to the weight of the homogeneous beam of cross-section A, $q = g\rho A$, per unit of length. At simple supports, the reaction moment vanishes, and the equilibrium of the parallel system of in-plane forces ($q\ dx$, P, A_v , B) is considered, $A_h = 0$,

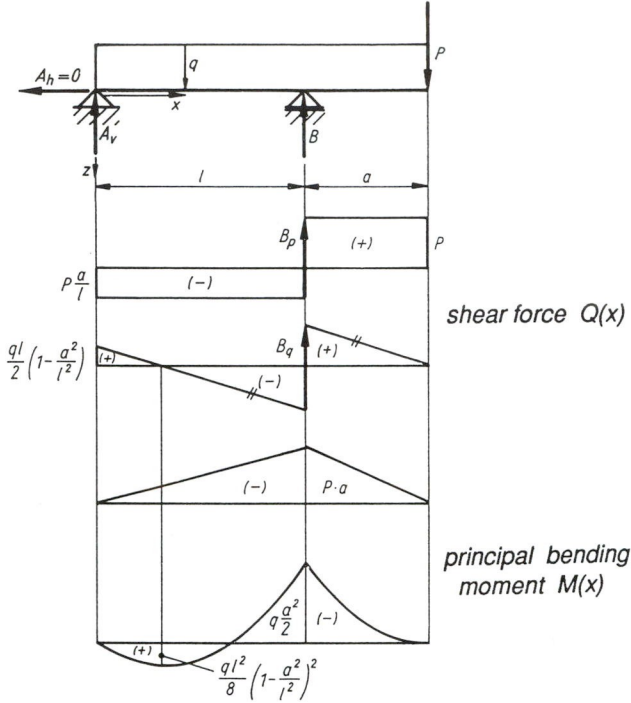

Fig. 2.37. A single-span beam with overhang, simply supported

$$P + R - A_v - B = 0 \ , \quad R = \int_0^{l+a} q \, dx = q \, (l + a) \ , \quad \text{at} \ \ x = (l + a)/2 \ .$$

With reference point $S(x = 0)$,

$$Bl - P(l + a) - R(l + a)/2 = 0 \ .$$

The latter is an explicit equation for the unknown supporting force B. Also, the reaction A_v can be directly determined when considering the equilibrium of moments about the reference point at the support $x = l$ of the simply supported beam

$$A_v l + P a - R [l - (l + a)/2] = 0 \ .$$

The vanishing sum of forces in the vertical direction proves the explicitly and independently derived results

$$A_v = q \, l \, [1 - (a/l)^2]/2 - P \, a/l \ , \quad B = q \, l \, (1 + a/l)^2/2 + P \, (1 + a/l) \ .$$

The consideration of the equilibrium of finite elements in the regions $0 < x < l$, from the left, and in $l < x < (l + a)$, from the right, renders the resultants of the internal forces $Q_z = Q(x)$ and $M_y = M(x)$ by solving

$$0 < x < l: \qquad A_v - q\,x - Q(x) = 0 ,$$

$$A_v\,x - qx\,x/2 - M(x) = 0 ,$$

$$l < x < (l + a): \qquad Q(x) - q(l + a - x) - P = 0 ,$$

$$M(x) + q(l + a - x)^2/2 + P(l + a - x) = 0 ,$$

A_v is to be substituted from above. The shear force varies linearly, $Q(x) = A_v - qx$ in $0 < x < l$, and with the same slope, $Q(x) = P + R - qx$ in the region of overhang, $l < x < (l + a)$. At the support $x = l$, where a single force B is applied to the beam, a jump occurs since

$$\lim_{x \to (l - 0)} Q(x) = A_v - q\,l , \qquad \lim_{x \to (l + 0)} Q(x) = A_v + B - q\,l ,$$

$$Q(l + 0) - Q(l - 0) = [Q(l)] = B .$$

<<At the point of application of a single force, the *shear force* has a *jump* whose value equals that of the external force.>> See Fig. 2.37. To the right of the load P, the shear drops suddenly to zero, according to the boundary condition of a free edge.

The bending moment diagram is steady and it consists of two parabolic arches $M(x) = A_v\,x - qx^2/2$ within the span, $0 \le x \le l$, and $M(x) = -[(q\,l/2)\,(1 + a/l)^2 + P(1 + a/l)]\,l + (P + R)\,x - qx^2/2$, in the adjacent overhang. It shows an analytical and a boundary extremum at those points where the shear goes through zero. The limit $a \to 0$ gives the uniformly loaded *single-span beam* ; the maximal moment at midspan $ql^2/8$ should be kept in mind.

(§) Illustrative Example: Eccentrical Axial Force. Figure 2.38 shows an unsteady moment diagram. A cantilever of length e is rigidly connected to a hinged-hinged beam of span l. At the free end, a single force of given strength F acts parallel to the straight beam axis. At the fixed support of the eccentrically loaded beam, $x = 0$, the reaction has two components $A_h = F$ and A_v. The reaction at the sliding support is B. The latter unknowns are determined by the remaining two conditions of equilibrium: The sum of moments about $S(x = 0)$ and $S(x = l)$ must vanish; $Bl - Fe = 0$ and $A_v\,l + Fe = 0$, respectively. Hence, $B = -A_v = Fe/l$ is a couple of reaction forces with the moment $-Fe$. The cross-sectional resultants are the normal force $N(x)$, the shear $Q(x)$, and the bending moment $M(x)$ and are to be determined from the relations,

$$0 \le x < a: \qquad N - A_h = 0 , \quad N(x) = A_h = F = \text{const}$$

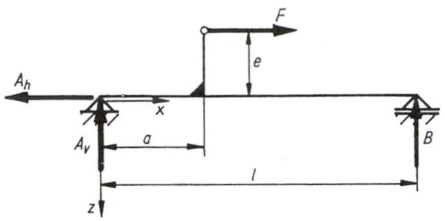

Fig. 2.38. Single-span beam loaded by a singular moment eF

$$Q - A_v = 0 \quad , \quad Q(x) = A_v = -Fe/l = const$$
$$M - A_vx = 0 \quad , \quad M(x) = A_vx = -Fe\,x/l$$
$a < x < l:$
$$N + F - A_h = 0 \quad , \quad N(x) = 0$$
$$Q - A_v = 0 \quad , \quad Q(x) = A_v = -Fe/l = const$$
$$M - A_vx - Fe = 0 \quad , \quad M(x) = A_vx + Fe = Fe(1 - x/l) \quad .$$

Since the limits taken from the left and from the right of the moment diagram are different, their difference renders a jump at a

$$\lim_{x \to (a-0)} M(x) = -Fe\,a/l \quad , \quad \lim_{x \to (a+0)} M(x) = Fe\,(1 - a/l) \quad ,$$
$$M\,(a + 0) - M(a - 0) = [M(a)] = Fe \quad .$$

[M(a)] equals the moment of the external force F with respect to $S\,(x = a)$. <<At the point of application of a couple of forces, the *bending moment* has a *jump* whose value equals that of the external

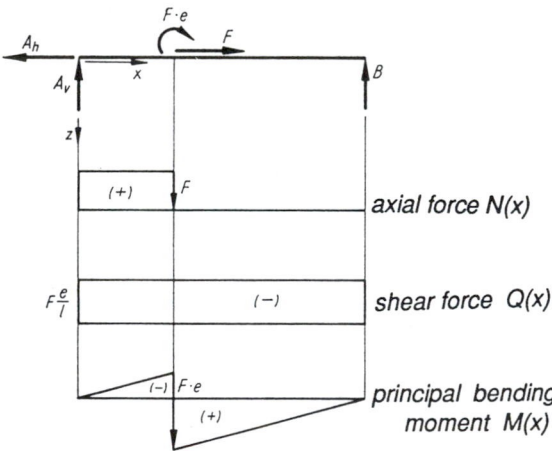

axial force $N(x)$

shear force $Q(x)$

principal bending
moment $M(x)$

Fig. 2.39. Statically equivalent loaded beam of Fig. 2.38 . Diagram of the constant shear and unsteady bending moment

moment.>> See Fig. 2.39, which shows the simply supported beam loaded in a statically equivalent manner.

(§)   The Graphic Solution by Means of the Force and Funicular Polygon. This solution is at first illustrated for a cantilever under single force loading $F = P$ (see Fig. 2.40). The scaled configuration is drawn as span l followed by the force P in the scale of the force plane with a pole C (see also Fig. 2.11).

The dummy forces 1 and 2 according to the polar rays are assumed to equilibrate the given load P; hence, the three forces have a common point of application along the line of action in the configurational plane. The actual forces acting on the beam (P, $A_v = A$, M_e) are supposed to be in equilibrium and, hence, are statically equivalent to the nonparallel plane system (P, 1, 2). Since P is a common force, the remaining systems are also statically equivalent, and the resultant A must equal the vector sum of the dummy forces, ie $A = P$. The force polygon is closed. The funicular polygon in the configurational plane is open and the clamping moment M_e results by considering the moments of the dummy forces with respect to a point of reference on the line of action of the resultant A, eg (minus the dummy force 2) times the normal distance n in Fig. 2.40 . The measurement of the distance η is preferable (here η_e), and the resulting moment M_e becomes equal to (minus the dummy force 2) times $\cos \alpha$ times η_e. In the force plane, the horizontal component of 2 is easily recognized to be the "universal" pole distance H, which is to be measured in the force scale. Thus, by measuring η_e,

$$- M_e = \mu_L \, \mu_F \, H \, \eta_e \quad , \tag{2.156}$$

where

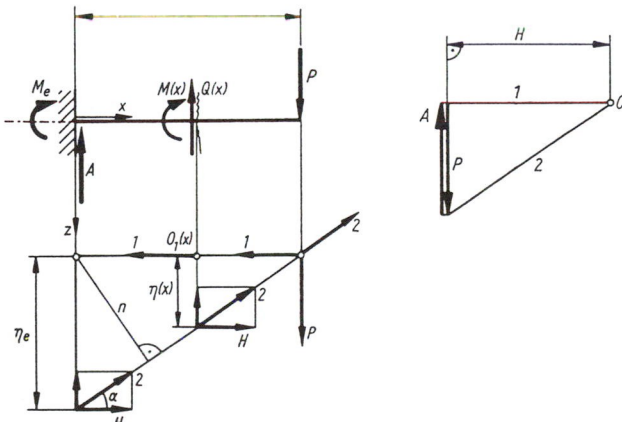

Fig. 2.40. Moment diagram of the cantilever beam and the constant shear, evaluated graphically in the configurational and in the force plane

μ_L = the length scale of the configuration.

μ_F = the force scale.

H = the pole distance in the force plane.

η_e = the lever measured along the line of action of $A_v = A$.

The sign of the moment is given by the equivalence condition to the dummy force couple. Note that only loading components perpendicular to the beam axis are to be considered, to keep H a universal constant in the case of multiple loads applied to the beam.

Analogously, the shear force $Q(x)$ and bending moment $M(x)$ in the field $0 < x < l$, are determined by considering the equilibrium of the (right) finite element shown in Fig. 2.40. Cutting also through the cord rays in the configurational plane renders the equivalent systems of forces: (P, Q, M) and $(P, 1, 2)$. Thus, the resultant of the dummy forces is $Q(x) = P$, but the moment is $-M(x) = \mu_L \mu_F H \eta(x)$. The moment pole must be a point on the line of action of the shear force.

Distributed loads $q_z = q(x)$ are equivalently substituted by piecewise uniform loadings by averaging in intervals of proper lengths and furthermore, by taking their partial resultants. A uniform loading $q = const$ of the *cantilever* of Fig. 2.41 may be equivalently replaced by the single force, the weight $G = q\,l$, attached at $x = l/2$, for consideration of the gross equilibrium. The graphic solution is $(A_v = A = G$ equals the resultant of the dummy forces *1* and *2*) the clamping moment $M_e = -\mu_L \mu_F H \eta_e = -G\,l/2$. With respect to internal forces, however, the single force G gives the resultant at $x = 0$ only. A better approximation of the shear and bending moment in the field is achieved by the consideration of two single forces, $P_1 = P_2 = G/2$, placed equivalently to the uniform load

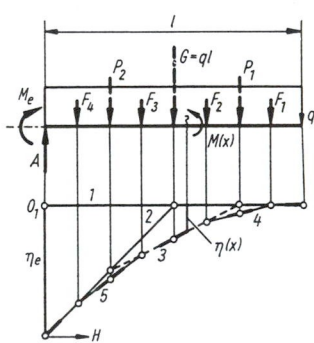 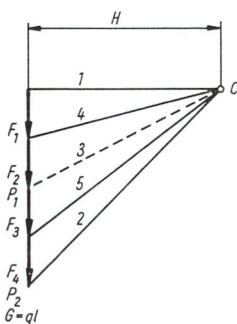

Fig. 2.41. Graphic approximation of the moment distribution of a uniformly loaded cantilever

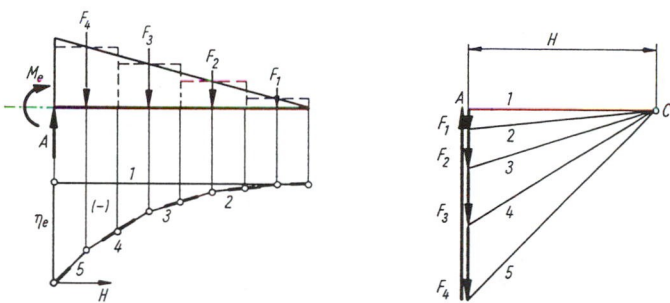

Fig. 2.42. Graphic approximation of the moment distribution of a cantilever loaded by a linearly varying force

at the quarter points of the span. In addition to the dummy forces 1 and 2, the force **3** is considered (it appears in dashed form in Fig. 2.41). Piecewise constant shear and piecewise linearly distributed moment follow according to the single loadings, but the actual line elements (point and tangent, $dM/dx = Q$) of the moment distribution coincide at the cross-section located at $x = l/2$, in addition to those of the previous approximation at $x = 0, x = l$. Further subdivisions (four intervals with $F_i = G/4$, $i = 1, 2, 3, 4$ are considered in Fig. 2.41) render more line elements at the end points of the subintervals (two more at $x = l/4$ and $x = 3 l/4$). The geometrical construction of a quadratic parabola in the configurational plane is formally the same. The moment distribution is completed by free-hand drafting,

$$M(x) = - \mu_L \mu_F H \eta(x) \quad . \tag{2.157}$$

The general case is demonstrated by the consideration of a linearly

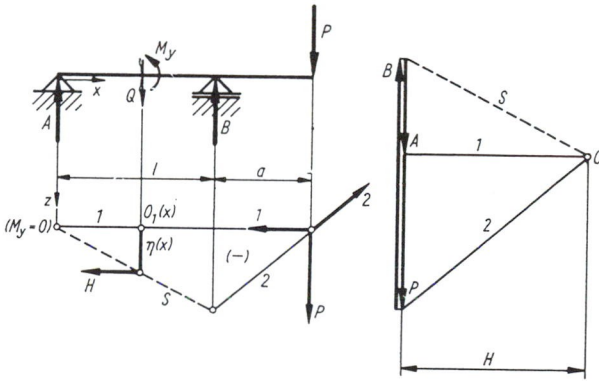

Fig. 2.43. Closing of the funicular polygon by the cord ray *s*

distributed loading without emphasizing the higher accuracy that is (easily) achievable by the exact statically equivalent reduction of the partial resultants of the parallel force loadings. Reaction moment M_e and the line elements shown in Fig. 2.42 are only approximate, since the single forces F_i, $i = 1, 2, 3, 4$ are not applied at the proper force centers. The supporting force $A_v = A$, however, is the sum of the forces F_i and has the proper value.

The *force* and *funicular polygon* renders the moment distribution and, thus, is considered a general graphic procedure to integrate the second-order linear differential equation (2.150),

$$d^2M/dx^2 = -q_z(x) \quad , \tag{2.158}$$

with two proper boundary conditions. The alternative method of replacing the distributed load of an interval by *two equivalent single forces* at the endpoints that is used with advantage in engineering is only mentioned here. The static idea is to discretize the given loading by means of intermediate beams, resting simply supported on the actual beam.

The hinged-hinged single-span beam with overhang loaded by the single force $F = P$ (Fig. 2.43) illustrates the *closing of the funicular polygon* in the case, where the gross moment vanishes in

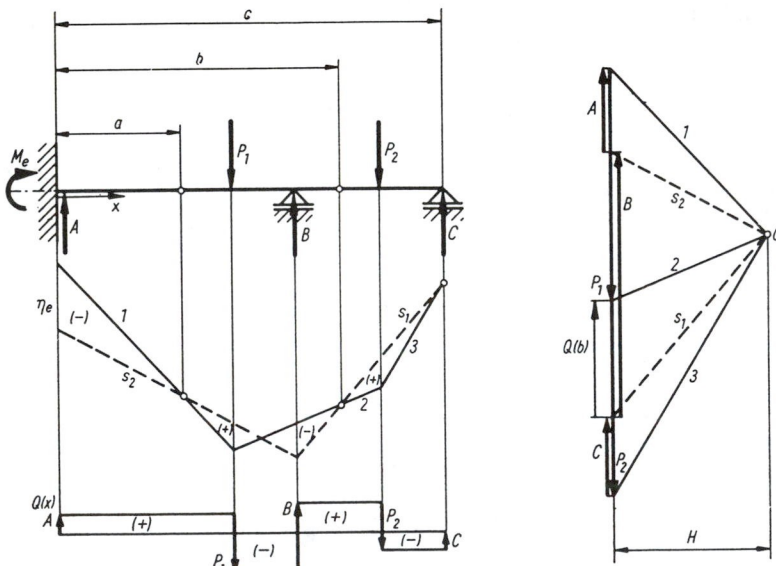

Fig. 2.44. Partial closing of the funicular polygon by the broken line of cord rays s_1, s_2, according to the boundary condition $M (x = c) = 0$, and the conditions at the joints, $M(x = a) = M(x = b) = 0$

equilibrium. The dummy forces 1 and 2 are statically equivalent to the support reactions $A_v = A$, B, and the bending moment is zero at $x = 0$. These conditions render the cord ray s in the configurational plane. Parallel translation through the pole C gives the proper third dummy force s. Otherwise, Fig. 2.43 is self-explanatory.

(§)   Continuous (Multispan) Beams. These beams, resting on three or more supports are sometimes made statically determinate by adding joints that allow for the transfer of axial and shear forces, but do not transfer the bending moment. Such a *double-span beam* with two *Gerber* joints when loaded by two single forces P_1 and P_2 is illustrated in Fig. 2.44.

2.5.1.3. Influence Lines

To represent both shear and bending moment at a selected cross-section at a station x for various loading cases, it is convenient to use the corresponding influence line. The latter considers a single

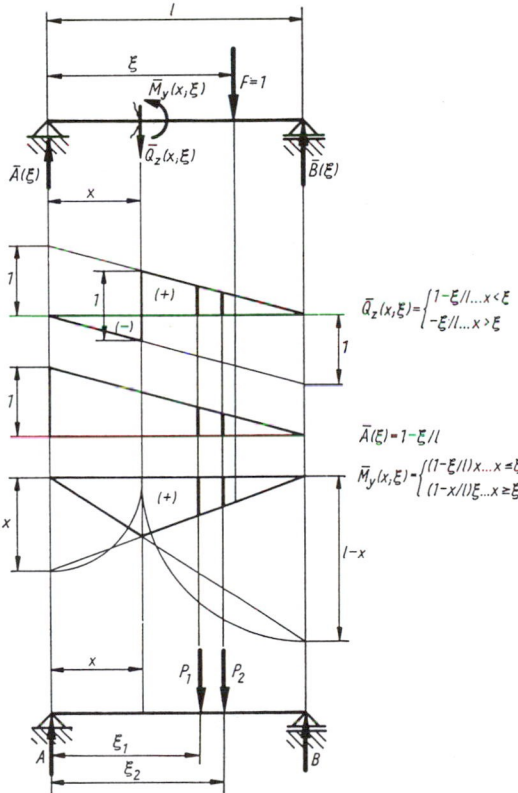

Fig. 2.45. Static influence lines of the single-span beam. See also Sec. 5.2

force loading $F = 1$ at a point in a distance ξ , which is considered variable, and thus gives

$$\overline{Q}_z(x;\xi) \text{ and } \overline{M}_y(x;\xi)$$

by the evaluation of the shear and moment at x , respectively. Boundary conditions are essential when considering these functions. For any load distribution $q_z(\xi)\, d\xi$, the resultants are found by superposition (integration)

$$Q_z(x) = \int_0^l \overline{Q}_z(x;\xi) q_z(\xi)\, d\xi \quad , \quad M_y(x) = \int_0^l \overline{M}_y(x;\xi) q_z(\xi)\, d\xi \quad .$$

$$(2.159)$$

The unit single force $F = 1$ is substituted by the loading intensity $q_z(\xi)\, d\xi$ at ξ. A single force loading $F = P$ at $\xi = \xi_1$ may be expressed by the pseudo-function $q_z(\xi) = P\, \delta(\xi_1 - \xi)$, where $\delta(\xi)$ is the *Dirac delta function* normalized to one, and with the essential property

$$\int_{-\varepsilon}^{+\varepsilon} \delta(\xi)\, d\xi = 1 \quad \text{and} \quad \int_{\xi_1 - \varepsilon}^{\xi_1 + \varepsilon} f(\xi)\, \delta(\xi_1 - \xi)\, d\xi = f(\xi_1) \quad , \quad \varepsilon > 0.$$

$$(2.160)$$

Substitution into the integrals (2.159) gives consequently, because of linearity,

$$Q_z(x) = P \int_0^l \overline{Q}_z(x;\xi)\, \delta(\xi_1 - \xi)\, d\xi = P\, \overline{Q}_z(x;\xi_1) \quad ,$$

$$M_y(x) = P \int_0^l \overline{M}_y(x;\xi)\, \delta(\xi_1 - \xi)\, d\xi = P\, \overline{M}_y(x;\xi_1) \quad .$$

$$(2.161)$$

$P = 1$ renders the influence line by definition. A loading by single forces $P_1, P_2, ..., P_n$ at $\xi_1, \xi_2, ..., \xi_n$ yields through superposition

$$Q_z(x) = \sum_{i=1}^n P_i \overline{Q}_z(x;\xi_i) \quad , \quad M_y(x) = \sum_{i=1}^n P_i \overline{M}_y(x;\xi_i) \quad .$$

$$(2.162)$$

This is a representation of the cross-sectional resultants, which may be used analogously to and as conveniently as the graphic solution, to approximate the integrals (2.159) for distributed loadings.

The application of Eq. (2.162) to the hinged-hinged beam of Fig. 2.45, loaded by the train of single forces P_1 and P_2 moving slowly over the span l, gives

$$Q_z(x) = P_1 \overline{Q}_z(x; \xi_1) + P_2 \overline{Q}_z(x; \xi_2) = P_1 + P_2 - (P_1\xi_1/l + P_2\xi_2/l) \ ,$$

$$M_y(x) = P_1 \overline{M}_y(x; \xi_1) + P_2 \overline{M}_y(x; \xi_2) = (P_1 + P_2) x - (P_1\xi_1/l + P_2\xi_2/l) x \ , \quad (2.163)$$

where $x < \xi_1 < \xi_2$. See also Exercise A 11.2. Putting $x = 0$ in the influence function of the shear renders the influence line of the reaction force $A_v = A$; see Fig. 2.45 for details.

In mathematical terms, the influence lines are *Green's* functions of the coupled system of first-order differential equations, (2.149) and (2.147),

$$\frac{d\overline{Q}_z(x, \xi)}{dx} = -\delta(x - \xi) \quad , \quad \frac{d\overline{M}_y(x, \xi)}{dx} = \overline{Q}_z(x, \xi) \ , \quad (2.164)$$

with the homogeneous boundary conditions

$$\overline{M}_y(x = 0) = \overline{M}_y(x = l) = 0 \ , \quad (2.165)$$

and the inhomogeneous ones

$$\overline{Q}_z(x = 0) = 1 - \frac{\xi}{l} \ , \quad \overline{Q}_z(x = l) = -\frac{\xi}{l} \ , \quad (2.166)$$

which apply for the simply supported beam considered above. Integration over x yields, when we note the subintervals $0 < x < \xi$, $\xi < x < l$,

$$\overline{Q}_z(x, \xi) = \begin{cases} \overline{Q}_z(0, \xi) = 1 - \dfrac{\xi}{l} \ , & x < \xi \\[2ex] -1 + \overline{Q}_z(0, \xi) = -\dfrac{\xi}{l} \ , & x > \xi \ . \end{cases}$$

A second integration gives

$$\overline{M}_y(x, \xi) = \begin{cases} \left(1 - \dfrac{\xi}{l}\right) x \ , & x < \xi \\[2ex] -\dfrac{\xi x}{l} + C \ , & x > \xi \end{cases} \quad , \quad \overline{M}_y(x = 0) = \overline{M}_y(x = l) = 0 \ \rightarrow \ C = \xi \ .$$

See also Sec. 5.2. An application is made in Sec. 6.2.3 (§). For elastic deflections see Exercise A 11.2.

2.5.1.4. Plane Frames and the Three-Hinged Arch

Geometrically interpreted, frames are beams with sharply bent axis or with bifurcating axes. A design goal is, however, to connect several straight beams in a common corner rigidly by vouting locally the cross-sections (thereby increasing the bending rigidity). At those rigid joints, which may be assumed to rotate rigidly during deformation of the frame, the internal forces are transmitted as well as the bending moments. Beam theory does not apply near the corner, where the cross-sectional area changes rapidly. Problem-oriented static analysis is illustrated by considering the planar one-story frame with statically indeterminate hinged supports, loaded by the single forces P and H according to Fig. 2.46. Such a frame is part of a three-dimensional structure, but may be isolated for analysis with respect to in-plane loadings. The planar system of forces (H, P given; A_v, B_v, A_h, B_h wanted) is in equilibrium if

$$A_h + B_h - H = 0 \quad , \quad A_v + B_v - P = 0 \quad ,$$
$$lB_v - eP - aH = 0 \quad \text{and} \quad lA_v + aH - (l - e)P = 0 \quad .$$

The moment conditions have the explicit solution

$$A_v = P(1 - e/l) - Ha/l \quad , \quad B_v = Pe/l + Ha/l \quad .$$

The sum of vertical forces $A_v + B_v = P$ merely verifies the solution. Thus, one of the horizontal support reactions remains statically indeterminate, eg $A_h = X$. Hence, $B_h = H - X$ is expressed by the given loadings and statically indeterminate force X. After this selection of the indeterminate force, the cross-sectional resultants are evaluated by considering the equilibrium of finite elements in the

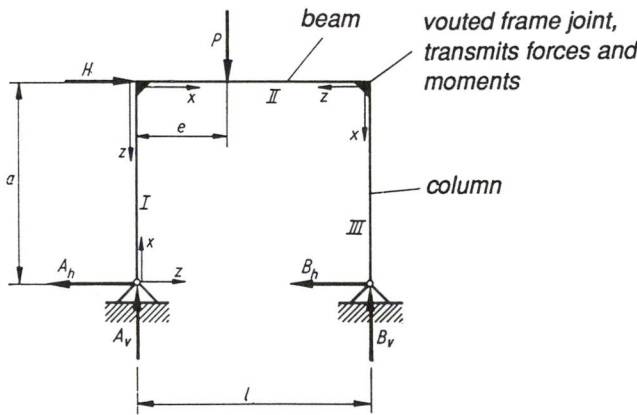

Fig. 2.46. Planar one-story frame with statically indeterminate supports

columns and in the beam (the *crossbar*). Free-body diagrams should be drawn before writing down the following conditions of equilibrium:

In column 1: $0 < x < a$,

$N + A_v = 0$, $N = -A_v$,
$Q - A_h = 0$, $Q = X$,
$M - xA_h = 0$, $M = xX$.

In the crossbar (beam) 11: $0 < x < e$,

$N + H - A_h = 0$, $N = X - H$,
$Q - A_v = 0$, $Q = A_v$,
$M - aA_h - xA_v = 0$, $M = aX + xA_v$,

For $e < x < l$,

$N + H - A_h = 0$, $N = X - H$,
$Q - A_v + P = 0$, $Q = A_v - P$,
$M - aA_h - xA_v + (x - e)P = 0$, $M = aX + eP + x(A_v - P)$.

In column 111: $0 < x < a$,

$N + B_v = 0$, $N = -B_v$,
$Q - B_h = 0$, $Q = B_h = H - X$,
$M + (a - x)B_h = 0$, $M = (a - x)(X - H)$.

In the case of a sliding (roller) support, the left hinge is free to move horizontally, $X = 0$; the frame becomes statically determinate. For the fixed support, the value of X depends on the stiffness distribution within the frame (see Fig. 6.24 and the analysis given there). A possible and practically useful design of a statically determinate frame, however, leaves the supports fixed, but adds a third hinged joint to the crossbar. Such a structure is, in general, called a *three-hinged arch*. Since the bending moment at the

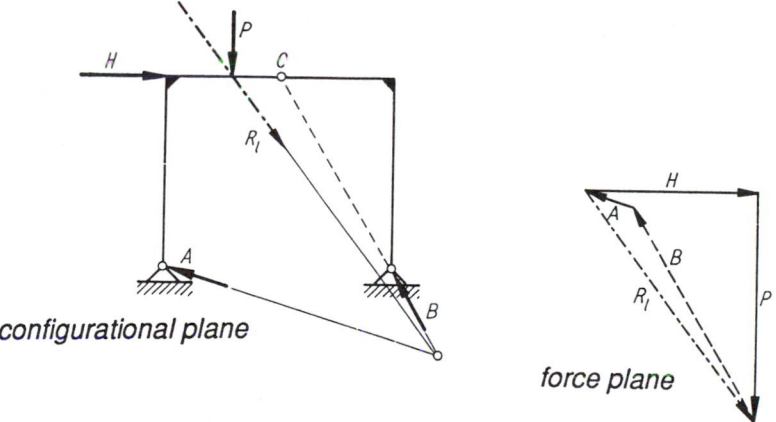

configurational plane

force plane

Fig. 2.47. A three-hinged arch loaded on the left-hand side. Graphic evaluation of the reaction forces in the hinged joints in the scaled configurational and force plane

additional hinge in the beam 11 vanishes, an additional static equation results, $X = A_h$,
In crossbar 11 , the hinge at $x = e$,

$$M(x = e) = aX + eA_v = 0 \rightarrow A_h = - A_v \, e/a = H \, e/l - P \, (1 - e/l) \, e/a \ .$$

Figure 2.47 shows the graphic evaluation of the support reactions of a three-hinged arch when loaded on the left-hand side only; C is located at $x = l/2 \neq e$. Since the part on the right hand side of the arch BC carries no loads, and the bending moment at C vanishes, the line of action of the supporting force $\boldsymbol{B}$ passes through C and is thus known in a scaled configurational plane. Three equilibrating forces in a plane, one is the resulting force of the given loading $\boldsymbol{R}_l$, $\boldsymbol{B}$ and $\boldsymbol{A}$, must be a central force system. This condition renders the direction of $\boldsymbol{A}$.

2.5.1.5. Two Statically Determinate Stress States

The inverse problem of Eq. (2.130), namely, the determination of the stress distribution in the cross-section, is, in general, statically not unique. A single moment vector, eg corresponds to an infinite number of equivalent couples. In thin-walled rods, however, the stress state may become statically determinate under additional approximating assumptions. Two, nontrivial examples are given below.
(§)   Bending Stresses in a Sandwich Cross-Section. Lightweight beams are designed having a sandwich cross-section, ie two girder plates (or surface layers of a material with high tensile and compressive strength, eg a proper metal) are kept in constant distance h by a core that eventually carries the shear stresses. The core material supports no normal stresses, $\sigma_{xx} = 0$. Interface bonding must transfer any shear stresses. According to Fig. 2.48, the normal bending stresses are assumed to be uniformly distributed in the thin-walled flanges, and the couple $\sigma_{xx}(x) \, tB \, h$ must be

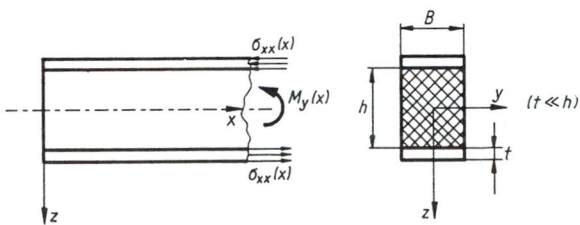

Fig. 2.48. Statically determinate bending stresses in a sandwich cross-section of a straight beam

statically equivalent to the given cross-sectional moment $M_y(x)$, at the station x,

$$M_y(x) = h\, t\, B\, \sigma_{xx}(x) \quad . \tag{2.167}$$

The tensile stress in the lower girder at a location x , thus, is given from the purely static consideration by

$$\sigma_{xx}(x) = M_y(x)/tBh > 0 \quad \text{if} \quad M_y(x) > 0 \quad , \tag{2.168}$$

and the compressive stress in the upper flange is

$$\sigma_{xx}(x) = -\,M_y(x)/tBh < 0 \quad \text{for} \quad M_y(x) > 0 \quad . \tag{2.169}$$

These stresses are the mean values when averaging over the flange thickness $t \ll h$. Equations (2.168) and (2.169) may be used also for dimensioning I-beams when the flanges are thin-walled, like above, and the web of height h is even thinner. Any shear stress is assumed to be distributed in the web of cross-section $A_S = s\,h$, and some simple design formulas assume a uniform distribution, thereby rendering also the mean shear statically determinate, $\sigma_{xz}(x) = \sigma_{zx} = Q_z(x)/A_s$.

   Beams of medium wide span are often discretized by means of a system of rods which are connected in nearly hinged joints, thus forming a two-dimensional truss. The latter may be statically determinate and the axial forces in the rods are statically equivalent to the shear force and bending moment; see Sec. 2.5.2.
(§)   Torsional Shear Stresses in a Thin-Walled Tube. The light-weight design of straight beams in torsion leads to closed thin-walled cross-sections. A rod in torsion is considered, loaded by a torque M_T , with a constant, single-cell cross-section of thickness $t(s)$, which may be variable with the arc-length s (see Fig. 2.49). The in-plane shear stress $\sigma_{xs}(s) = \sigma_{sx}(s)$ must be tangential to the traction-free surface and, thus, may be considered constant over the thickness t that is assumed small with respect to the other cross-sectional dimensions. The resultant force per unit of length $T = [t(s)\,\sigma_{xs}(s)]$ is called the *shear flow* . By consideration of the equilibrium of the wall element shown in Fig. 2.49, a constant shear flow T is recognized. Hence, static equivalence renders the differential contribution to the torque (again see Fig. 2.49)

$$dM_T = T\, ds\, p(s) = 2T\, dA \quad , \tag{2.170}$$

where $dA = p(s)\, ds/2$ is the area of the hatched triangular element, and, integration gives the first of *Bredt's* formulas,

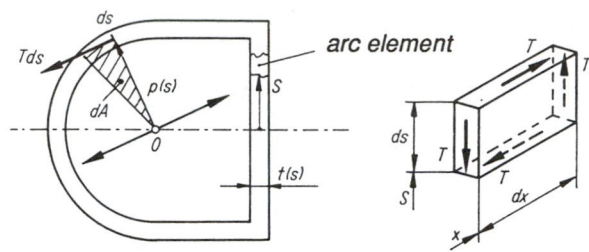

Fig. 2.49. Torsion of a thin-walled tube. The first of *Bredt´s* formulas

$$M_T = 2\,T\,A \; . \qquad\qquad (2.171)$$

$A = \int dA$ is the internal cell area, enclosed by the perimeter of the material wall; hence,

$$T = M_T / 2A \; , \qquad\qquad (2.172)$$

is the statically determinate shear flow and $\sigma_{xs}(s) = T/t(s)$, is the mean torsional shear. For a closed circular tube of radius a, $A = \pi a^2$, the shear is $\sigma_{xs} = M_T / 2\pi a^2 t$; for a quadratic box of base a, $A = a^2$, the torsional shear is $\sigma_{xs} = M_T / 2a^2 t$. The thin-walled closed circular shell is the ideal torsional hollow rod. See Sec. 6.2.4 for further considerations of elastic torsion.

2.5.2. Trusses

Contrary to beams and rigidly framed structures, where the elements are bent in addition to being stretched or compressed, a truss is a structural system of pin-jointed rods that are predominantly axially loaded. If all the member axes and the load are in a common plane, a planar truss is considered to be in equilibrium; otherwise, the three-dimensional structure of a space truss must be analyzed. Only the special case of a trussed beam renders a "one-dimensional-structure" in the sense of this chapter. But the members are, in general, one-dimensional slender rods. The static analysis of a truss is performed under the following idealizing assumptions of a proper design:

The members have perfectly straight axes.

The joints are ideally pinned (no bending, no friction).

The axes of all members having a common joint are crossing through this point.

The external forces are (statically equivalent) applied at the joints.

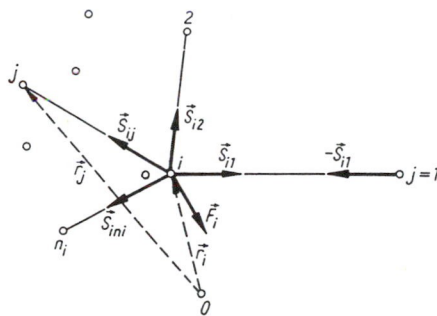

Fig. 2.50. Free-body diagram of an isolated idealized joint

The members of such an idealized truss are only axially loaded, both the shear force and bending moment vanish. In tension, the full load-carrying capacity of the cross-section can be used. In compression, slender rods may buckle: This failure mechanism has to be considered in the design by limiting the length of those compressional members. See Sec. 9.1.4 (*Euler* buckling).

Structural analysis is performed after the "geometric" design has been made in a first step. The nodes are fixed in space with respect to a global coordinate system, and the member axes connecting selected pairs of joints are chosen. Furthermore, the loadings are assumed to be given in the form of single forces attached to the joints. If the weight of the structure is considered, it is memberwise estimated and then reduced to the adjacent nodes in a statically equivalent manner. After the dimensioning of the truss members, such analysis is repeated with the true weight. The equilibrium of the gross system (a rigid body with hinged, fixed or sliding supports) may be considered separately to control the results derived from the free-body diagram of each pinned joint of the truss, where the members at the supports are included (see Fig. 1.8). The joints are numbered consecutively, such that the numbers of adjacent nodes show the least difference.

A general node of number i is considered in Fig. 2.50, where $\boldsymbol{S}_{ij}$ denotes the axial force in the member connecting the joint i with the node j and $\boldsymbol{F}_i$ is a given loading at this point. The systems i of central forces are in equilibrium if the vector conditions,

$$\boldsymbol{F}_i + \sum_{j=n_1}^{n_i} \boldsymbol{S}_{ij} = 0 \ , \quad \boldsymbol{S}_{ij} = -\boldsymbol{S}_{ji} \ ,$$

$$(2.173)$$

hold for all nodes $i = 1, ..., k$. The unit vectors

$$e_{ij} = (r_j - r_i)/l_{ij} \ , \tag{2.174}$$

determine the line of action of the internal forces $S_{ij} = S_{ij} \, e_{ij}$. Hence, Eq. (2.173) is rewritten in vector form

$$\sum_{j=n_1}^{n_i} (r_i - r_j) \, S_{ij} / l_{ij} = F_i \ , \quad i = 1, \dots, k \ . \tag{2.175}$$

The unknown internal forces S_{ij} (their number is equal to the number s of structural members plus the number of reaction forces at the external supports z) appear explicitly when taking the three components of Eq. (2.175) and, thus, when considering the $3k$ linear equilibrium conditions of a space truss or the $2k$ linear and inhomogeneous equations of a planar truss

$$\sum_{j=n_1}^{n_i} (x_i - x_j) \, S_{ij}/l_{ij} = X_i \ , \quad \sum_{j=n_1}^{n_i} (y_i - y_j) \, S_{ij}/l_{ij} = Y_i \ , \quad i = 1, \dots, k \ ,$$
$$\sum_{j=n_1}^{n_i} (z_i - z_j) \, S_{ij}/l_{ij} = Z_i \ , \quad l_{ij} = \sqrt{(x_i - x_j)^2 + (y_i - y_j)^2 + (z_i - z_j)^2} \ . \tag{2.176}$$

The $(3k \times s)$ coefficient matrix C contains the difference of the given nodal coordinates and should have a banded structure. In matrix notation,

$$C \, S = F \ . \tag{2.177}$$

S is the $(s \times 1)$ vector of the unknown internal forces, and F is the given $(3k \times 1)$ column matrix of the given loading and support reactions. In order for the truss to be a structure with load-carrying capacity in space (or in the plane), the number of unknowns (members) must be larger or equal to the number of linear equilibrium conditions

$$s + z \geq 3k \ (2k) \ . \tag{2.178}$$

If the truss is supported in a statically determinate manner, eg in three (or two) nodes, the number of reaction forces $z = 6$ (3), according to the degrees of freedom of the (restricted) motion of the associated rigid body. Hence,

$$s + 6 \ (3) \geq 3k \ (2k) \ . \tag{2.179}$$

In the case of external determinacy and the appearance of the inequality sign, the truss possesses redundant internal forces and is considered internally statically indeterminate. An equivalent number of conditions must be derived by considering the deformations, such that the redundant members fit into the deformed truss (see Sec. 5.4.1). These n redundant internal forces have to be properly selected and are added as additional elements to the column matrix F. The remaining

$$(s - n) + 6 \ (3) = 3k \ (2k) \tag{2.180}$$

equilibrium conditions may be solved to express the other internal forces as functions of the given loading and redundant forces X_l, $l = 1,..., n$. Some flexibility approaches of classical nature require this solution explicitly .

The externally and internally determinate truss, however, has a number of members according to the equation

$$s + 6 \ (3) = 3k \ (2k) \ . \tag{2.181}$$

The coefficient matrix C becomes quadratic in that case and must be nonsingular to solve the linear system of equations. The determinantal condition is

$$\det \{C\} \neq 0 \ . \tag{2.182}$$

Computer-oriented solution strategies for linear equations can be found, eg in
J. Dankert: Numerical Methods of Mechanics. Springer-Verlag, New York, 1977 (in German).
The singular case, however,

$$\det \{C\} = 0 \ , \tag{2.183}$$

renders a *shaky truss* with (little) mobility of the joints, which, of course, is an improper structure. These exemptional trusses, eg are regularly shaped with all the outer nodes on a sphere in space, or bounded by a conic section in the plane, respectively, when all these nodes are connected to each other by member paths. This phenomenon was already known to *Ritter* , since the graphic solution by means of the *Cremona* map fails in the case of a shaky truss. See
Ritter: Applications of Graphic Statics. 2nd Part. The Trusses. Meyer and Zeller, Zürich, 1890 (in German).
Those shaky trusses, as well as the trusses that become mobile at large, *(s + z) < 3k (2k)* , have no stable load-carrying capacity and are not considered structures in equilibrium. Such a design of a shaky truss must be changed by rearranging the nodes and not by just adding new member rods.

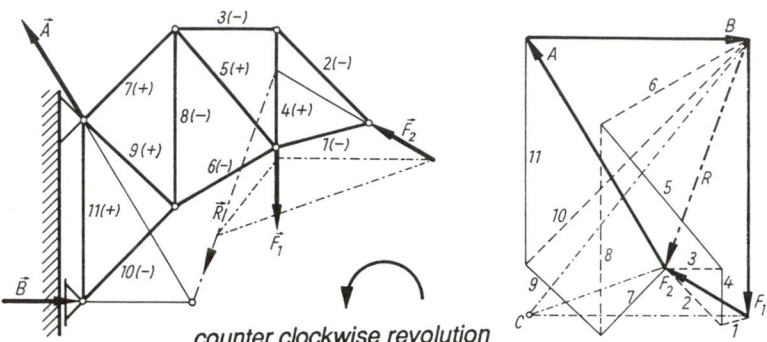

counter clockwise revolution

Fig. 2.51. Plane cantilever-type truss. *Cremona* map (must keep the counterclockwise direction)

2.5.2.1. Planar Trusses

Since the central force system at each of the nodes is plane, two conditions of equilibrium of Eq. (2.176) remain and are nontrivial. The graphic solution of these equations in a self-controlling manner is given by the *Cremona* map. Figure 2.51 illustrates the procedure for a cantilever-type plane truss, which may be part of a crane structure. Since the direction of B is known in advance, the line of action of A through the fixed support can be found by considering the equilibrium of three forces in a plane ($R = F_1 + F_2$ is the resultant of the given loading, B and A), which must be a central system in the scaled configurational plane. Summing the forces in a counterclockwise direction in the scaled force plane gives the values of the supporting forces. When we take a node where only two members are connected, the free-body diagram shows the given load F_2 and two unknown force values of members 1 and 2, and when we close the force polygon by keeping the counterclockwise arrangement of forces, F_2 is already present in the force plane, thereby initiating the *Cremona* map. Tensile forces point away from the node and, commonly, acquire a plus sign at the member; compressional internal forces are marked by a minus sign; no arrows are drawn in the force polygon. Cutting the next joint having no external nodal force, gives the forces 3 (−) and 4 (+) to equilibrate 2 (−), which has an arrow pointing to the node. Proceeding successively and keeping the counterclockwise direction should close the *Cremona* map without redrawing the reaction force, say, A. Such a self-controlling and convenient method is a cheap means for checking computational results.

If a plane truss can be divided into two substructures such that by the cut only three members are involved which are not

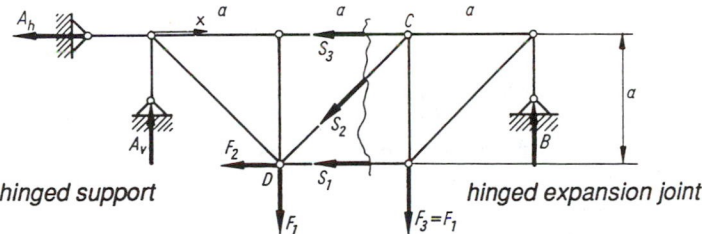

Fig. 2.52. Trussed single-span beam. *Ritter's* cut. Free-body diagram of the right substructure

attached to a common joint, the internal forces can be explicitly calculated. This cut is due to *Ritter* and is illustrated by the single-span beam structure of Fig. 2.52. The gross equilibrium of the plane force system (F_1, F_2, F_3 given; A_v, A_h, B wanted) renders three linear equations with the solution $A_v = F_1 - F_2/3$, $A_h = -F_2$, $B = F_1 + F_2/3$ for the special loading $F_1 = F_3$. When we consider the *Ritter* cut in Fig. 2.52 and the resulting free-body diagram of the right substructure, the force S_1 is explicitly calculated from the condition of vanishing moment about the node C, where the other two unknown forces S_2 and S_3 intersect. The force system to be considered is (F_3, B, S_1, S_2, S_3)

$$aS_1 - aB = 0 \quad \rightarrow \quad S_1 = B = F_1 + F_2/3 \ .$$

$S_1 > 0$ gives tension in the lower girder. Changing the moment pole to point D renders

$$aS_3 + 2aB - aF_3 = 0 \quad \rightarrow \quad S_3 = -2B + F_3 = -(F_1 + 2F_2/3) \ .$$

$S_3 < 0$, indicates compression in the upper girder. Since S_1 and S_3 are parallel, the force S_2 in the diagonal, which is a member of the web, is also explicitly calculated from

$$S_2/\sqrt{2} + F_3 - B = 0 \quad \rightarrow \quad S_2 = (B - F_3)\sqrt{2} = (\sqrt{2}) F_2/3 \ .$$

$S_2 > 0$ renders tension in the long diagonal member, which is favorable.

Any change in the design of the truss influences the internal forces. For example, consider the other diagonal member in Fig. 2.52. The forces are changed to

$$S_1' = F_1 + 2F_2/3 > S_1 > 0 \ , \quad F_1, F_2 > 0 \ ,$$
$$S_3' = -(F_1 + F_2/3) \ , \quad |S_3'| < |S_3| \ .$$

The compression in the upper girder is reduced, however,

$$S_2' = (-\sqrt{2})\, F_2/3 < 0$$

gives (unwanted) compression in the long diagonal.

Following the procedure outlined in Sec. 2.5.1.3, the *influence lines* of the forces in the members considered above are developed when restricting the loading to a parallel one, $F_2 = 0$. Following *Ritter* , the cross-section at x is taken through the joint C (Fig. 2.52), and the influence line of the bending moment according to the lower one in Fig. 2.53 is calculated (the single force $F = 1$ is assumed to be applicable at continuous ξ). Hence, the influence line of the lower girder force is proportional,

$$\overline{S}_1(\xi) = \overline{M}_C(\xi)/a \quad .$$

Taking x according to node D gives analogously

$$\overline{S}_3(\xi) = -\,\overline{M}_D(\xi)/a \quad .$$

The influence line of the force in the diagonal member is proportional to that of the shear force with x taken at nodes C or D, respectively,

$$\overline{S}_2(\xi) = -\sqrt{2}\,\overline{Q}_{C,D}(\xi) \quad ,$$

see Fig. 2.53.

According to the assumptions for the loading of trusses, ξ

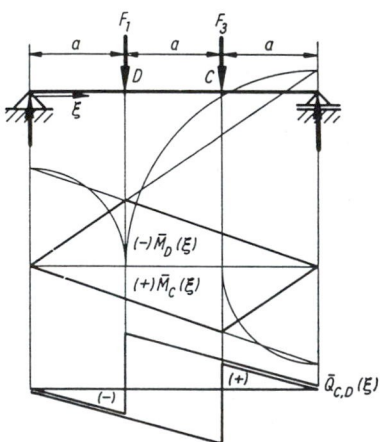

Fig. 2.53. Single-span beam laterally loaded like the truss of Fig. 2.52, with $F_2 = 0$. Influence lines of bending moments and of shear forces, respectively

takes on discrete values only, $\xi = 0, a, 2a, 3a$ in Fig. 2.53. Superposition renders the force in the lower girder for the given loadings F_1, F_3 [see Eq. (2.163)],

$$S_1 = F_1 \overline{S}_1(\xi = a) + F_3 \overline{S}_1(\xi = 2a) = \frac{F_1}{a}\left(1 - \frac{2a}{l}\right) a + \frac{F_3}{a}\left(1 - \frac{2a}{l}\right) 2a \ ,$$

finally to be $S_1 = F_1$, if $F_1 = F_3$.

2.5.3. Statics of Flexible Cables (and Chains)

Cables and chains are special one-dimensional structures of negligible bending stiffness, which carry tangential and lateral loads and thereby are being stressed in tension only. The complex stress states in the composite cross-section due to the contacting single wires of a cable and that in the individual links of the chain are not considered here. Our analysis is restricted to the determination of the resulting internal tensile force, which is tangent to the apriori unknown equilibrium shape. Any lateral load renders a curved configuration. Lateral loads, eg parallel to the vertical direction, give a deformation that is affined to the funicular curve (which is a polygon in the case of parallel single forces). For the single force loading by $-B$ of a suspension cable with span $(l + a)$, see the lower part of Fig. 2.43, where the dummy forces 2 and s are the cable tensions in the straight segments, respectively. Subsequently, the loading $q(s)$ is assumed to be distributed over the arc-length s of the cable. If we take the equilibrium of an element of length ds and of unknown position in space (Fig. 2.54), the sum of the three forces shown in the free-body diagram must vanish

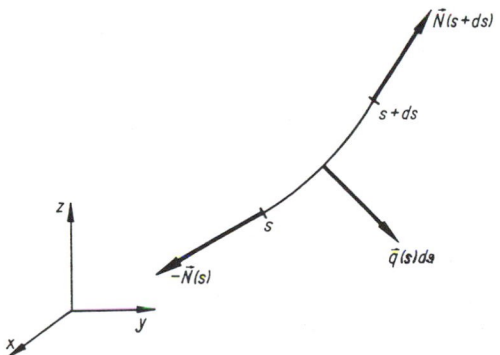

Fig. 2.54. Free-body diagram of a flexible cable element with tangential tension N and general loading $q\ ds$

$$N(s + ds) - N(s) + q(s)\, ds = 0 \quad .$$

It is sufficient to take a linear approximation of the difference of the cable tensions in closely spaced cross-sections to derive exactly the local equilibrium condition, with $ds \to 0$,

$$d\mathbf{N}/ds = -\mathbf{q}(s) \quad , \quad \mathbf{N} = X\, \mathbf{e}_x + Y\, \mathbf{e}_y + V\, \mathbf{e}_z \quad . \tag{2.184}$$

In components, where (x, y) is the horizontal plane and z points vertically upward

$$dX/ds = -q_x \quad , \quad dY/ds = -q_y \quad , \quad dV/ds = -q_z \quad . \tag{2.185}$$

These first-order differential equations may be formally integrated to render

$$X(s) = X_0 - \int_0^s q_x\, ds \quad , \quad Y(s) = Y_0 - \int_0^s q_y\, ds \quad , \quad V(s) = V_0 - \int_0^s q_z\, ds \quad .$$

It is convenient to consider the lowest point of the cable to have a horizontal tangent, even if the structure must be fictitiously extended, and to count the arc-length s from that point. Hence, with x pointing in the tangential direction, two of the integration constants must vanish: $V_0 = 0$, $Y_0 = 0$. The spatial curve has the canonical representation, $(ds)^2 = (dx)^2 + (dy)^2 + (dz)^2$,

$$dx : dy : dz = X : Y : V \quad . \tag{2.186}$$

For the *uniformly* distributed loading, $q_x = q_1$, $q_y = q_2$, $q_z = -q$, integration renders

$$X = X_0 - sq_1 \quad , \quad Y = -sq_2 \quad , \quad V = sq \quad , \quad dx : dy : dz = (X_0 - sq_1) : (-sq_2) : qs \quad ,$$

$$s = \int_0^s \left[1 + \left(\frac{dx}{dz}\right)^2 + \left(\frac{dy}{dz}\right)^2 \right]^{\frac{1}{2}} dz \quad .$$

A numerical solution of the nonlinear equations is to be performed by the finite-difference scheme. A simple analytical solution can be given for the plane curve, derived under the vertical loading condition, $q_2 = q_1 = 0$, $q = const$,

$$dz/dx = V/X_0 = q\, s/X_0 \quad . \tag{2.187}$$

$X = X_0$ becomes the constant horizontal tensile force component. Equation (2.187) is differentiated once, and $dz/dx = sinh\ u(x)$, as well as $d^2z/dx^2 = (du/dx)\ cosh\ u$, are substituted to render the linear first-order differential equation in the new variable $u(x)$

$$du/dx = q/X_0 = 1/a \quad , \quad [a] = meter,\ m \quad .$$

Integration yields $u(x) = x/a + b$, but $b = 0$, since $dz/dx = 0$ is assumed at $x = 0$. A second integration gives the final result of the in-plane equilibrium configuration

$$z(x) = a\ cosh\ (x/a) + C \quad . \tag{2.188}$$

By the vertical translation of the coordinate system to $z(0) = a$, the constant $C = 0$, and the parameter $a = X_0/q$ becomes the proper geometric and static interpretation. The *catenary* $z(x) = a\ cosh\ (x/a)$, for cables and "smeared" chains under uniform loading is of equal importance for the design of slender *arches in pure compression* ; see Sec. 6.4.1 and Eq. (2.190). By means of the arc-length $s = a\ sinh$ (x/a) , the vertical internal force component becomes

$$V(x) = q\ s = X_0\ sinh\ (x/a) \quad . \tag{2.189}$$

Note the definitions, $sinh\ z = (e^z - e^{-z})/2$, $cosh\ z = (e^z + e^{-z})/2$. The resulting tension is thus given by $N(x) = (X_0^2 + V^2)^{1/2} = X_0\ [1 + sinh^2(x/a)]^{1/2} = X_0\ cosh\ (x/a) = q\ z(x)$, $q = const$, and turns out to be proportional to the ordinates $z(x)$ in the specially oriented coordinate system, $z(0) = a$.

Cables with small sag, eg those in electrical high-voltage transmission lines, are highly tensioned and the parameter a becomes very large when compared to the span. Hence, $|x|/a \ll 1$, and

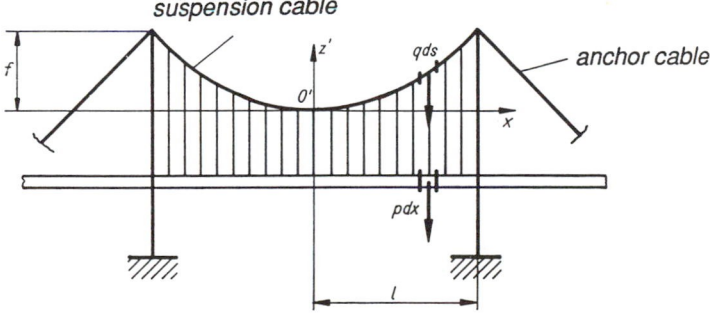

Fig. 2.55. Suspension bridge. Smeared loading of the suspension cable by $p = const$

the catenary when expanded in the *Taylor* series at $x = 0$ may be approximated up to the third order by the quadratic polynomial function (q is still assumed to be constant)

$$z(x) = a \cosh (x/a) \approx a + x^2/2a \quad , \quad z'(x) = z - a \approx q \, x^2 / 2X_0 \quad .$$

In this approximation, $N \approx X_0 = qa \gg V_{max}$ is nearly constant. The parabola becomes the exact configuration of the flexible cable for the constant loading p per unit of length of the horizontal projection x . For example, the suspension cable of the wide-span bridge of Fig. 2.55 is mainly loaded according to this assumption. Smearing the densely spaced loading of the cable according to

$$p \, dx = - q_z(s) \, ds \quad , \tag{2.190}$$

and inserting into Eq. (2.185)$_3$ give (p is assumed constant), $dV = [- q_z \, (s) \, ds] = p \, dx$. When we consider $V = V(x)$, integration renders the vertical component proportional to the horizontal distance from the lowest point, $V = p \, x$. Equation (2.187) becomes

$$dz'/dx = V/X_0 = px/X_0 \quad , \tag{2.191}$$

and integration renders the corresponding suspension curve exactly parabolic

$$z'(x) = p \, x^2 / 2X_0 \quad . \tag{2.192}$$

The allowable sag $f = p \, l^2/2X_0$ controls the tension X_0 ; the large parameter a is no longer useful.

Since the length of the cable and, hence, the sag change with

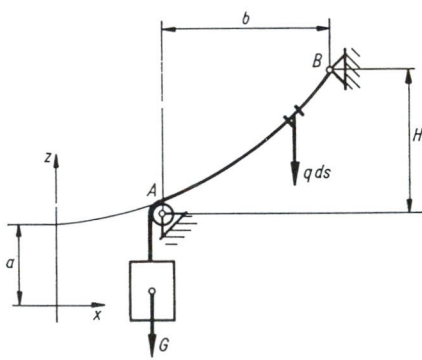

Fig. 2.56. Suspension cable with tensioning weight G . Fictitious extension shown. $G, q = const, b, H$ are given

temperature for fixed end supports, cables tensioned by a given weight G at a reel support are frequently used, eg to support cable railways (see Fig. 2.56) or to carry the upper supply line of an electrical railway. At the reel, point A, the resulting tangential force is approximately given by the weight , $N_A = qz_A \approx G$; the small and variable weight of the vertical part of the cable is neglected. This determines z_A in the special coordinate system shown. By means of the relation $z_B = z_A + H$, the maximal tension at the fixed point B is $N_B = qz_B$. The selection of the cable dimensions is already possible. The direction of the supporting force $-N_B$ may be found from the constant horizontal tension $X_0 = qa$, which is given after solving the nonlinear geometric condition for the parameter a by a proper numerical routine

$$z_B = z_A + H = a \cosh (x_A + b)/a = z_A \cosh (b/a) + (z_A^2 - a^2)^{1/2} \sinh (b/a) .$$

Additional vertical single force loadings are considered by joining the adjacent catenary arcs of the same parameter a properly. The graphic solution is approximately given by the funicular polygon.

2.6. Exercises A 2.1 to A 2.15 and Solutions

A 2.1: For the given area of Fig. A 2.1 with a circular hole, determine the location of the centroid C = S by a consideration of symmetry.

Solution: Since $x_S = 0$, the static moment about the x -axis is determined by

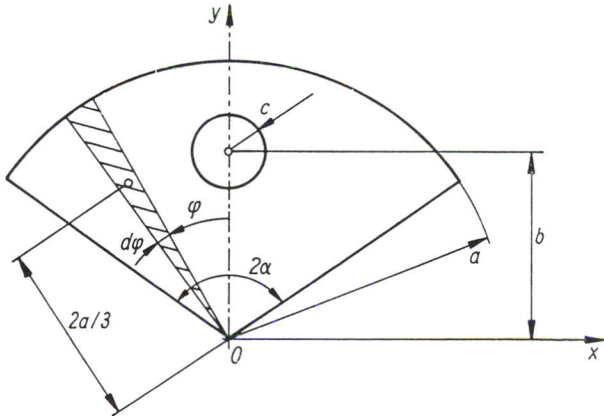

Fig. A 2.1.

$$Ays = \int_A y \, dA = A_1y_1 - A_2y_2 \quad , \quad A_1 = \alpha a^2 \quad , \quad A_2 = \pi c^2 \quad , \quad y_2 = b \; .$$

By considering the triangular area element in Fig. A 2.1, the central distance y_1 of the undisturbed area A_1 is derived from the static moment,

$$A_1y_1 = 2 \int_0^\alpha \left(\frac{2}{3} a \cos \varphi\right) \frac{a}{2} a \, d\varphi = \frac{2 \, a^3}{3} \sin \alpha \; ,$$

as $y_1 = (2a \sin \alpha)/3\alpha$. Hence, $y_S = a(2 \sin \alpha - 3\pi bc^2/a^3)/3(\alpha - \pi c^2/a^2)$. For $\alpha = \pi$, a circular area with an eccentric hole, the result is $y_S = -b/\pi \, [(a/c)^2 - 1]$.

A 2.2: The simple spatial truss of Fig. A 2.2 has three members connected at a pinned joint D and it is loaded by the single force $F = F_x \, e_x + F_y \, e_y + F_z \, e_z$. Determine the member forces S_i that are directly related to the reactions at the supports A, B, and C.

Solution: A spherical cut around D renders the central force system F , S_i , $i = 1, 2, 3$. The conditions of equilibrium contain the lengths of the members $l_1 = (a^2 + b_1^2)^{1/2}$, $l_2 = (a^2 + b_2^2)^{1/2}$, $l_3 = (a^2 + b_3^2 + c^2)^{1/2}$

$$\sum_{i=1}^4 X_i = 0 = F_x - S_1 a/l_1 - S_2 a/l_2 - S_3 a/l_3 \; ,$$

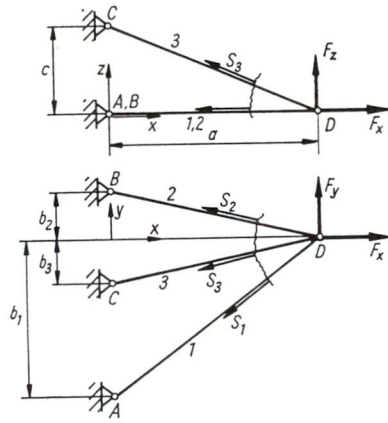

Fig. A 2.2

$$\sum_{i=1}^{4} Y_i = 0 = F_y - S_1 b_1/l_1 + S_2 b_2/l_2 - S_3 b_3/l_3 , \quad \sum_{i=1}^{4} Z_i = 0 = F_z + S_3 c/l_3 .$$

The solution of the inhomogeneous system of linear equations gives

$$S_1 = [b_2 F_x/a + F_y + (b_2 + b_3) F_z/c] l_1/(b_1 + b_2) ,$$

$$S_2 = [b_1 F_x/a - F_y + (b_1 - b_3) F_z/c] l_2/(b_1 + b_2) , \quad S_3 = -l_3 F_z/c .$$

A 2.3: A rigid platform carries single force loadings F_i , $i = 1, 2, 3$, at the corners, and it is supported by six doubly hinged columns according to Fig. A 2.3. The force center of the parallel loadings is to be determined first, followed by the statically equivalent reduction to the origin 0. Finally, the internal forces S_k , $k = 1, 2, ..,$ *6* , in the supporting rods are wanted.

Solution: The point of application of the load resultant $R = -R \, e_z = [- (F_1 + F_2 + F_3)] \, e_z$ (without moment) follows from the "static moment"

$$r R = \sum_{i=1}^{3} r_i F_i = l[F_1 e_x + F_2 (e_x + e_y) + F_3 e_y]$$

and has the radius vector *r* with components (coordinates) $r_x = (F_1 + F_2)l /R$, $r_y = (F_2 + F_3)l /R$, $r_z = 0$. Static equivalence renders *R* attached to 0 and the moment

$$M_O = \sum_{i=1}^{3} r_i \times F_i = r \times R = l[(F_1 + F_2) e_y - (F_2 + F_3) e_x] .$$

By cutting through all the supporting rods and considering the

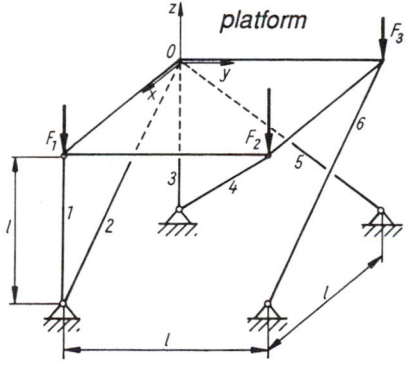

Fig. A 2.3.

unknown internal axial forces as tensions, the six conditions of equilibrium of the spatial force system $(F_1, F_2, F_3, S_1, S_2, \ldots, S_6)$

$$\sum_j X_j = 0 = \frac{\sqrt{2}}{2} S_2 - \frac{\sqrt{3}}{3} S_4 + \frac{\sqrt{2}}{2} S_6 \; , \quad \sum_j Y_j = 0 = -\frac{\sqrt{3}}{3} S_4 + \frac{\sqrt{2}}{2} S_5 \; ,$$

$$\sum_j Z_j = 0 = -S_1 - \frac{\sqrt{2}}{2} S_2 - S_3 - \frac{\sqrt{3}}{3} S_4 - \frac{\sqrt{2}}{2} S_5 - \frac{\sqrt{2}}{2} S_6 - R \; ,$$

$$\sum_j M_x^{(j)} = 0 = -l\frac{\sqrt{3}}{3} S_4 - l\frac{\sqrt{2}}{2} S_6 - lF_2 - lF_3 \; ,$$

$$\sum_j M_y^{(j)} = 0 = lS_1 + l\frac{\sqrt{3}}{3} S_4 + lF_1 + lF_2 \; , \quad \sum_j M_z^{(j)} = 0 = -l\frac{\sqrt{2}}{2} S_6 \; ,$$

render (a minus sign means the force is compressive) $S_1 = F_3 - F_1$, $S_2 = -(F_2 + F_3)\sqrt{2} = S_5$, $S_3 = 2F_2 + F_3$, $S_4 = -(F_2 + F_3)\sqrt{3}$, $S_6 = 0$. Supporting rod 6, under this loading conditions, is tensionless.

A 2.4:   By rolling up a strip of a (metal) sheet in a screw-shaped manner and by glueing together the edges, a thin-walled cylindrical shell of any length is obtained (see Fig. A 2.4). Used as a tensile rod, the sheet metal carries the normal stress σ_{xx} ; an admissible value is σ_{ad} . The interface bond is stressed by a normal (σ) and a shear-stress (τ) component. The latter is usually limited by $\tau_{ad} = \varepsilon\, \sigma_{ad}$, $\varepsilon <$ *1* . The pitch, expressed by the angle α , should be chosen such that the sheet metal and bond have equal strength.

Solution: A triangular infinitesimal element is cut out, where *ds* is part of the interface. The free-body diagram is shown in Fig. A 2.4. The equilibrium conditions render the stress vector in the bond to be statically determinate (see *Mohr's* circle) and, hence, $\tau = [(1/2)\,\sigma_{xx} \sin 2\alpha]$. Substituting the admissible stresses renders

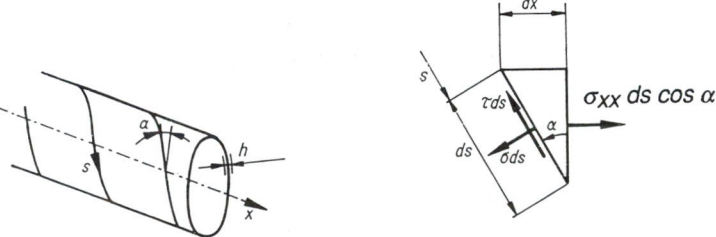

Fig. A 2.4.

$$\alpha = (1/2) \sin^{-1}(2\varepsilon) \quad \text{rad.}$$

Comment: In general, the stresses in the bond are statically indeterminate and the assumption of a constant stress distribution is a very rough one; see Exercise A 6.10 for an elastic and overlapping joint.

A 2.5: The cantilever of Fig. A 2.5 shows the alternative design of the web in comparison to the diagonal member of a plane truss, namely, the use of a thin panel, a sheet metal with edge beams. Due to the hinged joints and supports, the axial forces in the rods and the shear flow in the plate are statically determinate and are to be calculated. Loading F and the dimensions are given. *Note:* The edge stiffener is in welded contact with the web. If in the case of $H \leq L$, either one of the strong geometric inequalities, $H_G /H \ll 1$ and $(H \, t_{plate}) /A_G \ll 4$, holds, the assumption of axially stressed stiffeners and pure shear in the web becomes a sufficiently good approximation. The shear flow $T = \tau \, t_{plate}$ in the interface of any stiffener with the plate is constant. The shear force is carried by the web only.

Solution: The supporting reactions are $B_V = -F$, $B_H = -FL /H = -A_H = A$. The equilibrium of a differential stiffener element renders after integration $N_j(x_j) = -T_j \, x_j + N_j(x_j = 0)$, $j = 1, 2, 3, 4$. Circular cuts around the hinged joints of the stiffener yield $N_1(0) = N_3 (0) = F$, $N_2 (0) = -N_4 (L) = FL/H$, $N_1(H) = N_2 (l) = N_3 (H) = N_4 (0) = 0$. The shear flow is found after substitution and evaluation at $x_1 = x_3 = H$ and $x_2 = x_4 = L$ to be constant and given by $T_j = F/H = T$, $j = 1, 2, 3, 4$. The linear variation of the axial forces in the stiffener is shown in Fig. A 2.5 (b). See also Exercises A 6.8 and A 11.7 . See, eg *G Czerwenka and W. Schnell: Introduction to the Computational*

Fig. A 2.5. (a) Shear beam. (b) Free-body diagram of the edge beams. $T_j = T = const$, $i = 1, 2, .., 4$

Methods of Lightweight Structural Design. B. I. Hochschul-Taschenbücher Nr. 124. Bibliograghisches. Institut, Mannheim, 1967, (in German).

A 2.6: A crank cantilevered according to Fig. A 2.6 is loaded by an eccentrically attached axial force **F** . Determine the reaction forces and internal forces in the beam of span *l* .

Solution: The spatial force system (F, A, M_0) is in equilibrium under the conditions, $r_0 = l\,e_x + a\,e_y + b\,e_z$: $A_x + F_x = 0$, $A_y + F_y = 0$, $A_z + F_z = 0$, $M_x - b\,F_y + a\,F_z = 0$, $M_y - l\,F_z + b\,F_x = 0$, and $M_z - a\,F_x + l\,F_y = 0$. Considering the negative cross-sectional cut at x , note the minus sign of the internal resulting vectors in the free-body diagram of the forces $[F, -R(x), -M(x)]$ gives, $r(x) = (l - x)\,e_x + a\,e_y + b\,e_z$: axial force $N = F_x$; shear forces $Q_y = F_y$, $Q_z = F_z$; bending moments $M_y (x) = b\,F_x - (l - x)\,F_z$, $M_z (x) = (l - x)\,F_y - a\,F_x$; torque $M_x = a\,F_z - b\,F_y$.

A 2.7: The plane truss of Fig. A 2.7 may be part of the structural system of a roof; the loading is given by the forces F_1 , F_2 , F_3 . Find the internal forces S_i in members 1, 2, and 3 by means of a *Ritter* cut.

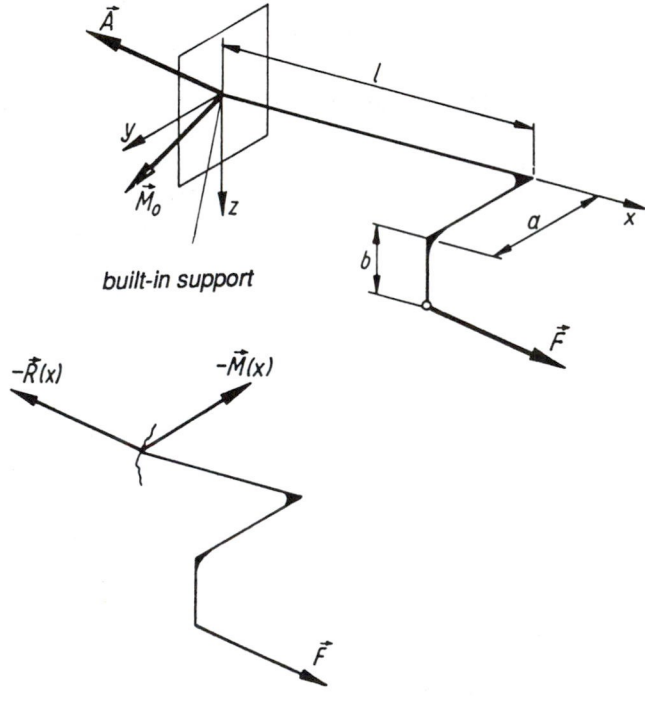

Fig. A 2.6.

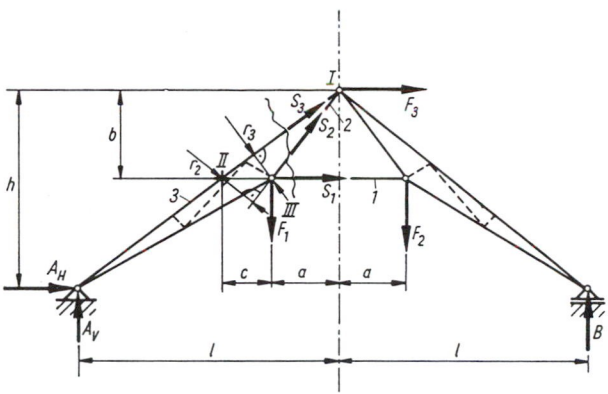

Fig. A 2.7.

Solution: Gross equilibrium conditions render the supporting reactions $A_H = - F_3$, $A_v = (1/2) [(1 + a/l)F_1 + (1 - a/l)F_2 - (h/l)F_3]$, $B = (1/2) [(1 - a/l)F_1 + (1 + a/l)F_2 + (h/l)F_3]$. The left subsystem of the *Ritter* cut is considered by application of the moment conditions about the poles 1, 11, and 111, respectively. Note the normal distances $c = -a + bl /h$, $r_2 = bc /(a^2 + b^2)^{1/2}$, $r_3 = hc /(h^2 + l^2)^{1/2}$ and find

$$bS_1 + hA_H - lA_v + aF_1 = 0 \ ,$$
$$r_2S_2 + (h - b)A_H - (l - a - c)A_v - cF_1 = 0 \ ,$$
$$r_3S_3 - (h - b)A_H + (l - a)A_v = 0 \ .$$

The truss is also internally statically determinate. The dashed lines in Fig. A 2.7 are stress-free members that are added to shorten the spans of the girders.

A 2.8: The three-hinged arch of Fig. A 2.8 consists of two subsystems: a plane truss and a "heavy" homogeneous beam of constant cross-section A_c and density ρ . Calculate the weight G of the beam and determine graphically the reaction forces **A** and **B** for the additional loadings $F_1 = G$ and $F_2 = G/2$. All member forces are to be determined by a *Cremona* map. The scaled configuration should be drawn first.

Solution: $G = g\rho (2\sqrt{2})l A_c$, and the *Cremona* map is drawn in the force plane of Fig. A 2.8 . Note the counterclockwise direction of arranging the forces.

A 2.9: Determine graphically the region of *self-locking* of a rigid ladder of length L when a weight G climbs a distance s according to Fig. A 2.9 . The coefficients of static friction are given by μ_1 at

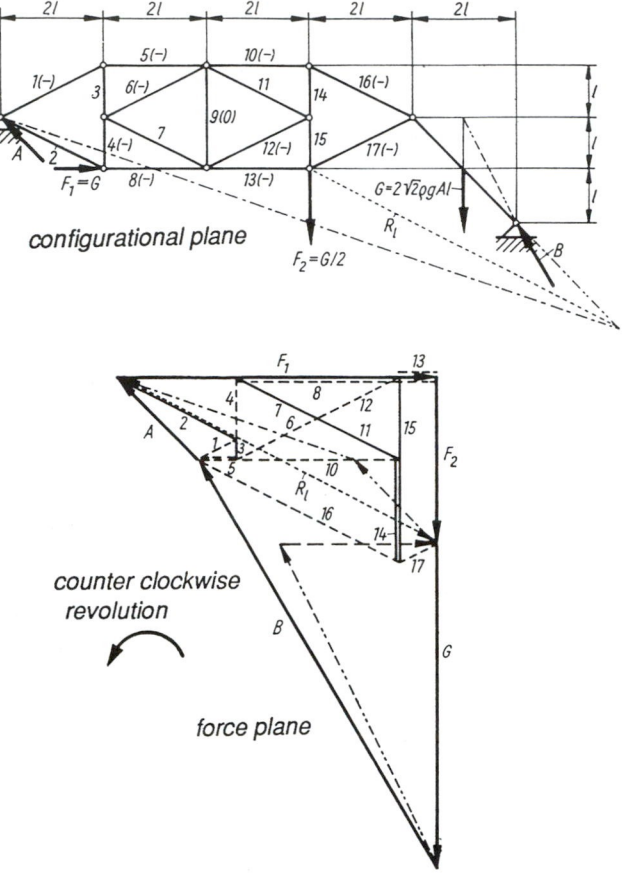

Fig. A 2.8.

the base and by μ_2 at the contact to the wall. What changes statically when the friction at the wall is neglected?

Solution: The lines of action of the three forces **G, A, B** must intersect at a common point, a necessary condition for the equilibrium of three forces in a plane. The cones of static friction at the base and at the wall have a common in-plane area that is hatched in Fig. A 2.9(a). For self-locking, the equilibrium must be independent of the value of **G** ; thus, the central point must fall into this region. The climbing weight reaches a critical position at the limit $s \to s_k$; hence, the region of self-locking is $0 \leq s \leq s_k$. The influence of the wall friction on the limit s_k is rather small in the common ladder configuration; compare (a) and (b) in Fig. A 2.9 . The three conditions of equilibrium, $T_1 - N_2 = 0$, $T_2 + N_1 - G = 0$, L (N_2

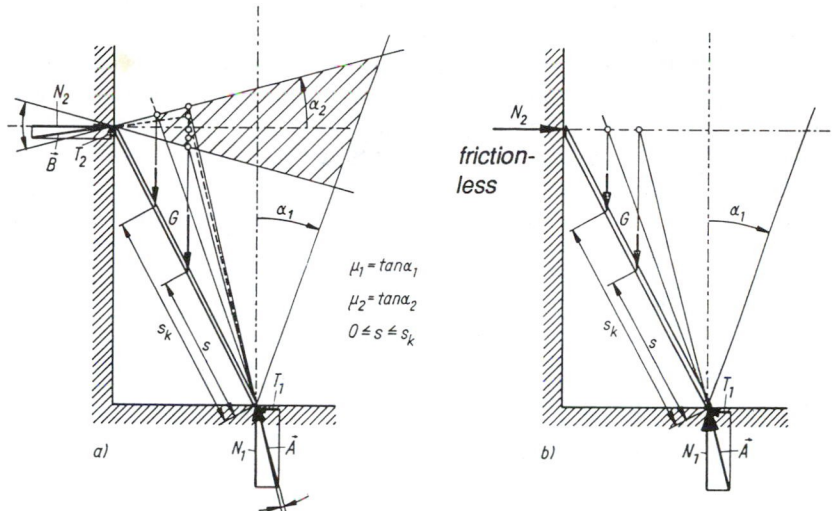

Fig. A 2.9.

$\sin \varphi + T_2 \cos \varphi) - s\, G \cos \varphi = 0$, contain four unknowns; the statically indeterminate force may be chosen as $X = T_2$. Due to the limiting conditions of static friction, the inequalities must hold for any precritical configuration, $|T_1| \le T_{F1} = \mu_1 N_1$, $|T_2| \le T_{F2} = \mu_2 N_2$, or when we consider the above equilibrium conditions, $N_2 = T_1 \le \mu_1 N_1$, $N_2 \ge (G - N_1)/\mu_2$, N_1 , $N_2 > 0$, $T_2 > 0$. The elimination of N_2 from $s/L = 1 - N_1/G + (N_2/G) \tan \varphi$ yields

$$(1 - N_1/G)\,[1 + (\tan \varphi)/\mu_2] \le s/L \le 1 - (N_1/G)\,(1 - \mu_1 \tan \varphi) > 0 .$$

Since, $1 - \mu_1 \tan \varphi > 0$, according to Fig. A 2.9(a), the inequality $N_1/G \ge 1/(1 + \mu_1 \mu_2)$ results. The elimination of N_1/G renders the geometric condition of self-locking by the limit (the system is statically determinate)

$$s_k/L = \mu_1\,(\mu_2 + \tan \varphi)/(1 + \mu_1 \mu_2) .$$

If $\mu_2 = 0$, all precritical configurations are also statically determinate, $X = T_2 = 0$. The limit is somewhat smaller, ie on the safe side,

$$\bar{s}_k/L = \mu_1 \tan \varphi .$$

A 2.10: A V-belt contacts a reel over an angle β and is stressed in the tangent straight parts by S_1 and S_2 , respectively. See Fig. A 2.10. Given the coefficient of static friction μ (there is always a slip over some part of the circumference due to the difference in strain in the incoming and outgoing part of the belt), determine the limit of the difference $|S_2 - S_1|$. Does a condition of self-locking exist? The limiting condition of static friction of a prismatic sledge on a straight wedge-shaped guiding rail when laterally loaded by a force F is to be derived as a special case.

Solution: The free-body diagram of a V-belt element shows the forces $[S(\varphi)$, $S(\varphi + d\varphi)$, dN , dT_φ , $dT_r]$. The sum of the force-components in the tangential and radial direction must vanish (2α is the opening angle of the wedge)

$$2 \, (dT_r \cos \alpha + dN \sin \alpha) - S \, d\varphi = 0 \quad , \quad dS - 2 \, dT_\varphi = 0 \quad .$$

Static friction limits the resultant tangential force

$$|dT| = \sqrt{dT_r^2 + dT_\varphi^2} \le dT_F = \mu \, dN \quad .$$

Just before slippage in the tangential direction occurs, the radial component dT_r vanishes and the first-order differential equation $dS/S = \mu' \, d\varphi$ results analogous to *Euler's* equation of ordinary belt friction on a reel (Eq. 7.67), but with a geometrically amplified static coefficient of friction of the V-belt $\mu' = \mu / \sin \alpha > \mu$. The solution is $S(\varphi) = S_1 \, exp \, (\pm \, \mu' \, \varphi)$. Thus,

$$S_2 \le S_1 \, exp \, (\mu'\beta) \quad \text{or} \quad S_1 \le S_2 \, exp \, (\mu'\beta) \quad .$$

Self-locking is not possible. The horizontal force H acting on a

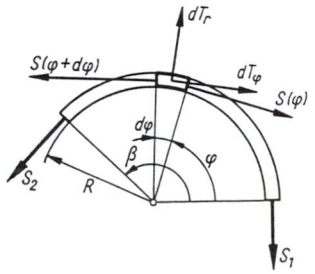
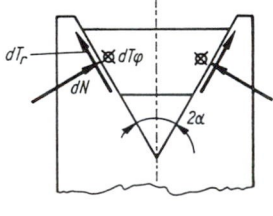

Fig. A 2.10.

prism at rest on a horizontal wedge-shaped guide is thus limited by $H \leq \mu' F$; compare this to the situation on a plane (see Fig. 2.4).

A 2.11: See Fig. A 2.11 for a diagram of a screw with its head in contact with a rigid plate with the following dimensions: a radius of screw head *a* , with a mean radius of the thread *r* , angle of pitch $\gamma(r)$, wedge angle α . It is pretensioned by the axial force *F* , the normal force *N* in the free shaft. Given the coefficient of static friction μ , which is assumed to have the same value in the contact zone of the head and in the thread, determine the upper and the lower limits of the external moment, where the screw rotation is initiated. Under what conditions does self-locking occur?

Solution: The forces on an element of the thread at *r* are projected onto the unit vectors according to the following natural coordinates:
$$\mathbf{e}_n = \kappa \ (\sin \alpha \cos \gamma \ \mathbf{e}_r + \cos \alpha \sin \gamma \ \mathbf{e}_\varphi - \cos \alpha \cos \gamma \ \mathbf{e}_z) , \ \mathbf{e}_u = -\cos \gamma \ \mathbf{e}_\varphi$$
$$- \sin \gamma \ \mathbf{e}_z , \ \mathbf{e}_f = \kappa \ (\cos \alpha \ \mathbf{e}_r - \sin \alpha \sin \gamma \cos \gamma \ \mathbf{e}_\varphi + \sin \alpha \cos^2\gamma \ \mathbf{e}_z) , \ \kappa =$$
$$(\cos^2\alpha + \sin^2\alpha \ \cos^2\gamma)^{-1/2} , \ \mathbf{e}_f = \mathbf{e}_u \times \mathbf{e}_n .$$
Hence, $d\mathbf{N} = -dN \ \mathbf{e}_n$, $d\mathbf{T} = dT_u \mathbf{e}_u + dT_f \mathbf{e}_f$. Equilibrium at the screw head requires $M - M_K - M_G = 0$, at the thread of total area *A* ,

$$- F + \kappa \cos \alpha \cos \gamma \int_A dN - \sin \gamma \int_A dT_u + \kappa \sin \alpha \cos^2 \gamma \int_A dT_f = 0 \quad ,$$

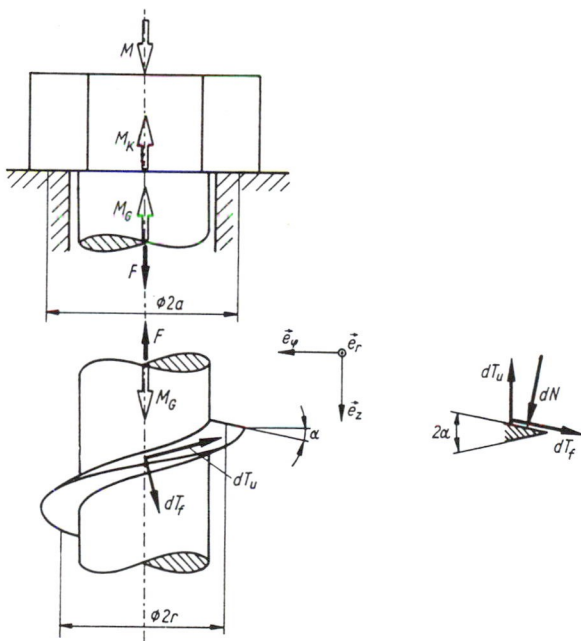

Fig. A 2.11.

$$M_G - r\left(\kappa \cos \alpha \sin \gamma \int_A dN + \cos \gamma \int_A dT_u + \kappa \sin \alpha \sin \gamma \cos \gamma \int_A dT_f\right) = 0 \ .$$

Elimination of the moment M_G related to the forces acting on the thread renders two equations for F and M. The inequalities

$$|M_K| \le M_{KF} = \mu \, a \, F \quad , \quad |dT| = (dT_u^2 + dT_f^2)^{1/2} \le dT_F = \mu \, dN$$

result from static friction. In the limit, dT_f vanishes, and the moment M is limited with respect to the tightening or loosening of the screw, respectively, by

$$M \, {\textstyle{\le \atop \ge}} \, \left(\pm \mu a + \frac{\kappa \cos \alpha \sin \gamma \pm \mu \cos \gamma}{\kappa \cos \alpha \cos \gamma - (\pm \mu \sin \gamma)} \, r \right) F \ .$$

Self-locking is characterized by a negative moment $M < 0$ for loosening the screw

$$(\mu - \kappa \cos \alpha \tan \gamma) \, r + \mu \, (\kappa \cos \alpha + \mu \tan \gamma) \, a > 0 \ ,$$

α may be put to zero for the small pitch of a fastening screw. If in addition also $\gamma \ll 1$, the function $\kappa \approx 1$.

A 2.12: The hydrostatic gage pressure in an incompressible fluid of density ρ produces a field of parallel forces on the trapezoidal area of an inclined plane retaining wall (see Fig. A 2.12). Calculate the statically equivalent resultant **R** and determine the proper point of application M.

Solution: The pressure is linearly distributed, $p = g \rho h$, and the resulting force follows by considering the in-plane coordinate of $h = \eta \cos \alpha$, the static moment $\eta_S A$, and the area $A = c \, (a + b)/2$ in the relation

$$R = \int_A p \, dA = (g\rho\eta_S \cos \alpha) \, A = g \, \rho \, h_S \, A \ ,$$

where specifically,

$$A\eta_S = \int_A \eta \, dA = \int_d^{d+c} \eta \left[b - (\eta - d)\frac{b-a}{c} \right] d\eta \, , \quad \eta_S = d + \frac{2a+b}{3(a+b)} \, c \, .$$

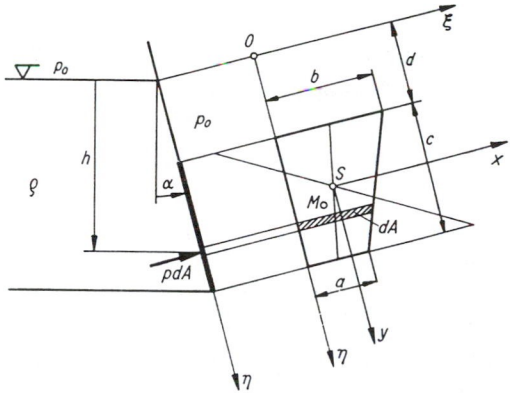

Fig. A 2.12.

Static equivalence requires

$$R\eta_M = \int_A \eta p\, dA = (g\rho \cos \alpha) J_\xi \ ,$$

$$J_\xi = \int_A \eta^2\, dA = \int_d^{d+c} \eta^2 \left[b - (\eta - d) \frac{b-a}{c} \right] d\eta$$

$$= \frac{b}{3} \left[1 + \frac{d}{b} \frac{b-a}{c} \right] \left[(d+c)^3 - d^3 \right] - \frac{b-a}{4c} \left[(d+c)^4 - d^4 \right] \ ,$$

$$\eta_M = \eta_S + i_x^2/\eta_S \ , \quad i_x^2 = c^2 (a^2 + 4ab + b^2) / 18(a+b)^2 \ ,$$

$$R\xi_M = \int_A \xi p\, dA = (g\rho \cos \alpha) J_{\xi\eta} \ ,$$

$$J_{\xi\eta} = \int_A \xi\eta\, dA = \frac{1}{2} \int_d^{d+c} \eta \left[b - (\eta - d) \frac{b-a}{c} \right]^2 d\eta$$

$$= \frac{b^2}{4} \left[1 + 2 \frac{d}{b} \frac{b-a}{c} + \left(\frac{d}{b} \right)^2 \left(\frac{b-a}{c} \right)^2 \right] \left[(d+c)^2 - d^2 \right]$$

$$- \frac{b}{3} \left[\frac{b-a}{c} + \frac{d}{b} \left(\frac{b-a}{c} \right)^2 \right] \left[(d+c)^3 - d^3 \right] + \frac{1}{2} \left(\frac{b-a}{2c} \right)^2 \left[(d+c)^4 - d^4 \right] \ ,$$

$$\xi_M = \xi_S - c (b^3 + 3ab^2 - 3a^2b - a^3) / 36 (a+b)^2 \eta_S \ ,$$

$$\xi_S = (a^2 + ab + b^2) / 3 (a+b) \ .$$

A 2.13: A three-hinged arch of a semicircular shape is loaded by the total hydrostatic pressure of an incompressible fluid of density ρ

from the outside and internally by a constant pressure p_c . Consider its own dead weight $q = g\rho_b s b$, where $s b$ is the cross-sectional area, when calculating the reaction forces in the three joints A, B, and C, and determine the internal resultants $N(\varphi)$, $Q(\varphi)$, $M(\varphi)$. See Fig. A 2.13.

Solution: The shear force at joint C vanishes due to the condition of symmetry, $C_V = 0$, $C_H = C$. The three loading cases are treated separately and may be superposed as a result of the linearity of the equilibrium conditions.

Consider case 1, uniform pressure loading by $p_1 = p_c - p_0$. Gross equilibrium renders $H_1 = 0$, $V_1 = C_1 = -ba\, p_1$. The free-body diagram of an arch element gives the normal force, $N_1 = ba\, p_1$, as constant and $Q_1 = 0$, $M_1 = 0$ [compare this with Eq. (2.98)].

Case 2, the hydrostatic gage pressure $p_2 = g\rho h = g\rho[t + a(1 - \sin\varphi)]$, remains to be considered. The constant atmospheric pressure p_0 is included in case 1. The central force field on one half of the arch has the resulting components $R_{2h} = g\rho h_s\,'A\,' = g\rho(t + a/2)ba$ and $R_{2v} = g\rho[t + a(1 - \pi/4)]ba$; the line of action of $\mathbf{R}$ passes through the center 0 of the circle. Gross equilibrium yields $H_2 = -g\rho ba^2(\pi - 2)/4$, $V_2 = C_2 = g\rho\,[t + a(1 - \pi/4)]ba$. Consideration of the arch element and its resultant loadings, $R_{2h}\,(\varphi) = g\rho ba\,[t + a(1 - (1/2)\sin\varphi)]\sin\varphi$, $R_{2v}\,(\varphi) = g\rho ba\,[t(1 - \cos\varphi) + a(1 - \cos\varphi + (1/2)\sin\varphi\cos\varphi - \varphi/2)]$, yields the cross-sectional resultants

$$N_2(\varphi) = -[H_2 + R_{2h}(\varphi)]\sin\varphi + [-V_2 + R_{2v}(\varphi)]\cos\varphi \ ,$$
$$Q_2(\varphi) = [H_2 + R_{2h}(\varphi)]\cos\varphi + [-V_2 + R_{2v}(\varphi)]\sin\varphi \ ,$$
$$M_2(\varphi)/a = [H_2 + R_{2h}(\varphi)]\sin\varphi - V_2\,(1 - \cos\varphi) - R_{2v}(\varphi)\cos\varphi \ .$$

In case 3, self-weight, $d\mathbf{G} = -qa\,d\varphi\,\mathbf{e}_z$, is a parallel system of

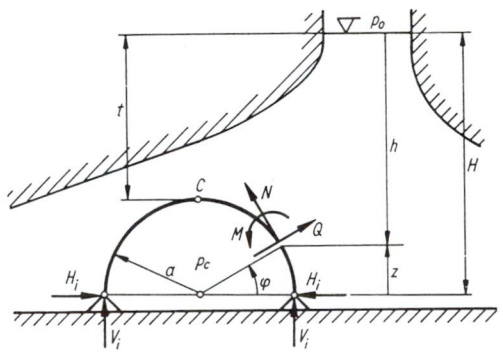

Fig. A 2.13.

loading forces with resultants $R_3 = \pi aq/2$ for each half of the arch in a normal distance of $2a/\pi$ from 0. Gross equilibrium yields the reaction forces $H_3 = C_3 = qa\,[(\pi/2) - 1]$, $V_3 = \pi aq/2$. By means of the force $R_3\,(\varphi) = \varphi aq$ in the normal distance $a\,\sin\,\varphi/\varphi$, the cross-sectional resultants become

$$N_3(\varphi) = -\,aq\,\{[(\pi/2) - 1]\,\sin\,\varphi + [(\pi/2) - \varphi]\,\cos\,\varphi\}\ ,$$
$$Q_3(\varphi) = aq\,\{[(\pi/2) - 1]\,\cos\,\varphi - [(\pi/2) - \varphi]\,\sin\,\varphi\}\ ,$$
$$M_3(\varphi)/a = -\,\pi aq\,[1 - \sin\,\varphi - (1 - 2\varphi/\pi)\,\cos\,\varphi]\,/\,2\ .$$

All the internal forces given above have been determined in a problem-oriented manner. They are the proper solutions of the differential equations (2.145) and (2.147); the loading enters there by means of the tangential and the radial (normal) components.

A 2.14: A homogeneous pendulum, a rod of length l, cross-section A, and density ρ_b, is partly submerged into an incompressible fluid of density ρ, $H < l$; see Fig. A 2.14 . Determine the equilibrium configurations and check the stability of the vertical configuration, $\varphi = 0$.

Solution: The sum of moments about the hinge 0 vanishes, $M_0 = [Gl/2 - A_S\,(l - H_F\,/2)]\,\sin\,\varphi = 0$, where the buoyancy is $A_S = g\rho A H_F$, and $H_F = l - (l - H)/\cos\,\varphi$. $\varphi = 0$ is an unconditional equilibrium configuration. Other configurations are given by the solution of the nonlinear equation, $\sin^2\varphi + (\rho_b\,/\rho)\,\cos^2\varphi = (H/l)\,(2 - H/l)$. The configuration $\varphi = 0$ is stable if a disturbance $|\delta\varphi| << 1$ yields a restoring moment. Linearization renders the inequality, $\delta M = Ag\,[\rho H (2\,l - H) - \rho_b\,l^2]\,\delta\varphi/2 < 0$, $\delta\varphi > 0$, and hence, the condition of stability, $\xi = H/l$,

$$\xi\,(2 - \xi) < \rho_b\,/\,\rho\ .$$

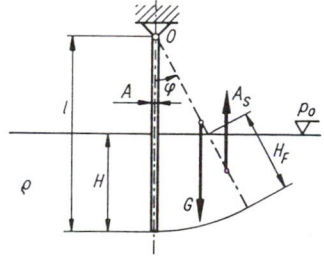

Fig. A 2.14.

Unconditional stability is given for $\rho_b > \rho$. Otherwise, the stability condition is $\xi < [1 - (1 - \rho_b / \rho)^{1/2}]$.

A 2.15: A rigid and homogeneous bridge deck (ρ, A, span l) is supported by a suspension cable of specific weight q per unit of length. The system is shown in Fig. A 2.15 . Given l, h, and b , determine the equation of the parameter a and calculate the length L of the suspension cable explicitly to keep the bridge horizontal.

Solution: The catenary curve is $z = a \cosh (x/a)$, $s = a \sinh (x/a)$. The tension $N = qz$ has the components $H = qa$ and $V = qs$. By means of the weight of the bridge $G = g\rho Al$, the vertical component at the joint becomes known, $V_1 = G/2$. Hence, $s_1 = V_1/q = G/2q = a \sinh (x_1/a)$ is the length of the fictitious part, dashed in Fig. A 2.15 and $x_1 = a \sinh^{-1}(G/2aq)$ determines the horizontal distance of the z-axis. Furthermore, the geometric relation holds (an addition formula is used below)

$$z_2 - z_1 = h$$
$$= a \cosh (x_2/a) - a \cosh (x_1/a) = a [\cosh ((x_1 + b)/a) - \cosh (x_1/a)]$$
$$= a [\cosh (x_1/a) \cosh (b/a) + \sinh (x_1/a) \sinh (b/a) - \cosh (x_1/a)] .$$

Substitution of x_1 yields the nonlinear equation for $\xi = h/a$,

$$\xi [1 - (G/2hq) \sinh (\xi b/h)] = [\cosh (\xi b/h) - 1] \cosh [\sinh^{-1}(\xi G/2hq)] ,$$

ready for numerical solution. Knowing a renders $L = s_2 - s_1 = s_2 - (G/2q)$, $s_2 = a \sinh [(x_1 + b)/a]$.

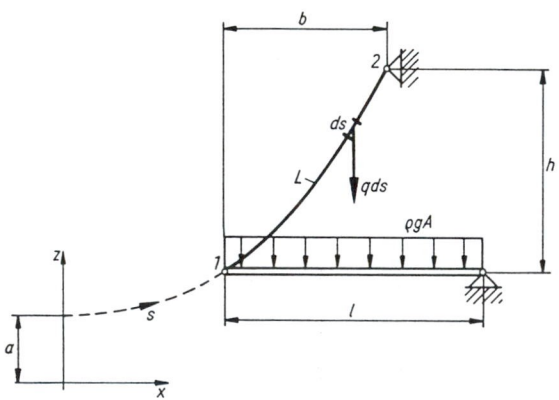

Fig. A 2.15.

3
Mechanical Work, Power, Potential Energy

3.1. Work and Power of Single Forces and Couples

A natural combination of motion (deformation) and force (stress) is given by the notion of mechanical work and power. The elementary work of a single force $F(t)$ by an infinitesimal displacement dr of its point of application at the body is given by the scalar product,

$$\delta W = F \cdot dr = X\,dx + Y\,dy + Z\,dz = |F|\,|dr|\cos\alpha \ . \tag{3.1}$$

Thus, see Fig. 3.1, the component of the displacement $|dr|\cos\alpha$, measured in the direction of the given force F, is crucial for the performance of work. In Eq. (3.1), δW is written instead of dW to distinguish the former from a total differential of a scalar function: The elementary work is, in general, not the analytical differential of the scalar function W. The dimension of work is force times length, $[\delta W] = Nm$, read Newtonmeter, with the newly derived SI-unit *1 Joule* $= 1\ J = 1\ Nm$. Following the pathline C of the point of application from $r_1 = r(t_1)$ to $r_2 = r(t_2)$, the finite work of the force F is the sum of the elementary work that is expressed by the line integral,

$$W_{1\to 2} = \int\limits_{\substack{\text{along C} \\ \text{from } r_1}}^{\text{to } r_2} \delta W = \int_{r_1}^{r_2} F \cdot dr = \int_{x_1}^{x_2} X\,dx + \int_{y_1}^{y_2} Y\,dy + \int_{z_1}^{z_2} Z\,dz \ .$$

$$\tag{3.2}$$

$W_{1\to 2}$, in general, depends on the shape and length of the curve C. This fact can be seen also by considering the ordinary integral $\int X\,dx$, since the component $X(x, y, z; t)$, in an *eulerian* representation, depends on C through the parametric equations $y = y(x), z = z(x)$ of the pathline, in addition to the nonstationary although explicit dependence on time t. However, there are forces, eg those that

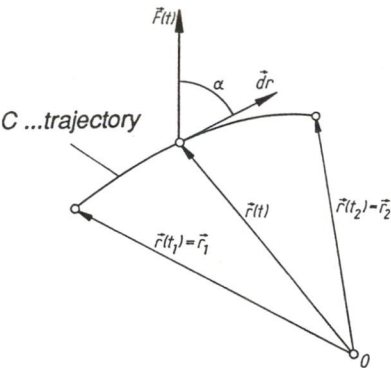

Fig. 3.1. Single force **F** and the trajectory C of the point of application

remain orthogonal to the path C, where all the elementary work $\delta W = 0$, and, hence, $W_{1 \to 2} = 0$. A group of forces may therefore be subdivided into those which do work and those that do not. The work per unit of time, the instantaneous rate, is also a scalar function of time and is called the power of the force **F**,

$$P = P(t) = \frac{\delta W}{dt} = \mathbf{F} \cdot \frac{d\mathbf{r}}{dt} = \mathbf{F} \cdot \mathbf{v} = X\,v_x + Y\,v_y + Z\,v_z = |\mathbf{F}||\mathbf{v}| \cos \alpha \ , \tag{3.3}$$

v(t) is the velocity of the point of application of **F**(t) . The dimension of P is work over time, [P] = Nm/s , the unit 1 Nm/s = 1 J/s = 1 W is called a Watt . Hence, 1 Ws = 1 J = 1 Nm is the unit of work and energy (a technical measure is, eg 1 KWh = 3.6 x 10⁶ J , read kilowatthours). The elementary work over the time interval dt is expressed by the power,

$$\delta W = P(t)\ dt \tag{3.4}$$

and

$$W_{1 \to 2} = \int_{t_1}^{t_2} P\ dt \ . \tag{3.5}$$

In general, the time integral also depends on the pathline C.

Equation (3.3) renders the power of a couple with a moment, **M** = **r** x **F** , for the case of the rigid rotation of the plane of action with the angular velocity **ω**. The point of application of $F_1 = F$ moves with velocity v_1 , the point of application of $F_2 = -F$ with v_2 . Due to the linear velocity distribution of any rigid rotation, $v_1 = v_2 + \omega$ x **r** , the expansion of the scalar triple product in Eq. (3.3) yields at once,

$$P = F_1 . v_1 + F_2 . v_2 = F . (v_1 - v_2) = (\omega \times r) . F = \omega . (r \times F) = M . \omega . \quad (3.6)$$

The elementary work of the moment M follows by multiplication with the time differential dt

$$\delta W = P \, dt = M . \omega \, dt = M . d\varphi \quad , \quad\quad\quad (3.7)$$

if the infinitesimal vector $d\varphi$ in the direction of the axis denotes the rotation of the plane of the couple through the small angle $d\varphi$. Hence, the component of $d\varphi$ in the direction of M, ie the small angle of rotation about an axis through M, determines the elementary work of the couple.

3.1.1. Example: The Work of Gravity Forces

The weight of the masses 1 and 2 in the guided motion according to Fig. 3.2 does elementary work, $\delta W = - (G_1 \, dz_1 + G_2 \, dz_2)$. The inextensibility of the connecting string gives the kinematic constraint $ds = R \, d\varphi = dz_1/\sin \alpha = - dz_2$, and, hence, if we do not consider any friction,

$$\delta W = (- G_1 \sin \alpha + G_2) \, ds \quad .$$

3.1.2. Example: The Work of a Couple

A crank with the radius of eccentricity R carries a tangential force F. The elementary work done during a small rotation $d\varphi$, ($ds = R \, d\varphi$ is the displacement in the direction of the force) follows from Eq. (3.1), as $\delta W = F \, ds = M \, d\varphi$. Reducing F to the center of the crank

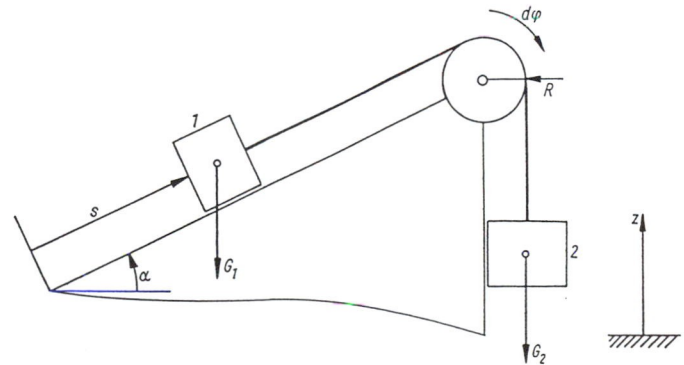

Fig. 3.2. Guided motion of the masses 1 and 2. Work of the gravity forces

renders the couple with the moment $M = FR$.

3.2. Power Density, Stationary and Irrotational Forces, Potential Energy

The single forces per unit of volume f of Eq. (2.18) are attached to the material points of any continuum and may be nonzero. Thus, the local density of power P' , measured in $[W/m^3]$, is given by the scalar product with velocity $v(r,t)$,

$$P' = P'(x, y, z; t) = (f \cdot v) \ . \tag{3.8}$$

The instantaneous power at time t of all the external and the internal forces (body-forces, surface tractions and stresses) acting on a body of material volume $V(t)$ follows by summation, ie by volume integration,

$$P(t) = \int_{V(t)} P' \, dt \ . \tag{3.9}$$

The total work done between two states at t_1 and t_2 is the time integral

$$W_{1 \to 2} = \int_{t_1}^{t_2} P \, dt = \int_{t_1}^{t_2} dt \left(\int_{V(t)} P' \, dV \right) \ . \tag{3.10}$$

The volume integral implies the splitting of the power density into two parts: The power of the stresses or internal forces, $P'^{(i)}$, and the power of the external forces, $P'^{(e)}$. Thus, $P' = P'^{(i)} + P'^{(e)}$.

A stationary density of forces, like any single force field, is recognized in the *eulerian* representation, $f = f(x, y, z; t)$, by $\partial f/\partial t = 0$, and, hence, by $f(x, y, z)$, showing no explicit dependence on time. For such a stationary field of forces, the work may become independent of the special paths of the material points. The necessary and sufficient conditions are subsequently developed for a single stationary force field $F(x, y, z)$, but apply as well to the force density $f(x, y, z)$. Requiring the elementary work to become equal to the total differential of a scalar function $E_p (x, y, z)$ has the following consequence:

$$\delta W = F \cdot dr = X \, dx + Y \, dy + Z \, dz = - dE_p = -\left(\frac{\partial E_p}{\partial x} \, dx + \frac{\partial E_p}{\partial y} \, dy + \frac{\partial E_p}{\partial z} \, dz \right) \to$$

$$X = -\frac{\partial E_p}{\partial x} \ , \quad Y = -\frac{\partial E_p}{\partial y} \ , \quad Z = -\frac{\partial E_p}{\partial z} \ . \tag{3.11}$$

In a vector notation, which becomes independent of the special choice of cartesian coordinates employed in the derivation of Eq. (3.11), this requirement reads,

$$\mathbf{F} = -\nabla E_p = -\operatorname{grad} E_p \quad . \tag{3.12}$$

<<The stationary force $\mathbf{F}$ is the gradient field of a scalar function, which is called the potential energy E_p .>>

A geometric interpretation of the gradient is given by considering Eq. (3.12) as the spatial derivative of the potential function E_p : At the surface of a small sphere centered at a point *(x, y, z)* , the central, radial vector-field E_p *(x₁ , y₁ , z₁)* $\mathbf{e}_n$, pointing outward, is constructed. The resultant is the vector-sum, expressed by the surface loop integral

$$\oint E_p(x_1, y_1, z_1) \, \mathbf{e}_n \, dS \quad .$$

(x₁ , y₁ , z₁) are the points at the surface of the sphere. Dividing through the spherical volume V_S and taking the limit render

$$\operatorname{grad} E_p(x, y, z) = \lim_{V_S \to 0} \frac{1}{V_S} \oint E_p(x_1, y_1, z_1) \, \mathbf{e}_n \, dS \quad . \tag{3.13}$$

<<The gradient is a vector which points in the direction of the strongest variation of the potential E_p .>>

Force in such a potential field is a *flux* in the sense of a mechanical driving agent. The vector lines of the force-field are the orthogonal trajectories of the equipotential surfaces, E_p *(x, y, z) = const* (compare this with the hydrostatic pressure field).

Taking the (steady) mixed second derivatives of the potential and noting the rule when interchanging the order yield these conditions:

$$\frac{\partial Y}{\partial z} = -\frac{\partial^2 E_p}{\partial y \partial z} = -\frac{\partial^2 E_p}{\partial z \partial y} = \frac{\partial Z}{\partial y} \quad , \text{hence,} \quad \frac{\partial Z}{\partial y} - \frac{\partial Y}{\partial z} = 0 \quad ;$$

$$\frac{\partial Z}{\partial x} = -\frac{\partial^2 E_p}{\partial z \partial x} = -\frac{\partial^2 E_p}{\partial x \partial z} = \frac{\partial X}{\partial z} \quad , \text{hence,} \quad \frac{\partial X}{\partial z} - \frac{\partial Z}{\partial x} = 0 \quad ;$$

$$\frac{\partial X}{\partial y} = -\frac{\partial^2 E_p}{\partial x \partial y} = -\frac{\partial^2 E_p}{\partial y \partial x} = \frac{\partial Y}{\partial x} \quad , \text{hence,} \quad \frac{\partial Y}{\partial x} - \frac{\partial X}{\partial y} = 0 \quad . \tag{3.14}$$

Independent of the special coordinates, the necessary and sufficient conditions for the existence of a potential function are summarized in the vector form

$$\nabla \times \mathbf{F} = \text{curl } \mathbf{F} = 0 \ . \tag{3.15}$$

<<Irrotational vector fields of forces $\mathbf{F}$ have an associated potential function E_p.>> This condition holds also for nonstationary forces; the potential function in that case depends explicitly on time, E_p *(x, y, z; t)* , but $\mathbf{F}(x, y, z; t) = -$ *grad* E_p is still valid.

Also the *curl* of a vector field is a spatial derivative. By construction of a field $\mathbf{F} \times \mathbf{e}_n$ that is tangential to the surface of a small sphere with its center at *(x, y, z)* and by consideration of the resultant vector, the *curl* becomes in the limit

$$\text{curl } \mathbf{F} = \lim_{V_S \to 0} \frac{1}{V_S} \oint \mathbf{F} \times \mathbf{e}_n \, dS \ . \tag{3.16}$$

Irrotational and stationary forces are derived from a potential, which is not explicitly dependent on time, and the work done by such a force is independent of the pathline C. The work of such a *conservative* system of forces equals the decrease of potential energy; the difference is taken between the values of the potential at the starting point and at the terminal point of the motion of the point of application of the force $\mathbf{F}$,

$$W_{1 \to 2} = \int_1^2 \mathbf{F} \cdot d\mathbf{r} = - \int_1^2 dE_p = E_p(\mathbf{r}_1) - E_p(\mathbf{r}_2) = E_1 - E_2 \ . \tag{3.17}$$

By considering the potential energy density $E_p{'}$ of the irrotational and stationary force density *f* in a continuum, the total work of the internal and external forces becomes,

$$W_{1 \to 2} = \int_{V(t)} W_{1 \to 2}{'} dV = \int_{V(t_1)} E_1{'} \, dV - \int_{V(t_2)} E_2{'} dV = E_1 - E_2 \ , \tag{3.18}$$

where $E_{1, 2}$ are the values of the total potential energy of the reference state and of the terminal state of the (elastically) deforming body, respectively. Accordingly, the potential energy is commonly split

$$E_p = E_p^{(e)} + E_p^{(i)} \ , \tag{3.19}$$

into the contribution of the conservative (external) loading *(e)* and that of the conservative (internal) stresses *(i)* .

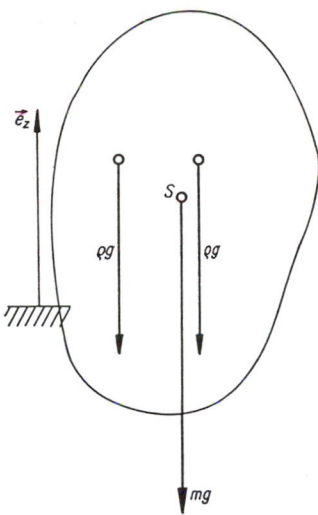

Fig. 3.3. Homogeneous and parallel gravity field, $-g\,\boldsymbol{e}_z = \boldsymbol{const}$

3.3.   Potential Energy of External Forces

The most important conservative forces are due to, approximately parallel fields of gravity, and due to central fields of gravity. Only the homogeneous parallel field of gravity is considered subsequently.

3.3.1. Homogeneous and Parallel Gravity, Potential of the Dead Weight

The vectors of the specific weight are assumed to be parallel to the vertical direction $\boldsymbol{e}_z$, Fig. 3.3. The nonvanishing force component is $Z = -g\rho$ attached to the material points. Hence, $\boldsymbol{k} = Z\,\boldsymbol{e}_z$ is an irrotational field and the potential density E' is the linear function of the height z over a horizontal reference plane,

$$E_g'(z) = -\int_0^z Z\,dz = g\rho z + C' \quad , \quad [\text{Ws/m}^3] \quad .$$

(3.20)

The total potential energy of the weight $G = mg$ is the volume integral; $z_S = z_M$ is the height of the center of gravity and mass,

$$E_g(z) = \int_{V(t)} E_g'(z) \, dV = g \int_{V(t)} \rho z \, dV + C = g \int_m z \, dm + C = mg \, z_S + C \ , \quad [Ws].$$

$$(3.21)$$

3.3.2. Central Force Field with Point Symmetry

With r , the radius vector, the force with unspecified intensity $f(r)$ is

$$\mathbf{F} = f(r) \, \mathbf{r}/r \ . \qquad (3.22)$$

Formally, the work done by this force is the sum of the elementary work, $(\mathbf{F} . d\mathbf{r}) = f(r) \, dr$, which, when we change its sign, equals the potential $E(r)$,

$$E(r) = - \int f(r) \, dr + C \ ,$$

since the central force field is irrotational. Taking the gradient, $grad \ E(r) = \nabla E = (dE/dr) \, \nabla r = - f(r) \, \nabla r$, reproduces the force $\mathbf{F}$ because $grad \ r = \nabla r = \mathbf{r}/r$, is the radial unit vector. The inhomogeneous central gravity field of a mass m_p attracts a sufficiently distant mass m by the inverse square law

$$f(r) = \mu \frac{m_p \, m}{r^2} \ , \qquad (3.23)$$

$\mu = 6.67 \times 10^{-11} \ m^3/kg \ s^2$, is the universal gravitational constant. Hence, for the planet Earth with mass $m_p \approx 5.97 \times 10^{24} \ kg$, average radius $r_0 \approx 6.371 \times 10^6 \ m$, the mean acceleration of gravity becomes $g = 9.81 \ m/s^2$. The variation of g with height follows by allowing $r = r_0 + z$ to be $g = 9.81 \times (1 + z/r_0)^{-2} \ [m/s^2]$. The acceleration of the free fall on the oblate and rotating Earth at a short distance from the surface is a function of the latitude $\varphi : g(\varphi, z) = 9.78049 \ (1 + 0.0052884 \ sin^2\varphi - 0.0000059 \ sin^2 2\varphi) - 0.0003086 \ z - 0.00011$, $[m/s^2]$.

The mass of the Sun is roughly given by $m_S \approx 1.99 \times 10^{30} \ kg$ and $r_0 \approx 6.96 \times 10^8 \ m$.

3.4. Potential Energy of Internal Forces

An elastic body is considered to illustrate the potential of internal forces (stresses). A thermodynamic definition according to *George Green* (1837) is given by the condition that every deformed configuration can be derived by a reversible (quasistatic and isothermal) process from the initially undeformed configuration that is assumed to be free of stresses. Thus, after sufficiently slow unloading, all the strains (deformations) are supposed to vanish and the elastic body recovers its original configuration. For such a sequence of states of equilibrium during loading or unloading, the force density $f = 0$ holds in all the material points, and, hence, the density of power $P' = f \cdot v = 0$, everywhere. According to Eq. (3.10), the total work must vanish

$$W_{1 \to 2} = W^{(e)}_{1 \to 2} - E^{(i)}_p = 0 \quad , \tag{3.24}$$

$$W^{(e)}_{1 \to 2} = E^{(i)}_p \equiv U \quad . \tag{3.25}$$

<<The work $W^{(e)}_{1 \to 2}$ done by the external forces which are acting on the elastic body equals the increase of the potential of the internal forces, the *elastic potential* $E_p^{(i)}$.>> The elastic potential, called the *strain energy* , under isothermal conditions is the internal energy U of the elastic body. Hence, in a closed-loop process of loading and unloading the elastic body, the work done by the external forces must vanish

$$W^{(e)}_{1 \to 2} + W^{(e)}_{2 \to 1} = U + (-U) = 0 \quad ,$$

and the points of application of the forces have closed pathlines.
The specific strain energy or the strain energy density,

$$U' = dU/dV \quad , \tag{3.26}$$

in each loading step is considered to be a function of the parameter, say, time t, assigned to each state of equilibrium. Hence, the velocity $v = dr/dt = du/dt$ determines the rate of deformation, and the power of the external forces can be compared to the rate of the strain energy in every state of equilibrium,

$$P^{(e)} = dU/dt \quad . \tag{3.27}$$

When we consider the free-body diagram of a finite volume V of the elastic body at some instant of time, this power becomes

$$P^{(e)} = \int_V k \cdot v \, dV + \oint_{\partial V} (\sigma_n \, dS \cdot v) = \frac{dU}{dt} \quad , \tag{3.28}$$

k is the given body force (specific weight), σ_n are the tractions at the surface ∂V of V. The surface integral can be changed to the volume integral by the application of the *Gauss* integral formula and by substituting Eq. (2.20),

$$\oint_{\partial V} (\sigma_n \, dS \cdot \mathbf{v}) = \oint_{\partial V} \left[n_x \, (\sigma_x \cdot \mathbf{v}) + n_y \, (\sigma_y \cdot \mathbf{v}) + n_z \, (\sigma_z \cdot \mathbf{v}) \right] dS$$

$$= \int_V \left[\frac{\partial}{\partial x} (\sigma_x \cdot \mathbf{v}) + \frac{\partial}{\partial y} (\sigma_y \cdot \mathbf{v}) + \frac{\partial}{\partial z} (\sigma_z \cdot \mathbf{v}) \right] dV .$$

Termwise differentiation of the products and the combination of the stress gradients with the given body force *k* give

$$P^{(e)} = \int_V \mathbf{f} \cdot \mathbf{v} \, dV + \int_V \left[(\sigma_x \cdot \frac{\partial \mathbf{v}}{\partial x}) + (\sigma_y \cdot \frac{\partial \mathbf{v}}{\partial y}) + (\sigma_z \cdot \frac{\partial \mathbf{v}}{\partial z}) \right] dV = \frac{dU}{dt} \ .$$

$$(3.29)$$

Since $\mathbf{f} = \mathbf{0}$ due to the balanced forces and Eq. (3.29) must hold for any partial volume V and also in the limit $V \to 0$, the specific rate of the strain energy becomes

$$- P^{\cdot(i)} = \frac{dU'}{dt} = \left[(\sigma_x \cdot \frac{\partial \mathbf{v}}{\partial x}) + (\sigma_y \cdot \frac{\partial \mathbf{v}}{\partial y}) + (\sigma_z \cdot \frac{\partial \mathbf{v}}{\partial z}) \right] \ ,$$

$$(3.30)$$

and $\mathbf{v}(x, y, z; t)$ is to be considered in the *eulerian* representation. In a subscript notation, the rate of internal energy is expressed by

$$\frac{dU'}{dt} = \sum_i \sum_j \sigma_{ij} \frac{\partial v_j}{\partial x_i} = \sum_i \sum_j \sigma_{ji} \frac{\partial v_i}{\partial x_j} \ .$$

$$(3.31)$$

By means of the symmetric tensor of the deformation rates in the *eulerian* representation,

$$V_{ij} = V_{ji} = (1/2)(\partial v_i/\partial x_j + \partial v_j/\partial x_i) \ ,$$

$$(3.32)$$

the specific rate of the strain energy is simply given by taking the scalar product with the symmetric stress tensor σ

$$\frac{dU'}{dt} = \sum_i \sum_j \sigma_{ij} V_{ij} \ .$$

$$(3.33)$$

The increment of internal energy during a small time interval *dt*, where the deformation $d\mathbf{u} = \mathbf{v} \, dt$ maps the instantaneous equilibrium configuration into the neighboring state, becomes by multiplication the expression that is free of the parameter time,

$$dU' = \sum_i \sum_j \sigma_{ij} \, V_{ij} \, dt = \sum_i \sum_j \sigma_{ij} \frac{1}{2} \, d \left[\frac{\partial u_i}{\partial x_j} + \frac{\partial u_j}{\partial x_i} \right] \, .$$

(3.34)

For the small increments of deformation, the linearized geometric relations apply, and, hence,

$$d\varepsilon_{ij} = \frac{1}{2} \, d \left[\frac{\partial u_i}{\partial x_j} + \frac{\partial u_j}{\partial x_i} \right] \, , \quad dU' = \sum_i \sum_j \sigma_{ij} \, d\varepsilon_{ij} \, .$$

(3.35)

The partial differentiation of Eq. (3.35) with respect to a strain component renders the stress $\partial U'/\partial \varepsilon_{ij} = \sigma_{ij}$. Thus, if the constitutive equation of the elastic body is given, $\sigma_{ij} (\varepsilon_{ij})$, termwise integration can be performed to yield the strain energy

$$U'(\varepsilon_{ij}) = \sum_i \sum_j \int \sigma_{ij} \, d\varepsilon_{ij} \, .$$

(3.36)

In the case of isotropy, this function depends only on the three invariants I_1 , I_2 , I_3 of the strain tensor.

3.4.1. The Elastic Potential of the Hookean Solid (Linear Spring)

Hooke´s law describes the linear elastic relation between stress and strain; see Sec. 4.1.1. In a tensile rod of constant cross-section A , the uniaxial stress, as well as the strain, are constant over the length l, and they are proportional to each other, $\sigma_{xx} = E \, \varepsilon_{xx}$, where E is *Young´s* modulus of elasticity. In that case, Eq. (3.36) becomes upon elimination of the normal stress component the constant strain energy density

$$U'(\varepsilon_{xx}) = E \, \frac{\varepsilon_{xx}^2}{2} \, .$$

(3.37)

Multiplication by the volume of the rod Al gives the spring potential (the totally stored elastic potential or strain energy) expressed by the strain or, alternatively, by the stress,

$$U(\varepsilon_{xx}) = \frac{1}{2} E \, A \, l \, \varepsilon_{xx}^2 = U(\sigma_{xx}) = \frac{1}{2} \frac{A \, l}{E} \, \sigma_{xx}^2 \, .$$

(3.38)

Substituting the axial force $N = A \, \sigma_{xx}$ renders the commonly encountered form, the complementary energy, to be

$$U(N) = \frac{1}{2} \frac{N^2}{EA} \, l = U^* \; .$$

(3.39)

For small and homogeneous strain, $\varepsilon_{xx} \ll 1$, the linearized geometric relation (1.21) becomes $\varepsilon_{xx} = du/dx = s/l$, where $s = u(x = l)$ is the elongation of the rod when fixed at $x = 0$,

$$U(s) = \frac{1}{2} \frac{EA}{l} s^2 \; ,$$

(3.39a)

and $EA/l = c$, measured in $[N/m]$, is the stiffness of the spring that is the force necessary to produce an unit elongation, $s = 1$. Hence, the gross *Hooke's* law of the spring $F = N = cs$ expresses proportionality between the axial load and the axial elongation. The work of the external force F during the deformation process is

$$W^{(e)}_{1 \to 2} = \int_0^s F(u) \, du = \int_0^s cu \, du = U(s) = \frac{1}{2} cs^2 = \frac{F^2}{2c} \; .$$

(3.40)

<<The work of the slowly increasing external force F is stored in the elastic rod in the form of recoverable elastic energy (of the internal forces N). During unloading, the work done amounts to exactly the same value (no dissipation).>> Thus, the strain energy depends on the stiffness c , a system parameter, and may be expressed by the elongation s or, alternatively, by the axial loading F . A property of general importance should be noted

$$dU/ds = F \quad \text{and} \quad dU/dF = s \; ,$$

(3.41)

where s is the deformation measured in the direction of the force $F = cs$.

The expression (3.39) of the strain energy applies to every linear elastic member of an *idealized truss* . Hence, if we assume a common *Young's* modulus E , summation over n members gives (S_i is the axial force in a member numbered i, which is expressed by the external loadings in case of static determinacy)

$$U = \frac{1}{2} \sum_{i=1}^{n} l_i S_i^2 / EA_i \; .$$

(3.42)

The partial differentiation of Eq. (3.42) with respect to an external force F renders the elastic displacement δ in the direction of that force at the point of application, a nodal displacement,

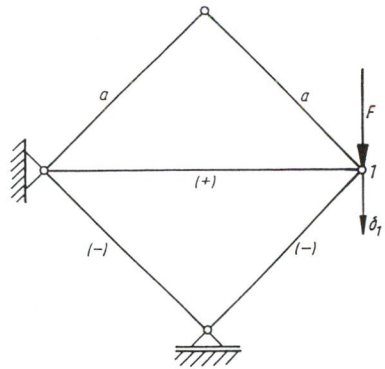

Fig. 3.4. Deformation of a linear elastic truss

$$\delta = \frac{\partial U}{\partial F} = \sum_{i=1}^{n} l_i (S_i / EA_i) \frac{\partial S_i}{\partial F} \ . \tag{3.43}$$

Noting that $\partial S_i/\partial F$ gives the relative change of the member force S_i with respect to the single force loading F , we see that it can be replaced in Eq. (3.43) by the member forces due to the individual loading by a dummy unit force $F = 1$. After that substitution, Eq. (3.43) becomes a convenient formula for engineering applications, eg for the control of deformations.

(§)   Example: A Simple Truss. Its design is shown in Fig. 3.4, with five members of equal cross-section $A = A_i = 6.25\ cm^2$, and the quadratic gross dimension $a = 0.25\ m$. The external loading $F = 50\ kN$ at node 1 gives rise to the normal forces $- F\sqrt{2}$ in the lower members, and to F in the horizontal diagonal (the upper members have zero stress). *Young's* modulus of steel, $E = 21 \times 10^6\ N/cm^2$, when inserted in Eq. (3.43) yields the lowering of joint 1 as $\delta_1 = (Fa/EA)[2(\sqrt{2})^2 + \sqrt{2}] = 0.516 \times 10^{-3}\ m$.

The uniaxial *Hooke's* law $F = c\ s$ can be generalized by considering the loading of a linear elastic body by several single forces and concentrated external moments, by means of matrix notation,

$$\mathbf{F} = \mathbf{k}\ \mathbf{u}_F \ , \tag{3.44}$$

where $\mathbf{k}$ is the stiffness matrix (the elements are the stiffness coefficients). Equation.(3.44) expresses *Hooke's* law by the linear vector transformation between the displacements u_i in the direction of the applied forces, sampled in the column matrix $\mathbf{u}_F$ and external loadings; the single forces F_i are the elements of the column matrix $\mathbf{F}$. In the case of concentrated moment loadings, the proper deformations are small rotations about the moment axis. Analogous to the tensile rod, where the stiffness c equals the applied force F for elongation $s = 1$, a column of the stiffness matrix takes on the values of the external forces that produce a unit deformation in the proper row of the matrix u_F when all other elements are kept at zero values,

$$F_i = \sum_j k_{ij} u_j \quad \rightarrow \ u_k = 1, \ u_j = 0 , \ j \neq k , \ F_i = k_{ik} , \ i = 1, 2 ,... \ .$$

The strain energy stored in the generally loaded linear elastic body becomes according to the matrix counterpart of Eq. (3.40)

$$U = \frac{1}{2} \ u_F^T \ k \ u_F = \frac{1}{2} \ F^T \ k^{-1} \ F .$$

$$(3.45)$$

A superscript T stands for the transpose of the matrix. The inverse k^{-1} of a nonsingular stiffness matrix is called the *flexibility* matrix. It exists when *det $\{k\} \neq 0$* . The elements are *influence numbers* , a generalization of the flexibility *$1/c = s/F$* of the uniaxial spring,

$$u_F = k^{-1} \ F .$$

$$(3.46)$$

A column of the flexibility matrix corresponds to the deformation u_F caused by the unit loading in the proper row of the loading vector, which otherwise has zero elements. Such a loading must be balanced to render the stiffness matrix nonsingular (eg all the degrees of freedom of rigid-body motions must be constrained by proper supports); see Eqs. (6.1), (6.2), (10.49) and (11.70).

The strain energy density of Eq. (3.36) is a quadratic function of the strains that for an *isotropic Hookean solid* must be of the form

$$U'(\varepsilon_{ij}) = C_1 \ e^2 + C_2 \ J_2 \ , \quad e = \varepsilon_{xx} + \varepsilon_{yy} + \varepsilon_{zz} \ ,$$

where *$C_{1, 2} = const$* and J_2 is the second invariant of the strain deviator, see Eqs. (1.35) and (2.36). The latter is independent of the dilatation e . When we take advance notice of the three-dimensional *Hooke's* law , Eq. (4.15),

$$\sigma_{ij} = 2G \left(\varepsilon_{ij} + \frac{\nu}{1 - 2\nu} \ e \ \delta_{ij} \right) \ ,$$

the integration in Eq. (3.36) can be performed to render

$$U'(\varepsilon_{ij}) = G\left[\frac{v}{1-2v}\, e^2 + \sum_i \sum_j \varepsilon_{ij}^2\right] = \frac{G(1+v)}{3(1-2v)}\, e^2 + 2G\, J_2 \ ,$$

$$(3.47)$$

a sum of two terms. Comparing coefficients gives $2C_1 = E/3(1-2v)$ $= K$, the *bulk modulus* , and $C_2 = 2G$, where $G = E/2(1+v)$ is the *shear modulus* and $-1 < v < 1/2$ is *Poisson's* ratio. The total elastic potential of a *Hookean* body is the volume integral

$$U = \int_V U'\, dV \ .$$

$$(3.47a)$$

In Eq. (3.47), the specific work of compression is given by the term K $e^2/2$, and $2G\, J_2$ represents the work of the configurational change of shape if the deformation gradients are small (the distortional strain energy). The latter becomes strain energy in the incompressible deformation process of an elastic solid.

3.4.2. The Barotropic Fluid

For the specific assumption of an ideal and barotropic fluid, the shear stresses vanish and the hydrostatic stress $\sigma_{ij} = -p\, \delta_{ij}$ has the given constitutive relationship to the density $p = p(\rho)$. Equation (3.33) in that case yields the power density of compression

$$dU'/dt = -p\ \mathrm{div}\ \mathbf{v} \ .$$

$$(3.48)$$

Substitution of the continuity equation (1.75) gives $dU'/dt = (p/\rho)$ $d\rho/dt$, and the increment of the internal potential becomes in a notation free of time

$$dU' = p\ (d\rho/\rho) \ .$$

$$(3.49)$$

3.5. *Lagrangean* Representation of the Work of Internal Forces, *Kirchhoff's* Stress Tensor

For large deformations, the nonlinear geometric relations (1.20) apply, and Eq. (3.33) should be transformed into *lagrangean* form. Formally, the material time derivative of Eq. (1.18) is

$$\frac{d}{dt}\left(dl^2 - dl_0^2\right) = 2\left[\dot{\varepsilon}_{xx}\, (dX)^2 + \dot{\varepsilon}_{yy}\, (dY)^2 + \dot{\varepsilon}_{zz}(dZ)^2\right.$$
$$\left. + 2\left(\dot{\varepsilon}_{xy}\, dX\, dY + \dot{\varepsilon}_{yz}\, dY\, dZ + \dot{\varepsilon}_{zx}\, dZ\, dX\right)\right] = 2\sum_k \sum_l \frac{d\varepsilon_{kl}}{dt}\, dX_k\, dX_l \ .$$

$$(3.50)$$

Alternatively, if we differentiate $dl^2 = dx^2 + dy^2 + dz^2$ and notice $d(dx)/dt = d(dx/dt) = dv_x$, etc., the derivative becomes

$$\frac{d}{dt}\left(dl^2 - dl_0^2\right) = 2\left[dx\,dv_x + dy\,dv_y + dz\,dv_z\right]$$

$$= 2\sum_i \sum_j V_{ij}\left(\frac{\partial x_i}{\partial X}dX + \frac{\partial x_i}{\partial Y}dY + \frac{\partial x_i}{\partial Z}dZ\right)\left(\frac{\partial x_j}{\partial X}dX + \frac{\partial x_j}{\partial Y}dY + \frac{\partial x_j}{\partial Z}dZ\right).$$

Termwise multiplication renders finally the quadruple sum

$$\frac{d}{dt}\left(dl^2 - dl_0^2\right) = 2\sum_i \sum_j \sum_k \sum_l V_{ij}\frac{\partial x_i}{\partial X_k}\frac{\partial x_j}{\partial X_l}dX_k\,dX_l.$$

(3.51)

Comparing coefficients and considering the deformation gradient $F_{ik} = \partial x_i/\partial X_k$ yield the strain rate in the form,

$$\frac{d\varepsilon_{kl}}{dt} = \sum_i \sum_j V_{ij}F_{ik}F_{jl} = \frac{\partial \varepsilon_{kl}}{\partial t}.$$

(3.52)

In the case of rigid-body motion, $d\varepsilon_{kl}/dt = 0$, as well as $V_{ij} = 0$.

Equation (3.52) may be solved for the elements of the symmetric tensor of the deformation rates

$$V_{ij} = \sum_k \sum_l \frac{d\varepsilon_{kl}}{dt}F_{ki}^{-1}F_{lj}^{-1},$$

and the divergence of the velocity field is thus given by summation

$$\sum_i V_{ii} = \text{div } \mathbf{v} = \sum_k \sum_l \frac{d\varepsilon_{kl}}{dt}\sum_i F_{ki}^{-1}F_{li}^{-1}.$$

The components

$$B_{kl} = \sum_i F_{ki}^{-1}F_{li}^{-1},$$

are the elements of *Finger's* strain tensor. (Late *J. Finger* was Professor of Mechanics at the TU-Vienna). The kinematic constraint of incompressibility, *div* $\mathbf{v} = 0$, can be expressed by

$$\sum_k \sum_l B_{kl}\dot{\varepsilon}_{kl} = 0.$$

By the multiplication of $d\varepsilon_{kl}/dt$ with a proper stress component S_{kl} and summation, the specific power of Eq. (3.33) per unit of mass is determined by

$$\frac{1}{\rho}\frac{dU'}{dt} = \frac{1}{\rho}\sum_i \sum_j \sigma_{ij} V_{ij} = \frac{1}{\rho_0}\sum_k \sum_l S_{kl} \frac{\partial \varepsilon_{kl}}{\partial t} \quad . \tag{3.53}$$

Comparing coefficients of the strain rates yields the relations between the *Cauchy* stresses of the *eulerian* representation and the *Kirchhoff* stresses

$$\sigma_{ij} = \frac{\rho}{\rho_0}\sum_k \sum_l S_{kl} F_{ik} F_{jl} \quad . \tag{3.54}$$

Solving for the stress components of the *lagrangean* form gives

$$S_{kl} = \frac{\rho_0}{\rho}\sum_i \sum_j \sigma_{ij} F_{ki}^{-1} F_{lj}^{-1} = S_{lk} \quad . \tag{3.55}$$

The newly defined stress components are the elements of the symmetric second *Piola-Kirchhoff* stress tensor. Most important, they produce the same power density per unit of mass when multiplied by *Green's* strain rates; ρ is the current density and ρ_0 the initial density in the undeformed reference state.

The stress vector or traction $\mathbf{S}_n$ according to the *Kirchhoff* stresses is constructed by relating the force to the undeformed area element with normal $\mathbf{e}_n$ in the undeformed reference configuration. The nonorthogonal components are given by projection into the obliquely angled triad that is the pointwise different output of the mapping of the initially orthogonal system $\mathbf{e}_x$, $\mathbf{e}_y$, $\mathbf{e}_z$ into the deformed state. The three base vectors that are not normalized are

$$\mathbf{g}_\alpha = \mathbf{e}_\alpha + \frac{\partial \mathbf{u}}{\partial X_\alpha} \quad , \quad \alpha = 1, 2, 3, \tag{3.56}$$

where

$$\mathbf{g}_\alpha \cdot \mathbf{g}_\beta = g_{\alpha\beta} = \delta_{\alpha\beta} + 2\varepsilon_{\alpha\beta} \quad , \tag{3.57}$$

are the elements of the metric tensor. Hence,

$$\mathbf{S}_n = \sum_{\alpha=1}^{3} S_{n\alpha} \mathbf{g}_\alpha \quad . \tag{3.58}$$

From these considerations of large deformations, it is already seen that a need exists for different stress tensors. Normalization of the base vectors $\mathbf{g}_\alpha$ renders the *engineering stress components* $\tau_{n\alpha}$ which are related to the undeformed area element

$$\mathbf{S}_n = \sum_{\alpha=1}^{3} \tau_{n\alpha} \, \mathbf{g}_\alpha / |\mathbf{g}_\alpha| \quad , \quad \tau_{n\alpha} = \mathbf{S}_{n\alpha} |\mathbf{g}_\alpha| \quad .$$

(3.59)

They do not obey the rules of transformation of tensor elements for rotated coordinate systems.

A second interpretation of the *Kirchhoff* stress vector $\mathbf{S}_n$ with components S_{ni} in the direction of the initial reference coordinates $\mathbf{e}_x$, $\mathbf{e}_y$, $\mathbf{e}_z$, say $i = x, y, z$, follows from the consideration of the deformed area element with the initial normal $\mathbf{e}_n$ in the undeformed configuration and by relating it to the force $d\mathbf{F}$ after a proper transformation into the undeformed state, $d\mathbf{F}_0$, with its force components and stresses given by

$$dF_{0i} = \sum_{j=1}^{3} \frac{\partial X_i}{\partial x_j} \, dF_j \quad , \quad \tilde{S}_{ij} = \frac{\rho_0}{\rho} \sum_k \sigma_{ik} F_{kj}^{-1} \quad .$$

(3.60)

In continuum mechanics, the associated nonsymmetric tensor is called first *Piola-Kirchhoff stress tensor* .

3.6. Exercises A 3.1 to A 3.2 and Solutions

A 3.1: Parallel and Series Connection of Springs. The effective stiffness of linear elastic springs arranged in series connection or, alternatively, in parallel connection, is to be determined by the addition of their elastic potentials. See Fig. A 3.1.

Solution: The total potential of *n* springs is the sum of the individually stored elastic energy

$$U = \sum_{i=1}^{n} U_i \quad , \quad n = 3 \text{ in Fig. A 3.1 .}$$

In the case of the parallel connection, the elongation *s* is common to all springs, hence, with $U_i = c_i \, s^2/2$,

$$U = \frac{s^2}{2} \sum_{i=1}^{n} c_i = \frac{1}{2} k_{eff} \, s^2 \quad , \quad k_{eff} = \frac{F}{s} = \sum_{i=1}^{n} c_i \quad , \quad F = \sum_{i=1}^{n} F_i \quad .$$

Each of the series-connected springs carries the same load *F* . Thus, the stored energy in the spring numbered *i* is $U_i = F^2/2c_i$, and

$$U = \frac{F^2}{2} \sum_{i=1}^{n} c_i^{-1} = \frac{F^2}{2 \, k_{eff}} \quad , \quad k_{eff} = \frac{F}{s} = 1 / \sum_{i=1}^{n} c_i^{-1} \quad , \quad s = \sum_{i=1}^{n} s_i \quad .$$

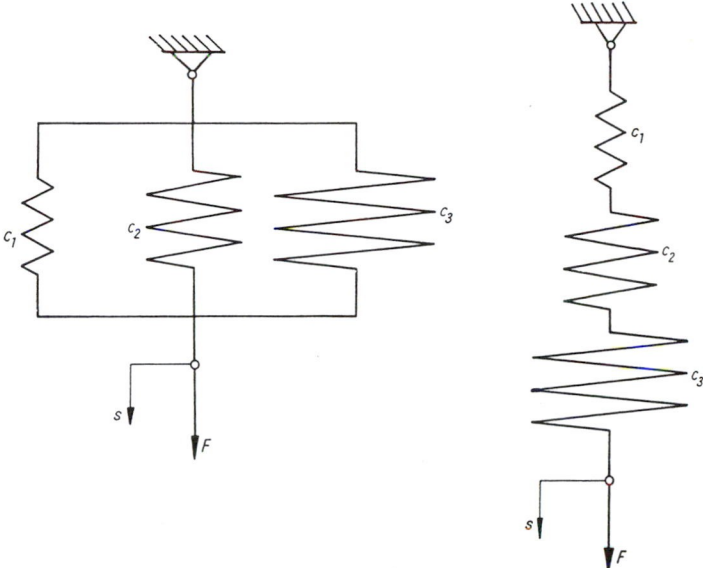

Fig. A 3.1. Parallel and series connection of linear elastic springs

See Sec. 4.2.2. and especially also the Exercises A 6.2, A 6.3 and Sec. 12.8.

A 3.2: Determine the moment small-angle relation of the spring-mounted foundation (Fig. A 3.2) by differentiating the elastic potential. The unsymmetric nonlinear constitutive relation of the distributed springs is given by a polynomial function of degree 3. The linear stiffness coefficient (per unit of length) *k´* is usually the common factor of the polynomial. For clockwise rotation, the elongation of an isolated spring is $s = -x\varphi$. Note the influence of the quadratic term and identify the linear rotational stiffness.

Solution: The force (per unit of length) displacement relation is

$$q = k´\left(s + \eta_1 s^2 + \eta_2 s^3\right) .$$

The specific energy equals the work of the external force per unit of length,

$$U´ = \int_0^s q(u)\, du ,$$

and the superposition yields the elastic potential

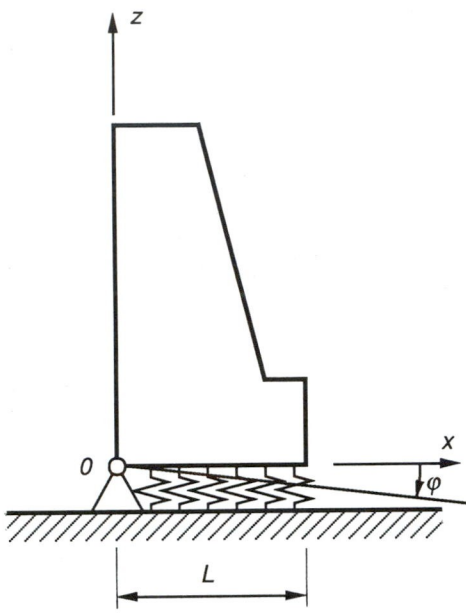

Fig. A 3.2. Physically nonlinear spring foundation

$$U = \int_0^L U' \, dx = k' \frac{L^3}{3} \left(\frac{\varphi^2}{2} - \eta_1 \frac{L}{4} \varphi^3 + \eta_2 \frac{3 L^2}{20} \varphi^4 \right) \ .$$

Differentiation renders the restoring moment of the compressed foundation (note the resulting linear torsional stiffness and the unsymmetry with respect to φ)

$$M = \frac{dU}{d\varphi} = k' \frac{L^3}{3} \left(\varphi - \eta_1 \frac{3 L}{4} \varphi^2 + \eta_2 \frac{3 L^2}{5} \varphi^3 \right) \ .$$

For the (linear) *Winkler* foundation see Sec. 9.6 and Fig. 11.7.

4
Constitutive Equations

"Nonlocal" material laws of solids or fluids are not considered in this textbook. Hence, the constitutive relations in every material point are either finite or time differential equations, which relate the stress tensor to the proper strains (the influence of temperature or other nonmechanical fields are excluded at this stage of consideration). Simple material laws are given for elastic, visco-elastic, elastic-plastic, and elastic-visco-plastic bodies. Especially with respect to the latter, the deformations are restricted to relatively small strains and strain rates, ie problems associated with plastic-forming techniques, like forging, are to be excluded. A loose guide of the subsequent considerations is given by the extensive and organized material testing that is performed in daily routine to render the numerical values of material parameters.

Except for single crystals, some fluids, and very pure (nontechnical) materials, a general theory is still lacking that relates the binding forces at the atomic or molecular level to the strength of a body. The latter is filled with all kinds of impurities, and thus, the smeared relations between stress and strain cannot be determined exactly. On the other hand, material science has developed theories and skills for the production of materials with optimal properties for special applications. The reader is referred to the vast literature on this important subject.

4.1. The Elastic Body, *Hooke´s* Law of Linear Elasticity

The general definition of nonlinear elasticity extends the thermodynamic considerations of the *hyperelastic* body by *George Green* ; see Eq. (3.35). The existence of an undeformed initial configuration, free of any stresses, in thermodynamical equilibrium is assumed. The elastic body returns to this state after unloading for any loading path and a unique finite relation exists between corresponding tensors of stress and strain. In the *eulerian* representation, the stress components of the *Cauchy* stress tensor are related to the *Almansi* strains

$$\sigma_{ij} = f(\varepsilon_{ij}) \ . \tag{4.1}$$

In the *lagrangean* description, *Kirchhoff´s* stress tensor is a function of *Green´s* strain tensor,

$$S_{ij} = G(\varepsilon_{ij}) \ . \tag{4.2}$$

Subsequently, linear and an example of a quadratic law of nonlinear elasticity are discussed.

4.1.1. The Linear Elastic Body, Hooke´s Law

The simplest and most frequently applied relation between stress and strain is the linear law of elasticity of *Hooke´s* material. *Robert Hooke* stated the law in 1678 in Latin, "Ut tensio sic vis." It means <<the force (of a spring) is proportional to its elongation>> or

$$F = cs \ . \tag{4.3}$$

The coefficient c is the stiffness of the spring and s the displacement in the direction of the applied force F ; see Fig. 4.1. Tensile rod-specimens (their design and size are specified in codes) are stressed in special testing machines to determine the stiffness c in the routine testing of elastic materials. The tension is increased in small steps or continuously and slowly, and the corresponding pair of force and elongation is precisely measured and registered in the diagram of the spring. The slope of the straight line is the stiffness c . The uniaxial linear elastic constitutive equation is derived from such a measurement by relating the axial force to the undeformed cross-sectional area A_0 , which renders the *Kirchhoff* stress component $S_{xx} = S$,

$$S = F / A_0 = (cl_0 / A_0) (s / l_0) = E\varepsilon \ , \tag{4.4}$$

where $\varepsilon = s/l_0$, is the homogeneous specific elongation and $E = (cl_0 /A_0)$, denotes *Young´s* modulus. For mild steel, $E = 2.06 \times 10^5$ N/mm^2 ; for other elastic materials, see the table in the appendix. In that part of the specimen where $\varepsilon = const$, the lateral dimensions d_0 are reduced by the factor

$$v = \varepsilon_q / \varepsilon \ , \quad \varepsilon_q = (d_0 - d) / d_0 \ , \tag{4.5}$$

and v is *Poisson´s* constant. In general, $v > 0$; for steel, $v \approx 0.3$. The admissible range is $-1 < v \le 1/2$; see, eg Eq. (4.11).

The measured specific elongation $\varepsilon \ll 1$ in the range of linear elasticity is approximately equal to the axial strain component, $\varepsilon \approx$

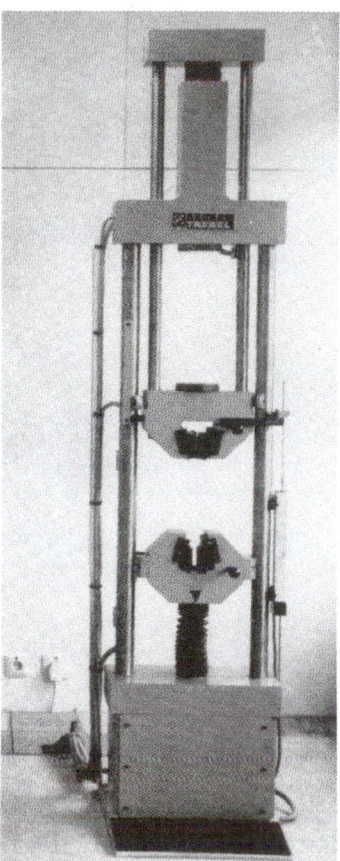

Fig. 4.1 (a). Hydraulic universal testing machine of Schenck-Trebel, Inc. Loading capacity *120 kN*

ε_{xx} ; analogously, $\varepsilon_q \approx -\varepsilon_{yy}$, since, eg for steel the limit of proportionality of stress and strain is about $\varepsilon \approx 0.5\%$. Consistently with the definition of internal work, the mathematical form of the uniaxial *Hooke's* law thus becomes

$$S = E\,\varepsilon_{xx} \ , \tag{4.6}$$

where ε_{xx} is the *Green* strain component. Equation (4.6) from a mathematical point of view holds also for large deformations, but the stress-specific elongation relation then becomes nonlinear! In a *eulerian* representation and also consistent with the definition of

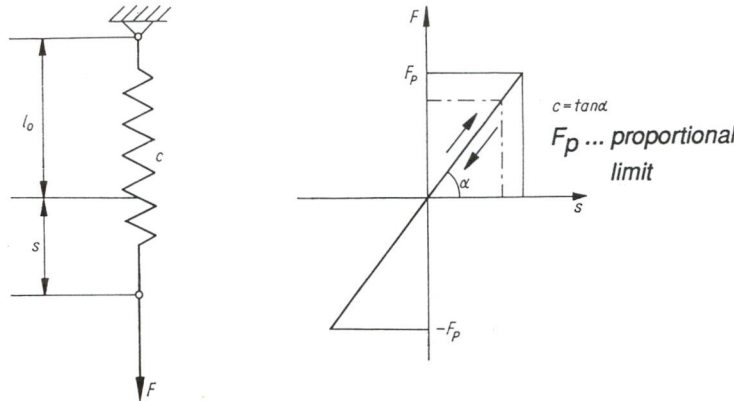

Fig. 4.1 (b). Symbolic model of *Hooke's* body. Force-displacement relation of the linear spring

internal work, another linear material law is found by referring the axial force to the current cross-sectional area A

$$\sigma_{xx} = F / A = E \, \varepsilon_{xx} \quad . \qquad (4.7)$$

ε_{xx} in the geometrically nonlinear case is the *Almansi* strain component. The stress-specific elongation relation is again nonlinear! A transformation of Eq. (4.7) to the *lagrangean* representation gives a nonlinear stress-strain law. Equations.(4.6) and (4.7) are the proper linear relations, which in the case of material isotropy allow the transition from a cross-sectional element to a rotated area element by transformation of the tensor elements, stress and strain; see Fig. 2.2. Equation (4.4), on the other hand, is merely a linear spring relation and not a constitutive (material) law.

A testing in *simple shear* , where all the strain components vanish except $\varepsilon_{xy} = \varepsilon_{yx}$, renders the (linear) relation between the shear stress and the shear angle from the measurement. The mathematical model, eg in the *eulerian* representation, is the linear constitutive law

$$\sigma_{xy} = 2G \, \varepsilon_{xy} \quad , \qquad (4.8)$$

where G is the shear modulus.

Young's modulus E and the shear modulus G are the pair of elastic constants of the homogeneous and isotropic *Hooke's* body (eg in the *eulerian* representation). In such a continuum, the principal stress and principal strain axes must be coincident in every material point. Each of the principal normal stresses of the three-dimensional stress state produces a lateral contraction; hence, when we generalize the uniaxial tension test, superposition renders in the three-dimensional case of a *eulerian* representation

$$\varepsilon_1 = \frac{1}{E}[\sigma_1 - v(\sigma_2 + \sigma_3)] \ , \ \varepsilon_2 = \frac{1}{E}[\sigma_2 - v(\sigma_3 + \sigma_1)] \ ,$$
$$\varepsilon_3 = \frac{1}{E}[\sigma_3 - v(\sigma_1 + \sigma_2)] \ .$$

(4.9)

When we take the mean stress $p = (\sigma_1 + \sigma_2 + \sigma_3)/3$ and the dilatation $e = \varepsilon_1 + \varepsilon_2 + \varepsilon_3$, the addition of Eqs. (4.9) yields the linear law of compressibility

$$e = [3(1 - 2v)/E]p \ .$$

(4.10)

For small strain, the dilatation e is the specific volume change, and the elastic constant

$$K = E/3(1 - 2v) \ ,$$

(4.11)

represents the bulk modulus. Under the condition of isotropic tension, $p > 0$ and $e > 0$. Hence, *Poisson's* ratio is limited to $0 \le v \le 1/2$. The range $-1 \le v < 0$ is generally not considered. The relation

$$p = Ke \ ,$$

(4.12)

gives the elasticity in compression and, thus, represents the first *Hooke's* law of the three-dimensional stress state. Both p and e are the first invariants of the corresponding tensors.

Compressibility of the material is thus separated, and consequently, the deviatoric principal strain components $\varepsilon_i' = \varepsilon_i - e/3$ [Eq. (1.35)] and the deviatoric stress components $\sigma_i' = \sigma_i - p$ [then Eq. (2.34) with the notation changed to $s_{ij} \equiv \sigma_{ij}'$] are linearly related by

$$\varepsilon_i' = [(1 + v)/E] \sigma_i' \ , \quad \sigma_i' = \sigma_i - p \ , \quad i = 1, 2, 3 \ .$$

(4.13)

When rotating the coordinate system with respect to the coinciding principal axes, the transformation formulas for tensor elements apply. See Eqs. (2.22) and (2.23) and change these to a principal axes representation; the deviatoric stresses are linearly related to the deviatoric strains by

$$\sigma_{ij}' = 2G \varepsilon_{ij}' \ , \quad i, j = 1, 2, 3 \ ,$$

(4.14)

if the relation for the shear modulus $G = E/2(1 + v)$ is introduced that is not in conflict with the definition from the test in pure shear. The relation (4.14) gives six equations in the case of the three-dimensional stress state and completes *Hooke´s* law. Elimination of the deviatoric components renders *Hooke´s* law in the form originally given by *Lame*

$$\sigma_{ij} = \lambda e\, \delta_{ij} + 2G\varepsilon_{ij} \quad , \quad \varepsilon_{ij} = \frac{1}{2G}\left(\sigma_{ij} - \frac{3v}{1+v}\, p\, \delta_{ij}\right) \quad , \quad i, j = 1, 2, 3 \ ,$$
$$\lambda = K - \frac{2}{3}G = vE/(1+v)(1-2v) \ .$$
$$(4.15)$$

For large deformations, the *Almansi* strain component has to be considered.

Under the conditions of *plane strain* , $\varepsilon_{zz} = \varepsilon_{zx} = \varepsilon_{zy} = 0$, the normal stress component that is orthogonal to the x, y plane separates

$$\sigma_{zz} = v\,(\,\sigma_{xx} + \sigma_{yy})\ .\qquad\qquad (4.16)$$

The strain components expressed by *Hooke´s* law (4.15) become, by substituting

$$3p = (1+v)\,(\sigma_{xx} + \sigma_{yy})\ ,\qquad\qquad (4.17)$$

$$\varepsilon_{xx} = (1/2G)\,[\sigma_{xx} - v\,(\,\sigma_{xx} + \sigma_{yy})] \quad , \quad \varepsilon_{yy} = (1/2G)\,[\sigma_{yy} - v\,(\,\sigma_{xx} + \sigma_{yy})] \quad ,$$

$$\varepsilon_{xy} = \sigma_{xy}/2G\ .\qquad\qquad (4.18)$$

In the *plane stress state* , the stress components $\sigma_{zz} = \sigma_{zx} = \sigma_{zy} = 0$, and the normal strain which is orthogonal to the x, y plane separates

$$\varepsilon_{zz} = -\,[v/(1-v\,)]\,(\varepsilon_{xx} + \varepsilon_{yy})\ .\qquad\qquad (4.19)$$

Thus, *Hooke´s* law renders under the conditions of plane stress

$$\varepsilon_{xx} = (1/E)\,[\sigma_{xx} - v\,\sigma_{yy}] \quad , \quad \varepsilon_{yy} = (1/E)\,[\sigma_{yy} - v\,\sigma_{xx}] \quad , \quad \varepsilon_{xy} = \sigma_{xy}/2G\ .\quad (4.20)$$

Such a stress state applies to thin plates of thickness h under in-plane loadings: The plane faces $z = \pm\, h/2$ are assumed traction-free. The internal forces per unit of length,

$$n_x = \int_{-h/2}^{h/2} \sigma_{xx}\, dz \quad , \quad n_y = \int_{-h/2}^{h/2} \sigma_{yy}\, dz \quad , \quad n_{xy} = \int_{-h/2}^{h/2} \sigma_{xy}\, dz \ ,$$
$$(4.21)$$

determine the mean stresses of the plane stress state, $\sigma_x = n_x/h$, $\sigma_y = n_y/h$, $\sigma_{xy} = n_{xy}/h$, which are independent of the lateral coordinate

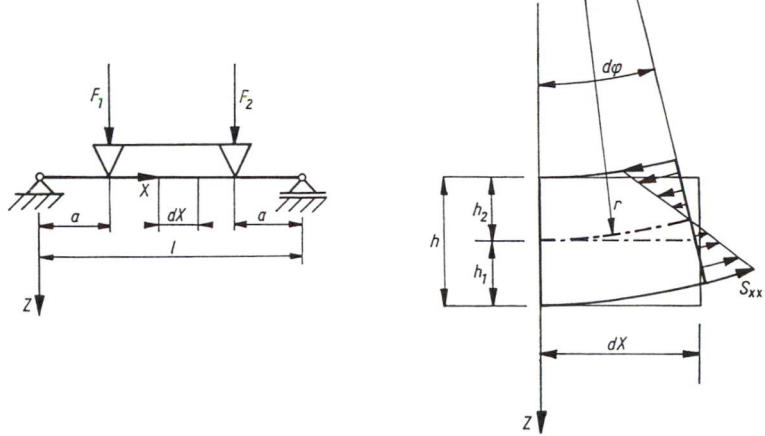

Fig. 4.2. Bending test. Stress distribution in pure bending

z and approximately substitute the actual stresses. The latter possibly show a slight variability with z. Equations (4.20) are rewritten for this state of *membrane stress* in thin plates and take on a form that depends on the midplane coordinates (x, y),

$$\varepsilon_x = (1/Eh)(n_x - \nu n_y) \ , \quad \varepsilon_y = (1/Eh)(n_y - \nu n_x) \ , \quad \varepsilon_{xy} = n_{xy}/2Gh \ . \quad (4.22)$$

Eh is the stretching stiffness of the linear elastic plate. Next, two additional tests are analyzed to determine the moduli E and G experimentally.

(§) The Bending Test. A single-span, simply supported beam is symmetrically loaded by two forces $F_1 = F_2 = F$ according to Fig. 4.2. Within the portion $(l - 2a)$ of the span, the shear forces vanish $(Q_z = 0)$ and the beam is loaded in pure bending by the constant moment $M_y = M = aF$. In this region, the originally straight beam axis is deformed to a circular arch; the z-axis is, eg an axis of symmetry of the cross-section.

In the case of a linear elastic material response, the experiment renders proportionality between the (negative) curvature of the deformed beam axis and the bending moment

$$1/r = -M/S_B = -aF/S_B \ , \quad (4.23)$$

where $S_B = EJ_y$ is the (constant) bending stiffness of the beam. Hence, measuring the radius r yields the bending stiffness

$$S_B = aF |r| \ , \quad (4.24)$$

and since the principal moment of cross-sectional inertia J_y of the probe is known, *Young's* modulus is given by

$$E = aF \, |r| \, / \, J_y \quad . \tag{4.25}$$

The curvature can be determined by measuring the elongations $\varepsilon_{1,\,2}$ of the fibers in the lower and upper surface, at $(r + h_1)$ and $(r - h_2)$, respectively. Hence (again see Fig. 4.2), $dX = r \, d\varphi$,

$$dX \, (1 + \varepsilon_1) = (r + h_1) \, d\varphi = (r + h_1) \, dX \, / \, r \, ,$$
$$dX \, (1 + \varepsilon_2) = (r - h_2) \, d\varphi = (r - h_2) \, dX \, / \, r \, .$$

Solving for the curvature renders

$$\left| \frac{1}{r} \right| = \frac{\varepsilon_1}{h_1} = -\frac{\varepsilon_2}{h_2} \quad , \tag{4.26}$$

or in the arithmetic mean that reduces the influence of errors in the measurement of the strains, $\varepsilon_2 < 0$,

$$\left| \frac{1}{r} \right| = \frac{\varepsilon_1 h_2 - \varepsilon_2 h_1}{2 \, h_1 h_2} \quad . \tag{4.27}$$

$h = h_1 + h_2$ is the height of the beam and h_1 the distance of the lower outermost fiber from the center of gravity in the cross-section. In the case of a doubly symmetric cross-section, $h_1 = h_2 = h/2$ and $|1/r| = (\varepsilon_1 - \varepsilon_2)/h$.

The elongations of fiber elements of length dX , initially in parallel orientation to the straight axis, are linearly distributed according to

$$\varepsilon_x = \frac{(r + Z) \, d\varphi - dX}{dX} = \frac{Z}{r} = Z \frac{\varepsilon_1}{h_1} \quad . \tag{4.28}$$

Hooke's law in a *lagrangean* representation

$$S_{xx} = E \, \varepsilon_{xx} = E \, \varepsilon_x \, (1 + \varepsilon_x / 2) \tag{4.29}$$

gives a nonlinear stress distribution over Z , the axis y is the line of vanishing stress, the *neutral axis* , and see Eq. (4.23),

$$M_y = \int_A Z S_{xx} \, dA = \frac{E \varepsilon_1}{h_1} \int_A Z^2 \, dA = E J_y \frac{\varepsilon_1}{h_1} = -S_B / r \quad . \tag{4.30}$$

Lateral to the X-axis, the specific elongations are $\varepsilon_y = \varepsilon_z = -\nu\,\varepsilon_x$. Elimination of *Young´s* modulus E renders under the assumption of small strains, $\varepsilon_x \ll 1$, the linear stress distribution of the structural beam theory

$$\sigma_{xx} = z\,M_y\,/J_y \;,\quad -h_2 \le z \le h_1 \;. \tag{4.31}$$

The extreme normal stresses are in the outer fibers and given by

$$\sigma_{1,2} = \pm\,\frac{M_y}{J_y}\,h_{1,2} \;. \tag{4.32}$$

(§) The Torsional Test. A circular cylindrical rod with a full or a hollow cross-section and of a given length l is loaded in such an experiment by a torque $M_x = M_T = M$. In the case of linear elasticity, the proportionality of the moment and the relative angular rotation of the terminal cross-sections are recognized from the measurements

$$\chi = \frac{M}{S_T}\,l \;, \tag{4.33}$$

where $S_T = GJ_T$ is the torsional stiffness. In case of the full circular cross-section, $J_T = J_p = \pi\,d^4/32$. The thin-walled circular cylindrical shell with wall-thickness $t \ll d$ has a torsional rigidity proportional to $J_T = \pi\,d^3\,t/4$. Hence, by measuring χ, the shear modulus can also be determined

$$S_T = \frac{M}{\chi}\,l \;,\quad \text{and}\quad G = \frac{M}{\chi J_T}\,l \;. \tag{4.34}$$

The cross-sections rotate rigidly during torsional deformation (see Fig. 4.3) and the warping becomes $u = \vartheta\,\varphi(Y, Z)$, independent of X. The constant angle per unit of length is $\vartheta = d\chi/dX = \chi/l$, and the warping function $\varphi(Y, Z)$ is determined in Sec. 6.2.4.

    When we consider two points of equal central distance to the x-axis in two cross-sections in the close neighborhood of dX, the rigid rotation gives the relations

$$v(X + dX) - v(X) = \frac{\partial v}{\partial X}\,dX = -Z\,d\chi\,,\quad w(X + dX) - w(X) = \frac{\partial w}{\partial X}\,dX = Y\,d\chi\,,$$

$$d\chi = \vartheta\,dX \;. \tag{4.35}$$

Hence, $dX \to 0$ renders the displacement gradients

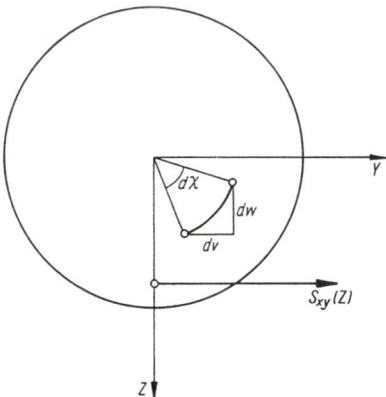

Fig. 4.3. Relative displacement components *dv* and *dw* are due to a rotation *dχ* of two
neighboring cross-sections and the shear stress S_{xy} in pure torsion

$$\frac{\partial v}{\partial X} = -\vartheta Z \;,\quad \frac{\partial w}{\partial X} = \vartheta Y \;.$$

(4.36)

The nonlinear geometric relation of ε_{xy} given by Eq. (1.19) renders,
without loss of generality it can be taken along the line $Y = 0$ where
$\partial/\partial Y = 0$ (rotational symmetry),

$$\varepsilon_{xy} = -\frac{\vartheta}{2} Z \;,$$

(4.37)

a linear distribution of the shear strain over the radial coordinate.
By means of *Hooke's* law,

$$S_{xy} = 2G\,\varepsilon_{xy} = -G\,\vartheta Z = S_{yx} \;,$$

(4.38)

a linear distribution of the torsional shear stresses occurs. The
static equivalence of the distributed couples and the torque *M*
yields

$$M = -\int_A r\, S_{x\varphi}\, dA \;=\; G\vartheta \int_A r^2\, dA \;=\; G\,\vartheta\, J_p \;.$$

(4.39)

Hence, *G* in the expression of the torsional rigidity S_T is identified
with shear modulus. The elimination of *G* gives

$$S_{xy} = -\frac{M}{J_p} Z \;,\quad Y = 0 \;,$$

(4.40)

and the maximal shear stress, tangential to the outer traction-free circle, is

$$\tau_{max} = S_{xy}\ (r = d/2) = \frac{M}{J_p}\ \frac{d}{2}\ .$$

(4.41)

See also Fig. 6.18.

4.1.2. A Note on Anisotropy

Many natural and engineering materials show linear elastic behavior within sufficiently small strain limits, but are also macroscopically *anisotropic* . Hence, if we still assume homogeneity, *Hooke's* law is generalized to include the directional dependence of elastic constants. Analogous to the isotropic case, a strain energy density $U'(\varepsilon_{ij})$ exists with the partial derivatives equal to the symmetric stresses

$$\sigma_{ij} = \frac{\partial U'}{\partial \varepsilon_{ij}} = \sigma_{ji}\ .$$

(4.42)

The latter are linear functions of the strain components. Thus, U' is a quadratic and positive-definite form of the symmetric strains,

$$U' = \frac{1}{2} \sum_i \sum_j \sum_k \sum_l C_{ijkl}\ \varepsilon_{ij}\ \varepsilon_{kl}\ ,\quad \text{where}\quad C_{ijkl} = C_{ijlk}\ ,\ C_{ijkl} = C_{jikl}\ ,\ C_{ijkl} = C_{klij}\ .$$

(4.43)

The tensor $\{C_{ijkl}\}$ under these conditions of symmetry has 21 mutually independent elastic constants. The general case is further discussed in the literature, see, eg
S. G. Lekhnitskii: Theory of Elasticity of an Anisotropic Elastic Body. Holden-Day, San Francisco, 1963.
Subsequently, the specific loadings of materials that exhibit further symmetries and, thus, are described by a reduced number of elastic constants are discussed.
(§)   Plane Stress State. Anisotropy with three planes of symmetry appears in orthotropic materials with nine elastic constants. In plane stress ($\sigma_{zz} = \sigma_{zx} = \sigma_{zy} = 0$) with a plane of symmetry in a thin plate, the number of elastic moduli is further reduced to four. If we take x, y in the principal directions of anisotropy, *Hooke's* law becomes

$$\varepsilon_{xx} = \frac{\sigma_{xx}}{E_x} - v_y \frac{\sigma_{yy}}{E_y}\ ,\quad \varepsilon_{yy} = \frac{\sigma_{yy}}{E_y} - v_x \frac{\sigma_{xx}}{E_x}\ ,\quad \varepsilon_{xy} = \frac{\sigma_{xy}}{2G_{xy}}\ .$$

(4.44)

E_x, E_y, v_x, v_y are the independent elastic parameters, which may be determined by tensile tests on rod specimens, taken out in the principal directions, respectively, and

$$G_{xy} = \sqrt{E_x E_y} \,/2(1 + \sqrt{v_x v_y}) \ . \tag{4.45}$$

(§) Transverse Isotropy Lateral to the x Axis. Such a case of cylindrical symmetry is, eg encountered in the material of a rod after drawing. Nonvanishing principal normal stresses are assumed in the radial σ_{rr} and axial σ_{xx} direction. Hooke's law becomes

$$\varepsilon_{xx} = \frac{\sigma_{xx}}{E_x} - v\frac{\sigma_{rr}}{E} \ , \quad \varepsilon_{rr} = \frac{\sigma_{rr}}{E} - v_x\frac{\sigma_{xx}}{E_x} \ , \quad \varepsilon_{xr} = 0 \ . \tag{4.46}$$

E and v are the elastic constants in the cross-section. For the tensile rod $\sigma_{rr} = 0$, such behavior is apparently isotropic.

4.1.3. A Note on Nonlinearity

A priori nonlinear elastic materials, like cast iron or concrete (in compression), where

$$\sigma_{ij} = f(\varepsilon_{kl}) \ , \tag{4.47}$$

are commonly assumed to be apparently isotropic. The incremental procedure due to time stepping or stepping the load factor is shown for the uniaxial stress state

$$\sigma_{xx} = f(\varepsilon_{xx}) \ , \tag{4.48}$$

with the stress increment

$$d\sigma_{xx} = \frac{\partial f}{\partial \varepsilon_{xx}} d\varepsilon_{xx} = E(\sigma_{xx}) \, d\varepsilon_{xx} \ . \tag{4.49}$$

In a numerical procedure, finite increments of stress are considered, and $E(\sigma_{xx})$ is approximated by the tangent or secant modulus, respectively See Fig. 4.4.

As an example, a cubic nonlinearity is considered, where we put $\sigma_{xx} = \sigma$, $\varepsilon_{xx} = \varepsilon$,

$$\sigma = E \, \varepsilon \, (\eta_0 + \eta_1 \, \varepsilon + \eta_2 \, \varepsilon^2) \ . \tag{4.50}$$

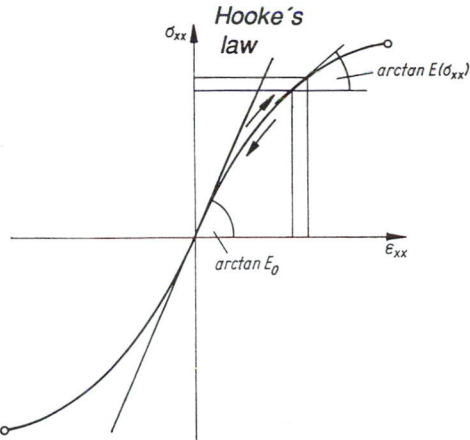

Fig. 4.4. Nonlinear elastic uniaxial stress-strain constitutive relation

η_k , $k = 0, 1, 2,$ are constants. Since $\eta_0 \geq 0$ and $\partial\sigma/\partial\varepsilon \geq 0$, the constitutive relation is valid within the limits of strain that follow from $E(\sigma) \geq 0$,

$$\eta_0 + 2\eta_1\varepsilon + 3\eta_2\varepsilon^2 \geq 0 , \quad \begin{cases} \text{if } \eta_1 \neq 0, \eta_2 = 0 : |\varepsilon| < \eta_0 / 2 |\eta_1| , \\ \text{if } \eta_2 \neq 0 : |\varepsilon| < \dfrac{1}{3\eta_2}\left(|\eta_1| - \sqrt{\eta_1^2 - 3\eta_0\eta_2}\right), \\ \text{when } \eta_1^2 - 3\eta_0\eta_2 \geq 0 , \end{cases}$$

otherwise, no strain limits apply if $\eta_2 > 0$. The coefficient $\eta_1 \neq 0$ renders the constitutive relation to be nonsymmetrical in tension and compression.

For *concrete* in compression, $\varepsilon < 0$, a quadratic relation is standardized ($\eta_2 = 0$), where $E = \sigma_p / \varepsilon_p$, $\eta_0 = 2$, $\eta_1 = \varepsilon_p^{-1}$, $\varepsilon_p = 0.2 \%$, and the value of σ_p , the maximum compressive stress, depends on the quality of the concrete (eg concrete standard B300, $\sigma_p = 2.25$ kN/cm^2). The tangent modulus according to Eq. (4.50) is

$$E(\varepsilon) = E \left(\eta_0 + 2\eta_1 \varepsilon + 3\eta_2 \varepsilon^2 \right) \geq 0 \quad , \tag{4.51}$$

hence, the tangent modulus of concrete is the linearly decreasing function of the compressive strain $\varepsilon < 0$, $|\varepsilon| < \varepsilon_p$,

$$E_{concr} = \frac{2\,\sigma_p}{\varepsilon_p} \left(1 + \frac{\varepsilon}{\varepsilon_p}\right) \quad , \quad E_{concr}(\varepsilon = 0) = \frac{2\,\sigma_p}{\varepsilon_p} = 2\,E \quad . \tag{4.52}$$

Poisson's constant of concrete in compression is assumed to be independent of strain in the range $0 \leq \nu_{concr} \leq 1/4$.

Nonlinear constitutive relations of an elastic and isotropic material, in general and in a multidimensional stress state are derived from the strain energy density, $U'(\varepsilon_{ij}) = U'(e, J_2, J_3)$, which is a function of the invariants of the strain tensor, by taking the derivative. When applying the chain rule, note that the invariant of the deviatoric strain tensor J_2 [see Eq. (2.36) and substitute the deviatoric strains ε_{ij}' for the stresses s_{ij}] and the dilatation e are independent, and that J_3 is the third invariant of the deviatoric strain tensor and, subsequently, is neglected (which is common practice)

$$\sigma_{ij} = \frac{\partial U'}{\partial \varepsilon_{ij}} = \frac{\partial U'}{\partial e}\frac{\partial e}{\partial \varepsilon_{ij}} + \frac{\partial U'}{\partial J_2}\frac{\partial J_2}{\partial \varepsilon_{ij}} + \frac{\partial U'}{\partial J_3}\frac{\partial J_3}{\partial \varepsilon_{ij}} \quad . \tag{4.53}$$

Hence, if we consider

$$e = \sum_i \varepsilon_{ii} \quad , \quad J_2 = \frac{1}{2}\sum_i \sum_j \varepsilon_{ij}'^2 \quad ,$$

and further,

$$\frac{\partial e}{\partial \varepsilon_{ij}} = \delta_{ij} \quad , \quad \frac{\partial J_2}{\partial \varepsilon_{ij}} = \frac{2}{3}e\,\delta_{ij} - \frac{\partial I_2}{\partial \varepsilon_{ij}} = \varepsilon_{ij} - \frac{e}{3}\delta_{ij} = \varepsilon_{ij}' \quad ,$$

the final form of the constitutive equation becomes

$$\sigma_{ij} = \frac{\partial U'}{\partial e}\delta_{ij} + \frac{\partial U'}{\partial J_2}\varepsilon_{ij}' \quad . \tag{4.54}$$

In the case of linear elasticity, *Hooke's* law applies

$$\frac{\partial U'}{\partial e} = K e \quad , \quad \frac{\partial U'}{\partial J_2} = 2 G \quad .$$

In engineering practice, the analytical constitutive relation has to be determined by the data fitting of the experimental output. Parameter identification is most easily done under the assumption of isotropy and homogeneity by choosing a positive-definite polynomial function, eg

$$U' = \frac{1}{2} K e^2 + 2G J_2 + D_1 e^3 + D_2 e J_2 + D_3 J_3 + E_1 e^4 +$$

$$E_2 e^2 J_2 + E_3 J_2^2 + E_4 e J_3 + O(\varepsilon^5) \tag{4.55}$$

that gives the first, quadratic correction of *Hooke's* law by means of Eq. (4.54) (J_3 is generally neglected)

$$\frac{\partial U'}{\partial e} = K e + 3 D_1 e^2 + D_2 J_2 \quad , \qquad \frac{\partial U'}{\partial J_2} = 2G + D_2 e \quad .$$

$$(4.56)$$

Splitting $\sigma_{ij} = \sigma_{ij}' + p \, \delta_{ij}$ renders the coupled equations

$$p = \frac{\partial U'}{\partial e} = (K + 3 D_1 e) e + D_2 J_2 \quad ,$$

$$(4.57)$$

$$\sigma_{ij}' = \frac{\partial U'}{\partial J_2} \, \varepsilon_{ij}' = (2G + D_2 e) \, \varepsilon_{ij}' \quad .$$

$$(4.58)$$

Taking small increments of Eqs. (4.57) and (4.58) with respect to the deformed state (e, ε_{ij}') gives

$$dp = (K + 6D_1 e) \, de + D_2 \, dJ_2 \quad , \qquad dJ_2 = \sum_i \sum_j \varepsilon_{ij}' \, d\varepsilon_{ij}' \quad ,$$

$$d\sigma_{ij}' = (2G + D_2 e) \, d\varepsilon_{ij}' + D_2 \, \varepsilon_{ij}' \, de \quad .$$

Two generalized tangent moduli, which depend on the state of deformation, and thus determine an inhomogeneous *Hooke's* law in terms of the "small" increments are determined by inspection

$$K_T = (K + 6D_1 e) \quad , \quad 2G_T = (2G + D_2 e) \quad .$$

Contrary to the uniaxial stress state, coupling terms remain in the incremental equations; these may be considered sources of self-stresses. The consideration of $D_3 \neq 0$ makes the coupling of the incremental deformations even stronger. Thus, the concept of equivalent stress must be employed that compares the strain energy density of the multidimensional stress state to the uniaxial one.

　　　Brittle materials are commonly considered elastic up to the limit load just before fracture occurs.

4.2. The Visco-Elastic Body

The deformations of some materials are not only related to the instantaneous state of stress but depend on the loading history. Those "*materials with memory* " are extensively discussed in the vast field of *rheology* . A simple theory has been developed for materials with *fading memory* . Subsequently, the internal friction in the flow of a viscous fluid is considered, followed by constitutive

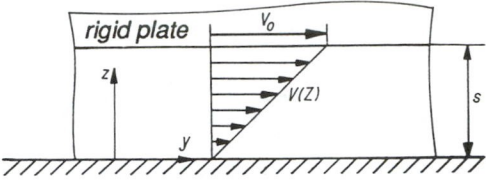

Fig. 4.5. Laminar shear flow of a viscous fluid (*Couette* flow)

relations of one-dimensional models. Three-dimensional relations
are derived through plausible arguments.

4.2.1. Newtonian Fluid

A typical experiment is the shear flow of a viscous fluid layer of
thickness s (see Fig. 4.5) by moving a rigid plate on top of the fluid
with a constant speed v_0. To keep the speed constant, a tangential
force has to be applied to the plate, which is transmitted to the
fluid by a constant shear-stress distribution in the interface (wet
friction). Measurements show that proportionality exists for many
fluids (liquids and gases) between the shear stress and the lateral
gradient of the velocity field of the fluid flow. Figure 4.5 illustrates
the linear variation of the velocity in a narrow slit. The "viscous"
law of the one-dimensional flow of such a *Newtonian* fluid is thus
the linear relation

$$\sigma_{zy} = \tau = \eta \frac{\partial v}{\partial z} = \sigma_{yz} \ .$$

(4.59)

For the linear velocity profile, $\partial v/\partial z = v_0/s$, and the shear stress
$\sigma_{zy} = \eta v_0/s$ is independent of z . The "inhomogeneous" parallel flow
is not irrotational, *curl* $v = (-\partial v/\partial z) \, e_x$. The material parameter η
is the *coefficient of dynamic or absolute viscosity* with the
dimension of stress over the strain rate, ie $[\eta] = N \, s \, m^{-2} = kg/s \, m$.
The value for water at a temperature $20°C$ is $\eta = 10^{-3} \, N \, s \, m^{-2}$. The
relatively strong dependence of η on temperature must be
emphasized, whereas the influence of the assigned pressure p
(normal stress σ_{zz}) on the shear flow and, hence, on the coefficient
of viscosity is negligible, except for very high pressure. For liquids,
η is decreasing with rising temperature; for gases, η is increasing.
Since viscosity is difficult to measure according to Eq. (4.59), in
industry specially designed apparatus is used. The time measured in
seconds for a liquid to flow out of a container of prescribed shape
and dimensions is determined. Empirical conversion to standard
units is performed. The derived unit of viscosity is named after *J. L.*

Poiseuille, 1 poise = *1 g /s cm* and its fraction *1 centipoise* = *1/100 poise* = 10 $^{-3}$ *N s m* $^{-2}$ is used. Since the viscosity of *Newtonian* fluids is small, shear stresses develop only for large lateral velocity gradients. Such gradients are encountered, eg in the boundary layer adjacent to a wall at rest; the outer stream may be deemed frictionless, ie without considering any shear stresses (see Chap. 13).

The shear-stress deformation-rate law (4.59) indicates the difference in the behavior of the fluid phase to that of the solid state: A fluid has no shear rigidity, a fluid at rest must be in the hydrostatic stress state, and the shear stresses vanish [see Eq. (2.80)]. Neglecting the volume viscosity of a compressible fluid renders in the *eulerian* representation, if we consider the deviatoric components, in "analogy" to *Hooke´s* law of linear elasticity

$$p = -K e \qquad (4.60)$$

(in the case of linear compressibility, *p* is the hydrostatic pressure, ie the negative mean normal stress),

$$\sigma_{ij}{}' = 2\eta\, V_{ij} - \frac{2}{3}\eta\, \delta_{ij} \sum_k V_{kk} \quad , \quad V_{ij} = \frac{1}{2}\left(\frac{\partial v_i}{\partial x_j} + \frac{\partial v_j}{\partial x_i}\right) . \qquad (4.61)$$

The constitutive relation is named after *G. G. Stokes* .

If the motion can be considered incompressible, the continuity equation (1.77) renders at once

$$\text{div } \mathbf{v} = \sum_k V_{kk} = 0 \quad \rightarrow \quad \sigma_{ij}{}' = 2\eta\, V_{ij} \ . \qquad (4.62)$$

$\sigma_{ij} = \sigma_{ij}{}' - p\, \delta_{ij}$ is related to the otherwise undetermined pressure *p* .

(§) Illustrative Example: The stationary and incompressible flow of a viscous fluid between parallel walls with a distance *2h* (see Fig. 4.6) is considered. Pressure must be constant in every cross-section and decreases in the direction of the inhomogeneous parallel flow *p* = *p(x)* in the *eulerian* representation. Due to incompressibility, the velocity is independent of the coordinate *x* ; thus, in all cross-sections, $v_x = u(z)$. Consideration of the free-body diagram of a control volume *dx dz* renders (acceleration is zero)

$$[p(x) - p(x + dx)]\, dz + [\sigma_{zx}(z + dz) - \sigma_{zx}(z)]\, dx = 0 \quad , \quad \sigma_{zx} = \sigma_{xz} \ .$$

Linear approximation gives exactly after we take the limit *dx* → *0*

$$-\frac{\partial p}{\partial x} + \frac{\partial \sigma_{zx}}{\partial z} = 0 \ .$$

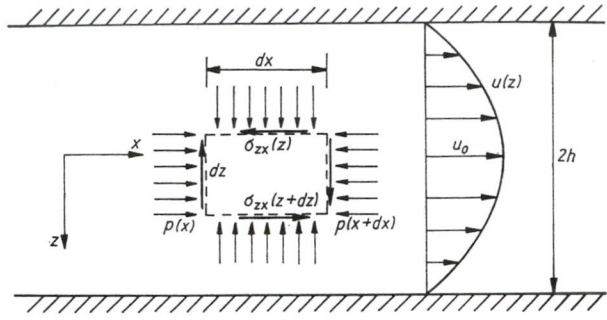

Fig. 4.6. Plane laminar viscous channel flow

Substituting $\sigma_{zx} = \eta\ \partial u/\partial z$ yields $\partial^2 u/\partial z^2 = \eta^{-1}\ \partial p/\partial x$, independent of the lateral coordinate z. Hence, integration is easily performed and the velocity profile becomes under the condition of symmetry, $\partial u/\partial z = 0$ at $z = 0$, $u(z = 0) = u_0$,

$$u(z) = u_0 + \frac{1}{2\eta}\frac{\partial p}{\partial x}\ z^2 \ .$$

The boundary conditions at $z = \pm h$, where the viscous fluid is attached to the wall (no-slip condition), are $u(z = \pm h) = 0$ and render the pressure gradient to be constant

$$\frac{\partial p}{\partial x} = -\frac{2\eta\ u_0}{h^2} \ ,$$

and therefore, the pressure decreases linearly with the distance x

$$p(x) = p_0 - \frac{2\eta\ u_0}{h}\frac{x}{h} > 0 \ . \tag{4.63}$$

The velocity distribution becomes the parabolic function

$$u(z) = u_0\left(1 - \frac{z^2}{h^2}\right) \ . \tag{4.64}$$

Measuring the difference of the pressure in some distance l may be used to identify the dynamic viscosity

$$\eta = \frac{p_0 - p(l)}{2\ l\ u_0}\ h^2 \ . \tag{4.65}$$

The maximal velocity u_0 can be expressed by the mass flow rate per unit of length

$$\dot{m} = \rho\ 2h\ \bar{u} = 2\rho \int_0^h u(z)\ dz = \rho\ 2h\ \frac{2u_0}{3}\ .$$

Thus, more convenient for evaluation, the *kinematic viscosity* is expressed by

$$v = \frac{\eta}{\rho} = \frac{2}{3}\ \frac{p_0 - p(l)}{\dot{m}\ l}\ h^3\ .$$

$$(4.66)$$

v is measured in m^2/s ; density is eliminated. Viscosity influences the flow of Fig. 4.6 mainly in two respects: The parallel flow becomes inhomogeneous in every cross-section, potential energy is dissipated, and pressure decreases linearly in the axial direction. A dimensionless hydraulic loss factor λ can be determined from the dimensionless drop in pressure

$$\frac{p_0 - p(l)}{\rho\bar{u}^2/2} = \frac{l}{2h}\ \frac{24v}{2h\bar{u}} = \frac{l}{2h}\ \lambda\ ,\quad \bar{u} = \dot{m}/2\rho h\ ,$$

$$(4.67)$$

in terms of the *Reynolds number Re*

$$\lambda = 24/Re\ ,\quad Re = 2h\ \bar{u}\ /\ v\ .$$

$$(4.68)$$

The hyperbolic law of the loss factor is typical for the laminar viscous flow of *Newtonian* fluids. The laminar flow is stable only for small *Reynolds* numbers. At a critical value of $Re = Re_{cr}$, the laminar parallel stream becomes unstable and changes to *turbulent flow* . The latter has, in the mean, an increased loss factor.

The *axisymmetric laminar flow* in a circular cylindrical pipe is named after *Hagen - Poiseuille* (see also Sec. 13.3.1), and the hydraulic loss factor becomes with the substitution of $h = a$ the cross-sectional radius

$$\lambda = 64/Re\ ,\quad Re < Re_{cr} = 2320\ .$$

$$(4.69)$$

(§) The One-Dimensional Viscous Model. A one-dimensional model of a "*Newtonian* fluid" as shown in Fig. 4.7. The generalized model body transmits a force proportional to the relative velocity between the "piston" and the "cylinder" of the dashpot; in Fig. 4.7, it is the speed of the point of application

$$F = \eta\ \dot{u}\ .$$

$$(4.70)$$

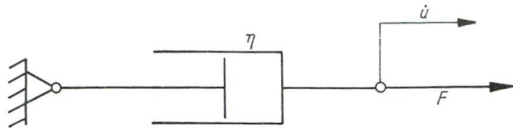

Fig. 4.7. Symbolic one-dimensional model of *Newtonian* fluid: The viscous damper
(dashpot)

The factor of proportionality is, eg linearly related to the viscosity coefficient of a *Newtonian* fluid when flowing in a narrow slit (Eq. 4.59).

4.2.2. Linear Visco-Elasticity

The deformations in an elastic solid are given by the instantaneous loading state; thus, the linear elastic body remembers only the undeformed reference state. All materials show, more or less, the deformation rate sensitivity of the stress distribution during the loading process. Internal friction, eg is activated and material damping changes the response to periodic loadings and free vibrations diminish. Also, under constant loading, irreversible deformations may develop in a state of creep of the visco-elastic body. Such permanent deformations may come to a halt in a finite time or grow indefinitely. The simplest linear models of these viscous effects are synthesized by the *Hookean* spring and the "*Newtonian* fluid." The coupling is shown subsequently, thereby forming the three basic models of linear visco-elasticity.

(§)   The *Kelvin-Voigt* Body. The one-dimensional model as shown in Fig. 4.8 consists of a parallel connection of a linear spring of stiffness c, and a *Newtonian* liquid phase, η is the damping constant of the dashpot. This two-phase solid model has the damping

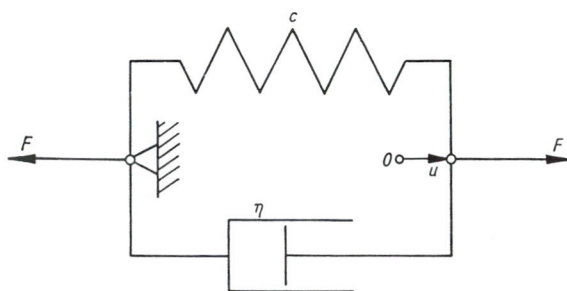

Fig. 4.8. Symbolic one-dimensional model of the *Kelvin-Voigt* solid

property mentioned above and exhibits the phenomenon known as *retarded elasticity* . A force F renders a joint deformation u , and both phases contribute

$$F = cu + \eta \dot{u} \quad , \quad u(t = 0) = 0 \quad . \tag{4.71}$$

The *creep function* $u(t)$ is the special solution of this first-order differential equation under the condition of a unit step loading. $F = H(t)$ is the *Heaviside* function, $H(t) = 1, t \geq 0$, $H(t) = 0, t < 0$. Equation (4.71) is linear and, thus, is solved by superposition, $u = u_h + u_p$. A particular solution for $t \geq 0$ is the constant $u_p = 1/c$, and the solution of the homogeneous equation is

$$u_h = A\,e^{-(c/\eta)t} \quad . \tag{4.72}$$

Consideration of the initial condition $u(0) = 0$ gives $A = -1/c$, and the creep function becomes

$$u(t) = \frac{1}{c}\left[1 - e^{-(c/\eta)t}\right] H(t) \quad , \quad \vartheta = \eta/c \quad . \tag{4.73}$$

ϑ is the *retardation time* of the displacement. The elastic deformation F/c is reached after keeping the loading $F = 1$ constant for a long time (in terms of the retardation time); elasticity is retarded.

The *relaxation function* of this model is the time force function under deformation control, by making the displacement equal to $u(t) = H(t)$ suddenly constant.

$$F(t) = c\,H(t) + \eta\,\delta(t) \quad , \quad \delta(t) = \frac{dH(t)}{dt} \quad , \tag{4.74}$$

is the *Dirac delta* function. Elasticity in the spring develops instantly; the viscous portion of the force becomes infinite at time zero and vanishes subsequently.

When we consider pure elasticity in compression, $p = K\,e$, and linearized geometric relations (small strains allow the superposition of the two material phases), the constitutive equations for a three-dimensional stress state in the homogeneous and isotropic *Kelvin-Voigt* body are

$$\sigma_{ij}' = 2G\left(\varepsilon_{ij}' + \vartheta\,\frac{\partial \varepsilon_{ij}'}{\partial t}\right) \quad , \quad \varepsilon_{ij}' = \frac{1}{2}\left(\frac{\partial u_i}{\partial x_j} + \frac{\partial u_j}{\partial x_i}\right) - \frac{e}{3}\,\delta_{ij} \quad , \quad e = \sum_i \frac{\partial u_i}{\partial x_i} \quad , \quad \vartheta = \eta/G \quad ,$$

$$\tag{4.75}$$

G is the shear modulus and K the bulk modulus of the *Hookean* phase; η is the viscosity coefficient of the viscous phase.

The *Kelvin-Voigt* model, Eq. (4.75), is well suited to describe the material damping in small amplitude time harmonic vibrations. The time function may be separated into $\sigma_{ij}'\,exp(i\omega t)$, $\varepsilon_{ij}'\,exp\,(i\omega t)$, the complex amplitudes are denoted likewise as the deviatoric stress and strain, respectively, for convenience, and Eq. (4.75) takes on the time reduced form

$$\sigma_{ij}' = 2G\left(1 + i\omega\vartheta\right)\varepsilon_{ij}' \ . \tag{4.76}$$

Formally, these relations are those of a *Hookean* solid with a *complex* shear modulus,

$$G^* = G + i\left(\omega\vartheta G\right) \ , \quad i = \sqrt{-1} \ . \tag{4.77}$$

Viscosity renders the stresses and the strains out of phase in their time behavior. More details are given in the discussion of the elastic-visco-elastic correspondence principle; see Sec. 6.10.

General loading renders a nonperiodic response and the *Laplace* transformation,

$$f^*(s) = \int_0^\infty e^{-st}\,f(t)\,dt \ , \tag{4.78}$$

when applied to Eq. (4.75) yields a linear relation of the transformed deviatoric stresses and strains. At $t = 0$, the undeformed reference state is considered [note with $f(0) = 0$]

$$s\,f^*(s) = \int_0^\infty e^{-st}\,\dot{f}(t)\,dt \ , \tag{4.79}$$

$$\sigma_{ij}'^* = 2G^*\,\varepsilon_{ij}'^* \ , \quad G^* = G\left(1 + s\vartheta\right) \ . \tag{4.80}$$

(§)   *Maxwell* Fluid. When we take the *Hookean* spring of stiffness c and the *Newtonian* fluid of viscosity η in a one-dimensional model in series connection (see Fig. 4.9), the two-phase material, called *Maxwell fluid* , exhibits the above-mentioned unlimited and irreversible creep property. Such a model explains the relaxation of

Fig. 4.9. Symbolic one-dimensional model of the *Maxwell* body: Spring and dashpot in series connection

stress that is experimentally observable. The force F is common to both phases in Fig. 4.9, and the deformation rate is superposed

$$\dot{u} = \frac{\dot{F}}{c} + \frac{F}{\eta} \quad , \quad u(0) = F(0)/c \quad .$$
(4.81)

The initial condition expresses the instantaneous elasticity of the *Maxwell* material under step loading. Hence, the *creep function* becomes, with $F(t) = H(t)$, $dF/dt = \delta(t)$,

$$u(t) = \frac{1}{c} \int \delta(t) \, dt + \frac{1}{\eta} \int H(t) \, dt = \left(c^{-1} + \eta^{-1} t\right) H(t) = \frac{1}{c} \left(1 + \frac{c}{\eta} t\right) H(t) \quad .$$
(4.82)

According to Eq. (4.82), creep is retarded against the instantaneous elastic response.

The *relaxation function* (force) $F(t)$ renders the deformation prescribed by $u(t) = H(t)$ [hence, $du/dt = \delta(t)$], and thus, it is the *Green's* function of the first-order differential equation

$$\dot{F} + \frac{c}{\eta} F = c \, \delta(t) \quad ,$$
(4.83)

$$F = c \, e^{-(c/\eta)t} \, H(t) \quad , \quad \vartheta = \eta/c \quad .$$
(4.84)

ϑ in this case is the *relaxation time*, the attenuation parameter of the stress.

If we assume pure elasticity in compression, $p = K \, e$, Eq. (4.81) is generalized to the three-dimensional stress-strain rate relation of the isotropic and homogeneous *Maxwell* fluid

$$\dot{\varepsilon}_{ij}{}' = \frac{1}{2G} \left(\dot{\sigma}_{ij}{}' + \frac{1}{\vartheta} \sigma_{ij}{}'\right) \quad , \quad \vartheta = \eta/G \quad .$$
(4.85)

When we consider the undeformed state at $t = 0$ for reference, the *Laplace* transformation, Eqs. (4.78) and (4.79), applied to the constitutive differential equation (4.85) gives

$$s \, \varepsilon_{ij}{}'^{*} = \frac{1}{2G} \left(s + \vartheta^{-1}\right) \sigma_{ij}{}'^{*} \quad .$$
(4.86)

The complex shear modulus of the *Maxwell* material renders formally with

$$G^{*} = G \frac{s}{s + \vartheta^{-1}} \quad ,$$
(4.87)

the linear relation between the *Laplace* transformed deviatoric stresses and strains according to *Hooke´s* law

$$\sigma_{ij}^{\prime *} = 2G^* \; \varepsilon_{ij}^{\prime *} \quad . \tag{4.88}$$

Refer to the elastic-visco-elastic correspondence; see Sec. 6.10.

(§)   The Multiple-Parameter Linear Visco-Elastic Body. The three-parameter *standard solid* has a one-dimensional model, which is the series connection of a *Hookean* spring of stiffness c_1 and the *Kelvin Voigt* phase (c_2 , η), or equivalently, the parallel arrangement of the *Hookean* spring c_1 and *Maxwell* fluid (c_2 , η). Such a three-phase material has a symmetric constitutive differential equation,

$$\dot{F} + p_0 F = q_1 \, \dot{u} + q_0 \, u \quad , \tag{4.89}$$

and may be determined by proper creep and relaxation functions. The three parameters p_0, q_0 and q_1 are expressed in terms of the elastic stiffness and viscosity coefficient in the case of the series connection by

$$p_0 = (c_1 + c_2) \, / \, \eta \quad , \quad q_0 = c_1 c_2 / \eta \quad , \quad q_1 = c_1 \quad . \tag{4.90}$$

The creep function in this series arrangement is the special solution

$$u(t) = \left[\frac{1}{c_1} + \frac{1}{c_2} \left(1 - e^{-t/\vartheta} \right) \right] H(t) \quad , \quad \vartheta = \eta/c_2 \quad . \tag{4.91}$$

Asymptotically, the finite deformation is reached

$$u(t \rightarrow \infty) = c_1^{-1} + c_2^{-1} \quad . \tag{4.92}$$

The parallel connection gives

$$p_0 = c_2 / \eta \quad , \quad q_0 = c_1 c_2 / \eta \quad , \quad q_1 = c_1 + c_2 \quad . \tag{4.93}$$

The relaxation function for this parallel arrangement is simply given by superposition

$$F(t) = \left(c_1 + c_2 \, e^{-t/\vartheta} \right) H(t) \quad . \tag{4.94}$$

Asymptotically, the stress decreases to the *elastic residual stress* $1 \; c_1$.

Substituting a deviatoric stress component for *F* and the corresponding strain for the displacement *u* gives the three-dimensional generalization, under the assumption of linearized geometric relations,

$$\dot{\sigma}_{ij}' + p_0\, \sigma_{ij}' = q_1\, \dot{\varepsilon}_{ij}' + q_0\, \varepsilon_{ij}' \quad . \tag{4.95}$$

The application of the *Laplace* transformation, Eqs. (4.78) and (4.79), renders

$$(s + p_0)\, \sigma_{ij}'^* = (q_1 s + q_0)\, \varepsilon_{ij}'^* \quad , \tag{4.96}$$

and the complex shear modulus

$$G^*(s) = \frac{q_1 s + q_0}{2(s + p_0)} \quad , \tag{4.97}$$

enters formally *Hooke's* law in the transformed domain

$$\sigma_{ij}'^* = 2G^* \,\varepsilon_{ij}'^* \quad . \tag{4.98}$$

Isotropy and homogeneity, as well as purely elastic compressibility, are understood.

The three-parameter *viscous* body is constructed by exchanging the *Hookean* phase of stiffness c_1 in the one-dimensional model with the *Newtonian* fluid of viscosity η_1. Hence, Eq. (4.89) is changed to

$$\dot{F} + p_0\, F = q_2\, \ddot{u} + q_1\, \dot{u} \quad . \tag{4.99}$$

The *creep function* of the series array is the special solution

$$u(t) = \left[\frac{1}{\eta_1} + \frac{1}{c_2}\left(1 - e^{-t/\vartheta}\right) \right] H(t) \quad . \tag{4.100}$$

The creep deformation grows beyond bounds with increasing loading time.

A first generalization of a *multiple parameter visco-elastic* constitutive relation is given by the differential equation of nth order with constant coefficients. Denoting the linear polynomial operator by $Q_n(D) = q_0 + q_1 D + q_2 D^2 + \ldots + q_n D^n$, $D = d/dt$, $P_m(D)$ analogously, the generalized differential relation between the deviatoric stress and strain becomes

$$P_m(D)\, \sigma_{ij}' = Q_n(D)\, \varepsilon_{ij}' \quad , \tag{4.101a}$$

pure elasticity is assumed in hydrostatic compression. The *Laplace* transformation, when we consider homogeneous initial conditions, renders formally *Hooke's* law in the transformed domain

$$\sigma_{ij}'^* = \frac{Q_n(s)}{P_m(s)} \, \varepsilon_{ij}'^* \quad , \quad \text{if } n \leq m \quad ,$$

(4.101b)

or the inverse relation in the case of $n \geq m$. The complex shear modulus G^* is found by inspection of the above equation in the form of a rational fraction

$$G^* = \frac{Q_n(s)}{2 \, P_m(s)} \quad .$$

(4.102)

The possibility of the decomposition into partial fractions is to be mentioned. Hence, G^* is synthesized by the superposition of "simple" (low-order) models.

(§)   General Linear Viscoelasticity. The complex shear modulus G^* may be taken as any proper function $G^*(s)$, instead of Eq. (4.102). Inversion of the product of two *Laplace* transforms [cf. Eq. (4.101b)] renders the convolution integral

$$\sigma_{ij}' = \int\limits_{-\infty}^{t} \sum_k \sum_l G_{ijkl}(x, y, z; t - \tau) \frac{\partial \varepsilon_{kl}'}{\partial \tau}(x, y, z; \tau) \, d\tau \quad .$$

(4.103)

Above even an inhomogeneous material's dependence on the spatial coordinates *(x, y, z)* and anisotropy may be considered, but the geometric relations are still linearized. G_{ijkl} is the fourth-order tensor of the relaxation functions of time. The inverse of the constitutive convolution integral has the kernel known as the tensor of the creep functions. Identification of these functions from (long duration) time testing is difficult and expensive and, thus, is restricted to special materials and specific applications. Multiple-parameter models are preferable. *Isotropy* simplifies the constitutive functional considerably when considering the undeformed reference state at $t = 0$

$$\sigma_{ij}' = \frac{\partial}{\partial t} \int\limits_{0}^{t} G(\tau) \, \varepsilon_{ij}'(x, y, z; t - \tau) \, d\tau \quad .$$

(4.104)

Also, the viscosity of the dilatational deformation can be included.
    All the visco-elastic constitutive equations discussed so far hold under isothermal conditions. For a thermal stress problem, see Sec. 6.10.2.

4.2.3. A Nonlinear Visco-Elastic Material

The one-dimensional linear *Maxwell* fluid has been generalized to include the nonlinear effects of stress on the creep deformation, mainly for metals, by *Bailey* and *Norton* . At the isothermal temperature T , the differential relation under the condition of uniaxial stress becomes a power law

$$\dot{\varepsilon} = \frac{\dot{\sigma}}{E} + \frac{1}{\vartheta}\left[\frac{\sigma}{\sigma_n(T)}\right]^{n(T)}, \quad \begin{cases} \vartheta \text{ is a characteristic time,} \\ \text{commonly } 10^9 \text{ seconds.} \end{cases} \tag{4.105}$$

When we denote the associated reference stress, $\sigma_n (T)/10^9 s = \sigma_{n9}$, the creep limit stress for another choice of the characteristic time is transformed according to

$$\sigma_n(T) = \left[\frac{10^9}{\vartheta}\right]^{1/n} \sigma_{n9} \quad . \tag{4.106}$$

The creep exponent $n(T)$ shows a weak dependence on temperature. In the case of an odd natural number n , the creep deformation in tension and compression is well described without further considerations of the signum function.

Generalization to the three-dimensional stress state is due to *F. K. G. Odqvist* and considers at first the uniaxial constitutive relation in the deviatoric stress components, $\sigma_{xx}' = (2/3)\,\sigma$, $\sigma_{yy}' = \sigma_{zz}' = -\,\sigma/3$, with the proper invariant J_2 to express the viscous strain rate portion

$$J_2 = \frac{1}{2}\sum_i \sum_j \sigma_{ij}'^2 = \sigma^2/3 \;\rightarrow\; \dot{\varepsilon}_v = f(J_2)\frac{2}{3}\,\sigma \quad . \tag{4.107}$$

Comparing with the second, viscous part of the strain rate (4.105) gives

$$f(J_2) = \frac{3}{2\vartheta\sigma_n{}^n(T)}\,(3J_2)^{(n-1)/2} \quad . \tag{4.108}$$

Due to the invariant coefficient in the uniaxial law (4.107), the three-dimensional constitutive equation of the strain deviation rate becomes simply

$$\dot{\varepsilon}_{ij}' = \frac{1}{2G}\,\dot{\sigma}_{ij}' + \frac{3}{2\vartheta\sigma_n{}^n(T)}\,(3J_2)^{(n-1)/2}\,\sigma_{ij}' \quad . \tag{4.109}$$

Purely linear elastic compressibility is still assumed. The introduction of the effective *equivalence stress*

$$\sigma_e = \sqrt{3J_2} \quad ,$$
(4.110)

which in the case of the uniaxial stress state of a tensile rod is the normal stress $\sigma_{xx} = \sigma$ [Eq. (4.107)], reduces the constitutive nonlinear differential equation to simpler form,

$$\dot{\varepsilon}_{ij}' = \frac{1}{2G} \dot{\sigma}_{ij}' + \frac{3}{2\vartheta\sigma_n^n(T)} \sigma_e^{(n-1)} \sigma_{ij}' \quad .$$
(4.111)

The specific dissipated power of deforming *Maxwell* type materials, in the case of isothermal creep, can easily be verified to be given in the following form (see Fig. 4.9):

$$D = \sum_i \sum_j \sigma_{ij}\, \dot{\varepsilon}_{ijv} = \frac{\sigma_e^{n+1}}{\vartheta\sigma_n^n(T)} \quad .$$
(4.112)

$d\varepsilon_{ijv}/dt$ is the viscous part of the strain rate, which is a unique function of the equivalence stress.

In the case of simple shear $\sigma_{xy} = \sigma_{yx} = \tau$, the equivalence stress is $\sigma_e = \tau\sqrt{3}$ and the rate of shear strain in that case is given by

$$\dot{\varepsilon}_{xy} = \frac{1}{2G} \dot{\tau} + \frac{3^{(n+1)/2}}{2\vartheta\,\sigma_n^n(T)} \tau^n \quad .$$
(4.113)

For a stationary creep process

$$\dot{\varepsilon}_{ijv}' = f(J_2)\,\sigma_{ij}' \quad ,$$

N. J. Hoff formulated an analogy between the viscous strain rate and the nonlinear elastic strains that would develop in an elastic body of the same shape under the same loading conditions

$$\varepsilon_{ij}' = f(J_2)\,\sigma_{ij}' \quad .$$

To illustrate the dimensions of the parameters in the law of *Bailey* and *Norton* , the values are given for a mild steel at the elevated temperature of *450°C* with $n = 5$ and $\sigma_{n7} = 70$ N/mm^2 . Details may be found in the monograph by
F. K. G. Odqvist and J Hult: Kriechfestigkeit metallischer Werkstoffe. Springer-Verlag, Berlin, 1962.

(§)   Example: The Creep Collapse of a Tensile Rod. To illustrate creep effects, a thin rod of *Norton* material is considered under constant tension by a force *F*. See, eg
H. Parkus: "On the lifetime of visco-elastic structures in a random temperature field." In Recent Progress in Applied Mechanics (The F. K. G. Odqvist Volume), B. Broberg, J. Hult and F. Niordson, Eds. Wiley, New York, 1967, pp. 391 - 397. (The late *H. Parkus* was Professor at TU-Vienna).
The viscous part of strain increases with loading time and outgrows the instantaneous elastic strain, which after a while becomes negligible. Creep deformation is considered under the condition of the constant volume of the rod

$$A \, dx = A_0 \, dX \quad , \tag{4.114}$$

where *A(t)* is the deformed cross-sectional area. According to Eq. (1.29), the logarithmic strain measure is employed for the large creep deformations

$$\varepsilon = \ln \frac{dx}{dX} = \ln (1 + \varepsilon_x) = - \ln \frac{A}{A_0} \quad , \tag{4.115}$$

and after differentiation, the rate is substituted in the constitutive relation,

$$\dot{\varepsilon} = \frac{\sigma^n}{\vartheta \, \sigma_n{}^n(T)} \quad , \tag{4.116}$$

to render the nonlinear first-order differential equation

$$\vartheta \dot{\alpha} + \left[\frac{\sigma_0}{\sigma_n(T)} \right]^n \alpha^{-n+1} = 0 \quad , \quad \alpha = A/A_0 \quad , \quad \dot{\alpha} = \dot{A}/A_0 \quad , \quad \sigma_0 = F/A_0 \quad . \tag{4.117}$$

The separation of variables and integration give the dimensionless time function of the decreasing cross-sectional area

$$\alpha^n(t) = 1 - \left[\frac{\sigma_0}{\sigma_n(T)} \right]^n \frac{t}{\vartheta} \quad . \tag{4.118}$$

The simplest criterion of a collapse of the steadily creeping rod to occur is $d\alpha/dt \to \infty$, or according to Eq. (4.117), equivalently $\alpha \to 0$. Thus, the nonlinear creeping homogeneous rod has a *finite lifetime*, $n > 1$,

$$\frac{t_L}{\vartheta} = \left[\frac{\sigma_n(T)}{\sigma_0} \right]^n \quad . \tag{4.119}$$

T is the isothermal temperature. A rod of mild steel at *500°C* under the constant load $\sigma_0 = 10^4 \ N/cm^2$ has the following parameters:

$$n(T) = 5 \ , \quad \vartheta\sigma_n^5 = 6.7 \times 10^{23} \ (N \ cm^{-2})^5 \ h \ ,$$

and its lifetime is about t_L = *6700 hours* . Note even small temperature fluctuations superposed on the mean value yield a strong reduction in expected lifetime. This sensitivity is important in practical engineering, as well as in designing creep experiments at elevated temperatures.

4.3. The Plastic Body

Besides the viscous deformations discussed above, irreversible deformations may develop in plastic bodies when loaded beyond the *yield limit* . The carrying capacity of such a structure is considerably reduced by the growth of the plastic deformations and by the spreading of the yielding zones over large parts of the structural volume. A plastic collapse may occur at the *limit load* of such a ductile structure. Large plastic deformations of ductile bodies that, eg occur in the course of remodelling by the forging and rolling of metals at elevated temperature are not considered in the course of this textbook. See,
H. Lippmann: Mechanik des Plastischen Fließens. Springer-Verlag, Berlin, 1981.
Structural plasticity, ie the ability of a ductile material to absorb energy, is favorable with respect to (short-time) overloadings. Also, yielding in a small plastic zone in the neighborhood of a load or stress concentration limits the stress peaks and renders a redistribution of stresses at much lower stress levels. The special design of structures enhances the positive effects of the built-in ductility of engineering materials.

Subsequently, an introduction into plasticity is presented by first considering the simple models in the one-dimensional representation, followed by the constitutive relation of the three-dimensional stress state. By means of the finite-element method, approximating solutions to the nonlinear problems of structural plasticity find their way into engineering practice.

4.3.1. The Rigid-Plastic Body

Coulomb's law of dry friction (see Fig. 2.4) renders the stick-slip, nonlinear stress-strain relation of the rigid-plastic, *Saint-Venant* phase. In the one-dimensional tensile rod model, no deformation develops if the force F is smaller than the yield limit, which is assumed to be equal in tension and compression; $|F| < F_Y$. Loading at

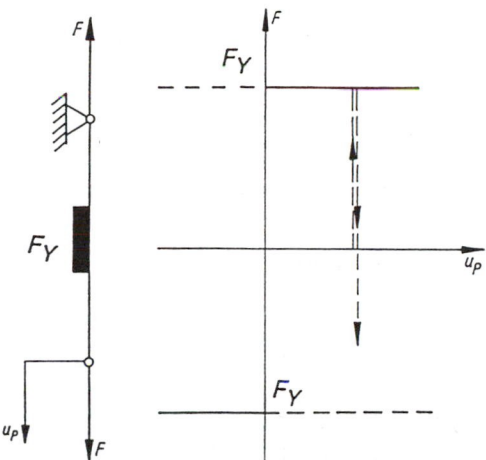

Fig. 4.10. Symbolic one-dimensional model of the *Saint-Venant* body. Loading cycle that reaches the yield limit F_Y

the yield limit, $|F| = F_Y$, produces stationary yielding without hardening, a permanent plastic deformation, $u_p(t)$. During unloading, the deformed rod reacts rigidly, $|F| < F_Y$. The work done during deformation, $F_Y u_p$, is totally dissipated. Symbolically, the model is shown in Fig. 4.10, together with the force displacement paths during loading and unloading.

   In the absence of any elastic deformation, the total axial strain equals the plastic portion, $\varepsilon_{xx} = \varepsilon_p$. The permanent lateral strain is given by the condition of incompressibility, $e = 0$, in the small strain approximation

$$\varepsilon_q = -\varepsilon_{yy} = -\varepsilon_{zz} = -\varepsilon_p/2 \ .$$

A simple generalization of the *Saint-Venant* model to a three-dimensional rigid-plastic body is given by adopting the *R. von Mises* yield criterion. An isochoric deformation process is still assumed, which is in good agreement with experimental results, and, if we assume the material remains isotropic, by means of the second invariant of the deviatoric stress tensor and a material constant k , the flow rule in the plastic zones becomes

$$J_2 - k^2 = 0 \ , \quad J_2 = \frac{1}{2}\sum_i \sum_j \sigma_{ij}'^2 \ .$$

$$(4.120)$$

In the case of the tensile rod, $J_2 = \sigma_{xx}^2/3$, where $\sigma_{xx} = F/A_0$ (linearized geometric relations are understood). Hence, unlimited yielding occurs if

$$\sigma_{xx} = k \sqrt{3} = \sigma_Y \; , \quad \sigma_Y = F_Y/A_0 \; , \tag{4.121}$$

takes on the value of the yield stress.

Similarly, in pure shearing, where $\sigma_{xy} = \sigma_{yx} = \tau$ and the second invariant $J_2 = \tau^2$, unlimited plastic shear deformation is encountered when

$$\tau = k = \tau_Y \; , \tag{4.122}$$

where τ_Y is the yield limit of the shear stress. Hence, the ratio of the yielding stresses in tension and shear is $\sqrt{3}$.

In the three-dimensional stress state, the second invariant becomes $J_2 = (3/2)\,\tau_0^2$. Thus, in the (bounded) plastic zones, it follows that

$$k = \tau_0 \sqrt{3/2} \; . \tag{4.123}$$

τ_0 is the octahedral shear stress, ie it is the deviatoric stress component in the surface element of an octahedron; see Eq. (2.36).

Under some specified loading conditions, the body consists of rigid parts of volume with undefined stress distribution (there only the equilibrium conditions apply) and bounded (enclosed) yielding zones of the ideal plastic material. In the plastically deformed regions, the *von Mises flow rule* is the set of (linear) stress-strain differential equations

$$\dot{\varepsilon}_{ij}{}' = \dot{\varepsilon}_{ij} = \lambda\,\sigma_{ij}{}' \quad . \tag{4.124}$$

Considering λ to be constant and independent of the stress state renders the linear viscous *Newtonian* fluid. To show the nonlinearity of the plastic flow law, the expression is substituted into the second invariant, and if we consider Eq. (4.120), gives

$$J_2 = \frac{1}{2}\sum_i \sum_j \sigma_{ij}{}'^2 = \lambda^{-2}\frac{1}{2}\sum_i \sum_j \dot{\varepsilon}_{ij}{}^2 = \lambda^{-2} J_{2\dot{\varepsilon}} = k^2 \; .$$

Hence, solving for λ yields

$$\lambda = \frac{\sqrt{J_{2\dot{\varepsilon}}}}{k} \; , \tag{4.125}$$

which depends on the second invariant of the current plastic strain rate tensor. The nonlinear *von Mises* flow rule, thus, becomes

$$\sigma_{ij}' = \frac{k}{\sqrt{J_{2\dot{\varepsilon}}}} \, \dot{\varepsilon}_{ij} \ .$$

$$(4.126)$$

If the strain rates are known, $e = 0$, the second invariant can be calculated, and the deviatoric stress tensor is explicitly given by Eq. (4.126) within the plastic zones. The terminal stresses at the end of the plastic deformation process are independent of the loading time, contrary to the stress state in a viscous body. Time is merely a parameter, and the rates can be replaced by the increments during every loading step.

From a historical point of view, it should be remarked that *St. Venant* did not apply the *von Mises* yield condition, but considered the one originally given by *Tresca* : Yielding takes place if the largest shear stress is constant and equal to the yield limit in shear, τ_Y , of the simple shear test. Hence, considering the principal normal stresses $\sigma_1 > \sigma_2 > \sigma_3$ yields

$$\max \{|\sigma_1 - \sigma_2|, |\sigma_2 - \sigma_3|, |\sigma_3 - \sigma_1|\} = 2k \ .$$

$$(4.127)$$

An invariant formulation of the *Tresca yield criterion* is rather involved and may be looked up in *W. Prager and P. G. Hodge: Theory of Perfectly Plastic Solids. Wiley, New York, 1951.*

Since several difficulties are encountered in the application of the rigid-plastic model to problems with localized plastic zones, which are the standard case in structural plasticity, elasticity is usually taken into account. The ideal elastic-plastic material is considered subsequently. The application of the above model is restricted to the technological plastic-forming process.

4.3.2. The Elastic-Plastic Body

The one-dimensional model of a perfectly *elastic-plastic body* consists of a *Hookean* and *Saint-Venant* phase in a series connection. The tensile rod shown in Fig. 4.11(a) exhibits a linear elastic response when loaded slowly from the undeformed state, where $F = cu$ holds, until the yield limit is reached the first time, $F = F_Y$. The yielding of this model becomes stationary, if the load is kept constant at the yield level. An actual rod of mild steel yields nonstationary due to the phenomenon of *necking* , ie locally a material instability occurs where the cross-sectional area is locally strongly decreased, and a three-dimensional stress state develops. The force displacement relation is assumed to be symmetric in Fig. 4.11(a). Also shown are some loading cycles and

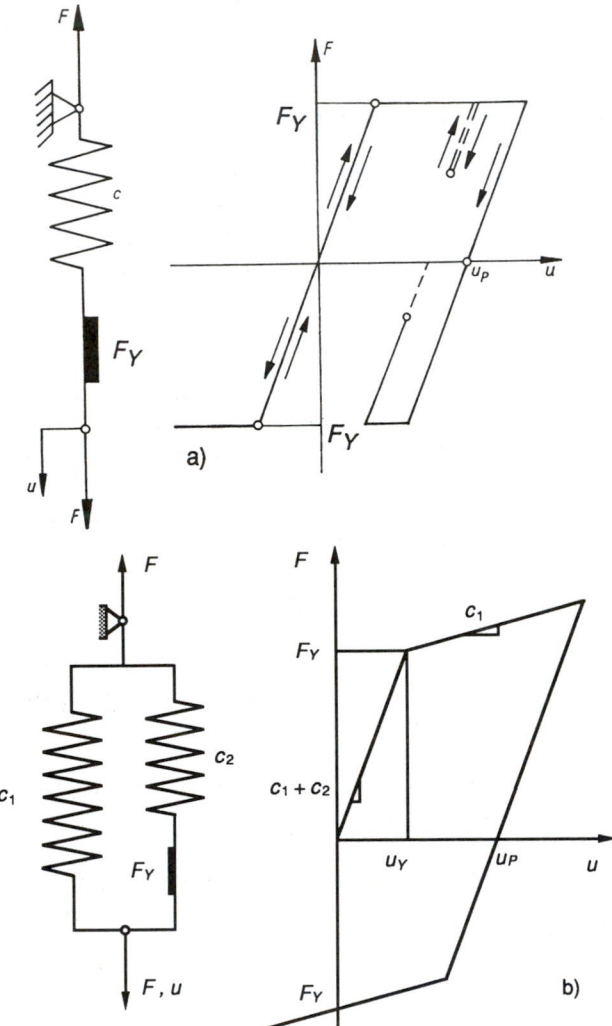

Fig. 4.11. Symbolic one-dimensional models: (a) The perfectly elastic-plastic solid with the yield limit F_Y . (b) The elastic-plastic body with linear strain hardening and the *Bauschinger* effect in cyclic plasticity

the permanent elongation u_p after unloading. The cycles follow the rules found by *Ludwig Prandtl* for the stress-strain diagrams of successive loadings and unloadings.

During *cyclic plastic deformations of metals* , it can be observed that under reloading conditions, plastic flow occurs at a lower stress level when compared to the yield limit of the initial loading path of the virgin material. This is referred to as the

Bauschinger effect (the original paper on this topic was published in 1886). It is one of the important effects of *strain hardening* . This phenomenon is represented by a one-dimensional bilinear model in Fig. 4.11(b), with a *Hookean* phase of stiffness c_1 in the parallel connection with the perfectly elastic-plastic body with parameters (c_2, F_Y). The tensile force in the plastic range is no longer constant, but is governed by the equation [see Fig. 4.11(b)]

$$F_p = F_Y + c_1 (u - u_Y) = c_1 u + \frac{c_2}{c_1 + c_2} F_Y \ , \quad u_Y = \frac{F_Y}{c_1 + c_2} \ .$$

Unloading follows the path of the *Hookean* double phase of stiffness $c_1 + c_2$. Reversed plastic deformations, however, do occur if

$$F_p = - F_Y + c_1 (u + u_Y) = c_1 u - \frac{c_2}{c_1 + c_2} F_Y \ .$$

Hence, the yield limit of any elastic path of reloading (in compression) that is initiated at the abscissa at a permanent displacement u_p is lowered with respect to the initial one F_Y, indicating the *Bauschinger* effect.

Another three-phase model limits the region of the linear strain hardening according to Fig. 4.12 by the parallel connection of two one-dimensional perfectly elastic-plastic bodies with different yield limits, $F_{1Y} < F_{2Y}$. Elongation u and strain ε_x are common. The elastic loading path, initiated at the undeformed configuration, is given by the linear relation

$$\sigma_{xx} = \frac{F}{A_0} = \frac{F_1 + F_2}{A_1 + A_2} = E \varepsilon_x \ , \quad (0 \le \varepsilon_x \le \varepsilon_{1Y} = \frac{F_{1Y}}{E A_1}) \ .$$

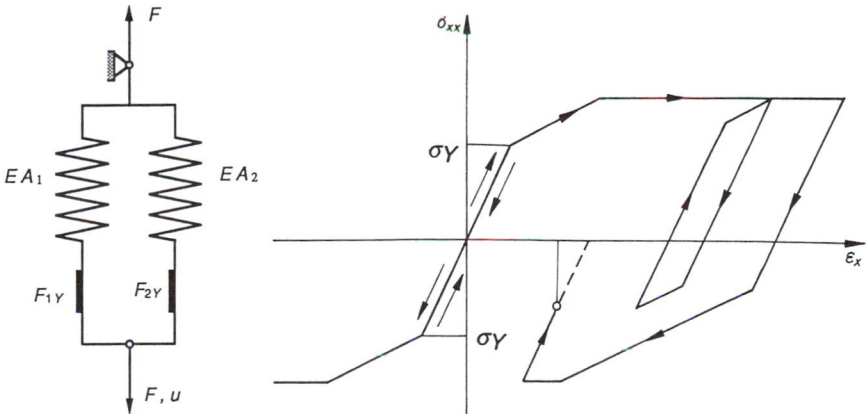

Fig. 4.12. Parallel connection of two perfectly elastic-plastic bodies. Limited linear strain hardening

For $\varepsilon_x > \varepsilon_{1Y}$, the phase one becomes perfectly plastic, $F_1 = F_{1Y} = const$, and

$$\sigma_{xx} = \frac{F}{A_0} = \frac{F_{1Y} + E\varepsilon_x A_2}{A_1 + A_2} \quad , \quad (\varepsilon_{1Y} \le \varepsilon_x \le \varepsilon_{2Y} = \frac{F_{2Y}}{E\,A_2}) \ .$$

A further increase in the tension renders the total strain $\varepsilon_x > \varepsilon_{2Y}$, both of the parallel phases become perfectly plastic, and

$$\sigma_{xx} = \frac{F}{A_0} = \frac{F_{1Y} + F_{2Y}}{A_1 + A_2} = const \quad , \quad (\varepsilon_x \ge \varepsilon_{2Y} = \frac{F_{2Y}}{E\,A_2}) \ .$$

A_1 and A_2 are the cross-sectional areas, F_1 and F_2 are the forces in the parallel connected two-phase models. The stress-strain relation of the virgin model is assumed to be symmetric in tension and compression according to Fig. 4.12, where some paths of cyclic loadings are shown.

The perfectly *elastic-plastic body* is considered in the three-dimensional stress state following the lines of the rigid-plastic model. Incompressible flow is assumed in the plastic zones, where the *von Mises* yield condition is taken into account [see Eq. (4.120)]

$$J_2 - k^2 = 0 \ , \tag{4.128}$$

and where the *von Mises* flow rule [see Eq. (4.124)] is to be replaced by the rate of the plastic strain portion $\varepsilon_{ij}{''}$

$$2G\,\dot{\varepsilon}_{ij}{''} = \lambda\,\sigma_{ij}{'} \ , \tag{4.129}$$

the proportionality factor λ is normalized by $2G$ for convenience. The consideration of the rate form of *Hooke's* law (this is important for the unloading path)

$$\dot{\sigma}_{ij}{'} = 2G\,\dot{\varepsilon}_{ij}{'} \ , \tag{4.130}$$

and of the deviatoric total strain components, by the superposition of the elastic portion $\varepsilon_{ij}{'}$ and the plastic strain $\varepsilon_{ij}{''}$,

$$\varepsilon_{ij} = \varepsilon_{ij}{'} + \varepsilon_{ij}{''} \ , \tag{4.131}$$

renders, when combined with Eq. (4.129), the stationary flow rule in the plastic zones

$$2G\,\dot{\varepsilon}_{ij} = \dot{\sigma}_{ij}{'} + \lambda\,\sigma_{ij}{'} \ , \tag{4.132}$$

If λ is considered constant, Eq. (4.132) is the constitutive relation of the *Maxwell* fluid. However, the plastic flow has a yielding condition, which becomes after differentiation

$$\dot{J}_2 = 0 = \sum_i \sum_j \sigma_{ij}' \dot{\sigma}_{ij}' \; .$$

(4.133)

Substitution of the stress rate gives

$$\dot{J}_2 = 0 = 2G \sum_i \sum_j \sigma_{ij}' \dot{\varepsilon}_{ij} - 2\lambda J_2 \; .$$

Inserting the yield condition $J_2 = k^2$ renders the factor λ not constant, but

$$\lambda = \frac{G}{k^2} P_C' \; ,$$

(4.134)

where

$$P_C' = \sum_i \sum_j \sigma_{ij}' \dot{\varepsilon}_{ij} \; ,$$

(4.135)

is the specific power of the distortional deformation. The deviatoric strain rate enters as a result of the assumption of incompressible plastic flow. The total strain is $\varepsilon_{ij}^{(t)} = \varepsilon_{ij} + e\,\delta_{ij}$, where the dilatation e is assumed purely elastic. The above power density is the total internal power of the stresses minus the power of elastic compression, taken per unit of volume: the rate of the specific distortional strain energy. For given stresses and strain rates in the plastic zones, P_C' and, hence, $\lambda > 0$ can be calculated. The substitution of λ renders the nonlinear *Prandtl-Reuss flow rule* according to

$$\dot{\sigma}_{ij}' = 2G \left(\dot{\varepsilon}_{ij} - \frac{P_C'}{2\,k^2} \sigma_{ij}' \right) \; .$$

(4.136)

The nonlinear differential equations are valid as long as the yield condition $J_2 = k^2$ holds, along with the condition that plastic flow is always accompanied by a dissipation of mechanical energy, $P_C' > 0$. They are supplemented by *Hooke's* law of purely elastic compressibility. In rate form,

$$\dot{p} = K\dot{e} \; , \quad \text{and,} \quad \dot{\sigma}_{ij} = \dot{\sigma}_{ij}' + \dot{p}\,\delta_{ij} \; ,$$

(4.137)

is the total stress rate.

In the elastic regions of the body, where $J_2 < k^2$, and in the case of unloading from the yield limit, $J_2 = k^2$ but $P_C' < 0$, all of *Hooke's* rate equations hold

$$\dot{\sigma}_{ij}' = 2G\,\dot{\varepsilon}_{ij} \ , \quad \dot{p} = K\,\dot{e} \ . \tag{4.138}$$

In many applications, compressibility has negligible influence and, thus, the limit $K \to \infty$ is considered to render approximately

$$e = 0 \ .$$

This additional equation covers the increase of the undetermined stress components by the unknown average stress p (contrary to the rigid-plastic model). The assumption of incompressibility reduces the numerical efforts drastically, but especially in problems with small plastic zones and, hence, elastic and plastic strains of the same (small) order, the stresses are possibly approximated with insufficient accuracy.

4.3.3. The Visco-Plastic Body

There are rheological materials (like enamel varnish) that exhibit a priori viscous effects in their plastic deformations, and there are engineering materials that show strong rate dependence in their deformations under severe dynamic loading conditions. The simple three-phase model of *Bingham* shows qualitatively those rate effects in plasticity. In Fig. 4.13, the one-dimensional model is shown to consist of a *Hookean* body of stiffness c in series connection with the parallel arrangement of a *Saint-Venant* stick-slip phase and a *Newtonian* fluid. In the range of loading $|F| < F_Y$, the rod is linear elastic. If, $|F| \geq F_Y$, plastic flow develops and exhibits a rate effect due to the viscosity η_p of the activated *Newtonian* fluid. In the visco-plastic range, the force results because of the common deformation

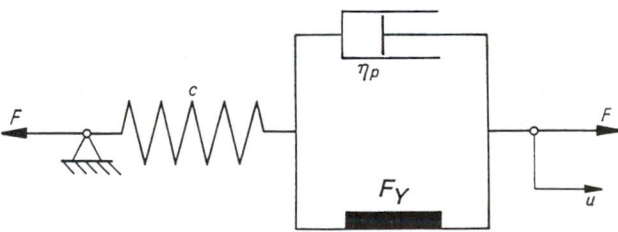

Fig. 4.13. Symbolic one-dimensional model of the elastic-visco-plastic *Bingham* body

$$F = F_Y + \eta_p \, \dot{u} = c \, u_e \quad ,$$

$$(4.139)$$

u_e is the elastic part of the elongation; η_p is called the plastic viscosity.

Considering the *von Mises* yield condition $J_2 = k^2$ in the plastic zone in a three-dimensional stress state renders the flow rule of the *Bingham* material in the generalized form [ε_{ij}'' is the deviatoric inelastic strain; see Eq. (4.126)]

$$\sigma_{ij}' = \sigma_{ij}'^P + \eta_p \, \dot{\varepsilon}_{ij}'' \quad , \quad \sigma_{ij}'^P = \frac{k}{\sqrt{J_{2\dot{\varepsilon}}}} \, \dot{\varepsilon}_{ij}'' \quad .$$

$$(4.140)$$

Exchanging the viscous phase in the *Bingham* model for a *Maxwell* fluid renders the *Schwedoff* material, a four-phase body. Many other generalized constitutive relations are discussed in the literature; also nonlinear viscosity is considered. Such a generalization is performed by solving Eq. (4.140) for the inelastic strain rate

$$\dot{\varepsilon}_{ij}'' = \frac{1}{\eta_p} \, \sigma_{ij}'^V \quad , \quad \text{where,} \quad \sigma_{ij}'^V = \left(\sigma_{ij}' - \sigma_{ij}'^P \right) \quad ,$$

$$(4.141)$$

is the deviatoric "overstress." Nonlinear viscous behavior is introduced by the extension

$$\dot{\varepsilon}_{ij}'' = \frac{1}{\eta_p} \, H(J_{2\sigma v}) \, \sigma_{ij}'^V \quad , \quad J_{2\sigma v} = \frac{1}{2} \sum_i \sum_j \sigma_{ij}'^V \sigma_{ij}'^V \quad .$$

$$(4.142)$$

Usually, σ_{ij}' and $\sigma_{ij}'^P$ can be assumed to be collinear. Hence, by means of the corresponding unit tensor n_{ij} , a scalar function of the second invariant of the overstress tensor can be factored out

$$\dot{\varepsilon}_{ij}'' = \frac{1}{\eta_p} \, [\sqrt{2J_{2\sigma v}} \, H(J_{2\sigma v})] \, n_{ij} \quad .$$

$$(4.143)$$

Instead of using the second invariant of the overstress, the difference of the invariants of the stress deviation and plastic stress portion may be employed. A function

$$\overline{\Phi}\left(\sqrt{J_2} - \sqrt{J_2^P}\right) = \frac{1}{\sqrt{2} \, k} \, \Phi(F) \quad , \quad F = \frac{\sqrt{J_2}}{k} - 1 \quad , \quad \sqrt{J_2^P} = k \quad ,$$

$$(4.144)$$

is substituted for the bracketed factor in Eq. (4.143) to render *Perzyna's* law of viscoplasticity

$$\dot{\varepsilon}_{ij}'' = \frac{1}{\eta_p} \, \{\Phi(F)\} \frac{\partial F}{\partial \sigma_{ij}} \quad .$$

$$(4.145)$$

The brace of Φ is an operational symbol: It indicates that the function is positive-semidefinite, $\Phi(F) \geq 0$ if $F \geq 0$, and $\Phi = 0$ if $F < 0$. This nonlinear visco-plastic flow rule has the advantage that the plastic stress deviation has been eliminated. Because of its simplicity, it can be easily programmed in incremental form, and due to its capability of describing the salient features of actual visco-plastic materials, *Perzyna's* constitutive relation is most frequently used in engineering and computational mechanics, including applications to cyclic structural plasticity. Refer to
P. Perzyna: *Fundamental Problems of Viscoplasticity. In Advances in Applied Mechanics, Vol. 9, 1966, pp. 243-377.*

4.4. Exercise A 4.1 and Solution

A 4.1: The ultimate bending moment $M_Y{}^*$ in a reinforced concrete (RC-) beam of rectangular composite cross-section $H \times B$ is wanted, such that the strains in the steel reinforcement and the compressional strain in the concrete take on their admissible limiting values $\varepsilon_e{}^*$ and ε_p, respectively. Employ the nonlinear elastic constitutive relation of concrete in compression [see Eq. (4.52)] under the assumption that the concrete phase is ruptured in the tensile region. See Fig. A 4.1 (state II). The reinforcing steel bars have a cross-sectional area summed to $A_e{}^*$.

Solution: The normal force is assumed to be zero; hence, $Z_e - D_b = 0$. The remaining couple of the internal forces is statically equivalent to the bending moment to be transmitted at the cross-section considered in Fig. A 4.1, $D_b z_b = M_Y(x)$, where

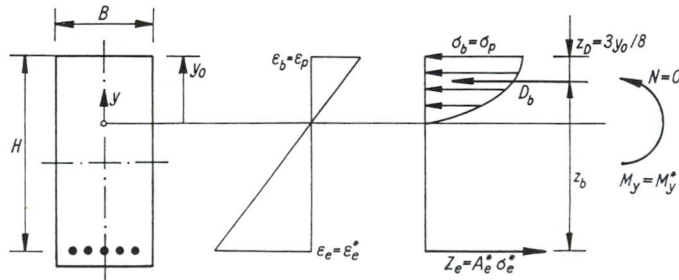

Fig. A 4.1. RC beam. Concrete is assumed ruptured in the tensile region, state II .
Subscript *b* stands for concrete, *e* refers to the steel reinforcement

$$D_b = B \int_0^{y_0} |\sigma_{xx}| \, dy \quad ,$$

is the resultant force transmitted in the unruptured concrete region. By means of Eq. (4.52) and the given linear distribution of the strains (cross-section is assumed to remain plane), $\varepsilon_{xx} = \varepsilon_p \, y/y_0$, where $y_0 = H \varepsilon_p /(\varepsilon_p + \varepsilon_e{}^*)$, the resultant becomes $D_b = (2/3) \, \sigma_p \, A_b$ with the unruptured concrete area $A_b = By_0$. The line of action of the resultant is found by the condition of static equivalence to pass through the point at a distance

$$z_D = \frac{B}{D_b} \int_0^{y_0} |\sigma_{xx}| (y_0 - y) \, dy = \frac{3}{8} y_0 \quad ,$$

from the upper face of the composite beam. With $z_b = H - z_D$, the ultimate bending moment of the cross-section in state II takes on the value

$$M_Y{}^* = \frac{2}{3} \, \sigma_p \, y_0 \, (A - \frac{3}{8} A_b) \quad , \quad \text{where } A = B H \quad .$$

Consideration of the stress $\sigma_e{}^*$ that accompanies the strain $\varepsilon_e{}^*$ in the reinforcing bars renders the required steel cross-section $A_e{}^* = D_b /\sigma_e{}^*$. Codes of RC structures limit the ratio A_e /A from above as well as from below.

5
Principle of Virtual Work

A structure loaded by external forces, like surface tractions and body forces, is considered to be in a state of equilibrium. The free-body diagram for such a set of conditions is given in Fig. 5.1. Necessarily, the resulting density of the force in every material point must vanish, $f = 0$. A material point with position vector r is given a *virtual displacement* δr , where $|\delta r| \ll l_{char}$ (characteristic length of the body), and the (elementary) virtual work per unit of volume vanishes consequently

$$f . \delta r = 0 \quad .$$

If we assume the virtual displacements are compatible with the integrity of the continuum, integration over the material volume causes the total virtual work of the internal and external forces to vanish

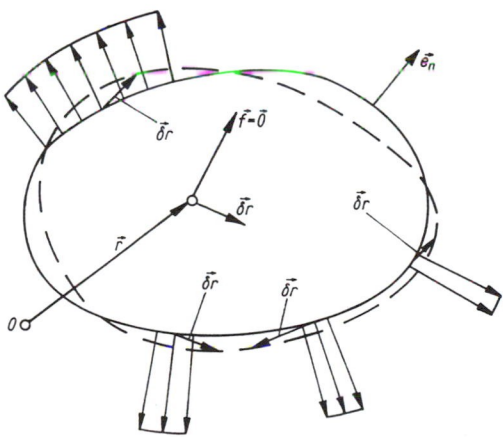

Fig. 5.1. Virtual displacements. Variation of the equilibrium configuration.
0 is a point of reference

$$\delta W = \int_V \left(\mathbf{f} \cdot \delta \mathbf{r} \right) dV = 0 \quad , \quad \mathbf{f} = \mathbf{k} + \sum_i \frac{\partial \boldsymbol{\sigma}_i}{\partial x_i} \quad ,$$
(5.1)

where $\int_V \mathbf{k} \cdot \delta \mathbf{r} \, dV$ is the virtual work of the (given) external body forces.

The second volume integral in Eq. (5.1) is changed by means of the *Gauss* integral theorem (the symmetry of the stress tensor is understood, and $\delta \mathbf{r} = \delta \mathbf{u}$)

$$\sum_i \int_V \left(\frac{\partial \boldsymbol{\sigma}_i}{\partial x_i} \cdot \delta \mathbf{r} \right) dV = \sum_i \int_V \frac{\partial}{\partial x_i} \left(\boldsymbol{\sigma}_i \cdot \delta \mathbf{r} \right) dV - \sum_i \int_V \boldsymbol{\sigma}_i \cdot \frac{\partial}{\partial x_i} \left(\delta \mathbf{r} \right) dV$$

$$= \oint_{\partial V} \left(\boldsymbol{\sigma}_n \cdot \delta \mathbf{r} \right) dS - \frac{1}{2} \sum_i \sum_j \int_V \sigma_{ji} \, \delta \!\left(\frac{\partial u_i}{\partial x_j} + \frac{\partial u_j}{\partial x_i} \right) dV \quad .$$

The surface integral

$$\oint_{\partial V} \left(\boldsymbol{\sigma}_n \cdot \delta \mathbf{r} \right) dS$$

is the virtual work done by the external surface tractions. Since the virtual displacements are assumed to be sufficiently small, the linearized geometric relations of Eq. (1.21) apply to the virtual variations of the strains $\delta \varepsilon_{ij}$,

$$\delta W^{(i)} = - \sum_i \sum_j \int_V \sigma_{ji} \, \delta \varepsilon_{ji} \, dV \quad .$$
(5.2)

Formally, a comparison with the differential increment of the elastic potential Eq. (3.36) in an elastic body is illustrative.

Hence, if we denote the virtual work of the external forces (body forces and surface tractions) by $\delta W^{(e)}$, the vanishing sum of the virtual work of the internal and external forces is rewritten in shorthand notation

$$\delta W = \delta W^{(i)} + \delta W^{(e)} = 0 \quad .$$
(5.3)

Equation (5.3) remains valid if on some parts of the body surface, the displacements are prescribed and, hence, are not varied virtually, $\delta \mathbf{r} = \mathbf{0}$ at these parts of the surface, ie the surface tractions (supporting forces) that are necessary to produce the prescribed displacements do not contribute to the virtual work of the external forces.

Iff Eq. (5.3) holds for all possible and admissible virtual displacement fields measured from a reference configuration, the principle of virtual work (in the special version, it is called the *principle of virtual displacements*) states that the body in the reference state is in equilibrium.

In the first application of the principle, it is proven that the six conditions of equilibrium of external forces [see Eq. (2.58)] are necessary for a body to remain at rest. The body in the deformed equilibrium configuration is considered in a free-body diagram, all the supports are replaced by the proper reaction forces, and by keeping the deformations frozen, the virtual displacements are chosen according to (small) rigid-body motion. Hence, Eq. (1.5) applies. A is a body-fixed point of reference, P is a general material point, $r_P = r$, $r_{PA} = r - r_A$,

$$\delta r = \delta r_A + \delta\alpha \times r_{PA} \quad , \tag{5.4}$$

and $\delta\alpha$ is the vector of the small virtual rigid-body rotation. The virtual work of the external forces (the given loadings and supporting forces) must vanish since the virtual work of the internal forces is identically zero due to rigid-body motion

$$\delta W = \delta W^{(e)} = \left[\int_V k \, dV + \oint_{\partial V} \sigma_n \, dS \right] \cdot \delta r_A$$

$$+ \left[\int_V (r_{PA} \times k) \, dV + \oint_{\partial V} (r_{PA} \times \sigma_n) \, dS \right] \cdot \delta\alpha = 0 \quad . \tag{5.5}$$

Since the virtual displacement of the material reference point δr_A and the virtual rotation $\delta\alpha$ are independent, Eq. (5.5) renders the two vector conditions of equilibrium of the external forces

$$R = 0 \quad , \quad M_A = 0 \quad . \tag{5.6}$$

The resulting force and resulting moment of the external forces must vanish. The conditions are necessary but not sufficient, since they have been derived for the special virtual displacement field of rigid-body motion. Only for the equilibrium of the rigid-body model are the conditions necessary and sufficient, since no other displacements are compatible in that special case.

The principle of virtual displacements is especially suitable to derive the conditions of equilibrium for mechanical systems with a finite number of degrees of freedom with respect to virtual motion (MDOF-systems). For example, the configuration of a solid body which is discretized by finite elements or a system of rigid bodies, interconnected in joints with or without elastic springs and considered at rest under the action of external forces, is described by a finite number of independent coordinates, q_1 , q_2 , ..., q_n . These are, eg the coordinates of the nodes of the finite elements or the independent coordinates of the individual rigid bodies (maximum six for a single body). All the radius vectors to the points of application of the forces become functions of these independent coordinates

$$\mathbf{r} = \mathbf{r}(q_1, q_2, ..., q_n) \ . \tag{5.7}$$

Hence, the virtual displacement can be expressed by the total derivative

$$\delta\mathbf{r} = \frac{\partial\mathbf{r}}{\partial q_1}\,\delta q_1 + \frac{\partial\mathbf{r}}{\partial q_2}\,\delta q_2 + ... + \frac{\partial\mathbf{r}}{\partial q_n}\,\delta q_n = \sum_{i=1}^{n} \frac{\partial\mathbf{r}}{\partial q_i}\,\delta q_i \ , \tag{5.8}$$

in terms of the independent virtual variations of the n configurational coordinates. If the internal and external forces can be represented by a system of single forces F_l, $l = 1, 2, ... k$, attached to the points $\mathbf{r}_l$, the virtual work is expressed by means of the generalized forces Q_i in the form

$$\delta W = \sum_{l=1}^{k} \mathbf{F}_l \cdot \delta\mathbf{r}_l = \sum_{l=1}^{k} \mathbf{F}_l \cdot \sum_{i=1}^{n} \frac{\partial\mathbf{r}_l}{\partial q_i}\,\delta q_i = \sum_{i=1}^{n} Q_i \cdot \delta q_i = 0 \ , \quad \text{where}$$

$$Q_i = \sum_{l=1}^{k} \mathbf{F}_l \cdot \frac{\partial\mathbf{r}_l}{\partial q_i} = 0 \ , \quad i = 1, 2, ..., n \quad , \tag{5.9}$$

gives exactly the number n of independent equations of equilibrium corresponding to the number n of the degrees of freedom. The vanishing generalized forces replace the actual forces in the sense that they produce the same virtual work. All those forces that do not contribute to the virtual work are eliminated from the conditions of equilibrium. Examples of those forces that do not enter are, eg the internal force in a frictionless hinge and the external reaction force at a frictionless sliding support. See also Eq. (10.7) for the generalized forces in dynamics.

5.1. Example: The Three-Hinged Arch

The statically determinate structure is loaded by the given external forces H and V and by the "internal" moment M_0 in the torsional spring (eg a prestressed linear elastic spring, $M_0 = k\varphi_0$); see Fig. 5.2. By the method of the selective partial free-body diagram, the horizontal component of the reaction force B_H at the simple support B is to be determined using Eq. (5.9). Adding this unknown force component to the external forces and opening the support B for a horizontal virtual displacement contribute to the total virtual work. All other reaction forces and the force in the hinge C, as well as the internal forces in the members of the three-hinged arch, which

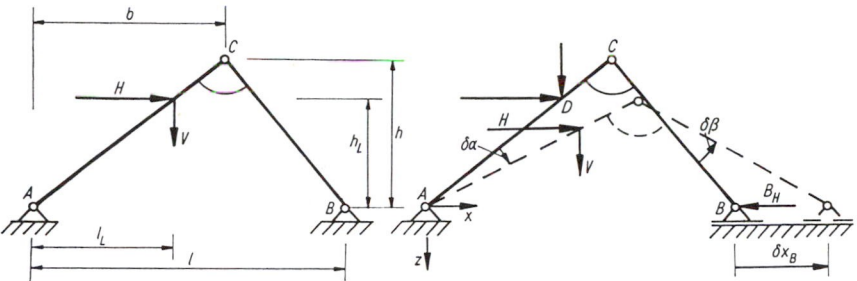

Fig. 5.2. Virtual displacement of a three-hinged arch from the equilibrium
configuration. Selective free-body diagram, SDOF-system

remain virtually undeformed, do not contribute in such a manner.
Hence, the virtual work of the external forces in the virtual rigid-
body motion of the two-bar mechanism of Fig. 5.2 becomes

$$\delta W^{(e)} = - B_H\, \delta x_B + H\, \delta x_D + V\, \delta z_D \quad,$$

and the internal moment contributes

$$\delta W^{(i)} = M_0\, (\delta\alpha + \delta\beta) \quad.$$

The mechanism in virtual motion has only a single degree of freedom
(SDOF-system), and the five virtual displacements are related to one
variation $\delta\alpha$

$$\delta x_D = h_L\, \delta\alpha \ , \quad \delta z_D = l_L\, \delta\alpha \ , \quad \delta z_C = b\, \delta\alpha = (l-b)\, \delta\beta \ ,$$
$$\delta x_B = \delta x_C + h\, \delta\beta = h\left(1 + \frac{b}{l-b}\right)\delta\alpha = h\,\frac{l}{l-b}\,\delta\alpha \ ,$$

and

$$\delta W = \left(- h B_H \frac{l}{l-b} + H\, h_L + V\, l_L + M_0 \frac{l}{l-b}\right)\delta\alpha = 0 \quad.$$

Since $\delta\alpha \neq 0$, the term in parenthesis must vanish

$$B_H = H\,\frac{h_L}{h}\frac{l-b}{l} + V\,\frac{l_L}{h}\frac{l-b}{l} + M_0/h \quad.$$

Analogously, the vertical component B_V can be determined
independently by considering the selective partial free-body
diagram in which the virtual motion of support B is controlled in the
vertical direction.

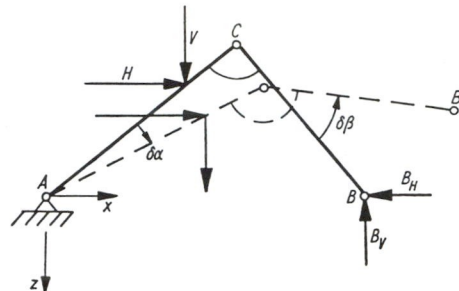

Fig. 5.3. Virtual displacement of a three-hinged arch from its equilibrium configuration. Selective free-body diagram for the determination of the reaction force **B**. MDOF-system with two degrees of freedom

If support B is set free to move virtually in the configurational plane, both components of the reaction force contribute to the virtual work, and the mechanism becomes a MDOF-system with two degrees of freedom. The virtual rotations $\delta\alpha \neq 0$ and $\delta\beta \neq 0$ become independent, and the virtual work of the forces shown in Fig. 5.3 is given by

$$\delta W = H\, \delta x_D + V\, \delta z_D + M_0\, (\delta\alpha + \delta\beta) - B_H\, \delta x_B - B_V\, \delta z_B = 0 \quad,$$

where

$$\delta x_D = h\, \delta\alpha \ , \quad \delta z_D = l_L\, \delta\alpha \ , \quad \delta x_B = h\, (\delta\alpha + \delta\beta) \ , \quad \delta z_B = b\, \delta\alpha - (l - b)\, \delta\beta \ .$$

The vanishing virtual work in that case is the sum of two terms

$$(H\, h_L + V\, l_L + M_0 - B_H\, h - B_V\, b)\, \delta\alpha + [M_0 - B_H\, h + B_V\, (l - b)]\, \delta\beta = 0 \quad,$$

which must be set to zero individually due to the independence of the variation of the configurational coordinates. Hence, the linear system of two equations results

$$B_H\, h + B_V\, b = H\, h_L + V\, l_L + M_0 \ ,$$
$$B_H\, h - B_V\, (l - b) = M_0 \ ,$$

with the solution B_H (unchanged) and $B_V = (H\, h_L + V\, l_L)/l$.

5.2. Influence Lines of Statically Determinate Structures

The method of the selective partial free-body diagram, combined with the principle of virtual work, renders directly the influence

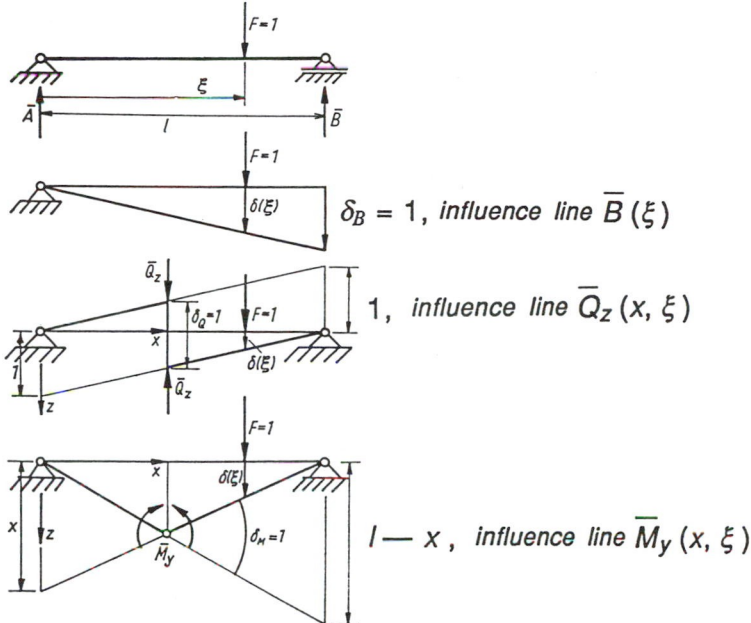

Fig. 5.4. Influence lines of reaction force B, shearing force Q and bending moment M_y, respectively. Evaluation by a kinematic method

lines of statically determinate structures when loaded by a single force $F = 1$. Such a kinematic procedure is illustrated by the influence lines of a single-span simply supported beam (see Fig. 5.4).

Replacing the sliding support by the force B allows a rigid virtual rotation about the hinge A. The virtual work $\delta W = F\,\delta(\xi) - B\,\delta_B = 0$ yields $B = F\,\delta(\xi)/\delta_B$. The influence line of the supporting force is by definition, see Fig. 5.4,

$$\overline{B} = \frac{B}{F} = \frac{\delta(\xi)}{\delta_B} = \xi/l \;,$$

(5.10)

The shear force in a cross-section at x contributes to the virtual work, if the beam is segmented and the two pieces are virtually rotated according to Fig. 5.4, whereas the bending moment, which is symmetrically attached to the positive and negative cross-sections, does not contribute. Hence, the virtual work becomes $\delta W = F\,\delta(\xi) - Q_z\,\delta_Q = 0$. The solution gives the shearing force $Q_z = F\,\delta(\xi)/\delta_Q$ and the influence line becomes

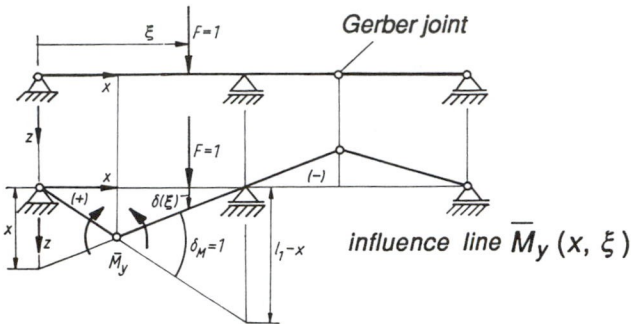

Fig. 5.5. Influence line of the bending moment M_y $(x; \xi)/F$ of a two-span beam
with a *Gerber* joint

$$\bar{Q}_z(x;\xi) = \frac{Q_z}{F} = \frac{\delta(\xi)}{\delta_Q} = \begin{cases} -\xi/l & , \quad 0 < \xi < x \quad , \\ (1 - \xi/l) & , \quad x < \xi < l \quad . \end{cases} \tag{5.11}$$

Assuming a hinged joint at x allows a rigid virtual rotation similar to that of a crank mechanism, and the bending moment contributes to the virtual work through the external double couples attached to the rigid bodies connected at that joint. The shear remains an internal force and does not contribute. Thus, $\delta W = F\,\delta(\xi) - M_y\,\delta_M = 0$ renders the bending moment, $M_y = F\,\delta(\xi)/\delta_M$, and δ_M is the relative angle of rotation (see Fig. 5.4). The influence line of the bending moment is given by

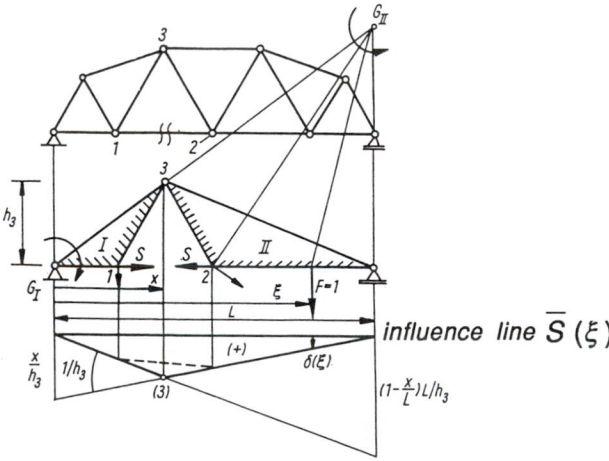

Fig. 5.6. Single-span truss. Influence line of the force S in member 12 evaluated by a
kinematic method

$$\overline{M}_y\,(x;\xi) = \frac{M_y}{F} = \frac{\delta(\xi)}{\delta_M} = \begin{cases} \xi\,(1 - x/l) \,, & 0 \le \xi \le x \,, \\ x\left(1 - \xi/l\right) \,, & x \le \xi \le l \,. \end{cases} \qquad (5.12)$$

In Fig. 5.5, a statically determinate double-span beam with a *Gerber* joint is considered and the influence line of the bending moment $M_y(x;\,\xi)/F = \delta(\xi)/\delta_M$ is evaluated by the kinematic method, ie the virtually deformed configuration is drawn with the relative angle of rotation $\delta_M = 1$.

The kinematic method can be generalized; as an example the influence line of the member force in the lower girder of a single-span truss is evaluated in Fig. 5.6. The point of application of the unit load F does not continuously move with ξ, but must jump from one node of the truss to the other. The influence line is valid only in these discrete points. The virtual motion follows that of a crank mechanism and the points 1 and 2 are moved apart horizontally by one unit. The centers of velocity $G_{I,\ II}$ are shown.

5.3. Conservative Mechanical Systems

The principle of virtual work is also an elegant method for establishing the differential, local conditions of equilibrium of continuous (nondiscretized) structures. The derivation is greatly simplified if the internal forces are irrotational and, hence, are determined by the (elastic) potential U. In that case, the virtual work of the internal forces becomes simply the variation of the internal energy $U(x, y, z)$ due to the variation of the state of equilibrium,

$$\delta W^{(i)} = -\,\delta U \,, \qquad (5.13)$$

and the principle takes on the simpler form

$$\delta W^{(e)} = \delta U \,. \qquad (5.14)$$

<<The virtual work of the external forces equals the virtual increase of the (elastic) potential which is due to the virtual displacements from the state of equilibrium.>>

Quite often, the external forces are also irrotational (eg when they are properly derived from dead weight loadings) and a potential function W_P exists in the neighborhood of the equilibrium configuration. Since the virtual work $\delta W^{(e)} = -\,\delta W_P$ [analogously to Eq. (5.13)], the principle of virtual work takes on the form given by *J. W. Gibbs* (the total potential energy of the internal and external forces is the sum $\Pi = U + W_P$)

$$\delta\Pi = \delta U + \delta W_P = 0 \quad . \tag{5.15}$$

<<The total potential energy of a conservative mechanical system in equilibrium has a stationary value; ie its first variation must vanish.>> Such a body at rest with no possibility of virtual rigid-body motion is in a stable equilibrium configuration if the potential energy Π has a local minimum. See Sec. 9.1.

   If the external forces are not related to the deformations, their potential is given by the volume integral of the (negative) work of the body forces k and the surface integral of the (negative) work of the prescribed tractions σ_n

$$W_P = -\int_V k \cdot u \, dV - \int_{\partial V_\sigma} \sigma_n \cdot u \, dS \; , \tag{5.16}$$

where u denotes the displacement vector and ∂V_σ is that part of the surface of the body where the tractions are prescribed.

5.3.1. Differential Equation of the Deflection of a Linear Elastic Beam

A linear elastic beam with a straight axis in the stress-free reference state carries a transverse load $q(x)$ per unit of length, and the deflection $w(x)$ determines the curved beam axis in the deformed state. The influence of shear on deflection will be neglected, and the *Bernoulli-Euler* hypothesis is assumed to hold. Plane cross-sections remain plane after deformation and are also assumed to remain orthogonal to the deformed beam axis. The strains are linearly distributed over the height of the beam in principal bending. If we assume small strains, the linearized geometric relations apply and also the normal stresses are linearly distributed according to *Hooke's* law; see Eq. (4.31) that is derived for pure bending. The resulting moment is, thus, given by

$$M_y(x) = -\int_A z \, E \frac{z}{r} \, dA = -\frac{E}{r} \int_A z^2 \, dA = -\frac{E \, J_y}{r(x)} \; ,$$

where $1/r$ is the curvature of the deformed beam axis. The work of bending an element of arc-length ds is [cf. the elastic spring, Eq. (3.25), and note Eq. (3.7)]

$$\frac{1}{2} M_y(x) \frac{ds}{r} = -\frac{E \, J_y}{2} \frac{ds}{r^2(x)} = -dU \quad .$$

The strain energy stored in the beam of length l becomes (the contribution of shear deformation is neglected

$$U = \int\limits_0^l \frac{EJ_y}{2} \frac{ds}{r^2(s)} \quad .$$

Considering the deflections to be sufficiently small renders approximately the linearized curvature, $1/r \approx d^2w/dx^2 = w''$, and putting $ds = dx$ gives the elastic potential of the beam according to the linearized bending theory,

$$U = \int\limits_0^l \frac{EJ_y}{2} (w'')^2 \, dx \quad . \tag{5.17}$$

The potential of the external forces, the dead weight loading $q(x)$, and the edge loadings Q_0 , M_0 and Q_1 , M_1 becomes under these assumptions

$$W_P = -\int\limits_0^l q(x) \, w(x) \, dx - M_0 \left(\frac{dw}{dx}\right)_{x=0} + M_1 \left(\frac{dw}{dx}\right)_{x=l} + Q_0 \, w_0 - Q_1 \, w_1 \quad . \tag{5.18}$$

Hence, the resulting potential is the sum of the bending energy and the negative work of the external forces

$$\Pi = U + W_P \quad . \tag{5.19}$$

The deformed state of equilibrium is derived when the potential energy takes on a stationary value. Thus, its first variation must vanish

$$\delta\Pi = \int\limits_0^l EJ_y \, w'' \, \delta(w'') \, dx - \int\limits_0^l q(x) \, \delta w(x) \, dx$$

$$- M_0 \, \delta\left(\frac{dw}{dx}\right)_{x=0} + M_1 \, \delta\left(\frac{dw}{dx}\right)_{x=l} + Q_0 \, \delta w_0 - Q_1 \, \delta w_1 = 0 \quad . \tag{5.20}$$

Twofold partial integration of the first term factors the variation δw of the deflection,

$$\delta\Pi = \int\limits_0^l \left[\frac{d^2}{dx^2} (EJ_y \, w'') - q(x)\right] \delta w \, dx$$

$$- [EJ_y \, (w'')_{x=0} + M_0] \, \delta\left(\frac{dw}{dx}\right)_{x=0} + [EJ_y \, (w'')_{x=l} + M_1] \, \delta\left(\frac{dw}{dx}\right)_{x=l}$$

$$+ \left[\frac{d}{dx} EJ_y (w'')_{x=0} + Q_0\right] \delta w_0 - \left[\frac{d}{dx} EJ_y (w'')_{x=1} + Q_1\right] \delta w_1 = 0 \quad . \tag{5.21}$$

Since the variation of the deflection δw is arbitrary in the interval of the span, $0 \leq x \leq l$, the coefficient in the integral must vanish

$$\frac{d^2}{dx^2} \left(EJ_y \frac{d^2 w}{dx^2}\right) - q = 0 \quad , \quad 0 \leq x \leq l \quad . \tag{5.22}$$

Equation (5.22) is the fourth-order differential equation of deflection of a slender, linear elastic beam in the state of principal bending according to the approximate *Bernoulli-Euler* beam theory.

Integrating twice and considering the expression for the bending moment (2.150) render the linear moment curvature relation (in linearized form)

$$\frac{d^2 w}{dx^2} = -\frac{M_y(x)}{EJ_y} \quad , \quad 0 \leq x \leq l \quad . \tag{5.22a}$$

The system of two second-order coupled differential equations, Eqs. (2.150) and (5.22a), is equivalent to the fourth-order Eq. (5.22). For the efficient integration procedure of *Mohr*, see Sec. 6.2.2.

The remaining terms in the variational Eq. (5.21) correspond to the virtual work of the edge forces that must also vanish. Table 5.1 gives sufficient conditions by the classical boundary conditions of engineering beam theory. At a rigidly clamped edge, the kinematic

Table 5.1. Classical kinematic (prescribed deflection and/or slope) or dynamic boundary conditions of a beam.

Kinematic   or   Dynamic
Boundary Conditions

$x = 0$:

$$\delta\left(\frac{dw}{dx}\right)_0 = 0 \qquad EJ_y \left(\frac{d^2 w}{dx^2}\right)_0 + M_0 = 0$$

$$\delta w_0 = 0 \qquad \frac{d}{dx}\left(EJ_y \frac{d^2 w}{dx^2}\right)_0 + Q_0 = 0$$

Kinematic   or   Dynamic
Boundary Conditions

$x = l$:

$$\delta\left(\frac{dw}{dx}\right)_1 = 0 \qquad EJ_y \left(\frac{d^2 w}{dx^2}\right)_1 + M_1 = 0$$

$$\delta w_1 = 0 \qquad \frac{d}{dx}\left(EJ_y \frac{d^2 w}{dx^2}\right)_1 + Q_1 = 0$$

boundary conditions $w = 0$ and $dw/dx = 0$ apply, ie the variations must vanish; see Table 5.1. At a free edge (eg of a cantilever), the dynamic boundary conditions $M = 0$ and $Q = 0$ apply that correspond according to Table 5.1 to the vanishing curvature $d^2w/dx^2 = 0$ and $d(EJ_y \, d^2w/dx^2)/dx = 0$. At a hinged support, the deflection is prescribed by $w = 0$ and the moment vanishes, $d^2w/dx^2 = 0$; see also Sec. 1.3.4.

5.3.2. The von Karman Plate Equations

A laterally loaded thin linear elastic and isotropic plate of constant thickness h , is considered, taking into account an in-plane prestress by membrane forces. The mid-plane $Z = 0$ coincides with the coordinate plane (X, Y) in the stress-free reference state. The displacements of its points are denoted $u_0 (X, Y)$, $v_0 (X, Y)$ within the plane and the deflection is $w(X, Y)$ in the *lagrangean* description. The displacements of a general material point are the functions $u(X, Y, Z)$, $v(X, Y, Z)$, and $w(X, Y, Z)$, and the nonlinear geometric relations [Eq. (1.20)] render the strains approximately by considering only the leading nonlinear terms in $\partial w/\partial X$ and $\partial w/\partial Y$, according to *von Karman* theory,

$$\varepsilon_{xx} = \frac{\partial u}{\partial X} + \frac{1}{2}\left(\frac{\partial w}{\partial X}\right)^2 \; , \; \varepsilon_{yy} = \frac{\partial v}{\partial Y} + \frac{1}{2}\left(\frac{\partial w}{\partial Y}\right)^2 \; , \; \varepsilon_{zz} = \frac{\partial w}{\partial Z} \; ,$$

$$2\varepsilon_{xy} = \frac{\partial v}{\partial X} + \frac{\partial u}{\partial Y} + \frac{\partial w}{\partial X}\frac{\partial w}{\partial Y} \; , \; 2\varepsilon_{yz} = \frac{\partial w}{\partial Y} + \frac{\partial v}{\partial Z} \; , \; 2\varepsilon_{zx} = \frac{\partial u}{\partial Z} + \frac{\partial w}{\partial X} \; . \tag{5.23}$$

The distribution of the in-plane displacements over Z is approximated by the plane strain condition $\varepsilon_{zz} = 0$, which is an inadmissible assumption with respect to *Hooke's* law. Hence, consequently, $\partial w/\partial Z = 0$, and $w(X, Y, Z) = w(X, Y)$ becomes independent of Z . If we assume that the external loading is free of any shear component and simply given by the pressure $p(X, Y)$, the shear stresses in the surfaces $Z = \pm h/2$ must vanish ($S_{yz} = S_{xz} = 0$), and due to *Hooke's* law also, the shear strains $\varepsilon_{yz} = \varepsilon_{xz} = 0$. In thin plates, those shear strains remain small in the interior ($-h/2 \leq Z \leq h/2$) and can be neglected. Within this approximation, the normal to the midplane remains a straight line after deformation and, in addition, is approximately normal to the deformed midsurface $w(X, Y)$. In analogy to the *Bernoulli* hypothesis of slender beams, this approximating *Kirchhoff* assumption is valid for thin plates. The integration of Eq. (5.23) yields

$$u(X, Y, Z) = u_0(X, Y) - Z\frac{\partial w}{\partial X} \; , \; v(X, Y, Z) = v_0(X, Y) - Z\frac{\partial w}{\partial Y} \; , \; w(X, Y, Z) = w(X, Y) \; .$$

and

$$\varepsilon_{xx}(X, Y, Z) = \bar{\varepsilon}_{xx}(X, Y) - Z\frac{\partial^2 w}{\partial X^2} \quad , \quad \varepsilon_{yy}(X, Y, Z) = \bar{\varepsilon}_{yy}(X, Y) - Z\frac{\partial^2 w}{\partial Y^2} \quad ,$$

$$2\,\varepsilon_{xy}(X, Y, Z) = 2\,\bar{\varepsilon}_{xy}(X, Y) - 2\,Z\frac{\partial^2 w}{\partial X \partial Y} \quad ,$$

where

$$\bar{\varepsilon}_{xx} = \frac{\partial u_0}{\partial X} + \frac{1}{2}\left(\frac{\partial w}{\partial X}\right)^2 , \ \bar{\varepsilon}_{yy} = \frac{\partial v_0}{\partial Y} + \frac{1}{2}\left(\frac{\partial w}{\partial Y}\right)^2 , \ 2\,\bar{\varepsilon}_{xy} = \frac{\partial v_0}{\partial X} + \frac{\partial u_0}{\partial Y} + \frac{\partial w}{\partial X}\frac{\partial w}{\partial Y} \ .$$

Since the normal stress component S_{zz} is of the order of the prescribed load p , it can be neglected and the stress state in the plate is approximately plane (S_{xx} , S_{yy} , and S_{xy} are the remaining stresses). The strain energy density of the *Hookean* plate is approximated by Eq. (3.47)

$$U' = \frac{G}{1-v}\left[\varepsilon_{xx}^2 + \varepsilon_{yy}^2 + 2\,v\,\varepsilon_{xx}\varepsilon_{yy} + 2\,(1-v)\,\varepsilon_{xy}^2\right] \ .$$

$$(5.24)$$

Volume integration gives the elastic potential of the plate. The substitution of the midplane strains and of the bending surface w and integrating over the plate thickness separate the strain energy into two terms

$$U = U_M + U_B \ .$$

$$(5.25)$$

The membrane energy depends linearly on h

$$U_M = \frac{Gh}{1-v}\iint_A \left[\bar{\varepsilon}_{xx}^2 + \bar{\varepsilon}_{yy}^2 + 2\,v\,\bar{\varepsilon}_{xx}\bar{\varepsilon}_{yy} + 2\,(1-v)\,\bar{\varepsilon}_{xy}^2\right]dX\,dY \ ,$$

$$(5.26)$$

and the bending energy is proportional to h^3

$$U_B = \frac{Gh^3}{12\,(1-v)}\iint_A \left[\left(\frac{\partial^2 w}{\partial X^2}\right)^2 + \left(\frac{\partial^2 w}{\partial Y^2}\right)^2 + 2v\left(\frac{\partial^2 w}{\partial X^2}\right)\left(\frac{\partial^2 w}{\partial Y^2}\right)\right.$$
$$\left. + 2(1-v)\left(\frac{\partial^2 w}{\partial X\,\partial Y}\right)^2\right]dX\,dY \ .$$

$$(5.27)$$

The potential of the pressure loading p is equal to the negative work

$$W_p = -\iint_A w\,p\,dX\,dY \ ,$$

$$(5.28)$$

no body forces or edge loads are considered. Finally, the potential energy is the superposition of all three contributions

$$\Pi = U_M + U_B + W_p \quad . \tag{5.29a}$$

The first variation must vanish to yield the conditions of equilibrium of the plate in the form of the *Euler-Lagrange* variational first-order differential equations. A system independent of Z separates

$$\delta\Pi = 0 \quad , \quad \text{gives} \rightarrow$$

$$\frac{\partial}{\partial X}\left(\bar{\varepsilon}_{xx} + v\bar{\varepsilon}_{yy}\right) + (1-v)\frac{\partial \bar{\varepsilon}_{xy}}{\partial Y} = 0 \;,\; (1-v)\frac{\partial \bar{\varepsilon}_{xy}}{\partial X} + \frac{\partial}{\partial Y}\left(\bar{\varepsilon}_{yy} + v\bar{\varepsilon}_{xx}\right) = 0 \;.\tag{5.29b}$$

Integrating over the plate thickness, or equivalently, multiplying by h, renders the membrane forces according to *Hooke's* law in plane stress, namely, the inverted Eq. (4.22) by inspection. Hence, the differential equations of equilibrium under plane stress conditions of the mean stresses are rederived; see Eq. (2.10). The third variational equation is the fourth-order plate equation that renders reliable results for deflections up to the order of the plate thickness

$$\nabla^2\nabla^2\, w \;=\; \frac{p}{K} + \frac{12}{h^2}\left[\left(\bar{\varepsilon}_{xx} + v\bar{\varepsilon}_{yy}\right)\frac{\partial^2 w}{\partial X^2} + \left(\bar{\varepsilon}_{yy} + v\bar{\varepsilon}_{xx}\right)\frac{\partial^2 w}{\partial Y^2}\right.$$

$$\left. + 2(1-v)\,\bar{\varepsilon}_{xy}\frac{\partial^2 w}{\partial X \partial Y}\right] \;,\quad K = \frac{E\,h^3}{12\,(1-v^2)} \quad . \tag{5.29c}$$

In the coupling terms of this equation, the midplane strains can be substituted by the membrane forces using *Hooke's* law (4.22), and the latter are derived from the *Airy* stress function by Eq. (2.11). Thus, the first of the *von Karman* plate equations takes on its final form

$$K\,\nabla^2\nabla^2\, w \;=\; p + \frac{\partial^2 F}{\partial Y^2}\frac{\partial^2 w}{\partial X^2} + \frac{\partial^2 F}{\partial X^2}\frac{\partial^2 w}{\partial Y^2} - 2\,\frac{\partial^2 F}{\partial X \partial Y}\frac{\partial^2 w}{\partial X \partial Y} \quad . \tag{5.30a}$$

The second of the *von Karman* plate equations is derived by enforcing the compatibility of strains through Eq. (1.22), where $i = X$ and $j = Y$. Hence, the coefficient of Z cancels, and a new relation between the midplane strains and the deflection results

$$\frac{\partial^2 \bar{\varepsilon}_{xx}}{\partial Y^2} + \frac{\partial^2 \bar{\varepsilon}_{yy}}{\partial X^2} - 2\,\frac{\partial^2 \bar{\varepsilon}_{xy}}{\partial X \partial Y} \;=\; \left(\frac{\partial^2 w}{\partial X \partial Y}\right)^2 - \frac{\partial^2 w}{\partial X^2}\frac{\partial^2 w}{\partial Y^2} \quad .$$

Elimination of the mean strain components by means of *Hooke´s* law [see Eq. (4.22)], substitutes the membrane stresses instead. The latter are expressed by the *Airy* stress function, $n_x = \partial^2 F / \partial X^2$, ... , [see Eq. (2.11)] to render also the second *von Karman* equation in its final form

$$\nabla^2 \nabla^2 F = -E\,h \left[\frac{\partial^2 w}{\partial X^2} \frac{\partial^2 w}{\partial Y^2} - \left(\frac{\partial^2 w}{\partial X \partial Y} \right)^2 \right] \ .$$

(5.30b)

Note the symmetry of the coupling nonlinear partial differential operator in Eqs. (5.30a) and (5.30b). The linearized biharmonic equations are further considered in Eqs. (6.225) and (6.269).

The potential of the edge forces is not taken into account in Eq. (5.28). The variation of this additional potential energy renders the proper boundary conditions of *von Karman* plate theory. In Sec. 6.6, classical boundary conditions are discussed by illustrative examples and by considering the linearized, first-order theory of elasticity of thin *Kirchhoff* plates. These results are derived by a consideration of the equilibrium of the undeformed plate element; hence, the shear force of these examples must be supplemented by terms that reflect the contribution of the membrane forces to the lateral equilibrium condition of the deformed plate element of *von Karman* theory; see, eg Eq. (9.60) and Exercises A 9.6 and A 11.8 for applications.

5.4. Principle of Complementary Virtual Work

In contrast to the previous more "natural" formulation of the principle of virtual work in which the variation of the equilibrium configuration is virtually performed, the variation of the equilibrating stress state is considered below. Thus, $f = 0$ has a first variation $\delta f = 0$, such that the virtual stresses are also in equilibrium. During this variation, the body force k is changed accordingly by δk and the external surface tractions $\sigma_n\,dS$ are virtually varied by $\delta\sigma_n\,dS$. The complementary virtual work per unit of volume of δf is determined by consideration of the actual displacement field u (measured from the undeformed reference configuration to the state of the body at rest)

$$\left(u \cdot \delta f \right) = 0 \ .$$

(5.31)

Integration over the body volume in the deformed configuration renders the complementary virtual work

$$\delta W^* = \int_V \left(\mathbf{u} \cdot \delta \mathbf{f}\right) dV = 0 \ .$$

$$(5.32)$$

The substitution of the virtual force density $\delta \mathbf{f} = \delta \mathbf{k} + \partial(\delta \boldsymbol{\sigma}_x)/\partial x + \partial(\delta \boldsymbol{\sigma}_y)/\partial y + \partial(\delta \boldsymbol{\sigma}_z)/\partial z$ and application of the *Gauss* integral theorem to the volume integral yield (the symmetry of the stress tensor is understood)

$$\delta W^* = \int_V (\mathbf{u} \cdot \delta \mathbf{k}) \, dV + \int_V \left[\frac{\partial}{\partial x} (\mathbf{u} \cdot \delta \boldsymbol{\sigma}_x) + \frac{\partial}{\partial y} (\mathbf{u} \cdot \delta \boldsymbol{\sigma}_y) + \frac{\partial}{\partial z} (\mathbf{u} \cdot \delta \boldsymbol{\sigma}_z)\right] dV$$

$$- \int_V \left[\left(\frac{\partial \mathbf{u}}{\partial x} \cdot \delta \boldsymbol{\sigma}_x\right) + \left(\frac{\partial \mathbf{u}}{\partial y} \cdot \delta \boldsymbol{\sigma}_y\right) + \left(\frac{\partial \mathbf{u}}{\partial z} \cdot \delta \boldsymbol{\sigma}_z\right)\right] dV$$

$$(5.33)$$

$$= \int_V (\mathbf{u} \cdot \delta \mathbf{k}) \, dV + \oint_{\partial V} (\mathbf{u} \cdot \delta \boldsymbol{\sigma}_n) \, dS - \frac{1}{2} \sum_i \sum_j \int_V \delta \sigma_{ij} \left(\frac{\partial u_i}{\partial x_j} + \frac{\partial u_j}{\partial x_i}\right) dV = 0 \ .$$

This complementary principle that is also called the *principle of virtual forces* is especially valuable for practical applications when the group of states of equilibrium of the virtual stresses is extended to include noncompatible ones. For example, the body forces are virtually not varied, $\delta \mathbf{k} = 0$, in the body volume, and also the variations of the tractions $\delta \boldsymbol{\sigma}_n = 0$ on those parts of the body surface where dynamic boundary conditions (tractions) are prescribed. Virtual stresses are applied over those parts of the surface only where kinematic boundary conditions are given (eg where the displacements $\mathbf{u}$ are enforced). Hence, Eq. (5.33) reduces under these assumptions to

$$\frac{1}{2} \sum_i \sum_j \int_V \delta \sigma_{ij} \left(\frac{\partial u_i}{\partial x_j} + \frac{\partial u_j}{\partial x_i}\right) dV - \int_{\partial V_u} (\mathbf{u} \cdot \delta \boldsymbol{\sigma}_n) \, dS = 0 \ .$$

$$(5.34)$$

∂V_u is that portion of the body surface where the stresses are not given by the surface-tractions. The elasticity of the body is not necessarily assumed; the complementary principle is in general valid if unloading is avoided.

If linearized geometric relations hold approximately, the actual strains can be substituted

$$\varepsilon_{ij} = \frac{1}{2} \left(\frac{\partial u_i}{\partial x_j} + \frac{\partial u_j}{\partial x_i}\right) \ .$$

If we assume the existence of the complementary energy density $U^{*'}$ (with respect to the strain energy U'), with the property

$$\varepsilon_{ij} = \frac{\partial U^{*\prime}}{\partial \sigma_{ij}} \quad , \tag{5.35}$$

the complementary potential is determined by

$$\Pi^* = \int_V U^{*\prime} dV - \int_{\partial V_u} \mathbf{u} \cdot \sigma_n \, dS \quad , \tag{5.36}$$

and the principle of complementary energy takes on the form, originally found by *Engesser* ,

$$\delta \Pi^* = 0 \quad . \tag{5.37}$$

<<Among all possible stress-states in equilibrium that are compatible with dynamic boundary conditions, that which makes the complementary energy stationary (to a minimum) is selected.>> The complementary principle is not limited to *Hookean* solids.

In the case of an isotropic linear elastic body at constant temperature, *Hooke's* law applies, and the complementary energy becomes

$$U^{*\prime} = \frac{1}{2K} p^2 + \frac{1}{2G} J_2 \, , \quad J_2 = \frac{1}{2} \sum_i \sum_j \sigma_{ij}^{\prime 2} \quad , \tag{5.38}$$

see Eq. (3.47).

Under the condition of linearized geometric relations, complementary energy is given by the *Legendre* transformation of the strain energy

$$U^{*\prime} = \sum_i \sum_j \sigma_{ij} \varepsilon_{ij} - U^{\prime}(\varepsilon_{ij}) \quad . \tag{5.39}$$

This is the common transformation of changing thermodynamic variables. $U^{*\prime}$ can be derived by substituting *Hooke's* law and, thus, by eliminating the strain in the elastic potential U only for the special case of the isothermal linear elastic solid.

5.4.1. Castigliano's Theorem and Menabrea's Theorem

If the loading of the body is given by a system of single forces, F_1, F_2, ... , F_n, the complementary principle takes on the form of *Castigliano's* theorem, which is very suitable for engineering applications. Even in computational mechanics, such an analysis can be built in by means of the new possibilities of the symbolic manipulation programs. The complementary energy is to be expressed by a function of these self-equilibrating forces

$$U^* = \int_V U^{*\prime}\, dV = U^*(F_1, ..., F_n) \quad .$$

(5.40)

Hence, the virtual variation becomes the finite sum of the total differential

$$\delta U^* = \sum_{i=1}^{n} \frac{\partial U^*}{\partial F_i}\, \delta F_i \quad ,$$

(5.41)

and the principle, Eq. (5.37), requires

$$\sum_{i=1}^{n} \frac{\partial U^*}{\partial F_i}\, \delta F_i = \sum_{k=1}^{n} u_k \cdot \delta F_k \quad .$$

(5.42)

Since the virtual forces in equilibrium are independent, *Castigliano's* theorem is derived in the form

$$\frac{\partial U^*}{\partial F_i} = u_i \quad .$$

(5.43)

<<The derivative of the complementary energy with respect to a force that is a member of a system in equilibrium renders the displacement of its point of application in the direction of that force.>> [see Eq. (3.41)].

Such a force in the self-equilibrating system may be statically indeterminate, $F_i = X_i$, and hence, the displacement in the direction of such a force must vanish

$$\frac{\partial U^*}{\partial X_i} = 0 \quad .$$

(5.44)

Eq. (5.43), specialized in this form that is proper for the analysis of redundant systems, is named *Menabrea's* theorem. Before taking the partial derivative of the complementary energy, all the reaction forces must be expressed by the given loading and properly selected statically indeterminate forces by means of the equilibrium conditions. Note that all participating forces must be named individually, especially those that have the same value (and are parallel) but have different points of application.

Both theorems remain valid if the derivative is performed with respect to a couple with the moment vector M_i ,

$$\frac{\partial U^*}{\partial M_i} = \alpha_i \quad ,$$

(5.45)

where α_i is the angle of rotation about the axis given by the line of action of that moment vector M_i. The couple must be an element of the group of forces $F_1, ..., F_n$, which ought to be in equilibrium.

(§)   The Linear Elastic, Thin, and Straight Rod. This rod has a rather simple expression of complementary energy. Since $\sigma_{yy} = \sigma_{zz} = \sigma_{zy} = 0$, the energy density becomes [x points in the axial direction; see Eq. (2.130)]

$$U^{*'} = \frac{1}{2E} \left[\sigma_{xx}^2 + 2(1 + v)(\sigma_{xy}^2 + \sigma_{xz}^2) \right] .$$

(5.46)

Integration of the first term renders, with the volume element $dV = dA\, dx$, the complementary energy contributed by the normal stress

$$\frac{1}{2E} \int_0^l \int_A \sigma_{xx}^2\, dA\, dx = \frac{1}{2E} \int_0^l \int_A \left(\frac{N}{A} + \frac{M_y}{J_y} z - \frac{M_z}{J_z} y \right)^2 dA\, dx .$$

(5.47)

Lateral y and z axes pass through the centroid of the cross-section A and are chosen to be the principal axes of inertia; see Eq. (2.125). Since the resulting internal forces $N(x)$, $M_y(x)$, and $M_z(x)$ are constants in the cross-sectional integral, the final expression of this contribution becomes

$$\frac{1}{2} \int_0^l \left(\frac{N^2}{EA} + \frac{M_y^2}{EJ_y} + \frac{M_z^2}{EJ_z} \right) dx .$$

(5.48)

Equation (5.48) remains valid for the *slightly curved beam* when the arc-element ds is substituted for dx.

The shear stresses in the second term of Eq. (5.46) are related only to the shear forces in the case of a loading which is free of any torsion. The distribution within the cross-section is determined in Sec. 6.2.1. Here, it suffices to consider the relation

$$\int_A \left(\sigma_{xy}^2 + \sigma_{xz}^2 \right) dA = \left(\kappa_y Q_y^2 + \kappa_z Q_z^2 \right) / A ,$$

(5.49)

where κ_y and κ_z are coefficients reflecting the distribution of the shear stress in the cross-section A that depends on the specific shape. For a circular cross-section, the value is $\kappa = 10/9$. Hence, the contribution of the shear force to the complementary energy is

$$\frac{1}{2} \int_0^l \left(\frac{\kappa_z Q_z^2}{GA} + \frac{\kappa_y Q_y^2}{GA} \right) dx .$$

(5.50)

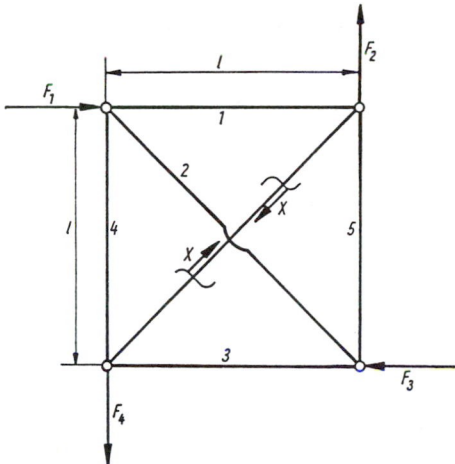

Fig. 5.7. Plane truss with single internal static indeterminacy

In the case of pure torsion, the sum of squares, $\sigma_{xy}^2 + \sigma_{xz}^2 = \tau^2$, equals the squared torsional shear stress with a distribution to be considered in Sec. 6.2.4. The field depends on the shape of the cross-section and enters the value of the torsional rigidity GJ_T . The contribution to the complementary energy can be expressed by the torque $M_x = M_T$

$$\frac{1}{2}\int_0^l \frac{M_T^2}{GJ_T} \, dx \quad .$$

(5.51)

The linear distribution of the torsional shear stress over the radius of a circular rod is derived in Eq. (4.40) where $J_T = J_p = 2J_x$, and the usefulness of Eq. (5.51) may be verified.

Subsequently, applications of the theorems of *Castigliano* and *Menabrea* to simple statically indeterminate structures are discussed, and the *dummy force* method is illustrated by considering the deflection of a linear elastic cantilever.

(§)   A Plane Truss With a Single Internal Static Indeterminacy (Fig. 5.7). The external loading is self-equilibrating if $F_1 = F_2 = F_3 = F_4 = F$. For simplicity, the stiffness of the members is assumed to be equal, $(EA)_1 = (EA)_2 = ... = (EA)_6 = EA$. The statically indeterminate force X in the diagonal member follows at once from *Menabrea's* theorem

$$\frac{\partial U^*}{\partial X} = 0 \quad , \quad U^* = \frac{1}{2} \sum_{i=1}^6 \frac{S_i^2}{EA} l_i \quad .$$

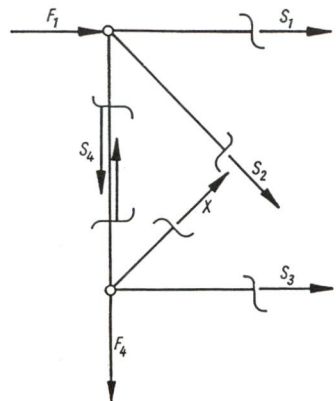

Fig. 5.8. Free-body diagrams of two isolated nodes

The member forces S_i are to be expressed by the loading and by the force X by means of the equilibrium conditions applied to each node; see Fig. 5.8 and consider the more general Fig. 2.50 in the plane,

$$S_1 = -X \frac{\sqrt{2}}{2} \;,\quad \frac{\partial S_1}{\partial X} = -\frac{\sqrt{2}}{2} \;,\quad S_2 = X - F\sqrt{2} \;,\quad \frac{\partial S_2}{\partial X} = 1 \;,$$

$$S_3 = -X \frac{\sqrt{2}}{2} \;,\quad \frac{\partial S_3}{\partial X} = -\frac{\sqrt{2}}{2} \;,\quad S_4 = F - X\frac{\sqrt{2}}{2} \;,\quad \frac{\partial S_4}{\partial X} = -\frac{\sqrt{2}}{2} \;,$$

$$S_5 = F - X\frac{\sqrt{2}}{2} \;,\quad \frac{\partial S_5}{\partial X} = -\frac{\sqrt{2}}{2} \;,\quad S_6 = X \;,\quad \frac{\partial S_6}{\partial X} = 1 \;.$$

Hence, $\dfrac{\partial U^*}{\partial X} = \dfrac{1}{EA}\displaystyle\sum_{i=1}^{6} S_i\, l_i \dfrac{\partial S_i}{\partial X} = 0$, yields, $X = F\dfrac{\sqrt{2}}{2}$.

(§) A Double-Span Beam Loaded According to Fig. 5.9. Two external edge moments are considered in addition to a spanwise uniform

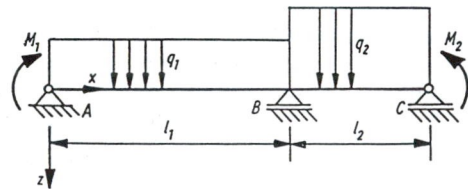

Fig. 5.9. A double-span beam with spanwise constant lateral loading and under the action of given edge moments

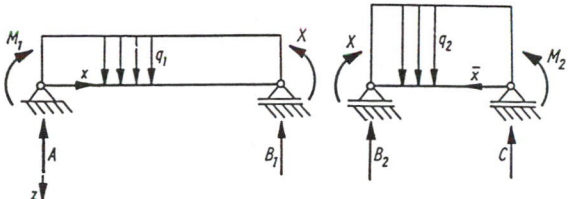

Fig. 5.10. A proper statically determinate system of the double-span beam of Fig. 5.9

transverse loading. The system has a single static indeterminacy and, as shown in Fig. 5.10, the bending moment at support B is selected as $X = M_y (x = l_1)$. The conditions of equilibrium are applied to the two statically determinate subsystems to render

$$A = \frac{1}{l_1} \left(X + \frac{q_1 l_1^2}{2} - M_1 \right) \quad , \quad M_y(x) = M_1 + Ax - \frac{q_1 x^2}{2} \quad , \quad 0 \le x < l_1 \ ,$$

$$C = \frac{1}{l_2} \left(X + \frac{q_2 l_2^2}{2} - M_2 \right) \quad , \quad M_y(\bar{x}) = M_2 + C\bar{x} - \frac{q_2 \bar{x}^2}{2} \quad , \quad 0 \le \bar{x} < l_2 \ .$$

The shear deformation is commonly neglected for slender beams and only the bending moment contributes to the complementary internal energy

$$U^* = \frac{1}{2} \int_0^{l_1 + l_2} \frac{M_y^2}{EJ_y} \, dx \ ,$$

which approaches *Menabrea's* theorem. Taking the derivative under the integral sign gives

$$\frac{\partial U^*}{\partial X} = \int_0^{l_1} \frac{1}{EJ_1} \left(M_1 + Ax - \frac{q_1 x^2}{2} \right) \frac{x}{l_1} \, dx + \int_0^{l_2} \frac{1}{EJ_2} \left(M_2 + C\bar{x} - \frac{q_2 \bar{x}^2}{2} \right) \frac{\bar{x}}{l_2} \, d\bar{x} = 0 \ .$$

The substitution of $A(X)$ and $C(X)$ (dependence on X has already been considered when taking the partial derivative) and integration yield the bending moment at B; bending stiffness is assumed to be constant spanwise

$$X = - \left[\frac{l_1}{EJ_1} \left(\frac{M_1}{2} + \frac{q_1 l_1^2}{8} \right) + \frac{l_2}{EJ_2} \left(\frac{M_2}{2} + \frac{q_2 l_2^2}{8} \right) \right] \bigg/ \left[\frac{l_1}{EJ_1} + \frac{l_2}{EJ_2} \right] \ .$$

(§) Deflection of a Uniformly Loaded Cantilever. The uniform loading of the linear elastic beam of constant bending rigidity $B = EJ$

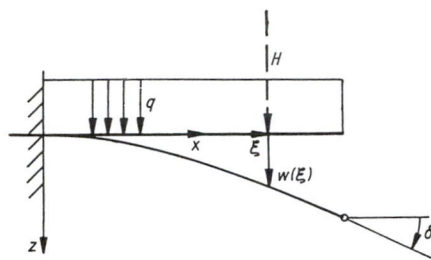

Fig. 5.11. Uniformly loaded cantilever. Dummy force H for the determination of the deflection $w(\xi)$

is considered in Fig. 5.11. The deflection of the point ξ of the axis is to be determined by the dummy force method and by means of *Castigliano's* theorem. The bending moment becomes with an additional single *dummy force* H at ξ in the region $\xi \leq x \leq l$, independent of H, $M_y(x) = -q(l-x)^2/2$, and in the interval $0 \leq x \leq \xi$, it is given by $M_y(x) = -q(l-x)^2/2 - H(\xi - x)$. If we neglect the shear deformation, the complementary internal energy becomes

$$U^*(H) = \frac{1}{2} \int_0^l \frac{M_y^2}{EJ_y} \, dx \quad .$$

Deflection in the lateral direction of the dummy force H is thus the partial derivative

$$w(\xi, H) = \frac{\partial U^*(H)}{\partial H} = \int_0^l \frac{M_y}{EJ_y} \frac{\partial M_y}{\partial H} \, dx \quad .$$

Taking the limit $H \to 0$ after differentiation renders the proper deflection

$$w(\xi) = \int_0^\xi \frac{q}{EJ} \frac{(l-x)^2}{2} (\xi - x) \, dx = \frac{q \, l^2}{8 \, B} \, 2\xi^2 \left(1 - \frac{2\xi}{3 \, l} + \frac{\xi^2}{6 \, l^2} \right) \quad .$$

By selecting a singular *dummy moment* H acting at ξ, instead of the aforementioned dummy force, the bending moment changes and the derivative of the proper complementary energy gives the slope at ξ

$$\lim_{H \to 0} \frac{\partial U^*(H)}{\partial H} = \frac{\partial w}{\partial x}\bigg|_{(x = \xi)} \quad .$$

The dummy couple H is assumed to act in a clockwise direction.

(§)   The Plane Snap Under the Action of Tip Forces. A shearing couple of forces F without moment causes bending and torsion in the circular snap shown in Fig. 5.12. The shear deformation is assumed negligible. *Castigliano's* theorem renders the tip opening δ. Bending moment $M = F R \sin \varphi$ and the torque $M_T = F R (1 - \cos \varphi)$ contribute to the internal complementary energy

$$U^* = \frac{1}{2} \int_0^{2R\pi} \frac{M^2}{EJ} \, ds + \frac{1}{2} \int_0^{2R\pi} \frac{M_T^2}{GJ_T} \, ds \ .$$

The displacement of the two points of application of the forces $F_1 = F_2 = F$ is the derivative

$$\delta = \frac{\partial U^*}{\partial F} = R \int_0^{2\pi} \frac{M}{EJ} \frac{\partial M}{\partial F} \, d\varphi + R \int_0^{2\pi} \frac{M_T}{GJ_T} \frac{\partial M_T}{\partial F} \, d\varphi$$

$$= F R \frac{A}{EJ} \left[1 + 6 (1 + v) \frac{J}{J_T} \right] \quad , \quad A = \pi R^2 \ .$$

For a snap with a circular cross-section of radius r , the torsional rigidity is proportional to the polar moment of inertia, $J_T = J_p$, and the ratio $J/J_p = 1/2$, where $J = \pi r^4/4$. The displacement of the tips is symmetrical to the mid-plane.

5.4.2. Betti's Method

Castigliano's theorem requires differentiation with respect to a

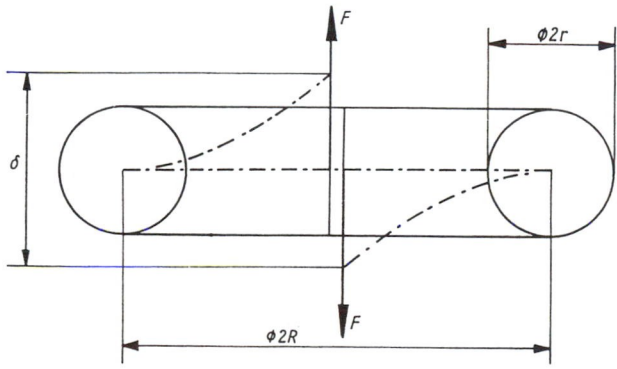

Fig. 5.12. Deformation of a plane snap by a couple without moment

(dummy) force or a statically redundant force of the analytically derived expression of the moment. This source of error and the consumption of manpower may be avoided by the direct application of the principle of virtual displacements with the choice of actual small linear elastic deformations for the virtual ones. The boundary conditions are already built in, but the geometric relations must be approximately linear. A deformation δ , a displacement or a rotation, at a selected material point is calculated by considering a proper generalized dummy load H , a force or a moment, and by expressing its virtual work $H\delta$, which in this case is the work of the external forces with respect to actual deformations. The dummy stresses due to the dummy loading H are denoted with an overbar in the expression of their virtual work by considering the actual (small) elastic strains. Note that the dummy stresses are constant during the virtual variation of the equilibrium configuration

$$-\sum_i \sum_j \int_V \bar{\sigma}_{ij}\, \varepsilon_{ij}\, dV \quad .$$

$$(5.52)$$

Hence, the principle of virtual work, Eq. (5.3), takes on the form of *Betti´s* method of the determination of δ

$$H\delta = \sum_i \sum_j \int_V \bar{\sigma}_{ij}\, \varepsilon_{ij}\, dV \quad .$$

$$(5.53)$$

(§) Thin-Walled Structures Free of Any Torsion. By considering such structures like beams and frames in tension, shear, and bending, and their internal resulting forces, integration over the cross-section can be performed to render Eq. (5.53) in a form that is convenient for solving practical problems

$$H\delta = \int_l \bar{N}\, \frac{N}{EA}\, dx + \int_l \left[\bar{M}_y\, \frac{M_y}{EJ_y} + \bar{M}_z\, \frac{M_z}{EJ_z} \right] dx$$

$$+ \int_l \left[\bar{Q}_y\, \frac{\kappa_y Q_y}{GA} + \bar{Q}_z\, \frac{\kappa_z Q_z}{GA} \right] dx \quad .$$

$$(5.54)$$

Putting the dummy loading at unity, $H = 1$, renders the deformation δ directly. Equation (5.54) holds also for slightly curved beams when $dx = ds$ is substituted.

(§) The Cantilever of Fig. 5.11. The rotation δ of the cross-section at the free end $x = l$ is to be calculated by selecting a proper clockwise dummy edge moment $H = 1$, and by considering the contribution of the bending curvature to Eq. (5.54) in *Betti´s* method only. Since the dummy bending moment is constant and equal to -1 , and the actual curvature due to the uniform loading is $M_y /EJ_y = -q\ (l$

$- x)^2/2EJ_y$, where the bending stiffness $B = EJ_y$ is assumed to be constant for convenience of integration,

$$1. \delta = \int_0^l \frac{q}{2EJ_y} (l-x)^2 \, dx = \frac{q \, l^3}{6EJ_y} = \frac{dw}{dx}\Big|_{(x=l)} \quad .$$

(§)   The Cantilever with an Additional Simple Support, Fig. 5.13 . *Betti's* method is applied to determine the redundant clamping moment $M_e = X$. The associated statically determinate system, thus, is the single-span hinged-hinged beam loaded uniformly and, in addition, by edge-moments. The rotation of the cross-section at $x = 0$ of the associated beam is determined for the actual uniform loading using a dummy moment $H = 1$ applied at the support A. Hence,

$$A_S = B_S = ql/2 \quad , \quad M(x)/EJ = (xA_S - qx^2/2)/EJ \quad ,$$

and

$$\overline{M} = 1. (l-x)/l \quad ,$$

when inserted into Eq. (5.54) (the subscript y is omitted for convenience), gives, $B = EJ = const$,

$$1. \delta_S = \int_0^l \frac{(l-x)}{l} \frac{\dfrac{ql}{2}x - \dfrac{qx^2}{2}}{EJ} \, dx = \frac{ql^3}{24 \, B} \quad .$$

Putting $q = 0$ and considering the rotation in the associated statically determinate system at $x = 0$ that is due to the dummy moment loading only renders

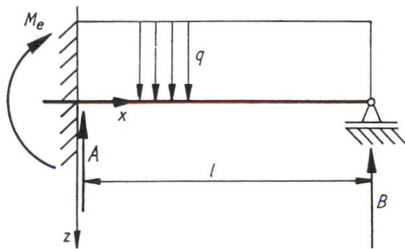

Fig. 5.13. Redundant beam, uniformly loaded. Clamping moment is considered statically indeterminate

$$\overline{M} = M(H = 1) \quad , \quad 1 . \delta_1 = \int_0^l \frac{(l - x)^2}{EJ \, l^2} \, dx = \frac{l}{3 B} \quad .$$

The kinematic condition of the rigidly clamped edge requires the resulting rotation to vanish

$$1 . \delta_S + X \, \delta_1 = 0 \quad , \quad X = M_e = - \delta_S / \delta_1 = - q \, l^2/8 \quad .$$

The reaction forces of the redundant beam are now easily calculated from conditions of equilibrium, eg $B \, l - q \, l^2/2 - M_e = 0$ renders

$$B = (q \, l^2/2 - q \, l^2/8)/l = 3 \, ql/8 \quad , \quad \text{and,} \quad A = B - ql = 5 \, ql/8 \quad .$$

The actual deformations δ of the redundant system can be determined directly from Eq. (5.54). The internal dummy forces due to the applied dummy force H may be efficiently determined in the associated statically determinate system, since the actual deformations of the redundant system are admissible virtual displacements of the statically determinate system with the external loading given. This is an outcome of the *reduction theorem* of the first-order theory of the elasticity of redundant structures.

5.4.3. Transformation of the Principles of Minimum Potential and Complementary Energy

The transformation is illustrated by considering the *Bernoulli-Euler* beam of Sec. 5.3.1. The linearized curvature $\kappa = w''$ in the potential energy of Eq. (5.19) is subsequently considered to be an independent function and the auxiliary conditions are multiplied by *Lagrangean factors* (generalized forces) and added thus

$$(\kappa - w'') \, M(x) = 0 \quad ,$$

and if kinematic boundary conditions apply, M_0 , M_1 , Q_0 , Q_1 are the constant *Lagrangean* factors of the associated terms, more auxiliary conditions are to be taken into account

$$M_0 \left(\frac{dw}{dx} \right)_{x = 0} = 0 \quad , \quad M_1 \left(\frac{dw}{dx} \right)_{x = l} = 0 \quad , \quad Q_0 \, w_0 = 0 \quad , \quad Q_1 \, w_1 = 0 \quad .$$

The resulting potential is unchanged when adding zeros and is expressed by

$$\Pi = \int_0^l \frac{EJ_y}{2} \kappa^2 \, dx - \int_0^l q(x) \, w(x) \, dx + \int_0^l (\kappa - w'') \, M_y(x) \, dx$$

$$- M_0 \left(\frac{dw}{dx}\right)_{x=0} + M_1 \left(\frac{dw}{dx}\right)_{x=1} + Q_0 w_0 - Q_1 w_1 \quad .$$

The functions and constants $\kappa, w, M, M_0, M_1, Q_0, Q_1$ may be varied independently and the first-order variation of the potential functional becomes accordingly

$$\delta\Pi = \int_0^1 \left[(M_y + EJ_y \,\kappa)\,\delta\kappa - (M_y'' + q)\,\delta w + (\kappa - w'')\,\delta M_y\right] dx$$

$$+ \left(Q_0 - M_y'(0)\right)\delta w_0 + \left(M_y(0) - M_0\right)\delta\left(\frac{dw}{dx}\right)_{x=0} - \left(Q_1 - M_y'(1)\right)\delta w_1$$

$$- \left(M_y(1) - M_1\right)\delta\left(\frac{dw}{dx}\right)_{x=1} + w_0\,\delta Q_0 - \left(\frac{dw}{dx}\right)_{x=0}\delta M_0$$

$$- w_1\,\delta Q_1 + \left(\frac{dw}{dx}\right)_{x=1}\delta M_1 = 0 \quad .$$

The coefficients of the independent variations must vanish individually to render the moment curvature relation, $\kappa = w'' = -(M_y /EJ_y)$, and the local equilibrium condition, $d^2 M_y /dx^2 = - q$; the *Lagrangean* coefficients are $Q_0 = M_y'(0)$, $M_0 = M_y(0)$, $Q_1 = M_y'(1)$, $M_1 = M_y(1)$.

Elimination of the curvature κ by the substitution of $-M_y/EJ_y$ and of w by the partial integration of the proper term in Π finally gives the result of the transformation, the complementary energy of the *Bernoulli-Euler* beam (the shear deformations remain neglected)

$$\Pi^* = \frac{1}{2}\int_0^1 \frac{M_y^{\,2}}{EJ_y}\, dx + w_0 Q_0 - \left(\frac{dw}{dx}\right)_{x=0} M_0 - w_1 Q_1 + \left(\frac{dw}{dx}\right)_{x=1} M_1 \quad . \tag{5.55}$$

In the complementary principle, the bending moment M_y and, in the case of kinematic boundary conditions, the proper forces Q_0, Q_1, M_0, M_1 are to be varied virtually, see, eg
K. Washizu: *Variational Methods in Elasticity and Plasticity.* Pergamon Press, Oxford, 1981, 3rd ed.

5.5. Exercises A 5.1 to A 5.4 and Solutions

A 5.1: A drawing table must be supported by a four-bar mechanism in such a way that its weight G is balanced by the counterweight Q in all practically required configurations; see Fig. A 5.1. The weight of the other rigid members should be negligible.

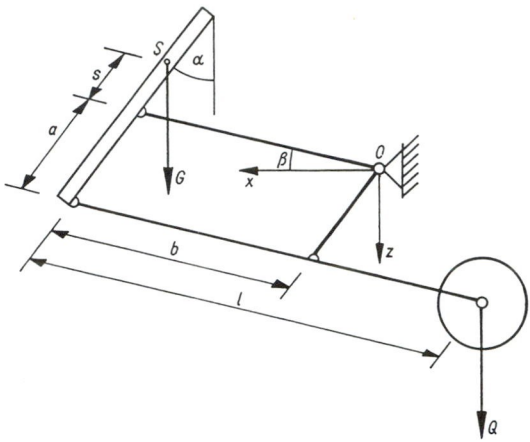

Fig. A 5.1.

Solution: The mechanism has two degrees of freedom; α and β are the independent configurational coordinates. The vertical coordinates of the mass centers are $z_S = - (b \sin \beta + s \cos \alpha)$, $z_Q = a \cos \alpha + (l - b) \sin \beta$, and the virtual variations are the total differentials $\delta z_S = - (b \cos \beta \, \delta\beta - s \sin \alpha \, \delta\alpha)$, $\delta z_Q = - a \sin \alpha \, \delta\alpha + (l - b) \cos \beta \, \delta\beta$. The principle of virtual displacements requires the virtual work to vanish: $\delta W = G \, \delta z_S + Q \, \delta z_Q = 0$. Hence, $[Gs - Qa] \sin \alpha \, \delta\alpha + [- Gb + Q(l - b)] \cos \beta \, \delta\beta = 0$. Since the virtual rotations are independent, two conditions of equilibrium are derived

$$\frac{G}{Q} = \frac{a}{s} \; , \quad \frac{G}{Q} = \left(\frac{l}{b} - 1\right) \; .$$

For a suitable design, both conditions are satisfied, and the table is at rest in all positions due to the independence of the condition from α and β.

A 5.2: A system of 15 heavy bars of equal length l and weight mg, simply connected according to Fig. A 5.2, is loaded by the horizontal forces F_1, F_2, and F_3. Determine the (deformed) equilibrium configuration by the application of the principle of virtual displacements.

Solution: All forces have potential, and the principle of virtual work takes on the form of the principle of minimum potential energy. According to Fig. A 5.2, the system has three degrees of freedom, and, the horizontal coordinates are $x_1 = 2l + l \sin \varphi_1$, $x_2 = x_1 + l (\sin \varphi_2 + \sin \varphi_3)$, $x_3 = l (\sin \varphi_1 + \sin \varphi_2)$. Similarly, the vertical

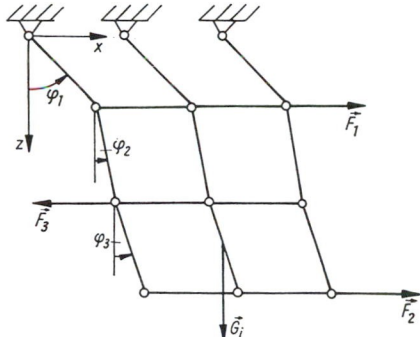

Fig. A 5.2.

distances of the mass centers are expressed by the independent configurational coordinates to render the potential energy:

$$W_p = W_G + W_F = -\, mgl\, [\, \tfrac{3}{2}\, \cos \varphi_1 + 2\, \cos \varphi_1 + 3 \left(\cos \varphi_1 + \tfrac{1}{2}\, \cos \varphi_2\right)$$

$$+\, 2\, (\cos \varphi_1 +\, \cos \varphi_2) + 3 \left(\cos \varphi_1 +\, \cos \varphi_2 + \tfrac{1}{2}\, \cos \varphi_3\right)$$

$$+\, 2\, (\cos \varphi_1 + \cos \varphi_2 + \cos \varphi_3)\,] - F_1\, x_1 - F_2\, x_2 - (-\, F_3)\, x_3\ .$$

Virtual rotations yield the first-order variation of the potential energy that is supposed to vanish

$$\delta W_p = \delta W_G + \delta W_F = -\, [(F_1 + F_2 - F_3)\, \cos \varphi_1 - 27\, G\, \sin \varphi_1\,]\, \delta \varphi_1$$

$$-\, [(F_2 - F_3)\, \cos \varphi_2 - 17\, G\, \sin \varphi_2\,]\, \delta \varphi_2$$

$$-\, [F_2\, \cos \varphi_3 - 7\, G\, \sin \varphi_3\,]\, \delta \varphi_3 = 0\quad ,\quad G = mg/2\ .$$

The coefficients of the angular variations must vanish individually to render the configuration of the system at rest

$$\varphi_1 = \tan^{-1} \frac{F_1 + F_2 - F_3}{27\, G}\quad ,$$

$$\varphi_2 = \tan^{-1} \frac{F_2 - F_3}{17\, G}\quad ,\quad \varphi_3 = \tan^{-1} \frac{F_2}{7\, G}\ .$$

A 5.3: Determine separately the reaction force at support B of the statically determinate triple-span beam loaded according to Fig. A 5.3 by means of a specially selected free-body diagram.

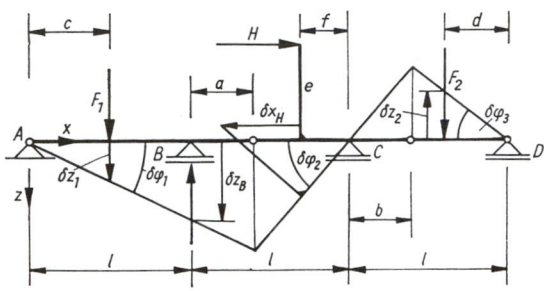

Fig. A 5.3.

Solution: Replacing the support by a vertical force B renders a SDOF-mechanism with the possibility of rigid-body virtual displacements. The virtual displacements are expressed by the virtual rotation $\delta\varphi_1$: $\delta z_1 = c\,\delta\varphi_1$, $\delta z_B = l\,\delta\varphi_1$, $\delta z_2 = -d\,[b/(l-b)][(l+a)/(l-a)]\,\delta\varphi_1$, $\delta x_H = -e[(l+a)/(l-a)]\,\delta\varphi_1$. The virtual work must vanish, $\delta W = F_1\,\delta z_1 + H\,\delta x_H + F_2\,\delta z_2 - B\,\delta z_B = 0$, and the reaction force follows

$$B = F_1 \frac{c}{l} - H \frac{e}{l}\frac{l+a}{l-a} - F_2\frac{d}{l}\frac{b}{l-b}\frac{l+a}{l-a} \; .$$

A 5.4: The member force S_1 of the statically determinate truss of Fig. 2.52 is to be determined by means of the principle of virtual displacements.

Solution: The virtual displacement field is properly selected if the given loads and the unknown double force without moment S_1 enter the principle. The reaction forces A and B should not contribute to the virtual work. Thus, removing the member in the lower girder makes it possible to rotate virtually the rigid subsystem ACD and the virtual work becomes,

$$\delta W = F_1\, a\delta\alpha + F_2\, a\delta\alpha + F_3\, 2a\delta\alpha - S_1\, a\delta\alpha - S_1\, a2\delta\alpha = 0.$$

Since $\delta\alpha \neq 0$, $S_1 = F_1 + F_2/3$.

6
Selected Topics
of Elastostatics

Only linearized problems are considered in this section, ie linearized geometric relations and *Hooke's* law are taken into account [see Eqs. (1.21) and (4.15)]. The consequences of these assumptions are illustrated by considering a linear elastic body loaded on its surface by a self-equilibrating system of single forces F_1, F_2, ..., F_n (reaction forces in the point supports are included). The loading is applied to the body by means of a common load factor λ , which is slowly increased from $0 \leq \lambda \leq 1$ to reach the terminal configuration by passing successively through states of equilibrium. The displacement u_i of a material point to the deformed state, the reference configuration is assumed free of stresses, is determined by the generalization of the linear law of a *Hookean* spring [see Eqs. (4.3) and (3.46)], and the common load factor λ cancels

$$u_i = \sum_{j=1}^{n} a_{ij} F_j \quad ,$$

$$(6.1)$$

where the influence coefficients a_{ij} are independent of the load intensity. They depend on the location and direction of the displacement and they are functions of the points of application of the forces F_j , as well as of their directions.

Equation (6.1) expresses the superposition principle of linear systems, and, thus, it is not restricted by the assumption of the common load factor: It is independent of the order of loading by the forces F_j . This is shown subsequently, without loss of generality, by considering two forces only and the displacement u_3 in a point number 3. The action of F_1 alone gives, say, $u_3 = a_{31} F_1$. Not changing the support conditions renders for a single force F_2 alone $u_3 = a_{32} F_2$. Increasing the combined loading $F_1 + F_2$ by a common load factor gives, say, $u_3 = a_{31}' F_1 + a_{32}' F_2$. Unloading at first by $F_1 \rightarrow 0$ changes that displacement to $u_3 = a_{31}' F_1 + a_{32}' F_2 - a_{31}'' F_1$. The body is still loaded by F_2 . Further unloading yields the undeformed reference configuration, and, hence,

$$u_3 = a_{31}' F_1 + a_{32}' F_2 - a_{31}'' F_1 - a_{32} F_2 = 0 \rightarrow$$

$$u_3 = (a_{31}' - a_{31}'') F_1 + (a_{32}' - a_{32}) F_2 = 0 \; .$$

The loadings F_1 and F_2 are assumed to be independent and, thus, $a_{32}' = a_{32}$ and furthermore, $a_{31}' = a_{31}'' = a_{31}$. <<The *influence coefficients* a_{ij} determine the superposition of displacements independently of the sequence of the loading.>>

The work of the external forces F_i is calculated under the assumption of a common load factor λ , and u_i is consequently the displacement component of the point of application of the force F_i measured in the force direction. With the increment $du_i = u_i \, d\lambda$,

$$\sum_{i=1}^{n} \int_0^1 \lambda F_i \, u_i \; d\lambda = \frac{1}{2} \sum_{i=1}^{n} F_i \, u_i = \frac{1}{2} \sum_{i=1}^{n} \sum_{j=1}^{n} a_{ij} \, F_j \, F_i = U^* \; , \tag{6.2}$$

the work is seen to be independent of the sequence of the loading and, hence, also any common load factor λ , and it equals the complementary energy [see Eq. (3.40)].

Maxwell's reciprocal theorem is illustrated by Eq. (6.2) since the influence coefficients of corresponding displacements and forces are symmetric, $a_{ij} = a_{ji}$. <<The displacement at a point numbered i due to a unit load at another point j is equal to the displacement at j due to a unit load at i, provided that the displacements of the points of application are measured in the directions of the applied forces, respectively.>> Symmetry is shown for two forces. The sequence of loadings is alternated and must render the same work

$$\frac{1}{2} F_1 u_1 + \frac{1}{2} F_2 u_2 = \frac{1}{2} a_{11} F_1^2 + a_{12} F_1 F_2 + \frac{1}{2} a_{22} F_2^2$$

$$= \frac{1}{2} a_{22} F_2^2 + a_{21} F_2 F_1 + \frac{1}{2} a_{11} F_1^2 \qquad . \tag{6.3}$$

Hence, $a_{12} = a_{21}$, ie the displacement component of the point of application of F_1 in the direction of F_1 due to the unit loading $F_2 = 1$ equals the component of the displacement of the point of application of the force F_2 measured in the direction of F_2 that is caused by the unit load $F_1 = 1$. A generalization is sometimes more convenient in applications and is stated below.

The *Betti-Rayleigh reciprocal theorem* considers two systems of forces having the same points of application at the linear elastic body and acting along the same group of (common) lines of action, and their fields of deformation given by the corresponding displacements of the points of application in the directions of the applied forces. The first system is given by F_1, F_2, ..., F_n with associated displacements u_1, u_2, ..., u_n and the second, primed system is denoted F_1', F_2', ..., F_n' with corresponding displacements u_1', u_2', ..., u_n'. Due to the symmetry of the influence coefficients, the work by the first system of forces on the primed displacements equals the work of the second system of forces on the unprimed displacements. The mathematical expression of the theorem is

$$\sum_{i=1}^{n} F_i u_i' = \sum_{k=1}^{n} F_k' u_k \quad .$$

(6.4)

In structural applications also the moments and corresponding angles of rotation of the plane of action of the couples are considered along with their contribution to the work (6.4); see Eq. (3.7).

The influence coefficients a_{ij} are the elements of the symmetric *flexibility matrix* . The matrix when inverted renders the symmetric *stiffness matrix* with elements k_{ij} ; see Eq. (3.44). Thus, the work of the external forces determines the complementary internal energy [see Eq. (6.2)] as well as the strain energy,

$$\frac{1}{2} \sum_{i=1}^{n} F_i u_i = U^* = \frac{1}{2} \sum_{i=1}^{n} \sum_{j=1}^{n} k_{ij} u_i u_j = U \quad ,$$

(6.5)

isothermal deformations are understood; see also Eq. (3.45).

6.1. Continuum Theory of Linearized Elastostatics

The symmetric stress tensor [see Eq. (2.24)] enters the local conditions of equilibrium at every material point, $f = 0$ [see Eq. (2.19)], and if we consider an isotropic and homogeneous linear elastic body, *Hooke's* law provides six linear stress-strain relations [see Eq. (4.15)]. Assuming the strains are sufficiently small, we see that the six linearized geometric relations of Eq. (1.21) apply as well, and all total the system of 15 basic linear equations of isothermal and linearized elastostatics are derived. The 15 unknowns are the field variables, the components of the displacement vector, and the elements of the stress and strain tensor. Elimination of the stresses and strains renders three coupled linear partial differential equations for the (small) displacement components, the celebrated *Navier's* equations,

$$\nabla^2 u_i + \frac{1}{1 - 2\nu} \frac{\partial e}{\partial x_i} = -\frac{k_i}{G} \quad , \quad e = \sum_{j=1}^{3} \frac{\partial u_j}{\partial x_j} \quad , \quad i = 1, 2, 3 \quad ,$$

(6.6)

$$\nabla^2 = \sum_{i=1}^{3} \frac{\partial^2}{\partial x_i^2} \quad , \quad \text{is the } Laplace \text{ operator .}$$

The equations are homogeneous in the absence of body forces, and when summed, they render the potential equation for the dilatation e, under isothermal conditions,

$$\nabla^2 e = 0 \quad . \tag{6.7}$$

By elimination of the displacements and strains in the basic system, six coupled partial differential equations are derived for the stress components. The *Beltrami-Michell* stress equations are also of second order

$$\nabla^2 \sigma_{ij} + \frac{3}{1+v} \frac{\partial^2 p}{\partial x_i \, \partial x_j} = -\left(\frac{\partial k_i}{\partial x_j} + \frac{\partial k_j}{\partial x_i}\right) - \frac{v}{1-v} \, \text{div } \mathbf{k} \, \delta_{ij} \, , \; p = \frac{1}{3} \sum_{k=1}^{3} \sigma_{kk} .$$
$$i, j = 1, 2, 3 \quad . \tag{6.8}$$

Considering a constant body force, $\mathbf{k} = const$, and putting $i = j$ and summing the Eqs. (6.8) give the potential equation for the mean normal stress

$$\nabla^2 p = 0 \quad . \tag{6.9a}$$

Both p and the dilatation e are solutions of the *Laplace* equation and, hence, harmonic functions of the coordinates. Equations (2.19), which are of first order, are to be taken into account as the auxiliary conditions of the stress equations (6.8). Furthermore, the homogeneous Eqs. (6.8) become, by application of the *Laplace* operator and by substitution of Eq. (6.9a), biharmonic equations

$$\nabla^2 \nabla^2 \sigma_{ij} = 0 \quad . \tag{6.9b}$$

In the mathematical theory of linear elasticity, analytic solution strategies are developed commonly by considering *Navier´s* equations: The method of complex functions by *E. Goursat* and *N. I. Muskhelishvili* in combination with conformal mapping techniques is a powerful tool for solving plane problems. See
N. I. Muskhelishvili: Some Basic Problems of the Theory of Elasticity. (transl. of the 3rd Russian ed. by J. R. M. Radok), Noordhoff, Groningen, 1953.
Methods relying on real functions are due to *Neuber-Papkovich* and *Westergaard-Galerkin* . Exact solutions are available for simple geometries and boundary conditions. Further details may be found in the vast literature, see, eg

A. E. H. Love: A Treatise on the Mathematical Theory of Elasticity. University Press, Cambridge, 1892, 1st ed. reprinted Dover Publ. New York, 1963 . Also,
R. Wm. Little: Elasticity. Prentice Hall, Englewood Cliffs, N.J.,1973.
V. D. Kupradze (ed.): Three-Dimensional Problems of the Mathematical Theory of Elasticity and Thermoelasticity. North-Holland, Amsterdam, 1979. H. Parkus: Thermoelasticity. 2nd ed., Springer-Verlag, New York, 1976.

(§)   The One-Dimensional Problems of Linear Elasticity. Solutions to that class of problems are considered subsequently. Depending on a parameter α , a unique solution is given for the simple case of uniaxial deformation, eg in the direction x (the uniaxial strain state is given by $\varepsilon_{xx} = \varepsilon_1 \neq 0$; all other strain components vanish), for the axisymmetric case of the radial deformation of a cylinder (under plane strain conditions), and for the point symmetric deformation of a sphere. The common differential equation for the nonvanishing displacement component u results by proper reduction of Eq. (6.6),

$$u'' + \alpha \left(\frac{u}{r}\right)' = 0 \ , \tag{6.10}$$

with the assumptions of homogeneity and vanishing body force. The parameter $\alpha = 0$ and $u' = \partial u/\partial x$ applies for the uniaxial deformation, and $\alpha = 1$, $\alpha = 2$ for the cylindrical and the spherical problem, where r in the latter cases is the radial coordinate and $u' = \partial u/\partial r$. The general solution is the superposition of two basic solutions

$$u(r) = A\,r + B\left(\frac{r_0}{r}\right)^{\alpha} \ . \tag{6.11}$$

The constants A and B must be determined according to two prescribed boundary conditions. The relevant normal stress is given by Hooke's   law, (4.15) or (4.18),

$$\sigma_{rr}(r) = \frac{2G}{1 - 2v}\left[(1 - v)\,u' + \alpha v\,\frac{u}{r}\right] = 2G\left[\frac{1 + (\alpha - 1)v}{1 - 2v}\,A - \alpha\,B\left(\frac{r_0}{r}\right)^{\alpha}\frac{1}{r}\right]. \tag{6.12}$$

The general solution is illustrated by considering a thick-walled cylinder and hollow sphere; the radii are denoted $R_{i,\,e}$, under the action of an inner and outer fluid pressure loading, $p_{i,\,e}$. The boundary conditions at the surfaces loaded by normal tractions render with Eq. (6.12)

$$\frac{p_{i,\,e}}{2G} = -\left[\frac{1 + (\alpha - 1)v}{1 - 2v}\,A - \alpha\,B\left(\frac{r_0}{R_{i,\,e}}\right)^{\alpha}\frac{1}{R_{i,\,e}}\right] \ . \tag{6.13}$$

Solving for A and B gives the radial stress component, $\rho = R_e / R_i$,

$$\sigma_{rr}(r) = -p_i \frac{(R_e/r)^{\alpha+1} - 1}{\rho^{\alpha+1} - 1} - p_e \frac{1 - (R_i/r)^{\alpha+1}}{1 - (1/\rho)^{\alpha+1}} \quad , \tag{6.14}$$

and the radial displacement, also superposed with respect to the inner and outer loadings,

$$2G\,\alpha\,\frac{u(r)}{r} = p_i \left[\alpha \frac{1 - 2v}{1 + (\alpha - 1)v} + (R_e/r)^{\alpha+1}\right] / \left[\rho^{\alpha+1} - 1\right]$$

$$- p_e \left[\alpha \frac{1 - 2v}{1 + (\alpha - 1)v} + (R_i/r)^{\alpha+1}\right] / \left[1 - (1/\rho)^{\alpha+1}\right] \quad . \tag{6.15}$$

The hoop stress is given by *Hooke's* law, Eq. (4.15),

$$\sigma_{\vartheta\vartheta} = \frac{2G\,v}{1 - 2v} \{u' + [1 + (\alpha - 2)]\frac{u}{r}\} = 2G \left[\frac{1 + (\alpha - 1)v}{1 - 2v} A - B (r_0/r)^{\alpha}(1/r)\right] , \tag{6.16}$$

and by substituting A and B,

$$\alpha\,\sigma_{\vartheta\vartheta} = p_i \frac{[\alpha + (R_e/r)^{\alpha+1}]}{[\rho^{\alpha+1} - 1]} - p_e \frac{[\alpha + (R_i/r)^{\alpha+1}]}{[1 - (1/\rho)^{\alpha+1}]} \quad . \tag{6.17}$$

In the sphere, the third principal normal stress equals the hoop stress when $\alpha = 2$ in Eq. (6.17). The parameter $\alpha = 1$ for the cylinder and the axial stress under plane strain condition becomes

$$\sigma_{zz} = v (\sigma_{rr} + \sigma_{\vartheta\vartheta}) \quad .$$

Note that the mean stress is independent of the radial coordinate r, since

$$\sigma_{rr} + \alpha\,\sigma_{\vartheta\vartheta} = \frac{(1 + \alpha)\,p_i}{\rho^{\alpha+1} - 1} - \frac{(1 + \alpha)\,p_e}{1 - (1/\rho)^{\alpha+1}} \quad . \tag{6.18}$$

(§)   Shrink Fit. The formulas also are still useful for calculating the elastic stresses in the case of technical importance. The elastic body in the interior is oversized in the stress-free reference configuration by $2R_i + h$. After shrinking, the bodies in contact take on the common constant reference temperature, and the fully bodied inclusion is in the state of constant radial stress $\sigma_{rr} = -p_i$. Due to the common interface, the kinematic condition applies

$$R_i + u(R_i) = R_i + \frac{h}{2} + u(R_i^{(-0)}) \quad ,$$

and $u(R_i^{(-0)})$ is the radial displacement of the inclusion surface. The case of an elastic shrink fit of two hollow bodies is somewhat more complicated and is not considered here. The solution becomes quite simple if the limit $R_e \to \infty$ applies, approximately for an extremely thick-walled cylinder, and the interface pressure in that case is given by

$$p_i = \frac{E\,h}{4\,(1-v^2)\,R_i} \; .$$

(6.19)

The shaft of such a simple shrink configuration is in a state of constant normal stresses

$$\sigma_{rr} = \sigma_{\vartheta\vartheta} = -p_i \quad , \quad \sigma_{zz} = -2v\,p_i \quad .$$

(6.20)

6.1.1. Thermoelastic Deformations

A homogeneous and isotropic linear elastic body has a constitutive relation given by *Hooke's* law [Eq. (4.15)]. A change of temperature with respect to a constant reference temperature, $\theta(x, y, z)$ denotes such a temperature field measured, eg in *centigrade* (°)(degrees *Celsius*), and that produces thermal strains, ie strains which are generally not compatible. They do not obey the compatibility conditions (1.22) and, thus, are sources of thermal stresses in the *Hookean* solid.

The unit of the *Celsius* temperature scale is equal to the unit of the *thermodynamic temperature*, named after *Kelvin* , $1°C = 1K$. It is suggested that the unit kelvin be used also to indicate a difference of temperature. It is defined by the fraction (1/273.16) of the thermodynamic (absolute) temperature of the triple point of water. The *Celsius* temperature is referred to as $T_0 = 273.16\ K$, the freezing point of water. The difference θ is to be measured with respect to the reference temperature of the body in the stress-free state.

The thermal strains are assumed to be normal strains that are linearly and isotropically related (proportional) to the temperature

$$\bar{\varepsilon}_{ii} = \alpha\,\theta \quad , \quad i = x, y, z \quad ,$$

(6.21)

where α is the *linear thermal expansion coefficient*, with a dimension of $[\alpha] = 1/K$. The superposition of the strains renders the generalized *Hooke's* law of thermoelasticity; compare this with Eqs. (4.12) and (4.14) for the dilatation

$$e = \frac{p}{K} + 3\alpha\theta \quad , \quad K = E/3(1 - 2\nu) \quad ,$$

(6.22)

3α is the isotropic *cubic expansion coefficient* . The linear relation for the deviatoric components remains unchanged, and the dependence of the elastic moduli on temperature is weak and is commonly neglected within the range of validity of linear elasticity

$$\varepsilon_{ij}' = \frac{1}{2G} \sigma_{ij}' \quad .$$

(6.23)

The elastic potential energy density of the isothermal state, Eq. (3.47), is increased by the thermal expansion of volume, and, hence, formally given by the same expression (any arbitrary function of temperature alone may be added)

$$U' = \frac{K}{2} e^2 + 2G J_2 \quad , \quad J_2 = \frac{1}{2}\sum_i \sum_j \varepsilon_{ij}'^2 \quad ,$$

(6.24)

when the total dilatation e of Eq. (6.22) is substituted.
The generalized complementary energy density, however, with the definition

$$\varepsilon_{ij} = \frac{\partial U^{*'}}{\partial \sigma_{ij}} \quad ,$$

(6.25)

becomes nontrivially

$$U^{*'} = \frac{1}{2K} p^2 + \frac{1}{4G}\sum_i \sum_j \sigma_{ij}'^2 + 3\alpha\theta\, p \quad .$$

(6.26)

It is derived by the *Legendre* transformation of the generalized strain energy U'

$$U^{*'}(\sigma_{ij}, \theta) = \sum_i \sum_j \sigma_{ij}\, \varepsilon_{ij} - U'(\varepsilon_{ij}, \theta) \quad ,$$

(6.27)

where $\varepsilon_{ij} = e\,\delta_{ij} + \varepsilon_{ij}'$ is eliminated by the substitution of the generalized *Hooke's* law, Eqs. (6.22) and (6.23).
The proof follows the common lines of thermodynamics: A small change of state is considered by taking the total differential of the complementary energy density on the left, and right-hand side of Eq. (6.27), respectively,

$$\sum_i \sum_j \frac{\partial U^{*'}(\sigma_{ij}, \theta)}{\partial \sigma_{ij}}\, d\sigma_{ij} + \frac{\partial U^{*'}(\sigma_{ij}, \theta)}{\partial \theta}\, d\theta = \sum_i \sum_j (\varepsilon_{ij}\, d\sigma_{ij} + \sigma_{ij}\, d\varepsilon_{ij})$$

$$-\sum_i \sum_j \frac{\partial U'(\varepsilon_{ij}, \theta)}{\partial \varepsilon_{ij}} \, d\varepsilon_{ij} + \frac{\partial U'(\varepsilon_{ij}, \theta)}{\partial \theta} \, d\theta \quad .$$

Comparing coefficients yields the relations that correspond to *Hooke's* law in its generalized form

$$\frac{\partial U^{*'}(\sigma_{ij}, \theta)}{\partial \theta} = -\frac{\partial U'(\varepsilon_{ij}, \theta)}{\partial \theta} \quad , \quad \sigma_{ij} = \frac{\partial U'(\varepsilon_{ij}, \theta)}{\partial \varepsilon_{ij}} \quad , \quad \varepsilon_{ij} = \frac{\partial U^{*'}(\sigma_{ij}, \theta)}{\partial \sigma_{ij}} \quad .$$

(6.28)

The partial derivative of Eq. (6.24) with respect to a strain component renders the generalized *Hooke's* law in its inverse form

$$\sigma_{ij} = 2G \left(\varepsilon_{ij} + \frac{v}{1 - 2v} \, e \, \delta_{ij} - \frac{1 + v}{1 - 2v} \, \alpha\theta \, \delta_{ij} \right) \quad .$$

(6.29)

(§)   The Complementary Energy of a Thermally Loaded Rod. A linear elastic rod with a straight axis in the reference configuration is considered. The torsional deformation is not coupled to the thermal loading in the linearized, first-order theory of elasticity. Neglecting the shear deformation gives the complementary internal energy of bending and stretching in the form of the volume integral of its density,

$$U^* = \int_0^l dx \int_A \left(\frac{1}{2E} \, \sigma_{xx}^2 + \alpha\theta \, \sigma_{xx} \right) dA \quad .$$

(6.30)

According to the *Bernoulli-Euler* hypothesis, the strains are linearly distributed in every cross-section. The latter is assumed to remain plane after loading and it rotates through the angle $\chi(x) = (-dw/dx)$ about the principal central y axis during the deflection of the beam. If we denote the strain of the beam axis at $z = 0$ by $\varepsilon_{xx}^{(0)}$, the uniaxial generalized *Hooke's* law becomes

$$\sigma_{xx}(x, z) = E \left(\varepsilon_{xx}^{(0)} + z \frac{d\chi}{dx} - \alpha\theta \right) \quad ,$$

(6.31)

where the thermal stress may become a nonlinear function of z in the case of a nonlinearly distributed temperature even in the linearized theory employed here. Integration over the cross-sectional area can be performed and yields the normal force

$$N(x) = \int_A \sigma_{xx} \, dA = EA \left(\varepsilon_{xx}^{(0)} - \alpha n_\theta \right) \quad ,$$

(6.32)

which depends on the mean temperature (the average over the cross-section)

$$n_\theta(x) = \frac{1}{A} \int_A \theta\, dA \quad,$$

(6.33)

and the principal bending moment

$$M_y(x) = \int_A z\, \sigma_{xx}\, dA = EJ_y \left(\frac{d\chi}{dx} - \alpha m_\theta \right) \quad,$$

(6.34)

which depends on the first cross-sectional moment of the temperature

$$m_\theta(x) = \frac{1}{J_y} \int_A z\, \theta\, dA \quad, \quad [m_\theta] = K/m \quad.$$

(6.35)

The substitution of the approximation $\chi = - \ dw/dx$ into Eq. (6.34) renders the linearized differential equation of the deflection, a superposition of the moment curvature and the thermal curvature; compare this with Eqs. (5.22a) and (6.70)

$$\frac{d^2w}{dx^2} = -\frac{d\chi}{dx} = -\left(\frac{M_y}{EJ_y} + \alpha m_\theta \right) \quad,$$

(6.36)

integration is discussed in the next section.

Elimination of all the deformation gradients gives the relation between the normal stress and the resulting forces and temperature

$$\sigma_{xx}(x, z) = \frac{N(x)}{A} + \frac{M_y(x)}{J_y} z + E\, \alpha\, (n_\theta + z\, m_\theta - \theta) \quad.$$

(6.37)

In addition to the linear distribution related to the internal resultants, a system of self-equilibrating thermal stresses is produced by any nonlinear deviation of the thermal strains from the linear distribution, according to the terms in parentheses in Eq. (6.37); see Sec. 2.5.1 and Eq. (4.31). The resultants of these thermal stresses vanish

$$\int_A E\, \alpha\, (n_\theta + z\, m_\theta - \theta)\, dA = 0 \quad, \quad \int_A E\, \alpha\, (n_\theta + z\, m_\theta - \theta)\, z\, dA = 0 \quad.$$

Taking Eq. (6.37) into account when performing the cross-sectional integration of Eq. (6.30) renders the complementary energy of a linear elastic rod of length l; any function of temperature alone can be omitted

$$U^*(N, M_y, n_\theta, m_\theta) = \int_0^l \left(\frac{N^2}{2EA} + \frac{M_y{}^2}{2EJ_y} + N\,\alpha n_\theta + M_y\,\alpha m_\theta \right) dx \ . \tag{6.38}$$

For a slightly curved beam, dx is to be replaced by the arc-element ds .

The *oblique bending* of an originally straight beam renders by the superposition of the two principal deflections

$$U^* = \int_0^l \left(\frac{N^2}{2EA} + \frac{M_y{}^2}{2EJ_y} + \frac{M_z{}^2}{2EJ_z} + N\,\alpha n_\theta + M_y\,\alpha m_{\theta y} + M_z\,\alpha m_{\theta z} \right) dx \ , \tag{6.39}$$

where

$$m_{\theta z}(x) = -\frac{1}{J_z} \int_A y\,\theta\,dA \ . \tag{6.40}$$

With the proper expressions (6.38) and (6.39), the theorems of *Castigliano* and *Menabrea* are extended to linearized thermo-elasticity.

(§)   Example: A Single-Span Redundant Beam. Such a beam of length *l* and of constant bending rigidity EJ_y , with clamped-clamped boundary conditions is loaded by a spanwise constant thermal curvature $\alpha m_{\theta y} = \alpha m_0$ and serves as an illustrative example for the application of *Menabrea's* theorem. The shear force is zero for reasons of symmetry, there is no lateral loading considered, and the bending moment is constant, $M_y(x) = M_0$. It equals the redundant clamping moment. The complementary energy is

$$U^* = \int_0^l \left(\frac{M_y{}^2}{2EJ_y} + M_y\,\alpha m_{\theta y} \right) dx \ = l \left(\frac{M_0{}^2}{2EJ_y} + M_0\,\alpha m_0 \right) \ .$$

The derivative with respect to M_0 must vanish,

$$\frac{\partial U^*}{\partial M_0} = l \left(\frac{M_0}{EJ_y} + \alpha m_0 \right) = 0 \quad ,$$

with the solution

$$M_0 = - EJ_y\,\alpha m_0 \ . \tag{6.41}$$

Thus, Eq. (6.36) renders the total curvature zero, and the built-in beam remains straight, $w(x) \equiv 0$; see also Eq. (6.96).

If we consider alternatively the hinged-hinged beam as the associated statically determinate system under the above given thermal loading conditions, the constant thermal curvature renders a circular deflection without bending moment [see also Eq. (6.89)]

$$w(x) = \alpha m_0 \, l^2 (x/2l)(1 - x/l) \ . \tag{6.41a}$$

Superposition of the circular deflection

$$w(x) = - (M_0/EJ_y) \, l^2(x/2l)(1 - x/l) \ , \tag{6.41b}$$

which is due to isothermal loading by the edge moment $M_0 = (- EJ_y \, \alpha m_0)$, gives the above result of total deformation as zero.

(§)   *Maysel´s*   Formula of Thermoelasticity. *Betti´s* method of linearized isothermal elasticity [see Eq. (5.53)] is generalized to include thermal strain contributions to the displacement δ_i at the point of application of a dummy loading H in the direction of this single unit force. *Maysel´s* formula is derived in two steps by applying the principle of virtual displacements twice in two different states of equilibrium of the elastic body.
1.   The elastic body is considered in the isothermal state, $\theta = 0$, at rest under the action of the dummy force H . Part of the material surface is free of dummy tractions; on the remaining portion, the displacements are prescribed (the body must be properly supported). The dummy loading, dummy reaction forces, and dummy stresses denoted by an overbar are in equilibrium; the dummy strains are also characterized by an overbar.
2.   The dummy loading is followed by properly raising the temperature to $\theta(x, y, z)$; thereby the quasistatic thermal stresses σ_{ij} and strains ε_{ij} are produced.
       Considering the dummy state of equilibrium and selecting the sufficiently small thermal deformations for the virtual variation of the configuration in Eq. (5.3) of the principle of virtual work renders

$$H \, \delta_i = \sum_k \sum_l \int_V \overline{\sigma}_{kl} \, \varepsilon_{kl} \, dV \ . \tag{6.42}$$

By means of the isothermal *Hooke´s* law [Eq. (4.15)] the dummy stresses are replaced by the dummy strains. The generalized *Hooke´s* law [Eqs. (6.22) and (6.23)] gives the total strains ε_{ij} in terms of the thermal stresses σ_{ij} and temperature θ . Substitution yields Eq. (6.42) in the form

$$H \, \delta_i = \sum_k \sum_l \int_V \sigma_{kl} \, \overline{\varepsilon}_{kl} \, dV + \int_V 3\alpha\theta \, \overline{p} \, dV \ . \tag{6.43}$$

In the actual state of the thermally loaded body, the thermal stresses σ_{ij} and reaction forces of the supports are in equilibrium. Hence, selecting now the dummy deformations for the virtual displacements and applying the principle of virtual work a second time render the work of the internal forces as zero, since the thermal reaction forces do not contribute to the virtual work of the external forces. Therefore, the double sum in Eq. (6.43) must vanish. Taking a unit dummy force $H = 1$, we derive *Maysel's* formula in the form

$$\delta_i = \int_V 3\alpha\theta\,\bar{p}\,dV \quad , \quad \bar{p} = \frac{1}{3}\sum_{j=1}^{3}\bar{\sigma}_{jj} \quad .$$

$$(6.44)$$

<<The thermal displacement at a point P, $\delta_i(P)$, is given by the weighted volume integral of the isotropic thermal dilatation. The kernel function is the isothermal dummy average normal stress considered in a point Q due to a dummy single force loading $H = 1$, at point P pointing in the direction e_i, in the same elastic body. The volume element dV_Q and temperature field $\theta(Q)$ are to be expressed in the coordinates of the variable point Q.>> Note the difference from the method of influence functions of isothermal statics; see, eg Eq. (2.159), where the summation is performed over the coordinates of the variable source point, which is the point of application of the dummy force. The gradient with respect to the coordinates of the source point can be taken under the integral sign, but care has to be taken about the integral that becomes singular. Formally, the thermal stresses are given through the generalized *Hooke's* law.

Equation (6.44) as it stands is the solution of the general three-dimensional problem of thermoelasticity. The temperature field is assumed known [to be calculated in advance by considering the heat conduction equation, (6.354)]. *Maysel's* formula may be adapted to the structural problems of thermally loaded beams, frames, plates, and shells as shown below. By means of the roots of this formula, ie the virtual work formulation, it is easily possible to consider continuum problems with axial or point symmetry. In such cases, the single dummy force is replaced by dummy ring loads or dummy tractions on the surface of a sphere. This specialization is illustrated for thermally loaded cylinders and spheres, respectively. (§)   A Hollow Sphere with Point Symmetry and a Thick-Walled Cylinder with Axial Symmetry, *Maysel's* Formula. A unit radial dummy traction $\sigma_H = 1$ at some internal radius R is considered to determine the radial thermal displacement with respect to a given temperature field $\theta(R)$. The dummy virtual work becomes

$$\sigma_H \, \delta_R \, 2\pi \, \overline{\alpha} \, R^{\overline{\alpha}} = \int_{R_i}^{R_e} 3\alpha \, \theta(R^*) \, \overline{p}(R^*, R) \, 2\pi \, \overline{\alpha} \, R^{*\overline{\alpha}} \, dR^* \quad ,$$

(6.45)

$\overline{\alpha} = 2$, for the sherical problem ,

$\overline{\alpha} = 1$, for the cylindrical problem .

If we put the dummy load $\sigma_H = 1$ and consider the isothermal influence function

$$3 \, \overline{p} \, (1 + v)^{(\overline{\alpha} - 2)} = \overline{\sigma}_{rr} + \overline{\alpha} \, \overline{\sigma}_{\vartheta\vartheta} \quad ,$$

the radial displacement is given by

$$\delta_R = R^{-\overline{\alpha}} \int_{R_i}^{R_e} 3\alpha \, \theta(R^*) \, \overline{p}(R^*, R) \, R^{*\overline{\alpha}} \, dR^* \quad .$$

(6.46)

The mean normal stress of the isothermal dummy load problem of the hollow bodies with traction-free surfaces is easily found by superposition of the solutions of two bodies with the common interface at the radius R , as given by Eqs. (6.14) and (6.15). The tractions at R are related by $p_i = p_e + \sigma_H$, $\sigma_H = 1$, and the radial displacement must be continuous

$$(1 - v) \, (1 + v)^{(\overline{\alpha} - 2)} \, [1 - \rho^{\overline{\alpha} + 1}] \, 3 \, \overline{p}(R^*, R)$$

$$= \overline{\alpha} \, (1 - 2v) \, (R/R_e)^{\overline{\alpha} + 1} + \begin{cases} [1 + (\overline{\alpha} - 1) \, v] & ...R^* < R \\ [1 + (\overline{\alpha} - 1) \, v] \, \rho^{\overline{\alpha} + 1} & ...R^* > R \end{cases} , \quad \rho = R_i/R_e \quad .$$

(6.47)

The analytical integration of Eq. (6.46) requires subdivision of the range of integration into the subintervals $[R_i, R]$ and $[R, R_e]$.

6.1.2. Saint Venant´s Principle

The possibility of finding special solutions of *Navier´s* equations [see Eq. (6.6)] depends strongly on the geometry of the elastic body and the complexity of the boundary conditions. Therefore, the tractions with an unknown distribution on those parts of the surface, where kinematic conditions are prescribed, are approximated by simpler, statically equivalent force systems to ease the problem. In many cases, the approximate stresses and the deformations in the far field (at a sufficiently large distance from the "small" part of the surface where the boundary conditions are approximate) approach the exact solution. The cantilever of Fig. 6.1 illustrates such an approach. In the near field, however, the stresses

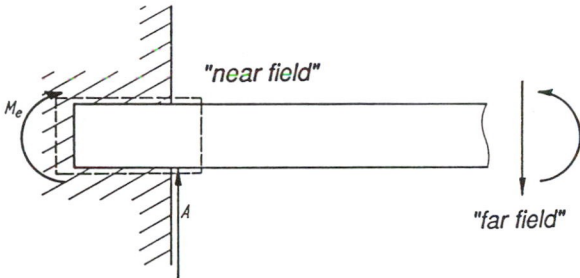

Fig. 6.1. Statically equivalent reaction forces and stress resultants in the far field. The dashed control surface indicates the near field

depend more or less exclusively on the exact distribution of the surface tractions. Approximations of that kind make many technical problems solvable that otherwise would be out of the scope of practical engineering and would become complex mathematical exercises.

In 1855, *Barre de Saint Venant* formulated a principle, with until now largely empirical justification: <<Statically equivalent forces, when applied to a small part of the body, render at a sufficiently large distance from this loading region approximately the same stresses and deformations.>>

Without discussion, the principle is many times applied in this book, see, eg the simple tension test, where the ends of the test specimen are clamped in the jaws of the testing machine, but the action over the control length of the probe is assumed to be nearly the same as if the forces were applied uniformly over the cross-sections at the ends.

There are some mathematical formulations of the principle given in theoretically oriented literature. The most important consider the loading to be self-equilibrating and concentrated at a small part of the body. Such a system may be found, eg at the free ends of a thermally loaded beam; see Eq. (6.37). The strain energy is introduced as a positive-definite measure of the intensity of the (self-equilibrating) stresses in the elastic body. However, the limits of applicability of the principle are not easily recognized. Care has to be taken when the principle is applied to thin-walled structures, like box-beams, shells, composite structures, etc., even in the elastic range of the loadings. If elastic-visco-plastic material behavior is to be considered, the limits of application of the principle become even less visible. For example, the formation of a plastic hinge in a statically indeterminate structure redistributes the stresses into the far field, a statically determinate structure becomes a mechanism.

6.1.3. Stress and Strain Hypotheses

The exertion of a material in a multiaxial stress state, ie the intensity of the stress state, is commonly measured by an equivalence stress. The elastic limits, eg the fracture of a brittle material or yielding of a visco-plastic body, are given by critical values of the equivalence stress according to a proper stress or strain hypothesis. The principal strain hypothesis is only mentioned here (it works similarly to the principal stress hypothesis), and the two simple stress hypotheses already useful for practical applications are discussed below.

(§) Principal Normal Stress Hypothesis. Brittle materials, in which sensitivity to the tensile component of the stress state predominates, may be described in their elastic range by *Hooke´s* law or a nonlinear law of elasticity until the limit given by the largest tensile principal normal stress is reached. The influence of the other two principal stresses on the elastic limit is neglected.

(§) *Hencky-von Mises* Energy Hypothesis. The exertion of an elastic-plastic solid can be measured by an elastic energy norm by taking the distortional strain energy density (energy of changing the configuration): the strain energy minus the work stored in compression. The elastic energy of dilatation must be ignored since the yielding of a visco-plastic material does not occur under even high hydrostatic pressure. Using deviatoric stress components and Eqs. (5.38) and (3.47) we see that the complementary energy of changing the configuration in the elastic material becomes

$$U_C^{*'} = \frac{1}{2G} J_2 \;,\quad J_2 = \frac{1}{2} \sum_i \sum_j \sigma_{ij}'^2 \;,\quad \sigma_{ij}' = \sigma_{ij} - p\delta_{ij} \;.$$

The energy density in a tension test probe of the same material stressed to σ_e is $J_2 = \sigma_e^2/3$,

$$U_C^{*'} = \frac{1}{6G} \sigma_e^2 \;.$$

The equivalence of the exertion in the multiaxial and uniaxial stress state renders the equivalence stress σ_e , and hence, with the elastic limit of the uniaxial stress state given by the yield stress σ_Y assumed to be equal in tension and compression, the elastic range is given by

$$\sigma_e = \frac{\sqrt{2}}{2} \left[(\sigma_1 - \sigma_2)^2 + (\sigma_2 - \sigma_3)^2 + (\sigma_3 - \sigma_1)^2 \right]^{1/2} \le \sigma_Y \;.$$

See also the *von Mises* yield condition of the ideal plastic solid [Eq. (4.127)]; σ_1 , σ_2 , and σ_3 denote the principal stresses in the material point of inspection. Under the conditions of plane stress, σ_3 = 0, the equivalence stress becomes

$$\sigma_e = \left[\sigma_{xx}^2 - \sigma_{xx}\sigma_{yy} + \sigma_{yy}^2 + 3\sigma_{xy}^2\right]^{1/2} \leq \sigma_Y .$$

(§) *Mohr-Coulomb* Stress Hypothesis. Many materials exhibit a transition from visco-plasticity to more brittle behavior when the deformation rate is increased. Under plane stress conditions, such behavior can be described by moving the center of *Mohr's* circle to increasing normal stresses and reducing its radius, the maximal allowable shear stress. The envelope in *Mohr's* hypothesis is given by two converging straight lines, in *Coulomb's* hypothesis by two parallel lines. An improvement is given in *Leon's* parabolic envelope, where the principal normal stress hypothesis of perfectly brittle material behavior is included. For details, see A. Nadai: *Theory of Flow and Fracture of Solids. 2nd ed., McGraw-Hill, New York, 1950; A Slattenschek: Plastic and brittle behavior of metallic materials under mechanical loadings. Schweißen und Schneiden Vol. 3 (1951), pp. 90-100, Special Issue (in German) and also, Basic theory of brittle fracture. Radex-Rundschau (1953), pp. 186-199.*
(§) The Concept of Allowable Stress. This concept is widely used in engineering in problems where stresses are proportional to the load factor. For example, considering a safety coefficient $v_Y > 1$ against reaching the yield stress of an elastic-plastic material gives an allowable stress (in the elastic range) $\sigma_a = \sigma_Y / v_Y$. The condition in terms of the equivalence stress becomes simply $\sigma_e \leq \sigma_a$.

6.2. Rods and Beams with Straight Axes

Rods are structural elements whose cross-sectional dimensions are small when compared to length or, in the case of slender beams, small with respect to span. Statics is discussed in Sec. 2.5.1. In general, the stress state is given by the normal stress σ_{xx} and the shear-stress components $\sigma_{xy} = \sigma_{yx}$ and $\sigma_{xz} = \sigma_{zx}$. x is the coordinate in the direction of the straight beam axis. If the deformations are small, linearized geometric relations apply, and no distinction is made between the undeformed reference state and the deformed configuration. The distribution of the stresses in the cross-sections is derived from the previously determined resultants: axial force $N(x)$, shear forces $Q_y(x)$, $Q_z(x)$, principal bending moments $M_y(x)$, $M_z(x)$, and torque $M_x = M_T$. The loading is commonly assumed to be given by single forces, the tangential $q_x(x)$, and the lateral $q_y(x)$, $q_z(x)$ load per unit of length. In addition, the boundary conditions are prescribed through idealized support conditions. The loading case where distributed external moments per unit of length are applied is not considered here.

(§)   Normal Force and Bending Moments. These resultants render within the limits of application of the *Bernoulli-Euler* hypothesis (cross-sections remain plane also in the case of nonvanishing shear) the linear distribution of the normal stress in the cross-section at x [see Eq. (4.31); dependence on that axial coordinate is understood]

$$\sigma_{xx}(y, z) = \frac{N}{A} + \frac{M_y}{J_y} z - \frac{M_z}{J_z} y \; .$$

(6.48)

The linear distribution becomes an exact solution of *Navier's* equations (6.6) for the pure bending and constant principal bending stiffness of a rectangular cross-section.

Extreme values of the normal stress occur always in points at the outer surfaces. If the normal force does not vanish, it can be reduced to a point of reference $A_N (y_N = b, z_N = c)$ at a distance from the cross-sectional centroid where the moment of the normal stress distribution vanishes [Eq. (2.135)]. Hence, the principal bending moments are given by the proper couples of the eccentrically acting axial force N

$$M_z = -N b \; , \quad M_y = N c \; .$$

(6.49)

The linear stress distribution (6.48) thus becomes

$$\sigma_{xx} = \frac{N}{A} \left(1 + \frac{c z}{i_y^2} + \frac{b y}{i_z^2} \right) \; , \quad J_{y, z} = A \, i_{y, z}^2 \; .$$

(6.50)

The expression is quite suitable for determining the straight line of vanishing normal stresses, the *neutral axis* ,

$$1 + \frac{c z}{i_y^2} + \frac{b y}{i_z^2} = 0 \; ,$$

(6.51)

which divides the cross-section into the sectors of tension and compression. Beams and columns made of materials with low stiffness in tension, like concrete or brick, when loaded in compression eccentrically by $N < 0$ should be designed such that the line of zero stress lies outside of the cross-section. The points of application A_N of the normal force N fill a subdomain of the cross-section around its areal centroid, which is called the *core of the cross-section*. Its shape is determined by putting the neutral axis tangential to the contour of the cross-section and solving Eq. (6.51) for the coordinates b and c . The latter are the coordinates of the antipoles of all tangents to the smallest convex envelope of the

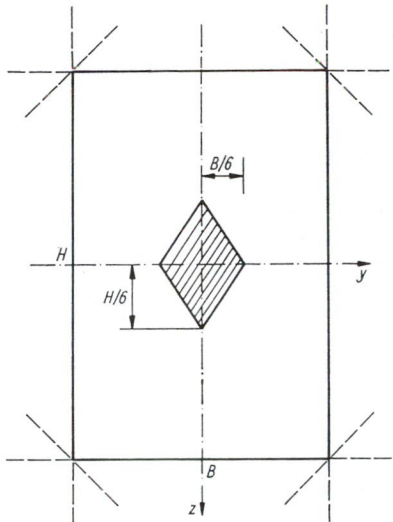

Fig. 6.2. Core of the rectangular cross section

cross-section. A straight segment of the envelope renders a single point of the circumferential line of the core, and vice versa.

The core of a rectangular cross-section *(H x B)*, with the squared principal radii of inertia $i_y^2 = H^2/12$ and $i_z^2 = B^2/12$, is a parallelogram. A point is given by inserting the straight segment $z = H/2$ into Eq. (6.51) and solving for $c = -H/6$, $b = 0$ for reasons of

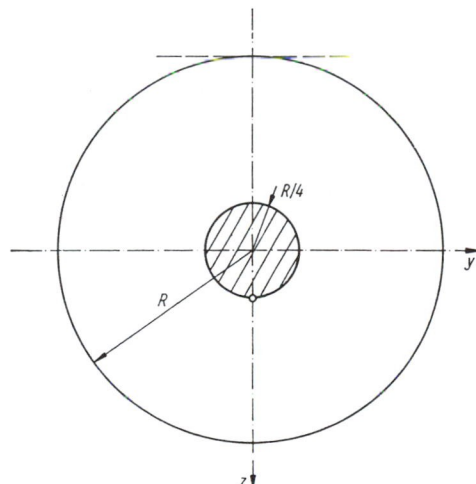

Fig. 6.3. Core of the circular cross-section

symmetry; see Fig. 6.2. Since $i^2 = R^2/4$ is isotropic for a circular cross-section, the antipoles are located on a concentric circle of radius $R/4$ and the points of application must lie within the hatched circular area of Fig. 6.3.

6.2.1. Shear Stresses and Deformations due to a Shear Force

The determination of the shear-stress distribution within a cross-section that is due to bending with nonvanishing shear force and torsion is quite involved. Only for thin-walled structures of thickness h in the cross-section, with surfaces free of traction, the distribution is a priori rather simple: At the free surfaces, the shear stress in the cross-section must be tangential due to the symmetry of the stress tensor. If we consider the mean shear stress over the thickness and, hence, the shear flow $T(s) = \tau(s)\, h(s)$, s is the arc-length along the thin-walled cross-section, and the equilibrium of an element (the free-body diagram is shown in Fig. 6.4) renders

$$\frac{\partial T}{\partial s} + h(s)\, \frac{\partial \sigma_{xx}}{\partial x} = 0 \ .$$

$$(6.52)$$

The substitution of the linear distribution of the normal stress in oblique bending,

$$\sigma_{xx} = \frac{M_y}{J_y}\, z - \frac{M_z}{J_z}\, y \ ,$$

$$(6.53)$$

yields with the relations (2.147), $ds = dx$, and (2.152) for the shear forces

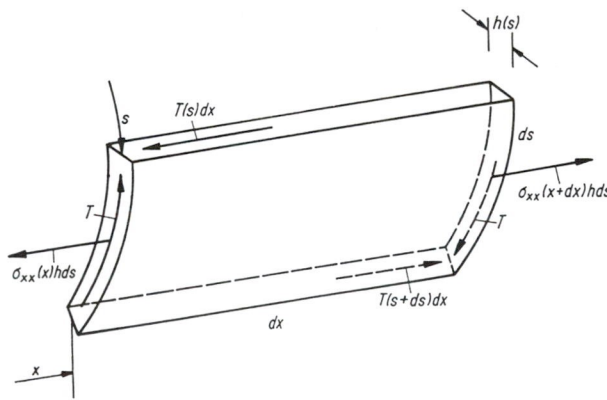

Fig. 6.4. Infinitesimal element of a rod with thin-walled cross-section. Equilibrium of the internal forces in shear bending

$$\frac{\partial T}{\partial s} + h(s) \left[\frac{Q_z}{J_y} z + \frac{Q_y}{J_z} y \right] = 0 \ .$$

(6.54)

The integration of the first-order differential equation gives the shear flow

$$T(s) = T_0 - \left[\frac{Q_z}{J_y} \int_0^s z \, h(s) \, ds + \frac{Q_y}{J_z} \int_0^s y \, h(s) \, ds \right] \ .$$

(6.55)

Hence, the mean shear stress $\tau = T/h$. The constant flow $T_0 = T(s = 0)$ vanishes for thin-walled and open cross-sections if $s = 0$ is an edge, free of any shear. It remains an undetermined constant for boxed (closed) cross-sections. The case of a single cell, where the constant shear flow due to a torque M_T becomes statically determinate, is considered in *Bredt's* formula (2.171). The integrals in Eq. (6.55) are the static moments of the area sections of arc-length s with respect to the principal central axes y and z ; they are denoted by

$$S_y(s) = \int_0^s z \, h(s) \, ds \quad \text{and} \quad S_z(s) = \int_0^s y \, h(s) \, ds \ .$$

(6.56)

Thus, the mean shear stress in the cross-section with the negative normal, which is due to the shear forces, becomes

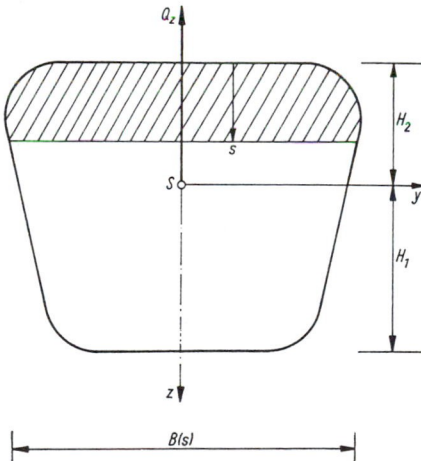

Fig. 6.5. Fully bodied cross-section loaded by the shear force $Q_z(x)$. The approximate shear-stress distribution is assumed to be proportional to the static moment of the hatched area and parallels z

$$\tau(s) = \frac{T}{h(s)} = \frac{T_0}{h(s)} - \left[\frac{Q_z S_y(s)}{J_y h(s)} + \frac{Q_y S_z(s)}{J_z h(s)}\right] . \tag{6.57}$$

The distribution of the shear stress in thick-walled cross-sections becomes complex. Thus, the engineering beam theory applies the same formula that holds for thin-walled beam structures to the fully bodied cross-section as a crude approximation. See Fig. 6.5 where $s = z + H_2$. Putting $Q_y = 0$ gives,

$$\sigma_{xz}(s) = \frac{T}{B(s)} = -\frac{Q_z S_y(s)}{J_y B(s)} . \tag{6.58}$$

Equation (6.58), eg does not take into account the boundary condition at the free surface of a beam with variable width $B(s)$, where τ must be parallel to the inclined contour as shown in Fig. 6.5.
(§)   Rectangular Cross-Section. In that case, $B = const$ and $A = B \times H$, the static moment is a quadratic function $(H = H_1 + H_2)$

$$S_y(s) = \frac{A}{2} s\left(1 - \frac{s}{H}\right) , \tag{6.59}$$

and the parabolically distributed shear stress is $(J_y = BH^3/12)$

$$\sigma_{xz}(s) = -\frac{Q_z}{A} \frac{6 s}{H} \left(1 - \frac{s}{H}\right) , \tag{6.60}$$

with the extremum at $s = H/2$, where $z = 0$ (see Fig. 6.6)

$$\text{Max}|\sigma_{xz}| = \frac{3}{2} \frac{Q_z}{A} . \tag{6.61}$$

The maximum outgrows the mean value Q_z/A by 50%.
(§)   Maximal Shear in an Elliptic or Circular Cross-Section. This shear in the cross-section of area is A when loaded by the shear force Q_z , it is analogously determined and given by the value

$$\text{Max}|\sigma_{xz}| = \frac{4}{3} \frac{Q_z}{A} . \tag{6.62}$$

(§)   Equation (6.58) when Applied to the T Cross-Section. The shear-stress distribution in the thin-walled fillet becomes (see Fig. 6.7)

$$\sigma_{xz}(s) = -\frac{B t H_2}{J_y h} Q_z - \frac{Q_z}{J_y} \frac{s}{H_2 - t} \left(1 - \frac{s}{2(H_2 - t)}\right) , \quad s \geq 0 . \tag{6.63}$$

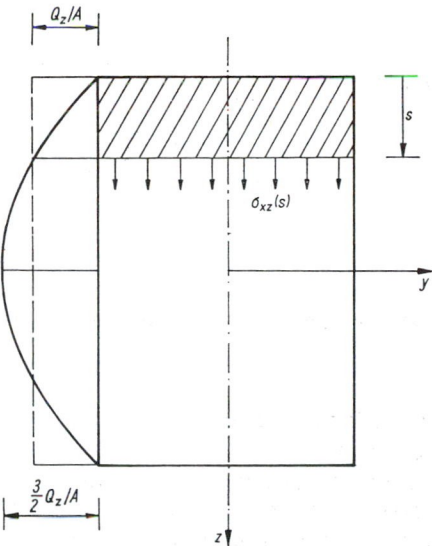

Fig. 6.6. Shear-stress distribution in the rectangular cross-section loaded centrically by the shear force Q_z free of any torsion

Equation (6.63) holds also for I-beams when substituting the proper cross-sectional moment of inertia J_y ; see Exercise A 6.14 for the shear center. Most important, the shear stress $\sigma_{xz} = \sigma_{zx}$ at $s = 0$; the constant first term of Eq. (6.63) is the interfacial shear that must be transmitted in the axial direction between the flange and

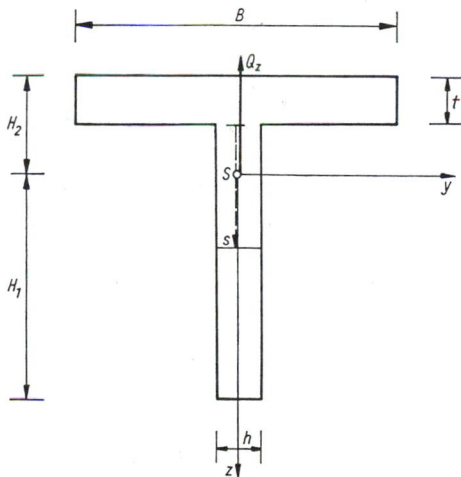

Fig. 6.7. Thin-walled T cross-section, loaded by the shear force Q_z

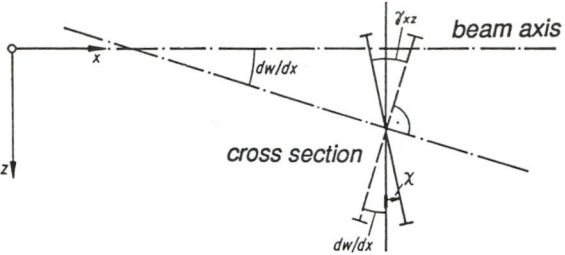

Fig. 6.8. Averaged rotation of the cross-section due to pure bending and the shear deformation

the fillet in bending. It loads the weld in a welded joint or the plugs in a plugged joint as well.

The *deformation in shear* contributes to the deflection an amount that is negligible for slender beams, but becomes of equal order for high beams. Within the approximation of the shear-stress distribution [Eq. (6.58)], the deflection in shear can be quantified. Assuming the shear stress σ_{xz} is constant in a cross-section, now with the normal pointing in the positive x direction, keeps the cross-section plane and renders an additional rotation due to shear

$$\gamma_{xz} = 2\varepsilon_{xz} = \frac{\sigma_{xz}}{G} = \frac{Q_z}{GA} \ . \tag{6.64}$$

As a consequence of this (approximate) deformation in shear, the plane cross-section remains no longer orthogonal to the deformed beam axis. When we consider the rotation due to pure bending [see Eqs. (5.22a) and (6.36); the principal bending moment is M_y]

$$\frac{d\chi}{dx} = \frac{M_y}{EJ_y} \ , \tag{6.65}$$

and take into account the variable distribution of the shear stress by replacing the shear deformation of Eq. (6.64) by an averaged rotation, the slope of the total deflection becomes (γ_{xz} denotes a small decrease of the right angle; see Fig. 6.8)

$$\frac{dw}{dx} = -\chi + \bar{\gamma}_{xz} \ . \tag{6.66}$$

The average rotation gives the same amount of work for the resulting shear force Q_z and the statically equivalent distributed shear stresses σ_{xz}

$$\frac{1}{2} \ Q_z \ dw = \frac{1}{2} \ \sigma_{xz} \frac{\partial w}{\partial x} \ dx \ dA = -\frac{1}{2} \ \chi \ dx \int_A \sigma_{xz} \ dA + dx \int_A \frac{\sigma_{xz}^2}{2G} \ dA$$

$$= \frac{dx}{2} \left[-\chi \ Q_z + \int_A \frac{\sigma_{xz}^2}{G} \ dA \right] \ .$$

(6.67)

The substitution of Eq. (6.58) (the minus sign does not apply at the positive cross-section) renders the work in the form

$$\frac{1}{2} \ Q_z \ dw = \frac{dx}{2} \left[-\chi + \frac{Q_z}{GJ_y^2} \int_A \frac{S_y^2}{B^2} \ dA \right] Q_z = \frac{dx}{2} \left[-\chi + \bar{\gamma}_{xz} \right] Q_z \ .$$

Comparing coefficients yields, with the corrected constant rotation of shear,

$$\bar{\gamma}_{xz} = \kappa_z \frac{Q_z}{GA} \ ,$$

(6.68)

[see Eq. (5.49)] the (geometric) correction coefficient

$$\kappa_z = \frac{A}{J_y^2} \int_A \left(\frac{S_y}{B}\right)^2 dA \ .$$

(6.69)

A second coefficient results for biaxial principal shear bending.

The correction coefficients are easily evaluated for simple cross-sections, eg for the rectangular shape of width B, $\kappa_z = 6/5$; for the circular cross-section the isotropic value is $\kappa = 10/9$; and for the thin-walled I-profile, $\kappa_z \approx A/A_{fillet}$.

Taking the derivative of Eq. (6.66) and inserting the lateral loading $q_z = - \ dQ_z /dx$ into the differentiated Eq. (6.68) render the linearized curvature by the superposition of the pure bending curvature and the curvature due to the shear deformation [compare this with Eq. (6.36)]

$$\frac{d^2w}{dx^2} = - \frac{M_y}{E \ J_y} - \frac{\kappa_z \ q_z}{G \ A} \ , \quad A = \text{const} \ .$$

(6.70)

The shear deformation influences also the boundary conditions, eg a rigidly clamped edge restricts the rotation of the cross-section due to pure bending $\chi = 0$, but due to the shear deformation, $dw/dx \neq 0$.

Equation (5.49) was derived by a mathematical argument. The contribution of shear to the complementary energy per unit of length can be expressed in terms of the correction coefficients [Eq. (6.69)] and the analogous expression, by relating the biaxial averaged rotations in shear to the mean values of the shear stresses using *Hooke´s* law

$$\int_A \frac{\sigma_{xy}^2 + \sigma_{xz}^2}{2G} \, dA = \frac{1}{2} \bar{\gamma}_{xy} \int_A \sigma_{xy} \, dA + \frac{1}{2} \bar{\gamma}_{xz} \int_A \sigma_{xz} \, dA$$

$$= \frac{1}{2} \bar{\gamma}_{xy} Q_y + \frac{1}{2} \bar{\gamma}_{xz} Q_z = \frac{1}{2} \left(\frac{\kappa_y Q_y^2}{G A} + \frac{\kappa_z Q_z^2}{G A} \right) \quad . \tag{6.71}$$

(§)   Example: Deflection of a Cantilever in Shear Bending. The linear elastic cantilever of length l is uniformly loaded by $q_z = q$ and has constant rigidity in bending ($B = EJ_y$) as well as in shear ($S = GA$), respectively. The differential equation (6.70), corrected for shear deformation, is easily integrated taking into account the parabolic bending moment $M_y (x) = - (q/2) (l - x)^2$ [note the shear force $Q_z (x) = dM_y(x)/dx = q(l - x)$]

$$w(x) = \frac{q}{24B} (l - x)^4 - \frac{\kappa_z \, q}{S} \frac{x^2}{2} + C_1 x + C_2 \quad .$$

The constants of integration are to be determined from the kinematic boundary conditions of the rigidly clamped edge at $x = 0$

$$w(0) = 0 \quad : \quad C_2 = - \frac{q \, l^4}{24 \, B}$$

$$\chi(0) = 0 \quad : \quad \frac{dw}{dx} (x = 0) = \bar{\gamma}_{xz} = \frac{\kappa_z}{S} Q_z(0) \quad , \quad C_1 = \frac{q \, l^3}{6 \, B} + \kappa_z \frac{q \, l}{S} \quad .$$

If we denote the contribution of pure bending to the deflection by w_B and the part due to the shear by w_S, the ratio taken at the tip of the cantilever takes on the value

$$\frac{w_S}{w_B} (x = l) = 8 (1 + v) \, \kappa_z \left(\frac{i_y}{l} \right)^2 \quad .$$

Since i_y is the principal radius of inertia of the cross-section and l the span, the factor $\kappa_z (i_y/l)^2 \ll 1$ for sufficiently slender beams. The portion w_S of the shear deformation is thus negligible for beams that are not too high. The order of magnitude of the ratio of deflections due to shear and pure bending is not changed by considering more sophisticated models of shear deformation.

6.2.2. Mohr's Method of Calculating Deflections

Originally, this method was developed by considering only the contribution of pure bending to the curvature of (slender) beams; hence, Eq. (5.22a) applies

$$\frac{d^2w}{dx^2} = - \frac{M_y}{EJ_y} \quad ,$$

(6.72)

which is a linear differential equation of the second order. Kinematic or dynamic boundary conditions render the deflection consistent with support conditions. Table 5.1 gives $w = 0$ and $dw/dx = 0$ at a rigidly clamped edge, $M_y = 0$ and $Q_y = dM_y/dx = 0$ at a free end, $w = 0$ and $M_y = 0$ (mixed conditions) at a simple support.

Analogous to the deflection w in Eq. (6.72), the bending moment M_y is the solution of the local condition of equilibrium that is also a linear second-order differential equation [see Eq. (2.152)] given the lateral loading q_z,

$$\frac{d^2M_y}{dx^2} = - q_z \quad .$$

(6.73)

The boundary conditions are given by the prescribed edge forces and couples: $M_y = M_0$, $Q_z = dM_y/dx = Q_0$.

Any solution strategy that solves Eq. (6.73) may be used to integrate Eq. (6.72), provided the boundary conditions correspond. The graphic solution technique of the static equation is the method of the force and funicular polygon. The bending moment in a beam with given loading is also found conveniently and problem-oriented by considering the equilibrium of a finite element of length x . Hence, comparing the right-hand sides of the above equations renders the conjugate load proportional to the given bending moment

$$\bar{q}_z = M_y \frac{J_0}{J_y} \quad ,$$

(6.74)

where J_0 is an inertia moment of reference. Comparing also the left-hand sides gives the correspondence between the actual deflection w and the generalized bending moment in the conjugate beam if it is properly loaded by the conjugate load

$$w = \frac{\overline{M}_y}{EJ_0} \quad , \quad \frac{dw}{dx} = \frac{\overline{Q}_z}{EJ_0} \quad .$$

(6.75)

The correspondence of the actual slope and conjugate shear force is also noted above; it is easily recognized by taking the derivative of the deflection and conjugate bending moment with respect to x . Hence, the analogy is to be completed by considering the proper boundary conditions of the conjugate beam through the correspondence of Eq. (6.75) and the given support conditions of the actual beam. Only a hinged support corresponds to a hinged support of the conjugate beam, since

$$w = 0 \rightarrow \overline{M}_y = 0 \quad , \quad \text{and} \quad \frac{dw}{dx} \neq 0 \rightarrow \overline{Q}_z \neq 0 \quad .$$

A free end of the actual beam (homogeneous dynamic boundary conditions apply) corresponds to a rigidly clamped edge of the conjugate beam (the actual dynamic conditions are transformed to homogeneous kinematic boundary conditions). An intermediate support of a continuous multispan beam corresponds to a *Gerber* joint of the conjugate beam, since

$$w = 0 \rightarrow \overline{M}_y = 0 \quad , \quad \text{and continuous} \quad \frac{dw}{dx} \rightarrow \overline{Q}_z \quad \text{continuous} \quad .$$

Note that the correspondence can be reversed. The analytic as well as the graphic analogy, called *Mohr's* method, is illustrated below by considering an elastic cantilever.

(§)   *Mohr's* Analytic Method Applied to the Cantilever of Fig. 6.9.
The loading is given by the single force F and the edge moment M_0, applied at the tip of the cantilever of constant bending stiffness $B = EJ_0$ and of length l. The bending moment is given by the linear function $M_y = M_0 - F(l - x)$. It is formally the lateral load of *Mohr's* conjugate beam that is also a cantilever but with exchanged boundary conditions due to Eq. (6.75). The conjugate system as shown in Fig. 6.9 is statically analyzed. If the actual tip deflection $w(x = l) = w_0$ and rotation of the terminal cross-section dw/dx ($x = l$) $= -\chi_0$ are to be determined, the analogy requires the determination of the conjugate support reactions only

$$\overline{Q}_z(x = l) = \frac{1}{2} F l^2 - M_0 l \quad , \quad \overline{M}_y(x = l) = \frac{1}{2} F l^2 \frac{2}{3} l - M_0 l \frac{1}{2} \quad .$$

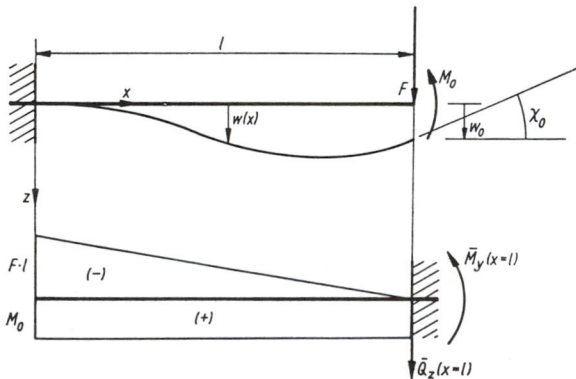

Fig. 6.9. Cantilever of constant bending rigidity, with given loading F and M_0. *Mohr's* conjugate cantilever laterally loaded by the bending moment $M_y(x)$

Equation (6.75) gives the following results

$$\chi_0 = -\frac{dw}{dx}(x = l) = -\frac{\overline{Q}_z(x = l)}{EJ_0} = \frac{l}{B}\,M_0 - \frac{l^2}{2B}\,F\ ,$$

$$w_0 = w(x = l) = \frac{\overline{M}_y(x = l)}{EJ_0} = -\frac{l^2}{2B}\,M_0 + \frac{l^3}{3B}\,F\ .$$

By sampling the local deformations in the column matrix [is transposed the row matrix $u_F{}^T = (\chi_0\,,\,w_0)$, and in the same sequence of the associated loading, the generalized force vector becomes $F^T = (M_0\,,\,F)$] a linear vector transformation results

$$u_F = f\,.\,F \quad \text{where} \quad f = \begin{pmatrix} \dfrac{l}{B} & -\dfrac{l^2}{2B} \\ -\dfrac{l^2}{2B} & \dfrac{l^3}{3B} \end{pmatrix},$$

$$(6.76)$$

is the symmetric 2x2 flexibility matrix of the cantilever of Fig. 6.9. The elements are the influence coefficients at $x = l$, ie the deformations due to individual unit loadings by $M_0 = 1$, $F = 0$ and $M_0 = 0$, $F = 1$, respectively.

The inversion of the flexibility matrix gives the stiffness matrix of the cantilever at $x = l$ and the linear relation is reversed

$$F = k\,.\,u_F \quad \text{where} \quad k = \begin{pmatrix} \dfrac{4\,B}{l} & \dfrac{6\,B}{l^2} \\ \dfrac{6\,B}{l^2} & \dfrac{12\,B}{l^3} \end{pmatrix}.$$

$$(6.77)$$

(§) *Mohr's* Graphic Method Applied to the Cantilever of Fig. 6.9.
The loading of the conjugate beam by the actual bending moment is assumed to be given and the graphic method is used for the determination of the conjugate shear force, in the force polygon, and the conjugate bending moment, by the funicular polygon; see Sec. 2.5.1.2. The distributed conjugate loading is approximated by a finite number of resultants; Fig. 6.10 shows the standard procedure using three subintervals. Alternatively, the resultants in each of the intervals may be equivalently reduced to the adjacent interval nodes. The values of the deflection *w* and of the slope *(−dw/dx)* at the intermediate points where two neighboring intervals meet are exact if the points of application of the resultants are properly chosen (in the sense of static equivalence). In Fig. 6.10, the midpoints are chosen for convenience's sake, and so the results are also approximate at the nodes. The funicular polygon approximates the actual deflection of the cantilever of Fig. 6.9.

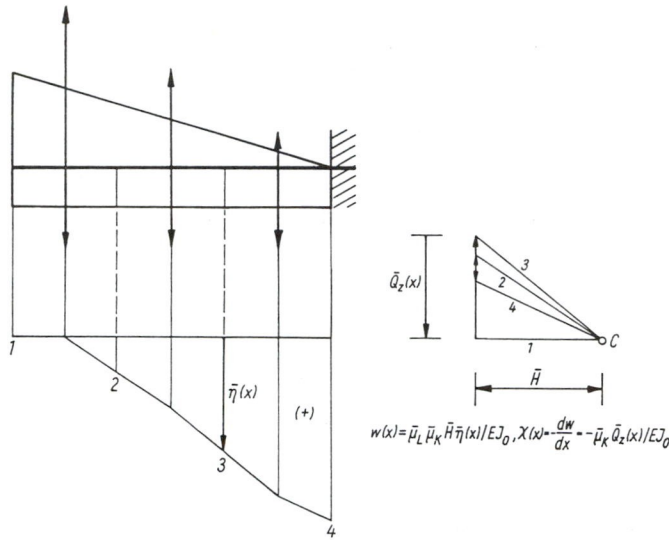

Fig. 6.10. *Mohr's* conjugate beam. Graphic solution of the actual deflection of the cantilever of Fig. 6.9 by the force and funicular polygon

(§)   Influence Lines of Deformations by *Mohr's* Method. The analogy provides a convenient means for the analysis of the influence lines of beams. The procedure is illustrated for the influence function of the deflection of a simply supported beam of constant bending rigidity $B = EJ_0$ according to Fig. 6.11. *Mohr's* conjugate beam has the same span l and identical support conditions, but it is loaded by the triangular moment distribution with the maximum $(1 - \xi/l) \xi$ under the unit load $F = 1$ of the actual problem. The conjugate reaction forces are

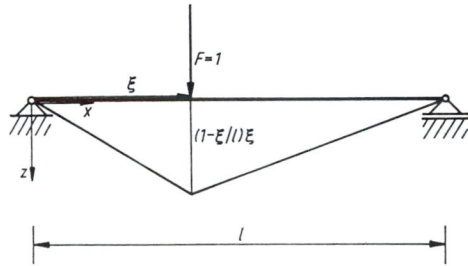

Fig. 6.11. Single-span beam and its conjugate beam. Actual unit force loading and conjugate lateral loading by the triangular distribution of the actual bending moment

$$\overline{A} = \frac{l^2}{3} \frac{\xi}{l} \left(1 - \frac{\xi}{l}\right)\left(1 - \frac{\xi}{2\,l}\right) \ , \quad \overline{B} = \frac{l^2}{6} \frac{\xi}{l}\left(1 - \frac{\xi^2}{l^2}\right) \ ,$$

and the influence line of the deflection is given by

$$\overline{w}(x;\xi) = \frac{\overline{M}_y(x)}{EJ_0} = \begin{cases} \dfrac{l^3}{6\,B}\left(1 - \dfrac{\xi}{l}\right)\left[\left(1 - \dfrac{x^2}{l^2}\right) - \left(1 - \dfrac{\xi}{l}\right)^2\right]\dfrac{x}{l} \ , & 0 \le x \le \xi \\[4mm] \dfrac{l^3}{6\,B} \dfrac{\xi}{l}\left[\left(1 - \dfrac{\xi^2}{l^2}\right) - \left(1 - \dfrac{x}{l}\right)^2\right]\left(1 - \dfrac{x}{l}\right) \ , & 0 \le \xi \le x \le l \end{cases}$$

(6.78)

(§)   *Mohr's* Method. This method also provides the deformation-related equations for the determination of statically indeterminate forces of redundant beams. The procedure is at first illustrated for the simple problem of a uniformly loaded cantilever with an additional simple support at the span l (see Fig. 6.12). If we consider the clamping moment $X = M_e$ as the statically indeterminate couple, the bending moment in the associated basic system, a hinged-hinged beam, is found by superposition, due to the given loading $q_z = q_0$ and edge moment X, $M_y(x) = (q_0\,x/2)(l - x) + (X/l)(l - x)$. *Mohr's* conjugate beam becomes a mechanism according to the analogy [Eq. (6.75)] and, hence, is kinematically indeterminate. If balanced,

$$\overline{M}(x = l) = 0 = \frac{2}{3} \frac{q_0\,l^2}{8}\,l\frac{l}{2} + X\frac{l}{2}\frac{2}{3}\,l \ ,$$

(6.79)

which is the additional equation for the unknown X. The solution is remarkable

$$X = M_e = - (q_0\,l^2)/8 \ .$$

(6.80)

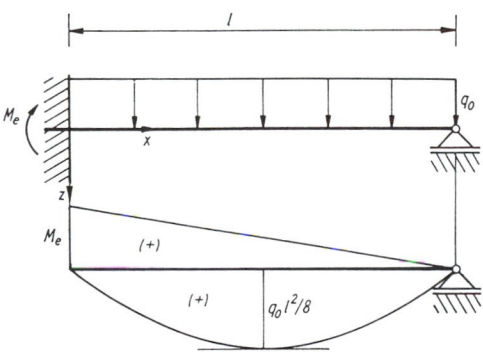

Fig. 6.12. Determination of the statically indeterminate couple $X = M_e$ by *Mohr's* method

The absolute value is larger than the analytic extreme of the bending moment in the field.

(§)   The Multispan Beam of Fig. 6.13. With given lateral load $q_z = q(x)$ it is a more complex redundant system and illustrates the power of *Mohr´s*   method for the determination of the statically unknown forces. Considering two out of the *n*   spans more closely, Fig. 6.13 shows the hinged-hinged associated subsystems connected at the supports by *Gerber*   hinges and additionally loaded by the unknown couples $X_k = M_k$, $k = 1, 2, ..., (n - 1)$; the actual bending moment is expressed in every field by the given loading and redundant edge couples. *Mohr´s*   conjugate beams have no supports at all except at the outer edges, and they are loaded by the resulting moments, reduced according to the bending rigidity that is assumed to be spanwise constant, as shown in Fig. 6.13. The resultant, ie the area under the moment line of the given loading, is denoted Φ_k and attached like a single force to the point of application a proper distance from the right support ξ_k according to static equivalence. Alternatively, a reduction to the adjacent nodes may be preferred. The kinematically indeterminate conjugate system is in equilibrium if (*k*   is the number of the intermediate support)

$$\overline{M}_{k-1} = 0 \text{ and } \overline{M}_{k+1} = 0 .$$

Hence, with $\overline{G}_k$ being the conjugate shear in the intermediate hinge, two independent equations result

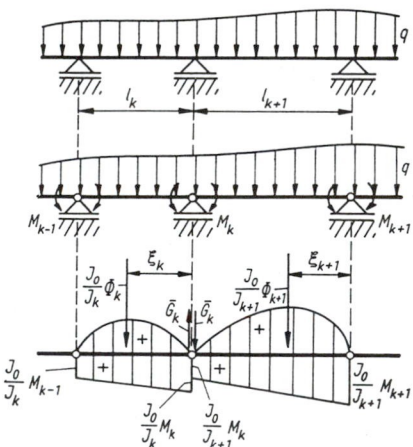

Fig. 6.13. Multispan beam with lateral loading *q(x)* . Associated statically determinate system loaded by the given load *q(x)* and by the unknown couples X_k. *Mohr´s* conjugate beam with *Gerber* joints loaded by the reduced moment distribution (a curvature)

$$\frac{J_0}{J_k} \, \Phi_k \, (l_k - \xi_k) + \frac{J_0}{J_k} \, M_{k-1} \, l_k \frac{l_k}{2}$$

$$+ \frac{1}{2} \left(\frac{J_0}{J_k} \, M_k - \frac{J_0}{J_k} \, M_{k-1} \right) l_k \frac{2 \, l_k}{3} - \overline{G}_k \, l_k = 0 \quad,$$

$$\frac{J_0}{J_{k+1}} \, \Phi_{k+1} \, \xi_{k+1} + \frac{J_0}{J_{k+1}} \, M_k \, l_{k+1} \frac{l_{k+1}}{2}$$

$$+ \frac{1}{2} \left(\frac{J_0}{J_{k+1}} \, M_{k+1} - \frac{J_0}{J_{k+1}} \, M_k \right) l_{k+1} \frac{l_{k+1}}{3} + \overline{G}_k \, l_{k+1} = 0 \quad.$$

The conjugate force at the intermediate node can be eliminated to render the *three moment equation* that is called *Clapeyron's* equation if $J_k = J_0$ is constant

$$\frac{l_k}{J_k} \, M_{k-1} + 2 \left(\frac{l_k}{J_k} + \frac{l_{k+1}}{J_{k+1}} \right) M_k + \frac{l_{k+1}}{J_{k+1}} \, M_{k+1}$$

$$= - 6 \left(\frac{l_k - \xi_k}{J_k \, l_k} \, \Phi_k + \frac{\xi_{k+1}}{J_{k+1} l_{k+1}} \, \Phi_{k+1} \right) \quad. \tag{6.81}$$

Writing down such an equation for each of the *(n − 1) Gerber* hinges of the conjugate beam yields an inhomogeneous system of linear equations with a banded structure that is readily solved.

Long beams of constant bending rigidity $B = EJ_0$ with many equi-distant supports, span l , may be considered in the limit to have infinite length. In the case of uniform lateral loading $q_z = q_0$, the bending moment at the support locations takes on the extreme value of the periodic distribution in the fields

$$M_k = - \frac{q_0 \, l^2}{12} \quad, \quad \text{since:} \quad M_k l = - \Phi \quad, \quad \Phi = \frac{2}{3} \, \frac{q_0 \, l^2}{8} \, l = \frac{q_0 \, l^3}{12} \quad. \tag{6.82}$$

In the case of multispan beams of finite length, the approximation becomes sufficiently accurate some distance from the ends.

6.2.3. Thermal Stresses in Beams

The temperature difference with respect to a constant reference temperature (of the stress-free state) in a beam may depend on the axial coordinate x and may be distributed over the cross-section as a possibly nonlinear function of (y, z) . In the transient heat conduction process, temperature varies with time. The temperature rate is assumed to be sufficiently small, such that every instant configuration can be considered to be in a state of equilibrium. Time is then only a parameter in the deformation process and inertia effects are negligible. Only thermal shock problems are truly nonstationary. Dependence on time in this quasistationary formulation is suppressed in the following formulas that are thus the stationary expressions where temperature is assumed to be a given function of the spatial coordinates. *Maysel's* formula (6.44) is applied and renders the thermoelastic displacements of the central points of the beam axis. Since $\sigma_{yy} = \sigma_{zz} = 0$, the hydrostatic influence stress is just

$$\bar{p} = \frac{\bar{\sigma}_{xx}}{3} \quad . \tag{6.83a}$$

That normal stress at a cross-section at ξ is due to a unit tensile force and a unit lateral single force loading (both forces are applied at x), and its linear distribution is related to the normal force and the principal bending moments [see Eq. (6.48)]

$$\bar{\sigma}_{xx}(\xi, y, z; x) = \frac{\bar{N}}{A} + \frac{\bar{M}_y}{J_y} z - \frac{\bar{M}_z}{J_z} y \quad . \tag{6.83b}$$

The mean influence stress is constant within the cross-section and the volume integration gives explicitly the thermoelastic elongation of the straight rod at some point x

$$\delta_x = \int_{\xi=0}^{\xi=1} \alpha \frac{\bar{N}(\xi, x)}{A} \, d\xi \int_A \theta(\xi, y, z) \, dA_{y,z} = \int_{\xi=0}^{\xi=1} \alpha n_\theta(\xi) \, \bar{N}(\xi, x) \, d\xi \quad , \tag{6.84}$$

where n_θ is the mean temperature at the cross-section at ξ

$$n_\theta(\xi) = \frac{1}{A} \int_A \theta(\xi, y, z) \, dA_{y,z} \quad . \tag{6.85}$$

The normal force at ξ, in general, equals the unit load $F = 1$ or it is zero.

The thermoelastic deflection in this first-order theory is independent of the axial deformation and, eg given by *Maysel's* formula by substituting the proper influence function

$$\delta = \int_{\xi=0}^{\xi=1} d\xi \int_A \left[\frac{\bar{M}_y(\xi, x)}{J_y} z \, \alpha\theta(\xi, y, z) - \frac{\bar{M}_z(\xi, x)}{J_z} y \, \alpha\theta(\xi, y, z) \right] dA_{y,z}$$

$$= \int_{\xi=0}^{\xi=1} \alpha \left[m_{\theta_y}(\xi) \, \bar{M}_y(\xi, x) + m_{\theta_z}(\xi) \, \bar{M}_z(\xi, x) \right] d\xi \quad . \tag{6.86}$$

The cross-sectional moments of first order of the temperature distribution are denoted by

$$m_{\theta y}(\xi) = \frac{1}{J_y} \int_A z\, \theta(\xi, y, z)\, dA_{y,z} \ , \quad m_{\theta z}(\xi) = -\frac{1}{J_z} \int_A y\, \theta(\xi, y, z)\, dA_{y,z} \ ,$$

(6.87)

and when multiplied by the linear thermal expansion coefficient α, they determine the principal thermal curvatures; see also Eqs. (6.33) and (6.35). The principal influence bending moments at ξ are due to a unit single lateral force loading $F = 1$ at x . By choosing the influence functions due to a unit external couple $M = 1$ at x instead of the force $F = 1$, *Maysel´s* formula (6.86) yields the thermoelastic rotation of the cross-section.

In conclusion, the thermoelastic displacements of the central points of the beam axis and the thermoelastic rotations of the cross-sections are given by a single integration over the span, provided the influence lines of the axial force and of the principal bending moments are known [see Sec. 2.5.1.2], and the mean temperature and principal thermal curvatures are given.

Substituting $w = \delta_z$ and $v = \delta_y$ into the generalized *Hooke´s* law (6.29) renders the thermal stress; the partial derivatives with respect to x may be taken under the integral sign of *Maysel´s* formula

$$\sigma_{xx}(x, y, z) = E\left[\frac{\partial \delta_x}{\partial x} - z\frac{\partial^2 \delta_z}{\partial x^2} - y\frac{\partial^2 \delta_y}{\partial x^2} - \alpha\theta(x, y, z)\right] \ .$$

(6.88)

A nonlinear temperature distribution over the cross-section renders the thermal stresses nonlinearly distributed. The nonlinear deviation from the linear function in *(y, z)* produces a system of self-equilibrating stresses; the resultants of these stresses are zero; see $\sigma_{xx}{}'''$ of Eq. (2.136). They are the only sources of thermal stresses in beams under statically determinate support conditions and of special importance, eg for bridges.

(§) The Single-Span, Simply Supported Beam. Bending rigidity $B = EJ_y$ and the load by the principal thermal curvature $\alpha m_{\theta y} = \alpha m_0$ are both given constants The influence line of the bending moment is given by Eq. (5.12), and *Maysel´s* formula (6.86) renders the thermal deflection free of any resulting (thermal) bending moment

$$w(x) = \int_{\xi=0}^{\xi=x} \alpha m_0\left(1 - \frac{x}{l}\right)\xi\, d\xi + \int_{\xi=x}^{\xi=l} \alpha m_0\, x\left(1 - \frac{\xi}{l}\right) d\xi = \frac{\alpha m_0\, l^2}{2}\,\frac{x}{l}\left(1 - \frac{x}{l}\right) \ .$$

(6.89)

Hence, the linearized curvature is constant,

$$d^2w/dx^2 = -\alpha m_0 \quad , \tag{6.90}$$

and *Hooke's* law renders the nonlinear thermal stresses

$$\sigma_{xx}(z) = E\alpha \left[n_\theta + z\, m_\theta - \theta(z) \right] \quad , \tag{6.91}$$

which are self-equilibrating, ie the resultants must vanish due to the statically determinate support conditions. At the traction-free ends of the beam, however, these stresses should vanish. If they do not vanish, *Saint Venant's* principle is assumed to hold and the solution becomes a good approximation at a sufficiently large distance from the ends. The stresses vanish altogether if the temperature is linearly distributed over the height H of the beam

$$\theta(z) - n_\theta = \frac{\theta_l - \theta_u}{H}\, z \quad , \tag{6.92}$$

where the difference of temperature of the lower and the upper surface enters. The first-order moment in a rectangular cross-section takes on the value

$$m_\theta = \frac{1}{J_y} \int_A z\, \theta(z)\, dA = \frac{\theta_l - \theta_u}{H} \quad , \tag{6.93}$$

and, hence, in this special case, it is easily verified that the term in brackets in Eq. (6.91) vanishes. A sliding support must allow the elongation of the beam due to the mean thermal strain αn_θ, which is given by

$$\alpha n_\theta = \alpha(\theta_l + \theta_u)/2 \quad , \tag{6.94}$$

for the rectangular cross-section.

(§) A Redundant Single-Span Beam. This beam with constant bending stiffness $B = EJ_y$, rigidly clamped on both ends, is loaded by a constant principal thermal curvature αm_0. The influence line of the bending moment is determined first under isothermal conditions

$$\overline{M}_y(\xi, x) = \left(1 - \frac{x}{l}\right)\left[\left(1 + \frac{x}{l} - 2\frac{x^2}{l^2}\right)\xi - \left(1 - \frac{x}{l}\right)x\right] \quad , \quad 0 \leq \xi \leq x \quad ,$$

$$\overline{M}_y(\xi, x) = \left(1 - \frac{x}{l}\right)\left[\left(1 + \frac{x}{l} - 2\frac{x^2}{l^2}\right)\xi - \left(1 - \frac{x}{l}\right)x\right] - \xi + x \quad , \quad x \leq \xi \leq l \quad . \tag{6.95}$$

Hence, *Maysel's* formula (6.86) causes the thermal deflection to vanish identically

$$w(x) = \alpha m_0 \int_0^l \overline{M}_y(\xi, x)\, d\xi \equiv 0 \quad .$$

(6.96)

The thermal stresses are

$$\sigma_{xx}(z) = -E\, \alpha \theta(z) \quad ,$$

(6.97)

and the resulting bending moment is constant (it equals the clamping moment)

$$M_y = \int_A z\sigma_{xx}(z)\, dA = -E\,\alpha \int_A z\theta(z)\, dA = -E J_y\, \alpha m_0 \quad .$$

(6.98)

In that case, the solution of the extended differential equation of the deflection [see Eq. (6.36)], together with the homogeneous kinematic constraints $w = w' = 0$ at $x = 0$ and $x = l$, is simpler than the application of *Maysel's* formula. Note that the reaction forces at the supports must be zero for reasons of symmetry.

6.2.4. Torsion

The torsion of a straight bar with a thin-walled single cell and constant cross-section gives a statically determinate shear flow, $T = \sigma_{xs} h$, which is related to the torque by the first of *Bredt's* formulas, Eq. (2.171). The linear elastic deformation is considered below, and its extension to statically redundant problems of multiple-cell cross-sections is further discussed. The introduction of the warping function allows for the analysis of the combined loading in torsion and shear that is of special importance for thin-walled, open cross-sections with small torsional rigidity. Hence, the center of twist gets special attention. Torsion with constrained warping is considered in some detail and illustrated for a C-profile. The twisting of fully bodied cross-sections, also with notched surfaces, and *Prandtl's* membrane analogy in torsion conclude this introduction to the engineering-oriented theory.

6.2.4.1. Thin-Walled, Single- and Multiple-Cell Cross-Sections

The constant shear flow in a single cell is related to the torque by the first of *Bredt's* formulas, Eq. (2.171), derived in the year 1896

$$M_T = 2 A T \quad ,$$

(6.99)

where A is the area within the contour of the cross-section. Pure torsion, ie *Saint Venant's* torsion, is the mechanical state of such a bar with constant cross-section under the action of a constant torque, if the warping deformation in the axial direction, if any, is not restricted. The (relative) *angle of twist* , $\Theta = d\chi/dx$, becomes a constant, and the relative rotation of two cross-sections at a distance x thus becomes the linear function $\chi = \Theta x$. Due to the rigid rotation of each cross-section, a point of the contour at a normal distance p from an arbitrary center is displaced in the tangential direction by $p\,\chi(x)$ and in the axial direction by $u = \Theta\,\varphi(s)$. s is the arc-length of the contour and $\varphi(s) = u(s)/\Theta$ the warping function that must be independent of x . The torsional angle χ is assumed to be sufficiently small, and the linearized geometric relations, Eqs. (1.21) and (1.31), render the shear strain

$$\gamma_{xs} = \frac{\partial(p\chi)}{\partial x} + \frac{\partial u}{\partial s} = \Theta\left(\frac{\partial\varphi}{\partial s} + p\right) .$$

(6.100)

Integration over the contour of the cell yields (the warping function is steady)

$$\oint \gamma_{xs}\, ds = \Theta \oint p\, ds = 2\,A\,\Theta ,$$

(6.101)

which is still independent of any constitutive relation. By means of *Hooke´s* law [see Eq. (4.8)],

$$\gamma_{xs} = \sigma_{xs}/G = T/G\,h = M_T/2\,G\,A\,h ,$$

(6.102)

Equation (6.101) becomes, when solved for the angle of twist, the *second of Bredt´s formulas* ; the wall thickness h may be variable

$$\Theta = \frac{M_T}{4G\,A^2} \oint \frac{ds}{h} .$$

(6.103)

Twist and torque are linearly related in linear elastic torsion, $\Theta = M_T / GJ_T$, and the torsional rigidity of the bar with a single-cell cross-section becomes

$$J_T = \frac{4A^2}{\oint \dfrac{ds}{h}} , \quad M_T = GJ_T\,\Theta .$$

(6.104)

$\oint \dfrac{ds}{h} = \dfrac{L}{h}$ for constant wall thickness. For a circular thin-walled tube, the perimeter is $L = 2\pi R$, and with $A = \pi R^2$ and $\Theta = M_T / GJ_T$, the torsional rigidity takes on the power law $J_T = 2\pi h\, R^3$.

Bredt´s formulas for a single cell remain valid also for a variable shear flow, eg in a multiple-cell cross-section where the individual cells are separated by intermediate webs; see Fig. 6.14 for a single web, C_3 . The shear flow must be constant in each part of the wall, but in a multiple-cell cross-section, the shear flow in the webs becomes statically indeterminate. However, the additional condition of a common deformation applies: The whole cross-section rotates rigidly and so do the single-cell subsystems. The shear flow at the intersections (nodes) is balanced analogously to the rate of mass flow of an incompressible fluid [see Eq. (1.86)]. Hence, according to Fig. 6.14, the incoming shear flow T_2 branches off into the shear flow T_1 and the shear flow T_3 in the web

$$T_2 = T_3 + T_1 \; . \tag{6.105a}$$

The same is true for the opposite node. The first of *Bredt´s* formulas is applied separately to each single-cell subsystem, and hence, the superposition of n cells yields the resulting torque

$$2 \sum_{i=1}^{n} A_i T_i = M_T \quad , \quad n = 2 \text{ in Fig. 6.14 .} \tag{6.105b}$$

A_i , $i = 1, 2, ..., n$ are the interior areas of the single-cell subsystems. Considering the shear flow to be constant in each individual cell, and clockwise in Fig. 6.14, renders in the web C_3 the shear flow by superposition and according to condition (6.105a). The second of *Bredt´s* formulas gives, according to the condition of a common angle of twist, for the two cells of Fig. 6.14,

$$2G\Theta = \frac{1}{A_1} \left(T_1 \int_{C_1} \frac{ds}{h} - T_3 \int_{C_3} \frac{ds}{h} \right) = \frac{1}{A_2} \left(T_2 \int_{C_2} \frac{ds}{h} + T_3 \int_{C_3} \frac{ds}{h} \right) . \tag{6.106}$$

Hence, with $n = 2$, Eqs. (6.105a, b) and (6.106) determine the three

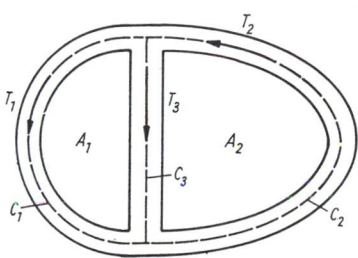

Fig. 6.14. Multiple-cell cross-section with one intermediate web. Branching of the shear flow

unknown shear flows T_1, T_2, T_3 , as well as the angle of twist Θ .

If the shear flow happens to vanish in a web, like $T_3 = 0$ in the case of a special geometry in Fig. 6.14, the web does not contribute to the torsional rigidity at all (it is blind).

The mean value of the shear stress is still given by $\sigma_{xs} = T_i /h$, but at the corners (the nodes) of the thin-walled cross-section, where the webs are attached to the contour, stress concentrations are encountered. At the inner surface of such a rounded shoulder, the peak of the shear stress depends on the ratio of the radius of curvature over the wall thickness. The geometric magnification factor, the peak value over the undisturbed mean stress, was calculated by *J. H. Huth in J. Appl. Mech. 17, ASME, New York, 1950, p. 388* . The results are reproduced in Fig. 6.15. Refer to the velocity distribution of the incompressible flow around a corner; see Fig. 13.12.

Combining the shear strain as expressed by Eq. (6.100) with *Hooke's* law (6.102) and integrating over the contour render the warping function; the shear flow may be variable in the sense discussed above,

$$\varphi(s) = \varphi_0 + \frac{1}{G\Theta} \int_0^s T \frac{ds}{h} - \int_0^s p \, ds \ .$$

$$(6.107)$$

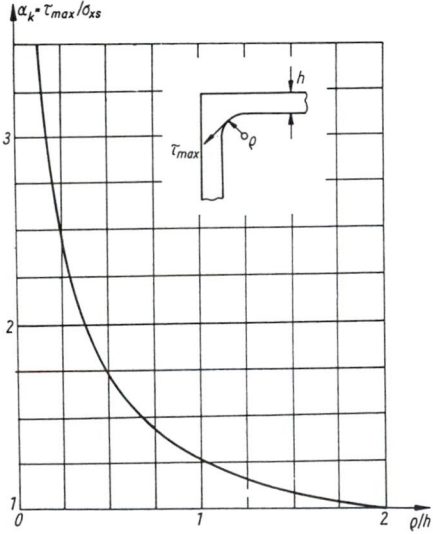

Fig. 6.15. Geometric stress magnification factor of the shear stress in a corner with radius of curvature ρ

Substituting the constant shear flow of pure torsion $T = M_T /2A$ and denoting the triangular area covered by a rotation of the normal p by $a(s)$ yield the special axial deformation $u(s)$, yet undetermined due to the integration constant

$$\varphi(s) = \varphi_0 + \frac{M_T}{2GA\Theta} \int_0^s \frac{ds}{h} - 2\,a(s) = \frac{u(s)}{\Theta} \quad , \quad a(s) = \frac{1}{2} \int_0^s p\,ds \quad .$$

If the center of shear is selected as the pole of rotation of the cross-section [see Eq. (6.118)], the axial displacement is fully determined. The substitution of the second of *Bredt's* formulas [Eq. (6.103)] renders, within the limits of the linear theory of elasticity the warping function related to the geometry of the cross-section

$$\varphi_0 - \varphi(s) = 2 \left[a(s) - \frac{A}{B}\, b(s) \right] \quad , \tag{6.108}$$

where

$$a(s) = \frac{1}{2} \int_0^s p\,ds \ , \ b(s) = \int_0^s \frac{ds}{h} \ \text{and} \ B = b(L) = \oint \frac{ds}{h} \ , \ L = \oint ds \ . \tag{6.109}$$

Equations (6.108) and (6.109) are evaluated for a thin-walled circular tube, where the center of shear is identical to the cross-sectional centroid, to render $a(s) = Rs/2, b(s) = s/h, A = \pi R^2, B = 2R\pi/h$, and hence, $\varphi(s) = \varphi_0$. The cross-section is free of any warping. For similar reasons of symmetry, other cross-sections are free of warping, eg a cell of constant wall thickness with a contour given by a closed polygon of tangents. The tube with a cross-section of triangular shape is free of warping also for the more general case where each side varies in thickness. The importance of such a property of no warping is emphasized with respect to structural applications.

More often actual loading conditions render a combination of torsion and shear bending; see Sec. 2.5.1 for the static considerations. The torque can be combined with the nonvanishing shear force to render the parallel resulting force without a moment. Hence, the location of the *center of shear* is determined through the condition of the equal axial moment of the shear force with that special point of application and that of the "given" shear flow T in the cross-section. The component Q_z of the shear force is considered first, $Q_y = 0$. Equation (6.55) reduces in that case to

$$T(s) = T_0 - \frac{Q_z}{J_y}\, S_y(s) \quad , \quad S_y(s) = \int_0^s z\, h(s)\, ds \quad , \tag{6.110}$$

y and z are the principal central axes of the cross-section. The condition of equal axial moment about the cross-sectional centroid renders (the condition of static equivalence is applied)

$$Q_z \, y_D = \oint T \, p(s) \, ds \quad,$$

(6.111)

and the central coordinate of the center of shear becomes

$$y_D = 2A \frac{T_0}{Q_z} - \frac{1}{J_y} \oint p(s) S_y(s) \, ds \quad.$$

(6.112)

If we note the differential static moment $dS_y = zh \, ds$ and $p \, ds = 2d[a(s)]$, a partial integration gives with $S_y(0) = S_y(L) = 0$

$$y_D = 2 \left[\frac{1}{J_y} \oint z \, a(s) \, h \, ds + A \frac{T_0}{Q_z} \right] \quad.$$

(6.113)

Analogously, the coordinate z_D of the center of shear is derived by considering the shear component Q_y and putting $Q_z = 0$, thus replacing J_y by J_z and the factor z by y and changing the sign in Eq. (6.113). The statically indeterminate shear flow T_0 follows from the condition that the angle of twist vanishes if the shear force Q_z is applied at the center of twist: Taking the whole loop of integration of Eq. (6.107) gives [the warping function is steady, $\varphi(L) = \varphi(0)$]

$$\frac{1}{G\Theta} \oint \frac{T}{h} \, ds = 2A \quad.$$

(6.114)

Substituting the shear flow of Eq. (6.110) yields, with $\Theta = 0$,

$$T_0 \oint \frac{ds}{h} - \frac{Q_z}{J_y} \oint S_y(s) \frac{ds}{h} = 0 \quad.$$

(6.115)

The equation, in general, becomes at least approximate if the shear force varies with the axial coordinate, $Q_z(x)$. Elimination of the constant shear flow T_0 in Eq. (6.113) and partial integration of the loop integral of the above equation give finally the coordinate of the center of shear related to the geometry of the cross-section (the expression holds within the limits of linear elasticity, but no elastic parameter is apparent)

$$y_D = \frac{2}{J_y} \oint z\left[a(s) - \frac{A}{B}\,b(s)\right] h\, ds \quad .$$

(6.116)

Analogously,

$$z_D = -\frac{2}{J_z} \oint y\left[a(s) - \frac{A}{B}\,b(s)\right] h\, ds \quad .$$

(6.117)

The term in brackets in the above integrals equals the warping function $[\varphi_0 - \varphi(s)]$, Eq. (6.108). Termwise integration renders the central coordinates of the center of shear; the static cross-sectional moments vanish,

$$y_D = -\frac{1}{J_y} \oint z\,\varphi(s)\, h\, ds \quad , \quad z_D = \frac{1}{J_z} \oint y\,\varphi(s)\, h\, ds \quad .$$

(6.118)

Since $h\, ds = dA$ is the area element of the thin-walled cross-section, the line integrals may be considered area integrals over the materialized area of the cross-section. The warping function is determined for all shapes of cross-sections, also for those of fully bodied rods. Thus, Eq. (6.118) when considered a cross-sectional area integral determines the center of shear in general; it is not restricted by the assumption of a thin-walled structure. See E. Trefftz, ZAMM **15**, (1935), p. 220 . The shear force should be constant. The formulas are free of any elastic constants but it should be kept in mind that they are restricted to linearized elasticity. Nonlinear material behavior, in general, influences the location of the center of shear.

6.2.4.2. Thin-Walled Open Cross-Sections

The torsional rigidity of a rod with an open cross-section is very low, and a torque produces large twists as well as large axial deformations. Shear stresses σ_{xs} must be tangential to the traction-free surfaces, but must be variable over the thickness h for the formation of couples. The shear flow of the torsional part of the shear stresses is zero. The warping function of Eq. (6.107) for pure torsion becomes with $T = 0$

$$\varphi(s) = \varphi_0 - 2\,a(s) \quad , \quad a(s) = \frac{1}{2}\int_0^s p(s)\, ds \quad .$$

(6.119)

For example, the ends of a circular tube (with a slit) of radius R are moved apart in the axial direction by the action of a torque $M_T = (GJ_T$

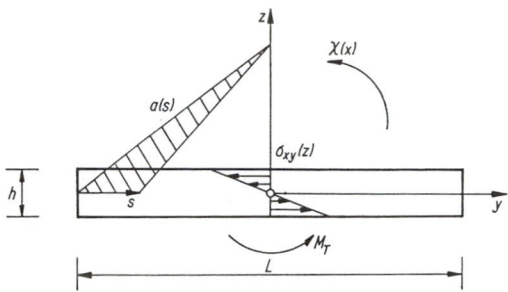

Fig. 6.16. Torsion of a bar with a slim rectangular cross-section, $L \gg h$

Θ) . The open cross-section in this case is symmetric and the areal centroid may be chosen for reference (any point of the axis of symmetry is suitable), $a(s) = R \, s/2$. And Eq. (6.119) gives when multiplied by Θ

$$u(s = 0) - u(s = 2R\pi) = \Theta \left[\varphi(0) - \varphi(2R\pi) \right] = 2\pi R^2 \, \Theta \quad , \tag{6.120}$$

unconstrained warping of the cross-sections is understood.
(§)   Torsion of a Thin-Walled Bar with Rectangular Cross-Section. The cross-section is doubly symmetric and the centroid coincides with the center of shear. Assuming $L \gg h$ in Fig. 6.16 and considering the area $a(s) = zs/2$ (z is chosen with respect to a point of reference at the axis of symmetry) render an approximation of the warping function by $\varphi(s) = \varphi_0 - zs$. Arbitrarily, $\varphi(s = L/2) = 0$, and putting $y = s - L/2$ renders the approximation in the form

$$\varphi(y, z) = -y \, z \quad . \tag{6.121}$$

y and z are a pair of central axes. Given an angle of twist per unit of length, the axial deformation is approximated by

$$u(y, z) = \Theta \, \varphi(y, z) = - \Theta \, y \, z \quad . \tag{6.122}$$

The other two displacement components are given for small torsional angles by (the cross-section rotates rigidly)

$$v(x, y, z) = - z \, \chi(x) = - z \, \Theta \, x \quad , \quad w(x, y, z) = y \, \chi(x) = y \, \Theta \, x \quad . \tag{6.123}$$

The deformation gradients, ie the derivatives of the above displacements, enter Hooke's law [Eq. (4.8)], and a linear distribution results

$$\sigma_{xy} = G\left(\frac{\partial u}{\partial y} + \frac{\partial v}{\partial x}\right) = G\left(-\Theta z - \Theta z\right) = -2G\Theta z \ ,$$

$$\sigma_{xz} = G\left(\frac{\partial u}{\partial z} + \frac{\partial w}{\partial x}\right) = G\left(-\Theta y + \Theta y\right) = 0 \ .$$

$$(6.124)$$

The maximal torsional shear stress appears at the mid-point of the longer side of the rectangle at $(y = 0, z = \pm h/2)$ and is given, sufficiently accurate, by the first of the approximations of Eq. (6.124)

$$\max |\sigma_{xy}| = G\,\Theta h \ . \qquad (6.125)$$

The shear stresses along the short side $h \ll L$ are neglected due to the second approximation in Eq. (6.124), and the condition of a traction-free surface is violated. *Lord Kelvin* already has addressed the fact that the couples of the shear-stress distribution of Eq. (6.124) do not even sum up to the given torque

$$M_x = \int_A \left(y\sigma_{xz} - z\sigma_{xy}\right) dA = 2G\Theta \int_A z^2\, dA = 2G\Theta\, J_y \neq M_T = G\Theta\, J_T \ . \qquad (6.126)$$

The magnified error in Eq. (6.124) is due to the differentiation of the approximate warping function. The resulting moment in Eq. (6.126) is related to the couples of the "large" components σ_{xy} with small lever arms of a length up to h and should contain the contributions that are of the same order of the "small" shear-stress components σ_{xz} (neglected above) that have large lever arms up to the length L. Hence, a better approximation of the torsional rigidity GJ_T is needed. Instead of the approximation of the warping function, the *stress function of torsion* is considered. See Eq. (2.11) and find the shear stresses by the derivatives

$$\sigma_{xy} = 2G\Theta\frac{\partial\Psi}{\partial z} \quad , \quad \sigma_{xz} = -2G\Theta\frac{\partial\Psi}{\partial y} \quad , \quad \Psi = \Psi(y, z) \ . \qquad (6.127)$$

Since $\sigma_{xx} = 0$, the local condition of equilibrium is identically satisfied

$$\frac{\partial\sigma_{xy}}{\partial y} + \frac{\partial\sigma_{xz}}{\partial z} = 2G\Theta\left(\frac{\partial^2\Psi}{\partial z\partial y} - \frac{\partial^2\Psi}{\partial y\partial z}\right) \equiv 0 \ . \qquad (6.128)$$

The integration of the approximate stress distribution of Eq. (6.124) yields the stress function, the errors are now somewhat reduced by smoothing, and the approximation is as good as that of the warping function (6.121)

$$\Psi(y, z) = -\int z \, dz = C - z^2/2 \quad . \tag{6.129}$$

The dynamic boundary condition of the traction-free surface with the contour $z = z(y)$ can be considered only at the long side $z = \pm\, h/2$. With Eq. (6.127), it becomes a condition for the stress function

$$\frac{dz}{dy} = \frac{\sigma_{xz}}{\sigma_{xy}} \to \sigma_{xy} \, dz - \sigma_{xz} \, dy = 2G\Theta \, d\Psi = 0 \to \Psi = \text{const} \quad .$$

There is only one closed contour; hence, the stress function is assumed to vanish at the free surface of the rectangle. The integration constant $C = h^2/8$ and

$$\Psi = \frac{h^2}{8}\left(1 - 4\,\frac{z^2}{h^2}\right) \quad . \tag{6.130}$$

The approximation does not depend on y; the boundary condition at $y = \pm\, L/2$ is not satisfied. The axial moment (6.126) is considered by substituting the stress function, Eq. (6.127), and performing an identical transformation and integration

$$M_x = \int_A (y\sigma_{xz} - z\sigma_{xy}) \, dA = -2G\Theta \int_A \left(y\,\frac{\partial\Psi}{\partial y} + z\,\frac{\partial\Psi}{\partial z}\right) dA$$

$$= -2G\Theta \left[-2 \int_A \Psi \, dA + \iint_A \frac{\partial(y\Psi)}{\partial y} \, dy \, dz + \iint_A \frac{\partial(z\Psi)}{\partial z} \, dz \, dy\right]$$

$$= 2G\Theta \left[2 \int_A \Psi \, dA - \int_{z_1}^{z_2} \{y_2(z) - y_1(z)\} \, \Psi \, dz + \int_{y_2}^{y_1} \{z_2(y) - z_1(y)\} \, \Psi \, dy\right]$$

$$= 2G\Theta \left[2 \int_A \Psi \, dA + \oint_{\partial A} \Psi(y, z) \, (z \, dy - y \, dz)\right] \quad . \tag{6.131}$$

The formula holds for general shapes of A. For a simply connected cross-section, the stress function is put to zero on the contour ∂A and the torque is simply given by

$$M_T = G\Theta \, 4 \int_A \Psi \, dA = G\Theta \, J_T \quad , \quad J_T = 4 \int_A \Psi \, dA \quad . \tag{6.132}$$

The substitution of Eq. (6.130) renders the torque and, thus, the torsional rigidity of the rectangle by integration of the approximate stress function; thereby the errors are further reduced by smoothing

$$M_T = G\Theta\, 4\, \frac{h^2}{8} \int_A \left(1 - 4\frac{z^2}{h^2}\right) dA = G\Theta\frac{h^2}{2}\left(A - \frac{4}{h^2} J_y\right) = G\Theta\, J_T \ .$$

(6.133)

The moment of inertia is $J_y = L\, h^3/12$ and the torsional stiffness of a slim rectangle is well approximated by the expression

$$J_T = \frac{L\, h^3}{3} \ .$$

(6.134)

By comparing Eqs. (6.133) and (6.126), it is easily seen that the couples of σ_{xy} are equivalent to the axial moment $M_x = M_T/2$. Thus, the missing half of the torque is contributed by the couples of σ_{xz} , the neglected component of the torsional shear stress. The stress distribution of Eq. (6.124), which is derived by taking the derivative of an approximation, is useful only in the neighborhood of the cross-sectional centroid, $|z| \le h/2$, $|y| \ll L/2$, but approximates the maximal shear sufficiently well.

(§) Generalization of Eq. (6.134). The approximations for the maximal shear stress and the torsional rigidity derived for the slim rectangle are generalized to include variable wall thickness $h(s)$ and curved contours of thin-walled open cross-sections

$$\max |\sigma_{xs}| = G\Theta\, h_{max} \quad , \quad J_T = \frac{1}{3} \int_0^L h^3\, ds \quad ,$$

(6.135)

where s is the arc-length measured along the contour. Especially, the torsional rigidity of an open profile consisting of rectangles joined together, eg by welding, is approximately proportional to the sum

$$J_T = \frac{1}{3} \sum_{i=1}^{n} h_i^3\, L_i \quad .$$

(6.136)

The rigidity of a circular tube with a slit of constant thickness is given by

$$J_T = \frac{1}{3} \int_0^{2\pi} h^3\, R\, d\theta = 2\pi R\, \frac{h^3}{3} \quad ,$$

(6.137)

and turns out to be much smaller than that of a circular pipe of closed cross-section

$$J_T = 2\pi h\, R^3 \quad .$$

(6.138)

The ratio of the angle of twist becomes large, namely, $3R^2/h^2 \gg 1$.

In conclusion, beams of open cross-sections should be kept free of torsion when laterally loaded. Their torsional rigidity is small and the angle of twist as well as the warping deformation becomes large even for small to moderate torques. Hence, the center of shear is of crucial importance as the point of application of the lateral loading. It is still given by Eq. (6.118)

$$y_D = -\frac{1}{J_y} \oint z\varphi(s)\, h\, ds \; , \quad z_D = \frac{1}{J_z} \oint y\varphi(s)\, h\, ds \; , \quad a(s) = \frac{1}{2} \int_0^s p\, ds \; ,$$

$$\varphi(s) = \varphi_0 - 2\, a(s) \; . \tag{6.139}$$

In that case, the shear force may be variable. The shear flow in open cross-sections due to shear bending and torsion is related only to the shear force [see Eq. (6.55)]

$$T(s) = -\frac{Q_z}{J_y} S_y(s) + \frac{Q_y}{J_z} S_z(s) \; . \tag{6.140}$$

The associated mean stress $T(s)/h$ is to be superposed to the torsional shear stress in case of combined loadings.

(§) Constrained Warping. Pure torsion needs freedom of deformation in the axial direction. In engineering practice, the warping is always more or less constrained, eg at the supports of the beam: The clamping of a cantilever in torsion may illustrate the constrained warping. Under such conditions, the stress state changes considerably also in the first order. The effect is pronounced in bars with open cross-sections. The angular rotation of the cross-sections must now be considered about a specified axis (which passes through the center of shear). To simplify the theory of torsion under the condition of constrained warping, the warping function of Eq. (6.119) is transformed and the center of shear is chosen to be the point of reference. The cross-sectional coordinates are changed to $y^* = y - y_D$, $z^* = z - z_D$; $u = \Theta\, \varphi$, $u = \Theta\varphi^*$, $v = -z\chi$, $w = y\chi$. And the shear stresses in pure torsion given by *Hooke's* law must be invariant under the coordinate translation

$$\sigma_{xy} = G\left(\frac{\partial u}{\partial y} + \frac{\partial v}{\partial x}\right) = G\Theta\left(\frac{\partial \varphi}{\partial y} - z\right) = G\Theta\left(\frac{\partial \varphi^*}{\partial y^*} - z^*\right) \; , \tag{6.141}$$

$$\sigma_{xz} = G\left(\frac{\partial u}{\partial z} + \frac{\partial w}{\partial x}\right) = G\Theta\left(\frac{\partial \varphi}{\partial z} + y\right) = G\Theta\left(\frac{\partial \varphi^*}{\partial z^*} + y^*\right) \; . \tag{6.142}$$

Integration and comparing the results give a linear transformation of the warping function

$$\varphi^* = \varphi - y\, z_D + z\, y_D \quad . \tag{6.143}$$

The basic assumption in the theory of constrained warping, in which the angle of twist becomes variable along the span $\Theta(x)$, puts the axial displacement in the form of separable functions: The product is analogous to pure torsion, but with the center of shear as the origin of the coordinate system

$$u(x, s) = \Theta(x)\varphi^*(s) \quad . \tag{6.144}$$

Since $\sigma_{yy} = \sigma_{zz} = 0$, according to the theory of thin rods, the warping stresses are derived from the uniaxial *Hooke´s* law

$$\sigma_{xx}^*(x, s) = E\frac{\partial u}{\partial x} = E\,\varphi^*(s)\,\frac{d\Theta(x)}{dx} \;\rightarrow\; \int_A \sigma_{xx}^*\, dA = \int_A y\sigma_{xx}^*\, dA = \int_A z\sigma_{xx}^*\, dA = 0 \,, \tag{6.145}$$

and are distributed as a system of self-equilibrating normal stresses in every cross-section. Hence, the constant of the warping function is determined through the equation

$$\int_A \varphi^*(s)\, dA = \int_A \varphi(s)\, dA = \varphi_0\, A - 2\int_A a(s)\, dA = 0 \,. \tag{6.146}$$

The remaining conditions of vanishing principal axial moments render the center of shear as the torsional center of rotation (the separable function approximation enters, however)

$$\int_A y\varphi^*(s)\, dA = \int_A y\varphi(s)\, dA \,- z_D\, J_z = 0 \,, \tag{6.147}$$

$$\int_A z\varphi^*(s)\, dA = \int_A z\varphi(s)\, dA \,+ y_D\, J_y = 0 \,. \tag{6.148}$$

The self-equilibrating warping normal stresses are variable in the axial direction and, thus, an additional shear flow is required for balancing the forces in the free-body diagram of an infinitesimal element of the thin-walled bar [refer to Eq. (6.52) and see Fig. 6.4]

$$h\frac{\partial \sigma_{xx}^*}{\partial x} + \frac{\partial T^*}{\partial s} = 0 \quad . \tag{6.149}$$

Integration yields, if we take into account the homogeneous boundary condition at the free end of the open cross-section $T^*(s = 0) = 0$,

$$T^*(x, s) = -\int_0^S z \frac{\partial \sigma_{xx}^*}{\partial x} \, ds = -E \frac{d^2\Theta}{dx^2} \int_0^S h\varphi^*(s) \, ds \; .$$

$$(6.150)$$

The additional warping shear stresses σ_{xs}^* and the corresponding shear flow T^* in the above given approximation do not produce any additional shear deformation. Measuring the normal distance of the area elements from the center of shear by p^* renders the in-plane displacements through the rigid rotation by $p^* \chi$, $(\partial\chi/\partial x = \Theta)$ and the transformed warping function. Hence, the shear strain vanishes

$$\varphi^*(s) = \varphi_0 - \int_0^S p^* \, ds \; , \quad \gamma_{xs}^* = \frac{\partial u}{\partial s} + \frac{\partial(p^*\chi)}{\partial x} = \Theta \frac{d\varphi^*}{ds} + p^*\Theta = 0 \; .$$

$$(6.151)$$

The warping shear flow T^*, which is tangential to the contour, contributes to the torque by the axial moment about the x axis that passes through the center of shear

$$M_x^*(x) = \int_0^L T^* p^* \, ds = -E \frac{d^2\Theta}{dx^2} \int_0^L p^*(s) \, ds \int_{\sigma=0}^{\sigma=s} h\varphi^*(\sigma) \, d\sigma \; .$$

Partial integration renders with the differential $d\varphi^* = -p^* \, ds$ and, by considering the cross-sectional integral of the warping function as vanishing,

$$M_x^*(x) = -E \frac{d^2\Theta}{dx^2} \int_0^L \varphi^{*2} h \, ds = -E C^* \frac{d^2\Theta}{dx^2} \; .$$

$$(6.152)$$

The constant that depends on the shape of the cross-section

$$C^* = \int_0^L \varphi^{*2} h \, ds = \int_A \varphi^{*2} \, dA \; , \quad [C^*] = m^6 \; ,$$

$$(6.153)$$

may be called the geometric coefficient of the warping rigidity. It has a dimension of length to the power of six. Hence, the torque is given by the superposition of the axial moment in pure torsion and the portion due to the constrained warping; J_T is given by Eq. (6.135)

$$M_T = GJ_T \Theta(x) + M_x^*(x) \; .$$

$$(6.154)$$

The substitution of Eq. (6.152) renders the linear second-order differential equation of the angle of twist, in torsion with constrained warping,

$$\frac{d^2\Theta}{dx^2} - a^{-2}\,\Theta = -\frac{M_T}{E\,C^*} \quad , \quad a^2 = 2(1+\nu)C^*/J_T \ .$$
(6.155)

Assuming the given torque, $M_T = const$ and considering the constant particular integral yield the general solution

$$\Theta(x) = \frac{M_T}{G\,J_T}\left(1 + D_1 \cosh\frac{x}{a} + D_2 \sinh\frac{x}{a}\right) \quad , \quad a = \sqrt{2(1+\nu)C^*/J_T} \ .$$
(6.156)

The constants of integration $D_{1,\,2}$ are determined by two boundary conditions: A rigidly clamped edge requires the axial displacements to vanish $u = 0$ and compatible with that condition, the angle of twist $\Theta = 0$. At the free end of unconstrained warping, the warping normal stress must vanish, $\sigma_{xx}^* = 0$, and hence, $d\Theta/dx = 0$.

(§)   The Cantilever with a C-Profile of Fig. 6.17. This cantilever is considered as an illustrative example of combined loading in shear bending and torsion with constrained warping. The single force loading F at the tip passes through the fillet (web). With the orientation of the coordinates shown in the figure, the shear force and principal bending moment become

$$Q_z = -F \quad , \quad M_y = F\,(l-x) \ .$$
(6.157)

Also, the shear force passes through the fillet. The torque can be

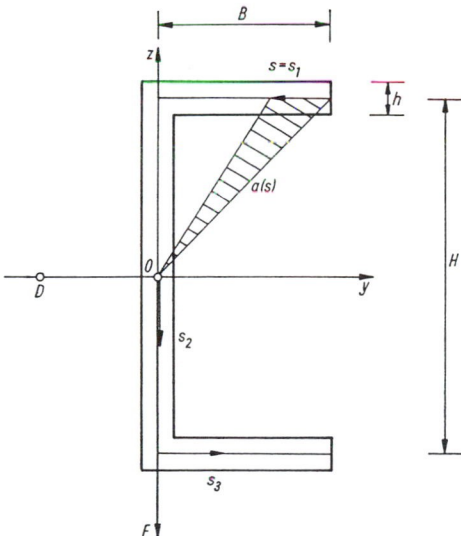

Fig. 6.17. A cantilever of length *l* with a C-profile loaded at the tip by F . The center of shear D

determined only if the center of shear is known. At first, the warping function (6.119) is to be determined and the constant is chosen such that Eq. (6.146) holds. The point of reference 0 is selected to lie on the symmetry axis for convenience (instead of the cross-sectional centroid) and s is counted counterclockwise:

Upper flange: $a(s_1) = \dfrac{1}{2}\dfrac{H}{2} s_1$, $\varphi = \varphi_0 - 2\, a(s_1)$, $\varphi^* = \dfrac{H}{2}(B - s_1 + y_D)$,

Web: $a(s_2) = \dfrac{1}{2}\dfrac{H}{2} B = const$, $\varphi = \varphi_0 - \dfrac{H}{2} B$, $\varphi^* = \left(\dfrac{H}{2} - s_2\right) y_D$,

Lower flange: $a(s_3) = \dfrac{1}{2}\dfrac{H}{2} B + \dfrac{1}{2}\dfrac{H}{2} s_3$, $\varphi = \varphi_0 - \dfrac{H}{2}(B + s_3)$, $\varphi^* = -\dfrac{H}{2}(s_3 + y_D)$.

Due to the symmetry of the C-profile, $z_D = 0$, and Eq. (6.143) reduces to $\varphi^* = \varphi + z\, y_D$. The static moment with respect to the central y axis vanishes and the integral of Eq. (6.146) equals the cross-sectional integral of the warping function φ . Hence, $\varphi_0 = HB/2$, and the wall thickness $h = const$. Equation (6.148) yields the center of shear at the distance

$$y_D = -\frac{1}{J_y} \int_A z\varphi(s)\, dA = -\frac{1}{J_y} \left[\int_0^B \frac{H}{2} \frac{H}{2} (B - s_1)\, h\, ds_1 \right.$$

$$\left. + \int_0^B \left(-\frac{H}{2}\right)\left(-\frac{H}{2} s_3\right) h\, ds_3 \right] = -\frac{H^2 B^2}{J_y} \frac{h}{4} \quad,$$

$$J_y \cong \frac{h\, H^3}{12} + 2\, Bh \left(\frac{H}{2}\right)^2 = \frac{H^2 B}{2}\left(1 + \frac{H}{6B}\right) h \quad.$$

Substitution of the above approximation of the inertia moment yields the result

$$y_D = -\frac{B}{2\,(1 + H/6B)} \quad. \tag{6.158}$$

The center is located opposite the flanges with respect to the fillet. Hence, the torque is given by $M_T = -\, y_D\, Q_z = F\, y_D = const$, and due to the rigidly clamped edge at $x = 0$, the torsion has constrained warping. The stresses at some cross-section at x are distributed according to the following:

(a) The bending moment M_y , $\sigma_{xx}(x, z) = \dfrac{M_y}{J_y} z$,

(b) The shear force Q_z , $\sigma_{xs}(s) = -\dfrac{Q_z S_y(s)}{h\, J_y}$ (in the web) ,

(c) Pure torsion $\overline{M}_T$, $\sigma_{xy}(x, s) = \dfrac{\overline{M}_T}{J_T}\left(\dfrac{\partial\varphi}{\partial y} - z\right)$, $\sigma_{xz} = 0$,

(in the flanges, $\partial/\partial y = -\,\partial/\partial s_1$ in the upper one, and $\partial/\partial y = \partial/\partial s_3$ in the lower one),

$$\sigma_{xz}(x, s) = \dfrac{\overline{M}_T}{J_T}\left(\dfrac{\partial\varphi}{\partial z} + y\right)\ ,\quad \sigma_{xy} = 0\ ,\ \text{(in the web, } \partial/\partial z = -\,\partial/\partial s_2)\ . \tag{6.159}$$

Equation (6.136) yields $J_T = (h^3/3)\,(2B + H) = (2/3)Bh^3\,(1 + H/2B)$.

The warping stresses are to be superposed. The warping rigidity of the C-profile is proportional to [see Eq. (6.153)]

$$C^* = \int_A \varphi^{*2}\, dA = \int_A \varphi^2\, dA - J_y\, y_D^2 = \dfrac{h\,H^2\,B^3\,[1 + (2H/3B)]}{24\,(1 + H/6B)}\ , \tag{6.160}$$

and the coefficient a^{-2} of Eq. (6.155) becomes

$$a^{-2} = \dfrac{J_T}{2\,(1 + v)\,C^*} = \dfrac{8\,h^2\,(1 + H/2B)(1 + H/6B)}{(1 + v)\,[1 + (2H/3B)]\,H^2\,B^2}\ . \tag{6.161}$$

By means of the boundary conditions of the cantilever of length l , the solution is

$$\Theta(x) = \dfrac{\overline{M}_T}{GJ_T}\left[1 - \cosh\dfrac{x}{a} + \tanh\left(\dfrac{l}{a}\right)\sinh\dfrac{x}{a}\right]\ , \tag{6.162}$$

and the angle of torsion under the condition of constrained warping is the nonlinear function

$$\chi(x) = \dfrac{a\,\overline{M}_T}{GJ_T}\left\{\dfrac{x}{a} - \sinh\dfrac{x}{a} + \tanh\left(\dfrac{l}{a}\right)\left[\cosh\left(\dfrac{x}{a}\right) - 1\right]\right\}\ . \tag{6.163}$$

Equation (6.145) renders the warping normal stresses

$$\sigma_{xx}^*(x, s) = E\,\varphi^*(s)\dfrac{\overline{M}_T}{aGJ_T}\left[\tanh\left(\dfrac{l}{a}\right)\cosh\left(\dfrac{x}{a}\right) - \sinh\dfrac{x}{a}\right]\ . \tag{6.164}$$

The distribution becomes maximal at $x = 0$. Critical points are $s_1 = 0$, $s_1 = B$ and $s_3 = 0$, $s_3 = B$. The warping shear flow is given through Eq. (6.150)

$$T^*(x, s) = -E \frac{d^2\Theta}{dx^2} \int_0^s h \, \varphi^*(s) \, ds \quad ,$$

(6.165)

where the integrals take on the values and the shear flow is steady

Upper flange: $\int_0^s \varphi^* \, ds = \frac{H}{2} \left(B - \frac{s_1}{2} + y_D\right) s_1$, max. at $s_1 = B + y_D$,

Web: $\int_0^s \varphi^* \, ds = \frac{y_D}{2} (H - s_2) s_2 + \frac{H}{2} \left(\frac{B}{2} + y_D\right) \frac{B}{2}$, max. at $s_2 = \frac{H}{2}$,

Lower flange: $\int_0^s \varphi^* \, ds = -\frac{H}{2} \left(y_D + \frac{s_3}{2}\right) s_3 + \frac{H}{2} \left(\frac{B}{2} + y_D\right) \frac{B}{2}$, max. at $s_3 = -y_D$,

and

$$\frac{d^2\Theta(x)}{dx^2} = -\frac{M_T}{a^2 G J_T} \left[\cosh \frac{x}{a} - \tanh \left(\frac{l}{a}\right) \sinh \frac{x}{a} \right] .$$

Hence, Eq. (6.152) determines the axial moment which is contributed to the torque by the constrained warping of the C-profile, C^* of Eq. (6.160),

$$M_x^*(x) = -EC^* \frac{d^2\Theta(x)}{dx^2} \quad , \quad M_T = \overline{M}_T(x) + M_x^*(x) \quad .$$

(6.166)

The maximal warping normal and shear stress are both in the clamped cross-section at $x = 0$.

At the free end $x = l$, the axial displacements take on maximal values at the corners of the flanges. Equation (6.144) gives

$$|u| = |\Theta| \frac{H B}{4} \frac{1 + H/3B}{1 + H/6B} \quad .$$

(6.167)

The lower edge moves forward, the upper one backward for $M_T > 0$. Another example is discussed in Exercise A 6.13.

6.2.4.3. Torsion of Elliptic and Circular, Full and Hollow Cylinders

The determination of the warping function φ or, equivalently, of the stress function Ψ of the torsion of rods with a full cross-section is a rather complicated exercise in the mathematical theory of elasticity and, thus, of potential-theory. After deriving the generally valid potential equations, their special solutions are discussed only for the elliptical shaped cross-section. An approximate solution for a rod of a rectangular or quadratic cross-section by *Galerkin´s* method is given in Sect. 11.3.3.

The cross-sections still rotate rigidly and the in-plane displacements are given for sufficiently small torsional angles, $\chi = \Theta\, x$, by

$$v = -z\chi \quad , \quad w = y\chi \quad ,$$

and the axial displacement is proportional to the warping function

$$u = \Theta\, \varphi(y, z) \quad , \quad \Theta = \frac{d\chi}{dx} \quad .$$

Navier´s equations (6.6) render at once the homogeneous potential equation for the warping function with the two-dimensional *Laplace* operator

$$\nabla^2 \varphi = 0 \quad , \quad \nabla^2 = \frac{\partial^2}{\partial y^2} + \frac{\partial^2}{\partial z^2} \quad . \tag{6.168}$$

The solutions are the harmonic functions. The linearized dilatation vanishes in torsion, $e = 0$. The dynamic boundary conditions at the traction-free surface become quite complex when expressed by the derivatives of the warping function. Hence, it is more convenient to consider the stress function $\Psi(y, z)$ of Eq. (6.127). In that case, the local conditions of equilibrium are identically satisfied, Eq. (6.128). An inhomogeneous partial differential equation, the *Poisson* equation, is the result of the elimination of the warping function from *Hooke´s* law

$$\sigma_{xy} = G\left(\frac{\partial u}{\partial y} + \frac{\partial v}{\partial x}\right) = G\Theta\left(\frac{\partial \varphi}{\partial y} - z\right) = 2G\Theta\frac{\partial \Psi}{\partial z} \quad ,$$

$$\sigma_{xz} = G\left(\frac{\partial u}{\partial z} + \frac{\partial w}{\partial x}\right) = G\Theta\left(\frac{\partial \varphi}{\partial z} + y\right) = -2G\Theta\frac{\partial \Psi}{\partial y} \quad . \tag{6.169}$$

With steady mixed derivatives of the warping function, by subtraction follows

$$-2G\Theta = 2G\Theta\left(\frac{\partial^2 \Psi}{\partial y^2} + \frac{\partial^2 \Psi}{\partial z^2}\right) \quad .$$

Hence, *Poisson´s* differential equation for the torsional stress function is derived in the form,

$$\nabla^2 \Psi = -1 \quad , \quad \nabla^2 = \frac{\partial^2}{\partial y^2} + \frac{\partial^2}{\partial z^2} \quad .$$

$$(6.170)$$

Shear stresses are parallel to the tangents of the contour of the cross-section at the surface and, thus, the boundary condition renders $\Psi = const$ along that contour; see also Eq. (6.130). For simply connected regions, the constant is put to zero. A hollow rod has at least a second closed free surface, and the stress function is constant and nonzero at the second contour. In a multiply connected region, the constants, in general, take on different values along the various contours. When given, the stress function for a specified cross-section determines the torsional rigidity by means of Eq. (6.131), where a cross-sectional and contour integral of the stress function is to be evaluated,

$$M_T = G\Theta \, J_T = G\Theta \left[4 \int_A \Psi \, dA + 2 \oint_{\partial A} \Psi \, (z \, dy - y \, dz) \right] \quad .$$

$$(6.171)$$

(§) Elliptic Cross-Section. Such a contour is given by

$$f(y, z) = \frac{y^2}{a^2} + \frac{z^2}{b^2} - 1 = 0 \quad ,$$

and allows an analytical solution of Eq. (6.170) by putting the stress function $\Psi = C \, f(y, z)$, since $\Psi = 0$ at the contour. The parameter C is found from *Poisson's* equation to be

$$C = \frac{a^2 \, b^2}{2 \left(a^2 + b^2 \right)} \quad .$$

The shear-stress components are exact and linearly distributed

$$\sigma_{xy} = -2G\Theta \, \frac{a^2}{\left(a^2 + b^2 \right)} \, z \quad , \quad \sigma_{xz} = 2G\Theta \, \frac{b^2}{\left(a^2 + b^2 \right)} \, y \quad .$$

$$(6.172)$$

The nonlinear distribution of the torsional shear stresses, hence, is given by

$$\tau = \sqrt{\sigma_{xy}^2 + \sigma_{xz}^2} = 2G\,\Theta \, \frac{b \, a^2}{\left(a^2 + b^2 \right)} \sqrt{\frac{z^2}{b^2} + \frac{y^2}{a^2} \frac{b^2}{a^2}} \quad .$$

$$(6.173)$$

The isostresses are the ellipses that are determined by putting $\tau = const$. Their semiaxes grow proportional to the shear stress τ with the parameter q , until the maximum is reached for $q = 1$

$$f(y, z; q) = \frac{y^2}{(qa^2/b)^2} + \frac{z^2}{(qb)^2} - 1 = 0 \ , \quad q = \frac{\tau}{2G} \frac{(a^2 + b^2)}{\Theta b a^2} \ . \tag{6.174}$$

Hence, the maximal torsional shear stress occurs at the surface point $(y = 0, z = \pm b)$, $b < a$. For slim ellipses, $b \ll a$, and the maximum equals that of the rectangle, Eq. (6.125), where $h = 2b$. The limit $(b/a) \rightarrow 0$, with $L = 2a$, gives a geometric interpretation of the approximation (6.130) of the torsional stress function of the slim rectangle, $h \ll L$.

Equation (6.171) renders with the cross-sectional area, $A = (\pi ab)$ of the ellipse, the principal moments of inertia, $J_y = \pi \, ab^3/4$, $J_z = \pi \, ba^3/4$, and, since $\Psi = 0$ on the contour, by areal integration

$$J_T = 2 \frac{a^2 b^2}{(a^2 + b^2)} \left(A - J_z /a^2 - J_y /b^2 \right) \ . \tag{6.175}$$

Hooke´s law (6.169) when integrated, yields the exact warping function of the elliptical cross-section

$$\varphi(y, z) = -\frac{1 - b^2/a^2}{1 + b^2/a^2} \, y \, z \ . \tag{6.176}$$

The axial displacements are proportional to this function, $u(y, z) = \Theta \, \varphi(y, z)$, and the cross-sections are deformed to surfaces according to a hyperbolic paraboloid (a saddle surface). The principal axes of inertia remain undeformed.

Putting $a = b = d/2$ in the solution for the elliptical cross-section renders specifically that of the torsion of a *circular cylinder* of diameter d. The isostresses are concentric circles and the linearly distributed torsional shear stresses $\tau = G\Theta r$ take on their maximum at the free surface, $max \, \tau = G\Theta \, (d/2)$. The torsional rigidity is proportional to the polar moment of inertia, $J_T = J_p = (\pi \, d^4/32)$, and the angle of twist becomes $\Theta = M_T/GJ_T = (32/\pi) \, M_T/Gd^4$. The circular cross-section is free of any warping, $\varphi \equiv 0$.

The torsional stress function of a hollow circular cylinder $R_i \leq r \leq R_e$, is simply given by

$$\Psi(r) = (R_e^2/4)(1 - r^2/R_e^2) \ , \quad \Psi(R_e) = 0 \ , \quad \Psi(R_i) = (R_e^2 - R_i^2)/4 \ . \tag{6.177}$$

The torsional rigidity is derived by the superposition of both integrals in Eq. (6.171)

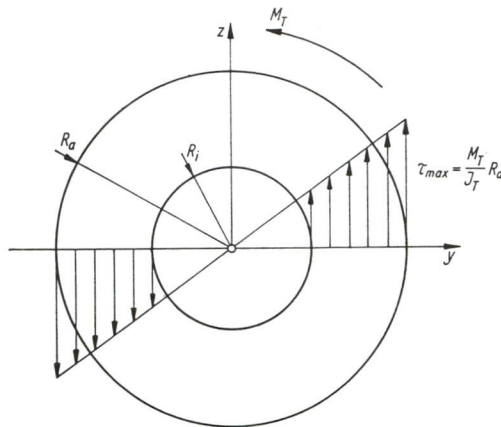

Fig. 6.18. Linear distribution of the torsional shear stresses in a thick walled cylinder

$$J_T = 4\int_A \Psi \, dA + 2\int_0^{2\pi} \Psi R_i^2 \, d\theta = \frac{\pi}{2}\left(R_e^4 - R_i^4\right) \ .$$

(6.178)

It is also given by taking the difference of the rigidity of the outer and inner circular cylinder, "the area of the inner circle is lost." The torsional shear stress remains linearly distributed (see Fig. 6.18)

$$\tau = 2G\Theta \left|\frac{\partial \Psi}{\partial r}\right| = G\Theta\, r \quad , \quad \max \tau = G\Theta\, R_e \ .$$

(6.179)

The circular ring cross-section remains free of any warping.

6.2.4.4. Torsion of a Notched Circular Shaft

The maximum of the torsional shear stress is expected to occur at the deepest point of a rounded notch. The solution for a circular cylindrical notch (Fig. 6.19) has been derived by *C. Weber, The theory of torsional rigidity. VDI-Forschungsheft, 1921, p. 249, (in German)*. If we consider the torsional stress function and its *Poisson* equation of definition (6.170) in polar coordinates *(r, α)* , with the origin at the center of the circular notch,

$$\nabla^2 \Psi = -1 \quad , \quad \nabla^2 = \frac{\partial^2}{\partial r^2} + \frac{1}{r}\frac{\partial}{\partial r} + \frac{\partial^2}{r^2 \partial \alpha^2} \ ,$$

(6.180)

together with the boundary condition $\Psi = 0$ at the contour $r = d\cos\alpha$ and $r = R$ suggests a solution in the form of the product

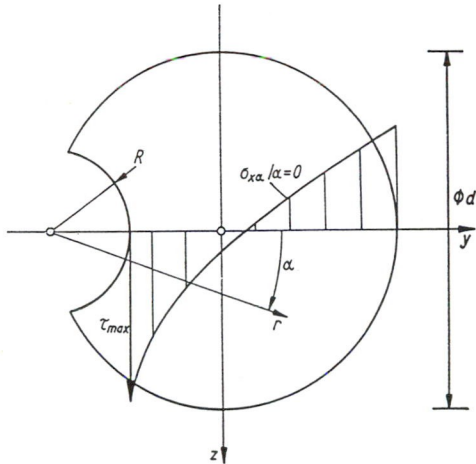

Fig. 6.19. Circular shaft with a circular notch, $R \ll d$. Maximal torsional shear stress

$$\Psi(r, \alpha) = (d \cos \alpha - r) f(r) \ , \quad f(r = R) = 0 \ . \tag{6.181}$$

Substitution into the partial differential equation (6.180) renders two ordinary differential equations of the second order with variable coefficients

$$r f'' + 3 f' + \frac{1}{r} f = 1 \ , \quad f' = \frac{df}{dr} \ , \tag{6.182}$$

and [analogous to Eq. (6.10) with $\alpha = 1$]

$$f'' + \left(\frac{1}{r} f\right)' = 0 \ . \tag{6.183}$$

Hence, the solution takes on the form [see Eq. (6.11)]

$$f(r) = C r + \frac{D}{r} \ . \tag{6.184}$$

The inhomogeneous equation (6.182) determines $C = 1/4$. The second constant follows from the boundary condition, $f(r = R) = R/4 + D/R = 0$, and the stress function finally becomes

$$\Psi(r, \alpha) = \frac{r}{4} (d \cos \alpha - r) \left[1 - (R/r)^2\right] \ . \tag{6.185}$$

Commonly, $R \ll d$, and the remaining area integral in Eq. (6.171) renders approximately $J_T < J_p = \pi d^4/32$,

$$J_T = 4 \int_A \Psi(r, \alpha) \, dA = J_p \left\{ 1 - 8 \, (R/d)^2 \, \left[1 + (R/d)^2 - (16 \, R/3\pi d) \right] \right\} . \tag{6.186}$$

The stress components are determined by Eq. (6.169), in polar coordinates,

$$\sigma_{xr} = 2G\Theta \frac{\partial \Psi}{r \, \partial \alpha} = -G\Theta \left[1 - (R/r)^2 \right] \frac{d}{2} \sin \alpha \ , \quad \Theta = M_T / G J_T \ ,$$

$$\sigma_{x\alpha} = -2G\Theta \frac{\partial \Psi}{\partial r} = -G\Theta \left\{ \left[1 + (R/r)^2 \right] \cos \alpha - 2r/d \right\} \frac{d}{2} . \tag{6.187}$$

The torsional shear stress along the line $\alpha = 0$ is hyperbolically distributed (Fig. 6.19)

$$\tau = G\Theta \left\{ 1 + (R/r)^2 - 2r/d \right\} \frac{d}{2} \ , \tag{6.188}$$

with the maximum at $r = R$,

$$\max \tau = \sigma_{x\alpha}(r = R, \alpha = 0) = G\Theta \, d \, (1 - R/d) = \alpha_k \frac{M_T}{J_p} \frac{d}{2} . \tag{6.189}$$

The stress magnification factor α_k when referred to the maximum of the torsional shear stress of the undisturbed circular shaft becomes

$$\alpha_k = \frac{2 \, (1 - \rho)}{1 - 8 \, \rho^2 \left(1 + \rho^2 - 16\rho/3\pi \right)} \ , \quad \rho = R/d . \tag{6.190}$$

In the limit $R \to 0$, the notch becomes merely a scratch of the

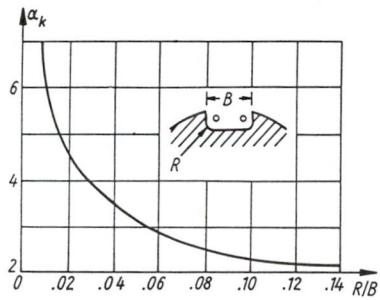

Fig. 6.20. Stress magnification factor of a shaft with a rectangular notch of width B with rounded shoulders of radius R

surface of the circular shaft, but still renders a magnification by a factor of 2   against the undisturbed shaft. This demonstrates the sensitivity of a shaft and the danger caused by even merely shallow surface breaking cracks. Technology of production responds by high-quality finish of the surface.

Figure 6.20 is a graph of the magnification factor of the torsional shear stress of a shaft with a rectangular notch with shoulders rounded with the radius R , taken from *H. Parkus, Österr. Ing.-Archiv 3, 1949, p. 336*. The analysis becomes quite involved due to the complex geometry.

6.2.4.5. *Prandtl´s* Membrane and an Electric Analogy

The (numerical) evaluation of the torsional rigidity and torsional shear-stress distribution of geometrically complex cross-sections becomes quite involved. *L. Prandtl* developed a convenient experimental method to circumvent the cumbersome calculations, which is based on the identity of *Poisson´s* differential equation of the torsional stress function Ψ and of the deflection $u(y, z)$ of a membrane loaded laterally by a uniform pressure $p = const$. A thin membrane, like a film of soap, has negligible bending stiffness and, like an idealized cable, takes on only (in-plane) tensile stresses when deflected. When we consider the principal curvatures to be linearized, the lateral condition of equilibrium of a prestressed membrane element renders with the isotropic tension $S = const$,

$$\nabla^2 u = -p/S \quad , \quad \nabla^2 = \frac{\partial^2}{\partial y^2} + \frac{\partial^2}{\partial z^2} \quad .$$

(6.191)

At the simply supported contour, the displacement of the membrane vanishes, $u = 0$. The relation to Eq. (6.170) is thus given by

$$\Psi = \frac{S}{p} u \quad .$$

(6.192)

The volume under the membrane surface, thus, is proportional to the torsional rigidity of a cross-section of the same shape and size of the projection [see Eq. (6.171)], and the gradient of the surface is related to the torsional shear stress in the cross-section when corresponding points are considered [see Eq. (6.169)]. In an inexpensive realization, a hole is produced in a thin metal plate of the shape of the cross-section and a soap-film is deformed to a surface by a slight one-sided increase of pressure. Using optical instruments for measurements renders the necessary data accurately without causing any disturbing mechanical contact with the membrane. Complications are encountered in the case of multiply connected regions.

An *electric analogy* gives the data more precisely. The cross-section is materialized by cutting a piece from an electrically conducting paper. A current of constant density per unit of area i flows in the region. In such a case, *Poisson's* equation applies to the voltage U, ρ is *Ohm's* resistance of the paper

$$\nabla^2 U = -\rho\, i \ .$$

(6.193)

The voltage U must be kept constant at the contour. The analogy is similar to Eq. (6.192). The lines of constant voltage are easily determined using a voltmeter. For an extensive discussion, see *Beadle and Conway: ASME Journal of Applied Mechanics 30, (1963), pp. 138-141, as well as Exp. Mechanics (1963), pp. 198-200* .

A *hydrodynamic analogy* is mentioned for its theoretical significance. *J. Boussinesq* gave a formulation by considering the laminar flow of a viscous fluid. *A. G. Greenhill* presented the analogy with respect to the ideal (inviscid) flow of a fluid with uniform circulation; see Eq. (13.1). The torsional stress function corresponds in that case to the stream function (velocity potential of the two-dimensional flow); see Eq. (13.48a).

6.3. Multispan Beams and Frames

Continuous beams on multiple supports and frames are (highly) redundant structures. Due to their importance in structural design, efficient methods have been developed for the analysis of stresses and elastic deformations: The force method and deformation method are illustrated below. The three-moment formula (6.81) is just one possibility of the force method applied to multispan beams.

(§) The Force Method of the Multispan Beam. A summary is given in Fig. 6.21, where the basic statically determinate system is shown for a single-span cut between joints (supports) of number 1 and 2. The statically indeterminate moments in the multispan beam are considered to be a positive bending moment M_1 and a negative one M_2 ; the angles of rotation of the cross-sections of the associated hinged-hinged single-span beam are accordingly denoted by φ_1 and φ_2 in the sense of contributing positive work. The force method determines the static unknowns M_1, M_2, ... from proper deformation conditions: The slope of the deflection at the supports must be a steady function. Hence, by superposition, the angles of rotation at the beginning and end of the field considered in Fig. 6.21 become

$$\varphi_1 = \varphi_{1L} + \varphi_{11} M_1 + \varphi_{12} M_2 \ , \quad \varphi_2 = \varphi_{2L} + \varphi_{21} M_1 + \varphi_{22} M_2 \ .$$

(6.194)

The first term is the deformation of the simply supported beam under the given loading $q_z(x)$; φ_{ij} are the influence numbers of this statically determinate single-span beam. *Mohr´s* method [see Sec. 6.2.2] may be applied also in the case of a nonconstant bending stiffness in the field. The same computation is applied to the left and right neighboring field and renders at support 1 the deformation $\varphi_1´$. At support 2 the angle is denoted $\varphi_2´$. The conditions of a steady slope at these joints yield two equations for the two unknowns

$$\varphi_1 - \varphi_1´ = 0 \quad , \quad \varphi_2 - \varphi_2´ = 0 \ . \tag{6.195}$$

The condition applies to each of the intermediate supports and the additional equations complete the system of equations for the determination of the unknowns in the multispan beam.

(§)   The Deformation Method. Contrary to the force method, a *kinematically determinate system* is associated with the multispan

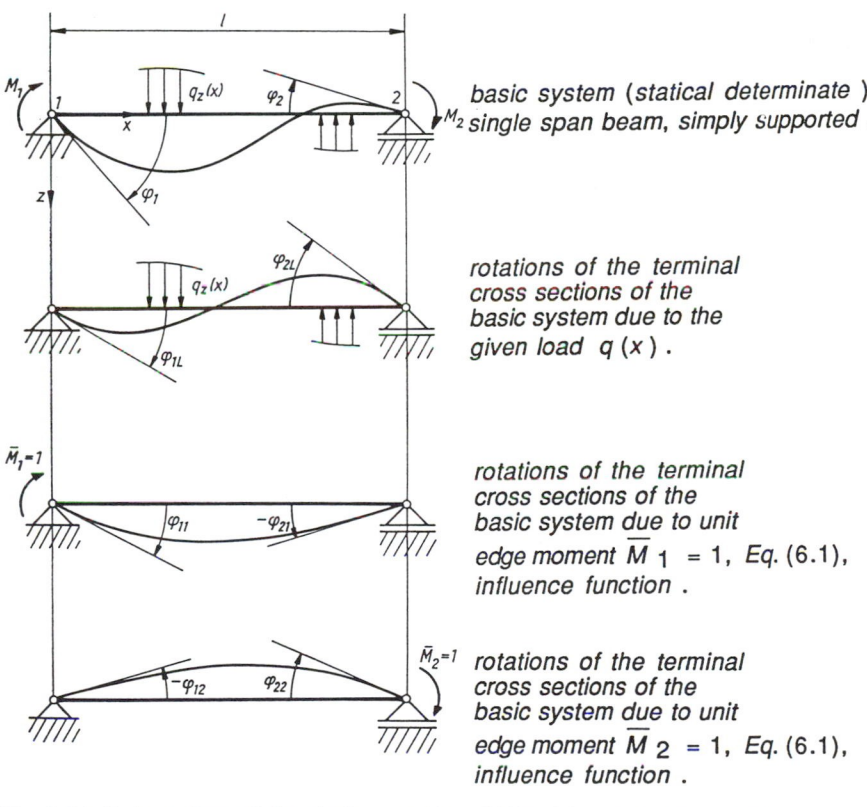

Fig. 6.21. Deformations of the single-span hinged-hinged beam of the statically determinate system associated with the continuous beam on multiple supports; see also Fig. 6.13. The force method

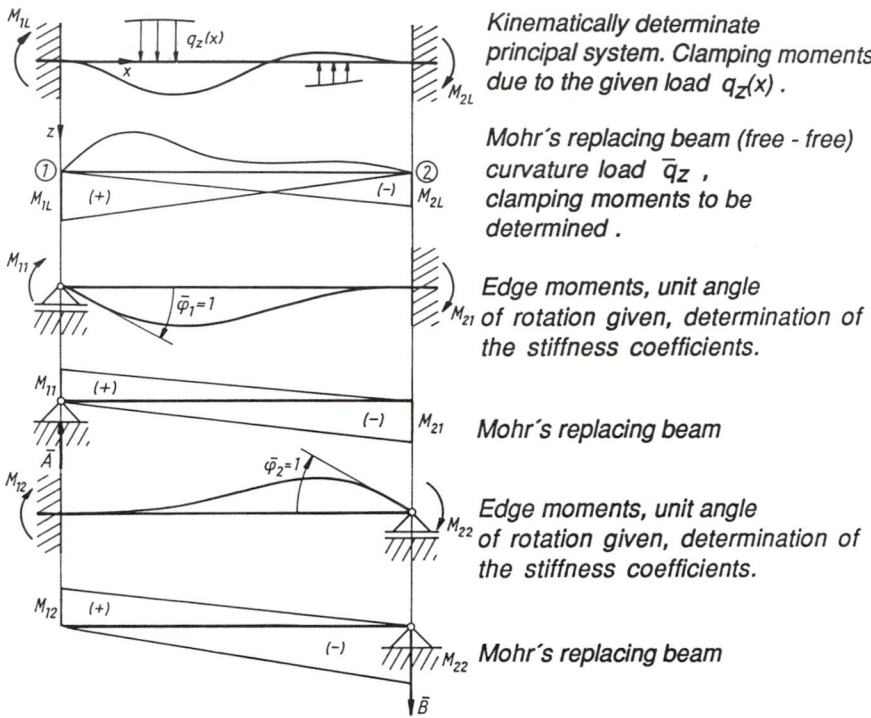

Fig. 6.22. Kinematically determinate clamped-clamped beam associated with a field of the multispan beam of Fig. 6.13. Deformation method. *Mohr´s* conjugate beams

beam. Instead of the statically unknown bending moments at the intermediate supports, the angles of rotation φ_1 and φ_2 are considered to be the unknowns in the single field. Hence, the rigidly clamped-clamped beam of Fig. 6.22 is considered (kinematically determinate) when loaded by the given lateral load $q_z(x)$. The analysis of this associated system that is statically redundant and of low order is conveniently performed by *Mohr´s* method, Eq. (6.22). *Mohr´s* conjugate free-free beam is shown in the Fig. 6.22. The superposition of the partial edge moments given by the clamping moments of the clamped-clamped beam under the given loading, M_{1L} and M_{2L}, respectively, and of the influence moments for prescribed rotations of the cross-sections renders the slope deflection equations

$$M_1 = M_{1L} + M_{11}\, \varphi_1 + M_{12}\, \varphi_2 \quad , \quad M_2 = M_{2L} + M_{21}\, \varphi_1 + M_{22}\, \varphi_2 \ . \quad (6.196)$$

M_{ij} are the torsional spring constants of the statically indeterminate systems also shown in Fig. 6.22. They are conveniently evaluated with prescribed unit rotations by *Mohr´s* method. The unknowns are given by the conditions of equilibrium of the moments at the joints when considering the adjacent fields

$$M_1 + M_1' = 0 \quad , \quad M_2 + M_2' = 0 \ . \tag{6.197}$$

In the case of an additional singular moment loading of the multispan beam just at an intermediate support, such an external moment has to be added to the above equilibrium condition at this joint. Due to the geometric constraints of the associated kinematically determinate system (Fig. 6.22), the stiffness in each field is increased with respect to the actual beam. The deflections produced by the actual lateral load q_z are reduced. These facts are contrary to the behavior of the subsystems of the force method (Fig. 6.21), where the stiffness is reduced and the deformations increased.

   If the intermediate supports are not rigid, eg in the case of spring supports (weak foundations), joints 1 and 2 are differently displaced, and an additional rigid rotation through an angle ψ occurs. It is counted positive counterclockwise against φ_k , ie $\psi > 0$ if joint 1 moves to a lower level than 2. Superposition renders in that more general case

$$M_1 = M_{1L} + M_{11} \varphi_1 + M_{12} \varphi_2 + (M_{11} + M_{12}) \psi \ ,$$
$$M_2 = M_{2L} + M_{21} \varphi_1 + M_{22} \varphi_2 + (M_{21} + M_{22}) \psi \ . \tag{6.198}$$

Additional conditions of equilibrium apply and the angle ψ is also easily determined.

(§) The Deformation Method Applied to Frames. Efficient procedures for the analysis of highly redundant frames are based on the deformation method in the special version where rotational angles are selected. The method is illustrated by considering the planar frame of Fig. 6.23 where several members are joined together at a "welded" and vouted joint of the frame. During the loading process, such a joint rotates rigidly, the configuration of the members at the joint is preserved, the angles between the members remain unaltered. Hence, in the deformation method, one unknown angle of rotation of the joint appears, whereas in the force method, the number of unknown bending moments at the joint is of the order of the number of the members connected at this point. In Fig. 6.23, the single force F gives an external moment Fl_1 when reduced to the joint. In addition, the member between joints 1 and 2 is uniformly loaded by q . The bending stiffness EJ is spanwise constant. The joint 1 is assumed to be fixed in space, its displacement is zero. Due to symmetry of the member 12, the clamping moments $M_{1L} = - M_{2L} = - ql^2/12$. Superposition gives the

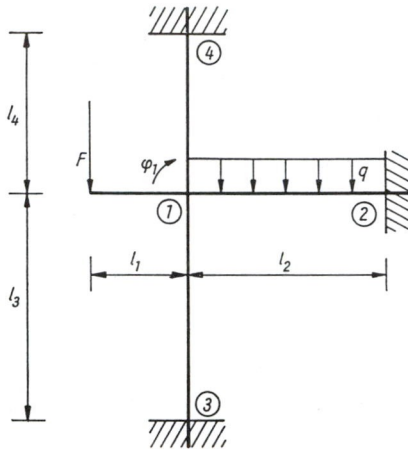

Fig. 6.23. Planar frame. Analysis by the deformation method

slope deflection equations; the angle of rotation φ_1 is common at the joint

$$M_{1,2} = M_{1L} + M_{11,2}\,\varphi_1 \; , \; M_{1,3} = M_{11,3}\,\varphi_1 \; , \; M_{1,4} = M_{11,4}\,\varphi_1 \; ,$$

$$M_{2,1} = M_{2L} + M_{21,1}\,\varphi_1 \; , \; M_{3,1} = M_{31,1}\,\varphi_1 \; , \; M_{4,1} = M_{41,1}\,\varphi_1 \; .$$

The rotational spring constants of the members are

$$M_{11,i} = 4\,\frac{(EJ)_i}{l_i} \; , \quad M_{i1,1} = 2\,\frac{(EJ)_i}{l_i} \; , \quad i = 2, 3, 4 \; .$$

The condition of vanishing resulting moment (the free-body diagram of joint 1 is considered) renders the equation

$$F\,l_1 + \sum_{i=2}^{4} M_{1,i} = 0 \; ,$$

which is solved for the angle of rotation

$$\varphi_1 = -\frac{(F\,l_1 + M_{1L})}{4\displaystyle\sum_{i=2}^{4}\frac{(EJ)_i}{l_i}} \; . \tag{6.199}$$

The deformation method with rotational angles is combined with the procedure of balancing the moments, an iterating procedure developed by *Cross* and *Kani* for the efficient analysis of highly redundant frames. See *H. Cross: Analysis of Continuous Frames by Distributing Fixed-End Moments. Transactions ASCE* **96**, *(1932), pp. 1-10. G. Kani: The Analysis of Multi-Story Frames. Wittwer, Stuttgart, 1949, (in German)*. An introduction to computer-oriented methods using matrix notation is given by *W. McGuire and R. H. Gallagher: Matrix Structural Analysis. Wiley, New York, 1979.*

6.3.1. *The Planar Single-Story Frame*

The frame of Fig. 6.24 has columns rigidly clamped to the base and, thus, is threefold redundant. The bending stiffness of the columns is assumed to be equal $(B = EJ)$, and the bending stiffness of the beam is $B_B = EJ_B$. Analysis of such a small system requires the same efforts, if done by the force or deformation method. Given the loading by the horizontal single force F and by the uniform lateral load q, the unknowns are selected to be the resulting internal forces at the midpoint of the cross-bar, N_0, Q_0, M_0.

Menabrea's theorem (5.44) is appropriate for the solution of such a small system. In the strain energy, for simplicity, only the bending curvatures are considered. However, deformations due to normal and shear forces are usually not negligible in such framed structures. The bending moment in the columns and in the beam must be expressed by the given loading and unknowns

$$-L \le x \le L: \quad M_y(x) = M_0 + x\,Q_0 - \frac{q\,x^2}{2}\ , \quad \frac{\partial M_y}{\partial Q_0} = x\ , \quad \frac{\partial M_y}{\partial M_0} = 1\ ,$$

$$\text{Column I}: \quad M_y(x_1) = M_0 - L\,Q_0 - \frac{q\,L^2}{2} - x_1\,(F + N_0)\ ,$$

$$\frac{\partial M_y}{\partial N_0} = -x_1\ , \quad \frac{\partial M_y}{\partial Q_0} = -L\ , \quad \frac{\partial M_y}{\partial M_0} = 1\ .$$

$$\text{Column II}: \quad M_y(x_2) = M_0 + L\,Q_0 - \frac{q\,L^2}{2} - x_2\,N_0\ ,$$

$$\frac{\partial M_y}{\partial N_0} = -x_2\ , \quad \frac{\partial M_y}{\partial Q_0} = L\ , \quad \frac{\partial M_y}{\partial M_0} = 1\ .$$

In *Menabrea's* theorem, the derivatives of the complementary energy with respect to the unknowns are taken under the integral sign

$$\frac{\partial U^*}{\partial N_0} = 0 = \int \frac{M_y}{B}\frac{\partial M_y}{\partial N_0}\,dx\ , \quad \frac{\partial U^*}{\partial Q_0} = 0 = \int \frac{M_y}{B}\frac{\partial M_y}{\partial Q_0}\,dx\ ,$$

$$\frac{\partial U^*}{\partial M_0} = 0 = \int \frac{M_y}{B}\frac{\partial M_y}{\partial M_0}\,dx\ .$$

Thus, the evaluation of the integrals that are supposed to vanish renders the unknowns explicitly

$$\frac{2}{B_B} \int_0^L Q_0 x^2 \, dx + \frac{1}{B} \int_0^H \left(2Q_0 L^2 + F x_1 L\right) dx_1 = 0 \;\; \rightarrow \;\; Q_0 = -\frac{F}{4}\frac{H}{L}\left(1 + \frac{L}{3H}\frac{B}{B_B}\right)^{-1} ,$$

(6.200)

and after solving the linear equations, we get

$$\int_0^H \left[-F x_1 + 2\left(M_0 - \frac{q}{2} L^2 N_0 x_1\right)\right](-x_1) \, dx_1 = 0 \;\; ,$$

$$\frac{2}{B_B} \int_0^L \left(M_0 - \frac{q}{2} x^2\right) dx + \frac{1}{B} \int_0^H \left[-F x_1 + 2\left(M_0 - \frac{q}{2} L^2 - N_0 x_1\right)\right] dx_1 = 0 \;\; \rightarrow$$

$$M_0 = \frac{qL^2}{6}\left(1 + \frac{3HB_B}{4LB}\right)\left(1 + \frac{HB_B}{4LB}\right)^{-1} , \;\; -N_0 = \frac{F}{2} + \frac{qL^2}{2H}\left(1 + \frac{HB_B}{4LB}\right)^{-1} . \;\; (6.201)$$

6.4. Plane-Curved Beams and Arches

The axis of the beam in the undeformed state is a plane curve and

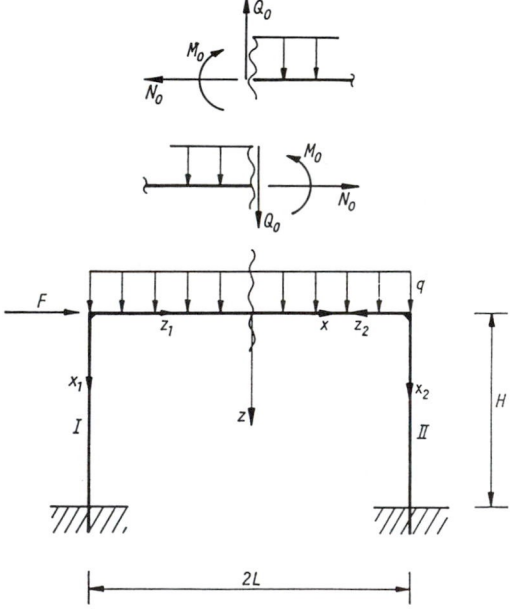

Fig. 6.24. Planar single-story frame. Application of *Menabrea's* theorem

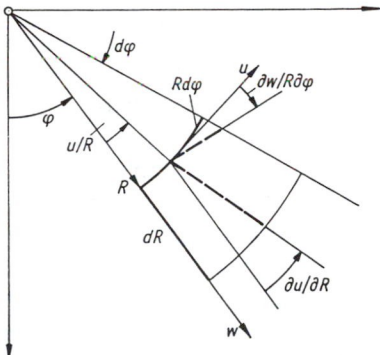

Fig. 6.25. Strains in polar coordinates. Linearized geometric relations

the in-plane loading is assumed to be such that the axis remains in the plane after deformation. Commonly, the cross-section is symmetric, which guarantees the state to be free of torsion. The local conditions of equilibrium are given by Eqs. (2.145), (2.146), and (2.147); see Fig. 2.31. The deformations and distribution of the stresses are considered in this section. The cross-section has a principal axis of inertia y, which must be orthogonal to the plane of loading. The cross-sections are assumed to remain plane in shear bending, and the kinematics of deformation in polar coordinates depends on the radial w and tangential displacement u of the central points. The radial strain becomes, according to Fig. 6.25 (the geometric relations are linearized)

$$\varepsilon_R = \frac{\partial w}{\partial R} \quad .$$

(6.202)

This relation is verified by considering the radial displacements of two neighboring points: $w(R)$ and $w(R + dR) = w(R) + (\partial w/\partial R)\, dR + \ldots$. The ratio of the fiber elongation $(\partial w/\partial R)\, dR$ to the undeformed length dR is the strain as given above. The hoop strain contains two contributions: The undeformed fiber $R\, d\varphi$ is radially displaced and its differential arc is thereby increased by $(R + w)\, d\varphi - R\, d\varphi$. The contribution to the strain is thus given by w/R ; the increment of the polar angle $d\varphi$ cancels. The second contribution is due to the difference in the tangential displacement of two points in a close distance $R\, d\varphi$. The fiber, oriented in the hoop direction is elongated by $u(\varphi + d\varphi) - u(\varphi) = (\partial u/\partial \varphi)\, d\varphi$. The ratio becomes ($d\varphi$ cancels) $(1/R)\, \partial u/\partial \varphi$ and must be added to find

$$\varepsilon_\varphi = \frac{w}{R} + \frac{\partial u}{R\,\partial \varphi} \quad .$$
(6.203)

The shear strain is derived from Fig. 6.25 by considering the decrease of the right angle between the two fibers. The hoop element rotates clockwise due to the difference in the radial displacements by $\partial w/R\partial \varphi$, and the radially oriented fiber rotates by $(\partial u/\partial R) - (u/R)$ due to the difference in the tangential displacements when the rigid rotation is eliminated

$$\gamma_{R\varphi} = 2\,\varepsilon_{R\varphi} = \frac{\partial w}{R\,\partial \varphi} + \frac{\partial u}{\partial R} - \frac{u}{R} \quad .$$
(6.204)

If we neglect the contribution of the shear deformation to the rotation of the beam axis, $\gamma_{R\varphi} = 0$ may be solved for the rotation of the tangent to the axis (which is the increment of the polar angle φ after deformation)

$$\chi = \frac{\partial u}{\partial R} = \frac{u_S}{R_S} - \frac{\partial w_S}{R_S\,\partial \varphi} \quad .$$
(6.205)

The displacement components at the cross-sectional centroid $C = S$ are denoted u_S and w_S ; $(1/R_S)$ is the initial curvature of the beam axis in the stress-free state. According to the theory of thin rods, the radial strain is assumed to vanish and Eq. (6.202) renders a uniform deflection of the material points in the cross-section, $w = w_S = const$. The latter is assumed to remain plane after deformation. Thus, consequently putting $u = R\,f(\varphi)$ yields the hoop strain in the form

$$\varepsilon_\varphi = \frac{w_S}{R} + f' \quad , \quad f' = \frac{df(\varphi)}{d\varphi} \quad .$$

Substitution into the uniaxial *Hooke´s* law renders the normal stresses to be hyperbolically distributed

$$\sigma_{\varphi\varphi} = E\,\varepsilon_\varphi = E\left(\frac{w_S}{z + R_S} + f'\right) \quad , \quad R = z + R_S \quad ,$$
(6.206)

with the resulting normal force given by the cross-sectional integral

$$N = \int_A \sigma_{\varphi\varphi}\,dA = E\,w_S \int_A \frac{dA}{(z + R_S)} + EA\,f' \quad ,$$
(6.207)

and the resulting moment (y is a central axis)

$$M_y = \int_A z\sigma_{\varphi\varphi}\, dA = E\, w_S \int_A \frac{z\, dA}{(z + R_S)} \quad . \tag{6.208}$$

The integral depends on the shape of the cross-section and it is related to the initial curvature of the undeformed beam axis. Commonly, this dependence is expressed by a new parameter κ with the relation (partial integration is applied)

$$-\frac{1}{A} \int_A \frac{z\, dA}{(z + R_S)} = \kappa \frac{J_y}{A\, R_S^2} \quad , \quad \kappa = 1 - \frac{1}{J_y} \int_A \frac{z^3\, dA}{(z + R_S)} \quad . \tag{6.209}$$

By means of this parameter, the bending moment becomes

$$M_y = -\kappa \frac{w_S}{R_S} \frac{E J_y}{R_S} \quad . \tag{6.210}$$

The relation may be solved for w_S/R_S , and the normal force is

$$N = EA \left(1 + \kappa \frac{J_y}{A\, R_S^2} \right) \frac{w_S}{R_S} + EA\, f' \quad . \tag{6.211}$$

Hence, f' in Eq. (6.206) depends on the normal force, and on the moment through Eq. (6.210),

$$f' = \frac{N}{EA} + \left(1 + \kappa \frac{J_y}{A\, R_S^2} \right) \frac{M_y}{\kappa E J_y} R_S \quad . \tag{6.212}$$

The nonlinear stress distribution (6.206) is finally determined by the resultants, the initial curvature $1/R_S$, and depends on the parameter κ

$$\sigma_{\varphi\varphi} = \frac{N}{A} + \left[\frac{J_y}{A\, R_S} + \frac{1}{\kappa} z \left(1 - \frac{z}{z + R_S} \right) \right] \frac{M_y}{J_y} \quad . \tag{6.213}$$

Note that the bending stress does not vanish at $z = 0$. The neutral axis is thus shifted also in the case of $N = 0$ to

$$z_0 = -\kappa \frac{J_y}{\left(1 + \kappa \dfrac{J_y}{A\, R_S^2} \right) A\, R_S} \quad . \tag{6.214}$$

Since the coordinate z is directed outward, the bending stresses are magnified at the inner surface and take on lower values at the outer surface when compared to the linear stress distribution $z M_y/J_y$ of the equivalent beam with a straight axis, Eq. (6.48). The latter is

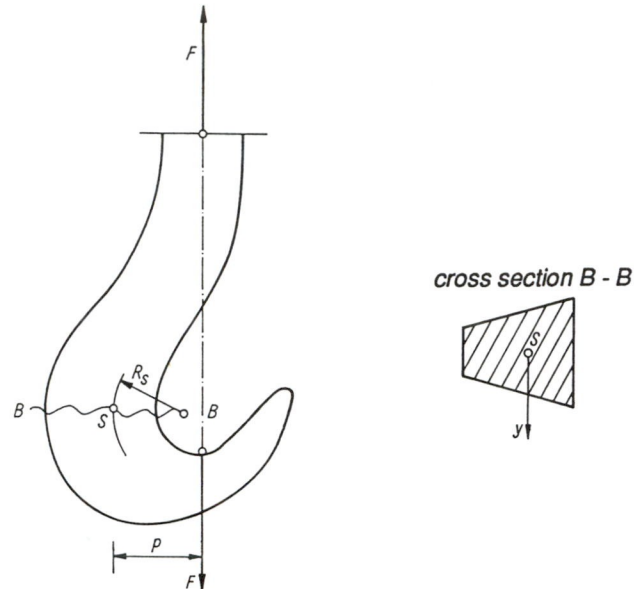

cross section B - B

Fig. 6.26. The hook loaded by F. Illustration of a strongly curved beam

derived in the limit of Eq. (6.213) when the initial curvature $1/R_S \rightarrow$ 0 and the parameter $\kappa \rightarrow 1$.

In the case of a rectangular cross-section $H \times B$, the principal moment of inertia is $J_y = BH^3/12$, and the parameter may be approximated

$$\kappa = \frac{AR_S^2}{J_y} \left[\frac{R_S}{H} \ln \frac{1 + H/2R_S}{1 - H/2R_S} - 1 \right] = 1 + \frac{3H^2}{20R_S^2} + \dots \quad . \tag{6.215}$$

It is a common practice to approximate the extreme stress distribution in the critical cross-section of a *hook*, loaded according to Fig. 6.26, by Eq. (6.213), despite the strongly variable cross-section of the curved beam. Actually, such a variation gives a three-dimensional stress state, similar to a vouted corner of a frame. The cross-section is orthogonal to the line of action of the load F, and hence, the normal force $N = F$. The moment is given by $M_y = -Fp$, where p is the normal distance of the centroid S ; see Fig. 6.26.

(§)   The Complementary Energy of the Curved Beam. The foregoing becomes an expression analogous to that of a straight beam [Eq (5.48)] by combining the normal force and moment times the initial curvature to construct a generalized force $N_H = N + M_y/R_S$ that substitutes the axial force

$$U^* = \frac{1}{2} \int_0^l \frac{N_H{}^2}{EA} \, ds + \frac{1}{2} \int_0^l \frac{M_y{}^2}{\kappa E J_y} \, ds \quad .$$

(6.216)

The expression is still independent of the sequence of the loading in tension and bending. Similarly, a formula analogous to Eq. (5.54) is derived by the application of the principle of virtual displacements in terms of this generalized normal force

$$1 . \delta = \int_0^l \overline{N}_H \frac{N_H}{EA} \, ds + \int_0^l \overline{M}_y \frac{M_y}{\kappa E J_y} \, ds \quad .$$

(6.217)

Thus, the theorems of *Castigliano* and *Menabrea* , previously discussed in Sec. 5.4.1, as well as Eq. (6.217), may be applied to solve the static problems of strongly curved beams. Finally, the coupled differential equations are listed that reflect the relations between the resulting forces and the deformations of the curved beam axis, $ds = R_S \, d\varphi$ is the arc element. Note the coupling terms and the variable coefficients,

$$\frac{du_S}{ds} + \frac{w_S}{R_S} = \frac{N_H}{EA} \quad , \quad \frac{d^2 w_S}{ds^2} + \frac{w_S}{R_S{}^2} - u_S \frac{d}{ds}\left(\frac{1}{R_S}\right) = -\frac{M_y}{\kappa E J_y} \quad .$$

(6.218)

6.4.1. Slightly Curved Beams and Arches

Small initial curvatures of slender beams render, in the limit of Eq. (6.215), $\kappa \to 1$ and the generalized normal force $N_H \to N$. The stress distribution (6.213) becomes approximately linear since $|z|/(R_S + |z|) \ll 1$. The neutral axis is assumed to pass through the cross-sectional centroid if $N = 0$. With the assumption of inextensibility of the beam axis, Eq. (6.218) gives

$$\frac{du_S}{ds} = -\frac{w_S}{R_S} \quad .$$

(6.218a)

(§)   The Slightly Curved Parabolic Arch of Fig. 6.27. The arch when uniformly loaded by a constant force per unit of the horizontal projection p illustrates the application of the simplified Eq. (6.217). The height of the shallow arch is assumed to be small with respect to the span, $H \ll L$, and the bending rigidity is taken to be constant, $B = E J_y = const$. Symmetry requires the reaction forces to be equal and given by $p \, L/2$. Due to the sliding support B, the horizontal components vanish and the bending moment is given by

$$M_y(x) = -p \, L \, x/2 + p \, x^2/2 \quad .$$

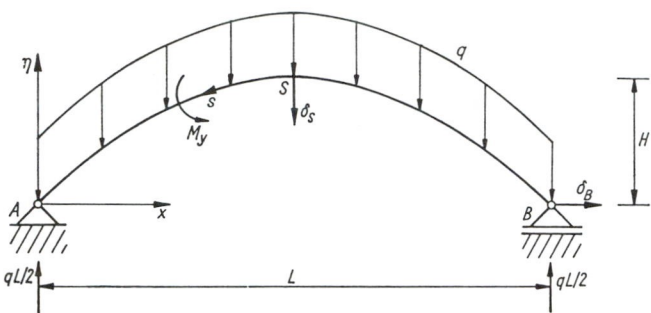

Fig. 6.27. Shallow parabolic arch, uniformly loaded by the constant force p per unit of length of the horizontal projection, $p\,dx = q\,ds$

The horizontal displacement δ_B of the support B is determined by means of the work of the dummy unit load $H_B = 1$, applied in the direction of the displacement. The dummy moment is distributed according to the equation of the parabola

$$\overline{M}_y(x) = -1.\eta \quad , \quad \eta = 4H \times (L-x)/L^2 \ .$$

The integration in Eq. (5.54) that is the limit of (6.217) can be extended over the span L instead of the curved axis, since $H \ll L$ (Fig. 6.27)

$$\delta_B = \frac{1}{B} \int_0^L \frac{p}{2L^2} (L\,x - x^2)\, 4H \times (L-x)\, dx = \frac{pHL^3}{15\,B} \ , \quad B = EJ_y \ , \quad p\,dx = q\,ds \ .$$

The deflection at apex S is determined analogously by considering a dummy unit force loading pointing vertical downward; integration is performed over half the span and then doubled

$$\delta_S = \frac{2}{B} \int_0^{L/2} \frac{p}{2}(L\,x - x^2)\frac{x}{2}\, dx = \frac{5\,pL^4}{384\,B} \ , \quad B = EJ_y \ .$$

Considering the hinged support at B to be fixed in space (a common design goal) renders the horizontal components of the reaction forces in A and B to be opposite and equal, and the system becomes singly redundant, say, by the unknown X. The determination of the horizontal displacement at the sliding support B of the arch of Fig. 6.27 when loaded by a unit force $H_B = 1$,

$$\delta_1 = \frac{1}{B} \int_0^L \eta^2 \, dx = \frac{8 \, L H^2}{15 \, B} \, , \quad B = E J_y \, ,$$

together with the geometric condition that the actual displacement at B due to the given loading p and unknown reaction force X must vanish,

$$\delta_B + X \delta_1 = 0 \, ,$$

yields $X = - \delta_B / \delta_1 = - p L^2 / 8H$. Hence, the bending moment in the redundant arch vanishes identically: $M_y (x) \equiv 0$. The parabolic arch is the *funicular curve* of the uniform loading p with respect to the horizontal projection; see Eq. (2.192). The result holds also in the case of rigidly clamped supports and remains valid even for supports at different levels. The parabolic arch with both supports fixed is thus an important structural design element (eg in concrete, that has high compressive strength).

(§)　The Slightly Curved Ring. This ring may be analyzed along the same lines. In general, such a beam with an axis forming a closed curve in the plane is threefold redundant (N_0 , Q_0 , M_0) . The *circular ring* of Fig. 6.28 if symmetrically loaded, eg by diametrically opposed forces F , is considered in detail. Similar problems are encountered with frames having a contour in the shape of a closed polygon. Considering the free-body diagram of one of the symmetric

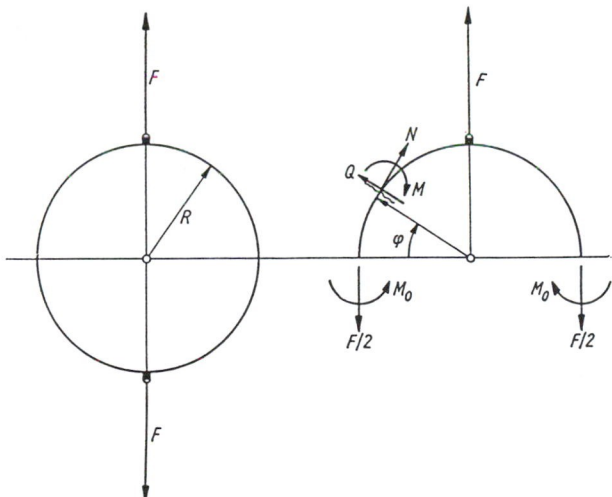

Fig. 6.28. Slightly curved circular ring loaded by diametrically opposed forces F (a double force without moment). The symmetric cut and free-body diagram of the symmetric half are shown

halves renders the shear force $Q_0 = 0$ at once, and also the normal force becomes statically determinate, $N_0 = F/2$, due to reasons of symmetry. Only the bending moment $X = M_0$ remains unknown. The field moment is expressed by the given loading and that unknown, M_y $(\varphi) = X + (FR/2) (1 - \cos \varphi)$, $0 \leq \varphi \leq \pi/2$. *Menabrea's* theorem (5.44) yields (the influence of the normal as well as shear force is neglected, and the bending rigidity is assumed to be constant)

$$\frac{\partial U^*}{\partial X} = \frac{4}{EJ_y} \int_0^{\pi/2} M_y \frac{\partial M_y}{\partial X} R \, d\varphi = 0 \quad .$$

The redundant moment is the solution

$$X = - \frac{FR}{2} \left(1 - \frac{2}{\pi} \right) \quad .$$

The field moment becomes

$$M_y = \frac{FR}{2} \left(\frac{2}{\pi} - \cos \varphi \right) \; , \; 0 \leq \varphi \leq \frac{\pi}{2} \; , \; \max |M_y| = \frac{FR}{\pi} \; \text{at} \; \varphi = \frac{\pi}{2} \; . \quad (6.219)$$

(§) Spinning Rings. These rings are often supported by, say, *n* spokes. An important mechanical device for storing energy is a *flywheel* . Figure 6.29 shows a section of a wheel in stationary rotation with angular speed ω ; $v = R \omega$ is the circumferential velocity of the rim. The latter is radially loaded by $q = \rho A v^2/R = const$, whereas the spoke carries the variable load $q_S = \rho A_S r \omega^2$. The rim segment between cuts of symmetry is approximated by a

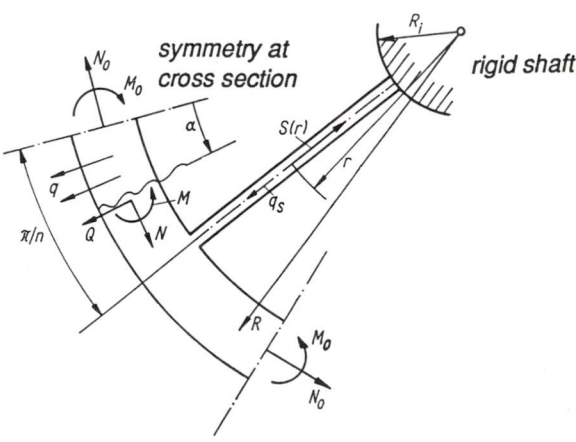

Fig. 6.29. Segment of a flywheel with spokes. Symmetric cuts shown

slightly curved beam. The edge loads are the static unknown normal force N_0 and bending moment M_0 ; the shear force Q_0 vanishes for reasons of symmetry. The resultants in the field are thus expressed by

$$M_y(\alpha) = M_0 + RN_0 (1 - \cos \alpha) - qR^2 \, 2 \sin^2 (\alpha/2) \ , \quad Q(\alpha) = (N_0 - qR) \sin \alpha \ ,$$

$$N(\alpha) = N_0 \cos \alpha + qR \, 2 \sin^2 (\alpha/2) \ , \quad 0 \le \alpha \le \pi/n \ . \tag{6.220}$$

The tension in the spoke, at a radial distance r from the axis, is simply given by the sum

$$S(r) = -2Q(\alpha = \pi/n) + \int_r^R q_S(\bar{r}) \, d\bar{r} = 2(qR - N_0) \sin \pi/n + \frac{qR \, A_S}{2A} \left(1 - r^2/R^2\right) .$$

$$\tag{6.221}$$

Substitution into the complementary energy (the shear deformation is neglected)

$$U^* = \frac{2n}{2} \int_0^{\pi/n} \frac{M_y^2}{EJ_y} R \, d\alpha + \frac{2n}{2} \int_0^{\pi/n} \frac{N^2}{EA} R \, d\alpha + \frac{n}{2} \int_{R_i}^R \frac{S^2}{EA_S} \, dr \ ,$$

and the application of *Menabrea's* theorem (5.44) (the partial differentiation with respect to the unknowns N_0 and M_0 is performed under the integral sign) yields

$$M_0 = -(N_0 - qR) \, R \left[1 - \frac{n}{\pi} \sin (\pi/n)\right] \ , \quad f(n) = \cos (\pi/n) + \frac{\pi}{n \sin (\pi/n)} \ , \tag{6.222}$$

$$N_0 - qR = \frac{-4 \, qR/3}{f(n) + 4(A/A_S) \sin (\pi/n) + (AR^2/J_y) \left[f(n) - 2(n/\pi) \sin (\pi/n)\right]} \ . \tag{6.223}$$

A ring without spokes if radially loaded by q is in a state of pure tension: $N_0 = qR$ and $M_0 = 0$. Hence, Eq. (6.220) renders the bending moment $M_y (\alpha) \equiv 0$ and $N(\alpha) = N_0 = const$. The radial extension of the ring in that case becomes $u(R) = R \, \varepsilon_{\alpha\alpha} = qR^2/EA$. The normal force in the rim contributes significantly to the deformation, also in the wheel design with spokes.

6.5. In-Plane Loaded Plates

Thin plates are plane two-dimensional structures. They remain plane after deformation under the action of the in-plane loadings (see Sec. 9.1.6 for the phenomenon of the buckling of plates under dominant compressive loads). Short beams of a rectangular cross-section, or the fillets (webs) of high I-beams, shear walls, etc., are illustrative examples, as well as the rotating disk. The stresses are integrated over the thickness h to account for the membrane stresses (forces per unit of length) n_x, n_y, n_{xy} in the plane middle surface. The mean stresses follow by dividing through h and render the plate in the state of plane stress. The faces are assumed to be free of tractions. Variations of the actual stresses over the plate thickness are small and, thus, are usually neglected. The introduction of *Airy's* stress function, Eq. (2.11), solves the local conditions of equilibrium for constant or vanishing body forces identically. The partial differential equation of this stress function is derived by eliminating the strains in the compatibility equation (1.22) by means of *Hooke's* law, Eq. (4.44). The set of equations is rewritten in terms of the membrane stresses

$$2 \frac{\partial^2 \varepsilon_{xy}}{\partial x \partial y} = \frac{\partial^2 \varepsilon_{xx}}{\partial y^2} + \frac{\partial^2 \varepsilon_{yy}}{\partial x^2} \;,\; \varepsilon_{x,y} = (n_{x,y} - \nu n_{y,x})/Eh \;,$$

$$\gamma_{xy} = 2 \varepsilon_{xy} = n_{xy}/Gh \;,\; \varepsilon_z = -\nu(n_x + n_y)/Eh \;,\; n_x = \frac{\partial^2 F}{\partial y^2} \;,$$

$$n_y = \frac{\partial^2 F}{\partial x^2} \;,\; n_{xy} = -\frac{\partial^2 F}{\partial x \partial y} \;\rightarrow\; \frac{\partial^4 F}{\partial x^4} + 2 \frac{\partial^4 F}{\partial x^2 \partial y^2} + \frac{\partial^4 F}{\partial y^4} = 0 \;.$$

$$(6.224)$$

The biharmonic differential equation is of the fourth order and becomes, in a notation free of the special coordinate system of reference, by means of the *Laplace* operator,

$$\text{e. g., } \nabla^2 = \frac{\partial^2}{\partial x^2} + \frac{\partial^2}{\partial y^2} \;,\; \text{or} \;,\; \nabla^2 = \frac{\partial^2}{\partial r^2} + \frac{1}{r} \frac{\partial}{\partial r} + \frac{1}{r^2} \frac{\partial^2}{\partial \theta^2} \;\rightarrow\; \nabla^2 \nabla^2 F = 0 \;.$$

$$(6.225)$$

In addition, the *Laplace* operator in polar coordinates is shown above.

The complementary energy, Eqs. (5.40) and (5.38), can be expressed by the stress function (integration is performed over A), which is the area of the midplane

$$U^* = \frac{1}{2Eh} \int_A \left\{ (\nabla^2 F)^2 - 2(1+\nu) \left[\frac{\partial^2 F}{\partial x^2} \frac{\partial^2 F}{\partial y^2} - \left(\frac{\partial^2 F}{\partial x \partial y} \right)^2 \right] \right\} dA \;.$$

$$(6.226)$$

The elementary boundary value problems of a rectangular plate are encountered, if two opposite edges, $x = \pm a$, are uniformly loaded by n_0 , while the other two edges remain free of tractions.

The quadratic function $F_1 = C y^2$ is a special solution of the biharmonic equation and renders the membrane stress to be constant

$$n_x = \frac{\partial^2 F}{\partial y^2} = 2C \quad , \quad n_y = n_{xy} = 0 \quad : \quad C = n_0/2 \ .$$

$$(6.227)$$

Analogously, the exact solution is found if the pair of edges $y = \pm b$ is uniformly loaded. Hence, superposition renders *Airy's* stress function of a plate in equilibrium when loaded at all four edges

$$F_2 = \frac{1}{2} \left(n_0 \, y^2 + n_1 \, x^2 \right) \ .$$

$$(6.228)$$

Isotropic tension is included by putting $n_0 = n_1$. Similarly, the special case of pure shear follows from the loading $n_0 = - n_1$ and it is illustrated by considering *Mohr's* circle (see Fig. 2.7). Note the maximal shear stress $\tau_{max} = n_0/h$.

Pure bending in the plane of the plate is produced by the application of a moment M_0 at the edges $x = \pm a$; if the forces are linearly distributed, $n_0 = 12 \, M_0 \, y/(2b)^3$. Integrating $n_x = - n_0$ at the boundary twice gives the stress function

$$F = - M_0 \, y^3 / 4 \, b^3 \ .$$

$$(6.229)$$

Since all the boundary conditions are satisfied, Eq. (6.229) is an exact solution and $n_x = - 3M_0 \, y/2b^3$, $n_y = n_{xy} = 0$. The cross-sections remain plane after deformation, as in the pure bending case of a slender beam.

6.5.1. The Semiinfinite Plate

The stresses in the near field of a concentrated load are of considerable importance with respect to a proper design. Possibly, the boundary conditions prescribed at the edges in the far field are not very influential; hence, a semiinfinite plate may be considered (see Fig. 6.30), loaded by the tractions n_0 . The latter are assumed to be uniformly distributed in the interval $|x| \leq c$. The load is represented by the *Fourier* integral; see, eg *Bronstein-Semendjajew, (1980), Table 4.4.2.2*

$$n_0(x) = \frac{2}{\pi} \, n_0 \int\limits_0^\infty \frac{\sin{(c\xi)}}{\xi} \cos{(x\xi)} \, d\xi \quad , \quad -\infty < x < \infty \ .$$

$$(6.230)$$

Asymptotically, the stresses must vanish if $y \to \infty$. They are related to the stress function by second derivatives, and a solution may possibly take on the form

$$F(x, y) = \int_0^\infty \frac{A(\xi) + y\,\xi\,B(\xi)}{\xi^2}\; e^{-y\xi} \cos(x\xi)\, d\xi\; .$$

(6.231)

Equation (6.224) gives (the second derivatives of that trial function are taken under the integral sign)

$$n_x(x, y) = \frac{\partial^2 F}{\partial y^2} = \int_0^\infty \left[A(\xi) - 2\,B(\xi) + y\,\xi\,B(\xi) \right] e^{-y\xi} \cos(x\xi)\, d\xi\; ,$$

$$n_y(x, y) = \frac{\partial^2 F}{\partial x^2} = -\int_0^\infty \left[A(\xi) + y\,\xi\,B(\xi) \right] e^{-y\xi} \cos(x\xi)\, d\xi\; ,$$

$$n_{xy}(x, y) = -\frac{\partial^2 F}{\partial x\, \partial y} = -\int_0^\infty \left[A(\xi) - B(\xi) + y\,\xi\,B(\xi) \right] e^{-y\xi} \sin(x\xi)\, d\xi\; .$$

(6.232)

At the boundary $y = 0$, the condition holds

$$\frac{2}{\pi}\, n_0 \int_0^\infty \frac{\sin(c\xi)}{\xi}\, \cos(x\xi)\, d\xi = n_y(y = 0) = -\int_0^\infty A(\xi) \cos(x\xi)\, d\xi\; ,$$

(6.233)

and by comparing coefficients, we get

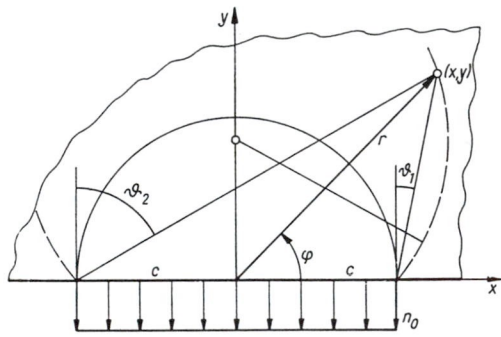

Fig. 6.30. Semiinfinite plate of thickness h with surface tractions n_0. Semicircle $r = c$ of isostresses of maximal shear $\tau_{max} = n_0/\pi h$

$$A(\xi) = -\frac{2}{\pi} n_0 \frac{\sin(c\xi)}{\xi} \ .$$

$$(6.233a)$$

The homogeneous dynamic boundary condition of vanishing shear, n_{xy} $(y = 0) = 0$, yields $A(\xi) = B(\xi)$. All the semiinfinite integrals in Eq. (6.232) have closed-form solutions to be found in tables of integrals. By means of the two angular coordinates shown in Fig. 6.30, the stress distribution is given independently of any of the elastic moduli but holds within the limits of applicability of the linearized theory of elasticity

$$n\binom{x}{y} = \frac{n_0}{2\pi} \left[2\left(\vartheta_2 - \vartheta_1\right) \pm \left(\sin 2\vartheta_1 - \sin 2\vartheta_2\right) \right] \ , \quad n_{xy} = \frac{n_0}{2\pi} \left(\cos 2\vartheta_1 - \cos 2\vartheta_2\right) \ .$$

$$(6.234)$$

The lines of constant principal shear stress are determined by means of *Mohr's* circle (Fig. 2.7) and with Eq. (6.234)

$$\tau = \frac{n_0}{\pi h} \sin\left(\vartheta_2 - \vartheta_1\right) = \text{const} \ ,$$

$$(6.235)$$

and are circular arches through the surface points at the ends of the given tractions $(x = \pm c, y = 0)$. Centers are located at the axis of symmetry. The maximal shear $\tau_{max} = n_0 / \pi h$ is reached along the semicircle with the radius $r = c$.

(§) The *Boussinesq* Problem. This problem is derived by increasing the uniform loading n_0 beyond any bounds but keeping the resultant $F = 2cn_0$ finite through the limit of proper order $c \rightarrow 0$. Asymptotically, the stresses due to a single force loading $\mathbf{F} = -F\mathbf{e}_y$, applied at the surface point $(0, 0)$, result. Thus, taking these limits in Eq. (6.234) and considering the polar coordinate system of Fig. 6.31 yield the cartesian components

$$n_x = \frac{2}{\pi} F x^2 y/r^4 \ , \quad n_y = \frac{2}{\pi} F y^3/r^4 \ , \quad n_{xy} = \frac{2}{\pi} F x y^2/r^4 \ , \quad r^2 = x^2 + y^2 \ .$$

$$(6.236)$$

The principal normal stresses are aligned to the polar coordinates and form the central force system, according to *J. Boussinesq* ,

$$n_r = \frac{2}{\pi} \frac{F}{r} \sin \varphi \ , \quad n_\varphi = n_{r\varphi} = 0 \ .$$

$$(6.237)$$

The lines of constant radial stress, the isostresses, are circles through the point of application of the single force loading and with the center located at the axis of symmetry (Fig. 6.31)

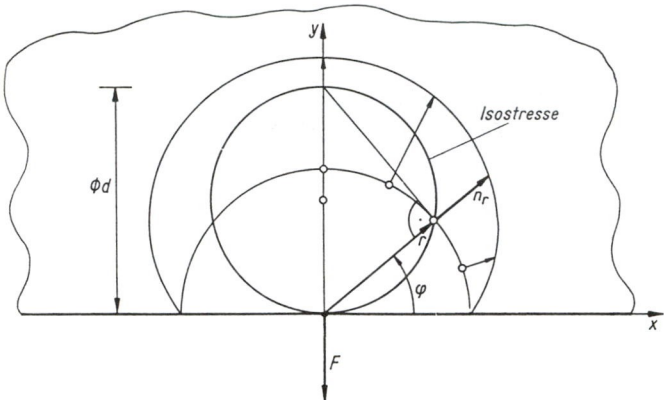

Fig. 6.31. Semiinfinite plate loaded by a single force F. Isostresses of constant radial stress

$$\frac{\sin\varphi}{r} = 1/d = \text{const} \quad , \quad n_r = \frac{2}{\pi}\frac{F}{d} \; . \tag{6.238}$$

Along these circles, the principal shear stress $\tau = F/\pi h d$ is also constant. The cartesian components of the membrane stresses are

$$n_x(r,\varphi) = n_r \cos^2\varphi \;, \quad n_y(r,\varphi) = n_r \sin^2\varphi \;, \quad n_{xy}(r,\varphi) = \frac{n_r}{2}\sin 2\varphi \; . \tag{6.239}$$

Also remaining in the *Boussinesq* problem are the stresses independent of the special choice of the elastic parameters.
(§) The Stress Function in the Case of a Tangential Single Force. With the load at the surface $F = -F_t\,\boldsymbol{e}_x$ this function is derived analogously

$$F(x,y) = -\frac{F_t}{\pi}\int_0^\infty y\,\frac{\sin(x\xi)}{\xi}\,e^{-y\xi}\,d\xi \; . \tag{6.240}$$

The principal radial normal stress has a cosine distribution in the polar coordinate system of Fig. 6.31; compare with Eq. (6.237)

$$n_r(r,\varphi) = \frac{2}{\pi}\frac{F_t}{r}\cos\varphi \;, \quad n_\varphi = n_{r\varphi} = 0 \; . \tag{6.241}$$

Solutions to other, more complex loadings are sampled in the literature; see, eg *K. Girkmann: Flächentragwerke. 6th ed., Springer-*

Verlag, Wien, 1963, (in German). (The late *K. Girkmann* was Professor at TU Vienna).

6.5.2. Stationary Spinning Disks

If we consider the angular velocity $\omega = const$, the problem becomes axisymmetric. Figure 6.32 shows the free-body diagram of an element, and the condition of equilibrium in the radial direction renders, in the limit of contraction of the element to a material point, the relation between the principal normal stresses and the loading q [see also q_x in Exercise A 11.6 and Eq. (1.16)]

$$r \frac{dn_r}{dr} + n_r - n_\varphi = -rq \ , \ q = \rho h \, r\omega^2 \ .$$
(6.242a)

The choice of a stress function $f(r)$ according to $n_r = f/r$, $n_\varphi = \rho h \, r^2\omega^2 + df/dr$ identically solves the local equilibrium condition. Hence, the compatibility condition of the principal strains, $\varepsilon_r - d(r \, \varepsilon_\varphi)/dr = 0$, derived from Eqs. (6.202) and (6.203) with $u = 0$, where $\varepsilon_r = du/dr$, $\varepsilon_\varphi = u/r$, renders on the elimination of the strains by means of *Hooke's* law [see Eq. (6.224)] the ordinary differential equation, $f' = df/dr$,

$$r f'' + \left(1 - \frac{r}{h}\frac{dh}{dr}\right) f' - \left(\frac{1}{r} - \frac{v}{h}\frac{dh}{dr}\right) f = -(3 + v)\rho h \, r^2\omega^2 \ .$$
(6.242b)

A closed-form solution exists if the thickness of the disk varies according to the potential function $h = H r^n$ (the constant thickness H is included by putting $n = 0$)

$$f(r) = C \, r^{\alpha_1} + D \, r^{\alpha_2} - \frac{(3 + v)\rho H\omega^2}{8 + (3 + v)n} r^{3+n}, \ \alpha_{1,2} = \frac{n}{2} \pm \sqrt{\frac{n^2}{4} + 1 - v \, n} \ .$$
(6.243)

The constants of integration, C and D, are determined in the case of

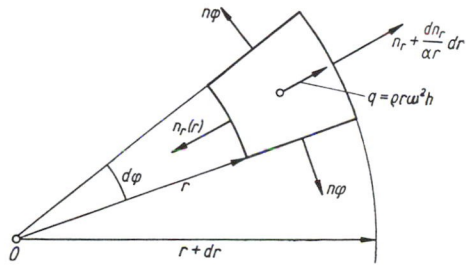

Fig. 6.32. Free-body diagram of a plate element under radial loading conditions

a hollow circular plate by the boundary conditions prescribed at the inner and outer surface, $r = R_{i, e}$. However, in the case of a full disk, the condition at $r = R_e$ is supplemented by the requirement of finite stresses at $r = 0$. The principal normal stresses in a hollow rotating disk of constant thickness with traction-free surfaces are collected below

$$n_r = N \left(\frac{r^2}{R_i^2} - 1 \right) \left(1 - \frac{r^2}{R_e^2} \right) \frac{R_e^2}{r^2} \quad , \quad N = K R_i^2 \quad , \quad K = (3 + v) \, \rho H \, \omega^2 / 8 \quad ,$$

$$n_\varphi = N \left[1 + \left(\frac{R_e^2}{R_i^2} + 1 \right) \frac{r^2}{R_e^2} - \frac{1 + 3v}{3 + v} \frac{r^4}{(R_i R_e)^2} \right] \frac{R_e^2}{r^2} \quad . \tag{6.244}$$

The hoop stress becomes maximal at the inner boundary; see Fig. 6.33

$$\max \left(n_\varphi \right) = 2N \left[R_e^2 / R_i^2 + (1 - v) / (3 + v) \right] \quad . \tag{6.245}$$

Even in the limit $R_i \rightarrow 0$ of the cylindrical notch, the hoop stress remains magnified by a factor of 2 with respect to the undisturbed full disk spinning with the same angular speed

$$\lim_{R_i \rightarrow 0} \left(n_\varphi \right) = \alpha \, n \quad , \quad n = \bar{n}_r = \bar{n}_\varphi = K R_e^2 \quad , \quad \alpha = 2 \quad , \tag{6.246}$$

n is the isotropic tension at the center of the full disk (Fig. 6.33).
The plane stress state is considered above in thin-walled plates. The solutions are applicable to the plane strain state in *rotating thick-walled shafts* as well, if *Poisson´s* constant v is replaced by the function $v/(1 - v)$.

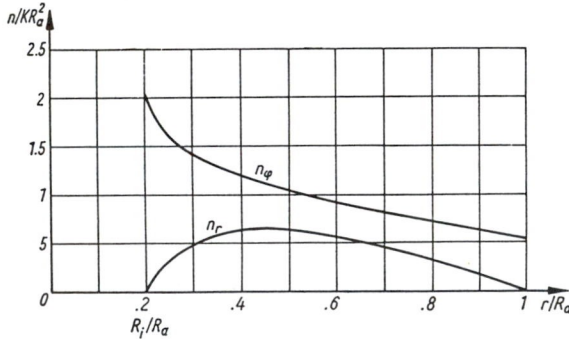

Fig. 6.33. Dimensionless membrane stresses in stationary rotating hollow disks with traction-free surfaces. $n = K R_e^2$ is the isotropic tension at the center of the full plate

6.5.3. The Infinite Plate with a Circular Hole: Kirsch's Problem

An infinite plate with a circular hole is considered with uniform tension applied at infinity. The stresses are magnified with respect to the undisturbed plate, and a solution was derived independently by *G. Kirsch* and *A. Leon* (The late *A. Leon* was Professor at the TU-Vienna). Following the lines outlined for the tensile rod [see Eq. (2.6)], the loading at infinity $n_x = n_0 = const$, $n_y = n_{xy} = 0$ is projected onto the polar coordinates shown in Fig. 6.34

$$n_r = n_0 \cos^2 \varphi = \frac{n_0}{2} + \frac{n_0}{2} \cos 2\varphi \ , \ n_{r\varphi} = -\frac{n_0}{2} \sin 2\varphi \ . \tag{6.247}$$

The stresses due to a constant radial tension $n_r = n_0 /2 = - hp_e$ can be taken from Eqs. (6.14) and (6.17), where $\alpha = 1$ and $R_e \to \infty$

$$\bar{n}_r = \frac{n_0}{2} \left[1 - (R/r)^2\right] \ , \ \bar{n}_\varphi = \frac{n_0}{2} \left[1 + (R/r)^2\right] \ . \tag{6.248}$$

Hence, the stresses due to the load which is distributed over the double angle 2φ in Eq. (6.247) must be determined and superposed. The stress function is assumed to have the same azimuthal dependence and is separated

$$\bar{\bar{F}}(r, \varphi) = f(r) \cos 2\varphi \ . \tag{6.249}$$

Thus, Eq. (6.225) in polar coordinates reduces to an ordinary differential equation with the general solution

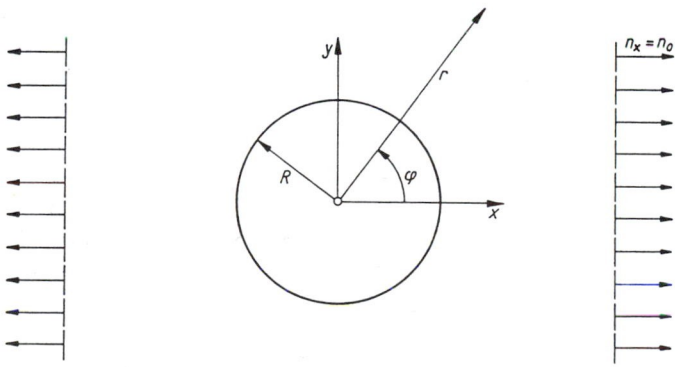

Fig. 6.34. The infinite plate with a circular hole: *Kirsch's* problem

$$f(r) = A\, r^2 + B\, r^4 + C\, r^{-2} + D \; . \qquad (6.250)$$

The stresses remain finite at infinity if the limit $r \to \infty$ of the second derivative of Eq. (6.250) is finite; hence, $B = 0$. The remaining three constants follow from the two homogeneous dynamic boundary conditions at the traction-free surface $r = R$, $\sigma_{rr} = \sigma_{r\varphi} = 0$, and from the condition of equilibrium at a cylindrical surface in the far field, $r \gg R$,

$$A = -n_0/4 \;\; , \;\; C = -n_0\, R^4/4 \;\; , \;\; D = n_0\, R^2/2 \; . \qquad (6.251)$$

The superposition of the principal normal stresses that are derived from the resulting stress function $F(r, \varphi)$ leaves the surface of the hole free of any tractions, due to the individually applied boundary conditions at $r = R$

$$n_r = \frac{1}{r}\frac{\partial F}{\partial r} + \frac{1}{r^2}\frac{\partial^2 F}{\partial r^2} = \frac{n_0}{2}\left[1 - \frac{R^2}{r^2} + \left(1 - 4\frac{R^2}{r^2} + 3\frac{R^4}{r^4}\right)\cos 2\varphi\right],$$

$$n_\varphi = \frac{\partial^2 F}{\partial r^2} = \frac{n_0}{2}\left[1 + \frac{R^2}{r^2} - \left(1 + 3\frac{R^4}{r^4}\right)\cos 2\varphi\right],$$

$$n_{r\varphi} = -\frac{\partial}{\partial r}\left(\frac{1}{r}\frac{\partial F}{\partial \varphi}\right) = -\frac{n_0}{2}\left(1 + 2\frac{R^2}{r^2} - 3\frac{R^4}{r^4}\right)\sin 2\varphi \; . \qquad (6.252)$$

Asymptotically, the stresses at infinity are given by

$$\lim_{r \to \infty} n_r = \frac{n_0}{2}\left[1 + \cos 2\varphi\right] \;\; , \;\; \lim_{r \to \infty} n_\varphi = \frac{n_0}{2}\left[1 - \cos 2\varphi\right] \; ,$$

$$\lim_{r \to \infty} n_{r\varphi} = -\frac{n_0}{2}\sin 2\varphi \; . \qquad (6.253)$$

This limit proves that the principal stresses at infinity take on the proper values of the given loading, $n_x = n_0 = const$ and $n_y = 0$, for all values of the angle φ. Note *Mohr's* stress circle.

The critical hoop stress $n_\varphi = 3\, n_0$ at the internal boundary at $\varphi = \pm \pi/2$ [Eq. (6.252)] turns out to be independent of the radius of the hole R. Thus, the magnification factor of *3* as derived for the circular notch is merely a shape factor. It remains constant even in the limit $R \to 0$. Similar problems of engineering importance were collected by *G. N. Sawin: Stress Concentrations at the Surface of Holes. VEB-Verlag Technik, Berlin, 1956, (in German)*. Note such stress magnifications when ordering drilling of holes to save weight in light-weight structural design.

6.5.4. Thermal Membrane Stresses in Plates

The temperature field in thin plates produces in-plane stresses, if the average over the thickness h does not vanish. *Maysel´s* formula (6.44) is quite suitable for the determination of the thermoelastic displacements: The dummy force $H = 1$ acts in the direction e_i in the middle plane and produces the work

$$1.\ \delta_i = \iint_A d\xi\, d\eta \int_{-h/2}^{h/2} 3\alpha\theta(\xi,\eta,\zeta)\, \overline{p}(\xi,\eta,\zeta; x,y)\, d\zeta\ ,\quad 3\overline{p} = (\overline{n}_x + \overline{n}_y)/h\ .$$

Hence, integration over the thickness yields by means of the mean temperature $n_\theta\,(x,\,y)$

$$\delta_i = \int_A \alpha n_\theta(\xi,\eta)\, \overline{n}(\xi,\eta; x,y)\, dA_{\xi,\eta}\ ,\quad n_\theta = \frac{1}{h}\int_{-h/2}^{h/2} \theta(\xi,\eta,\zeta)\, d\zeta\ . \tag{6.254}$$

The isothermal influence function of an infinite plate, due to the load $H = 1$ applied at the origin and pointing in the negative x direction, is quite simple and it is reported here for convenience

$$\overline{n}\,(\xi,\eta; 0,0) = \frac{1+\nu}{2\pi}\, \frac{\xi}{r^2}\ ,\quad r^2 = \xi^2 + \eta^2\ . \tag{6.255}$$

A translation of the coordinate system renders the general form. Equation (6.254), when supplemented by a proper loop integral, becomes the starting point of the *boundary element method (BEM)*, if the kernel function of the infinite domain is substituted. An unknown distribution of forces along the actual boundary of the finite plate has to be determined by solving (numerically) an integral equation. This procedure requires a discretization of the boundary only.

6.6. Flexure of Plates

Thin plates are considered under the action of transverse loads. The originally plane middle surface is deflected to a surface $w(x,\,y)$. Like a straight beam, the deformation depends on the bending rigidity and is mainly controlled by the boundary conditions. Analogous to the linear elastic beam, the normal stresses σ_{xx} and σ_{yy} are assumed to be linearly distributed over the thickness h. Their moments per unit of length are the resultants. Note that the subscript refers rather to the spatial orientation of the cross-sectional element and, thus, does not give the direction of the vector component

$$m_x = \int_{-h/2}^{h/2} z\, \sigma_{xx}\, dz \quad , \quad m_y = \int_{-h/2}^{h/2} z\, \sigma_{yy}\, dz \quad .$$

$$(6.256)$$

The torsional moment per unit of length results by summing the couples of the shear stresses σ_{xy}

$$m_{xy} = \int_{-h/2}^{h/2} z\, \sigma_{xy}\, dz \quad .$$

$$(6.257)$$

Since the stresses are the elements of the symmetric 2 x 2 stress tensor and the integration is a linear operation, the tensor property is conserved for the symmetric matrix of the moments. The rotation of the coordinate system to principal axes follows the rules of *Mohr´s* circle (Fig. 2.7). Equations (2.13) and (2.14) render the principal moments $m_{1,\,2}$ and, hence, the bending stresses

$$\sigma_{11} = \frac{12\, m_1}{h^3}\, z \quad , \quad \sigma_{22} = \frac{12\, m_2}{h^3}\, z \quad .$$

$$(6.258)$$

They take on their extreme values at $z = \pm\, h/2$. *Bernoulli´s* hypothesis that the cross-sections remain plane after deflection of a beam is generalized for thin plates to *Kirchhoff´s hypothesis* . <<The material normal of the midplane of a plate remains normal to the deflected midplane and keeps its length unaltered, ie the plate is assumed inextensible in the lateral direction.>> Hence, in this approximation, $\varepsilon_{zz} = \varepsilon_{xz} = \varepsilon_{yz} = 0$. Furthermore, the normal stress σ_{zz} that is only of the order of the lateral pressure loading is neglected. In the case of very small deflections, $|w| \ll h$, fibers in the midplane are assumed inextensible; see also Sec. 5.3.2, where the nonlinear coupling terms of Eq. (5.30) become negligible quantities. However, the deformation conditions are more restrictive when compared to the *Bernoulli-Euler* beam theory. Consequently, the slopes $\partial w/\partial x$ and $\partial w/\partial y$ (see Fig. 6.35) in this first-order theory of *Kirchhoff* plates are set equal to the rotations of the normal during deformation, and the displacement components in a distance z from the midplane are proportional

$$u(z) = -z\, \frac{\partial w}{\partial x} \quad , \quad v(z) = -z\, \frac{\partial w}{\partial y} \quad .$$

$$(6.259)$$

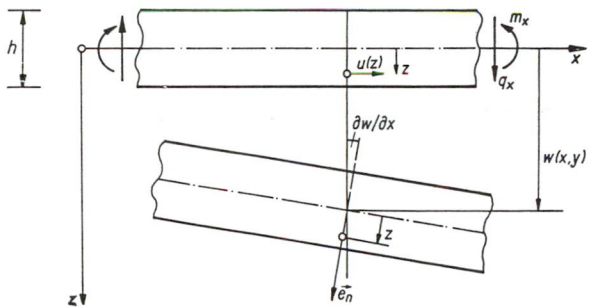

Fig. 6.35. Displacements, slopes, and resultants in a laterally loaded plate

The linearized geometric relations render the strains proportional to the linearized curvatures (only their dependence on z is explicitly shown)

$$\varepsilon_x(z) = \frac{\partial u}{\partial x} = -z\frac{\partial^2 w}{\partial x^2} \quad , \quad \varepsilon_y(z) = \frac{\partial v}{\partial y} = -z\frac{\partial^2 w}{\partial y^2} \ . \tag{6.260}$$

The shear strain is proportional to the mixed derivative of the deflection

$$\gamma_{xy} = 2\,\varepsilon_{xy}(z) = \frac{\partial u}{\partial y} + \frac{\partial v}{\partial x} = -2\,z\frac{\partial^2 w}{\partial x\partial y} \ . \tag{6.261}$$

Hooke's law (4.20), putting $\sigma_{zz} = 0$, when multiplied by z and integrated over the plate thickness h gives the moment curvature relations and a relation for the torsional moment

$$m_x = -K\left(\frac{\partial^2 w}{\partial x^2} + v\frac{\partial^2 w}{\partial y^2}\right) , \ m_y = -K\left(\frac{\partial^2 w}{\partial y^2} + v\frac{\partial^2 w}{\partial x^2}\right) , \ m_{xy} = -(1-v)\,K\frac{\partial^2 w}{\partial x\,\partial y} \ ,$$

$$K = \frac{Eh^3}{12\,(1-v^2)} \quad , J = \frac{1.\,h^3}{12} \ . \tag{6.262}$$

K in Eq. (6.262) is the bending rigidity and J is thus the moment of inertia of a strip of width one. Hence, putting *Poisson's* ratio $v = 0$ renders the bending stiffness of a beam of unit width, $B = EJ$. The equations are exact for small curvatures for a plate in pure bending, ie if the plate is loaded only by self-equilibrating edge moments, m_1 and m_2. Analogous to the beam [Eq. (4.23)] where the curvature $1/R$ $= -M/EJ$ in pure bending, an extension to larger principal curvatures of the plate, $1/R_1$ and $1/R_2$, is easily shown to hold. Superposition still applies

$$\frac{1}{R_1} = \frac{12}{Eh^3}(m_1 - v\, m_2) \quad , \quad \frac{1}{R_2} = \frac{12}{Eh^3}(m_2 - v\, m_1) \quad .$$
$$(6.263)$$

Inversion of the linear equations yields the desired relation of pure bending

$$m_1 = K\left(\frac{1}{R_1} + v\frac{1}{R_2}\right) \quad , \quad m_2 = K\left(\frac{1}{R_2} + v\frac{1}{R_1}\right) \quad .$$
$$(6.264)$$

Linearization of the curvatures gives the principal moment relations of Eq. (6.262), $m_{12} = 0$.

The deflection of the midplane to circular cylindrical and spherical surfaces, respectively, is of practical importance for the special cases of a plate strip and of a circular plate under hinged support conditions. In the first case, $1/R_2 = 0$; in the second case, the principal curvatures are equal, $1/R_1 = 1/R_2 = 1/R$, hence, $m_1 = m_2$.

Shear deformations are not considered in the *Kirchhoff* approximation of thin plates, analogous to the *Bernoulli-Euler* theory of slender beams. The *shear forces* per unit of length enter the conditions of equilibrium of a plate element (Fig. 6.36) and are related to the field moments by

$$q_x = \frac{\partial m_x}{\partial x} + \frac{\partial m_{xy}}{\partial y} \quad , \quad q_y = \frac{\partial m_y}{\partial y} + \frac{\partial m_{xy}}{\partial x} \quad .$$
$$(6.265)$$

The substitution of the moment curvature relations (6.262) renders the shear forces proportional to the third derivatives of the deflection

$$q_x = -K\frac{\partial(\Delta w)}{\partial x} \quad , \quad q_y = -K\frac{\partial(\Delta w)}{\partial y} \quad , \quad \Delta = \nabla^2 = \frac{\partial^2}{\partial x^2} + \frac{\partial^2}{\partial y^2} \quad .$$
$$(6.266)$$

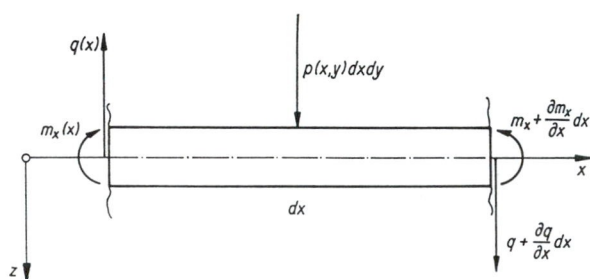

Fig. 6.36. Free-body diagram of a plate element

The equilibrium of the forces in the lateral z direction of Fig. 6.36 relates the gradients to the given pressure loading p

$$\frac{\partial q_x}{\partial x} + \frac{\partial q_y}{\partial y} = -p(x, y) \quad .$$

(6.267)

Elimination of the shear forces by combining the two equations yields the second-order differential equation

$$\frac{\partial^2 m_x}{\partial x^2} + 2\frac{\partial^2 m_{xy}}{\partial x \partial y} + \frac{\partial^2 m_y}{\partial y^2} = -p(x, y) \quad .$$

(6.268)

The substitution of Eq. (6.262) eliminates the moments and renders the inhomogeneous *biharmonic plate equation*

$$K \Delta\Delta w = p(x, y) \quad , \quad \Delta = \frac{\partial^2}{\partial x^2} + \frac{\partial^2}{\partial y^2} \quad .$$

(6.269)

The first invariant of the moment tensor $m = m_x + m_y$ is formed by summing Eq. (6.262). The result is a first *Poisson´s* differential equation for the deflection

$$\Delta w = -\frac{1}{K(1+v)} m(x, y) \quad , \quad \Delta = \frac{\partial^2}{\partial x^2} + \frac{\partial^2}{\partial y^2} \quad .$$

(6.270)

Inserting into Eq. (6.269) yields a second, coupled *Poisson´s* equation for the sum of moments

$$\Delta m = -(1+v)\, p(x, y) \quad , \quad \Delta = \frac{\partial^2}{\partial x^2} + \frac{\partial^2}{\partial y^2} \quad .$$

(6.271)

In those cases where the boundary value of the sum of moments is known, eg at a straight and simply supported edge of a (polygonal) plate where *Navier´s limit condition* applies, namely, $m = 0$, the solution is more easily found by solving the system of two second-order equations (6.270) and (6.271) instead of integrating the equivalent biharmonic equation. At first, Eq. (6.271) is integrated independently to render the distribution of the sum of moments that enters *Poisson´s* equation (6.270) as the forcing function. Solving for the deflection and the linearized curvatures determines the individual moments by substitution into Eq. (6.262). A low-order system of equations is also preferred in numerical integration algorithms, eg in the finite difference scheme.

The *complementary energy* of the *Kirchhoff* plate is given by the integral over the area of the midplane

$$U^* = \frac{1}{2} \int_A \frac{12}{Eh^3} \left[m^2 + 2(1+v)(m_{xy}^2 - m_x\, m_y) \right]\, dA \ , \quad m = m_x + m_y \ .$$

(6.272)

Castigliano's theorem (5.43) renders the deflection at the point of application of a single force. The generalization of *Betti's* method (5.54) to thin plates is straightforward if the shear deformations are negligible.

It will be shown for the special planform of a circular plate that, in general, two boundary conditions ought to be prescribed at the edge of a plate in order to select the proper solution from the general one. At the free edge of a plate with the normal e_n and with the tangent unit vector e_s , eg the tractions are supposed to vanish, but the three resultants m_n , m_{ns} , and q_n cannot be put to zero individually. Therefore, *Thomson* and *Tait* suggest a combination that renders two independent conditions in accordance with the plate theory: They replace the torsional moment by a statically equivalent couple of shear forces. At an infinitesimal lever *ds* , the force $(\partial m_{ns}/\partial s)\, ds$ remains in the *z* direction, which leads to *Kirchhoff's effective shear force* per unit of length

$$\bar{q}_n = q_n + \frac{\partial m_{ns}}{\partial s} \ .$$

(6.273a)

It replaces the actual shear force q_n in the formulation of the dynamic boundary conditions, ie at the free boundary, the two dynamic effective conditions, being compatible with the *Kirchhoff* plate theory, are

$$m_n = 0 \ , \quad \bar{q}_n = q_n + \frac{\partial m_{ns}}{\partial s} = 0 \ .$$

(6.273b)

In addition, a single force appears where the torsional moment has a finite jump accompanied by an infinite derivative. At those points of the boundary, the plate needs a special support. Not properly anchored corners are likely to uplift.

6.6.1. Axisymmetric Flexure of Circular Kirchhoff Plates

Under the conditions of polar symmetry, the *Laplace* operator becomes an ordinary second-order differential operator in the field variable *r* , and the fourth-order plate equation (6.269) has a general solution in the form of a multiple integral of the weighted given pressure loading *p(r)* ,

$$\Delta\Delta w = \frac{p(r)}{K} \ , \quad \Delta = \frac{1}{r}\frac{\partial}{\partial r}\left(r\frac{\partial}{\partial r}\right) \ , \quad w_p(r) = \int\frac{dr}{r} \int r\, dr \int\frac{dr}{r} \int\frac{p(r)}{K}\, r\, dr \ ,$$

$$w(r) = A + B\,r^2 + C\,\ln\frac{r}{R_e} + D\,r^2\,\ln\frac{r}{R_e} + w_p \ .\tag{6.274a}$$

The four constants of integration are to be determined by the boundary conditions at the outer and inner (concentric) circular edge of a hollow plate. The deflection at the center of a full circular plate, $w(0)$, must be finite: $C = 0$ as well as the curvature, $D = 0$. For singular loading by a single force F applied at the center, see Exercise A 6.19, where $D = F/8\pi K \neq 0$. The particular solution becomes for the uniform loading $p = p_0 = const$

$$w_p(r) \;=\; \frac{p_0}{64\,K}\,r^4 \ .\tag{6.274b}$$

The constants A and B of the deflection of a uniformly loaded full plate are easily determined by the boundary conditions at the outer edge, $r = R_e$.

The geometric conditions of a rigidly clamped circular edge are $w = \partial w/\partial r = 0$ and thus render $A = p_0\,R_e^4/64K$, $B = -\,p_0\,R_e^2/32K$

At the simply supported boundary, $w = 0$ is supplemented by the dynamic boundary condition $m_r = 0$ (r points in the direction of the normal of the circular edge), and the constants become in relation to the above values: $A_h = (5 + v)A/(1 + v)$, $B_h = (3 + v)B/(1 + v)$.

6.6.2. The Infinite Plate Strip

The parallel edges $x = 0$ and $x = a$ of a very long strip are supported by hinges, and the pressure loading $p = p(x)$ is assumed to vary over the width only. The cylindrical deflection $w = w(x)$ is a solution of the reduced plate equation

$$\frac{d^4w}{dx^4} = \frac{p(x)}{K} \ .\tag{6.275}$$

It corresponds to the ordinary differential equation of the beam deflection (5.22). A comparison of the bending rigidities renders the deflection of the plate strip as $(1 - v^2)$ times that of a beam of rectangular cross-section $1 \times h$. In addition to the principal bending moment m_x that is the same in both structures, the plate carries the second principal component $m_y = v\,m_x$. Uniform loading $p = p_0 = const$ gives the solution

$$w(x) \;=\; \frac{p_0\,a^4}{24\,K}\left(\frac{x^3}{a^3} - 2\,\frac{x^2}{a^2} + 1\right)\frac{x}{a} \ , \quad m_x = -K\frac{\partial^2 w}{\partial x^2} = \frac{p_0\,a^2}{2}\left(1 - \frac{x}{a}\right)\frac{x}{a} \ .\tag{6.276}$$

6.6.3. The Rectangular Plate with Four Edges Simply Supported

The boundary conditions $w = 0$ and $m = 0$ [see Eq. (6.270)] at $x = 0, a$ and $y = 0, b$ suggest a general solution in the form of a *Fourier* double series. It was originally given by *Navier*

$$w(x, y) = \sum_j \sum_k w_{jk} \sin \frac{j\pi x}{a} \sin \frac{k\pi y}{b} \quad . \tag{6.277}$$

The coefficients w_{jk} are determined by comparing coefficients in the plate equation (6.269) after a periodic extension of the given pressure loading $p(x, y)$ is performed, such that it can be expanded into an odd double sum with the periods $2a$ and $2b$

$$p(x, y) = \sum_j \sum_k p_{jk} \sin \frac{j\pi x}{a} \sin \frac{k\pi y}{b} \quad . \tag{6.278}$$

The result is (the expansion coefficients p_{jk} are assumed to be known)

$$w_{jk} = p_{jk} / \pi^4 K \, (j^2/a^2 + k^2/b^2)^2 \quad . \tag{6.279}$$

The expansion coefficients of the uniform loading $p = p_0 = const$ are odd-numbered,

$$p_{jk} = 16 \, p_0 / \pi^4 j \, k \quad , \quad j, k = 1, 3, 5, \ldots \tag{6.280}$$

A single force loading F at the general point of application $x = \xi$, $y = \eta$ has expansion coefficients, given by the products

$$p_{jk} = \frac{4 \, F}{a \, b} \sin \frac{j\pi\xi}{a} \sin \frac{k\pi\eta}{b} \quad , \quad j, k = 1, 2, 3, \ldots \tag{6.281}$$

Hence, the influence function of the sum of moments becomes with $F = 1$ and the summation of one of the *Fourier* series

$$m(x, y; \xi, \eta) = 2 \, (1 + v) \sum_j \sin \alpha_j x \sin \alpha_j \xi \sinh \alpha_j (b - y) \sinh \alpha_j \eta \, / \, \alpha_j a \sinh \alpha_j b,$$

$$0 \le \eta \le y \quad , \quad \alpha_j = j\pi/a \quad , \quad \sinh z = (e^z - e^{-z})/2 \quad . \tag{6.282}$$

The exponentially growing term should be eliminated in the numerical evaluation by substituting the *sinh*-function. In the region $y \le \eta \le b$, the coordinates y and η are to be interchanged. The single-series representation exhibits fast convergence.

6.6.4. Thermal Deflection of Plates

Given the temperature difference $\theta(x, y, z)$ with respect to a constant reference temperature, *Maysel's* formula (6.44) is an efficient tool for analyzing the plate deflections and thermal stresses. The sum of the isothermal normal stresses due to a single unit force loading of the plate is expressed by the sum of the moments [the coordinates in the influence function (6.282) ought to be interchanged]

$$3\,\overline{p} = \frac{12}{h^3}\,(\overline{m}_x + \overline{m}_y)\,\zeta = \frac{12}{h^3}\,\overline{m}\,\zeta \ .$$

$$(6.283)$$

Integration over the plate thickness h can be performed. The thermal deflection thus becomes an integral over the area of the midplane,

$$w(x, y) = \iint_A dA \left[\int_{-h/2}^{h/2} \alpha\theta(\xi, \eta, \zeta)\,\frac{12}{h^3}\,\overline{m}\,\zeta\,d\zeta \right] = \iint_A \alpha m_\theta(\xi, \eta)\,\overline{m}\,(\xi, \eta; x, y)\,dA_{\xi, \eta} \ .$$

$$(6.284)$$

Analogous to Eq. (6.87) of the beam, the first-order thickness moment of the temperature enters

$$m_\theta(\xi, \eta) = \frac{12}{h^3} \int_{-h/2}^{h/2} \zeta\,\theta(\xi, \eta, \zeta)\,d\zeta \ .$$

$$(6.285)$$

Multiplication by the linear coefficient of thermal expansion determines the thermal curvature that is to be superposed on the bending curvatures. Inversion renders the generalized moments of the moment tensor [compare this with Eq. (6.262)]

$$m_x = -K\left(\frac{\partial^2 w}{\partial x^2} + v\frac{\partial^2 w}{\partial y^2} + (1+v)\,\alpha m_\theta\right) , \quad m_y = -K\left(\frac{\partial^2 w}{\partial y^2} + v\frac{\partial^2 w}{\partial x^2} + (1+v)\,\alpha m_\theta\right) ,$$

$$m_{xy} = -(1-v)\,K\,\frac{\partial^2 w}{\partial x\,\partial y} \ , \quad K = \frac{Eh^3}{12\,(1-v^2)} \ , \quad J = \frac{1.\,h^3}{12} \ .$$

$$(6.286)$$

The possibly nonlinear thermal stresses are then given by

$$\sigma_{xx} = \frac{12\,m_x}{h^3}\,z + \frac{E\alpha}{1-v}\,(n_\theta + zm_\theta - \theta) \ , \quad \sigma_{xy} = \frac{12\,m_{xy}}{h^3}\,z \ ,$$

$$\sigma_{yy} = \frac{12\,m_y}{h^3}\,z + \frac{E\alpha}{1-v}\,(n_\theta + zm_\theta - \theta) \ .$$

$$(6.287)$$

In the first-order theory, the mean stresses n_x/h, n_y/h, n_{xy}/h are decoupled and eventually superposed.

(§)   A Plate of Quadratic Planform. Assuming all four edges hinged, such a plate serves as an illustrative example of the effects of constant thermal curvature loading. The principal moments take on their maximal values at the center

$$m_x = m_y = -\frac{1-v^2}{2} K \alpha C \ , \quad m_\theta = C = const \ . \tag{6.288a}$$

At arbitrary field points, the sum of moments remains proportional to the thermal curvature, also for simply supported plates with more general polygonal planforms,

$$m = m_x + m_y = -(1 - v^2) K \alpha m_\theta \ . \tag{6.288b}$$

(§)   The Infinite Plate. Such a plate is studied for generating particular solutions. Associated is the isothermal, logarithmically singular influence function of the sum of moments,

$$\overline{m} = \overline{m}_x + \overline{m}_y = -\frac{(1+v)}{2\pi} \ln r \ , \quad r^2 = (x-\xi)^2 + (y-\eta)^2 \ , \tag{6.289}$$

For example, a constant thermal curvature loading in the interior of a circle $r \le R$ renders the principal moments according to a centered polar coordinate system

$$m\left\{ \begin{matrix} r \\ \varphi \end{matrix} \right\} = -\frac{1-v^2}{2} K \alpha C \left\{ \begin{matrix} (\pm)(R/r)^2 \ , \ r > R \\ 1 \quad , \ r \le R \end{matrix} \right. \ , \quad m_\theta = C = const \ . \tag{6.290}$$

Hence, the hoop moment jumps at the thermally insulated "boundary", at $r = R$, by the amount

$$[m_\varphi] = -(1 - v^2) K \alpha C \ . \tag{6.291}$$

The above value of the jump is independent of the special geometry of the thermal boundary. If we take the limit $A = \pi R^2 \to 0$, but $C \to \infty$, such that the product, the strength of the resulting *bending hot spot*, remains finite, the singular solution becomes with Eq. (6.289)

$$w(r) = -\alpha \kappa \overline{m} \ , \quad \kappa = \lim_{R \to 0} \pi R^2 C \ . \tag{6.292}$$

Eq. (6.292) defines a thermal influence function of considerable structural importance.

6.7. Thin Shells of Revolution

Contrary to a plate that is flat in the undeformed state, a shell has a curved middle surface already in the reference configuration. At this point, a comparison with the load-carrying capacity of the one-dimensional structure given by a straight beam and its curved counterpart, the arch, is quite illustrative. A shallow shell can be considered within the context of plate theory by substituting the (linearized) initial curvature on the right-hand side of the first *von Karman* equation (5.30a). Thin shells of revolution of arbitrary curvatures are considered subsequently under the action of axisymmetric loadings, including support reactions. The points of the middle surface are located by their angle of longitude ϑ and latitudinal angle φ . A general material point is further determined by the normal distance z from the midsurface (see Fig. 6.37). The shell is considered free of torsion, $\sigma_{\varphi\vartheta} = 0$, and the normal stress σ_{zz} , which is of the order of the normal pressure loading, is neglected. For the pressurized cylindrical and spherical shell, see Fig. 2.21. Analogous to plate theory, the remaining stresses are integrated over the thickness h of the shell, the results are indicated in Fig. 6.37

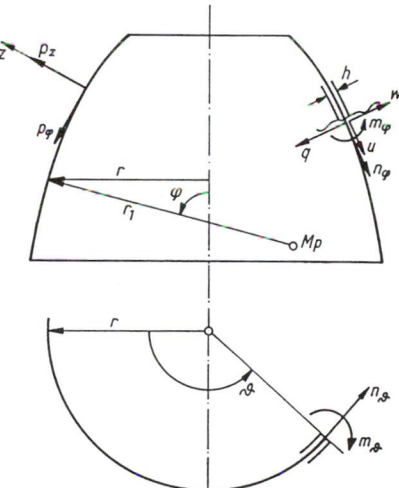

Fig. 6.37. Thin shell of revolution. Axisymmetric loading. Supports not shown. Displacements and stress resultants. See also Fig. 2.21

Membrane stresses, $n_\varphi = \int_{-h/2}^{h/2} \sigma_{\varphi\varphi}\, dz$, $n_\vartheta = \int_{-h/2}^{h/2} \sigma_{\vartheta\vartheta}\, dz$,

$$(6.293)$$

Bending moments, $m_\varphi = -\int_{-h/2}^{h/2} z\, \sigma_{\varphi\varphi}\, dz$, $m_\vartheta = -\int_{-h/2}^{h/2} z\, \sigma_{\vartheta\vartheta}\, dz$,

$$(6.294)$$

Shear force, $q = -\int_{-h/2}^{h/2} \sigma_{\varphi z}\, dz$.

$$(6.295)$$

In the case of smoothly distributed surface tractions, they are resolved into their components p_z and p_φ (see Fig. 6.37), steady curvatures of the midsurface, and constant or slightly varying thickness, the shear force changes slowly and does not contribute considerably to the local conditions of equilibrium. Thus, the *membrane stresses* approximately equilibrate the given tractions in the field. See also the membrane state of stresses of the pressurized cylindrical and spherical shell, Eq. (2.98).

(§) Membrane Stresses. The balance of forces in the axial direction of a part of the shell derived by a cut at some latitude $\varphi = const$ (the parallel circle cut of the middle surface) yields the internal statically determinate tangential membrane stress per unit of length (note the axial symmetry)

$$2\pi r\, n_\varphi \sin\varphi = 2\pi \int_{\varphi_0}^{\varphi} r\left(p_z \cos\varphi - p_\varphi \sin\varphi\right) r_1\, d\varphi + C \ .$$

$$(6.296)$$

The constant of integration C is the resultant force of the given loading at the upper edge of the open shell at $\varphi = \varphi_0$, which is necessarily parallel to the shell axis or it must be zero. For a dome-shaped shell, $C \neq 0$ corresponds to a single axial force (tensile loading). The membrane state of stresses of a closed shell in that case becomes singular at the top. In the near field of the single force, the bending stresses dominate. Considering the balance of forces in the normal z direction of a shell element, $r_1\, d\varphi\, r\, d\vartheta$, renders, after division through the product of differentials $d\varphi\, d\vartheta$ in the limit of a vanishing area element, the exact expression for the determination of the internal statically determinate (second) circumferential membrane stress

$$n_\vartheta\, r_1 \sin\varphi = r\, r_1\, p_z - r\, n_\varphi \ .$$

$$(6.297)$$

The membrane stresses may still be externally statically indeterminate. The elastic deformations of the midsurface associated with the membrane stresses are, in general, not compatible with the actual boundary conditions. Hence, the membrane stress state is (it is hoped only in the near field of the boundary, which is a design goal) perturbed by bending stresses. The bending perturbation is related to the deformations of the membrane state. *Hooke´s* law gives the membrane strains in the midsurface [see Eq. (6.224); h is the thickness of the shell]

$$\varepsilon_\varphi = \frac{1}{Eh}\,(n_\varphi - v n_\vartheta) \;\;,\;\; \varepsilon_\vartheta = \frac{1}{Eh}\,(n_\vartheta - v n_\varphi) \;.$$

(6.298)

The linearized geometric relations are determined similarly to the curved beam (Fig. 6.25). u is the tangential and w the normal displacement component of the points on the extensional mid-surface

$$\varepsilon_\varphi = \frac{1}{r_1}\left(\frac{du}{d\varphi} + w\right) \;\;,\;\; \varepsilon_\vartheta = \frac{\Delta r}{r} = \frac{1}{r}\,(u \cos \varphi + w \sin \varphi) \;\;,$$

(6.299)

$1/r_1$ is the initial curvature of the meridian $\vartheta = const.$

(§) Bending Perturbation of the Membrane State. After the considerations above, the shell is now assumed to be free of any smoothly distributed tractions. The free-body diagram of the shell at the parallel circle cut contains the additional normal stresses denoted $n_\varphi{}^*$, the shear force q, and the bending moment m_φ. Equilibrium requires (Fig. 6.37 in the absence of external loads)

$$2\pi r \left(n_\varphi{}^* \sin \varphi + q \cos \varphi\right) = 0 \;\;,\;\; n_\varphi{}^* = -q \cot \varphi \;.$$

(6.300)

Equation (6.300) enters also the jump condition of the unsteady normal stress and unsteady shear force at any parallel circle of application of an external ring load. The local condition of equilibrium in the normal z direction of a traction-free shell element, $r_1\,d\varphi\,r\,d\vartheta$, renders ($n_\vartheta{}^*$ is the additional circumferential normal stress)

$$\left[\frac{d(rq)}{d\varphi} + r n_\varphi{}^* + r_1 n_\vartheta{}^* \sin \varphi\right] d\varphi\,d\vartheta = 0 \;.$$

(6.301)

The elimination of $n_\varphi{}^*$, by means of Eq. (6.300), also gives $n_\vartheta{}^*$ explicitly in terms of the shear force

$$(r_1 \sin \varphi)\,n_\vartheta{}^* = -\left[\frac{d(rq)}{d\varphi} - rq \cot \varphi\right] \;.$$

(6.302)

Thus, knowing the shear force, determines the additional normal stresses in the region of any bending perturbation of the membrane state, which are to be superposed on the (statically determinate) membrane stresses. Also, the additional strains of the midsurface in the bending zones are thus implicitly related to the shear force through *Hooke's* law

$$\varepsilon_\varphi^* = \frac{1}{Eh}\left(n_\varphi^* - vn_\vartheta^*\right) \; , \; \varepsilon_\vartheta^* = \frac{1}{Eh}\left(n_\vartheta^* - vn_\varphi^*\right) \; .$$

(6.303)

The meridional tangent rotates during deformation through the angle χ , positive in the sense of increasing the latitude φ [see Eq. (6.205)]

$$\chi = \frac{u}{r_1} - \frac{dw}{r_1\,d\varphi} \; ,$$

(6.304)

and thus couples the tangential u , and the normal w , displacement of the midsurface. Hence, the compatibility of the strains (6.299) also requires

$$\chi = \frac{1}{\sin\varphi}\left[\left(\varepsilon_\varphi - \varepsilon_\vartheta\right)\cos\varphi - \frac{r}{r_1}\frac{d\varepsilon_\vartheta}{d\varphi}\right] \; .$$

(6.305)

During the small additional rotation in the zone of the bending perturbation, denoted χ^* , the fiber element $r_1\,d\varphi$ in a normal distance z is elongated by $z\,d\chi^*$. Furthermore, the radius $(r + z\,sin\,\varphi)$ of the parallel circle is increased to $r + z\,sin\,(\varphi + \chi)$. Since, $\chi^* \ll 1$ and $|z| \ll r$, a fiber element $r\,d\vartheta$ is elongated by $z\,\chi^*\,cos\,\varphi\,d\vartheta$. The strains in the bending zone are thus approximated by the linear distribution over the shell thickness

$$\varepsilon_\varphi(z) = \varepsilon_\varphi^* + \frac{z}{r_1}\frac{d\chi^*}{d\varphi} \; , \; \varepsilon_\vartheta(z) = \varepsilon_\vartheta^* + \frac{z}{r}\chi^*\cos\varphi \; , \; -h/2 \le z \le h/2 \; .$$

(6.306)

Equation (6.305), when considered in terms of these bending disturbances yields the desired relation between the shear force q and the angle χ^* . A substitution of the linearly distributed strains into *Hooke's* law (4.20), multiplication by z , and integration over the shell thickness render the principal bending moments related to χ^*

$$m_\varphi = -K\left(\frac{1}{r_1}\frac{d\chi^*}{d\varphi} + \frac{v}{r}\chi^*\cos\varphi\right) \; , \; K = \frac{Eh^3}{12\left(1 - v^2\right)} \; ,$$

$$m_\vartheta = -K\left(\frac{\chi^*}{r}\cos\varphi + \frac{v}{r_1}\frac{d\chi^*}{d\varphi}\right) \; .$$

(6.307)

The equilibrium of the axial moments about the tangent of the parallel circle of the traction-free element, $r_1 \, d\varphi \, r \, d\vartheta$, gives the static relation between the bending moments and shear force

$$\frac{d(r \, m_\varphi)}{d\varphi} - r_1 \, m_\vartheta \cos\varphi - r \, r_1 \, q = 0 \; .$$
(6.308)

The integration of the system of four coupled differential equations, (6.305) to (6.308), taking into account the boundary conditions, poses a rather complex mathematical problem. Exact solutions of the moments m_φ , m_ϑ , the shear q , and the rotation χ^* , and experience with several shell geometries indicate the possibility of drastically simplifying the equations, since the bending disturbance is, in general, limited to a boundary layer. That is, the bending moments decrease rapidly with the distance from any source of perturbation of the membrane state of stresses. As long as the radius of the parallel circle r does not become too small in such a perturbed zone (only the highest-order derivative is the leading term), the unknowns and their lower-order derivatives can be neglected. Elimination of the bending moments yields in such a boundary-layer approximation

$$\frac{d^2 q}{d\varphi^2} = Eh \left(\frac{r_1 \sin\varphi}{r}\right)^2 \chi^* \; , \quad \frac{d^2 \chi^*}{d\varphi^2} = -\frac{q}{K} \, r_1^2 \; .$$
(6.309)

The further elimination of χ^* renders the approximate fourth-order differential equation of the shear force

$$\frac{d^4 q}{d\varphi^4} + 4 \kappa^4 q = 0 \; , \quad \kappa^4 = 3 \left(1 - v^2\right) \frac{r_1^4}{h^2 \, r^2} \sin^2\varphi \; .$$
(6.310)

The shell design should limit the bending perturbation Hence, $|\kappa(\varphi)| \gg 1$, and under that condition, the approximation also becomes reliable. In general, a numerical procedure is required to solve Eq. (6.310). If, however, $\kappa = const$, eg for a *spherical shell* , where $\kappa_s^4 = 3(1 - v^2) \, (R/h)^2$, the shear force has functional the form of a "damped vibration." Time is replaced by the latitude

$$q = e^{\kappa\varphi} \left(C_1 \cos\kappa\varphi + C_2 \sin\kappa\varphi\right) + e^{-\kappa\varphi} \left(C_3 \cos\kappa\varphi + C_4 \sin\kappa\varphi\right) \; .$$
(6.311)

The constants of integration are determined by the boundary conditions at the lower, $\varphi = \varphi_1$, and at the upper, $\varphi = \varphi_0$, edge of the open shell. If $(\varphi_1 - \varphi_0)$ and the latitude φ_0 are both sufficiently large, the zones of perturbation of the membrane state do not overlap. In that case, approximately, $C_3 = C_4 = 0$ at the lower parallel circle and $C_1 = C_2 = 0$ at the upper edge of the open shell.

In general, the solution of Eq. (6.310) gives the shear force, and when we integrate twice, it renders χ^*. Substitution into Eqs. (6.300) and (6.302) yields the additional membrane stresses. The bending moments are determined by Eq. (6.307). The total normal stresses are given by the superposition

$$\sigma_{\varphi\varphi} = \frac{n_\varphi + n_\varphi{}^*}{h} - \frac{12\, m_\varphi}{h^3}\, z \quad , \quad \sigma_{\vartheta\vartheta} = \frac{n_\vartheta + n_\vartheta{}^*}{h} - \frac{12\, m_\vartheta}{h^3}\, z \quad . \tag{6.312}$$

The shear force q renders shear stresses that are assumed to be parabolically distributed over the shell thickness, as in the case of beams or plates,

$$\sigma_{\varphi z} = -\frac{3\, q}{2\, h} \left[1 - \left(\frac{2z}{h}\right)^2 \right] \quad . \tag{6.313}$$

6.7.1. Thin Circular Cylindrical Shells

The limit $r_1 \to \infty$, $d\varphi \to 0$, $r_1\, d\varphi = dx$ and putting $\varphi = \pi/2$, $r = R = const$ render simply

$$\chi = -\frac{dw}{dx} \quad . \tag{6.314}$$

The approximate equations of the bending perturbation become

$$\frac{d^2\chi^*}{dx^2} = -\frac{q}{K} \quad , \quad \frac{d^2q}{dx^2} = \frac{Eh}{R^2}\, \chi^* \quad \to \quad \frac{d^4q}{dx^4} + 4\, \frac{\kappa^4}{R^4}\, q = 0 \quad , \tag{6.315}$$

where κ equals a constant. The latter is formally given by the expression κ_S of the sphere. Equation (6.311) is still the solution of the shear force if x/R is substituted for the latitude φ. The perturbed resultants in the boundary layer are

$$n_x{}^* = 0 \quad , \quad n_\vartheta{}^* = -R\, \frac{dq}{dx} = \frac{Eh}{R}\, w^* \quad , \quad m_x = -K\, \frac{d\chi^*}{dx} = K\, \frac{d^2w}{dx^2} \quad , \quad m_\vartheta = v\, m_x \quad . \tag{6.316}$$

Hence,

$$\frac{du^*}{dx} = -v\, \frac{w^*}{R} \quad , \tag{6.317}$$

and from Eq. (6.316),

$$w^* = - \frac{R^2}{E h} \frac{dq}{dx} \quad . \tag{6.318}$$

The membrane stresses are still statically determinate and in the limit of Eqs. (6.296) and (6.297) they become decoupled

$$n_\vartheta = R \, p_z \quad , \quad n_x = - \int_0^x p_x \, dx + n_0 \quad . \tag{6.319}$$

Hooke´s law determines the strains in the midsurface

$$\varepsilon_x = \frac{1}{E h} (n_x - v n_\vartheta) \quad , \quad \varepsilon_\vartheta = \frac{1}{E h} (n_\vartheta - v n_x) \quad , \tag{6.320}$$

and by means of the linearized geometric relations, the displacements can be found

$$\varepsilon_x = \frac{du}{dx} \quad , \quad \varepsilon_\vartheta = \frac{w}{R} \quad . \tag{6.321}$$

(§)   The Open Cylindrical Storage Tank. This tank filled with a fluid of density ρ_F serves as an illustrative example of the theory of thin cylindrical shells. The density of the homogeneous shell material is denoted ρ , and the distributed loadings are given by its own dead weight $p_x = \rho g h$ and hydrostatic pressure $p_z = g \, \rho_F \, x$. The upper edge of the shell, $x = 0$, is also the free surface of the fluid; the tank is filled to the rim. See Fig. 6.38. The principal membrane stresses are thus linear functions of the axial coordinate, $n_x = (- \rho \, g \, h \, x)$ and $n_\vartheta = \rho_F \, g \, R \, x$; see also Eqs. (2.95) and (2.98) of the pressurized shell. At the lower edge of the shell, say, at $x = L$, the boundary condition $u = 0$ can be considered within the membrane theory, but the radial displacement becomes $w_0 = [(R/Eh) \, (\rho_F \, R + v \, \rho h) \, gL]$ and the rotation is $\chi_0 = - \, dw/dx = - \, w_0 /L$. In general, a bending perturbation is to be superposed. If the wall is rigidly clamped to the plate at the bottom, the total displacement as well as the total rotation must vanish: $w_0 + w^* = 0$, $\chi_0 + \chi^* = 0$. Equation (6.311) gives in the limit $\varphi \to (L - x)/R$ the shear force

$$q = \left[A \cos \left(\kappa \frac{L - x}{R} \right) + B \sin \left(\kappa \frac{L - x}{R} \right) \right] e^{- \kappa (L - x)/R} \quad ,$$

and by means of Eqs. (6.318) and (6.315), the radial bending deflection is expressed in the form of the exponentially decaying function

$$w^* = - \frac{\kappa R}{Eh} \left\{ q + \left[A \sin \left(\kappa \frac{L - x}{R} \right) - B \cos \left(\kappa \frac{L - x}{R} \right) \right] e^{- \kappa (L - x)/R} \right\} \quad ,$$

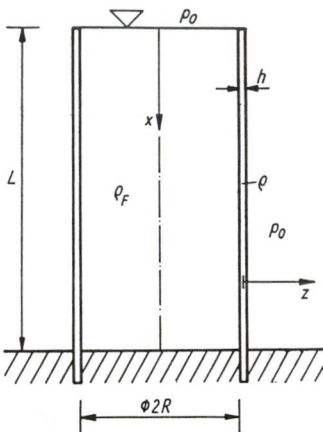

Fig. 6.38. Circular cylindrical shell filled with a fluid to the rim. The lower edge is rigidly clamped to the foundation

and further, the additional rotation in the boundary layer becomes

$$\chi^* = 2 \frac{\kappa^2}{Eh} \left[A \sin\left(\kappa \frac{L-x}{R}\right) - B \cos\left(\kappa \frac{L-x}{R}\right) \right] e^{-\kappa(L-x)/R} \quad .$$

The boundary conditions at $x = L$ render two linear equations for A and B, which are readily solved,

$$\chi_0 = -\frac{w_0}{L} = 2 \frac{\kappa^2}{Eh} B \quad , \quad w_0 = \frac{\kappa R}{Eh} (A - B) \quad .$$

The bending moment

$$m_x = -K \frac{d\chi^*}{dx} \quad ,$$

is extreme at the lower edge and takes on the value $2K \kappa \chi_0 (1 - \kappa L/R)/R$. Since $m_\vartheta = v\, m_x$ and $n_\vartheta^* = Eh\, w^*/R$, all the resultants are determined. The analysis of a sufficiently long cylindrical pressure vessel closed by end plates follows the same lines.

6.7.2. The Semispherical Dome of Fig. 6.39

The thin-walled spherical shell of Fig. 6.39 is loaded by its own dead weight, and the corresponding pressure components become $p_z = -\rho g h \cos \varphi$ and $p_\varphi = \rho g h \sin \varphi$. Equations (6.296) and (6.297) render

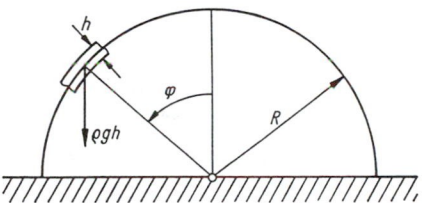

Fig. 6.39. Spherical dome under its own dead weight load. $\rho g h = const$

the membrane stresses as statically determinate, $C = 0$, $r_1 = R = const$,

$$n_\varphi = -\frac{\rho g h R}{1 + \cos \varphi} \quad , \quad n_\vartheta = -\left(1 - \cos \varphi - \cos^2 \varphi\right) n_\varphi \quad .$$

At the edge, the membrane stresses take on the values $n_\varphi = - n_\vartheta = (-\rho g h R)$. Thus, the deformations at the largest parallel circle are, under membrane stress conditions, $w_0 = R \, \varepsilon_\vartheta = R \, (n_\vartheta - \nu \, n_\varphi)/Eh = (1 + \nu) \, \rho g \, R^2/E$ and $\chi_0 = -(2 + \nu) \, \rho g \, R/E$.

The boundary conditions of a freely sliding edge, $\varphi = \pi/2$, $q = 0$, $m_\varphi = 0$, are compatible with that membrane solution. The shear force vanishes identically.

If, however, the edge of the dome is kept inextensible by a rigid bandage, the radial displacement is supposed to vanish, $w_0 + w^* = 0$, and according to a simply supported edge, $m_\varphi = 0$ at $\varphi = \pi/2$. These conditions render a bending perturbation of the membrane state. Since $\kappa = const$, Eqs. (6.311), (6.309), (6.307), (6.300), and (6.302) take on the special forms that hold for spherical shells of constant thickness

$$q = e^{\kappa\varphi}\left(C_1 \cos \kappa\varphi + C_2 \sin \kappa\varphi\right) \, , \, \chi^* = \frac{2\kappa^2}{Eh} \, e^{\kappa\varphi}\left(- C_1 \sin \kappa\varphi + C_2 \cos \kappa\varphi\right) \, ,$$

$$m_\varphi = -\frac{K}{R}\left[\left(\kappa + \nu \cot \varphi\right) \chi^* - \frac{2\kappa^3}{Eh} \, q\right] \, , \, m_\vartheta = \nu \, m_\varphi - \left(1 - \nu^2\right) \frac{K}{R} \, \chi^* \cot \varphi \, ,$$

$$n_\varphi^* = - q \cot \varphi \, , \, n_\vartheta^* = -\left(\kappa q + \frac{Eh}{2\kappa} \, \chi^*\right) \, .$$

A system of two linear equations results at $\varphi = \pi/2$. $m_\varphi = 0$ gives the relation

$$\chi^* = \frac{2\kappa^2}{Eh} \, q \quad \rightarrow \quad C_1\left(\cos \kappa\frac{\pi}{2} + \sin \kappa\frac{\pi}{2}\right) + C_2\left(\sin \kappa\frac{\pi}{2} - \cos \kappa\frac{\pi}{2}\right) = 0 \quad .$$

Furthermore, $w^* = R\,\varepsilon_\vartheta^* = -\,w_0 = -\,(1 + v)\,\rho g\,R^2/E$. The bending perturbation is completely determined by the constants $C_{1,\,2}$

$$q = \frac{1+v}{2\,\kappa}\,\rho gh\,R\,e^{-\kappa\psi}\,(\cos\kappa\psi - \sin\kappa\psi)\ ,\quad \psi = \frac{\pi}{2} - \varphi\ ,$$

$$\chi^* = \frac{1+v}{E}\,\rho g\,R\,\kappa\,e^{-\kappa\psi}\,(\cos\kappa\psi + \sin\kappa\psi)\ .$$

In the case of an edge rigidly clamped to the foundation, the boundary conditions $w_0 + w^* = 0$ and $\chi_0 + \chi^* = 0$ apply and render an exponentially decreasing bending perturbation of the membrane state given by

$$q = \frac{1+v}{\kappa}\,\rho gh\,R\,e^{-\kappa\psi}\,[\cos\kappa\psi - \mu\,(\cos\kappa\psi + \sin\kappa\psi)]\ ,\quad \mu = \frac{2+v}{2(1+v)\kappa}\ ,$$

$$\chi^* = \frac{2\,(1+v)}{E}\,\rho g\,R\,\kappa\,e^{-\kappa\psi}\,[\sin\kappa\psi + \mu\,(\cos\kappa\psi - \sin\kappa\psi)]\ ,\quad \psi = \frac{\pi}{2} - \varphi\ .$$

6.7.3. Thermal Stresses in Thin Shells of Revolution

The temperature field is assumed to be axisymmetric and *Maysel´s* formula (6.44) is applied. The dummy load is accordingly chosen to be a uniform ring force $q_0 = 1$ with an angle of inclination Ψ against the local normal z direction. The membrane forces as well as the dummy bending perturbation contribute to the isothermal influence function

$$3\,\bar p = \bar\sigma_{\varphi\varphi} + \bar\sigma_{\vartheta\vartheta} = \frac{\bar n_\varphi + \bar n_\vartheta}{h} - 12\,\frac{\bar m_\varphi + \bar m_\vartheta}{h^3}\,z\ . \tag{6.322}$$

Integration over the shell thickness h in the Eq. (6.44) is performed and the mean temperature

$$n_\theta = \frac{1}{h}\int_{-h/2}^{h/2}\theta(\varphi, z)\,dz\ , \tag{6.323}$$

as well as the first-order thickness moment,

$$m_\theta = \frac{12}{h^3}\int_{-h/2}^{h/2} z\,\theta(\varphi, z)\,dz\ , \tag{6.324}$$

enter the remaining integral over the shell surface

$$\left(u \sin \Psi + w \cos \Psi\right) = \frac{1}{r} \int_{\varphi_0}^{\varphi} \left[\alpha n_\theta(\varphi^*) \, \overline{n}(\varphi^*, \varphi) - \alpha m_\theta(\varphi^*) \, \overline{m}(\varphi^*, \varphi)\right] r^* \, r_1^* \, d\varphi^* \quad ,$$

$$\overline{n} = \overline{n}_\varphi + \overline{n}_\vartheta \; , \quad \overline{m} = \overline{m}_\varphi + \overline{m}_\vartheta \; . \tag{6.325}$$

$u(\varphi)$ and $w(\varphi)$ are the thermal displacements of the midsurface; the geometry is shown in Fig 6.37.

The thermal stresses are derived from the resultants analogous to those in a plate

$$n_\varphi = \frac{E\,h}{1-v^2} \left[\frac{1}{r_1}\left(\frac{du}{d\varphi} + w\right) + \frac{v}{r}\left(u \cos\varphi + w \sin\varphi\right) - (1+v)\,\alpha n_\theta \right],$$

$$n_\vartheta = \frac{E\,h}{1-v^2} \left[\frac{v}{r_1}\left(\frac{du}{d\varphi} + w\right) + \frac{1}{r}\left(u \cos\varphi + w \sin\varphi\right) - (1+v)\,\alpha n_\theta \right], \tag{6.326}$$

$$m_\varphi = -K\left[\frac{1}{r_1}\left(u - \frac{dw}{d\varphi}\right)\frac{d(r_1^{-1})}{d\varphi} + \frac{1}{r_1^2}\left(\frac{du}{d\varphi} - \frac{d^2w}{d\varphi^2}\right) \right.$$
$$\left. + \frac{v}{r\,r_1}\left(u - \frac{dw}{d\varphi}\right)\cos\varphi - (1+v)\,\alpha m_\theta \right],$$

$$m_\vartheta = -K\left[\frac{v}{r_1}\left(u - \frac{dw}{d\varphi}\right)\frac{d(r_1^{-1})}{d\varphi} + \frac{v}{r_1^2}\left(\frac{du}{d\varphi} - \frac{d^2w}{d\varphi^2}\right) \right.$$
$$\left. + \frac{1}{r\,r_1}\left(u - \frac{dw}{d\varphi}\right)\cos\varphi - (1+v)\,\alpha m_\theta \right]. \tag{6.327}$$

(§)   The Radial Thermal Expansion of a Circular Cylindrical Shell. Such expansion is determined by means of the axial integral

$$1.\; w(x) = \alpha \int_0^L \left[n_\theta \, \overline{n}_\vartheta(\xi, x) - m_\theta \, \overline{m}(\xi, x) \right] d\xi \; . \tag{6.328a}$$

The isothermal influence functions of a radial "tensile" unit ring load applied at x are to be substituted in the above integral. By means of proper superposition, it is possible to produce the special solutions from the influence function of the infinitely long circular cylindrical shell

$$\overline{w}(\xi, x = 0) = \frac{1 \cdot R^3}{8\,K\,\kappa^3}\left(\cos\frac{\kappa\xi}{R} + \sin\frac{\kappa\xi}{R}\right) e^{-\kappa\xi/R} \; . \tag{6.328b}$$

Thermal stresses at the inlet of a semiinfinite circular shell and other solutions are reported in *F. Ziegler and H. Irschik: "Thermal*

Stress Analysis Based on Maysel's Formula," in Thermal Stresses, Vol.2 (ed. R. B. Hetnarski), Elsevier Science Publ. Amsterdam, 1987. Isothermal solutions are sampled in *S. Timoshenko and S. Woinowski-Krieger: Theory of Plates and Shells. 2nd ed., McGraw-Hill, New York, 1959, p. 471.*

6.8. Contact Problems (The *Hertz* Theory)

If two linear elastic bodies are brought into contact at a single point of their smooth surfaces, or in the case of cylinders along a single straight line, a common tangent plane exists. Hence, a common normal vector determines its orientation. The application of a collinear double force without moment *F* produces deformations such that the pressure remains finite in the (small) contact zone. The first step in the analysis is to find the size and shape of this contact area under the simplifying assumption that the shear tractions, if any, are negligibly small. *Hertz* considered the bodies to be sufficiently flat in the neighborhood of the plane contact surface to approximate the stresses by the stress state in the half-space. *J. Boussinesq* derived the three-dimensional solution of the elastic half-space when loaded by a single compressive force *F* at the point *(r = 0, z = 0)* of the otherwise traction-free surface. In a centered cylindrical coordinate system, the nonvanishing and axisymmetric stress components are

$$\sigma_{rr} = \frac{F}{2\pi}\left[(1-2v)\frac{(1-z/R)}{r^2} - 3\frac{z\,r^2}{R^5}\right] \ , \quad \sigma_{\theta\theta} = -\frac{F}{2\pi}(1-2v)\left[\frac{(1-z/R)}{r^2} - \frac{z}{R^3}\right] \ ,$$

$$\sigma_{rz} = -\frac{3F}{2\pi}\frac{r\,z^2}{R^5} \ , \quad \sigma_{zz} = \frac{z}{r}\,\sigma_{rz} \ , \quad R^2 = r^2 + z^2 \ .$$

$$(6.329)$$

The singular traction acting on a buried area element with normal e_z at a depth *z* is a vector pointing to the origin where the load is applied

$$\sigma_z = \sigma_{zr}\,e_r + \sigma_{zz}\,e_z = \frac{1}{r}\,\sigma_{rz}(r\,e_r + z\,e_z) \ , \tag{6.330}$$

ie as in the two-dimensional case (see Fig. 6.31), a central system of internal forces balances the loading. The absolute value of the traction is constant at the spherical surface, with the northern pole at the point of application of the load *F*

$$|\sigma_z| = \frac{3F}{2\pi R^2}\cos^2\varphi \ , \quad \cos\varphi = z/R \ . \tag{6.331}$$

Its extreme value is constant on the sphere of the diameter $d = R/\cos \varphi$. The associated displacements are given by *Hooke's* law

$$u = r \, \varepsilon_{\theta\theta} = \frac{r}{E} \left[\sigma_{\theta\theta} - \nu \left(\sigma_{rr} + \sigma_{zz} \right) \right] = - \frac{1-2\nu}{G} \, \frac{F}{4\pi r} \left[1 - \frac{z}{R} - \frac{1}{1-2\nu} \, \frac{z \, r^2}{R^3} \right] , \tag{6.332}$$

and by integrating the derivatives

$$\frac{\partial w}{\partial z} = \varepsilon_{zz} = \frac{1}{E} \left[\sigma_{zz} - \nu \left(\sigma_{rr} + \sigma_{zz} \right) \right] \quad , \quad \frac{\partial w}{\partial r} = \gamma_{rz} - \frac{\partial u}{\partial z} = \frac{\sigma_{rz}}{G} - \frac{\partial u}{\partial z} . \tag{6.333}$$

Thus, the vertical displacement becomes

$$w = \frac{F}{2\pi R \, G} \left(1 - \nu + \frac{z^2}{2R^2} \right) . \tag{6.334}$$

The plane surface $z = 0$ is deformed to a surface of revolution that is translated in the z direction by any constant displacement

$$w(r, z = 0) = \frac{(1-\nu)F}{2\pi \, r \, G} , \tag{6.335}$$

and the radial displacement becomes

$$u(r, z = 0) = - \frac{(1-2\nu)F}{4\pi \, r \, G} . \tag{6.336}$$

The solution approximates the frictionless contact zone of two spherical bodies. This is the special case of *Hertz's* theory with a circular contact area. The undeformed convex surfaces of the bodies have the constant positive curvatures $1/R_1$ and $1/R_2$ at the point of contact, respectively. Taking the radial coordinate in the common tangent plane thus renders the surface points of the undeformed bodies 1 and 2 in the distances

$$Z_1 = \frac{r^2}{2 \, R_1} \quad , \quad Z_2 = \frac{r^2}{2 \, R_2} . \tag{6.337}$$

The values are sufficiently accurate in a small surrounding of the origin (according to the theorem of the height r in a rectangular triangle with hypotenuse $2 \, R_i$, $i = 1, 2$). Hence, the undeformed distance of two points before contact becomes the sum

$$Z_1 + Z_2 = \frac{R_1 + R_2}{2 \, R_1 \, R_2} \, r^2 . \tag{6.338}$$

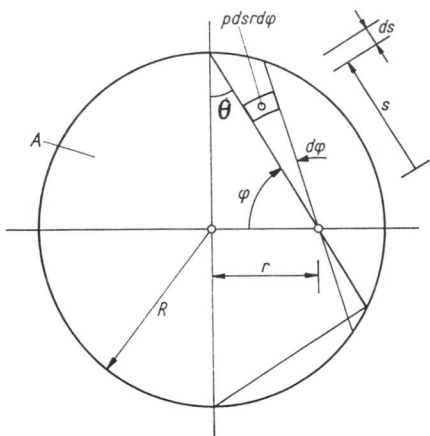

Fig. 6.40. Circular contact area of the axisymmetric *Hertz* problem

If we denote the displacements at the surfaces by $w_i\,(r)$, positive in the directions of Z_i , $i = 1, 2$, and consider the reduction of the distance of two points, located in the far field on the axis of symmetry, by w_0 , the condition of contact of two surface points within $r \leq R$ becomes

$$w_0 - (w_1 + w_2) = Z_1 + Z_2 = \kappa\, r^2 \;, \quad \kappa = \frac{R_1 + R_2}{2\,R_1\,R_2} \;. \tag{6.339}$$

Equation (6.335) approximates these displacements; $p(s,\,\varphi)$ denotes the pressure in the contact area A (see Fig. 6.40)

$$w_i(r) = \frac{1 - \nu_i}{2\pi\,G_i}\, \iint\limits_{A} \frac{p(s,\,\varphi)\,ds\,s\,d\varphi}{s} = k_i \iint\limits_{A} p\,ds\,d\varphi \;,\quad k_i = \frac{1 - \nu_i}{2\,\pi G_i} \;,\quad -\pi/2 \leq \varphi \leq \pi/2 \;,$$

$$\tag{6.340}$$

and ds is the differential of the chord $2R\,\cos\,\theta$ of the contact circle. Equation (6.339), after the substitution of the vertical displacements, becomes an integral equation of the yet unknown pressure distribution $p(r)$ in the contact zone

$$(k_1 + k_2) \iint\limits_{A} p\,ds\,d\varphi = w_0 - \kappa\, r^2 \;. \tag{6.341}$$

A rough approximation of the solution is given by the assumption of a semiellipsoidal pressure distribution with the maximum $p(r = 0) = p_0 = k\,R$. The integration of such a function with respect to s renders

$$\int p\, ds = k \frac{\pi}{2} \left(R^2 - r^2 \sin^2 \varphi\right) ,$$
(6.342)

and the definite integration of the remaining symmetric function gives a quadratic polynomial

$$2 \int_0^{\pi/2} \left(R^2 - r^2 \sin^2 \varphi\right) d\varphi = A - \frac{\pi r^2}{2} , \quad A = \pi R^2 .$$
(6.343)

Comparing the coefficients in Eq. (6.341) reduces the number of parameters from three to one; p_0 remains to be determined

$$w_0 = (k_1 + k_2)\, \pi A\, \frac{p_0}{2R} ,$$
(6.344)

and solving

$$\kappa = (k_1 + k_2)\, \pi^2 \frac{p_0}{4R} ,$$
(6.345)

gives the radius of the contact area R in terms of p_0

$$R = (k_1 + k_2)\, \pi^2 \frac{p_0}{4\kappa} .$$
(6.346)

The resultant of the pressure balances the load $F = 2p_0 A/3$ and the maximal contact pressure $p_0 = (3/2)\, F/A$ turns out to be magnified by a factor of $3/2$ with respect to the mean pressure F/A. Substitution renders the diameter of the contact area in terms of the geometry, the elastic parameters, and the force F

$$2R = \left[3F\left(\frac{1-\nu_1}{G_1} + \frac{1-\nu_2}{G_2}\right)\frac{R_1 R_2}{(R_1 + R_2)}\right]^{1/3} .$$
(6.347)

The assumptions made in the *Hertz* theory on the condition of contact and the distribution of the pressure are *not* compatible.

The stresses in the near field of the frictionless contact area are approximated by the stresses that would develop in a half-space when uniformly loaded over a circular area of radius R at the otherwise traction-free surface. The solution is found by a proper integration of the singular stress state. The distribution along the axis of symmetry is given by ($\zeta = z/R$)

$$\frac{\sigma_{rr}}{p_0} = \frac{\sigma_{\varphi\varphi}}{p_0} = -(1+\nu)\left[1 - \zeta \tan^{-1}\left(1/\zeta\right)\right] - \frac{\sigma_{zz}}{2p_0} , \quad \frac{\sigma_{zz}}{p_0} = -\left(1+\zeta^2\right)^{-1} .$$
(6.348)

Tables of additional data are presented by *G. Lundberg and F. K. G. Odqvist, Ing. Vetenskaps Akad´Handl.116 (1932), p.64.*

Poisson´s ratio v is the only material parameter left. It is to be substituted in Eq. (6.348) according to the material of body 1 or 2. The absolute maximum of the shear stress is reached at a depth ζ = 0.47 if $v = 0.3$. That value is $\tau_{max} = 0.31\ p_0$. The largest tensile stress, however, is observed at the boundary of the contact area at $r = R$, $z = 0$, and is given by

$$\frac{\sigma_{rr}\,(r = R, z = 0)}{p_0} = \frac{(1 - 2v)}{3} \ .$$

(6.349)

For a more general elliptic contact area and for a generalization of the loading to include a tangential force, see the classic by *S. P. Timoshenko and J. N. Goodier: Theory of Elasticity. McGraw-Hill, New York, 1970.* Rolling contact, as well as frictional effects, are still under theoretical investigations, see, eg *Proceedings of IUTAM-Symposium on "The Mechanics of the Contact Between Deformable Bodies," (eds. A. D. de Pater and J. J. Kalker), Delft University Press, The Netherlands, 1975.*

6.9. Stress-Free Temperature Fields, *Fourier´s* Law of Heat Conduction

A structure under statically determinate support conditions is considered. The question in which circumstances a temperature field does not produce any thermal stresses (such internal forces would necessarily be self-equilibrating) is of considerable technical importance. On the other hand, the conditions of a temperature distribution not able to produce thermal deformations may be of equal importance. In that latter case see Exercise A 6.22 and for further details and a discussion of nondeflecting beams refer to *F. Ziegler and H. Irschik: "Thermal Stress Analysis Based on Maysel´s Formula," in Thermal Stresses, Vol.2 (ed. R. B. Hetnarski) Elsevier Science Publishers, Amsterdam, 1987.*

A simply connected linear elastic region free to expand is considered to remain free of thermal stresses , $\sigma_{ij} = 0$. Only the thermal strains remain nonzero if a temperature field $\theta(x, y, z)$ is applied

$$\varepsilon_{ii} = \alpha\theta \quad , \quad \varepsilon_{ij} = 0, \quad i \neq j = 1, 2, 3 \ .$$

(6.350)

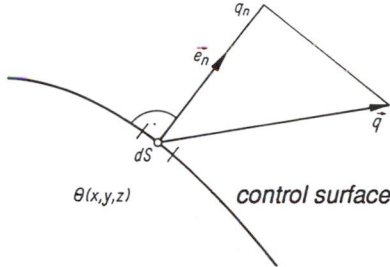

Fig. 6.41. Heat flux through a surface element, $(q \cdot e_n)$ dS

That thermal strains must be compatible [see Eq. (1.22)]; otherwise, stresses are produced. The conditions are

$$\frac{\partial^2(\alpha\theta)}{\partial x_j^2} + \frac{\partial^2(\alpha\theta)}{\partial x_i^2} = 0 , \text{ and } , \frac{\partial^2(\alpha\theta)}{\partial x_i \partial x_j} = 0 , i \neq j = 1, 2, 3 .$$

(6.351)

A solution common to all the differential equations is the linear temperature distribution in a cartesian coordinate system

$$\alpha\theta = a_0 + a_1 x + a_2 y + a_3 z ,$$

(6.352)

with arbitrary coefficients that may also be slowly varying in time.

Since the transport of heat energy by radiation is negligible in the range of temperatures admissible for a solid to remain elastic, the conduction of heat is considered. *Fourier's* law states that for a homogeneous and isotropic solid, the heat flux q_n in the direction e_n per unit of area and per unit of time is proportional to the (negative) temperature gradient (Fig. 6.41 shows a control surface element with the normal e_n)

$$q_n = -k \frac{\partial\theta}{\partial n} .$$

(6.353)

k with the dimension $[k]$ = W/mK is the isotropic *thermal conductivity* . It is assumed constant in the small range of admissible temperature variations. Analogous to the equation of continuity (1.74), the local heat conduction equation is derived in the limit $dV \rightarrow 0$ of balancing the heat flux of a control volume element $dV = dx \, dy \, dz$, similarly to the mass flow rate. The result is

$$\frac{\partial \theta}{\partial t} = a \, \nabla^2 \theta + \frac{S}{\rho c} \ , \quad a = k/\rho c \ ,$$
(6.354)

where the coefficient a with the dimension $[a] = m^2/s$ is the *thermal diffusivity* . The product ρc is the (isochoric) specific heat capacity per unit of volume. The initial and boundary value problems of the diffusivity equation (6.354) are extensively discussed in *H. S. Carslaw and J. C. Jaeger: Conduction of Heat in Solids, 2nd ed., Clarendon Press, Oxford, 1959* . See also *H. Schuh: Heat Transfer in Structures. Pergamon Press, London, 1965.*

The equation is inhomogeneous in the case of the action of external heat sources of capacity S in the interior of the solid. It also contains the heat generated or absorbed during deformations. The coupling of stress and temperature fields is usually neglected in linear elasticity and for sufficiently slow deformation rates. The substitution of the stress-free temperature field (6.352) renders a restrictive condition on the distribution of the external heat sources

$$\dot{a}_0 + \dot{a}_1 x + \dot{a}_2 y + \dot{a}_3 z = S/\rho c \, \alpha \ .$$
(6.355)

In the case of $S = 0$, there are no heat sources, the coefficients turn out to be $a_k = const$, and the linear temperature field must be stationary. Thus, statically determinate beams remain free of thermal stresses if the temperature is linear in the cross-sectional coordinates [see Eq. (6.91)].

In multiply connected bodies, Eq. (6.355) is only a necessary condition of integrability. Further details may be found in *B. A. Boley and J. Weiner: Theory of Thermal Stresses. Wiley, New York, 1960* , pp. 92-95.

6.10.   The Elastic-Visco-Elastic Analogy

In Sec. 4.2.2, the linear constitutive relations of visco-elastic materials are considered and complex moduli discussed. *T. Alfrey* (1944) and later *E. H. Lee* (1955), developed the correspondence of elastic and visco-elastic solutions. A rather general formulation of the analogy is derived by applying the *Laplace* transformation (4.78) to the local conditions of equilibrium (2.19), to the linearized geometric relations (1.21), and to the linear rate-dependent material law [Eqs. (4.101)], as well as to the boundary conditions. The transformed equations are formally those of a linear elastic solid of the same configuration and, in the separable case, with the same boundary conditions. Thus, it is assumed that any time-dependent loading can be expanded into a finite series where the time factor is separated. The correspondence in that case is complete and the complex moduli are then given by [see Eq. (4.102)]

$$G^*(s) = \frac{1}{2} \frac{Q_n(s)}{P_m(s)} \quad, \quad K^*(s) = \frac{\overline{Q}_q(s)}{\overline{P}_p(s)} \quad, \quad n \le m \quad, \quad q \le p \ .$$

(6.356)

The complex *Poisson´s* ratio is expressed by

$$v^*(s) = \frac{1}{2} \frac{K^*(s) - 2\,G^*(s)}{K^*(s) + G^*(s)} \ .$$

(6.357)

s is the (complex) transform variable. The homogeneous initial conditions of these quasistationary problems are understood. In the case of pure elastic dilatation (compressibility), $K^* = K = 2\dot{G}\,(1 + v)/(1 - 2v)$ is real and Eq. (4.12) holds. The remaining complex parameters of the *Maxwell* fluid are, eg [see Eq. (4.87)]

$$G^*(s) = G \frac{s}{\left(s + \vartheta^{-1}\right)} \quad, \quad v^*(s) = v \frac{s + 1/2v\vartheta''}{s + 1/\vartheta''} \quad, \quad \vartheta'' = \frac{3}{2\,(1 + v)}\,\vartheta \ .$$

(6.358)

This analogy is of a special practical value if the linear elastic solution is already known and dependence on the elastic parameters G and v is expressed in analytical form. The *Laplace* -transformed visco-elastic response of the same structure after switching on the load at time $t = 0$ is found by multiplication with $1/s$ and by substituting the complex moduli $G^*(s)$ and $v^*(s)$. The inversion of this *Laplace* image function is performed by contour integration in the complex s plane or by inspection; see, eg G. Doetsch: *Guide to the Practical Use of the Laplace Transformation, 2nd ed.,* Oldenbourg, Munich, 1961, (in German) . For tables see A. Erdelyi, et. al.: *Tables of Integral Transforms (Bateman Project). Mc-Graw Hill, New York, 1954* . The results are the time-dependent creep deformations and the possibly redistributed nonstationary stresses. Two simple examples illustrate the direct application of the correspondence principle.

6.10.1. The Creeping Simply Supported Single-Span Beam

The linear elastic deflection of a simply supported beam under uniform lateral loading q_0 and with constant principal bending stiffness $B = EJ = 2(1 + v)\,GJ$ is easily derived, eg by properly integrating Eq. (6.72)

$$w(x) = \frac{q_0\,l^4}{24\,B} \frac{x}{l} \left[1 - 2\left(\frac{x}{l}\right)^2 + \left(\frac{x}{l}\right)^3\right] \ .$$

(6.359)

The quasistatic elastic response to the load jump $q_0\, H(t)$, where $H(t) = 1$ for $t > 0$, is the *Heaviside* function and has the image

$$w^*(x, s) = \frac{1}{s}\, w(x) \; .$$

(6.360)

The complex *Young's* modulus, say, of the *Maxwell* fluid

$$E^*(s) = E\, \frac{s}{s + 1/\vartheta''} \quad ,$$

(6.361)

is substituted for E in the bending rigidity, and the image of the creep deflection becomes by this direct application of the correspondence principle

$$w_c^*(x, s) = \frac{s + 1/\vartheta''}{s^2}\, w(x) \; .$$

(6.362)

The inversion of the Laplace transformation is termwise performed and renders the *Heaviside* time function according to the instantaneous elastic response of a *Maxwell* body and its time integral

$$w_c(x, t) = H(t) \left(1 + t/\vartheta''\right) w(x) \quad , \quad t \geq 0 .$$

(6.363a)

The creep deformation grows "beyond any bounds" affine to the elastic deflection in the time scale

$$\vartheta'' = 3\, \vartheta/2\, (1 + v) \quad ,$$

(6.363b)

which is proportional to the relaxation time of the *Maxwell* material, Eq. (4.84). The unloading of the *Maxwell* beam after a finite duration leaves behind an irreversible deflection. Other material laws, eg those discussed in Sec. 4.2.2 should be substituted above to gain experience of the various effects of creep.

6.10.2. The Heated Thick-Walled Pipe (Fig. 6.42)

The temperature at the internal surface $r = R_i$ may be fixed by $\theta = \theta_i = const$ in a pipe conveying the flow of a heated fluid. The outer surface $r = R_e$ is kept at zero temperature. The proper axisymmetric and stationary solution of the heat conduction equation (6.354) is easily derived

$$\theta(r) = \frac{\theta_i}{\ln \beta} \ \ln (R_e/r) \ , \ \ \beta = R_e/R_i \ .$$

(6.364)

Under the conditions of plane strain that apply to a long and linear elastic pipe, the thermal stresses are easily calculated by the superposition of a particular solution of the inhomogeneous thermal problem and the isothermal solutions of the pipe under internal and external pressure [see Eq. (6.14)], thus rendering the boundaries free of any tractions. The elastic principal stress state due to *E. Melan* (The late *E. Melan* was Professor at the TU-Vienna), is taken from *H. Parkus: Thermal Stresses in Handbook of Engineering Mechanics. (W. Flügge, ed.), Mc-Graw Hill, New York, 1962, Chap. 43,*

$$\sigma_{rr} = -G \ \frac{1+\nu}{1-\nu} \ \alpha\theta_i \ \left(\frac{\ln (R_e/r)}{\ln \beta} - \frac{(R_e/r)^2 - 1}{\beta^2 - 1} \right) ,$$

$$\sigma_{\varphi\varphi} = -G \ \frac{1+\nu}{1-\nu} \ \alpha\theta_i \ \left(\frac{\ln (R_e/r) - 1}{\ln \beta} + \frac{(R_e/r)^2 + 1}{\beta^2 - 1} \right) ,$$

(6.364)

$$\sigma_{zz} = -G \ \frac{1+\nu}{1-\nu} \ \alpha\theta_i \ \left(\frac{2 \ln (R_e/r)}{\ln \beta} + 2\nu \left[\frac{1}{\beta^2 - 1} - \frac{1}{2 \ln \beta} \right] \right) .$$

(6.365)

The state of stress in the viscous material is derived by multiplication with *(1/s)* and by substituting the complex moduli $G^*(s)$ and $\nu^*(s)$ for the elastic parameters G and ν . It is easily recognized by the inspection of Eq. (6.365) that the axial stress consists of two parts with different time behavior and thus will be properly redistributed by the inversion of the *Laplace* transform. Choosing the *Maxwell* material for the illustration of the finite creep effects specifies the substitutions. The first factor is

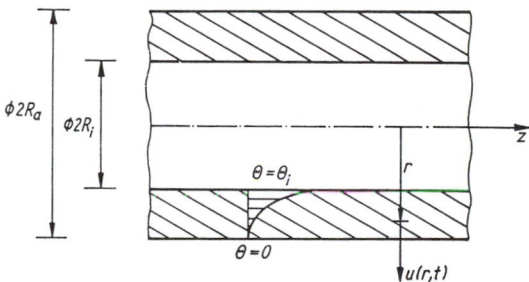

Fig. 6.42. Internally heated thick-walled visco-elastic pipe

$$\frac{1}{s}\,G\,\frac{1+v}{1-v}\,\alpha\theta_i \;\rightarrow\; \frac{1}{s}\,G^*(s)\,\frac{1+v^*(s)}{1-v^*(s)}\,\alpha\theta_i \;=\; G\,\frac{1+v}{1-v}\,\frac{1}{s+1/\vartheta'}\,\alpha\theta_i \;,$$

$$\vartheta' = 3\,\frac{1-v}{1+v}\,\vartheta\;.$$

(6.366)

And the second coefficient becomes,

$$\frac{1}{s}\,G\,\frac{v(1+v)}{1-v}\,\alpha\theta_i \;\rightarrow\; \frac{1}{s}\,G^*(s)\,\frac{v^*(s)\,[1+v^*(s)]}{1-v^*(s)}\,\alpha\theta_i$$

$$= G\,\frac{v(1+v)}{1-v}\,\frac{1}{s+1/\vartheta'}\,\frac{s+1/2v\vartheta''}{s+1/\vartheta''}\,\alpha\theta_i \;,\quad \vartheta'' = \frac{3\vartheta}{2(1+v)}\;.$$

(6.367)

The inversion of the *Laplace* transforms results in the relaxation of the thermal stresses in the *Maxwell* material in the cross-section of the pipe

$$\sigma_{rr}^{(c)} = e^{-t/\vartheta'}\sigma_{rr}\;,\quad \sigma_{\varphi\varphi}^{(c)} = e^{-t/\vartheta'}\sigma_{\varphi\varphi}\;,$$

(6.368)

and the axial stresses (they are redistributed over r)

$$\sigma_{zz}^{(c)} = -G\,\frac{1+v}{1-v}\,\alpha\theta_i\left\{\left[\frac{2\ln(R_e/r)-1}{\ln\beta} - \frac{2}{1-\beta^2}\right]e^{-t/\vartheta'}\right.$$

$$\left. +(1-v)\left[\frac{1}{\ln\beta} + \frac{2}{1-\beta^2}\right]e^{-t/\vartheta''}\right\}\;.$$

(6.369)

The linear elastic radial displacement is given by

$$u(r) = \frac{r}{4}\,\frac{1+v}{1-v}\,\frac{\alpha\theta_i}{\ln\beta}\left[1+2\ln(R_e/r) + \frac{1+(R_i/r)^2}{1-1/\beta^2} - \left(1+2\ln\beta\right)\frac{1+(R_e/r)^2}{\beta^2-1}\right.$$

$$\left. +2v\,\frac{1+2\ln\beta-\beta^2}{\beta^2-1}\right]\;.$$

(6.370)

The application of the correspondence to that analytic expression and the inversion of the *Laplace* transforms render the viscous growth of the cross-section (note the two time scales)

$$u^{(c)}(r,t) = \frac{r}{4}\,\frac{1+v}{1-v}\,\frac{\alpha\theta_i}{\ln\beta}\left\{\left[1+2\ln(R_e/r) + \frac{1+(R_i/r)^2}{1-\beta^{-2}}\right.\right.$$

$$- \left(1 + 2 \ln \beta\right) \frac{1 + (R_e / r)^2}{\beta^2 - 1} \Bigg] \frac{3\,(1 - v) - 2\,(1 - 2v)\,e^{-t/\vartheta'}}{1 + v}$$

$$+ \frac{1 + 2 \ln \beta - \beta^2}{\beta^2 - 1} \left[1 - (1 - 2v)\,e^{-t/\vartheta''}\right] \Bigg\} \; .$$

$$(6.371)$$

6.11. Exercises A 6.1 to A 6.22 and Solutions

A 6.1: A cantilever with a symmetrically L-shaped cross-section is loaded by the edge moment $\boldsymbol{M} = M_\eta\,\boldsymbol{e}_\eta + M_\zeta\,\boldsymbol{e}_\zeta$ according to Fig. A 6.1. Under those biaxial bending conditions, calculate the linear elastic stress distribution and verify that the vector deflection is not orthogonal to the vector of the bending moment. In addition, determine the core of the L-shaped cross-section.

Solution: Since the axis of symmetry y is also a principal axis of the cross-section, the area is $A = 2bt - t^2$. Only the location of the centroid S must be determined to fix the second, orthogonal principal axis z (see Fig. A 6.1)

$$e = \frac{1}{2}\left(t + b\,\frac{b - t}{2b - t}\right) \; .$$

The moments of inertia with respect to the η, ζ axes are easily calculated

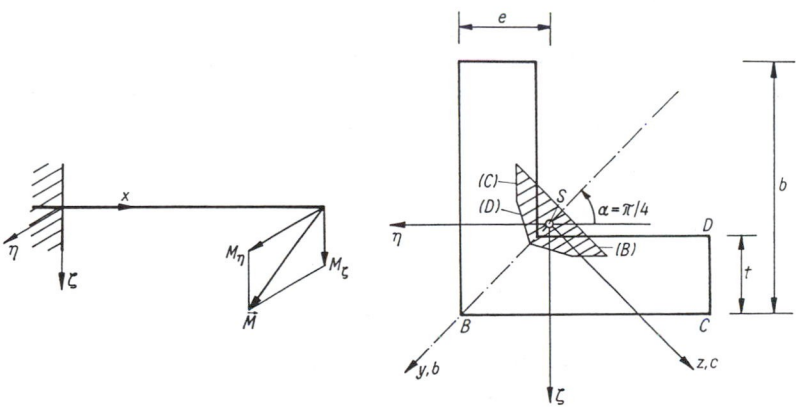

Fig. A 6.1.

$$J_\eta = J_\zeta = \frac{t}{12}\left[b^3 + bt^2 - t^3 + 3b\frac{(b-t)^3}{2b-t}\right] \ , \quad J_{\eta\zeta} = \frac{tb^2}{4}\frac{(b-t)^2}{2b-t} \ ,$$

and transformed into the rotated coordinate system of principal axes y, z

$$J_1 = J_y = \frac{t}{12}\left[4b^3 + 4bt^2 - t^3 - 6tb^2\right] \ , \quad i_y^2 = J_y/A \ ,$$

$$J_2 = J_z = \frac{t}{12}\left[4b^3 + 4bt^2 - t^3 - 6\frac{b^4}{2b-t}\right] \ , \quad i_z^2 = J_z/A \ .$$

The loading is projected to render axial moments with respect to the principal axes

$$M_y = M_\eta \cos\alpha + M_\zeta \sin\alpha \ , \quad M_z = -M_\eta \sin\alpha + M_\zeta \cos\alpha \ ,$$

and the result of the superposition of the two stress states of the principal bending cases is (the field moments are constant)

$$\sigma_{xx} = \frac{M_y}{J_y}z - \frac{M_z}{J_z}y \ .$$

Extreme normal stresses are to be observed in the points B, C, and D. Deflections in the planes of principal bending are the integrals of the linearized second-order differential equation (5.22a) of w, and its counterpart for v,

$$w(x) = -\frac{M_y}{EJ_y}x^2 \ , \quad v(x) = \frac{M_z}{EJ_z}x^2 \ .$$

Since the principal bending rigidities are unequal, the vector $u = v\,e_y + w\,e_z$ is not orthogonal to $M = const$. Inserting the coordinates of the points B, C, and D into the equation of the neutral axis (6.51) renders the straight lines of the polygonal area of the core shown in the figure in successive order. The b, c coordinates are summarized in the following table:

Point	B	C	D
y	$e\sqrt{2}$	$e\sqrt{2} - b/\sqrt{2}$	$e\sqrt{2} - (b+t)/\sqrt{2}$
z	0	$b\sqrt{2}$	$(b-t)/\sqrt{2}$

A 6.2: The rigid cylinder head is pressed against the rigid block with an intermediate elastic seal by means of n tension bolts, see Fig. A 6.2. Assuming the initial prestressing of a bolt is equal to S, how does the normal force vary if the gauge pressure p changes, eg according to the strokes of a piston? How large must the allowable maximal pressure be to keep the pressure vessel sealed? Geometry and stiffness parameters are given.

Solution: The balance of forces acting on the lid requires $F_p + F_D - nF_S = 0$, $F_p = pA_i$. The seal of initial thickness h_0 is assumed to be linear elastic in compression, $F_D = k_D (h_0 - h)$. The stiffness is approximated by $k_D = E_D A_D /h_0$; thus, we assume a lubricated contact with no shear. The linear elastic bolt of unstressed length l_0 has stiffness $k_S = E_S A_S /l_0$ and *Hooke's* law gives $F_S = k_S (l - l_0)$. The geometric condition of contact of the seal is $l - l_1 = h - h_1$, if l_1 and h_1 are the lengths in the prestressed state with $p = 0$. Equilibrium with no pressure present, $p = 0$, renders the compression of the seal $D = nS$. With reference to the initial lengths $(l - l_0) - (l_1 - l_0) = (h_0 - h_1) - (h_0 - h)$ and by elimination of the elongations due to the spring forces, the following relation results:

$$k_S^{-1} (F_S - S) = k_D^{-1} (D - F_D) \quad .$$

Further elimination of F_D yields the current force in the bolts

$$F_S = S + \frac{F_p}{n (1 + k_D/nk_S)} \quad .$$

Since the fatigue strength of the bolts strongly depends on the fluctuation of the normal force, the design rule $k_D \gg k_S$ results, which keeps the amplitude small with respect to the pressure

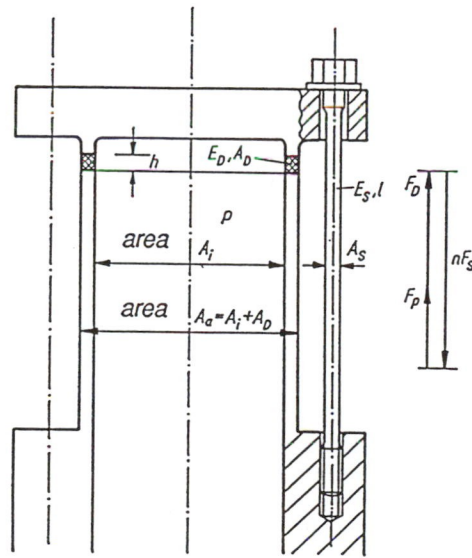

Fig. A 6.2.

variations. In practice, a stiff seal and long (soft) bolts are preferred. The maximal allowable pressure follows from the inequality $s_D = F_D /A_D \geq p$, such that no leakage occurs, by using the equality sign

$$p_{max} = nS \frac{1 + k_D /nk_S}{A_D + A_a k_D /nk_S} \quad , \quad p_{max} \approx nS/A_a \ .$$

The approximation holds under the design condition $k_D /nk_S >> 1$. Usually, the area of the cylinder is given as A_i, and since $A_a = A_i + A_D$, the cross-sectional area of the seal should be small to maximize the pressure. Thus, the high stiffness of the seal requires a large *Young's* modulus E_D; the thickness h_o is also very limited.

A 6.3: A force F is distributed by a rigid cross-bar to n linear elastic rods of variable stiffness $(EA)_i$, $i = 1, 2, ..., n$ (see Fig. A 6.3). The normal forces N_i and the displacement s of the rigid cross-bar are to be calculated with the additional thermal loading given by the memberwise constant temperature θ_i.

Solution: The balance of forces acting on the rigid cross-bar renders

$$F = \sum_{i=1}^{n} N_i \ .$$

Castigliano's theorem yields with the condition of a common elongation of all the member rods

$$s = \frac{\partial U^*}{\partial N_i} = l \sum_{i=1}^{n} \frac{N_i}{(EA)_i} + (\alpha\theta)_i \ ,$$

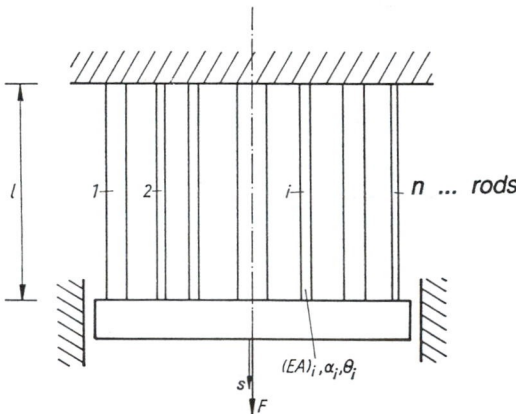

Fig. A 6.3.

and after the elimination of s , the $(n - 1)$ linear equations complete the system for the n unknowns

$$\frac{N_{i+1}}{(EA)_{i+1}} - \frac{N_i}{(EA)_i} = (\alpha\theta)_i - (\alpha\theta)_{i+1} \quad , \quad i = 1, 2, ..., (n-1) .$$

Thermal stresses depend essentially on the temperature gradient lateral to the direction of free expansion. If all the n rods are equal, $(EA \, \alpha)_i = EA \, \alpha = const$, the solution clearly indicates this fact since the thermal stresses are proportional to the deviation of the local temperature from the mean value

$$N_i = F/n + EA \, \alpha \left(\frac{1}{n} \sum_{j=1}^{n} \theta_j - \theta_i \right) .$$

The common strain in that special case is given by

$$\varepsilon = s/l = F/(nEA) + \alpha \frac{1}{n} \sum_{j=1}^{n} \theta_j .$$

A 6.4: Each of the two parallel cantilevers carries a tip load F . For design reasons, a rod AB couples the cantilevers to form a framed structure according to Fig. A 6.4. Show that even the uniform heating of the whole structure renders a redistribution of the load stresses. The connecting rod is built-in free of stress at the reference temperature.

Solution: Putting the shear force at A, $Q_z \, (x = l) = Q_A$, and that at B

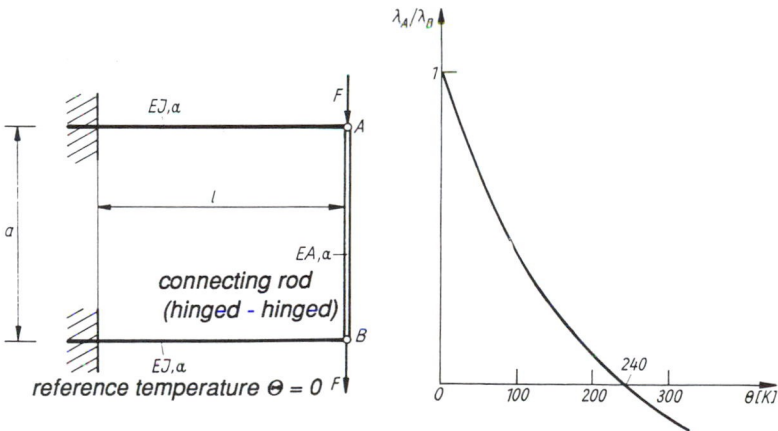

Fig. A 6.4.

equal to $Q_z (x = l) = Q_B = 2F - Q_A$ determines the linear bending moments in the field, and the complementary energy becomes (the shear deformation is neglected as well as the thermal expansion of the cantilevers)

$$U^* = \frac{1}{2} \left[\frac{Q_A^2 \, l^3}{3B} + \frac{(2F - Q_A)^2 \, l^3}{3B} + \frac{(Q_A - F)^2 \, a}{EA} \right] + (Q_A - F) \, a \, \alpha\theta \quad , \quad B = EJ \quad .$$

Menabrea's theorem $\partial U^*/\partial Q_A = 0$ renders the redundant force in the heated frame Q_A and, hence, also Q_B to be proportional to the given loading F

$$Q_A = F \lambda_A \quad , \quad Q_B = F \lambda_B \quad ,$$

but with differing thermal load factors,

$$\lambda_A = 1 - \beta E \alpha \theta \, l^2/F \quad , \quad \lambda_B = 1 + \beta E \alpha \theta \, l^2/F \quad , \quad \beta = \frac{3 \, a \, J}{2 \, l^5 \left(1 + 3 \, a \, J/2 \, A \, l^3 \right)} \quad .$$

Figure A 6.4(b) shows the ratio of these load factors as a function of the increasing temperature and thus indicates the redistribution of the loading F and of the associated load stresses in the cantilevers.

A 6.5: The simply supported frame of Fig. A 6.5 is loaded by a thermal curvature $\alpha m_\theta = \kappa_{th}$ that is assumed to be constant in all three member bars. The mean temperature is assigned as zero. Calculate the internal forces in the cross-bar if the (principal) bending rigidity of the columns and of the beam are equal, $B = EJ$.

Solution: The balance of the external forces renders the vertical components $A = B = 0$, and the statically indeterminate force remains $H_1 = H_2 = X$. If we neglect in a crude approximation the axial and shear deformations, *Menabrea's* theorem considers the

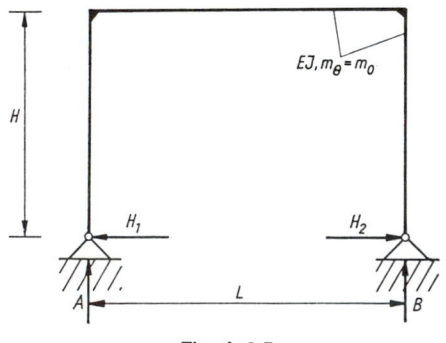

Fig. A 6.5.

derivative of the complementary energy

$$U^* = 2 \int_0^H \left(\frac{M_1^2}{2B} + M_1 \kappa_{th} \right) dx + \int_0^L \left(\frac{M_2^2}{2B} + M_2 \kappa_{th} \right) dx \quad ,$$

$$M_1 = x\, X \quad , \quad M_2 = H\, X \quad , \quad \frac{\partial U^*}{\partial X} = 0 \quad .$$

Solving for X gives

$$X = -3\, B\, \kappa_{th} \frac{1 + L/H}{3L + 2H} \quad .$$

The normal force $N = X$ and bending moment $M_2 = H\, X$ are constant in the cross-bar, and the shear force is zero.

A 6.6: Determine the portion of the shear deformation of the total deflection of the single-span beam laterally loaded by the distributed force $q_z\,(x)$ and by a single force F at a general point ξ. The rigidities of the simply supported beam of Fig. A 6.6 are constant.

Solution: The bending moment is statically determinate. Hence, the curvature due to shear becomes $w_S{''} = -\kappa\,[q_z + F\,\delta(x - \xi)]\,/GA$. $\delta(x)$ is the *Dirac* delta function. The hinged-boundary conditions refer to the *Bernoulli-Euler* beam theory, and *Mohr's* analogy becomes applicable to the shear curvature. *Mohr's* conjugate beam has the same supports and span, and it is loaded by the shear curvature to render the shear deflection equal to the "bending moment"

$$\bar{q}_z = \kappa \left[q_z + F\delta(x - \xi) \right] /GA \quad , \quad w_S = \overline{M}_y(x) \quad .$$

The *influence line* of the shear deflection thus becomes ($q_z = 0$)

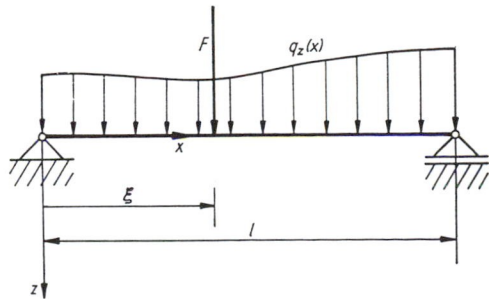

Fig. A 6.6.

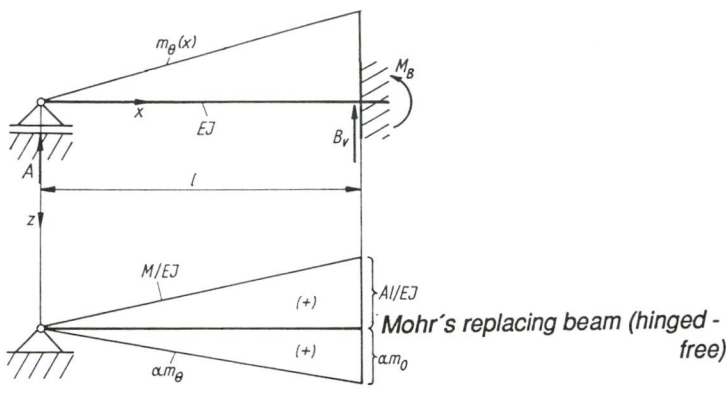

Fig. A 6.7.

$$w_S = \overline{M}_y(x) = \begin{cases} \kappa F\, x(1 - \xi/l)/GA \ , & 0 \le x \le \xi \\ \kappa F\, \xi(1 - x/l)/GA \ , & \xi \le x \le l \end{cases} \ , \ F = 1 \ .$$

A 6.7: The redundant beam of Fig. A 6.7 is loaded by a linearly increasing thermal curvature $\alpha m_\theta = (x/l)\, \kappa_0$. Determine the deflection by properly generalizing *Mohr's* analogy [Eq. (6.74)] to include the thermal loading [see Eq. (6.36)].

Solution: Mohr's conjugate beam is loaded by the moment curvature *M/EJ* and thermal curvature αm_θ , and rigid rotation is prevented by the balance equation of their moments about the hinged support. This renders the statically indeterminate force *A* , and the conditions of equilibrium of the actual structure give

$$A = -B = -EJ\, \kappa_0/l \ , \ M = lA \ .$$

Hence, the total deflection vanishes identically; see Eq. (6.96) and the reference given there and refer to Sec. 6.9. Note that Eq. (6.86) may be considered an orthogonality condition.

A 6.8: The shear beam of Fig. A 2.5 is considered to be linear elastic, the tensile stiffness of the edge beams is given as $D = E_G A_G$ and the shear modulus of the fillet is G_{Bl} . Calculate the displacement of point D in the direction of load *F* by means of *Castigliano's* theorem.

Solution: The normal forces in the edge beams contribute to the complementary energy the portion, $T = F/H$,

$$U_N^* = \frac{1}{2D}\left[2\int_0^L (xT)^2\, dx + 2\int_0^H (xT)^2\, dx\right] \quad\rightarrow\quad \frac{\partial U_N^*}{\partial F} = \frac{2\,HF}{3\,D}\left[1 + (L/H)^3\right] \quad.$$

The shear force in the fillet contributes approximately, if we consider the mean stress,

$$U_Q^* = \frac{1}{2\,G_{Bl}\,(tH)_{Bl}}\int_0^L (HT)^2\, dx \quad\rightarrow\quad \frac{\partial U_Q^*}{\partial F} = \frac{L\,F}{G_{Bl}\,(tH)_{Bl}} \quad.$$

The superposition of the deformations renders in the case of a *stable* structure

$$w_D = \frac{\partial U_N^*}{\partial F} + \frac{\partial U_Q^*}{\partial F} \quad.$$

A 6.9:   The sandwich beam of constant height H of Fig. A 6.9 is laterally loaded and considered to be linear elastic in the parallel girder plates, as well as in the shear of the core. Adapt Eq. (6.70) as the differential equation of deflection of the composite beam under the assumption that the cross-section remains plane after deformation. The effective bending rigidity should be deduced from the equivalence of the bending portion of the elastic energy. For further readings, see *V. Dundrova, V. Kovarik, and P. Slapak: Bending Theory of Sandwich Plates. Springer-Verlag, Wien-New York, 1970, (in German); K. Stamm and H. Witte: Sandwich-Structures. Springer-Verlag, Wien-New York, 1974, (in German).*

Solution: The shear stresses in the core are constant since $\sigma_{xx} \equiv 0$ is assumed, and $\partial\sigma_{xx}/\partial x + \partial\sigma_{xz}/\partial z = 0$ holds locally. Hence, in Eq. (6.70), $\kappa_z = 1$, and the shear rigidity is that of the core alone, $G A = G_K A_K$. The bending stiffness EJ_y is to be replaced by an effective

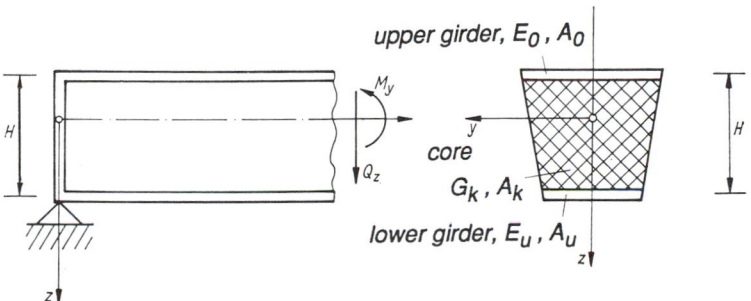

Fig. A 6.9.

rigidity: The equivalence of the potential energy of bending (only the girder plates contribute) gives

$$\int_0^l \frac{M_y^2}{2(EJ_y)_{eff}} \, dx = \int_0^l \left[\frac{\sigma_{xx}^{(0)2}}{2E_0} A_0 + \frac{\sigma_{xx}^{(u)2}}{2E_u} A_u \right] dx \quad .$$

Hence, by substituting the normal stresses

$$\sigma_{xx}^{(0)} = - \frac{M_y}{H A_0} \quad , \quad \sigma_{xx}^{(u)} = \frac{M_y}{H A_u} \quad ,$$

and comparing coefficients we get

$$(EJ_y)_{eff} = \frac{(EA)_0 (EA)_u}{(EA)_0 + (EA)_u} H^2 \quad .$$

A 6.10: The tensile force F is transmitted through a welded or glued joint of length l according to Fig. A 6.10. Thickness h of the material interface (the width is b) is small with respect to l. Determine the mean shear stress $\tau(x)$ or, equivalently, the shear flow $T = \tau b$ by considering the normal force $N_2(x)$ statically indeterminate and by applying *Menabrea's* theorem. The tensile stiffness EA of the two members is a given constant.

Solution: A cut at x through the joint and the free-body diagram shown in Fig. A 6.10 give $F - (N_1 + N_2) = 0$. The balance of the forces acting on an infinitesimal element renders in the limit $dx \to 0 : T + dN_2/dx = 0$. The complementary energy stored in the joint is given by the contributions of the normal forces and of the shear in the material interface (shear modulus G_S); the latter density is $(\tau^2/2G_S)hb$

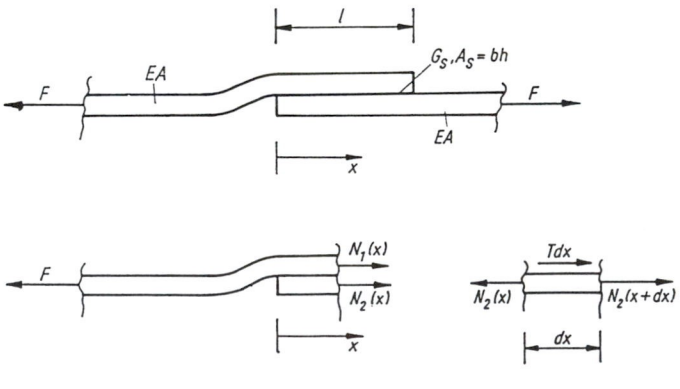

Fig. A 6.10.

$$U^* = \int\limits_0^l \left(\frac{N_1^2}{2EA} + \frac{N_2^2}{2EA} + \frac{T^2 h}{2G_S b} \right) dx \quad .$$

The application of *Menabrea´s* theorem, $\partial U^*/\partial N_2 = 0$, where $(\partial N_1 /\partial N_2) = -1$, $\partial T/\partial N_2 = (\partial T/\partial x) (\partial x/\partial N_2) = -(1/T) \, d^2 N_2 /dx^2$, yields the second-order differential equation

$$\frac{d^2 N_2}{dx^2} - \alpha^2 N_2 = -\frac{\alpha^2}{2} F \quad , \quad \alpha^2 = \frac{2 G_S b}{E A h} \quad .$$

Its solution according to the boundary conditions $N_2 (0) = 0$, $N_2 (l) = F$ is

$$N_2 = \frac{F}{2} (1 - \cosh \alpha x) + \frac{F(1 + \cosh \alpha l)}{2 \sinh \alpha l} \sinh \alpha x \quad .$$

The derivative with respect to the coordinate x renders the negative shear flow that takes on extreme values at the very ends of the joint.

A 6.11: A helical spring of mean diameter $2R$ has n windings within the length l . The wire has a circular cross-section with a diameter $d \ll 2R$ (Fig. A 6.11). The spring is axially loaded by a force F . Determine the stiffness k and maximal shear stress in the wire.

Solution: The axial displacement under the load F is easily determined by means of *Castigliano´s* theorem. The main contribution to the complementary energy of the helical spring with a small slope, $\alpha \ll 1$, is due to torsion. At a cross-section (x, φ) , the bending moment of the slightly curved wire is $M = FR \sin \alpha$ and the torque $M_T = FR \cos \alpha$. Hence, $|M| \ll |M_T| \approx FR$ and

$$U^* \approx \int\limits_0^{2n\pi} \frac{M_T^2}{2G \, J_T} R \, d\varphi = \frac{\pi n R^3}{G J_T} F^2 \quad , \quad J_T = J_p = \frac{\pi d^4}{32} \quad .$$

Also, the contributions of the normal force and shear have been neglected. The axial displacement is $u_0 = \partial U^*/\partial F$, and the stiffness k becomes by definition

$$k = \frac{F}{u_0} = \frac{G J_T}{2\pi n R^3} = \frac{(EA)_{eff}}{l} \quad .$$

The effective stiffness $(EA)_{eff}$ of a fictitious solid tensile rod is introduced above for further reference. The shear force is

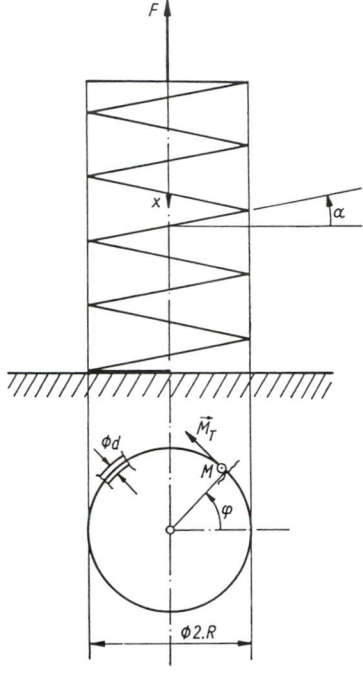

Fig. A 6.11.

approximately $Q \approx F$ and the maximal shear stress thus becomes by superposition

$$\tau_{max} = \frac{4\,F}{3A} + \frac{d\,M_T}{2\,J_T} = \frac{16\,F}{3\pi\,d^2}\left(1 + \frac{3R}{d}\right) .$$

A 6.12: The torsional rod has a doubly symmetric cross-section, consisting of three cells like those shown in Fig. A 6.12. Determine the shear flows and twist by means of *Bredt´s* formulas.

Solution: Due to the rigid rotation of the cross-section,

$$2\,G\,\vartheta = \frac{4\,T_1}{bh_1} - \frac{8\,T_3}{\pi bh_3} = \frac{2\,T_2}{bh_2} + \frac{2\,T_3}{ah_3} .$$

Static equivalence and the flow condition render two more equations

$$M_T = \frac{\pi b^2}{2}\,T_1 + 2ab\,T_2 \quad , \quad T_1 + T_3 = T_2 \quad .$$

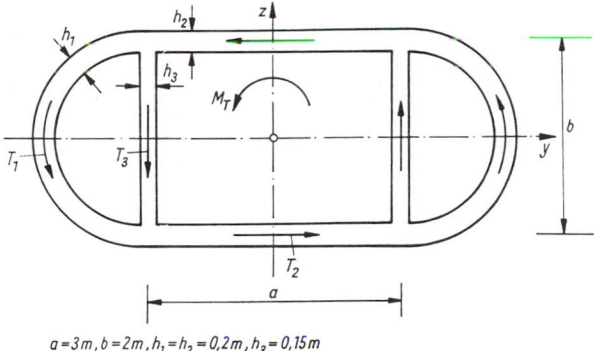

$a=3m, b=2m, h_1=h_2=0,2m, h_3=0,15m$

Fig. A 6.12.

The numerical solution according to the given dimensions of the cross-section becomes $[M_T] = Nm$, $[G] = Nm^{-2}$, $(T_1, T_2, T_3) = (46, 59, 13) \times 10^{-3} \times M_T$, and in Eq. (6.104), $J_T = 5.66\ m^4$.

A 6.13: A thin-walled open circular cross-section of a torsional rod is given in Fig. A 6.13. Determine the shear center, torsional stiffness, warping function, and warping rigidity. The wall thickness $h \ll R$ is constant.

Solution: The shear center is located on the axis of symmetry. The principal moment of inertia $(J_y = \pi h R^3$, $z = R \sin \alpha)$ and the area $(a(s) = \alpha R^2/2$, $ds = R\ d\alpha)$ are substituted into the formula valid for open sections

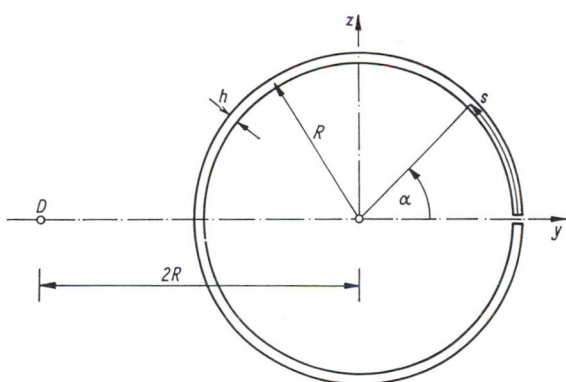

Fig. A 6.13.

$$y_D = \frac{2}{J_y} \int_0^{2\pi R} z \, a(s) \, h \, ds = -2R \quad , \quad z_D = 0 \quad .$$

The warping function, $\varphi(s) = \varphi_0 - 2a(s)$, is determined by enforcing the condition of symmetry, $\varphi(s = \pi R) = 0$, and hence, $\varphi_0 = \pi R^2$. With reference to the center of shear, $\varphi^* = \varphi + z \, y_D = (\pi - \alpha - 2 \sin \alpha) \, R^2$. Equation (6.146) should be verified. Torsional rigidity is proportional to $J_T = Lh^3/3 = (2\pi/3) \, Rh^3$, and the warping rigidity becomes [see Eq. (6.153)]

$$C_w = \int_0^{2\pi R} \varphi^{*2} \, h \, ds = \pi h \left(\frac{2 \pi^2}{3} - 4\right) R^5 \quad .$$

A 6.14: Determine the center of shear and torsional rigidity of the nondoubly symmetric I cross-section of Fig. A 6.14.

Solution: The point of reference 0 is selected on the axis of symmetry and the radius vector $r(s)$ must run smoothly along the contour to determine the area $a(s)$. For example, in the lower girder $- b_1/2 \le s_1 \le b_1/2$, $s_1 = 0$ before entering the web.

Lower girder: $a = Hs_1/4$, $\varphi = \varphi_0 - 2a(s_1) = \varphi_0 - Hs_1/2$,

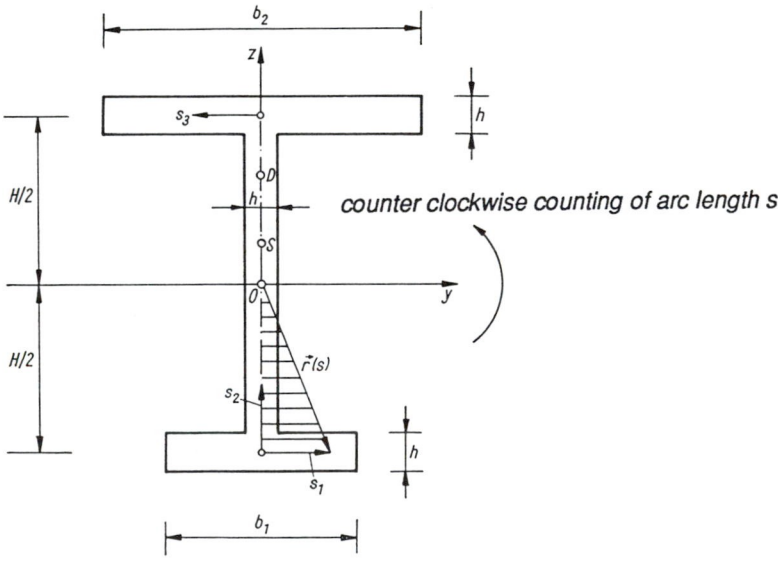

counter clockwise counting of arc length s

Fig. A 6.14.

Web: $a = 0$, $\varphi = \varphi_0 - 2a(s_2) = \varphi_0 = 0$, $\varphi(s)$ skew symmetric,

Upper girder: $a = Hs_3/4$, $\varphi = \varphi_0 - 2a(s_3) = \varphi_0 - Hs_3/2$.

Equation (6.146) holds with $\varphi_0 = 0$. Substituting $J_z = h\,(b_1{}^3 + b_2{}^3)/12$ renders

$$z_D = \frac{1}{J_z} \int_0^L y\,\varphi(s)\,h\,ds = \frac{H}{2}\frac{b_2{}^3 - b_1{}^3}{b_2{}^3 + b_1{}^3} > z_S = \frac{H}{2}\frac{b_2 - b_1}{b_2 + b_1} , \quad b_2 > b_1 .$$

"The center of shear is beyond the areal centroid, with respect to material concentration." The torsional rigidity is given by Eq. (6.136)

$$J_T = \frac{1}{3}\sum_{i=1}^{3} L_i\,h_i{}^3 = \frac{h^3}{3}\,(b_1 + b_2 + H) .$$

A 6.15: A thin circular disk of thickness h is loaded by an axisymmetric temperature field $\theta(r)$. Find the inhomogeneous *Navier's* differential equation of the radial thermal displacement $u(r)$ and its general solution.

Solution: The elimination of the membrane stresses, namely, radial stress n_r and hoop stress n_φ , from the local equilibrium condition (no body forces), $r\,dn_r/dr + n_r - n_\varphi = 0$, by means of *Hooke's* law (plane stress conditions apply)

$$n_r = \frac{Eh}{1-v^2}\left(\frac{\partial u}{\partial r} + v\frac{u}{r} - (1+v)\,\alpha n_\theta\right) , \quad n_\theta = \frac{1}{h}\int_{-h/2}^{h/2}\theta\,dz ,$$

$$n_\varphi = \frac{Eh}{1-v^2}\left(\frac{u}{r} + v\frac{\partial u}{\partial r} - (1+v)\,\alpha n_\theta\right) ,$$

renders a differential equation of the *Euler* type and inhomogeneous

$$\frac{\partial^2 u}{\partial r^2} + \frac{1}{r}\frac{\partial u}{\partial r} - \frac{u}{r^2} = (1+v)\,\alpha\,\frac{\partial n_\theta}{\partial r} .$$

The general solution combines a particular integral with the solution of the homogeneous equation

$$u = C_1 r + \frac{C_2}{r} + \frac{1+v}{r}\,\alpha\int r\,n_\theta\,dr .$$

The constants of integration are to be determined from the boundary conditions. The full disk requires $C_2 = 0$, and if the boundary $r = R$ is free of any traction, the self-equilibrating thermal stresses become

$$n_r = \alpha Eh\left[f(R) - f(r)\right] \quad , \quad n_\varphi = \alpha Eh\left[f(R) + f(r) - n_\theta\right] \quad , \quad f(r) = \frac{1}{r^2}\int_0^r \rho\, n_\theta(\rho)\, d\rho \quad .$$

A 6.16: In Sec. 6.5.3, the membrane stresses are given for a plate with a circular hole. The stress distribution changes if the hole is elliptically shaped according to Fig. A 6.16. The solution has been derived by *N. I. Muschelishwili*, see, eg *G. N. Sawin: Stress Concentration at the Boundary of Holes. VEB Verlag Technik, Berlin, 1956, p. 86. (in German).*

The largest hoop stress is observed at $x = a > b , y = 0$ to be *max* $n_\varphi = n_0 \,(1 + 2a/b) = n_0 \,[1 + 2\,\sqrt{(a/R_1)}]$, where the principal curvature of the hole is $1/R_1 = a/b^2$. The (compressive) hoop stress at $x = 0 , y = b$ is, independent of the geometry, given by $n_\varphi = - n_0$. The load is applied in the y direction. With $a = b$ (the circular hole), the magnification factor becomes 3 . If $b << a$, the maximum is approximately *max* $n_\varphi \approx 2n_0\,\sqrt{(a/R_1)}$. In the limit $b \to 0$, the *Griffith crack* of length $2a$, oriented orthogonal to the loading axis, is derived. A crack that parallels the external tension does not influence the constant stresses $n_y = n_0$. The singular stress distributions of the *Griffith crack* along $y = 0$ for $|x| \geq a$ are derived by the proper limiting process from the stress field around the elliptic hole according to *Eshelby*

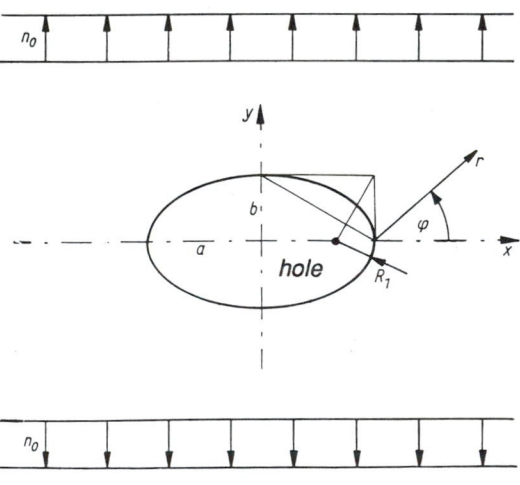

Fig. A 6.16.

$$n_x = n_0 \left(\frac{\xi}{\sqrt{\xi^2 - 1}} - 1 \right) \quad , \quad n_y = n_0 \frac{\xi}{\sqrt{\xi^2 - 1}} \quad , \quad n_{xy} = 0 \quad , \quad \xi = x/a \; .$$

Determine the near field of the stresses at the crack tip by a transformation to the local polar coordinates (r, φ) for $\varphi = 0$ and $r \ll a$. Find the $1/\sqrt{r}$ singularity of the plane problems of linear Fracture Mechanics and determine the stress intensity factor K_1 (fracture mode 1) by comparing the coefficients in $n_i = K_1/\sqrt{(2\pi r)}$. Also find the principal (out-of-plane) shear stress. For a survey, see *H. Liebowitz (ed.): Fracture. An Advanced Treatise. Vol.I to VII, Academic Press, New York, 1968-1972.* A short introduction is given in *H. Rossmanith (ed.): Foundations of Fracture Mechanics. Springer-Verlag, Wien-New York, 1982, (in German).*

Solution: Substituting $\xi = 1 + r/a$ and taking the limit give the leading terms in the near field

$$n_r(r, \varphi = 0) = n_0 \sqrt{a/2r} = K_1 / \sqrt{2\pi r} \quad , \qquad K_1 = n_0 \sqrt{\pi a} \; ,$$

$$n_\varphi(r, \varphi = 0) = n_0 \sqrt{a/2r} = K_1 / \sqrt{2\pi r} \quad , \quad n_{r\varphi}(r, \varphi = 0) = 0 \; .$$

With h, the constant thickness of the plate, the principal out-of-plane shear stress becomes

$$\max |\tau| = K_1 / (2h \sqrt{2\pi \, r}) \; .$$

A 6.17: An infinite plate strip of width a is simply supported and loaded by a pressure $p(x)$ (Fig. A 6.17). Determine the cylindrical surface of deflection by its *Fourier* series.

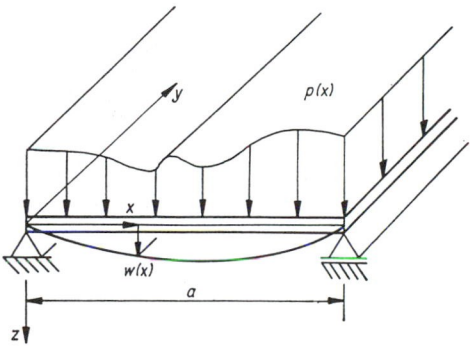

Fig. A 6.17.

Solution: The simply supported edge requires an odd periodic extension, and the *Fourier* series

$$w(x) = \frac{1}{K} \sum_{n=1}^{\infty} \frac{p_n}{\alpha_n^4} \sin \alpha_n x \quad ,$$

is a fast converging solution of the plate equation (6.275) if the factors p_n are the *Fourier* coefficients of the given loading

$$p_n = \frac{2}{a} \int_0^a p(x) \sin \alpha_n x \, dx \quad , \quad \alpha_n = n\pi/a \quad .$$

The bending moments still show a rather fast convergence

$$m_x(x) = \sum_{n=1}^{\infty} \frac{p_n}{\alpha_n^2} \sin \alpha_n x \quad , \quad m_y(x) = v\, m_x(x) \quad , \quad m_{xy} = 0 \quad .$$

The result is of special importance as a particular solution of a finite rectangular plate with a pair of parallel and simply supported edges; see Exercise A 6.18.

A 6.18: A rectangular plate is simply supported at the edges $x = 0$ and $x = a$, but rigidly clamped at the remaining boundaries at $y = -b/2$ and $y = b/2$. Calculate the deflection if the loading $p(x)$ does not depend on y . The results derived for the plate strip of Exercise A 6.17 should be used in the analysis.

Solution: Separable functions are applied in the solution of the homogeneous plate equation under the following conditions: Even functions of y must be found and the boundary conditions at the hinged edges should be builtin individually. The result is an infinite series (the terms are quite obvious)

$$w_h(x, y) = \sum_{n=1}^{\infty} \frac{1}{\alpha_n^2} (A_n \cosh \alpha_n y + B_n \alpha_n y \sinh \alpha_n y) \sin \alpha_n x \quad , \quad \alpha_n = n\pi/a \quad .$$

Superposition of the particular solution given by the strip deflection of Exercise A 6.17 and considering the clamping conditions $w(x, y = b/2) = 0$, $\partial w(x, y = b/2)/\partial x = 0$ of the total solution render, by comparing coefficients,

$$A_n = -\frac{p_n}{N_n} \left(2 \sinh \frac{\alpha_n b}{2} + \alpha_n b \cosh \frac{\alpha_n b}{2}\right) \quad , \quad B_n = \frac{2 p_n}{N_n} \sinh \frac{\alpha_n b}{2} \quad ,$$
$$N_n = K \alpha_n^2 (\alpha_n b + \sinh \alpha_n b) \quad .$$

The total deflection is given with the above coefficients by superposition

$$w(x, y) = w_h(x, y) + \frac{1}{K} \sum_{n=1}^{\infty} \frac{p_n}{\alpha_n^4} \sin \alpha_n x \quad .$$

A 6.19: The axisymmetric deflection of a full circular plate with a central single force loading can be deduced from Eq. (6.273a). By a limiting procedure, derive the *Green's* function of the infinite plate domain; put $F = 1$. By means of *Maxwell's* theorem, give the influence function of a couple, a double force with moment $M = 1$, applied at (ξ, η) in the infinite plate, as an extension of the above solution. See Fig. A 6.19.

Solution: The homogeneous part of the solution of the plate equation is $w(r) = A + B r^2 + C \ln (r/R_e) + D r^2 \ln (r/R_e)$. Deflection at $r = 0$ must be finite, $C = 0$. The shear force is related to the given single load F through the condition of equilibrium of a finite plate element, $2\pi r \, q_r + F = 0$. Since $q_r = - K \, \partial(\Delta w)/\partial r$, the constant $D = F/8\pi K$. The remaining coefficients A and B are to be determined by the boundary conditions prescribed at $r = R_e$. The limit $R_e \to \infty$ renders the infinite plate, and a coordinate transformation gives the force *Green's* function, $F = 1$,

$$w^F(x, y; \xi, \eta) = \frac{r^2}{8\pi K} \ln r \; , \quad r^2 = (x - \xi)^2 + (y - \eta)^2 \; .$$

Equation (6.3) renders with Fig. A 6.19 (the moment produces work by a rotation through a small angle)

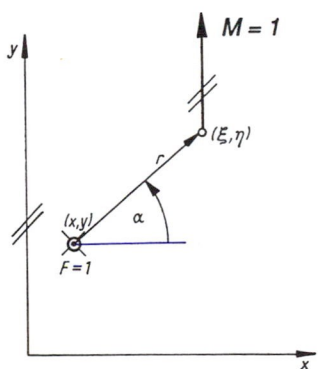

Fig. A 6.19.

$$F \, w^M(x, y; \xi, \eta) = M \, \frac{\partial w^F(\xi, \eta; x, y)}{\partial \xi} \quad .$$

The unit load yields the influence function of a moment acting in an infinite plate

$$w^M(x, y; \xi, \eta) = \frac{r \cos \alpha}{4\pi K} \ln r \ , \quad \alpha = \tan^{-1} \frac{\eta - y}{\xi - x} \quad .$$

A 6.20: A shell of revolution in the form of a truncated cone is hanging from the ceiling by its upper fixed edge and carries a uniform ring loading q_0 at the lower boundary (see Fig. A 6.20). Such welded-steel roof structures are found in industrial buildings and civic centers. Considering the membrane stresses and deformations only, determine the axial stiffness coefficient.

Solution: Since the distributed loadings $p_z = p_\varphi = 0$, the equations (6.296) and (6.297) render at once the membrane stresses ($\varphi = \pi/2 - \alpha$, $r = a + s \sin \alpha$)

$$n_\varphi \equiv n_s = \frac{q_0}{[1 + (s/a) \sin \alpha] \cos \alpha} \ , \quad n_\vartheta \equiv 0 \quad \text{since } r_1 \to \infty \ .$$

The complementary energy of the membrane stress state is

$$U^* = \int_0^{H/\cos \alpha} \frac{n_\varphi{}^2}{2 \, Eh} \, dA \ , \quad dA = 2\pi \, (a + s \sin \alpha) \, ds \quad .$$

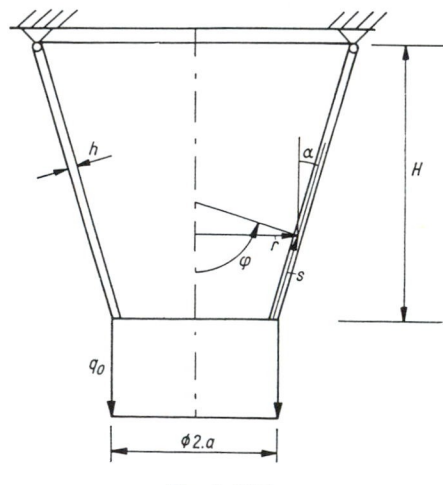

Fig. A 6.20.

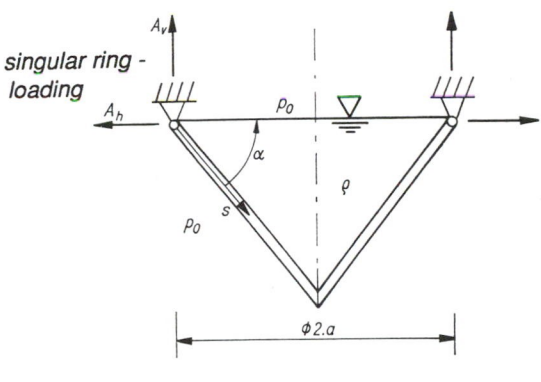

Fig. A 6.21.

Castigliano's theorem renders the tip displacement

$$w_0 = \frac{1}{2\pi a} \frac{\partial U^*}{\partial q_0} = 4 \frac{aq_0}{Eh} \frac{\sin \alpha}{\sin^2 2\alpha} \ln\left(1 + \frac{H}{a} \tan \alpha\right) .$$

Thus, the stiffness coefficient is easily determined: $c = 2\pi a q_0 / w_0$. The stiffness of a cylindrical shell of the same length follows by working out the limit $\alpha \to 0$ by means of *l'Hospital's* rule, $w_0^* = (q_0 / Eh) H$.

A 6.21: A hanging conical shell roof is filled to the rim with a fluid (Fig. A 6.21). Determine the membrane stresses due to the hydrostatic pressure loading.

Solution: The vertical supporting ring force per unit of length due to the weight of the fluid filling is $A_v = (g\rho a^2/6) \tan \alpha$. Since $\varphi = \pi - \alpha$ = *const* and $r = a - s \sin \alpha$, Eqs. (6.296) and (6.297) render, with the hydrostatic pressure loading $p_z = g\rho s \sin \alpha$, the membrane stresses $(r_1 \to \infty)$

$$n_\varphi = \frac{g\rho}{6} (a - s \cos \alpha)(2 s + a/\cos \alpha) , \quad n_\vartheta = g\rho (a - s \cos \alpha) s .$$

The vertical component of $n_\varphi(s)$ may be determined by considering the equilibrium of a finite truncated cone.

A 6.22: In the case of slender beams which are exposed to a temperature environment the question arises: What are the most general distributions of thermal curvature that keep the deflections zero? Interpret *Maysel´s* formula for some simple redundant beams to remain undeflected.

Solution: The thermal curvature $\alpha m_\theta(\xi)$ enters the domain integral. The latter determines the inner product with respect to the isothermal influence function of the linear elastic beam. It vanishes if the functions are orthogonal. A first example of the clamped-clamped beam is given by Eq. (6.96) if the thermal curvature is a constant. The clamped-simply supported beam of Fig. 6.12 remains undeflected if the thermal curvature increases linearly towards the clamped edge. Consequently, the thermal deflection vanishes, ie the thermal bending moment becomes proportional to the cross-sectional temperature moment in the case of a double-span beam, Fig. 5.9, or for a symmetric triple-span beam on simple supports, if the imposed thermal curvature has a triangular or a trapezoidal distribution, respectively. The statically determinate system should be selected according to Fig. 5.10, or for multispan beams according to Fig. 6.13, to recognize the answer at once. For details of the force method see *Thermal Stresses. Vol. II, (ed. R. B. Hetnarski), Elsevier Science Publisher, Amsterdam, 1987, chap. 3, p. 152.*

7
Dynamics of Solids and Fluids, Conservation of Momentum of Material and Control Volumes

Dynamics is understood in the narrow sense of kinetics, which means that the inertia of accelerated masses is of crucial importance when considering the stresses and deformations in moving bodies. *Newton's* law of motion is applicable if the velocities remain small with respect to the speed of light. In vacuum, the speed of light is the natural constant *c = 299 792 458 m/s* . Since 1983, the definition of the unit of length, *1 meter* , is based on the above derivable fraction of the distance traveled by light in vacuum within *1 second* . The basic dynamic law within the limits of *newtonian mechanics* in the formulation of *Euler-Cauchy* states that proportionality exists in every material point of a continuum between the force density *f* and the absolute acceleration *a* . The latter is to be measured with respect to an inertial reference system, point 0 of Fig. 7.1 [see Eq. (2.18)]

$$\mathbf{f} = \rho \, \mathbf{a} \; . \tag{7.1}$$

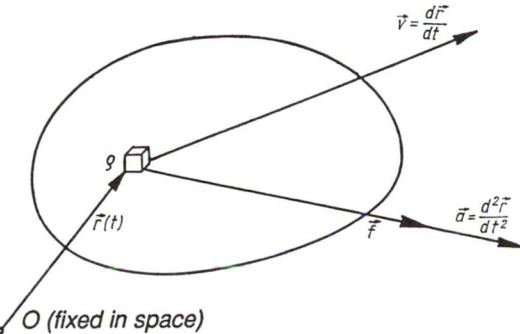

Fig. 7.1. The *Euler-Cauchy* equation of motion in *newtonian mechanics*

The scalar factor of proportionality is the mass density $\rho(x, y, z, t)$ with the dimension mass per unit of volume, kg/m^3 , that should not explicitly depend on the acceleration, velocity, or force. One consequence of high speeds is the definition of mass at rest, ρ_0 ; ρ in Eq. (7.1) in such a case is to be replaced by $\rho_0 /\sqrt{(1 - v^2/c^2)}$. Details are given in textbooks on special relativity theory.

Newton's Law of Inertia (1687), is derived by integrating Eq. (7.1) over the volume of a rigid body that is in pure translational motion with respect to a reference system that is fixed in space. The acceleration a of all the mass elements $dm = \rho\, dV$ is a uniform parallel vector field; hence,

$$\int_V \mathbf{f}\, dV = \mathbf{R} = \int_m \mathbf{a}\, dm = m\, \mathbf{a} \quad , \quad m = \int_V \rho\, dV \ . \tag{7.2}$$

The uniform acceleration a is produced by the colinear resulting external force $\mathbf{R}$ and m is the total mass of the rigid body. Equation (7.2) may be used for the definition of inertial reference systems: <<If a rigid body is in translational motion in an arbitrary direction, $v = v_0$, and the external forces are permanently self-equilibrating or zero, $R = 0$, all the material points move with constant speed along parallel straight lines with respect to any other proper reference system.>> This, Galilei's Law of Inertia , determines the inertial reference systems by these rigid bodies in translational and constant motion. It follows that one inertial frame cannot be accelerating with respect to another, the position vector of a particle in parallel inertial frames 1 and 2, eg is related by the linear Galilean transformation, $r_2 = r_1 - v_{21}\, t$. The fixed stars of our firmament define an empirical inertial system. The Earth is in rather complicated motion, but the daily rotation about the axis through the poles [the mean angular speed is, of course, $(2\pi/24$ hours)] is of the greatest importance with respect to the property that makes it actually a noninertial system. Its angular velocity in common units is therefore $\omega = 7.27 \times 10^{-5}$ rad/s . The rotation causes the difference between the acceleration of the free fall and the gravitational acceleration and, most important, allows for north-south orientation by means of the gyro-compass. Many geophysical phenomena are caused by the eigen rotation of the Earth. The weight of a mass in a parallel gravity field is given by Eq. (7.2) if the acceleration of the free fall is substituted

$$G = mg \ . \tag{7.3}$$

Due to the eigenrotation of the Earth, the weight is not a constant; it varies with the latitude on the surface (see Secs. 2.1 and 3.3.2). In many applications, however, it suffices to consider the mean acceleration $g = 9.81$ m/s^2 constant.

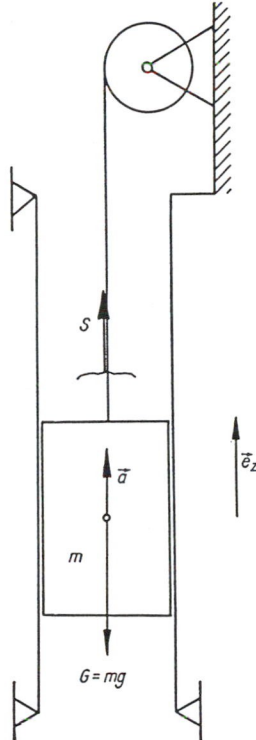

Fig. 7.2. Free-body diagram of an elevator cabin

The dynamics of simple motions sometimes may be considered the generalization of statics through the inclusion of inertial force in a balance of the given forces. Consider, eg the translational motion of the cabin of an elevator (Fig. 7.2). The cable tension S equilibrates the weight of the cabin $G = mg$ during any stationary phase of the motion, ie if the speed up or down is a constant. Approximately, the Earth can be considered the reference system: $S_0 - mg = 0$ or $S_0 = mg$. If the tension in the cable is changed by pulling stronger, the cabin accelerates. *Newton's* law (7.2) gives, with the current value of the cable tension, $S - mg = ma$ or $S = m(g + a)$.

It is common practice, to express the acceleration in terms of fractions of $g = 9.81 \ m/s^2$; hence, $a = \lambda g$ and $S = mg (1 + \lambda)$. The maximal acceleration of an elevator has to be limited, and $(1 + \lambda_{max})$ is the *dynamic load factor* in the above quasistatic relation between the cable tension and the weight of the cabin. Analogously, also distributed inertial forces are introduced if it comes to the calculation of stress fields in moving bodies; see, eg Sec. 6.5.2 and Exercise A 11.4.

The vector equation (7.1) renders three relations between the cartesian components of absolute acceleration and the stress gradients (the body forces per unit of volume **k** are assumed to be given)

$$f_i = \sum_j \frac{\partial \sigma_{ji}}{\partial x_j} + k_i = \rho\, a_i \;\;,\;\; i, j = x, y, z \;\;,\;\; a = \frac{dv}{dt} = \frac{d^2 u}{dt^2} \;\;.$$

(7.4)

Since the stresses and strains (deformation gradients) are related, the constitutive equations of the deformable body must be known before any time integration of the above local differential equations becomes visible. See Eq. (7.161) of the isotropic, homogeneous linear elastic solid and Chap. 11. The gross conservation laws of motion are derived subsequently; they are independent of the special material laws.

7.1.   Conservation of Momentum

Equation (7.1) may be integrated over the material volume $V(t)$ at any time, the mass m is considered to be constant within the material volume, $r(t)$ is the radius vector of Fig. 7.1

$$\int_{V(t)} f\, dV = \int_{V(t)} k\, dV + \int_{V(t)} \sum_j \frac{\partial \sigma_j}{\partial x_j} dV = \int_m a\, dm = \frac{d^2}{dt^2} \int_m r\, dm$$

$$= m \frac{d^2 r_M}{dt^2} = m\, a_M \;\;,\;\; \int_m r\, dm = m\, r_M \;\;.$$

(7.5)

Integration over the constant mass m and the time derivative have been interchanged and the static moment of the mass distribution with respect to the origin 0 can be expressed by the mass times the radius vector to the mass center M; see Eq. (2.72). Hence, the resulting inertial force is proportional to the absolute acceleration of the mass center, a_M. The volume integral of the given body force renders the resultant **K** , eg the weight of the mass. The *Gauss* integral theorem is applied to the remaining volume integral and renders termwise the surface integrals

$$\int_{V(t)} \frac{\partial \sigma_j}{\partial x_j} dV = \oint_{\partial V(t)} \sigma_j\, n_j\, dS \;\;.$$

(7.6)

Summation of the integrand over j gives the surface traction according to Eq. (2.20), and the surface integral over the forces σ_n dS yields the resultant of the external surface forces Σ . Since **K** +

$\Sigma = R$ is the resultant of all the external forces, the integral law of motion results

$$m\, a_M = R \quad . \tag{7.7}$$

<<The acceleration of the center of mass of any deformable body with constant mass is proportional to the resultant of the external forces.>> The internal forces have no influence on the motion of the center of mass (or the center of gravity). An illustrative system is a gun and the bullet before it leaves the barrel. The high internal gas pressure accelerates the bullet, but the mass center of the total and deforming system remains at rest. Thus, the barrel must move backward. This renders the well-known recoil of a gun after firing. Equation (7.7) represents also the vector equation of motion of a point mass m under the action of a force R .

The momentum of a mass element dm is defined by the infinitesimal vector $dl = v\, dm$ in the direction of its velocity v . Thus, the momentum of a mass distribution is the resulting vector [mass m in the material volume $V(t)$ is constant]

$$l = \int_m v\, dm = \frac{d}{dt} \int_m r\, dm = m\, v_M \quad . \tag{7.8}$$

The internationally recommended symbol of momentum, $p = m\, v_M$, is not used here to avoid any possible confusion with the hydrostatic pressure p . With the absolute velocity of the mass center v_M and if we consider $m = const$, Eq. (7.7) becomes the rate equation of the momentum vector

$$\frac{dl}{dt} = m\frac{dv_M}{dt} = R \quad . \tag{7.9}$$

<<The rate of the absolute momentum of a body with constant mass equals the resultant of the external forces.>> Time integration renders the law of conservation of momentum

$$l(t_2) - l(t_1) = \int_{t_1}^{t_2} R\, dt \quad . \tag{7.10}$$

<<The momentum remains constant if the impulse of the external forces (their time integral) vanishes.>> Momentum of a mass and impulse of force have the same dimension, [kg m/s = Ns] . For the impulse during impact, see Eq. (12.3). The notation I is justified.

There are several important applications in which it is more convenient to consider a *control volume V* (quite often fixed in space) with mass flowing through the closed *control surface ∂V* , (see Sec. 1.6), instead of considering the material volume $V(t)$ where the mass, enclosed in the material surface $\partial V(t)$, is constant. The increase of the momentum $I(t)$ of a mass $m(t)$ that occupies the control volume at time t , which is then necessarily nonstationary, ie time dependent, is caused by two contributions, the so called production terms:

1)   The mass elements within the control volume are accelerated by external forces.

2)   Mass flows through the control surface, which results in a transport of momentum. Since the normal of the surface is pointing outward, the net flow outward is counted as positive in Eq. (1.71). Hence, the contributions are

$$1) \dots \int_V \frac{d\mathbf{v}}{dt} \rho \, dV = \int_V \mathbf{a} \, dm \ , \quad 2) \dots -\oint_{\partial V} \mu\mathbf{v} \, dS \ .$$

The nonstationary rate of the momentum equals the superposed production terms,

$$\frac{d \, I(t)}{dt} = \frac{d\,[m(t)\mathbf{v}_M(t)]}{dt} = \int_V \frac{\partial(\rho\mathbf{v})}{\partial t} \, dV = \int_V \mathbf{a} \, dm - \oint_{\partial V} \mu\mathbf{v} \, dS \ ,$$

$$I(t) = \int_V \rho\mathbf{v} \, dV = m(t) \, \mathbf{v}_M(t) \ . \tag{7.11}$$

Since Eq. (7.1) still applies to all material points in the control volume at some time instant t , integration at a fixed time renders

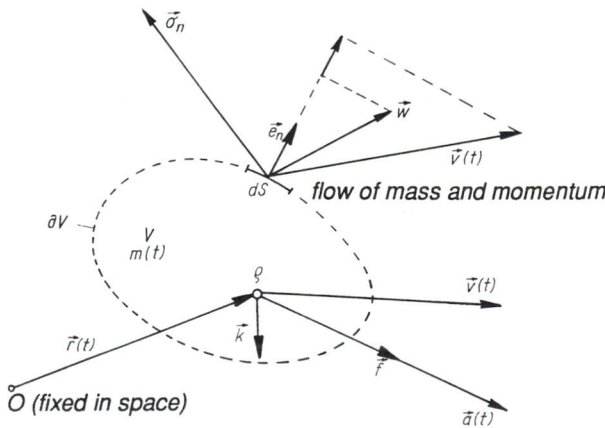

Fig. 7.3. Control surface ∂V fixed in space, $w = 0$, or moving in a prescribed motion with velocity $\mathbf{w}$. Conservation of momentum of a control volume with flow of mass through the control surface

the first production term of the impulse rate

$$\int_V \mathbf{a} \, dm = \int_V \mathbf{f} \, dV = \mathbf{R}(t) \ ,$$

(7.12)

to be equal to $\mathbf{R}$, the resultant of the external forces acting on the mass in the control volume (the resulting body force plus the resulting traction on the control surface). Thus, the equivalent to Eq. (7.9), if a flow of mass is considered, becomes

$$\frac{d\mathbf{I}(t)}{dt} + \oint_{\partial V} \mu \mathbf{v} \, dS = \mathbf{R} \ , \quad \mu = \rho \, (\mathbf{v} \cdot \mathbf{e}_n) \ .$$

(7.13)

The integrand of the surface integral $(\mu \, \mathbf{v})$ is denoted as the density of the momentum flux rate.

If the control volume V fixed in space is chosen so it coincides instantly with the material volume $V(t)$, the left-hand side becomes the *Reynolds* transport theorem of momentum vector (the three-dimensional generalization of *Leibniz´s* rule of differentiation); compare this with Eq. (7.9) (see Sec. 1.6)

$$\frac{d}{dt} \int_{V(t)} \rho \mathbf{v} \, dV = \int_V \frac{\partial (\rho \mathbf{v})}{\partial t} \, dV + \oint_{\partial V} \mu \mathbf{v} \, dS = \mathbf{R} \ .$$

In many cases of stationary flow problems, the momentum vector remains constant, despite the continuous exchange of particles in the control volume: Its rate vanishes and the resultant external force equals the net outflow of momentum through the control surface. Illustrative applications are discussed in Sec. 7.3.

Formally, Eq. (7.13) remains valid for moving control surfaces ∂V^* with prescribed velocities $\mathbf{w}$ [see Eqs. (1.80) and (1.82) and Fig. 7.3], provided the rate of the mass flow density of Eq. (1.81) is substituted. If the control volume is changed during the motion, the nonstationary rate of the momentum vector takes on this form, which is generally valid,

$$\frac{d\mathbf{I}}{dt} = \frac{\partial}{\partial t} \int_{V^*(t)} (\rho \mathbf{v}) \, dV \ , \quad \mu = \rho \, (\mathbf{v} - \mathbf{w}) \cdot \mathbf{e}_n^* \ .$$

(7.14)

Equation (7.9) results if $\mathbf{v} \equiv \mathbf{w}$ is chosen on the control surface, $\mu = 0$; the moving control volume becomes identical to the material volume and the enclosed mass remains constant.

In the practically important case of the prescribed rigid-body motion of the control volume, Eq. (1.83) applies to a scalar integrand. The momentum flux density $\rho \mathbf{v}(x', y', z', t)$, however, is a vector function, and its rate includes a term due to any rigid-body rotation of the control volume. See Eq. (7.59) for taking the proper time derivative.

7.2.   Conservation of Angular Momentum

Summing the forces $f\,dV$ in Sec. 7.1 was analogously performed in the static case of Eq. (2.45). Hence, a second vector equation results if, as in Eq. (2.56), the moments are considered. To be sufficiently general, an intermediate point of reference A′ is considered that moves against the inertial frame with velocity v_A and acceleration a_A . The moments on the left- and right-hand side of Eq. (7.1) with respect to A′ are calculated with the radius vector $r(t)$ of Fig. 7.4 [integration is over the material volume $V(t)$ with constant mass m]

$$\int_{V(t)}(\mathbf{r} \times \mathbf{f})\,dV = \int_{m}(\mathbf{r} \times \mathbf{a})\,dm \quad .$$

$$(7.15)$$

Analogously to Eq. (7.5), the volume integral on the left is transformed to a surface integral by the *Gauss* integral theorem. Termwise it follows, with an identical transformation taken into account, that, $j = 1, 2, 3$,

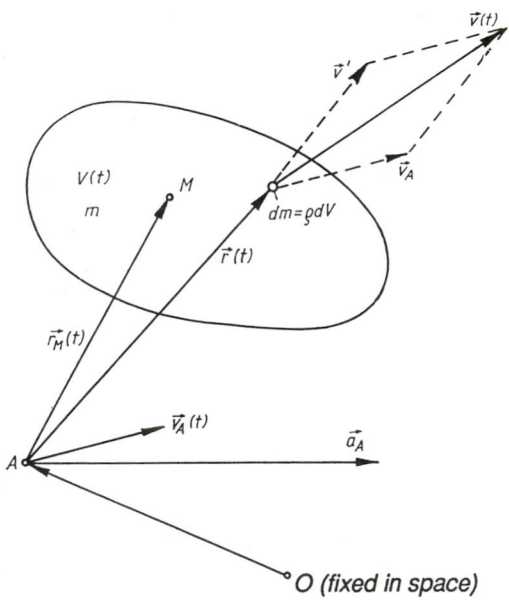

Fig. 7.4. The relative angular momentum with respect to the intermediate reference point A′. The mass center M. Origin 0 of the inertial frame

$$\int_{V(t)} \left(\mathbf{r} \times \frac{\partial \sigma_j}{\partial x_j} \right) dV = \int_{V(t)} \frac{\partial}{\partial x_j} \left(\mathbf{r} \times \sigma_j \right) dV - \int_{V(t)} \left(\mathbf{e}_j \times \sigma_j \right) dV = \oint_{\partial V(t)} \left(\mathbf{r} \times \sigma_j \right) n_j \, dS$$

$$- \int_{V(t)} \left(\mathbf{e}_j \times \sigma_j \right) dV , \quad \frac{\partial \mathbf{r}}{\partial x_j} = \mathbf{e}_j .$$

$$(7.16)$$

The summation of the force density $\mathbf{f}$, according to Eq. (2.18), renders the moment of the surface tractions $\mathbf{r} \times \sigma_n$ in the completed surface integral and thus, the moment of the external forces with respect to point A′ , $\mathbf{M}_A$, results

$$\mathbf{M}_A = \int_{V(t)} \left(\mathbf{r} \times \mathbf{k} \right) dV + \oint_{\partial V(t)} \left(\mathbf{r} \times \sigma_n \right) dS = \int_{V(t)} \left(\mathbf{r} \times \mathbf{a} \right) dm$$

$$+ \int_{V(t)} \left[(\sigma_{xy} - \sigma_{yx}) \, \mathbf{e}_z + (\sigma_{yz} - \sigma_{zy}) \, \mathbf{e}_x + (\sigma_{zx} - \sigma_{xz}) \, \mathbf{e}_y \right] dV .$$

$$(7.17)$$

Boltzmann's axiom about the stress tensor of a point continuum remaining symmetric also in dynamics is expressed in Eq. (7.17) by a volume integral: The internal forces do not contribute to the resulting moment of the inertial forces. Hence, Eq. (7.17) reduces to the "moment equation"

$$\mathbf{M}_A = \int_{V(t)} \left(\mathbf{r} \times \mathbf{a} \right) dm , \quad m = \int_{V(t)} \rho \, dV .$$

$$(7.18)$$

The remaining volume integral is manipulated to render Eq. (7.18) complementary to the momentum rate equation (7.9). Due to the intermediate reference point A′, the total acceleration of a mass element is split into $\mathbf{a} = \mathbf{a}_A + \mathbf{a}'$, where the relative acceleration is measured in the moving frame, $\mathbf{a}' = d\mathbf{v}'/dt = d^2\mathbf{r}/dt^2$, and the relative velocity is $\mathbf{v}' = d\mathbf{r}/dt = \mathbf{v} - \mathbf{v}_A$. See Fig. 7.4. Substitution and considering $m = const$ in the material volume $V(t)$ give

$$\mathbf{M}_A = \int_{V(t)} \mathbf{r} \, dm \times \mathbf{a}_A + \int_{V(t)} \left(\mathbf{r} \times \frac{d\mathbf{v}'}{dt} \right) dm = m \, \mathbf{r}_M \times \mathbf{a}_A + \frac{d}{dt} \int_{V(t)} \left(\mathbf{r} \times \mathbf{v}' \right) dm .$$

$$(7.19)$$

By definition of the relative momentum of a mass element with respect to the moving reference point A′, $d\mathbf{I}_A = \mathbf{v}' dm$, the moment of the relative momentum becomes $d\mathbf{H}_A = \mathbf{r} \times d\mathbf{I}_A$, and by performing the volume integration, the rate of the relative *angular momentum* (or moment of momentum) $\mathbf{H}_A$ enters the equation

$$\frac{dH_A}{dt} + m\,r_M \times a_A = M_A \quad,\quad H_A = \int_{V(t)}(r \times v')\,dm \quad,\quad v' = v - v_A \;.$$

$$(7.20)$$

Most important, the equation is further reduced and considerably simplified, if the center of mass M is chosen as the intermediate reference point $A' = M$. The static mass moment $m\,r_M = 0$ and, furthermore, since

$$H_M = \int_{V(t)}[r \times (v - v_M)]\,dm = \int_{V(t)}(r \times v)\,dm \quad,\quad v' = v - v_M \;,$$

the relative angular momentum of a mass distribution about its mass center equals the absolute moment of momentum with respect to M . The standard form of the rate equation becomes therefore (there is no restriction with respect to the deformations of the body)

$$\frac{dH_M}{dt} = M_M \quad.$$

$$(7.21)$$

Also, a reference point fixed within the inertial frame $A' = 0$ puts the additional term in Eq. (7.20) at zero, $a_A = 0$, and the rate of the absolute angular momentum equals the moment of the external forces taken about this point

$$\frac{dH_0}{dt} = M_0 \quad,\quad H_0 = \int_{V(t)}(r \times v)\,dm \quad.$$

$$(7.22)$$

A point of reference A' that moves with constant speed on a straight line gives the same result. The selection of the intermediate point of reference, such that its acceleration is permanently parallel to the point vector of the mass center, also drops the additional term in the general equation (7.20), but its use is not recommended for further application.

The internationally accepted symbols for the angular momentum of a moving particle with momentum p are $L = r \times p$ or $J = r \times p$. It is more convenient in continuum mechanics to use the symbol H recommended for the angular impulse instead.

The law of conservation of the (absolute) angular momentum about the center of the mass distribution M is given by the time integration of Eq. (7.21)

$$H_M(t_2) - H_M(t_1) = \int_{t_1}^{t_2} M_M(t)\,dt \quad.$$

$$(7.23)$$

<<The increase of the moment of momentum of a body with constant mass m equals the angular impulse, which is the time integral of the external moments.>> Notation H is also justified.

The laws of conservation of momentum and angular momentum, Eqs. (7.10) and (7.23), or their rate forms (7.9) and (7.22), respectively, reflect the conditions of equilibrium, $R = M_0 = 0$, that are necessary but not sufficient for a body to remain at rest.

Analogous to the momentum of Sec. 7.1, the angular momentum of a mass $m(t)$ that is instantly contained in a control volume V * with the control surface ∂V * (possibly moving in space with prescribed velocity w) is considered subsequently. By the inspection of Eq. (7.11), the nonstationary rate of the moment of momentum about an intermediate (moving) point of reference A' is obviously given by

$$\frac{dH_A}{dt} = \frac{\partial}{\partial t} \int_{V^*(t)} (r \times \rho v') \, dV = \int_{V^*(t)} \frac{d}{dt} (r \times v') \, dm - \oint_{\partial V^*(t)} (r \times \mu v') \, dS \ ,$$

$$\mu = \rho (v - w) . e_n^* \ .$$

By establishing the absolute acceleration as $a = a_A + dv'/dt$, the production term, the volume integral, is eliminated by the moment equation (7.18); see also Eq. (7.20). The result is

$$\frac{dH_A}{dt} + \oint_{\partial V^*(t)} (r \times \mu v') \, dS + m \, r_M \times a_A = M_A \ , \quad \mu = \rho (v - w) . e_n^* \ .$$
(7.25)

The rate of the flux of the moment of the relative momentum enters the surface integral. For the rigid-body motion of the control volume, see Sec. 7.1. Putting $A' = 0$ as fixed in space and selecting the control surface at rest render the equation complementary to Eq. (7.13), in the reduced form,

$$\frac{dH_0}{dt} + \oint_{\partial V} (r \times \mu v) \, dS = M_0 \ , \quad \mu = \rho \, v . e_n \ .$$
(7.26)

<<The nonstationary rate of the angular momentum plus the net rate of outflow of the moment of absolute momentum equal the resulting moment of the external forces acting on the mass in the control volume.>> Since the control surface is fixed in space [see also Eqs. (1.68) and (7.14)]

$$\frac{dH_0}{dt} = \int_V \frac{\partial}{\partial t} (r \times \rho v) \, dV \ .$$
(7.27)

In stationary flow, the angular momentum vector is constant and its rate vanishes in Eq. (7.26).

It is sometimes more convenient to calculate the angular momentum with respect to the mass center at first, and subsequently, a transformation is applied for changing that point of reference to an arbitrarily selected intermediate (moving) point of reference A'. By definition it follows,

$$\mathbf{r} = \mathbf{r}_{MA} + \mathbf{r'} \quad , \quad \mathbf{v'} = \mathbf{v}_{MA} + \mathbf{v''} \quad , \quad \mathbf{v}_{MA} = \mathbf{v}_M - \mathbf{v}_A \quad \rightarrow \quad \mathbf{l}_A = m\, \mathbf{v}_{MA} \quad .$$

$$\mathbf{H}_A = \int_{V(t)} (\mathbf{r} \times \mathbf{v'})\, dm = \int_{V(t)} (\mathbf{r}_{MA} + \mathbf{r'}) \times (\mathbf{v}_{MA} + \mathbf{v''})\, dm = \mathbf{H}_M + \mathbf{r}_{MA} \times \mathbf{l}_A \quad .$$

$$\tag{7.28}$$

Restriction of that transformation to hold for a material volume $V(t)$ is a sufficient condition for its validity; see Fig. 7.4.

7.3. Applications of Control Volumes

To illustrate the transport of momentum and of its moment, the stationary flow of mass through the control surface is considered next.

7.3.1. Stationary Flow Through an Elbow

As indicated in Fig. 7.5 by the dashed line, the control surface consists, in a natural way, of the fluid structure interface and is completed by fictitious cross-sections A_1 and A_2 in the adjacent straight parts of the pipeline. If we consider averaged velocities, the mass flow rate to be constant requires

$$\dot{m} = \rho_1\, v_1\, A_1 = \rho_2\, v_2\, A_2 \quad .$$

$$\tag{7.29a}$$

Since the momentum vector $\mathbf{l}$ of the mass in the control volume does not depend on time in the stationary flow, Eq. (7.13) becomes independent of time. By substituting the nonvanishing mass flow rate densities

$$\mu_1 = \rho_1\, (\mathbf{v}_1 . \mathbf{n}_1) = -\rho_1\, v_1 \quad , \quad \mu_2 = \rho_2\, (\mathbf{v}_2 . \mathbf{n}_2) = \rho_2\, v_2 \quad ,$$

$$\tag{7.29b}$$

the surface integral is reduced to two simple cross-sectional area integrals and, by considering the uniform pressures at the inlet and outlet of the control surface, only the resultant of the surface tractions at the wall enters the sum of the external forces as the unknown (no body forces are assumed to be present)

$$\oint_{\partial V} \mu \mathbf{v} \, dS = \int_{A_1} \mu_1 \, \mathbf{v}_1 \, dA + \int_{A_2} \mu_2 \, \mathbf{v}_2 \, dA = \mathbf{R} = -p_1 A_1 \, \mathbf{n}_1 - p_2 A_2 \, \mathbf{n}_2 - \mathbf{F}_W \quad ,$$

$$-\mathbf{F}_W = \int_{\partial V_W} \boldsymbol{\sigma}_n \, dS \quad .$$

$$(7.29c)$$

The tractions $\boldsymbol{\sigma}_n$ act on the fluid surface, and hence, for technical reasons the resultant $\mathbf{F}_W$ of the reactions, ie of the pressure and shear applied at the wall ∂V_W , is considered. The shear stresses in the boundary layer at the wall are small when compared to the pressure $p = -\sigma_{nn}$, the normal stresses. Since elbows are rather short, the loss in pressure height is commonly negligible. Nevertheless, the effects of viscosity are included in the assigned pressure p_2 . Thus, the resulting force on the elbow, the dynamic reaction, is given by the sum of the two vectors

$$\mathbf{F}_W = -(p_1 + \rho_1 v_1{}^2) A_1 \, \mathbf{n}_1 - (p_2 + \rho_2 v_2{}^2) A_2 \, \mathbf{n}_2 \quad . \qquad (7.30)$$

It is the superposition of the pressure force and rate of momentum flux at the inlet and outlet of the control surface. The stationary rate of the flux of momentum apparently causes an increase in pressure by $\rho_k v_k{}^2$, $k = 1, 2$ in the quasistatic expression. In an incompressible flow, $\rho_k = \rho = const$. The force $\mathbf{F}_W$ acts as an external loading of the elbow. To avoid permanent stresses and deformations of the adjacent pipeline segments, elbows should be properly anchored. In addition, rapid changes of flow rates render further nonstationary loadings due to water hammer; see also Sec. 12.8.

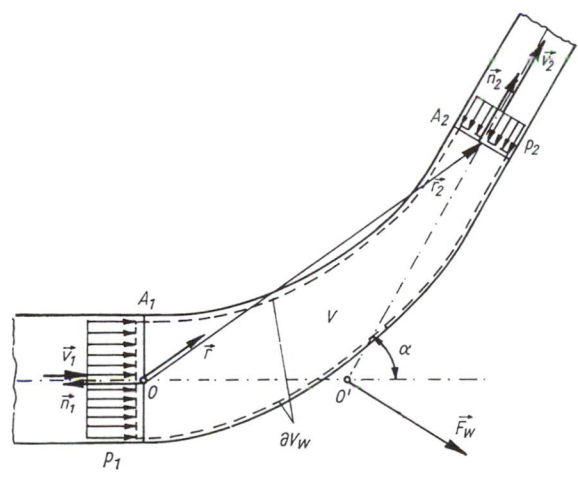

Fig. 7.5. Force acting on a plane elbow conveying a stationary flow of mass

The center of the inlet area 0 is chosen for reference of the moments in Eq. (7.26). The rate term of the angular momentum vanishes, and if we denote the reaction of the moment of the tractions on the fluid interface to the wall by

$$\mathbf{M_W} = -\int_{\partial V_W} (\mathbf{r} \times \boldsymbol{\sigma}_n) \, dS \ ,$$

(7.31a)

the net stationary rate of the flux of angular momentum equals (body forces are absent; see Fig. 7.5)

$$\oint_{\partial V} (\mathbf{r} \times \mu \mathbf{v}) \, dS = \int_{A_2} (\mathbf{r} \times \mu_2 \mathbf{v}_2) \, dA \ = \mathbf{M}_0 = \int_{A_2} \mathbf{r} \times (-p_2 \mathbf{n}_2) \, dA - \mathbf{M_W} \ .$$

(7.31b)

Thus, the external moment acting at 0 on the elbow, the dynamic reaction, becomes

$$\mathbf{M_W} = (-\mathbf{r}_2 \times \mathbf{n}_2) \, (p_2 + \rho_2 v_2^2) \, A_2 \ .$$

(7.32)

The total dynamic reaction on the elbow caused by changing the direction of the mass flow and possibly by accelerating the flow by changing the cross-sections from the inlet to the outlet is statically equivalent to the force $\mathbf{F_W}$ at point 0 with a couple $\mathbf{M_W}$. The subscript W stands for the tractions acting on the wall. The result is derived without knowing the details of the flow in the interior of the elbow.

(§)   The Plane Elbow. This elbow has already been determined by Eq. (7.30) since the lines of action of the two generalized forces are in the plane and have a common point 0′ where the resultant $\mathbf{F_W}$ is attached with the moment zero (see Fig. 7.5). The reduction of the force $\mathbf{F_W}$ with $\mathbf{M_W}$ from the origin 0 to 0′ can be performed by static considerations since $\mathbf{M_W}$ is orthogonal to $\mathbf{F_W}$. See Fig. 7.5 for the distance 00′. Two special configurations are mentioned for their practical importance.

(§)   A Nozzle with Straight Axis. Such a nozzle ($\alpha = 0$ in Fig. 7.5) is loading by an axial force, $\mathbf{e}_x$ points in the flow direction,

$$\mathbf{F_W} = \left[(p_1 + \rho_1 v_1^2) \, A_1 - (p_2 + \rho_2 v_2^2) \, A_2 \right] \mathbf{e}_x \ , \quad \mathbf{M_W} = 0 \ .$$

(7.33)

(§)   Plane U-Shaped Elbow. For this elbow $\alpha = 180°$, the diameter of the semicircle is a. A reduction of the forces to point 0, the center of the inlet, gives in that configuration

$$\mathbf{F_W} = -\left[(p_1 + \rho_1 v_1^2) \, A_1 + (p_2 + \rho_2 v_2^2) \, A_2 \right] \mathbf{n}_1 \ , \quad \mathbf{M_W} = (p_2 + \rho_2 v_2^2) \, a \, A_2 \ .$$ (7.34)

The point of application 0′ of the axial force without a couple is located in the plane between the parallel axes of the attached pipeline segments. If the cross-sections are equal, the force F_W passes through the center of the semicircle. The same force is excited by a jet that when hitting the bucket of a *Pelton* wheel (an impulse type of hydraulic turbine but held at rest), is divided by a splitter and is nearly reversed in its flow direction. The pressure in the impinging free jet after leaving the nozzle is enforced by the surrounding air pressure, as well as the pressure in the two emerging free jets. Hence, $p_1 = p_2 = p_0$, $\rho_1 = \rho_2 = \rho$, $A_1 = 2A_2/2$, $v_1 = v_2 = v$.

7.3.2. Thrust of a Propulsion Engine

Contrary to a rocket, the air-breathing propulsion engines take the oxygen for continuously burning the fuel from the surrounding air, and thus, a flux of momentum at the intake of the airmass also is to be considered. In addition, rockets can be propelled in vacuum. The control surface as indicated in Fig. 7.6 is moving in the flight direction; it is fixed to the engine. Assuming a complete expansion of the ejected (hot) gas jet to atmospheric pressure determines the axial component of the resultant of the external forces by $R = F\,n_2$ (containing the air resistance of the flying object, the drag, and a weight component in an inclined flight).
Equation (7.13), generalized by (7.14),

$$\frac{d\mathbf{l}}{dt} + \oint_{\partial V}\mu \mathbf{v}\,dS = \mathbf{R} \quad , \quad \begin{aligned}\mu_1 &= \rho_1\,(\mathbf{v}_1 - \mathbf{w}) \cdot \mathbf{n}_1 = -\rho_1 u_1 \quad , \\ \mu_2 &= \rho_2\,(\mathbf{v}_2 - \mathbf{w}) \cdot \mathbf{n}_2 = \rho_2 u_2 \quad , \end{aligned}$$

together with the momentum of the mass within the control volume, which is approximated by the simplifying assumption that the speed of its mass center equals the flight speed,

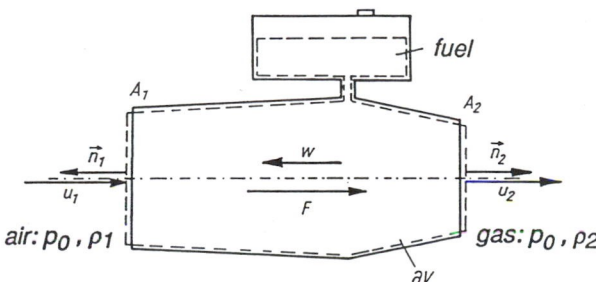

Fig. 7.6. Air-breathing propulsion engine or rocket. Control surface ∂V moves with the speed of flight w in the flight direction. Fuel is burned at a constant rate

$$I = \int_V \rho \mathbf{v}\, dV = m\, \mathbf{v_M} \approx m w\, \mathbf{n_1} \quad,$$

renders in the $\mathbf{n_2}$ direction (note that the control volume is accelerated)

$$-\frac{d(mw)}{dt} - \rho_1 u_1 (u_1 - w) A_1 + \rho_2 u_2 (-w + u_2) A_2 = F \quad. \tag{7.35}$$

u_1 and u_2 are the relative speeds of the jet at the intake of air and at the outlet of hot gas with the assigned cross-sections A_1 and A_2, respectively. Hence, the absolute velocities of the jets are to be substituted

$$\mathbf{v_1} = \mathbf{w} + \mathbf{u_1} = (u_1 - w)\,\mathbf{n_2} \quad, \quad \mathbf{v_2} = \mathbf{w} + \mathbf{u_2} = (-w + u_2)\,\mathbf{n_2} \quad.$$

A further substitution of the nonstationary rate of mass $dm/dt = \rho_1 u_1 A_1 - \rho_2 u_2 A_2$, and considering the $\mathbf{n_1}$ direction change Eq. (7.35) to the "flight equation of motion"

$$m(t)\frac{dw}{dt} = \rho_2 u_2^2 A_2 - \rho_1 u_1^2 A_1 - F \quad. \tag{7.36}$$

The generalized force $\rho_2 u_2{}^2 A_2$ acts in an accelerating manner and is called the *thrust* of the engine. Considering the air intake at A_1 and the flow of hot gas through A_2,

$$\dot{m}_1 = \rho_1 u_1 A_1 \quad, \quad \dot{m}_2 = \rho_2 u_2 A_2 \quad,$$

respectively, and the flight as stationary ($w = const$) yields the quasistatic condition

$$F = \dot{m}_2\, u_2 - \dot{m}_1\, u_1 \quad, \quad w = const \quad. \tag{7.37}$$

For a rocket, $u_1 \equiv 0$. In the case of a ramjet engine (air-breathing, but without a compressor), $u_1 = w$. The turbojet engine is also air-breathing, but the air passes through a compressor before reaching the combustion chamber, $u_1 > w$.

A rocket in the ignition phase before take-off is clamped to the ground in a vertical position

$$\dot{m}_2\, u_2 = F > m(t)g \quad. \tag{7.38}$$

The external force F is the sum of the instant weight and the reaction of the supporting tower.

Equation (7.36) is easily integrated for a vertically ascending flight after take-off in a homogeneous field of gravity by neglecting air resistance: The separation of the variables w and m and with $u_2 = u = const$ and $u_1 = 0$ yields

$$w(t) = u \ln \frac{m_L}{m(t)} - g\,t \quad , \quad \dot{m}_2 = \dot{m} = \frac{dm}{dt} = const \quad .$$

(7.39)

m_L is the mass at the instant of take-off. A second time integration of the logarithmic law yields the theoretical altitude time relation for constant thrust

$$h(t) = u\left[t - \frac{m(t)}{\dot{m}} \ln \frac{m_L}{m(t)}\right] - \frac{g\,t^2}{2} \quad .$$

(7.40)

The height at burn-off is given by substituting the mass consumption $m_B = m_L - m_0$, totally burned at the constant rate, and substituting the time elapsed since take-off

$$t_B = m_B/\dot{m} \quad .$$

The ideal instant vertical velocity (after a loss-less flight in the homogeneous gravity field in vacuum) of the remaining mass m_0 at burn-off is thus given by

$$w_B = u\left[\ln \frac{m_L}{m_0} + \frac{m_0}{m_L} - 1\right] \quad .$$

(7.41)

A high speed at burn-off requires a large ratio of the rocket masses m_L/m_0 : Given the payload, an optimum is derived by the design of a multistage rocket.

7.3.3. Euler's Turbine Equation

The wheel of a radial (reaction-type) turbine with a horizontal or, alternatively, a vertical axis, is spinning with $\Omega = const$. The stationary rate of flow of fluid mass from the interior is assumed incompressible and is guided properly by the blade arrangement at rest toward the inlet of the spinning system of blades. The cylindrical control surface is shown in Fig. 7.7 by dashed circles, and the moment Eq. (7.26) with 0 as the point of reference fixed in space gives (the rate of the angular momentum vector is zero)

$$\oint_{\partial V} (\mathbf{r} \times \mu\mathbf{v})\,dS = \mathbf{M}_0 = -\mathbf{M}_W^* \quad .$$

(7.42)

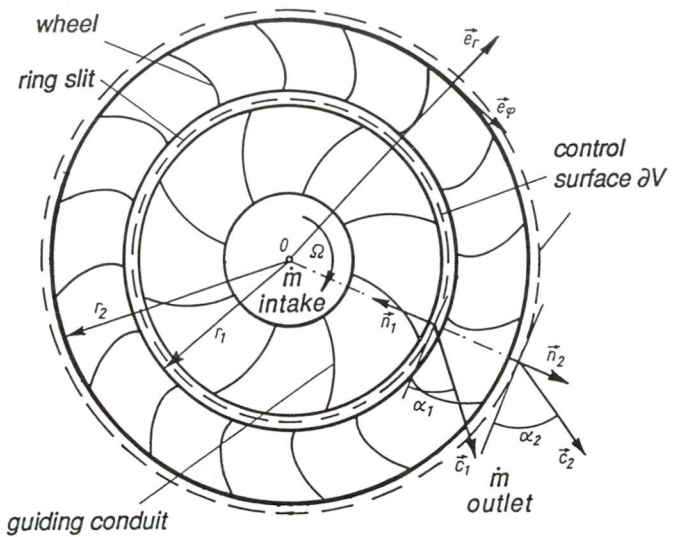

Fig. 7.7. Cross-section of a radial turbine with mass flow from the interior. Control surface fixed in space with bladesurface tractions smeared as fictitious body forces with the same reaction moment $-\boldsymbol{M_W}^*$

In fact, the spinning control surface must cutout every spinning blade to put the surface tractions into action as the true external forces of the mass flow with the reaction moment $-\boldsymbol{M}_W$ with respect to 0. By replacing these tractions on the blades by a system of fictitious external body forces acting on the mass that is instantly enclosed in the volume of the spinning wheel without blades and having the same moment about 0, $\boldsymbol{M}_W = \boldsymbol{M}_W^*$, the control surface of Eq. (7.42) may be assumed fixed in space; see also Eq. (8.32). Since $\boldsymbol{\Omega}$ is parallel $\boldsymbol{H}$, even the nonstationary rate is not influenced by this assumption. Also the mass flow rate μ is unchanged. For the lift force of a single blade in a parallel arrangement, see Sec. 13.2.

     A polar coordinate system is considered in the axisymmetric flow, and the absolute velocities of the fluid at the intake $\boldsymbol{c}_1$ and outlet $\boldsymbol{c}_2$ are projected. With a , the axial extension of the channel, the mass flow rates are

$$\mu_1 = -\dot{m}/A_1 \ , \ \ \mu_2 = \dot{m}/A_2 \ , \ \ A_k = 2\pi\, r_k a \ , \ \ k = 1, 2 \ .$$

The surface integral of Eq. (7.42) is easily evaluated to render

$$r_1\, \boldsymbol{e}_r \times \mu_1 \left(c_1 \sin \alpha_1\, \boldsymbol{e}_r + c_1 \cos \alpha_1\, \boldsymbol{e}_\varphi\right) A_1$$

$$+ r_2\, \mathbf{e}_r \times \mu_2 \left(c_2 \sin \alpha_2\, \mathbf{e}_r + c_2 \cos \alpha_2\, \mathbf{e}_\varphi\right) A_2 = -\, \mathbf{M_W^*} \; . \tag{7.43}$$

By putting $\mathbf{M_W^*} = M_W\, \mathbf{e}_z$, the axial moment excited by the stationary flow on the spinning wheel becomes (the tangential components of the absolute fluid velocities are abbreviated by $c_{kt} = c_k \cos \alpha_k$, $k = 1, 2$)

$$M_W = \dot{m} \left(r_1\, c_{1t} - r_2\, c_{2t}\right) \; . \tag{7.44}$$

The formula is the celebrated *Euler's turbine equation* . With Eq. (3.6), the theoretical power is given by the scalar product

$$P = \mathbf{M_W} \cdot \mathbf{\Omega} \; . \tag{7.45}$$

If you consider the hoop speeds of the inner and outer edges of the spinning blades, $u_k = \Omega\, r_k$, $k = 1, 2,$ the theoretical power of the radial turbine in a loss-free stationary state is thus given by the simple formula

$$P = \dot{m} \left(u_1\, c_{1t} - u_2\, c_{2t}\right) \; . \tag{7.46}$$

It should be maximized by a proper design of the blade configurations. See also the reaction wheel of *Segner* , Fig. 8.12.

7.3.4. Water Hammer in a Straight Pipeline

A stationary and incompressible flow of fluid with the speed $v_0 = const$ is considered in a straight (horizontal) pipeline of constant cross-sectional area A . At time $t = 0$, a control gate valve at a distance L from a large reservoir begins to shut down the flow. The mass flow thus becomes time-dependent, $\rho A v(t)$, $t \geq 0$. The momentum equation (7.13) in its nonstationary form may be applied to the straight cylindrical control volume indicated by the dashed control surface in Fig. 7.8

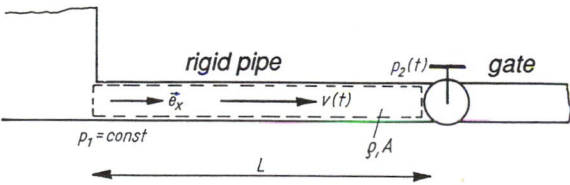

Fig. 7.8. Closing of a gate in a straight pipeline. The structure is assumed to be rigid and the flow incompressible. The control surface is indicated by a dashed line

$$\frac{d\mathbf{l}}{dt} + \int_{\partial V} \mu \mathbf{v} \, dS = \mathbf{R} \quad , \quad \mu_1 = -\rho v(t) \quad , \quad \mu_2 = \rho v(t) \quad , \quad A_1 = A_2 = A \ .$$

The net flux of momentum vanishes, and the rate of the time-dependent momentum $\mathbf{l} = \rho A L v(t) \, \mathbf{e}_x$ in the axial direction gives [$p_2(t)$ is the pressure at the closing valve and $p_1 = const$ is the pressure in the reservoir assigned at the inlet; viscous effects (shear stresses) at the wall are neglected]

$$\rho A L \frac{dv}{dt} = [p_1 - p_2(t)] A \ . \tag{7.47}$$

The given time control of the gate determines the nonstationary rate dv/dt , quite often it is assumed constant and the speed of the flow decreases then linearly with time. Hence, the generally time-dependent pressure at the gate becomes

$$p_2(t) = p_1 - \rho L \frac{dv(t)}{dt} \ . \tag{7.48}$$

During the time of closing the gate, $dv/dt < 0$, and hence, the pressure p_2 is increased over its stationary value p_1 , $p_2 > p_1$. Opening the gate initiates a flow from the state at rest, $dv/dt > 0$, and the pressure p_2 is decreased with respect to the pressure p_1 of the quiet fluid, $p_2 < p_1$. High rates of control render strong pressure fluctuations in the flow and, thus, give rise to high-stress fluctuations in the structure conveying that flow. Also the rapid opening of the gate is dangerous, since the low pressure may cause the fluid to evaporate, which initiates the phenomenon of *cavitation* (ie it gives rise to impact loadings and corrosion). The assumption of incompressibility, however, limits the applicability of the formula (7.48) to low to moderate control rates. For high rates and compressible flow, the wave propagation of the disturbances is to be considered; see Sec. 12.8. For further information, see
J. Parmakian: Waterhammer Analysis. Dover, New York, 1963 .
        Equation (8.35) should be applied in the case of curved pipes and elbows.

7.3.5. Carnot's Loss of Pressure Head

A stationary and incompressible flow through a straight pipe with an abruptly increasing cross-section is examined, and Eq. (7.13) is applied to the control volume indicated in Fig. 7.9. The rate of the momentum vector is zero. Downstream of the unsteady increase of the cross-sectional area, a nearly uniform distribution of the speed of flow v_2 can be reached only through the action of the viscous shear stresses. An inviscid jet would not change its initial cross-

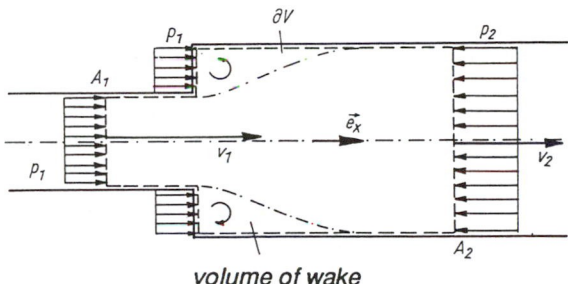

volume of wake

Fig. 7.9. Separation of a viscous flow at an abruptly increasing cross-section. *Carnot´s* pressure loss. The control surface is shown by a dashed line

section A_1 and its speed v_1 . Thus, analogous to a free jet, the mean pressure p_1 must enter also the wake of the flow, ie the dead end of the corner filled with the fluid , where A_1 jumps to $A_2 > A_1$. With that assumption and if we neglect the viscous shear stresses at the wall, the stationary flux of momentum becomes

$$\oint_{\partial V} \mu \mathbf{v}\, dS = (p_1 - p_2)\, A_2\, \mathbf{e}_x \ , \quad \mu_1 = -\rho v_1 \ , \quad \mu_2 = \rho v_2 \ , \quad \dot{m} = \rho A_1 v_1 = \rho A_2 v_2 \ .$$

The surface integral is easily evaluated and the difference in pressure of the viscous flow is given independently of the viscosity parameter by

$$p_2 - p_1 = \frac{\rho v_1^2}{2}\, (1 - A_1/A_2)\, \frac{2\, A_1}{A_2} \ , \quad A_1/A_2 < 1 \ . \tag{7.49}$$

The loss in pressure head with respect to an ideal flow is calculated in Sec. 8.5.9 and the dimensionless loss factor is given by Eq. (8.60). The pressure difference in the latter idealized case is much larger. Hence, *diffusors* (in subsonic flows) are designed with a smooth and only slightly increasing cross-sectional area for high efficiency. The length of such a diffusor, on the other hand, is limited, to keep the loss of pressure head due to the viscous shear at the wall sufficiently small.

The work of the frictional forces in the wake is converted into the internal energy of the fluid. That portion, which is converted into heat, is not fully recoverable in the form of the mechanical energy of the flow: A loss of potential energy of the internal forces (the pressure head) results. The first law of thermodynamics is derived in Sec. 8.6.

7.4.   Applications to Rigid-Body Dynamics

The rate equation of the angular momentum (7.20) is drastically simplified by the assumption of an undeformable body (with constant mass). The point of reference A' for the moment of momentum is chosen to be fixed to the rigid body and, consequently, the relative velocity of any mass element is linearly distributed according to Eq. (1.4). If we consider the angular velocity ω and the material vector of constant length, $|r| = const$, the expression

$$\mathbf{v}' = \boldsymbol{\omega} \times \mathbf{r} \ , \tag{7.50}$$

is substituted and the volume integral is expanded by means of the formula of the vector triple product

$$\mathbf{H}_A = \int_m (\mathbf{r} \times \mathbf{v}') \, dm = \int_m \mathbf{r} \times (\boldsymbol{\omega} \times \mathbf{r}) \, dm = \int_m [r^2 \boldsymbol{\omega} - (\mathbf{r} \cdot \boldsymbol{\omega}) \mathbf{r}] \, dm \ . \tag{7.51}$$

In a cartesian frame of origin A', the expressions $r^2 = x^2 + y^2 + z^2$ and $(\mathbf{r} \cdot \boldsymbol{\omega}) = x \omega_x + y \omega_y + z \omega_z$. They are substituted to delineate the integrand in the form of the three vector components $[(r^2 - x^2) \omega_x - xy \omega_y - xz \omega_z] \mathbf{e}_x$, $[(r^2 - y^2) \omega_y - yz \omega_z - yx \omega_x] \mathbf{e}_y$, $[(r^2 - z^2) \omega_z - zx \omega_x - zy \omega_y] \mathbf{e}_z$. Integration over the mass distribution at a fixed time t factors the constant angular velocity and renders through the definition of axial mass moments of inertia (see Sec. 2.4)

$$I_x = \int_m (y^2 + z^2) \, dm \quad , \quad I_y = \int_m (z^2 + x^2) \, dm \quad , \quad I_z = \int_m (x^2 + y^2) \, dm \quad , \tag{7.52}$$

and the centrifugal moments of the rigid mass distribution [see Eq. (2.117)]

$$I_{xy} = \int_m x \, y \, dm \quad , \quad I_{yz} = \int_m y \, z \, dm \quad , \quad I_{zx} = \int_m z \, x \, dm \ . \tag{7.53}$$

They are the elements of the 3×3 symmetric tensor of the moments of inertia of the rigid body which are evaluated at a material point of reference A'

$$I = \begin{pmatrix} I_x & -I_{xy} & -I_{xz} \\ -I_{yx} & I_y & -I_{yz} \\ -I_{zx} & -I_{zy} & I_z \end{pmatrix} \ . \tag{7.54}$$

The tensor becomes time-invariant if the coordinate frame is fixed to the rigid body. The angular momentum is the matrix product, a linear vector transformation ($\boldsymbol{\omega}$ is a column matrix)

$$H_A = I \cdot \omega \ . \tag{7.55}$$

Choosing the principal axes of inertia at A' renders [Eq. (7.54) is in diagonal form]

$$H_A = I_1 \, \omega_1 \, e_1 + I_2 \, \omega_2 \, e_2 + I_3 \, \omega_3 \, e_3 \ , \tag{7.56}$$

where $e_k(t)$, $k = 1, 2, 3$ are mutually orthogonal material unit vectors fixed to the rigid body in motion. Their directional time dependence must be taken into account in the rate of the angular momentum. In many applications, there is an axis of symmetry and the inertia-moment-tensor does not change if the rigid body is allowed to rotate about this axis. Hence, the rate of the angular momentum is calculated under the more general assumption that the frame rotates with an assigned angular speed Ω, whereas the rigid body still rotates with ω. Formal differentiation yields with Eq. (1.6) and substituting Ω

$$\frac{dH_A}{dt} = \dot{H}_x \, e_x + \dot{H}_y \, e_y + \dot{H}_z \, e_z + H_x \, \dot{e}_x + H_y \, \dot{e}_y + H_z \, \dot{e}_z \ , \quad \dot{e}_i = \Omega \times e_i \ , \tag{7.57}$$
$$i = x, y, z \ .$$

By denoting the rate of the angular momentum relative to the rotating frame by

$$\frac{d'H_A}{dt} = \sum_i \dot{H}_i \, e_i \ , \tag{7.58}$$

by superposition, the total rate is the sum of the two vectors

$$\frac{dH_A}{dt} = \frac{d'H_A}{dt} + \Omega \times H_A \ . \tag{7.59}$$

Equation (7.59) holds for the rate of any vector if an intermediate rotating frame is used for observational purpose: Especially the nonstationary rate of the momentum and moment of momentum vectors in Eqs. (7.14) and (7.25) is easily calculated if the control volume $V = const$ is selected to move as a rigid body, ie if Eq. (1.4) holds for the velocity field w .

When we consider the mass center M, which is fixed in a rigid body, for reference, Eq. (7.21) takes on the vector form

$$\frac{d'H_M}{dt} + \Omega \times H_M = M_M \ . \tag{7.60}$$

If the rigid body is in a pure rotational state of motion about a point fixed in space, $A' = 0$, the point is also fixed in the body; it is the ideal point of reference. Considering the frame with the origin M or 0 fixed to the body, $\Omega \equiv \omega$, and selecting the principal axes of inertia render the three nonlinear differential equations with constant coefficients, known *Euler's gyroscopic equations*

$$I_1 \dot{\omega}_1 - (I_2 - I_3) \omega_2 \omega_3 = M_1 \quad ,$$

$$I_2 \dot{\omega}_2 - (I_3 - I_1) \omega_3 \omega_1 = M_2 \quad ,$$

$$I_3 \dot{\omega}_3 - (I_1 - I_2) \omega_1 \omega_2 = M_3 \quad . \tag{7.61}$$

If one of the principal axes is also an axis of symmetry, say, number 1, the principal moment of inertia $I_1 = const$, $I_2 = I_3 = I$, even if the rigid body is allowed to rotate about this axis with the angular velocity σ . The latter is called the *spin* , measured with respect to the intermediate reference system. Hence, $\omega_1 = \Omega_1 + \sigma$, $\omega_k = \Omega_k$, $k = 2, 3$. Equation (7.60) takes on the component form

$$I_1 \dot{\omega}_1 = M_1 \quad , \quad I \dot{\omega}_2 + [(I_1 - I) \omega_1 + I \sigma] \omega_3 = M_2 \quad ,$$

$$I \dot{\omega}_3 - [(I_1 - I) \omega_1 + I \sigma] \omega_2 = M_3 \quad . \tag{7.62}$$

Simple illustrative examples indicate the way, how these equations are to be applied to solve actual problems.

7.4.1. The Rolling Rigid Wheel

A wheel spinning with the angular velocity ω_0 is at time $t = 0$ put into a sliding contact with a rigid plane. The contact forces with components N and T and the weight mg , shown in Fig. 7.10, are applied subsequently. The nonstationary motion is subject to *Coulomb's* law of dry friction

$$T_R = -\mu N \, sgn \, (v') \quad , \tag{7.63}$$

where μ is the coefficient of dry friction, N the normal contact force, and v' the relative velocity of sliding, the argument of the signum function. At time t , momentum and angular momentum about the mass center have built up (see Fig. 7.10)

$$I = m \dot{x}_S \, e_x \quad , \quad H_S = I_S \omega \, e_y \quad , \quad I_S = m i^2 \quad .$$

Equations (7.9) and (7.21) in vector and component form are applied to this plane motion

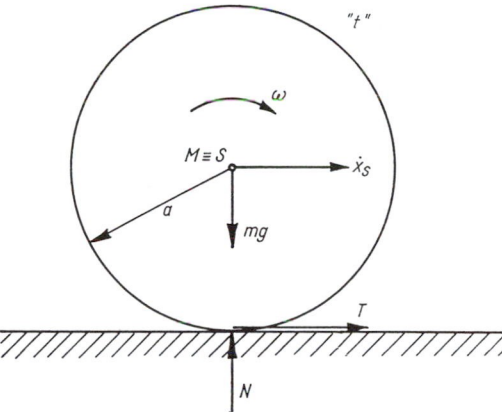

Fig. 7.10. Nonstationary rolling of a wheel with slip

$$\frac{d\mathbf{I}}{dt} = m\ddot{x}_S\,\mathbf{e}_x = \mathbf{R} \; , \; \frac{d\mathbf{H}_S}{dt} = I_S\dot{\omega}\,\mathbf{e}_y = -aT\,\mathbf{e}_y \; , \; m\ddot{x}_S = T \; , \; 0 = (-mg + N) \; .$$

Any resisting couple of "rolling friction" that is due to the finite extension of the contact zone is neglected in this phase of motion. With $N = mg$, the remaining differential equations are easily integrated. During the slip phase of the wheel, Eq. (7.63) holds. $T = T_R = \mu mg = const$ is the accelerating external force component in the time interval $0 \le t \le t_0$, and

$$\dot{x}_S = \mu g\,t + C_1 \; , \quad x_S = \mu g\,\frac{t^2}{2} + C_1 t + C_2 \; , \quad \omega = -\mu g\,\frac{a\,t}{i^2} + C_3 \; .$$

The constants of integration are determined by the initial conditions at $t = 0 : x_S = dx_S/dt = 0$, $\omega = \omega_0$, to be $C_1 = C_2 = 0$, $C_3 = \omega_0$. The condition of pure rolling contact without slip is reached the first time at t_0 if the contacting point becomes the velocity pole (its speed v' is zero)

$$\dot{x}_S = a\omega : \quad \mu g\,t_0 = -\mu g\left(\frac{a}{i}\right)^2 t_0 + a\omega_0 \; , \; t_0 = \frac{a\omega_0}{\mu g\left(1 + a^2/i^2\right)} \; . \tag{7.64}$$

The force T no longer equals T_R , and for $t \ge t_0$ no slip occurs. Pure rolling contact requires $dx_S/dt = a\,\omega$, and if we still neglect the couple of rolling friction, renders

$$m\ddot{x}_S = T \text{ and } I_S\dot{\omega} = -aT \rightarrow m\ddot{x}_S = -\left(\frac{a}{i}\right)^2 T :$$

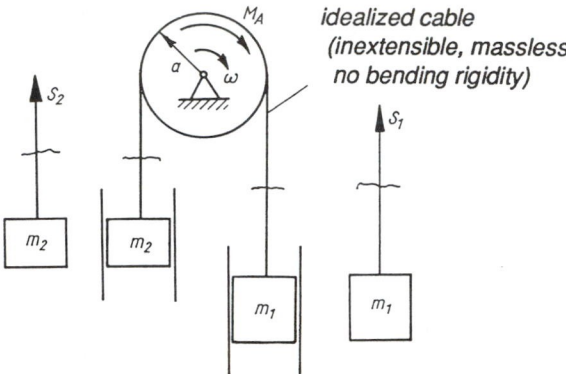

Fig. 7.11. Cable drive. Three rigid bodies in guided motion with a single degree of freedom. The mass-less cable is assumed to be inextensible. M_A is a given external couple

Only the trivial solution of these incompatible equations is possible, $T \equiv 0$, and, hence, in stationary motion, $T < \mu N$ is the no-slip condition

$$\ddot{x}_S \equiv 0 \quad \text{and} \quad \dot{x}_S \equiv \dot{x}_S(t = t_0) \quad .$$

7.4.2. Cable Drive

The cable pulley is driven by an external couple of forces M_A with the angular speed $\omega(t)$. The masses m_1 and m_2 are in vertical translational motion and attached to the cable that is idealized to be inextensible and massless (Fig. 7.11). All losses due to friction and slip in the contact zone of the cable to the wheel are neglected. The cable tensions are denoted S_k , $k = 1, 2$. (Note that there is an unavoidable slip of the strained cable when it moves over the pulley from a state of strain due to S_2 to that of S_1 which is different.) Since the cable is assumed to be in perfect adhesive contact to avoid any excessive wear, the system of three rigid bodies has only one degree of freedom. Each body is considered in a free-body diagram

$$m_1 \left(a\dot{\omega} \right) = m_1 g - S_1 \quad , \quad \dot{x}_1 = \dot{x}_2 = a\,\omega \quad , \quad m_2 \left(a\dot{\omega} \right) = -m_2 g + S_2 \quad ,$$

$$I\,\dot{\omega} = a\left(S_1 - S_2\right) + M_A \quad , \quad I = m\,i^2 \quad .$$

The elimination of the cable tensions gives the equation of motion (the angular acceleration as)

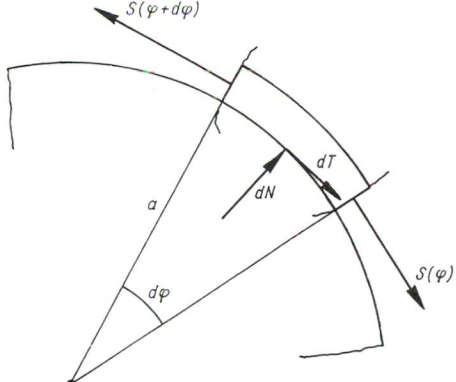

Fig. 7.12. Mass-less cable element in no-slip (adhesive) contact. Balance of forces
(inertia neglected)

$$\dot{\omega} = \frac{a\,(m_1 - m_2)\,g + M_A}{(m_1 + m_2)\,a^2 + m\,i^2} \ . \tag{7.65}$$

The extreme of the external couple M_A is to be determined in the
limit of the adhesive contact of the cable that is characterized by
the initiation of an overall slip: *Coulomb's* law of dry friction
applies in this limit. When we consider Fig. 7.12, the balance of the
cable tensions and contact forces at the rim gives the quasistatic
relations (φ is the polar angle and the inertia of the cable element
is neglected)

$$\frac{dS}{d\varphi}\,d\varphi - dT = 0 \ , \quad dT \le dT_R \ , \quad dN - S\,d\varphi = 0 \ . \tag{7.66}$$

In the limit of adhesive contact, $dT = dT_R = \mu\,dN$, and the system is
statically determinate. The elimination of dN yields in the limit $d\varphi$
$\rightarrow 0$ *Euler's* equation of cable friction

$$\frac{dS}{d\varphi} - \mu S = 0 \ \rightarrow \ \frac{dS}{S} = \mu\,d\varphi \ . \tag{7.67}$$

The separation of the variables and integration, if we take into
account the boundary condition $S(\varphi = 0) = S_1$, render the practically
important exponential growth of the cable tension with the
circumferential angle

$$S(\varphi) = S_1\,e^{\mu\varphi} \ . \tag{7.68}$$

With $\alpha = \pi$ in Fig. 7.11, the inequality applies

$$S_2 \leq S_1 \, e^{\mu\pi} \quad . \tag{7.69}$$

Equality applies in the limit of the adhesive contact. With Eq. (7.65), the upper limit of the external couple is found. Since the direction of slip may be reversed, $\mu \to -\mu$, a lower bound also results. The external moment is thus bounded from above and below by (equality gives the directional extreme)

$$g \, \frac{\left(m_1 e^{-\mu\pi} - m_2\right) m i^2 + 2 \left(e^{-\mu\pi} - 1\right) m_1 m_2 a^2}{\left(m_1 e^{-\mu\pi} + m_2\right) a} \leq$$

$$M_A \leq g \, \frac{\left(m_1 e^{\mu\pi} - m_2\right) m i^2 + 2 \left(e^{\mu\pi} - 1\right) m_1 m_2 a^2}{\left(m_1 e^{\mu\pi} + m_2\right) a} \quad . \tag{7.70}$$

7.4.3. Dynamics of the Crushing Roller (Fig. 1.3)

A heavy rigid wheel of radius r and with the moment of inertia $I_1 = m i^2$, is rolling on a circular path of radius R as a symmetric gyroscope in guided and stationary motion. The reference system of origin 0, a point fixed in space and to the rigid body, rotates about the vertical principal axis 2 such that the spin of the wheel about the 1 axis is σ. Pure rolling with no slip gives the geometric condition for the angular speeds, $\sigma = (R/r) \, v$. Equation (7.62) is well suited to be applied with $\omega_1 = \sigma$, $\Omega_1 = 0$, $\omega_2 = \Omega_2 = v$, $\omega_3 = \Omega_3 = 0$

$$M_1 = M_2 = 0 \quad , \quad M_3 = -m i^2 \, \sigma v = -m i^2 \, \frac{R}{r} \, v^2 \quad . \tag{7.71}$$

Weight and contact force N contribute to the moment $M_3 = (mg - N) R$, and the dynamically magnified contact force becomes

$$N = mg \left(1 + \frac{i^2}{r \, g} v^2\right) \quad . \tag{7.72}$$

Any couple of the rolling resistance must be balanced by a driving moment in the condition $M_1 = 0$. Similarly, the contact forces of the wheels of a turning vehicle on a circular path are dynamically changed (Fig. 1.4 may serve as a guide) but note the inertia of the vehicle body that must be taken into account.

7.4.4. Swing Crane with a Boom

According to Fig. 7.13, the dynamics of the following stationary motion is considered: The boom is lowered in a controlled motion with constant angular speed ω_0 and the tower, with the boom

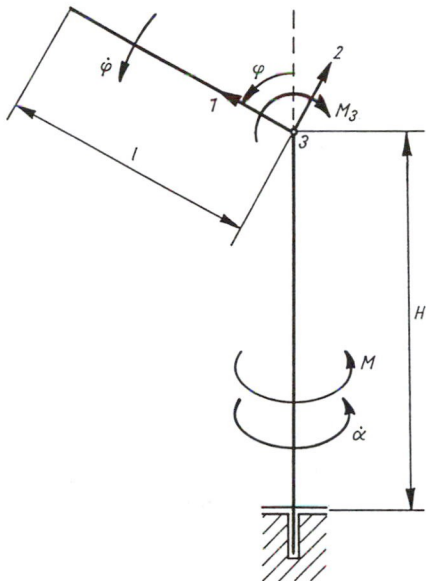

Fig. 7.13. Swing crane with a boom in stationary rotational motion. Overturning moment
$M_2 \cos \varphi$

properly attached, is independently rotated by another motor also with constant angular speed Ω_0. The reference frame with the origin fixed in space and at the joint is fixed to the boom and identified by the principal axes of inertia 1, 2, 3.

In Fig. 7.13, the angular speeds of the boom are given by

$$\omega_1 = \Omega_0 \cos \varphi \ , \quad \omega_2 = \Omega_0 \sin \varphi \ , \quad \omega_3 = -\omega_0 \ , \quad \omega_0 = \dot{\varphi} \ , \quad \Omega_0 = \dot{\alpha} \ . \quad (7.73)$$

Booms are slender cantilevers, $I_1 \approx 0$, $I_2 = I_3 = I_0 = m i_0{}^2$, and *Euler's* equations (7.61) render the external moment

$$M_1 \approx 0 \ , \quad M_2 = (\dot{\omega}_2 - \omega_1 \omega_3) \, m i_0{}^2 \ , \quad M_3 = (\dot{\omega}_3 + \omega_1 \omega_2) \, m i_0{}^2 \ . \quad (7.74)$$

The weight and motor of the boom contribute the moment

$$M_3 = (\dot{\alpha}^2 \sin 2\varphi) \, \frac{m i_0{}^2}{2} \ , \quad (7.75)$$

and the drive of the rotational motion of the whole system requires a couple about the vertical axis

$$M = M_2 \sin \varphi = (\dot{\alpha} \, \dot{\varphi} \sin 2\varphi) \, m i_0{}^2 \ . \quad (7.76)$$

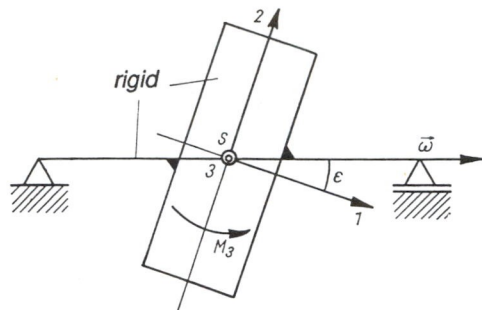

Fig. 7.14. Dynamically unbalanced rigid rotor

The out-of-plane component $M_2 \cos \alpha$ gives a couple that stresses the joint of the boom and that is a bending moment of the crane tower, as well as a possibly unexpected overturning moment.

7.4.5. Balancing of Rotors

Any eccentricity of the mass center S from the axis of rotation is assumed to be already removed by the procedure of static balancing, ie by properly removing or adding mass. But according to Fig. 7.14, the principal axis of inertia 1 is misaligned with respect to the axis of rotation of the rigid rotor as shown. To keep the supports free of dynamic forces requires dynamic balancing with the output $\varepsilon = 0$. Otherwise, the moment M_3 of Fig 7.14 is nonzero and causes a couple of reaction forces in the bearings that are generally not tolerable. *Euler´s* equations (7.61) render with the components of the angular velocity $\omega_1 = \omega \cos \varepsilon$, $\omega_2 = \omega \sin \varepsilon$, and $\omega_3 = 0$

$$M_1 = M_2 = 0 \quad , \quad M_3 = \frac{m}{2} \left(i_2^2 - i_1^2 \right) \omega^2 \sin 2\varepsilon \; , \quad I_k = m \, i_k^2 \; . \tag{7.77}$$

The plane of action of the couple M_3 rotates with n, the given rotations per minute, *rpm.*, where $\omega = \pi n/30$. See, *K. Federn: Auswuchttechnik. Springer-Verlag, Berlin-New York, 1977, (in German)* for comprehensive information on all aspects of the balancing of rotors.

7.4.6. The Gyro-Compass

The basic idea of a gyro-compass goes back to *Foucault* (1852). The spin axis 1 of a cylindrical gyroscope of high revolutionary speed (the spin is denoted σ) is fixed to the local horizontal plane on

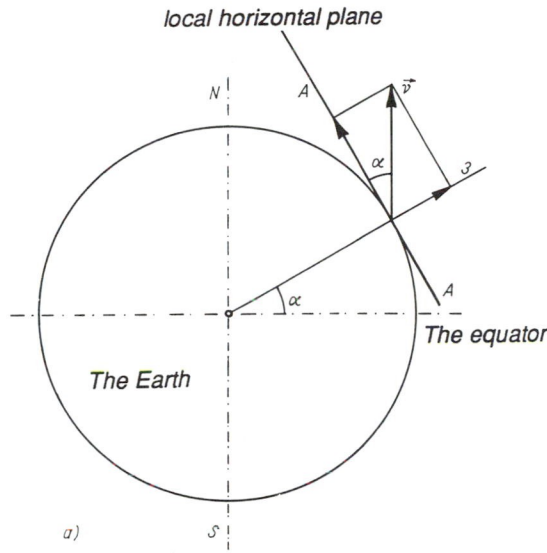

Fig. 7.15(a). Gyro-compass at latitude α

Earth, such that it performs a rotational motion about the vertical axis 3. The latter passes through the center of mass.

Figure 7.15(a) depicts the geographic situation at latitude α with the angular speed of the (noninertial) Earth about the N-S axis denoted v. The deviation $\varphi(t)$ of the spin axis 1 from the N-S pointing tangent of the meridian is shown in Fig. 7.15(b). The angular velocity of the intermediate reference system is selected so that it has these components in the directions of the principal axes: $\Omega_1 = v \cos \alpha \cos \varphi$, $\Omega_2 = -v \cos \alpha \sin \varphi$, $\Omega_3 = d\varphi/dt + v \sin \alpha$. Equations (7.62) are applied with the mass center as the point of reference to render, with $M_1 = M_3 = 0$,

$$I_1 \dot{\omega}_1 = 0, \quad \omega_1 = const, \quad I_2 = I_3 = I, \quad I \dot{\omega}_2 + [I_1\omega_1 - I(\omega_1 - \sigma)] \omega_3 = M_2,$$

$$I \dot{\omega}_3 - [I_1\omega_1 - I(\omega_1 - \sigma)] \omega_2 = 0. \tag{7.78}$$

The couple M_2 produces reaction forces in the bearings. The third equation contains the compass effect. With the assumptions $v \ll \sigma$ and $\alpha = const$, the plane motion of the axis of symmetry is approximated by the solution of the "pendulum" equation with nonlinear restoring moment [see Eq. (7.115)]

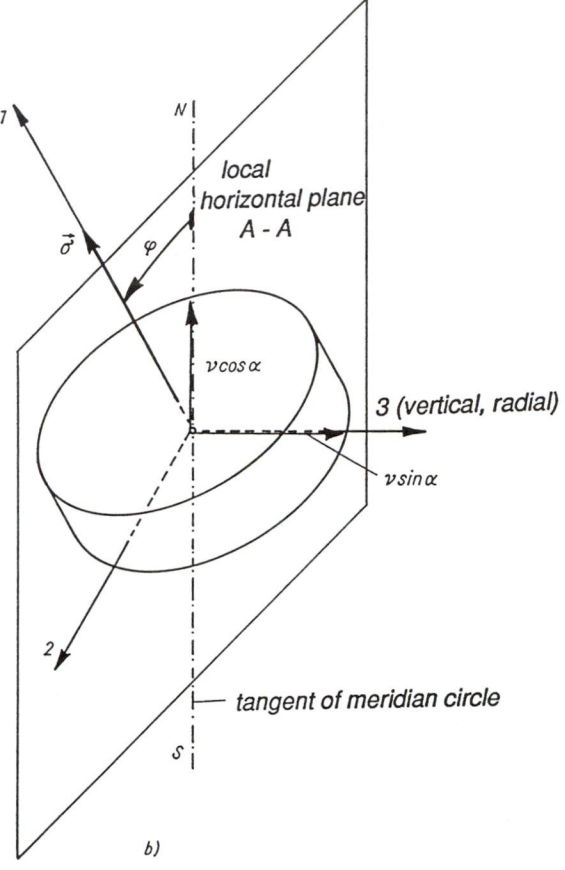

Fig. 7.15(b). The spin axis of symmetry 1 moves in the local horizontal plane

$$\ddot{\varphi} + p^2 \sin \varphi = 0 \quad , \quad p^2 = \left(\frac{i_1}{i}\right)^2 \sigma v \cos \alpha \quad , \quad I_1 = m i_1^2 \quad , \quad I = m i^2 \; . \tag{7.79}$$

At the Poles, the compass effect vanishes, *cos* $\alpha = 0$. The solution, subject to any initial disturbance, determines the undamped natural vibration about $\varphi = 0$, which is in the desired N-S direction. With damping present, the spin axis returns to the N-S direction after a while. This phenomenon is called <<the tendency toward equally oriented parallelism of the angular velocity vectors σ and v.>>

7.4.7. The Linear Oscillator

Simple, linear elastic dynamic systems with a single degree of freedom are symbolically modeled by the translational linear

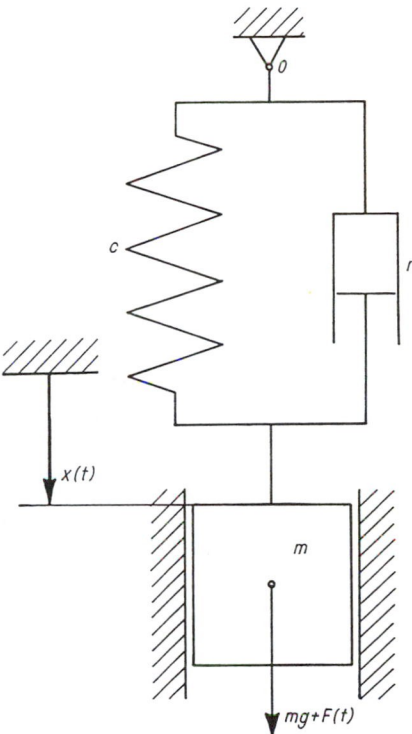

Fig. 7.16. Forced spring-mass-dashpot system, symbolic model of the linear oscillator. $x(t)$, vertical coordinate measured from the unstressed configuration. $\xi(t)$, Eq. (7.81), is a translated coordinate measured from the state of equilibrium in any direction of motion in space.

oscillator shown in Fig. 7.16. A *Hookean* spring of stiffness c supplies the linear restoring force to a mass m . Light material damping is mathematically approximated by a linear viscous damper [see Eq. (4.70)] in a parallel arrangement. The combination is called the (mass-less) *Kelvin-Voigt* body; see Sec. 4.2.2(§). A hysteretic type of material damping that is closer to reality can be considered by assuming the stiffness of the spring to be complex, $c^* = c (1 + i \varepsilon)$, $\varepsilon \ll 1$, $i = \sqrt{(-1)}$. The external force of excitation is assumed to be given, say, by $F(t)$. If we consider the coordinate $x(t)$ of the (vertical) motion to be measured from the undeformed state, *Newton's* law renders with the external forces acting in the free-body diagram of the mass

$$m \, \ddot{x} = m \, g + F(t) - c \, x - r \, \dot{x} \; . \tag{7.80a}$$

All parameters are assumed constant; the system is thus time-invariant. Dividing by m yields the normal form of the second-order time differential equation of the linear oscillator. It is inhomogeneous due to the forcing $F(t)$, and since it was assumed that the weight of the mass suddenly loads the untensioned spring,

$$\ddot{x} + 2\zeta\omega_0\,\dot{x} + \omega_0^2\,x = g + \frac{F(t)}{m} \quad , \quad \zeta = \frac{r}{2\sqrt{mc}} \quad , \quad \omega_0 = \sqrt{\frac{c}{m}} = \sqrt{\frac{g}{a_s}} \quad , \quad a_s = \frac{mg}{c} \; .$$
$$\text{(7.80b)}$$

In horizontal motion, the constant forcing by g is absent. A simple coordinate translation always removes any constant force: Superposition applies. Hence, putting $\xi = x - x_0$ transforms Eq. (7.80b) to

$$\ddot{\xi} + 2\zeta\omega_0\,\dot{\xi} + \omega_0^2\,\xi = \frac{F(t)}{m} = \omega_0^2\,\xi_e(t) \quad , \quad \xi_e = \frac{F(t)}{c} \quad , \quad x_0 = a_s = \frac{mg}{c} \; . \quad \text{(7.81)}$$

The coordinate $\xi(t)$ is to be measured from the state of equilibrium of the mass m under its own weight. The vibration of the mass is not influenced by the homogeneous parallel gravity field. The system is determined by two parameters: the damping coefficient $\zeta \ll 1$ representing the light material damping and the circular natural frequency ω_0. The numerical values of the latter are tabulated in Table 7.1 in relation to the static elongation of the spring when stressed by a force equal to the weight of the mass. That static deformation, a system property, does not depend on the actual orientation of the oscillator in space. The general solution of the linear equation (7.81) is given by the superposition of a particular solution of the inhomogeneous equation to the general solution of the homogeneous differential equation (when forcing is absent)

$$\xi(t) = \xi_h(t) + \xi_p(t) \; . \quad\quad\quad\quad\quad\quad (7.82)$$

The homogeneous equation describes the free vibration, ie the motion after a disturbance of the state of equilibrium,

$$\ddot{\xi}_h + 2\zeta\omega_0\,\dot{\xi}_h + \omega_0^2\,\xi_h = 0 \; , \quad\quad\quad\quad\quad (7.83a)$$

Table 7.1. Natural frequency in relation to the static deformation a_s of the spring loaded by a force $F_s = mg$ in the direction of motion. Circular frequency ω_0, linear frequency f, and natural period T.

$a_s = mg/c$	cm	0.01	0.05	0.1	0.5	1.0	10	20
$\omega_0 = \sqrt{g/a_s}$	rad/s	313.2	140.1	99.05	44.29	31.32	9.90	7.00
$f = \omega_0/2\pi$	Hz	49.85	22.29	15.76	7.05	4.98	1.58	1.11
$T = f^{-1}$	s	0.020	0.045	0.063	0.142	0.201	0.63	0.90

with the general complex solution (note the characteristic equation and its conjugate complex roots)

$$\alpha^2 + 2\zeta\omega_0\,\alpha + \omega_0{}^2 = 0 \;\rightarrow\; \alpha_{1,2} = \omega_0\!\left(-\zeta \pm i\,\sqrt{1-\zeta^2}\right),$$

$$\xi_h = C_1\,e^{\alpha_1 t} + C_2\,e^{\alpha_2 t}\;,\;\zeta < 1\;,\;i = \sqrt{-1}\;. \tag{7.83b}$$

The root of double multiplicity, $\alpha = -\zeta\omega_0$, refers to the special limiting case of an aperiodic motion where the mass creeps back into the state of equilibrium, possibly with a single zero-crossing. The remaining exponential function in that case is to be multiplied by a linear time function (a polynomial of degree "multiplicity minus 1")

$$\xi_h{}^{(a)} = (C_1 + C_2\,t)\,e^{-\omega_0 t}\;,\;\zeta = 1\;. \tag{7.84}$$

Aperiodic motions exist for $\zeta > 1$ that are not discussed any further. The critical value of the damping coefficient, $\zeta = 1$, separates the vibrational free motion from the creeping one. In the case of modeling the material damping $\zeta \le 20\%$, and the approximation $\sqrt{(1-\zeta^2)} \approx 1$ holds with sufficient accuracy. Factoring out the exponential damping in Eq. (7.83b) and applying the cartesian representation of the unimodular complex numbers,

$$e^{\pm i\varphi} = \cos\varphi \pm i\,\sin\varphi\;,$$

render exactly for $\zeta < 1$

$$\xi_h(t) = (C_1\cos\omega_D t + C_2\sin\omega_D t)\,e^{-\zeta\omega_0 t}\;,\;\omega_D = \omega_0\,\sqrt{1-\zeta^2} \approx \omega_0\,,\zeta \le 0.2\;. \tag{7.85}$$

Changing the integration constants to the amplitude, $a = \sqrt{(C_1{}^2 + C_2{}^2)}$ and to the phase angle, ε , to be related by $cos\;\varepsilon = C_1/a$ and $sin\;\varepsilon = C_2/a$, and, furthermore, applying the addition theorem of trigonometric functions yield the free damped vibration in a form that exhibits the time-dependent, exponentially decaying amplitude

$$\xi_h(t) = a\,e^{-\zeta\omega_0 t}\cos(\omega_D t - \varepsilon)\;. \tag{7.86}$$

The logarithmic decrement δ is defined by the natural logarithm of the ratio of any two successive maxima of the displacement and thus is given by $\delta = 2\pi\zeta / \sqrt{(1-\zeta^2)}$. If the amplitude diminishes to 1/2 of its initial value after n cycles, the factor of light damping is determined by the approximate relation $\zeta \approx 0.110/n$. Equidistant zero-crossings still have the period $T = 2\pi/\omega_D$ that, for light damping, is approximately the natural period of the undamped system. Equation (7.86) for $\zeta = 0$ becomes the time-harmonic motion of the undamped oscillator with the constant amplitude a and the natural frequency $f = \omega_0 / 2\pi$ measured in *Hertz* .

(§)   Periodic Forcing Function , $F(t) = F(t + T_e)$. Periodic functions can be expanded in *Fourier* time series. Excitation is then considered termwise, and finally, superposition is applied. One term of the forcing function is, eg the time harmonic with an assigned circular excitation frequency v , $F(t) = F_0 \cos vt$. A special particular solution of the inhomogeneous Eq. (7.81) for this special forcing must be time-harmonic. The constants in

$$\xi_p(t) = A \cos vt + B \sin vt \quad , \tag{7.87}$$

are determined by comparing coefficients in Eq. (7.81). A linear set of two equations results

$$\begin{pmatrix} \omega_0^2 - v^2 & 2\zeta\omega_0 v \\ -2\zeta\omega_0 v & \omega_0^2 - v^2 \end{pmatrix} \begin{pmatrix} A \\ B \end{pmatrix} = \begin{pmatrix} F_0/m \\ 0 \end{pmatrix} . \tag{7.88}$$

Since the determinant of the coefficient matrix

$$\Delta = \left(\omega_0^2 - v^2 \right)^2 + 4\zeta^2\omega_0^2 v^2 \neq 0 \ \rightarrow \ \zeta \neq 0 \quad , \tag{7.89}$$

the particular solution of the damped system is determined for the whole spectrum of the forcing frequencies by

$$\begin{pmatrix} A \\ B \end{pmatrix} = \begin{pmatrix} \omega_0^2 - v^2 \\ 2\zeta\omega_0 v \end{pmatrix} \frac{F_0}{m \Delta} \quad . \tag{7.90}$$

The amplitude of the particular solution may be determined analogously to the homogeneous one by

$$a_p = \sqrt{A^2 + B^2} = \frac{F_0}{m \sqrt{\Delta}} \quad , \tag{7.91a}$$

and the phase angle φ is related to A and B by

$$\cos \varphi = A/a_p = \left(\omega_0^2 - v^2 \right) / \sqrt{\Delta} \quad , \quad \sin \varphi = B/a_p = 2\zeta\omega_0 v / \sqrt{\Delta} \ . \tag{7.91b}$$

The addition theorem applied to Eq. (7.87) thus renders the forced time-harmonic vibration term by

$$\xi_p(t) = a_p \cos \left(vt - \varphi \right) \ . \tag{7.92}$$

Compare to the natural vibration of the undamped system, Eq. (7.86), with $\zeta = 0$. The total solution, subject to incorporate any initial conditions, becomes by the superposition of Eqs. (7.86) and (7.92) (*a* and ε are the constants of integration)

$$\xi(t) = = a\, e^{-\zeta\omega_0 t}\cos{(\omega_0 t - \varepsilon)} + a_p \cos{(\nu t - \varphi)} \ . \tag{7.93}$$

The continuously forced and damped oscillator has a *fading memory* with respect to the initial conditions prescribed at $t = 0$ due to the exponentially decaying amplitude of the free vibration term. After a sufficiently long time, the vibration becomes stationary and, consequently, is approximated by the forced vibration term only. If we relate the amplitude a_p of the steady-state solution to the static deformation $a_0 = F_0/c$, the dynamic magnification factor of the harmonic displacements is determined by the dimensionless function of the dimensionless forcing frequency and the damping parameter

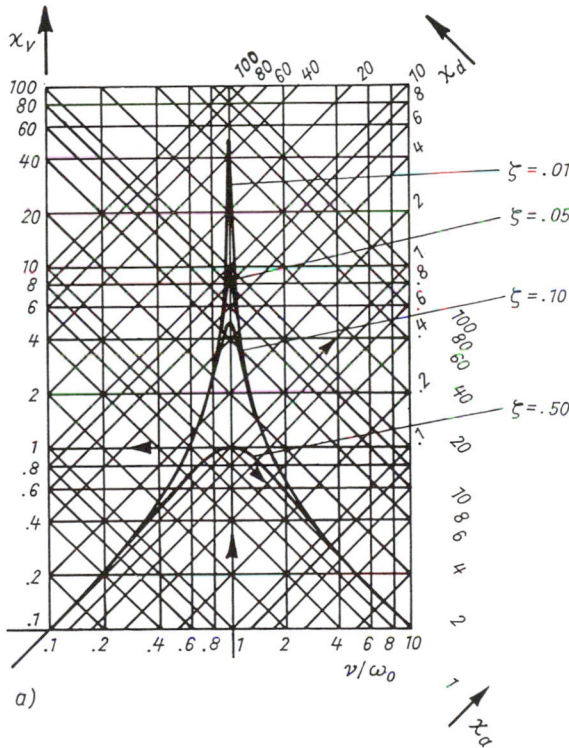

a)

Fig. 7.17(a). *Blake's* diagram. Amplitude frequency response functions of displacement, velocity, and acceleration

$$\chi_d(v, \zeta) = a_p/a_0 = 1 / \sqrt{\left[1 - (v/\omega_0)^2\right]^2 + 4\,\zeta^2 (v/\omega_0)^2} \ . \qquad (7.94)$$

By sweeping the forcing frequency slowly, the vibrational response is altered to be quasistationary. The displacement magnification factor takes on its maximal value at the critical forcing frequency v_c , which depends only on and is limited by the damping coefficient

$$\max \chi_d(\zeta) = \frac{1}{2\zeta \sqrt{1 - \zeta^2}} \approx \frac{1}{2\zeta} \rightarrow \zeta \le 0.2 \,, \ v_c = \omega_0 \sqrt{1 - \zeta^2} \approx \omega_0.$$

$$(7.95)$$

The resonance of a slightly damped system appears close to the

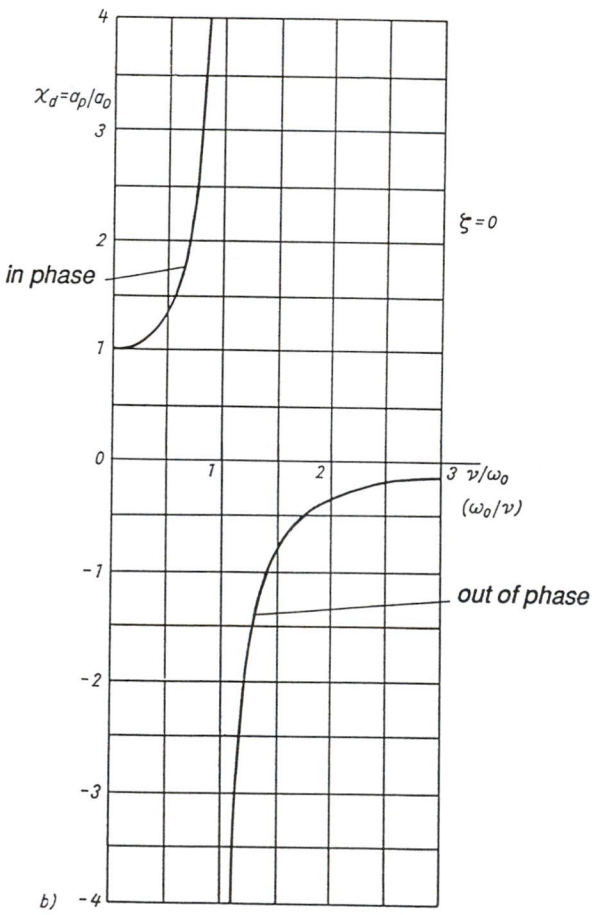

Fig. 7.17(b). Resonance curve of the undamped linear oscillator

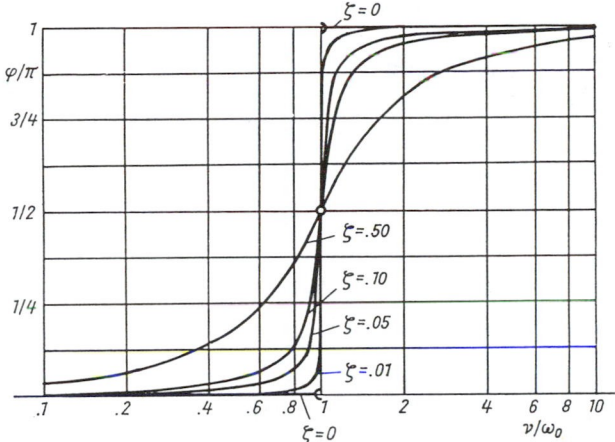

Fig. 7.18. Phase frequency response function of the steady-state vibrations of viscously damped linear oscillators. Note the phase jump of the undamped response

natural frequency of the undamped oscillator. Of equal importance are the resonance curves of the velocity, $\chi_v (v, \zeta) = v \chi_d (v, \zeta)$ and of the acceleration, $\chi_a (v, \zeta) = v \chi_v (v, \zeta) = v^2 \chi_d (v, \zeta)$. The resonance curves are represented parametrically with respect to the damping coefficient, in a fourfold logarithmic diagram named after *Blake* [see Fig. 7.17(a)]. Note the dimensionless forcing frequency in the logarithmic scale of the abscissa. Other authors use the forcing period instead. Putting $\zeta = 0$ in Eq. (7.94) renders the response curve of the undamped oscillator which is singular at the natural frequency [see Fig. 7.17(b)]. In that case, the stationary motion is in phase with the harmonic force in the subcritical range of forcing frequencies, $v < \omega_0$, like the static loading of the spring, but it is in counterphase in the supercritical range, $v > \omega_0$. The latter is a solely dynamic phenomenon.

The phase shift of the steady-state motion of the damped oscillator with respect to the driving force changes continuously with the forcing frequency, Eq. (7.91b). The frequency response function of the phase angle $\varphi(v, \zeta)$ in fractions of π is shown in Fig. 7.18. It increases monotonically from the static value zero (at zero frequency) to $\pi/2$ at "resonance," where the forcing frequency equals the natural frequency of the undamped system, for all admissible values of the damping coefficient, and reaches asymptotically the value π of the counterphase. The phase jump of the undamped steady-state vibration is also shown. The force transmitted to the support in steady-state motion has an amplitude of $F_T(v, \zeta) = \chi_d F_0 \sqrt{[1 + 4 \zeta^2 (v/\omega_0)^2]}$. Vibration isolation is thus achieved for $v/\omega_0 > \sqrt{2}$, since the ratio $F_T/F_0 < 1$. The high efficiency of isolation requires very low values of viscous damping.

Amplitude and phase frequency responses may be combined with the complex frequency response function of the linear oscillator. The transfer function is defined by the ratio of the complex amplitude of the output A_0 to that of the input. The excitation is of complex form

$$F(t)/m = A_e \, e^{ivt} \; . \tag{7.96}$$

If we put the stationary response equal to

$$\xi_p = A_0 \, e^{ivt} \; , \tag{7.97}$$

the complex frequency response function results

$$H(v, \zeta) = \frac{A_0}{A_e} = \frac{1}{\omega_0^2 - v^2 + 2i\,\zeta\omega_0 v} = \frac{1}{\Delta}\left(\omega_0^2 - v^2 - 2i\,\zeta\omega_0 v\right) \equiv \frac{1}{\omega_0^2}\,\chi_d\,e^{-i\varphi} \; . \tag{7.98}$$

Mass times the reciprocal of the transfer function is the *mechanical impedance* $Z(v)$. In the complex plane, H is plotted as a function of the forcing frequency and, thereby, combines the resonance curves of amplitude and phase into a single curve.

Given a periodic excitation by its complex *Fourier* series,

$$F(t)/m = F(t + T_e)/m = \sum_{n=-\infty}^{\infty} c_n \, e^{in\omega t} \; , \quad \omega = 2\pi/T_e \; , \tag{7.99}$$

the stationary response is still given by superposition and, hence, by the (infinite) complex series

$$\xi_p(t) = \sum_{n=-\infty}^{\infty} c_n \, H_n(\zeta) \, e^{in\omega t} \; , \quad H_n(\zeta) \equiv H(v = n\omega, \zeta) \; . \tag{7.100}$$

In practice, the forcing is considered band-limited, and only a finite number of terms is examined in the above series representations.

(§) Excitation by a Nonperiodic Forcing Function. This kind of excitation that, contrary to the discrete spectrum of the periodic force (7.99), has a continuous spectrum according to the *Fourier* integral.

$$c(\omega) = \frac{1}{m} \int\limits_{t\,=\,-\infty}^{\infty} F(t)\, e^{-i\omega t}\, dt \ ,$$

(7.101)

renders the response by the weighted integration in the frequency domain, which is the continuous form of the superposition principle,

$$\xi_p(t) = \frac{1}{2\pi} \int\limits_{\omega\,=\,-\infty}^{\infty} H(\omega,\, \zeta)\, c(\omega)\, e^{i\omega t}\, d\omega \ .$$

(7.102)

Numerical evaluations of these integrals are efficiently performed by means of the computer routines of the *fast Fourier transformation (FFT)* that formally go back to the (finite) series of Eq. (7.100) in combination with a binary representation. See, eg *R. W. Clough and J. Penzien: Dynamics of Structures. McGraw-Hill, New York, 1975, p. 114.*

In the time domain, however, the general solution for $t > 0$ is given by *Duhamel's convolution integral* , if we assume homogeneous initial conditions (the oscillator is at rest at $t = 0$; the transient phase of the vibration is included)

$$\xi(t) = \frac{1}{m\omega_D} \int\limits_0^t F(\tau)\, e^{-\zeta\omega_0(t\,-\,\tau)} \sin\left[\omega_D(t - \tau)\right] d\tau \ .$$

(7.103)

The application of an addition theorem to the trigonometric function leads to the representation

$$\xi(t) = A(t) \sin \omega_D t - B(t) \cos \omega_D t \ .$$

(7.104)

The time-dependent amplitudes are given by the numerically stable integrals

$$A(t) = \frac{1}{m\omega_D} \int\limits_0^t F(\tau)\, \frac{e^{\zeta\omega_0\tau}}{e^{\zeta\omega_0 t}} \cos \omega_D\tau \ d\tau \ , \quad B(t) = \frac{1}{m\omega_D} \int\limits_0^t F(\tau)\, \frac{e^{\zeta\omega_0\tau}}{e^{\zeta\omega_0 t}} \sin \omega_D\tau \ d\tau \ .$$

(7.105)

With respect to the oscillatory weighting function in the integrand, the time step in a numerical integration scheme must be selected so it is shorter than $T/10$.

(§) Representation of the Motion in the Phase Plane $(\xi,\, d\xi/dt)$. Such a representation is illustrative for the linear oscillator, and the field of isoklinic lines is the basis of an efficient numerical integration scheme that is subject to generalization to nonlinear problems of vibrations. The phase curves have constant slope when crossing an isoklinic line. Equation (7.81) is transformed into a system of two first-order differential equations by the trivial substitution

$$\xi = \xi_1 \ , \quad \dot{\xi} = \dot{\xi}_1 = \xi_2 \quad . \tag{7.106}$$

Displacement and velocity are considered independent variables

$$\dot{\xi}_1 = \xi_2 \ ,$$
$$\dot{\xi}_2 = -\omega_0^2 \, \xi_1 - 2\zeta\omega_0 \, \xi_2 + F(t)/m \quad . \tag{7.107}$$

The state vector, a column matrix, of the single degree of freedom (SDOF) vibrational system is the radius vector of an image point that moves according to the time-mechanical state of motion in the phase plane along the phase curve. The row vector is transposed

$$\xi^T = (\xi_1, \xi_2) \quad . \tag{7.108}$$

Equation (7.107), by means of the definition (7.108), becomes the linear vector transformation according to the rules of matrix multiplication, plus the given forcing vector b ,

$$\dot{\xi} = A\xi + b \ , \quad b^T = (0, F(t)/m) \quad . \tag{7.109}$$

The system matrix is given by inspection

$$A = \begin{pmatrix} 0 & 1 \\ -\omega_0^2 & -2\zeta\omega_0 \end{pmatrix} \quad . \tag{7.110}$$

Many computer routines of systems analysis are based on such state equations.

Using matrix functional analysis generalizes the simple solutions of the single first-order differential equation to the solution of the state equations. The homogeneous part, the free motion in the phase plane, is formally given by the matrix exponential function

$$\xi_h = \xi_0 \, e^{At} \quad . \tag{7.111a}$$

In general, for all types of system matrices, the convergent infinite series exists that represents the solution

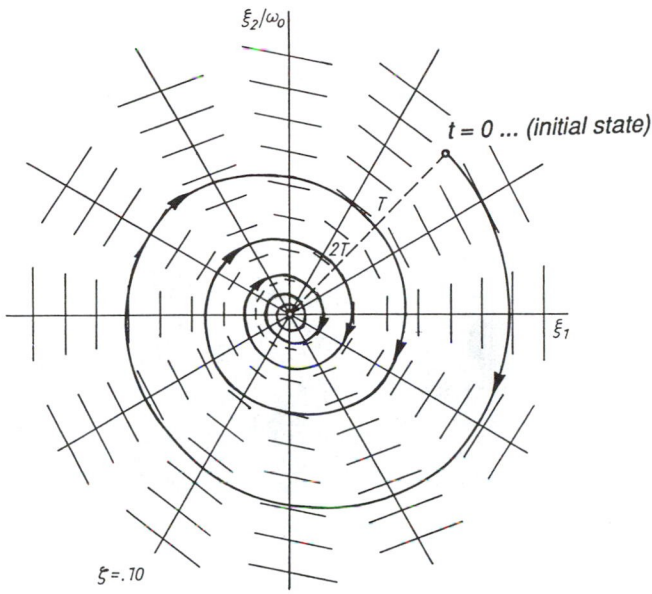

Fig. 7.19. Phase curve of the free and damped vibrational motion and the field of isoklines in the phase plane

$$e^{At} = \sum_{k=0}^{\infty} \frac{(At)^k}{k!} \quad .$$

(7.111b)

But time can be eliminated by the ratio of the homogeneous components of Eq. (7.107)

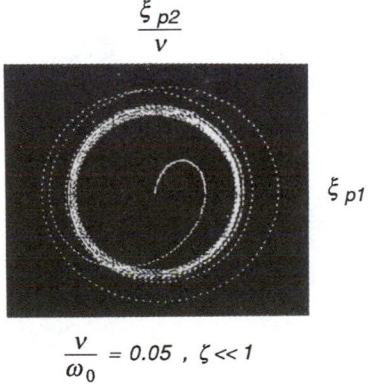

$$\frac{v}{\omega_0} = 0.05 \ , \ \zeta \ll 1$$

Fig. 7.20. Limit cycle of the steady state motion. Transient vibration starting at rest

$$\left(\frac{\dot{\xi}_2}{\dot{\xi}_1}\right)_h = \left(\frac{d\xi_2}{d\xi_1}\right)_h = -\omega_0\left[2\zeta + \omega_0\left(\frac{\xi_1}{\xi_2}\right)_h\right] \quad .$$

(7.112a)

Figure 7.19 shows the field of isoklinic lines of the damped free motion, which are the semiinfinite radial lines of constant slope,

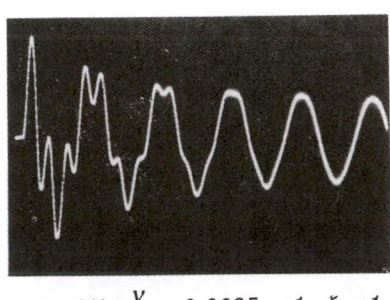

a) $\xi(t)$: $\dfrac{v}{\omega_0} = 0.0625 \ll 1$, $\zeta \ll 1$

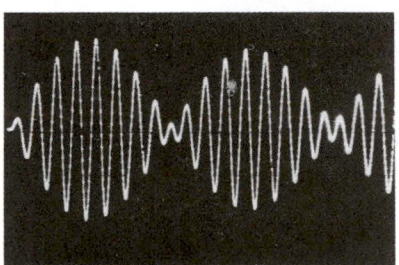

b) $\xi(t)$: $\dfrac{v}{\omega_0} = 0.8125 \approx 1$ (beat)

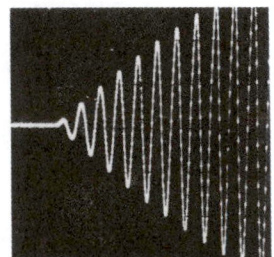

c) $\xi(t)$: $\dfrac{v}{\omega_0} = 1$, $\zeta \approx 0$

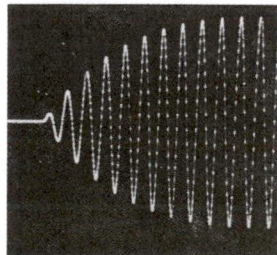

d) $\xi(t)$: $\dfrac{v}{\omega_0} = 1$, $\zeta \ll 1$

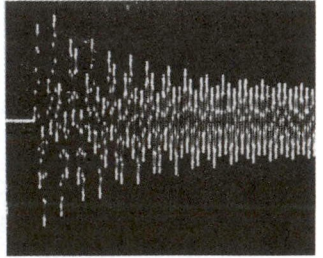

e) $\xi(t)$: $\dfrac{v}{\omega_0} = 12.5 \gg 1$, $\zeta \ll 1$

Fig. 7.21. Transient vibrations under harmonic excitation with forcing frequency v assigned. Simulations by means of an analog computer

$$\left(\frac{d\xi_2}{d\xi_1}\right)_h = -\omega_0\left[2\zeta + \omega_0\left(\frac{\xi_1}{\xi_2}\right)_h\right] = \text{const} \ .$$

$$(7.112b)$$

Given the initial conditions, the motion of the image point is easily drawn and renders the logarithmic spiral line of Fig. 7.19, thus indicating the diminishing motion of the damped system. Note the orthogonal crossings of the abscissa.

The steady-state motion of the harmonically forced oscillator is given by Eq. (7.92). Time differentiation renders the velocity and, hence, the *limit cycle* , determined by the equation of an ellipse in the phase plane

$$\dot{\xi}_p = -a_p v \sin(vt - \varphi) \ \rightarrow \ \xi_{p1}^2 + \left(\xi_{p2}/v\right)^2 = a_p^2 \ .$$

$$(7.113)$$

Time is apparently eliminated and Fig. 7.20 shows the limit circle reached asymptotically after starting the oscillator at rest: The ordinate is changed to velocity over the forcing frequency. The image point passes one cycle of the phase curve in the time $T_e = 2\pi/v$.

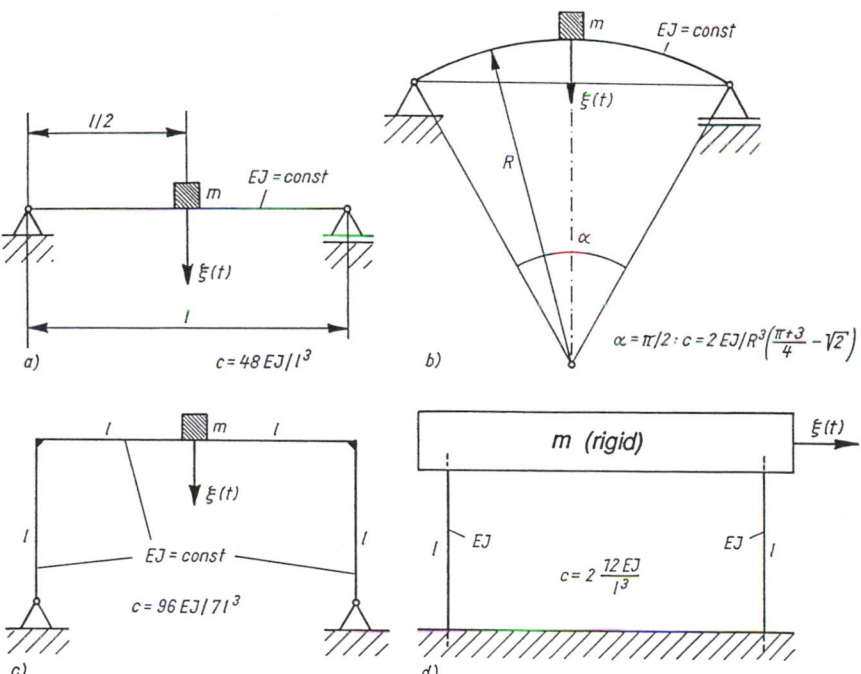

Fig. 7.22. Structural lumped mass models of the linear oscillator. Stiffness of the restoring spring *c*

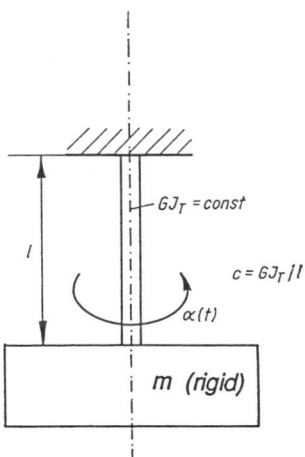

Fig. 7.23. Torsional vibrations. Angular coordinate $\alpha(t)$. Stiffness of the restoring torsional spring c

Also the natural, but undamped motion gives such a circle, but the time of passing one cycle changes to the natural period $T = 2\pi/\omega_0$. Its radius is determined by initial conditions. In the case of a more general periodic excitation, the limit cycle is no longer an ellipse but remains a closed line in the phase plane.

Figure 7.21 samples some time functions of the transient motion phase, Eq. (7.93), of a harmonically forced oscillator when starting at rest. Note the beat phenomenon of (b) where the forcing frequency is close to the resonance frequency and (c) and (d) that depict resonance excitation. These photos were taken from the screen of an analog computer simulation at the Technical University of Vienna. These experiments can be repeated on a personal computer with a graphics display.

(§)   Some Structural Models of the Linear Oscillator. These are illustrated in Fig. 7.22 where lumped masses are assumed to be "known" a priori: These systems are of the "heavy concentrated mass soft spring" type. The stiffness of the spring in the model (Fig. 7.16) is related to the static structural and distributed rigidity. Models of that type are extensively discussed in Sec. 11.2.

(§)   Linear Torsional Vibrations. Such vibrations are illustrated by the system shown in Fig. 7.23. The free and undamped torsional motion of the rigid disk of mass m and with the given radius of gyration i_S is the solution of one component of Eq. (7.21) (the inertia of the linear elastic shaft is assumed to be negligible)

$$m i_S^2 \, \ddot{\alpha} = -c\alpha \;\; , \;\; \ddot{\alpha} + \omega_0^2 \alpha = 0 \;\; , \;\; i_S = \sqrt{I_S/m} \;\; , \;\; \omega_0 = \sqrt{c/m i_S^2} \;\; . \quad (7.114)$$

Angular coordinates are changed to $i_S \, \alpha$ in coupled vibrations.

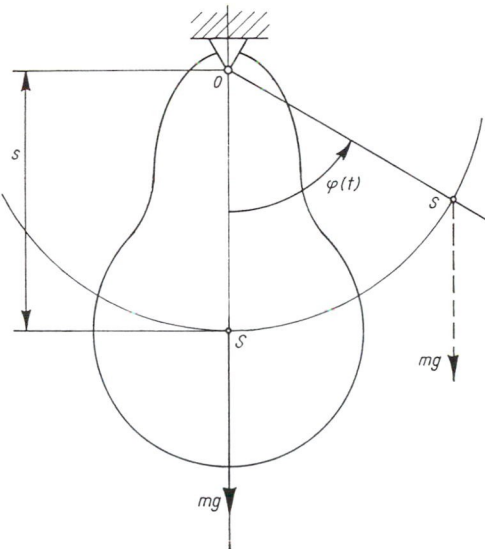

Fig. 7.24. Plane vibration or rotation of a pendulum in a homogeneous and parallel gravity field

7.4.8. Nonlinear Vibrations

A general theory of mechanical systems with nonlinear restoring forces and nonlinear (eg hysteretic) damping cannot be given here, but reference can be given to a vast specialized literature on this important subject. Illustrations of the nonlinear effects are given below exemplarily by consideration of the free motion of the planar pendulum in a homogeneous parallel field of gravity, and by taking into account the frictional damping of an oscillator. The dependence of the natural period on the amplitude of free motion is encountered in the first case. Approximate solutions are discussed further in Chap. 11 including the *Duffing* oscillator.

(§) Motion of a Planar Pendulum. A rigid body when supported by an axle that allows rotational motion in the plane represents a physical planar pendulum. At rest, the center of gravity S is located vertically below the simple support 0 at a given distance s ; see Fig. 7.24. Any disturbance of the state of equilibrium renders the restoring moment of the gravity force about the point 0 fixed in space. Equation (7.22) applies and renders, if damping is neglected, the nonlinear equation of free motion $\varphi(t)$

$$m i_0^2 \, \ddot{\varphi} = - mgs \sin \varphi \;\; \rightarrow \;\; \ddot{\varphi} + \omega_0^2 \sin \varphi = 0 \;\; , \;\; i_0^2 = i_s^2 + s^2 \;\; , \;\; \omega_0 = \sqrt{gs/i_0^2} \;\; .$$

$$(7.115)$$

Comparison with a mathematical pendulum, where a concentrated mass *m* is assumed to be guided by a mass-less rigid rod of (constant) length *l* , renders the same period. To make the linear natural frequency ω_0 equal suffices when $l = i_0^2/s$.

Assuming small amplitude oscillations, $\alpha \ll 1$, allows the linearization of the restoring force. A linear *Taylor* approximation of the trigonometric function at $\varphi_0 = 0$, the equilibrium position, renders the linearized equation of motion [compare this with Eq. (7.114)]

$$\ddot{\varphi} + \omega_0^2 \, \varphi = 0 \;\; .$$

$$(7.116)$$

The approximating solution is time-harmonic, $\varphi(t) = a \cos (\omega_0 t - \varepsilon)$, and the linear period is constant, $T = 2\pi/\omega_0$. If we consider the initial conditions at $t = 0$ to be $\varphi = \alpha \ll 1$, $d\varphi/dt = 0$, the free motion is simply given by $\varphi(t) = \alpha \cos \omega_0 t$.

The first integral of the nonlinear equation (7.115) is derived by integrating the equation, premultiplied by the angular velocity, and noting the identity

$$\ddot{\varphi} \, \dot{\varphi} = \frac{d}{dt} \left(\frac{\dot{\varphi}^2}{2} \right) \;\; \rightarrow \;\; \frac{\dot{\varphi}^2}{2} + \omega_0^2 \int \sin \varphi \, d\varphi = C \;\; .$$

$$(7.117)$$

Taking into account the special initial conditions yields the nonlinear first-order equation

$$\dot{\varphi}^2 = 2\omega_0^2 \left(\cos \varphi - \cos \alpha \right) \;\; .$$

$$(7.118)$$

A second integration can be performed after a simplifying transformation to half-angles by considering the relation $1 - \cos \varphi = 2 \sin^2 \varphi/2$. With the new variable $\xi = (1/k) \sin \varphi/2$, $k = \sin \alpha/2$, the first-order differential equation takes on the integrable form

$$\dot{\xi}^2 = \omega_0^2 \left(1 - \xi^2 \right) \left[1 - \left(k\xi \right)^2 \right] \;\; , \;\; 0 \le k \le 1 \;\; .$$

$$(7.119)$$

The separation of variables gives

$$\frac{d\xi}{\sqrt{\left(1 - \xi^2 \right) \left[1 - \left(k\xi \right)^2 \right]}} = \omega_0 \, dt \;\; ,$$

$$(7.120)$$

and integrating both sides renders (the elliptic integral of the first kind and its inverse function are tabulated and even stored in computer libraries)

$$\int_0^{\xi} \frac{d\eta}{\sqrt{(1 - \eta^2)[1 - (k\eta)^2]}} = \omega_0 t + D \ .$$

(7.121)

The inverse is called the *Jacobi* elliptical sinus-amplitudinis function

$$\xi = sn (\omega_0 t + D) \ .$$

(7.122)

Characteristic is its periodicity with the period of *4K* ; *sn K = 0* and *ξ = 1* correspond to a quarter-period

$$K(k = \sin \alpha/2) = \int_0^1 \frac{d\eta}{\sqrt{(1 - \eta^2)[1 - (k\eta)^2]}} \ , \quad T = 4K/\omega_0 \ .$$

(7.123)

The limit of small amplitudes α or $k \to 0$ corresponds to $K \to \pi/2$ and $T \to 2\pi/\omega_0$ and renders the constant parameters of the linearized solution. The nonlinear free motion is characterized by the dependence of the period on the amplitude, $T(\alpha)$. A limit cycle exists in the phase plane for $\alpha < \pi$, indicating vibrational motion. The solution for $\alpha = \pi$ plus an initial angular velocity renders rotations; the phase curve is no longer closed but is extended to infinity. A separatrix, the phase curve for $\alpha = \pi$, is still given by Eq. (7.118) and separates the two kinds of motion in the phase plane.

The support reactions are found from the momentum equation (7.7). The acceleration of the mass center is derived by taking the second time derivatives of the coordinates $x_S = s \sin \varphi$ and $z_S = (-s \cos \varphi)$

$$\ddot{x}_S = s\ddot{\varphi} \cos \varphi - s\dot{\varphi}^2 \sin \varphi = - s\omega_0^2 (3 \cos \varphi - 2 \cos \alpha) \sin \varphi \ ,$$
$$\ddot{z}_S = s\ddot{\varphi} \sin \varphi + s\dot{\varphi}^2 \cos \varphi = - s\omega_0^2 + s\omega_0^2 (3 \cos \varphi - 2 \cos \alpha) \cos \varphi \ .$$

(7.124)

The horizontal and vertical reactions, acting on the body at 0, both are given as functions of the rotational angle $\varphi (t)$ (time is eliminated)

$$H(\varphi) = m \ddot{x}_S = m \left(s\ddot{\varphi} \cos \varphi - s\dot{\varphi}^2 \sin \varphi\right) = - ms\omega_0^2 (3 \cos \varphi - 2 \cos \alpha) \sin \varphi \ ,$$

$$V(\varphi) = m (g + \ddot{z}_S) = m (g - s\omega_0^2) + ms\omega_0^2 (3 \cos \varphi - 2 \cos \alpha) \cos \varphi \ .$$

(7.125)

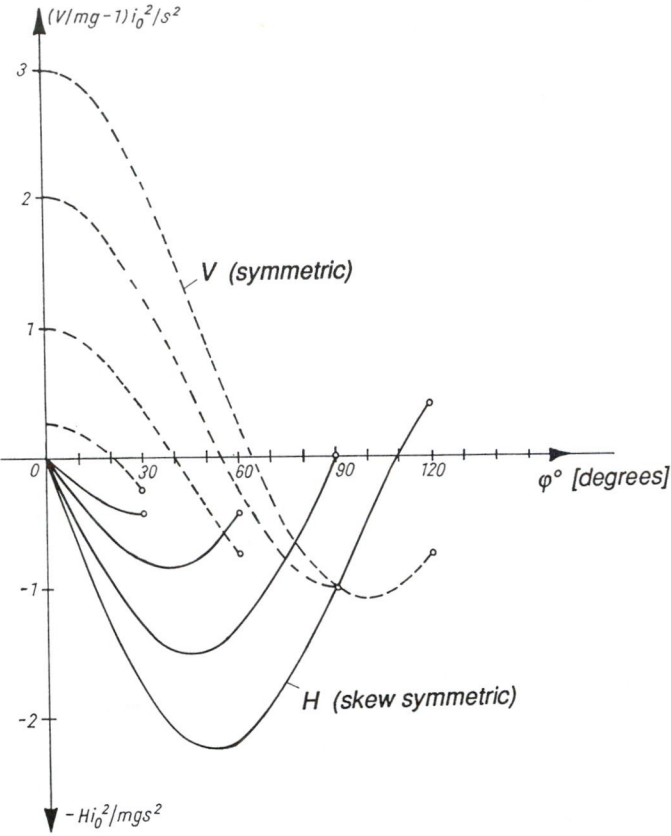

Fig. 7.25. Support reactions of the natural vibrations of a pendulum, $-\alpha \leq \varphi(t) \leq \alpha$

The forces are periodic but nonharmonic, and the higher-order terms in a *Fourier* time series are not small. Their dynamic portions are shown graphically in Fig. 7.25. The vertical component is an even function of the angle, the horizontal component an odd function. The reaction forces are the source of vibration of slender bell towers.

(§) SDOF-System with Dry Friction. The model of a vibrational system with *Coulomb´s* damping force consists of a mass attached to a *Hookean* spring in parallel connection to a *St.-Venant* body; see Fig. 4.11 for the series connection. Qualitatively, both the effects of dry friction in mechanical devices, as well as the free motion of a mass connected to a rigid-plastic structure with linear hardening, are illustrated in the model of Fig. 7.26. The stick-slip motion of the mass and the possibility of its rendering self-excited vibrations can be studied by moving the support in the axial direction according to

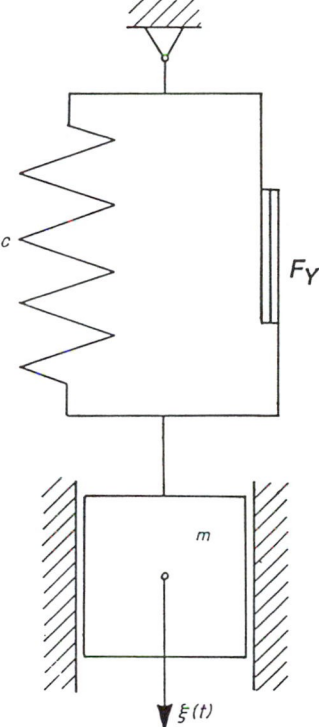

Fig. 7.26. Model of the oscillator with dry friction. Free vibrations

a prescribed law. Such problems, eg are encountered in positioning mechanisms. Subsequently, only free motion in the absence of any permanent forcing is studied. The equation of such translational motion is simply given by *Newton´s* law

$$m\ddot{\xi} = -c\xi + T_R \quad , \quad T_R = -F_y \, \text{sgn} \, \dot{\xi} \ . \tag{7.126a}$$

It is a nonlinear differential equation due to the sign function that changes its sign in reversed motions, but it is piecewise linear

$$\ddot{\xi} + \omega_0^2 \left(\xi + s \, \text{sgn} \, \dot{\xi} \right) = 0 \quad , \quad \omega_0^2 = c/m \quad , \quad F_y/m = s\omega_0^2 \ . \tag{7.126b}$$

The length *s* characterizes the frictional force by the limit of the spring force $cs = F_y$, which can be balanced by adhesion (the coefficients of dry friction and that at the limit of adhesion are assumed to be equal). The transformation

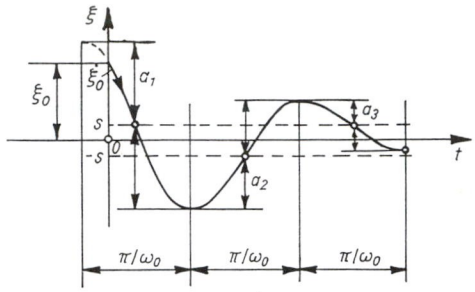

Fig. 7.27. Free motion when damped by dry friction

$$x = \xi + s \, \text{sgn} \, \dot{\xi} \ , \tag{7.127a}$$

permanently renders the equation to be linear

$$\ddot{x} + \omega_0^2 x = 0 \ . \tag{7.127b}$$

Hence, the solution of the linear undamped oscillator (7.86), $\zeta = 0$ applies,

$$x(t) = a \cos(\omega_0 t - \varepsilon) \ . \tag{7.128}$$

Free motion results by the successive superposition of harmonic halfwaves that are shifted by the drift $\pm s$, but leave the linear frequency unaltered (see Fig. 7.27)

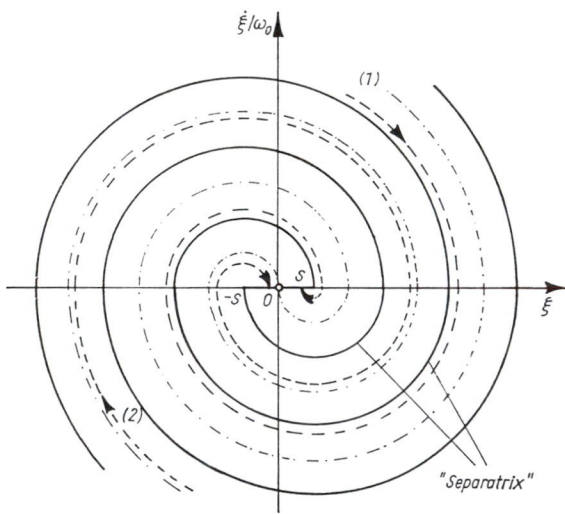

Fig. 7.28. Phase curves and separatrix of free frictional vibrations

$$\xi(t) = -s \, \text{sgn} \, \dot{\xi} + a \cos(\omega_0 t - \varepsilon) \ . \tag{7.129}$$

Figure 7.28 shows the free motions according to different initial conditions by dashed lines and the separatrix in the phase plane. Note the orthogonal crossings of the abscissa and the positions of the mass at rest.

7.4.9. Linear Elastic Chain of Oscillators

Simple, multiple-degree-of-freedom systems (MDOF-systems) can be represented by a nonbranched chain of linear oscillators in a series connection. Figure 7.29 shows the arrangement with two degrees of freedom under the condition of support excitation, ie $x_0 = x_e(t)$ is prescribed. If we consider the absolute coordinates shown in the figure and the free-body diagram of each spring mass system, *Newton's* law on the translational motion of the mass number one gives (note the motion of both ends of the springs in Fig. 7.29)

$$m_1 \ddot{x}_1 = -c_1(x_1 - x_0) + c_2(x_2 - x_1) \ . \tag{7.130}$$

Analogously, the equation for an intermediate mass number k results

$$m_k \ddot{x}_k = -c_k(x_k - x_{k-1}) + c_{k+1}(x_{k+1} - x_k) \ . \tag{7.131}$$

If the chain has an open end with mass m_n, m_2 in Fig. 7.29, the restoring force is supplied by the single connecting spring only

$$m_n \ddot{x}_n = -c_n(x_n - x_{n-1}) \ . \tag{7.132}$$

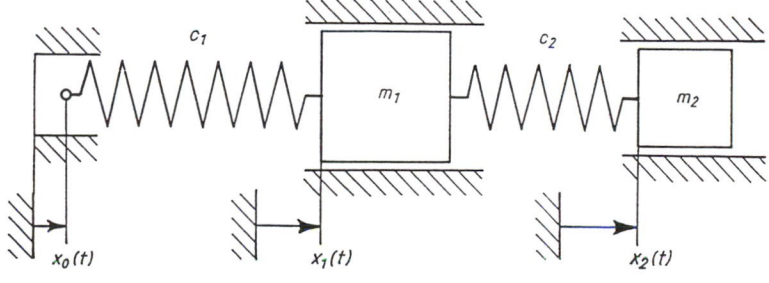

Fig. 7.29. The kinematically forced 2 degree-of-freedom linear oscillator

The linear system of second-order differential equations, inhomogeneous due to support excitation and with constant coefficients, is coupled by the underscored spring forces. The coupling of m_k is very local, namely, to the adjacent masses m_{k-1} and m_{k+1} only. The chains related to more complicated systems may have additional branches with far-reaching couplings. Changing the coordinates to the relative ones that are given by the deformations of the springs renders an inertia type of coupling.

The forcing is assumed to be time-harmonic with assigned frequency ω

$$x_e(t) = A_0\, e^{i\omega t} \ . \tag{7.133}$$

The steady-state responses of the masses are also time-harmonic and, therefore, the proper particular solutions are

$$x_k(t) = A_k\, e^{i\omega t} \ . \tag{7.134}$$

Substitution renders each of the equations (7.131) reduced in time, and a linear inhomogeneous system of equations results; $k = 1, 2, ...,$ n. It is real, since no damping is considered

$$-m_k\, \omega^2\, A_k = c_k\, A_{k-1} - (c_k + c_{k+1})\, A_k + c_{k+1}\, A_{k+1} \ . \tag{7.135}$$

(§) The Residual Method of *Holzer* and *Tolle* . The natural frequencies and associated modes of free vibrations can be found by stepping the forcing frequency ω and selecting the amplitude $A_n = A$ of the mass m_n at the free end of the chain. The zero crossings of $A_0(\omega)$ render the natural frequencies. The intermediate amplitudes are also registered and finally determine the modes of free vibrations. In that context, the equations are solved successively beginning at the free end

$$A_{n-1} = \left(1 - m_n\omega^2/c_n\right) A \ , \tag{7.136}$$

and the tension of the spring number n is $c_n\,(A - A_{n-1}) = m_n\, \omega^2 A$. The $(n-1)$ th equation gives, with A_{n-1} known,

$$A_{n-2} = A_{n-1} - m_{n-1}\omega^2 A_{n-1}/c_{n-1} + c_n A_{n-1}/c_{n-1} - A c_n/c_{n-1} \ . \tag{7.137}$$

The force in the spring number $n - 1$ is thus $c_{n-1}\,(A_{n-1} - A_{n-2}) = (Am_n + m_{n-1}\,A_{n-1})\,\omega^2$, etc. Hence, the tension in the first spring that is forced becomes

$$c_1 (A_1 - A_0) = \omega^2 \sum_{k=1}^{n} m_k A_k \quad \rightarrow \quad A_0(\omega) = A_1 - \frac{\omega^2}{c_1} \sum_{k=1}^{n} m_k A_k \quad .$$

$$(7.138)$$

Equation (7.138) may be related to (7.136) by means of the *transfer matrix* that is a product of the field matrices (without inertia forces) and the point matrices (connecting the state, displacement, and force, before and after a concentrated mass). Combination with a root-searching technique forms an easily programmable routine. The signs of the amplitudes must be registered to assign the order of the mode and not to miss a natural frequency. For two masses, $n = 2$, it follows explicitly:

$$A_1 = \left(1 - \frac{m_2 \omega^2}{c_2}\right) A \quad , \quad A_0(\omega) = A_1 - \frac{\omega^2}{c_1}(m_2 A + m_1 A_1) =$$

$$\left[1 - \omega^2 \left(\frac{m_1 + m_2}{c_1} + \frac{m_2}{c_2}\right) + \frac{m_1 m_2}{c_1 c_2} \omega^4\right] A \quad .$$

$$(7.139)$$

$A_0 (\omega) = 0$ is the frequency equation of the second order in the variable ω^2. In that explicit form, it equals the characteristic equation of the eigenvalue problem, that results when putting the coefficient determinant of the homogeneous system of equations of A_1 and A_2 (with $A_0 = 0$) to zero. The roots are the positive, squared, circular natural frequencies

$$\omega_{2,1}^2 = \frac{1}{2}\left(\Omega_1^2 + \Omega_2^2\right) \pm \frac{1}{2} \sqrt{\left(\Omega_1^2 - \Omega_2^2\right)^2 + (2k_1 k_2)^2} \quad , \quad \Omega_1^2 = (c_1 + c_2)/m_1 \quad ,$$

$$\Omega_2^2 = c_2/m_2 \quad , \quad k_1^2 = c_2/m_1 \quad , \quad k_2^2 = c_2/m_2 \quad .$$

$$(7.140)$$

The basic frequency ω_1 and the first superharmonic frequency ω_2 are subject to the self-explanatory inequalities (Fig. 7.30),

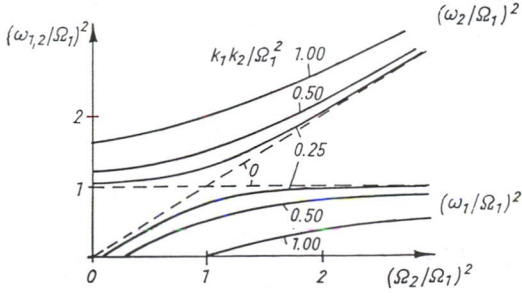

Fig. 7.30. The natural frequencies of the 2 degree-of-freedom oscillator, damping neglected

$$\omega_1 \le \min(\Omega_1, \Omega_2) \quad , \quad \omega_2 \ge \max(\Omega_1, \Omega_2) \quad .$$

(§)   *Dunkerly's* Formula. This formula provides an approximation from below for the natural frequencies: For the 2 degree-of-freedom vibrations, two associated frequencies are found to be the natural ones when putting $m_1 = 0$

$$\omega_{m_2}^2 = [(1/c_1 + 1/c_2)\, m_2]^{-1} \quad , \quad m_1 = 0 \quad ,$$

and by putting $m_2 = 0$,

$$\omega_{m_1}^2 = c_1/m_1 \quad , \quad m_2 = 0 \quad .$$

An equivalent frequency is given by the formula

$$\frac{1}{\overline{\omega}^2} = \frac{1}{\omega_{m_1}^2} + \frac{1}{\omega_{m_2}^2} \quad , \quad \overline{\omega}^2 = \frac{c_1\, c_2}{(c_1 + c_2)\, m_2 + c_2\, m_1} \quad .$$

The summation of the reciprocals of the squared natural frequencies also renders

$$\frac{1}{\overline{\omega}^2} = \frac{1}{\omega_1^2} + \frac{1}{\omega_2^2} \quad .$$

Hence, the inequality applies to the reciprocals

$$\frac{1}{\omega_1^2} < \frac{1}{\overline{\omega}^2} \quad ,$$

and thus gives an estimation of the basic frequency

$$\overline{\omega}^2 < \omega_1^2 \quad .$$

This result illustrates *Dunkerly's* formula: When we consider a vibrational system with n masses and take away all the masses except one, the equivalent frequency becomes

$$\overline{\omega} = \left(\sum_{j=1}^{n} \omega_{m_j}^{-2} \right)^{-1/2} \quad ,$$

which turns out to be a lower bound for the basic frequency ω_1 of the actual lumped system. The quality of the lower bound depends on how rapidly the terms of the sum are decreasing in regard to their increasing order. For an upper bound, see the *Ritz* procedure (Sec. 11.1).

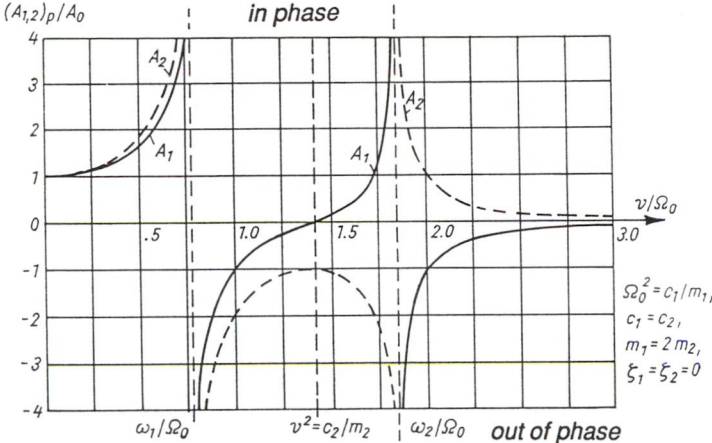

Fig. 7.31. Resonance curves of the 2 degree-of-freedom undamped oscillator.
A negative amplitude indicates a phase angle of 180°

(§)   Natural Modes. The amplitudes of the steady-state vibration
are all positive in the frequency range $\omega < \omega_1$, eg A_0 , A_1 , $A_2 = 1$
positive. Within the bandwidth $\omega_1 < \omega < \omega_2$, one sign changes, and so
on. The free vibrations for $n = 2$ are given by the real part of the
complex solution, $\omega_1 \neq \omega_2$,

$$(x_1)_h = A_{11} e^{i\omega_1 t} + A_{12} e^{i\omega_2 t} \quad , \quad (x_2)_h = A_{21} e^{i\omega_1 t} + A_{22} e^{i\omega_2 t} \quad . \quad (7.141)$$

More details are given in Sec. 10.1, and the alternative method of
Stodola and von Mises is discussed in Exercise A 11.11, further in
R. W. Clough and J. Penzien: Dynamics of Structures. McGraw Hill,
New York, 1975.

(§)   The Amplitude Frequency Response Functions of the Two-Mass
System. These functions are shown in Fig. 7.31, where $v = \omega \neq \omega_1$, ω_2
is the forcing frequency, and $x_e(t) = A_0 \cos vt$ is the real excitation

$$(A_1)_p = A_0 (c_2 - m_2 v^2) \frac{c_1}{\Delta(v)} \quad , \quad (A_2)_p = A_0 \frac{c_1 c_2}{\Delta(v)} \quad ,$$

$$\Delta(v) = m_1 m_2 v^4 - [(c_1 + c_2) m_2 + c_2 m_1] v^2 + c_1 c_2 \neq 0 \quad . \quad (7.142)$$

Considering the zero crossing of $(A_1)_p$ at the forcing frequency $v = \sqrt{(c_2/m_2)}$ renders the two-mass system so that it functions as an
undamped dynamic absorber : The principal mass m_1 remains at rest
due to the tuned motion of the absorber mass m_2 for that single

frequency. Damping somewhat destroys the ideal picture of the tuned vibration absorber. For the two fixed points of the principal resonance curves with variable damping and an optimality criterion for tuning with damping present, see *J. P. Den Hartog: Mechanical Vibrations. McGraw-Hill, New York, 1956.* The "lossless" dynamic absorber has many applications in mechanical and civil engineering (eg it is applied to reduce the basic and uncomfortable vibration of high-rise buildings under wind or earthquake loadings).

7.5. Bending Vibrations of Linear Elastic Beams

The beam may be considered a continuous vibrating chain of mass elements $\rho A \, dx$. The free-body diagram of such an infinitesimal element is shown in Fig. 7.32; y is a principal axis and $q(x,t)$ an external loading. The axial force is assumed to be zero, no torsion is considered, and conservation of momentum in the z direction renders

$$\rho A \, dx \, \frac{\partial^2 w}{\partial t^2} = \frac{\partial Q_z}{\partial x} \, dx + q \, dx + ... \quad .$$

$$(7.143)$$

Conservation of the angular momentum about the y axis gives

$$\rho J_y \, dx \, \frac{\partial^2 \varphi}{\partial t^2} = \frac{\partial M_y}{\partial x} \, dx - Q_z \, dx + ... \quad .$$

$$(7.144)$$

Cancelling the common factor dx and taking the limit $dx \to 0$ render

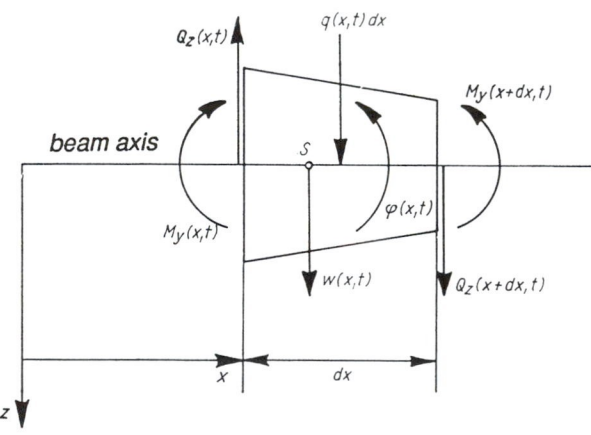

Fig. 7.32. Mass element of a beam in principal bending motion. Shear deformation and rotational inertia of the cross-section are incorporated according to *Timoshenko* theory

the dynamic equivalent to Eq. (2.147)

$$\frac{\partial Q_z}{\partial x} = -q + \rho A \frac{\partial^2 w}{\partial t^2} \quad , \quad \frac{\partial M_y}{\partial x} = Q_z + \rho J_y \frac{\partial^2 \varphi}{\partial t^2} \quad .$$

(7.145)

The cross-section rotates through the angle φ and is assumed to remain plane. Equations (6.65) and (6.66) include the shear deformation

$$\frac{\partial \varphi}{\partial x} = \frac{M_y}{EJ_y} \quad , \quad -\varphi = \frac{\partial w}{\partial x} - \frac{\kappa Q_z}{GA} \quad .$$

(7.146)

Hence, four equations result for the four unknowns, M_y (x, t) , Q_z (x, t) , $\varphi(x, t)$, and for the deflection, $w(x, t)$. Elimination of the former variables renders a fourth-order partial differential equation for the deflection w that is named after *Timoshenko* . For additional information, see *H. N. Abramson et al: "Stress Wave Propagation in Rods and Beams," in Advances in Applied Mechanics, Vol. 5. New York, 1958.*

In the case of slender beams with a sufficiently smooth variation of the cross-section and a distributed loading varying slowly over the span, and for small deflections, the simplified *Bernoulli-Euler* theory applies. In that case, the shear deformation is neglected and $\varphi = -\partial w/\partial x$. The drastically simplified fourth-order partial differential equation results upon elimination

$$\frac{\partial^2}{\partial x^2} \left(EJ_y \frac{\partial^2 w}{\partial x^2} \right) + \rho A \frac{\partial^2 w}{\partial t^2} = q(x, t) \quad .$$

(7.147)

With the initial and boundary conditions prescribed, the solution yields time-dependent small deflections. The resulting bending moment and shear force are

$$M_y = -EJ_y \frac{\partial^2 w}{\partial x^2} \quad , \quad Q_z = \frac{\partial M_y}{\partial x} \quad .$$

(7.148)

Practically, the resultants of the internal forces are determined by static considerations of the given loading, plus the inertia forces with time t kept fixed

$$q^* = q(x, t) - \rho A \ddot{w}(x, t) \quad .$$

The solution becomes independent of time in the steady state for a time-harmonic excitation and the generalized loading in that case is frequency-dependent.

(§)   Natural and Forced Vibrations of a Slender, Hinged-Hinged, Single-Span Beam. Within the *Bernoulli-Euler* approximation such vibrations are considered subsequently. With a constant cross-section, the total mass is $m = \rho Al$, and the bending rigidity $B = EJ_y = const$. The homogeneous partial differential equation (7.147) with $q = 0$ is solved by means of the separable function (*Bernoulli's* trial)

$$w(x, t) = Y(t) \, f(x) \ .\tag{7.149}$$

Substitution and dividing through w render two ordinary differential equations that must be equal to a common constant ω^2

$$k^2 \frac{f^{iv}}{f} = -\frac{\ddot{Y}}{Y} = \omega^2 \ , \quad k^2 = B/\rho A \ , \quad f^{iv} = \frac{d^4 f}{dx^4} \ .\tag{7.150}$$

Thus, the time equation of the linear oscillator results with the solution

$$\ddot{Y} + \omega^2 Y = 0 \quad \rightarrow \quad Y(t) = a \cos(\omega t - \varepsilon) \ .\tag{7.151}$$

The spatial, fourth-order equation with the constant coefficient $\kappa^2 = \omega/k$ has a general real solution in the form of a combination of trigonometric and hyperbolic functions

$$f^{iv} - \kappa^4 f = 0 \rightarrow f(x) = C_1 \sin \kappa x + C_2 \cos \kappa x + C_3 \sinh \kappa x + C_4 \cosh \kappa x \ .\tag{7.152}$$

The spectrum of the natural frequencies depends on the boundary conditions. The simply supported beam of span l requires at $x = 0, l$: $w = w'' = 0 \rightarrow f = f'' = 0$. Hence, the coefficients of the even functions must vanish in this case and the two linear equations remain to be solved

$$C_1 \sin \kappa l + C_3 \sinh \kappa l = 0 \ , \quad -C_1 \sin \kappa l + C_3 \sinh \kappa l = 0 \ .$$

A nontrivial solution exists if $C_3 = 0$, and by taking into account the transcendental frequency equation

$$\sin \kappa l = 0 \ , \quad \kappa^2 = \omega/k \ \rightarrow \ \kappa = \pi/l \, , \, 2\pi/l \, , \, \dots \, , \, n\pi/l \, , \, \dots\tag{7.153}$$

The squared natural frequencies are thus given by substituting the eigenvalues: $\omega_n^2 = (n\pi)^4 \, B/m \, l^3$, $n = 1, 2, 3, \dots$ The associated natural bending modes are the eigenfunctions $f_n(x) = \sin(n\pi x/l)$. The general solution of the dynamic deflection can be expressed by the infinite modal sum, since the superposition principle applies and the eigenfunctions are linearly independent (they form a system of orthogonal functions)

$$w(x, t) = \sum_{n=1}^{\infty} Y_n(t) \, f_n(x) = \sum_{n=1}^{\infty} a_n \sin{(n\pi x/l)} \cos{(\omega_n t - \varepsilon_n)} \ ,$$

$$\int_0^l f_n(x) \, f_m(x) \, dx = N_n \delta_{mn} \ . \tag{7.154}$$

Analogous to a *Fourier* expansion, any given loading that is represented by separable functions, $q(x, t) = h(t) \, g(x)$, can be at first projected on the mode of free vibrations and then represented by an infinite series

$$A_n = \frac{1}{N_n} \int_0^l g(x) \, f_n(x) \, dx \quad \rightarrow \quad g(x) = \sum_{n=1}^{\infty} A_n \, f_n(x) \ . \tag{7.155}$$

Especially, the steady-state vibration that results from a time-harmonic excitation, $h(t) = \cos{vt}$, has the modal series representation

$$w_p(x, t) = \sum_{n=1}^{\infty} B_n \, f_n(x) \cos{vt} \ . \tag{7.156}$$

The coefficients B_n are determined by substitution into Eq. (7.147). The simply supported single-span beam gives, by comparing coefficients (damping is neglected)

$$\rho A B_n = \frac{A_n}{\omega_n^2 - v^2} \quad , \quad v \neq \omega_n = k \, (n\pi/l)^2 \ . \tag{7.157}$$

For example, the dynamic reaction at the hinged support B can be determined by means of the conservation of the angular momentum about the joint B

$$l \, B - \int_0^l xq \, dx = - \int_0^l x \ddot{w} \rho A \, dx \ .$$

In the steady state, the force becomes by means of Eq. (7.156)

$$B(t) = \frac{1}{l} \int_0^l x \, (q - \rho A \ddot{w}_p) \, dx \ . \tag{7.158}$$

Putting

$$B(t) = \sum_{n=1}^{\infty} b_n \cos vt \quad , \tag{7.159}$$

renders, by comparing coefficients, for a constant mass per unit of length, $\rho A = const$,

$$b_n = \frac{1}{l} \left(A_n + \rho A\, B_n v^2\right) \int_0^l x\, f_n(x)\, dx \; = \; \frac{l\, A_n}{n\,\pi} \frac{(-1)^{n+1}}{1-(v/\omega_n)^2} \quad , n = 1, 2, .. \tag{7.160}$$

Analogously, the internal dynamic reactions are determined by a similar infinite series.

The separation shown in Eq. (7.149) in combination with a proper selection of an admissible function $f(x)$ is the basic ingredient of the *Ritz* approximation; see Sec. 11.1.

7.6. Body Waves in the Linear Elastic Solid

Supplementing the static *Navier´s* equations (6.6) by the inertia force per unit of volume according to Eq. (7.1) renders the local equations of motion of a homogeneous and isotropic *Hookean* body in vector form (k is subsequently the body force per unit of mass)

$$(\lambda + G)\, \nabla \left(\nabla . \, \mathbf{u}\right) + G\, \nabla^2 \mathbf{u} + \rho\, \mathbf{k} = \rho\, \frac{\partial^2 \mathbf{u}}{\partial t^2} \quad . \tag{7.161}$$

G is the shear modulus and $\lambda = 2G\, v/(1 - 2v)$ the second of *Lame´s* constants; see also Eq. (4.15). Their dynamic (adiabatic) values may differ from the static (isothermal) ones. The *Helmholtz* decomposition of the displacement vector

$$\mathbf{u} = \nabla\Phi + \nabla \times \mathbf{\Psi} \equiv grad\; \Phi + curl\; \mathbf{\Psi} \quad , \tag{7.162}$$

allows a distinction between two systems of body waves with different speeds of propagation. The fast longitudinal P-waves (unda prima) and the slower S-waves (unda secunda) separate, if the body force is absent or has an analogous representation by means of the force potentials $k = grad\; b + curl\; B$. The speeds are, eg determined by the eigenvalues of plane waves. The substitution of the *Helmholtz* decomposition into Eq. (7.161) renders a scalar wave equation for the irrotational potential Φ and a vector wave equation for the vector potential $\mathbf{\Psi}$

$$c_P^2 \nabla^2 \Phi + b = \ddot{\Phi} \quad , \quad c_S^2 \nabla^2 \Psi + B = \ddot{\Psi} \quad , \quad (\nabla . \Psi) \equiv \operatorname{div} \Psi = 0 \quad . \quad (7.163)$$

The constant coefficients

$$c_P = \sqrt{(\lambda + 2G)/\rho} \quad , \quad c_S = \sqrt{G/\rho} \quad , \qquad (7.164)$$

have the dimension of speed, c_P is recognized as the speed of sound. The potentials Φ and Ψ refer to wave systems that propagate independently in the (infinite) solid. No losses are considered, and thus, the deformations are according to a thermodynamic viewpoint, isentropic changes of state. Hence, it was already mentioned above that the elastic constants λ or E and G are to be measured under adiabatic conditions (contrary to the static ones that refer to isothermal conditions). Since *curl grad* $\Phi \equiv 0$, the associated deformation is irrotational and simply corresponds to a compressional wave like a sound wave (a propagating small disturbance) in a fluid, where $G = 0$ and $\lambda = K$ is the modulus of linear compressibility [see Eq. (2.87)], and, therefore, the speed of sound is $c = \sqrt{(K/\rho)}$. The propagating dilatation is $e = \operatorname{div} \boldsymbol{u} = \nabla^2 \Phi$, and the second wave system corresponds to an isochoric deformation: *div curl* $\Psi \equiv 0$. The coupling of the wave systems occurs, however, during any refraction at an interface between two elastic solids or at a free surface by reflection.

Plane waves are the *D´Alembert´s* solutions f of the equations of motion in the absence of body forces ($k = 0$)

$$\boldsymbol{u} = \boldsymbol{A} \, f\!\left(t - \frac{(\boldsymbol{r}.\boldsymbol{e}_n)}{c}\right) \quad . \qquad (7.165)$$

$(\boldsymbol{r} . \boldsymbol{e}_n) = const$ is the equation of the plane wavefront with the normal vector $\boldsymbol{e}_n$. Substitution into Eq. (7.161) renders the homogeneous equations of the components of the amplitude vector

$$(\lambda + G)(\boldsymbol{A}.\boldsymbol{e}_n)\boldsymbol{e}_n + (G - \rho c^2)\boldsymbol{A} = 0 \quad . \qquad (7.166)$$

Nontrivial solutions $A \neq 0$ exist only for selected eigenvalues c^2 .

(§) The Longitudinal Wave. This wave with a displacement in the direction of propagation, $\boldsymbol{A} = A\,\boldsymbol{e}_n , A \neq 0$, has the characteristic equation

$$(\lambda + 2G - \rho c^2)\,A = 0 \quad , \quad c^2 = c_P^2 = (\lambda + 2G)/\rho \quad , \quad A \neq 0 \quad . \qquad (7.167)$$

The scalar potential and its gradient, the displacement vector of the plane P-wave are

$$\Phi = \Phi_0 \, g\left(t - \frac{(\mathbf{r}.\mathbf{e}_n)}{c_P}\right) \ , \quad \mathbf{u} = \nabla\Phi = -\Phi_0 \, \frac{\mathbf{e}_n}{c_P} \, \dot{g}\left(t - \frac{(\mathbf{r}.\mathbf{e}_n)}{c_P}\right) \ , \tag{7.168}$$

$$\Psi = 0 \ , \quad f = -\dot{g} \ .$$

(§) The Shear Wave. This wave has a displacement that is orthogonal to the direction of propagation, $(\mathbf{A}.\mathbf{e}_n) = 0$. Equation (7.166) reduces to

$$(G - \rho c^2) \, \mathbf{A} = 0 \ , \quad c^2 = c_S^2 = G/\rho < c_P^2, \ |\mathbf{A}| \neq 0 \ . \tag{7.169}$$

The vector potential takes on the form

$$\Psi = \Psi_0 \, g\left(t - \frac{(\mathbf{r}.\mathbf{e}_n)}{c_S}\right) \ , \quad \left(\Psi_0 . \mathbf{e}_n\right) = 0 \ . \tag{7.170}$$

By means of the unit polarization vector $\mathbf{e}_p$, the amplitude vector is represented by the vector product, $\Psi_0 = \Psi_0 \, (\mathbf{e}_p \times \mathbf{e}_n)$, and the displacement vector, ie the *curl* of the potential, becomes

$$\mathbf{u} = \nabla \times \Psi = -\Psi_0 \, \frac{\mathbf{e}_p}{c_S} \, \dot{g}\left(t - \frac{(\mathbf{r}.\mathbf{e}_n)}{c_S}\right) \ , \quad (\mathbf{e}_p . \mathbf{e}_n) = 0 \ . \tag{7.171}$$

With respect to a characteristic direction $\mathbf{e}_3$ the shear wave is termed horizontally polarized if $\mathbf{e}_p = \mathbf{e}_3$. The propagation vector $\mathbf{e}_n$ of such a SH-wave is lying in a plane that is orthogonal to $\mathbf{e}_3$. If, however, $(\mathbf{e}_p . \mathbf{e}_3) = 0$, ie $\mathbf{e}_p$ as well as $\mathbf{e}_n$ (the latter is orthogonal to $\mathbf{e}_p$) are spanning a single plane, a vertically polarized shear wave, SV-wave, is propagating with speed c_S .

An inhomogeneous plane wave, the *Rayleigh* surface wave, is discussed in Exercise A 7.12. Note its decomposition into body waves with common wave number. Such a representation clearly exhibits the interaction of the plane P- and SV-waves in the course of time. For details see, *W. Scheidl and F. Ziegler, in Proceedings IUTAM-Symposium on Modern Problems in Elastic Wave Propagation (eds. J. Achenbach and J. Miklowitz), Wiley, New York, 1978, pp. 145-149*

7.7. Exercises A 7.1 to A 7.12 and Solutions

A 7.1: A railed vehicle of mass m on a straight track moves with constant speed v northbound on Earth. Calculate the lateral guiding force by considering the eigenrotation v of the Earth.

Solution: Only the *Coriolis* acceleration acts in the horizontal plane; according to Eq. (8.62), $\mathbf{a}_C = 2 \, (\mathbf{v} \times \mathbf{v}_{rel}) = a_\varphi \, \mathbf{e}_\varphi = -2 \, v \, v \sin \alpha \, \mathbf{e}_\varphi$; α is the northern latitude; and $\mathbf{e}_\varphi$ points eastward. See Fig. 7.15.

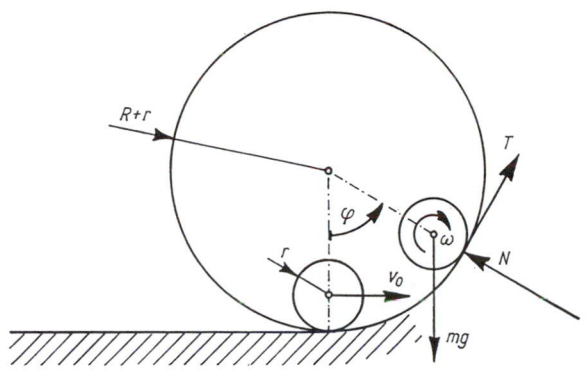

Fig. A 7.2.

Hence, Eq. (7.7) when applied in the lateral direction gives $m\, a_\varphi\, \boldsymbol{e}_\varphi =$ $R_h\, \boldsymbol{e}_\varphi$ and the guiding force is $R_h = -\,(2\,v\,v/g)\,mg\,\sin\alpha$, which is the reaction of the eastern rail, $\alpha > 0, v > 0$. The force is orthogonal to $\boldsymbol{v}_{rel}$ and does not change the magnitude if the direction of motion is altered.

A 7.2: A wheel of radius r in rolling contact enters a circular guiding track of radius $R + r$ with the initial speed v_0 . Determine the minimum value of v_0 to secure contact (see Fig. A 7.2).

Solution: Polar coordinates are convenient and the condition of rolling (no slip) is expressed by $v = R\,d\varphi/dt = r\,\omega$. Hence, the conservation of momentum and angular momentum render

$$m\dot{v} = mR\ddot{\varphi} = T - mg\sin\varphi \;,\;\; mv^2/R = mR\dot{\varphi}^2 = N - mg\cos\varphi \;,$$

$$mi^2\,\dot{\omega} = mi^2\,R\ddot{\varphi}/r = -rT \;.$$

Elimination of the tangential force T yields the nonlinear equation of motion that is analogous to that of a pendulum

$$\ddot{\varphi} + \left[g\,/R\!\left(1 + i^2/r^2\right)\right]\sin\varphi = 0 \;\; \rightarrow \;\; \dot{\varphi}(\varphi) \;.$$

The condition of contact is

$$N = mg\cos\varphi + mR\,\dot{\varphi}^2$$

$$= mg\left[v_0^2\left(1 + i^2/r^2\right)/gR - 2 + \left(3 + i^2/r^2\right)\cos\varphi\right]/\left(1 + i^2/r^2\right) > 0 \;.$$

The inequality yields

$$\frac{v_0^2}{gR} > \left[2 - \left(3 + i^2/r^2\right) \cos \varphi \right] / \left(1 + i^2/r^2\right) .$$

It holds for all angles φ if the initial speed

$$v_0 > v_{crit} = \left[\left(5 + i^2/r^2\right) gR / \left(1 + i^2/r^2\right)\right]^{1/2} .$$

Putting the radius of inertia to zero , $i = 0$, gives the idealized condition of permanent contact during the frictionless motion of a point mass on a circular track in a vertical plane ($T = 0$).

A 7.3: Two rigid disks, I_1 and I_2 , are connected to each other and to infinite masses at both ends of the chain by linear elastic and mass-less shafts according to Fig. A 7.3. The upper end is rotated by α_0 and suddenly released (by disengaging a clutch or after fracture in torsion). Determine the free torsional vibrations of the 2 degree-of-freedom system.

Solution: In an instant deformed configuration given by the angles $\varphi_1(t)$, $\varphi_2(t)$, the free-body diagrams of the disks are considered. Conservation of angular momentum renders with the proper restoring moments

$$I_1 \ddot{\varphi}_1 = - k_1\varphi_1 + k_2 \left(\varphi_2 - \varphi_1\right) , \quad I_2 \ddot{\varphi}_2 = - k_2 \left(\varphi_2 - \varphi_1\right) - k_3\left[\varphi_2 - \alpha(t)\right] ,$$

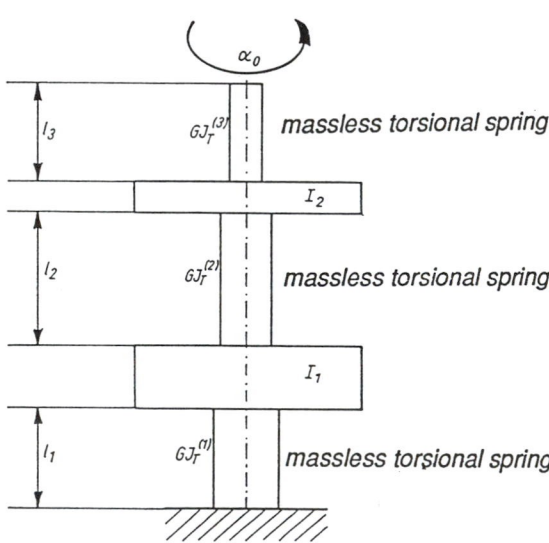

Fig. A 7.3. Torsional vibrations. Torsional rigidities of the massless springs:
$$k_i = GJ_T^{(i)}/l_i , \quad i = 1, 2, 3$$

where $\alpha(t)$ is still a general angular excitation. Putting $\alpha(t) = \alpha_0$ gives the equilibrium condition with all the angular accelerations set equal to zero

$$\varphi_1^{(0)} = \mu\Omega_2^2\Omega_3^2\,\alpha_0/\Delta_0 \ , \quad \varphi_2^{(0)} = \Omega_1^2\Omega_3^2\,\alpha_0/\Delta_0 \ , \quad \mu = l_2/l_1 \ , \quad \Omega_i^2 = k_i\,/l_2 \ , i = 2,\,3,$$

$$\Delta_0 = \Omega_1^2\Omega_2^2 + \Omega_1^2\Omega_3^2 - \mu\,\Omega_2^4 \ , \quad \Omega_1^2 = (k_1 + k_2)\,/l_1 \ .$$

The free-end condition is $k_3 = 0$. Solving the linear eigenvalue problem renders the squared natural frequencies as

$$\omega_{2,\,1}^2 = \frac{\Omega_1^2 + \Omega_2^2}{2}\left\{1 \pm \left[1 + 4\,\mu\Omega_2^4\,/\left(\Omega_1^2 + \Omega_2^2\right)^2\right]^{1/2}\right\} \ .$$

By means of the abbreviations

$$\kappa_1 = \Omega_2^2/\left(\Omega_2^2 - \omega_1^2\right) \ , \quad \kappa_2 = \Omega_2^2/\left(\Omega_2^2 - \omega_2^2\right) \ ,$$

the general solution of the homogeneous system of linear differential equations becomes

$$\varphi_1 = A_1 \cos \omega_1 t + A_2 \cos \omega_2 t \ , \quad \varphi_2 = \kappa_1 A_1 \cos \omega_1 t + \kappa_2 A_2 \cos \omega_2 t \ .$$

The initial conditions determine the constants

$$A_1 = \left[\left(\Omega_1^2 - \omega_2^2\right)\varphi_1^{(0)} - \mu\,\Omega_2^2\varphi_2^{(0)}\right]/\left(\omega_2^2 - \omega_1^2\right) \ ,$$

$$A_2 = \left[-\left(\Omega_1^2 - \omega_1^2\right)\varphi_1^{(0)} - \mu\,\Omega_2^2\varphi_2^{(0)}\right]/\left(\omega_2^2 - \omega_1^2\right) \ .$$

A 7.4: A mechanical device for measuring dynamic states has an indicator of mass m attached through a *Kelvin-Voigt* , spring-dashpot model to a casing of mass M (see Fig. A 7.4). The instrument is fixed to a vibratory system of mass M_1 . Excitation is by ground displacement $u(t)$. Set up the equations of motion in absolute and relative coordinates, respectively. The transformation to the reading $s(t)$ makes it possible to tune the instrument so that it is more sensitive to acceleration or displacement.

Solution: Free-body diagrams of the masses m and $m_1 = M + M_1$ in translational motions are considered and *Newton's* law renders in absolute coordinates (the reference state is free of stresses)

$$m\,\ddot{x} = -\,c\,(x - x_1) - r\,(\dot{x} - \dot{x}_1) - mg \ ,$$

$$m_1\,\ddot{x}_1 = c\,(x - x_1) + r\,(\dot{x} - \dot{x}_1) - c_1\,(x_1 - u) - r_1\,(\dot{x}_1 - \dot{u}) - m_1\,g \ .$$

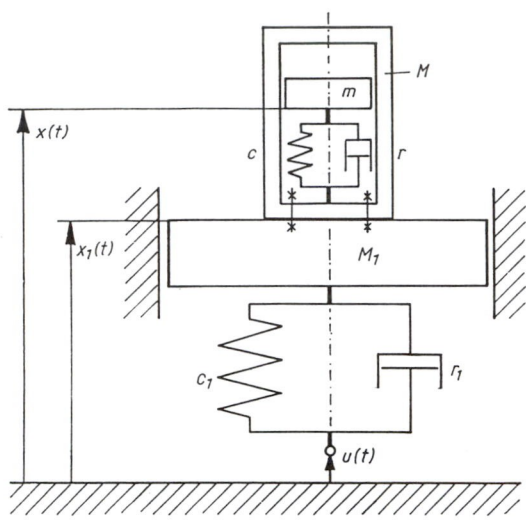

Fig. A 7.4.

Transformation to relative coordinates (static equilibrium is considered in the reference state) gives a system of coupled and inhomogeneous linear differential equations

$$m \frac{d^2\xi}{dt^2} + r \frac{d\xi}{dt} + k \, \xi = F(t) \quad , \quad \xi = \begin{Bmatrix} \xi \\ \xi_1 \end{Bmatrix} \quad , \quad F(t) = \begin{Bmatrix} -m\ddot{u} \\ -m_1\ddot{u} \end{Bmatrix} \quad ,$$

$$\xi = x - u + \left(\frac{m_1 + m}{c_1} + \frac{m}{c} \right) g \quad , \quad \xi_1 = x_1 - u + (m_1 + m) \, g/c_1 \quad ,$$

$$m = \begin{pmatrix} m & 0 \\ 0 & m_1 \end{pmatrix} \quad , \quad r = \begin{pmatrix} r & -r \\ -r & r+r_1 \end{pmatrix} \quad , \quad k = \begin{pmatrix} c & -c \\ -c & c+c_1 \end{pmatrix} \quad .$$

Putting $\xi_1 = u_1$ and substituting the reading $s(t) = \xi - \xi_1$ leave the forcing vector unchanged, but render the mass matrix nondiagonal and the stiffness and damping matrices nonsymmetric

$$M = \begin{pmatrix} m & m \\ 0 & m_1 \end{pmatrix} \quad , \quad R = \begin{pmatrix} r & 0 \\ -r & r_1 \end{pmatrix} \quad , \quad K = \begin{pmatrix} c & 0 \\ -c & c_1 \end{pmatrix} \quad , \quad X = \begin{Bmatrix} s(t) \\ u_1(t) \end{Bmatrix} \quad ,$$

$$M \, \ddot{X} + R \, \dot{X} + K \, X = F(t) \quad .$$

The standard form of the equations of motion becomes

$$\ddot{s} + \ddot{u}_1 + 2\zeta\Omega_s \dot{s} + \Omega_s^2 \, s = -\ddot{u} \quad ,$$

$$\ddot{u}_1 + 2\zeta_1\Omega_1\dot{u}_1 - 2\zeta_2\Omega_2\dot{s} + \Omega_1^2 u_1 - \Omega_1^2 u_1 = -\ddot{u} \ ,$$

where

$$\Omega_s^2 = c/m \ , \quad \Omega_1^2 = c_1/m_1 \ , \quad \Omega_2^2 = c/m_1 \ , \quad \zeta = r/2\sqrt{mc} \ ,$$

$$\zeta_1 = r_1/2\sqrt{m_1c_1} \ , \quad \zeta_2 = r/2\sqrt{m_1c_1} \ .$$

The measurement of the base acceleration requires tuning by

$$\Omega_s^2 \gg \left(\Omega_1^2, \Omega_2^2\right) \ \rightarrow \ \Omega_s^2 s \cong -(\ddot{u} + \ddot{u}_1) = -\ddot{x}_1 \ , \quad s \cong -\ddot{x}_1 / \Omega_s^2 \ .$$

If furthermore,

$$\Omega_1^2 \gg \Omega_2^2 \ \rightarrow \ \dot{u}_1 = -\frac{\Omega_1}{2\zeta_1} u_1 + \frac{\Omega_s^2}{2\zeta_1\Omega_1} s \ ,$$

$$u_1 = u_1(0) \, e^{-\lambda t} + \frac{\Omega_s^2}{2\zeta_1\Omega_1} \int_0^t e^{-\lambda(t-\tau)} s(\tau) \, d\tau \ , \quad \lambda = \Omega_1/2\zeta_1 \ .$$

Tuning with respect to displacement sensitivity requires $\Omega_s^2 \ll \Omega_1^2$ ($\Omega_s <$ maximal forcing frequency of u) together with the additional assumption

$$\Omega_s^2 \gg \Omega_2^2 \ \rightarrow \ \ddot{s} \cong -(\ddot{u} + \ddot{u}_1) = -\ddot{x}_1 \ , \quad s = -x_1 + x_{1\text{stat}} \ .$$

The second equation gives in that case, $\lambda = \Omega_1/2\zeta_1$,

$$\dot{u}_1 = -\frac{\Omega_1}{2\zeta_1} u_1 + \frac{1}{2\zeta_1\Omega_1} \ddot{s} \ , \quad u_1 = u_1(0) \, e^{-\lambda t} + \frac{1}{2\zeta_1\Omega_1} \int_0^t e^{-\lambda(t-\tau)} \ddot{s}(\tau) \, d\tau \ .$$

A 7.5: The (viscous) flow round an immersed body (eg the rigid cylinder of Fig. A 7.5, a bridge pier) is considered in the far field of the wake. The velocities at the inlet and also approximately at the outlet of the control surface are parallel, but the flow becomes inhomogeneous as shown in Fig. A 7.5. By measuring the velocity profile of the wake, $v_2 (y)$ of the two-dimensional flow, the drag F_D = F_w can be calculated from the conservation of momentum, if the pressure at A_2 is assumed to be constant and unchanged. Note the deficit of the mass flow rate in the direction e_x .

Solution: The mass flow rate through A_3 is determined by the stationary part of Eq. (1.82) and the deficit at the outlet

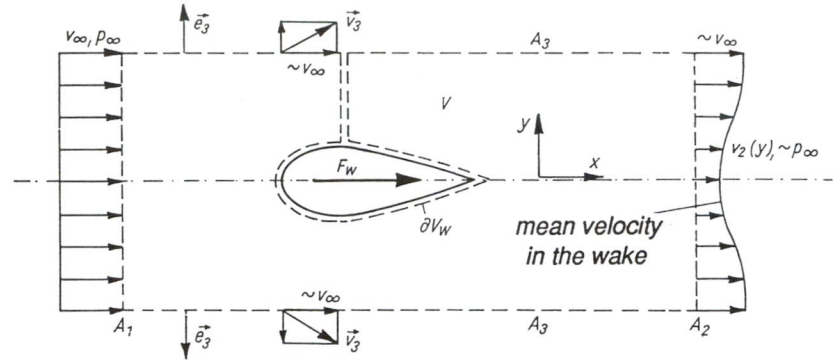

Fig. A 7.5. The drag of an immersed cylindrical body

$$- \rho_\infty v_\infty A_1 + \int_{A_2} \mu_2\, dA + \int_{A_3} \mu_3\, dA = 0 \quad , \quad \mu_2 = \rho\, v_2(y) \quad , \quad \mu_3 = \rho\,(\mathbf{v}_3 \cdot \mathbf{e}_3) \ .$$

The conservation of the x component of momentum yields

$$- \rho_\infty v_\infty^2\, A_1 + \int_{A_2} \mu_2 v_2\, dA + v_\infty \int_{A_3} \mu_3\, dA = -\, F_w \ ,$$

and hence, the drag is determined by the integral; see also Sec. 13.4.7,

$$F_w = \int_{A_2} \mu_2 \left[v_\infty - v_2(y) \right] dA = \frac{\rho_\infty v_\infty^2}{2}\, 2 \int_{A_2} \frac{v_2}{v_\infty} \left(1 - \frac{v_2}{v_\infty} \right) dA \quad , \quad dA = b\, dy \ .$$

A 7.6:   Determine the propelling force of an air-screw fixed in space, which is in a stationary state of rotation or, equivalently, the effective drag of a wind turbine in an incident parallel stream (the angular speed is kept constant by a power control) by the (measured) velocity profile of the wake. See Fig. A 7.6 and compare it with A 7.5.

Solution: The circular cylindrical control surface spins with the stationary speed of the rotor. Conservation of mass renders the flow rate through A_3 , which is the cylindrical part of the control surface

$$- \rho_\infty v_\infty A_1 + \int_{A_2} \mu_2\, dA + \int_{A_3} \mu_3\, dA = 0 \quad , \quad \mu_2 = \rho\, v_2(r) \ ,$$

$$\mu_3 = \rho\,(\mathbf{v}_3 \cdot \mathbf{e}_3) \begin{cases} < 0 \ \ \text{propeller} \\ > 0 \ \ \text{wind turbine} \end{cases} .$$

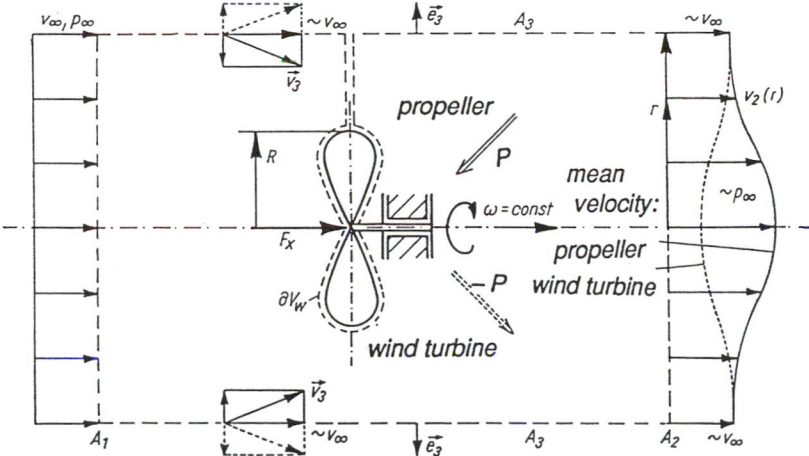

Fig. A 7.6. The axial force of a propeller and the effective drag of a wind turbine

Conservation of momentum yields the force $F_w = F_x \, e_x$ (the pressure at A_1 and A_2 is assumed to be approximately equal)

$$- \rho_\infty v_\infty^2 \, A_1 + \int_{A_2} \mu_2 v_2 \, dA + v_\infty \int_{A_3} \mu_3 \, dA = - F_x \ .$$

Analogous to the expression derived in Exercise A 7.5, the integral becomes

$$F_x = \int_{A_2} \mu_2 \left[v_\infty - v_2(r) \right] dA = \frac{\rho_\infty v_\infty^2}{2} \, 4\pi \int_0^{R_2} \frac{v_2}{v_\infty} \left(1 - \frac{v_2}{v_\infty} \right) r \, dr \ .$$

In simplified propeller theory, an engineering formula is derived by considering a mean velocity in the shadow behind the wheel that is given by the mass flow rate

$$\dot{m} = \int_{A_3} \mu_3 \, dA = \rho_\infty \pi R^2 \, \bar{v}_2 \ ,$$

$2R$ is the outer diameter of the wheel. Hence, the flow through the ring area portion $(A_2 - \pi R^2)$ seems to be replaced by the undisturbed parallel flow with speed v_∞ and

$$F_x = \frac{\rho_\infty v_\infty^2}{2} \left(1 - \frac{\bar{v}_2}{v_\infty} \right) \frac{\bar{v}_2}{v_\infty} \, 4\pi \int_0^R r \, dr = \dot{m} \left(v_\infty - \bar{v}_2 \right) \quad \begin{cases} F_x < 0 \ \text{... propeller} \\ F_x > 0 \ \text{... wind turbine} \end{cases} .$$

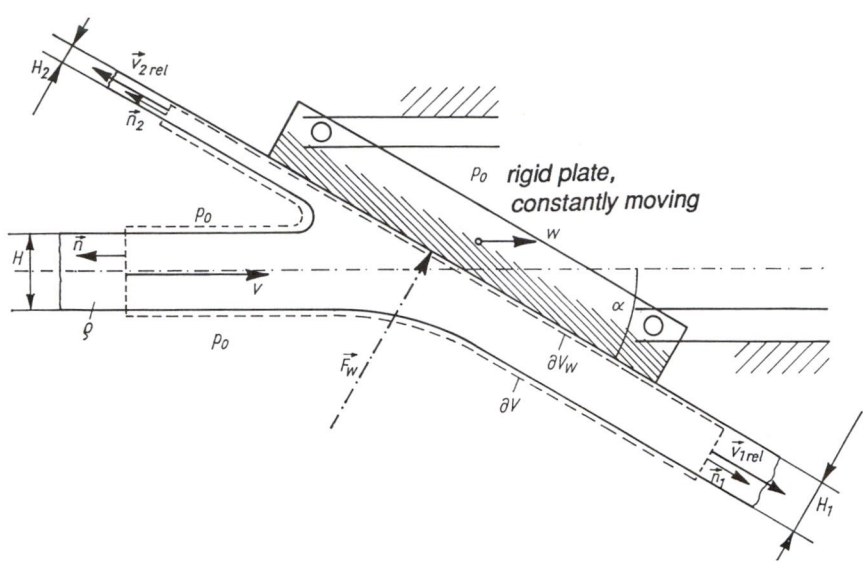

Fig. A 7.7.

A 7.7: A planar jet of rectangular cross-section $B \times H$ and with a constant mass flow rate is guided between two parallel walls and impinges on an inclined rigid plate. The latter is moving with an assigned constant speed w in the direction of the flow (see Fig. A 7.7). Neglecting any viscosity effects, determine the resulting force F_w of the pressure distribution acting on the moving plate, as well as its power that is produced by the stationary and controlled translational motion. Also, associate an effective lift and drag coefficient with the proper force components.

Solution: At the free surface, the atmospheric pressure p_0 is assigned. The same pressure acts in the cross-sections of the free jets that are thus sufficiently distant from the plate. The control volume V moves rigidly with constant speed w and is permanently attached to the plate. The branching of the jets in that case becomes stationary. Since $p_1 = p_2 = p_0$, the relative speeds of the mass stream become equal, $v - w = v_{1rel} = v_{2rel}$. *Bernoulli's* equation (8.36) applies also to the relative streamlines since $w = const$, and the moving frame is not distinguishable from an inertial frame. The relative mass flow rate densities are $\mu = \rho(v - w) \cdot n = -\rho(v - w)$, $\mu_1 = \rho(v_{1rel} + w - w) \cdot n_1 = \rho(v - w)$, $\mu_2 = \rho(v_{2rel} + w - w) \cdot n_2 = \rho(v - w)$, and the conservation of momentum gives

$$\oint_{\partial V} \mu v \, dS = -\rho(v - w) BH \, v + \rho(v - w) BH_1 (v_{1rel} + w)$$

$$+ \rho\,(v - w)\, BH_2\,(v_{2rel} + w) = -\,\mathbf{F_w}\,.$$

The resulting force is orthogonal to the moving plate surface

$$F_w = \rho\,(v - w)\, B\,(vH - wH_1 - wH_2)\sin\alpha\,,$$

and in the tangential direction, it follows that, since the external force component is zero, and shear stresses are neglected,

$$\rho\,(v - w)\, B\left[\, vH \cos\alpha - (v_{1rel} + w \cos\alpha)\, H_1 - (-\,v_{2rel} + w \cos\alpha)\, H_2 \right] = 0\,.$$

A second condition for the branched jet thicknesses is derived from the balance of mass flow, $B = const$,

$$\oint_{\partial V}\mu\,dS = -\rho\,(v - w)\, BH + \rho\,(v - w)\, BH_1 + \rho\,(v - w)\, BH_2 = 0 \;\;\rightarrow$$

$$H_1 + H_2 = H\,.$$

Thus, solving for $H_{1,\,2}$ renders

$$H_{1,\,2} = \frac{H}{2}\left(1 \pm \cos\alpha\right)\,.$$

The final expression of the force acting on the plate becomes

$$F_w = \rho\,(v - w)^2\, BH \sin\alpha\;,\;\;v > w\;.$$

Its theoretical power is given by

$$P = \rho\,w\,(v - w)^2\, BH \sin^2\alpha\;,\;\;v > w\;.$$

Considering the component of the force $\mathbf{F_w}$ in the direction of the impinging jet and the lateral component allows an interpretation in terms of effective drag and effective lift, respectively,

$$F_D = c_D\,\frac{\rho\,(v - w)^2}{2}\, BH \;\rightarrow\; c_D = (1 - \cos 2\alpha)\;,$$

$$F_L = c_L\,\frac{\rho\,(v - w)^2}{2}\, BH \;\rightarrow\; c_L = \sin 2\alpha\;.$$

For details, see Sec. 13.2. α is the "angle of attack" of the main stream.

A 7.8: A pipe conveying a stationary flow of mass is attached to a rigid wall of the container and supported by a linear spring (see Fig. A 7.8). The hinged support and restoring force allow a vibration like

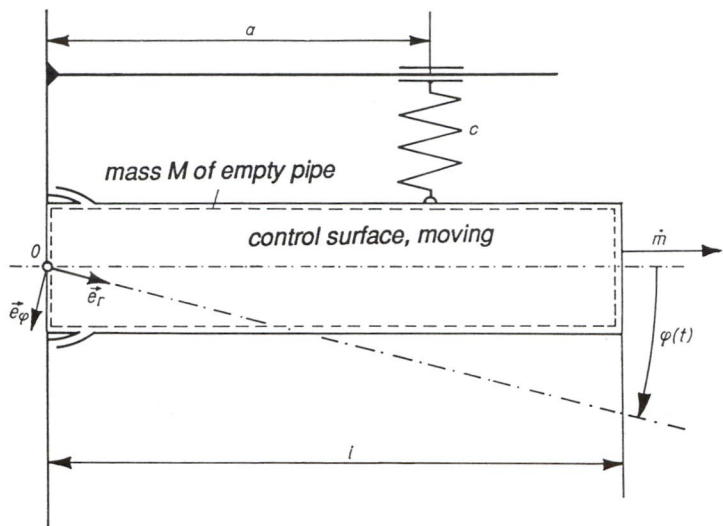

Fig. A 7.8. Vibrational motion in the horizontal plane

that of a planar pendulum. Determine the equation of motion if the inviscid fluid flow enters atmospheric conditions of assigned pressure p_0.

Solution: The angular momentum vector of the fluid becomes directionally time-dependent

$$\mathbf{H}_0 = \int_0^l (\mathbf{r} \times \mathbf{v})\, \rho A\, dr = \frac{\rho A l}{3}\, l^2\, \dot\varphi\, \mathbf{e}_z, \quad \mathbf{r} = r\, \mathbf{e}_r, \quad \mathbf{v} = \frac{\dot m}{\rho A}\, \mathbf{e}_r + r\dot\varphi\, \mathbf{e}_\varphi, \quad \dot{\mathbf{H}}_0 = \frac{\rho A l}{3}\, l^2\, \ddot\varphi\, \mathbf{e}_z.$$

Equation (7.26) renders for the fluid

$$\frac{\rho A l}{3}\, l^2\, \ddot\varphi + \dot m l^2\, \dot\varphi = -M_w.$$

A second equation is derived by considering the pendulum without fluid mass but loaded by the resulting moment M_w, and the restoring moment of the spring

$$\frac{M l^2}{3}\, \ddot\varphi = M_w - \frac{c a^2}{2}\, \sin 2\varphi.$$

Thus, M_w can be eliminated by adding the equations, and the result is a damped vibration, which is the solution of

$$\ddot{\varphi} + R\dot{\varphi} + \frac{\omega_0^2}{2} \sin 2\varphi = 0 \quad , \quad R = \frac{3\dot{m}}{M + \rho Al} \quad , \quad \omega_0^2 = \frac{3c}{M + \rho Al} \left(\frac{a}{l}\right)^2 .$$

A 7.9: Determine the natural torsional vibrations of a homogeneous linear elastic cantilever with a circular cross-section of diameter *d* and of a given length *l* . The density is *ρ* and shear modulus of elasticity is *G* .

Solution: The conservation of angular momentum of an infinitesimal element renders the wave equation in terms of the angle of twist

$$\frac{\partial^2 \chi}{\partial t^2} = c^2 \frac{\partial^2 \chi}{\partial x^2} \quad , \quad \vartheta = \frac{\partial \chi}{\partial x} \quad , \quad c^2 = \frac{GJ_T}{I} = \frac{G}{\rho} \quad , \quad I = \rho J_T \quad , \quad J_T = \frac{\pi d^4}{32} .$$

Solution in the form of separable functions, $\chi(x, t) = Y(t) \, \varphi(x)$, yields

$$\frac{\ddot{Y}}{Y} = c^2 \frac{\varphi''}{\varphi} = -\omega^2 \quad \rightarrow \quad \left(\begin{array}{l} Y(t) = A \cos \omega t + B \sin \omega t \\[2mm] \varphi(x) = C \cos \left(\frac{\omega}{c} x\right) + B \sin \left(\frac{\omega}{c} x\right) \end{array} \right.$$

The geometric boundary condition at the clamped edge, $\chi(x = 0, t) = 0$, and the dynamic condition at the free end, $M_T (x = l, t) = GJ_T \, \partial\chi/\partial x = 0$, both render $C = 0$ and determine the natural frequencies as well. A discrete spectrum results

$$\omega_n = \frac{c}{l} \frac{2n - 1}{2} \pi \quad , \quad n = 1, 2, 3, \dots$$

A 7.10: Does a both statically and dynamically balanced shaft still show critical rotational speeds? In case it does, then the deflected shaft (bending stiffness *B = EJ*) also must rotate with the same stationary speed *ω* as the (undeformed) shaft with a straight axis.

Solution: The rotating deflected shaft is loaded by the inertia forces per unit of length $q = \rho A \, \omega^2 w$. Thus, the stationary differential equation of principal bending, $B \, d^4w/dx^4 = q$, becomes

$$\frac{d^4 w}{dx^4} - \omega^2 \frac{\rho A}{B} w = 0 .$$

It is identical to Eq. (7.152). If the boundary conditions are also identical, the same frequency equation and, hence, the same discrete spectrum of circular frequencies of bending vibrations result. By definition, they determine the critical angular speeds of the shaft. The latter allow a rotation of the straight shaft, as well as of the deflected (frozen) configuration. As a consequence, the rotational speed is chosen as $\omega \neq \omega_n$, $n = 1, 2, 3, \ldots$, unequal to the critical speeds.

A 7.11: Consider the free vibrations of isotropically stressed membranes of a rectangular and circular planform. S is the given in-plane hydrostatic force per unit of length; see Fig. A 7.11.

Solution: The lateral load p of Eq. (6.191) is substituted by the inertial force $- \rho h \, d^2w/dt^2$, and the resulting equation of motion is reduced in time in the case of harmonic vibrations, $w = W \exp (i\omega t)$, to render the *Helmholtz* differential equation for the distribution of the amplitude W

$$\Delta W + \lambda^2 W = 0 \quad , \quad \lambda^2 = \rho h \omega^2 / S \quad , \quad \omega = \lambda \sqrt{S/\rho h} \; .$$

Under *Dirichlet* boundary conditions, $W = 0$, a real and discrete spectrum of natural frequencies corresponding to the eigenvalues λ_n, $n = 1, 2, \ldots$ exists with the associated modes of vibration W_n . In the case of the rectangular planform, the solution is separable, $W(x, y) = F(x) \, G(y)$, and two ordinary differential equations result, each one corresponding to the vibrations of a string. The solutions become compatible with the choice of the eigenvalues and associated modes

$$\lambda_{mn}^2 = \pi^2 \left(\frac{m^2}{a^2} + \frac{n^2}{b^2} \right) \quad , \quad W_{mn} = A_{mn} \sin \frac{m\pi x}{a} \sin \frac{n\pi y}{b} \quad , \quad m, n = 1, 2, \ldots$$

The circular membrane requires the application of polar coordinates to make the solution separable, $W(r, \varphi) = F(r) \, G(\varphi)$, and with n^2 as a natural number, two differential equations result

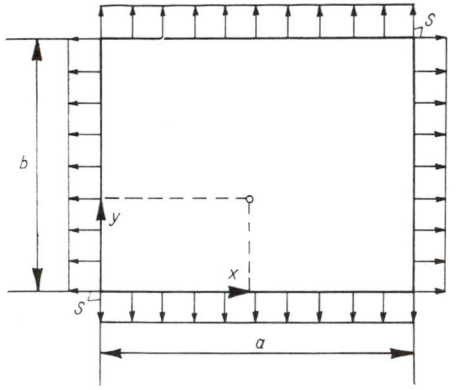
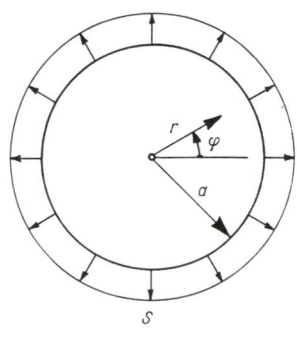

Fig. A 7.11. Lateral vibrations of prestressed membranes

$$\frac{d^2G}{d\varphi^2} + n^2 G = 0 \quad , \quad F'' + \frac{1}{r} F' + \left(\lambda^2 - \frac{n^2}{r}\right) F = 0 \ .$$

The first one is of the vibrational type, the second one is the *Bessel* differential equation. If we denote the *Bessel* functions of the first kind and of n th order by $J_n (z)$, the modes are

$$W_{nm}(r, \varphi) = A_{nm} J_n(\lambda_{nm} r) \left\{ \begin{array}{c} \cos n\varphi \\ \sin n\varphi \end{array} \right. \ , \quad n = 0, 1, 2, ..., \quad m = 1, 2, ...$$

The eigenvalues are the well-documented roots of $J_n(\lambda_{nm} a) = 0$, $(W = 0, r = a)$. The first seven roots are smaller than 6 and listed here for convenience: $\lambda_{01} a = 0.8936$, $\lambda_{11} a = 2.1971$, $\lambda_{21} a = 3.3842$, $\lambda_{02} a = 3.9577.$, $\lambda_{31} a = 4.5270.$, $\lambda_{12} a = 5.4297.$, $\lambda_{41} a = 5.6452$.

A 7.12: Show by means of the *Helmholtz* decomposition of the displacement vector $\boldsymbol{u} = \nabla \Phi + \nabla \times \boldsymbol{\Psi}, \ \boldsymbol{\Psi} = \Psi \boldsymbol{e}_y , \ \partial\Psi/\partial y = 0$, the existence of an inhomogeneous plane surface wave that propagates without dispersion with a constant speed $c_R < c_S < c_P$ along the traction-free surface $z = 0$ of a linear elastic half-space (in the direction x). See Fig. A 7.12. For simplicity, a monochromatic wave with assigned frequency ω , (ie varying harmonically in time) is assumed.

Solution: The nonvanishing displacement components are given by

$$u = \frac{\partial\Phi}{\partial x} - \frac{\partial\Psi}{\partial z} \quad , \quad w = \frac{\partial\Phi}{\partial z} + \frac{\partial\Psi}{\partial x} \ .$$

Inhomogeneous plane waves have the z dependence of the amplitudes

$$\Phi = \varphi(z) \, e^{i\omega(t - x/c)} \quad , \quad \Psi = \psi(z) \, e^{i\omega(t - x/c)} \ ,$$

and the body waves must have a common phase velocity c , which is the speed of the plane $x = const$ of constant phase. Substitution into the homogeneous wave equations (7.163) renders two ordinary differential equations with the solutions $\varphi(z) = C \, e^{-b_1 z}$ and $\psi(z) = D \, e^{-b_2 z}$. The latter are properly decaying when z goes to infinity, if the positive roots of $b_{1, 2}{}^2 = k^2 (1 - c^2/c_{P, S}{}^2)$ are selected. The common wave number is $k = \omega/c$. At the surface, the tractions must vanish, $\sigma_{zz} = \sigma_{zx} = 0$ at $z = 0$. Hence, the phase velocity is the eigen-value of the resulting two linear and homogeneous equations. *Hooke's* law with linearized geometric relations must be considered

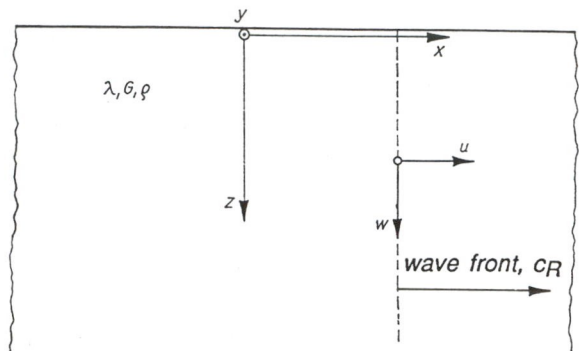

Fig. A 7.12. The *Rayleigh-* surface wave. Retrograde motion of the points at the free surface

$$\sigma_{zz} = \left[\lambda \frac{\partial^2 \Phi}{\partial x^2} + (\lambda + 2G) \frac{\partial^2 \Phi}{\partial z^2} + 2G \frac{\partial^2 \Psi}{\partial z \partial x} \right]_{z=0}$$

$$= \left[(c_P^2 - 2c_S^2)(-k^2) + c_P^2 b_1^2 \right] C + 2i\, k\, c_S^2 b_2\, D = 0 \;,$$

$$\sigma_{zx} = G \left[2 \frac{\partial^2 \Phi}{\partial z \partial x} - \frac{\partial^2 \Psi}{\partial z^2} + \frac{\partial^2 \Psi}{\partial x^2} \right]_{z=0} = 2i\, k\, b_1\, C + (b_2^2 + k^2)\, D = 0 \;.$$

The existence of a nontrivial solution requires the determinant of the coefficients, the *Rayleigh* determinant, to vanish: The characteristic equation has an approximate solution that is commonly of sufficient numerical accuracy

$$\left(2 - \frac{c^2}{c_S^2} \right)^2 - 4 \sqrt{1 - \frac{c^2}{c_P^2}} \sqrt{1 - \frac{c^2}{c_S^2}} = 0 \;\;\rightarrow$$

$$c_R \cong \frac{0.862 + 1.14\, v}{1 + v} \quad c_S < c_S < c_P \;.$$

The amplitude vector of the *Rayleigh* surface wave is real and is determined by the ratio

$$\frac{C}{D} = -\frac{1 - c_R^2/2c_S^2}{\sqrt{1 - c_R^2/c_P^2}} \;.$$

This representation illustrates that the surface wave is a special superposition of the P- and SV-waves with a common wave number

$$\Phi = C\, e^{-b_1 z}\, e^{i\omega(t - x/c_R)} \;\;, \quad \Psi = D(C)\, e^{-b_2 z}\, e^{i\omega(t - x/c_R)} \;.$$

8
First Integrals of the Equations of Motion, Kinetic Energy

The equations of motion are second-order time differential equations. By a single integration, their order can be reduced by one. Simple time integrations are discussed that render a general theorem on work related to the increase of kinetic energy and possibly, for idealized systems, to the conservation of mechanical energy. Integration along a streamline is performed by keeping the time fixed, with the result known as the generalized *Bernoulli* equation. The latter becomes a first integral as well in the case of steady flow.

8.1. The Power Theorem and Kinetic Energy

The *Euler-Cauchy* vector equation of motion (7.1) is the starting point of constructing the power density per unit of volume in each material point of a moving body by scalar multiplication, with the absolute velocity $v(t, r)$

$$\frac{dP}{dV} = P' = f \cdot v = \rho a \cdot v \quad , \quad a = \frac{dv}{dt} \quad . \tag{8.1}$$

Integration (summation) over the material volume $V(t)$ of the body (time is kept constant) renders the instant value of the total power of the internal and the external forces $P(t)$

$$\int_{V(t)} P' dV = \frac{\delta W}{dt} = P(t) = \int_{V(t)} \rho \frac{d}{dt} \left(\frac{v^2}{2}\right) dV = \frac{d}{dt} \left[\frac{1}{2} \int_{V(t)} v^2 \, dm\right] \quad ,$$

$$a \cdot v = \frac{1}{2} \frac{d}{dt} (v \cdot v) \quad . \tag{8.2}$$

The positive semidefinite expression that solely depends on the distribution of the squared speed of the mass elements,

$$E_k \equiv T(t) = \frac{1}{2} \int_{V(t)} v^2 \, dm \geq 0 \quad , \tag{8.3}$$

is the *kinetic energy* of the body with the dimension of work and, hence, of energy $[T] = 1 \ kg \ (m/s)^2 = 1 \ Nm = 1 \ Ws = 1 \ J$. For a rigid body of mass m in translational motion, the kinetic energy is simply given by

$$E_k \equiv T(t) = \frac{mv^2}{2} \quad . \tag{8.4}$$

The same expression holds for a point mass with instant velocity $v(t)$ in arbitrary motion.

Thus, Eq. (8.2) renders the rate of kinetic energy of a body of constant mass m in a material volume $V(t)$ equal to the total power of the internal and external forces

$$\frac{dE_k}{dt} \equiv P(t) = \frac{\delta W}{dt} \quad . \tag{8.5}$$

This power theorem is considered in addition to the first law of thermodynamics. A time integration renders the law of conservation of kinetic energy

$$E_k(t_2) - E_k(t_1) \equiv T_2 - T_1 = \int_{t_1}^{t_2} P \, dt = W_{1 \to 2} \quad . \tag{8.6}$$

<<The finite increase of the kinetic energy of a body equals the total work of the internal and external forces in the time interval elapsed.>> If the total amount of work vanishes, the kinetic energy is conserved. Equation (8.6) contains only the velocities and, thus, is a first integral of the second-order equations of motion of general validity. It is of special value for checking the (numerical) solution of dynamic problems.

8.2. Conservation of Mechanical Energy

If the internal and external forces are stationary and irrotational, the associated potentials U and W_P are not explicitly time-dependent, and the total work is path-independent [given by the difference of the total potential energy, Eq. (3.18)]

$$W_{1 \to 2} = E_p(t_1) - E_p(t_2) \equiv V_1 - V_2 \quad , \quad E_p \equiv V = U + W_P \quad .$$

Substitution into Eq. (8.6) renders

$$T_2 - T_1 = V_1 - V_2 \quad , \tag{8.7}$$

<<The increase of kinetic energy equals the decrease of the total potential energy.>> Rearrangement expresses the conservation of mechanical energy

$$T_1 + V_1 = T_2 + V_2 \quad \text{or} \quad E_k(t) + E_p(t) = E_0 = \text{const} \quad , \tag{8.8}$$

<<The sum of mechanical energy (kinetic plus total potential energy) of a body in motion through a conservative field of (internal and external) forces remains constant.>> The sum of mechanical energy is conserved if the forces are stationary and irrotational: Thus, the field of forces is called *conservative* , as well as synonymous with such a mechanical system in motion. Equation (8.8) is applied for checking the solution of the equations of motion of conservative mechanical systems, and especially, it can be used for the determination of stability of a state of equilibrium in energy norm; see Sec. 9.1. For conservative SDOF-systems, Eq. (8.8) becomes the first-order (integrated) equation of motion. It is always nonlinear; see also Eq. (7.118) on a pendulum in plane motion. Equation (8.8) taken per unit of mass or volume, and considered along a streamline of an ideal (inviscid) fluid in stationary flow, becomes the classical *Bernoulli* equation [see Eq. (8.36)].

Before any further applications are considered, kinetic energy for rigid bodies is evaluated and put into a convenient form.

8.3. Kinetic Energy of a Rigid Body

The velocity field of an undeformable body is linearly distributed according to Eq. (1.4) and, by putting the material vector $r_{PA} = r'$, it may be substituted into the definition

$$E_k(t) = \int_m \frac{v^2}{2} \, dm \quad , \tag{8.9a}$$

to render three terms. A' is the point of reference fixed in the rigid body, ω is the angular velocity and the scalar triple product is reformulated according to the rules of vector algebra, $v \cdot v = (v_A + \omega \times r')^2$,

$$E_k(t) = \frac{m \, v_A^2}{2} + (v_A \times \omega) \cdot \int_m r' \, dm + \frac{1}{2} \int_m |\omega \times r'|^2 \, dm \quad . \tag{8.9b}$$

The choice of the mass center M as the point of reference A′ renders the static mass moment as zero and the second term vanishes identically. The remaining volume integral is evaluated by considering a cartesian frame with its origin at the mass center: $r' = x\,e_x + y\,e_y + z\,e_z$, and $\omega = \omega_x\,e_x + \omega_y\,e_y + \omega_z\,e_z$. The integrand becomes $|\omega \times r'|^2 = (z\omega_y - y\omega_z)^2 + (x\omega_z - z\omega_x)^2 + (y\omega_x - x\omega_y)^2$, and the integrals, by definition, are the moments of inertia about the central axis, Eqs. (7.52) through (7.54). In matrix notation with the inertia tensor I , the kinetic energy of a rigid body is given by

$$E_K(t) = \frac{mv_M^2}{2} + \frac{1}{2}\,\omega^T \cdot I \cdot \omega \quad .$$

(8.10)

With the mass center as the point of reference, the kinetic energy becomes the sum of the translational kinetic energy (mass is considered to be concentrated in the mass center) and of the rotatory energy. With respect to principal axes, the latter is given by (the inertia tensor is diagonal)

$$\frac{1}{2}\,\omega^T \cdot I \cdot \omega = \frac{1}{2}\left(I_1\omega_1^2 + I_2\omega_2^2 + I_3\omega_3^2\right) \quad .$$

(8.11a)

Two important special motions are considered subsequently.

8.3.1. Pure Rotation of the Rigid Body About a Fixed Point 0

The point of reference, A′ = 0, is fixed to the body as well as in space, $v_A \equiv 0$, and its kinetic (rotatory) energy can be expressed formally by Eq. (8.11a), with the inertia tensor evaluated with respect to the point 0

$$E_k = \frac{1}{2}\,\omega^T \cdot I_O \cdot \omega = \frac{1}{2}\left(I_1^{(O)}\omega_1^2 + I_2^{(O)}\omega_2^2 + I_3^{(O)}\omega_3^2\right) \quad .$$

(8.11b)

The principal axes of inertia now pass through the fixed point 0.

8.3.2. Rotation About an Axis e_a Fixed in Space

If we consider the axis passing through A′ = 0, again $v_A \equiv 0$, and, in addition, the angular velocity has a constant line of action, ie $\omega = \omega(t)\,e_a$. The kinetic energy becomes with the axial moment of inertia I_a

$$E_k = \frac{1}{2}\,I_a\,\omega^2 \quad .$$

(8.12)

8.4. Conservation of Energy in SDOF-Systems

The time-integrated power theorem of nonconservative systems and conservation of mechanical energy both render first-order equations of motion, and as is quite important for applications, the characteristic deformations are determined by finite relations.

8.4.1. Motion of a Linear Oscillator After Impact (Fig. 8.1)

A mass m is attached to a linear spring of stiffness c and moves (after impact) with the initial velocity v_0 . The first turning point of the oscillatory motion is to be determined; light damping is of such little influence that it is negligible. Hence, the conservation of mechanical energy applies especially between the initial state and the turning point in a distance a (where the kinetic energy vanishes the first time)

$$E_0 = \frac{1}{2} m v_0^2 = \frac{1}{2} c a^2 \ \rightarrow a = v_0 \sqrt{m/c} \ . \tag{8.13}$$

Since the natural circular frequency is determined by the reciprocal of the above root, Eq. (7.80), the amplitude of the undamped free vibration about the state of equilibrium becomes $a = v_0 /\omega_0$, and $x(t) = a \sin \omega_0 t$.

Conservation of energy applies at any instant of time and gravity has no influence on the natural vibrations; see Fig. 8.1

$$T + U = E_0 = \frac{1}{2} m v_0^2 \ .$$

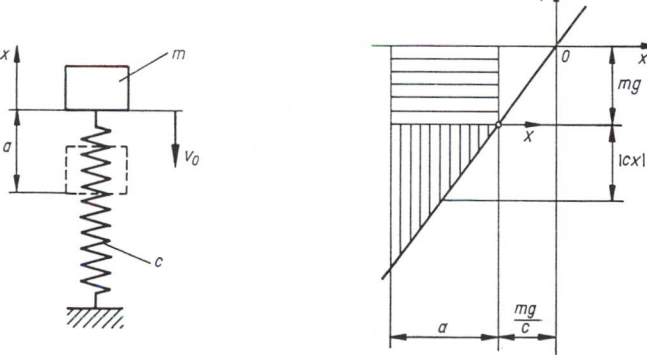

Fig. 8.1. Turning point of a linear oscillator after impact and the force displacement diagram showing the relative elastic energy stored in the spring (prestressing occurs through the self-weight of the mass)

The nonlinear first-order equation of motion results

$$\frac{1}{2}\left(m\dot{x}^2 + cx^2 = mv_0^2\right) \;\rightarrow\; \dot{x}^2 = v_0^2 - \omega_0^2\, x^2 \;.$$

Substitution of the time-harmonic functions gives

$$\frac{m}{2}\, a^2\omega_0^2 \cos^2 \omega_0 t + \frac{c}{2}\, a^2 \sin^2 \omega_0 t = E_0 \;.$$

When we consider the time instants where the kinetic energy vanishes, $T = 0$, and that the elastic potential $U = U_{max} = ca^2/2 = E_0$ is maximal, $t = \pi/2\omega_0$, $3\pi/2\omega_0$, ..., and contrary to this, the time instants $t = 0,\ \pi/\omega_0,\ 2\pi/\omega_0,\ ...$, where the kinetic energy takes on the maximal value $T = T_{max} = ma^2\omega_0^2/2 = E_0$ and the strain energy is zero, $U = 0$, the natural circular frequency ω_0 follows also from the condition of energy conservation. *Rayleigh's quotient* renders

$$T_{max} = \frac{m}{2}\, a^2\omega_0^2 = E_0 = U_{max} = \frac{c}{2}\, a^2 \;\rightarrow\; \omega_0^2 = \frac{U_{max}}{\left(1/\omega_0^2\right) T_{max}} \;. \qquad (8.14)$$

The relation is independent of the initial conditions and gives $\omega_0 = \sqrt{(c/m)}$. Nontrivially, the denominator T_{max}/ω_0^2 remains independent of the natural frequency also in the case of a vibrational MDOF-system.

8.4.2. The Basic Vibrational Mode of a Linear Elastic Beam

Again, light material damping may be neglected and the *Rayleigh* quotient (8.14) follows from the conservation of mechanical energy taken over one-half period

$$T_{max} = E_0 = U_{max} \;. \qquad (8.15)$$

An approximate value of the basic natural frequency results if the mode shape $\varphi(x)$ is assumed to be affine to the static deflection due to the beam's own weight: $q_0 = \rho g A$. The maximal elastic potential

$$U_{max} = \frac{1}{2} \int_0^L \rho g A\, \varphi(x)\, dx \;, \qquad (8.16)$$

is substituted together with the kinetic energy of the time-harmonic vibration $w(x, t) = a\, \varphi(x) \cos \omega_0 t$ ($a = 1$ is assumed, and its maximum is arrived at $t_1 = \pi/2\omega_0$)

$$T_{max} = \frac{1}{2} \int_0^L \rho A \, \dot{w}^2(t_1, x) \, dx = \frac{\omega_0^2}{2} \int_0^L \rho A \, \varphi^2(x) \, dx \; .$$

$$(8.17)$$

Rayleigh's quotient in the first approximation yields the result (the integrals may be evaluated numerically)

$$\omega_0^2 \cong g \, \frac{\int_0^L \rho A \, \varphi(x) \, dx}{\int_0^L \rho A \, \varphi^2(x) \, dx} \; .$$

$$(8.18)$$

The ratio (8.18) becomes exact if the first mode $\varphi = \varphi_0 \, (x)$ is substituted. Any admissible shape function renders the approximation of the natural frequency from above. The error is reduced by an iterative procedure: The deflection due to the inertia loading $q_1 = \rho g A \, \omega_0^2 \, \varphi(x)$ is $\omega_0^2 a_1 \, \varphi_1 \, (x)$ and is closer to the basic mode $\varphi_0 \, (x)$. Analogous to Eq. (8.16), the maximal energies are expressed by the integrals over the span to get

$$U_{max} = \frac{1}{2} \int_0^L q_1 \omega_0^2 a_1 \varphi_1(x) \, dx = \frac{a_1}{2} \, \omega_0^4 \int_0^L \rho g A \, \varphi(x) \, \varphi_1(x) \, dx \; ,$$

$$(8.19)$$

and

$$T_{max} = \frac{a_1^2}{2} \int_0^L \rho A \left[\omega_0^3 \, \varphi_1(x) \right]^2 dx \; .$$

$$(8.20)$$

Hence, the first step in the iteration gives an approximation of the squared natural frequency ($U_{max} = T_{max}$)

$$\omega_0^2 = g \, \frac{\int_0^L \rho A \, \varphi(x) \varphi_1(x) \, dx}{a_1 \int_0^L \rho A \, \varphi_1(x)^2 \, dx} \; .$$

$$(8.21)$$

It may be considered a generalized *Rayleigh* quotient, and any admissible shape function $a_0 \, \varphi(x)$ can be substituted for the static deflection $g \, \varphi(x)$. Essential for any shape function to be admissible are the kinematic boundary conditions of the beam; see also the *Ritz* approximation in Sec. 11.1. Equation (8.21) is not restricted to the bending vibrations of beams. The integrals can be extended over the domain of any structure, eg to frames, plates, and shells. For an

application of the generalized *Rayleigh* quotient within the matrix
iteration of natural vibrations, see Exercise A 11.11.

8.4.3. Acceleration of a Motorized Vehicle

The power law (8.5) renders explicitly the acceleration *a* of any
powered vehicle, eg with four wheels, that moves uphill (slope α)
with speed *v* . The nonconservative system of total mass *m* with the
(external) moment *M* acting on the wheels under a perfect rolling
condition, $\omega = v/R$, has kinetic energy (the body plus four wheels are
considered; the rotatory energy of the transmission can easily be
included)

$$E_k = \frac{m\,v^2}{2} + 4\frac{I}{2}\,(v/R)^2 \ .$$

(8.22)

I is the axial moment of inertia of a single wheel (possibly with the
portion of the effective moment of inertia of the rotating
transmission lines included). Hence, the rate of kinetic energy
becomes

$$\frac{dE_k}{dt} = P = M\frac{v}{R} - (D + mg\sin\alpha)v \ .$$

(8.23)

D is the total drag force. Taking the time derivative of Eq. (8.22)
renders (the speed *v* cancels)

$$a = \frac{dv}{dt} = \frac{M/R - D - mg\sin\alpha}{m + 4\,I/R^2} \ .$$

(8.24)

8.4.4. The Turning Points of a Nonlinear, Dry-Friction Oscillator

The model according to Fig. 7.26, with *Coulomb´s* law of resistance
$T_R = -\mu mgv/|v|$, is considered in free motion and the turning points
are calculated by means of the power law. The coordinate s_k is
counted from the individual successive turning points, and the time-
integrated power law renders

$$0 = W_{1\to2} = \int_0^{s_1} T_R\,dx + U_1 - U_2 \ , \ U_1 = cs_0^2/2 \ , \ U_2 = c\,(s_0 - s_1)^2/2 \ .$$

(8.25)

The distance to the next turning point is the nonvanishing solution
of the factored equation

$$s_1 = 2\,(s_0 - s) > 0 \ , \ cs_0 > cs \equiv \mu mg \ .$$

(8.26)

The work of the friction force $|T_R| \, s_1 = U_1 - U_2$ is dissipated in the form of heat. The next turning point is reached under the condition

$$cs_1 > cs \equiv \mu mg \, , \, 0 = W_{2 \to 3} = \int_0^{S_2} T_R \, dx + U_2 - U_3 \, , \, U_3 = c \, (s_0 - s_1 + s_2)^2 \, /2 \, ,$$

(8.27)

which reduces to $s_2 \, [s_2 - 2(s_1 - s_0 - s)] = 0$. The nontrivial solution is

$$s_2 = (2s_0 - 6s) \quad , \quad \text{etc.} \tag{8.28}$$

The amplitude decays with the decrement $2s$ and the successive values are

$$|x_0| = s_0 \, , \, |x_1| = s_0 - 2s \, , \, |x_2| = s_2 - x_1 = s_0 - 4s \, , \, \ldots \tag{8.29}$$

8.5. *Bernoulli's* Equation of Fluid Mechanics

The relationship of pressure and (subsonic) velocity in inviscid flow is of crucial importance in a *eulerian* representation. The vector equation of motion in the absence of any shear stresses

$$\rho \mathbf{a} = \mathbf{k} - \text{grad } p \quad , \tag{8.30}$$

may be integrated along a selected streamline, keeping the time constant (see Fig. 8.2). A projection of Eq. (8.30) in the tangential direction $\mathbf{e}_t$ and integration over the arc-length s give

$$\int_{s_1}^{S_2} \mathbf{a} \cdot \mathbf{e}_t \, ds = \int_{s_1}^{S_2} \frac{1}{\rho} k_t \, ds - \int_{s_1}^{S_2} \frac{1}{\rho} \frac{\partial p}{\partial s} \, ds \, , \, t = \text{const} \, .$$

(8.31)

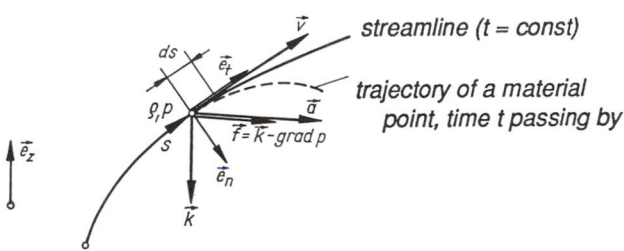

streamline (t = const)

trajectory of a material point, time t passing by

Fig. 8.2. Streamline of a nonstationary flow at $t = const$. The dashed curve is the pathline of an individual particle

Locally, the differential $v\, dt = e_t\, ds$ of the pathline of the material point considered in Eqs. (8.1) and (8.6) is replaced by $e_t\, ds$ of the streamline that, at time t, is the material line of an infinite number of particles. The tangential component of the body force $(k \cdot e_t) = k_t$ and the projection of the pressure gradient give simply $(grad\ p) \cdot e_t = \partial p/\partial s$. The tangential component of acceleration in a *eulerian* description of nonstationary flow is given by $a \cdot e_t = \partial v/\partial t + \partial(v^2/2)/\partial s$; see Eq. (1.42). Hence, integration of the left-hand side yields $[v(s_j, t) = v_j, j = 1, 2]$

$$\int_{s_1}^{s_2} \frac{\partial v}{\partial t}\, ds + \frac{1}{2}\left(v_2^2 - v_1^2\right) .$$

(8.32)

Local acceleration and, hence, the remaining integral in (8.32) are to be considered in nonstationary flows. If the body force is due to gravity, a parallel force field is assumed, $k = -\rho g\, e_z$, and the tangential component per unit of mass becomes $(1/\rho)k_t = -g\,(e_z \cdot e_t)$. However, the direction cosine renders simply $(e_z \cdot e_t)\, ds = dz$, and integration yields the difference of the potential energy per unit of mass according to the difference in the geodesic height of the two points of the streamline with respect to a common horizontal reference plane

$$\int_{s_1}^{s_2} \frac{1}{\rho} k_t\, ds = -g \int_{s_1}^{s_2} dz = -g\,(z_2 - z_1) .$$

(8.33)

The constitutive relation of a barotropic fluid is the (nonlinear) density pressure relation, $\rho = \rho(p)$; temperature has been eliminated. And in that case, the remaining integral in Eq. (8.31), the pressure potential, can be evaluated. In the case of an incompressible flow of a homogeneous fluid, $\rho = const$, and

$$\int_{s_1}^{s_2} \frac{1}{\rho} \frac{\partial p}{\partial s}\, ds = \frac{p_2 - p_1}{\rho} , \quad p_j = p(s_j, t) , \quad j = 1, 2 .$$

(8.34)

Thus, Eq. (8.31) when considered along a streamline of an ideal fluid in a homogeneous gravity field takes on the simple form

$$\int_{s_1}^{s_2} \frac{\partial v}{\partial t}\, ds + \frac{1}{2}\left(v_2^2 - v_1^2\right) = -g\,(z_2 - z_1) - \frac{p_2 - p_1}{\rho} .$$

(8.35)

The left-hand side determines the increase in kinetic energy per unit of mass along the streamline of the nonstationary flow at a constant time. The right-hand side represents the negative difference of potential energy per unit of mass of the gravity field and of the internal forces, respectively. These work terms are fictitious since the particles, in general, do not move along the streamline in a nonstationary flow.

In stationary flow, all the state variables do not explicitly depend on time, $\partial v/\partial t \equiv 0$, and Eq. (8.35) takes on the form of conservation of mechanical energy per unit of mass since the material points sitting on the streamline move along, time may pass, and the pathlines coincide with the streamline

$$\frac{v_1^2}{2} + \frac{p_1}{\rho} + gz_1 = \frac{v_2^2}{2} + \frac{p_2}{\rho} + gz_2 \ .$$

$$(8.36)$$

The conservation of mechanical energy per unit of mass holds for all points of a single streamline (and pathline), and the *Bernoulli* equation takes on one of its classical forms

$$\frac{v^2}{2} + \frac{p}{\rho} + gz = \text{const} \ .$$

$$(8.37)$$

The total energy per unit of mass may change from streamline to streamline. According to Eq. (8.8) the terms per unit of mass are the kinetic energy, the pressure potential (the potential of the internal forces), and the gravity potential (the potential of the external forces). Dividing by g renders all energy terms with a dimension of length or height, and the technical form of *Bernoulli's* equation is given by

$$\frac{v^2}{2g} + \frac{p}{\rho g} + z = h_E = \text{const} \ .$$

$$(8.38)$$

The terms are illustrated in Fig. 8.3 and z is recognized to be the geodesic height above a constant horizontal reference plane, positive upward, $p/\rho g$ is the pressure head [see also Eq. (2.85)], and $v^2/2g$ is the remaining velocity head to the constant energy level h_E , assigned to this streamline of the ideal flow.

Gravity can be eliminated by considering the static pressure in the fluid body at rest (see Sec. 2.3.1),

$$p_s + \rho gz = \text{const} \ ,$$

$$(8.39)$$

and by subtraction from the *Bernoulli* equation that is premultiplied by the density ρ (all terms are of the dimension of stress)

$$\frac{\rho v^2}{2} + (p - p_s) = \text{const} \ .$$

$$(8.40)$$

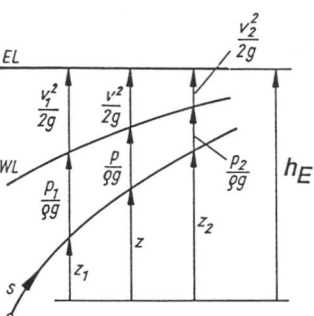

Fig. 8.3. Streamline of a stationary, incompressible, and inviscid flow
with the superposition of partial energy heads

The sum of the kinetic energy per unit of volume (the stagnation pressure) and the dynamic pressure, $p_d = p - p_s$, is a constant of the streamline, and the geodesic height is eliminated as long as no boundary conditions are to be expressed in terms of absolute pressure (eg if there is no free surface).

If the stationary flow is also irrotational, *curl v = 0* , the (convective) acceleration expressed by Eq. (1.44) becomes a gradient field, and *Euler´s* equation of the inviscid flow relates the two gradients

$$\rho \, \text{grad} \, \frac{v^2}{2} = - \text{grad} \, (W' + p) \ , \ W' = \rho g z \ .$$

(8.41)

Projection in any direction in space and integration render the conservation of mechanical energy per unit of volume with a universal *Bernoulli* constant (which is then equal for all the streamlines). Assuming the flow to be irrotational is a sufficient condition

$$\frac{\rho v^2}{2} + p + W' = \text{Const} \ .$$

(8.42a)

Elimination of gravity yields the sum of the stagnation pressure and dynamic pressure universal constant

$$\frac{\rho v^2}{2} + p_d = \text{Const} \ .$$

(8.42b)

The constant is universal, eg if all the streamlines of the inviscid flow originate from some large basin of liquid in which the velocities are so small that their squares can be neglected.

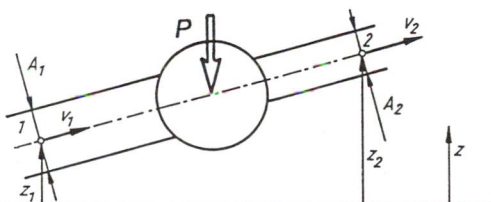

Fig. 8.4. Stationary incompressible flow with
power charging, $P > 0$, or discharging, $P < 0$

Equation (8.40) is applied to the stationary flow through pipes. In lifelines, the viscous loss of pressure head must be taken into account (see the subsequently performed generalization).

8.5.1. Stationary Flow with Power Charging or Discharging

The one-dimensional flow through a pipe is considered in which through a pump the power is supplied to the stream of mass (see Fig. 8.4). The flow is assumed to be stationary and incompressible. Without going into the details of the force distributions at the surfaces of the blades of the (rotating) pump, let us say that fictitious body forces of density k^* act in the interior of the pump supplying the additional work. Since a streamline is also a pathline, the generalized *Bernoulli* equation can be derived according to Eq. (8.6) that is taken per unit of mass

$$\frac{1}{2} \left(v_2^2 - v_1^2 \right) = -g \left(z_2 - z_1 \right) - \frac{1}{\rho} \left(p_2 - p_1 \right) + \int_{S_1}^{S_2} \frac{1}{\rho} \, k_t^* \, ds \; .$$

(8.43)

The mass flow rate

$$\dot{m} = \rho v_1 A_1 = \rho v_2 A_2 \; ,$$

is constant, and, in a time differential dt, the mass element dm is charged with the work $P \, dt$ in the pump. Hence, the work per unit of mass is

$$\frac{\delta W}{dm} = \frac{P \, dt}{\dot{m} \, dt} = \frac{P}{\dot{m}} = \int_{S_1}^{S_2} \frac{1}{\rho} \, k_t^* \, ds \; .$$

(8.44)

Substitution into Eq. (8.43) renders the increase of kinetic energy density equal to the total work of the internal and external forces [$P > 0$ charges the flow with the power (through pumping), $P < 0$ discharges the flow, eg in a turbine]

$$\frac{1}{2}\left(v_2{}^2 - v_1{}^2\right) = -g\left(z_2 - z_1\right) - \frac{1}{\rho}\left(p_2 - p_1\right) + \frac{P}{\dot{m}} \quad , \quad \dot{m} = \text{const} .$$
(8.45a)

The equation is also suitable for considering the energy loss in stationary viscous flow through the pipe. Dividing through g renders (the loss must be a pressure head h_l due to the continuity of stationary mass flow)

$$\frac{1}{2g}\left(v_2{}^2 - v_1{}^2\right) = -\left(z_2 - z_1\right) - \frac{1}{\rho g}\left(p_2 - p_1\right) - h_l \quad , \quad \dot{m} = \text{const} .$$
(8.45b)

For a discussion of *Newtonian* fluid viscosity, see Eq. (4.67) and Sec. 13.3.1. Equation (8.45b) holds also for a narrow streamtube in a general (viscous) flow if it remains stationary.

In hydraulic systems with power charging, Eq. (8.45a) is multiplied by ρ and, with the definition of total dynamic pressure, the kinetic energy per unit of volume is eliminated

$$p_t = \frac{\rho v^2}{2} + p_d \quad , \quad p_d = p - p_s \quad , \quad p_s + \rho g z = \text{const} .$$
(8.46)

The work supplied per unit of volume flow through the pump increases the total pressure by

$$p_{t2} - p_{t1} = \frac{\rho P}{\dot{m}} .$$
(8.47)

Cross-section 1 is the intake of flow into the machine and 2 is its outlet.

8.5.2. Velocity of Efflux from a Small Aperture in an Open Vessel or a Pressurized Tank (Fig. 8.5)

The fluid is emitted in the form of a free jet of cross-section $A_2 << A_1$. Atmospheric pressure at the free surface 1 in an open vessel and acting also on the free jet is assumed to be the same, p_0 . It extends into the interior of the jet. *Bernoulli´s* equation (8.38) holds approximately for the streamline shown in Fig. 8.5, where the velocity v_1 at the free surface can be assumed to be zero. With $H = const$, the flow is approximately stationary and

$$\frac{p_1}{\rho g} + H = \frac{p_2}{\rho g} + \frac{v^2}{2g} .$$
(8.48a)

By putting $p_1 = p_2 = p_0$, *Torricelli´s* theorem (who has been a student of *Galileo*) is derived

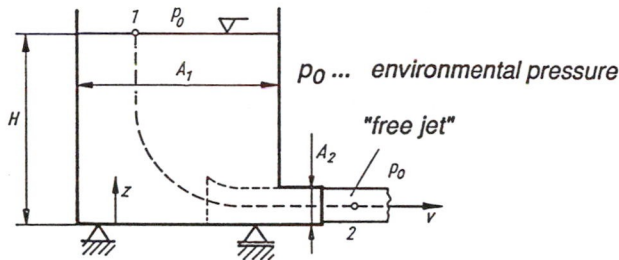

Fig. 8.5. Stationary efflux from an orifice.
Atmospheric pressure is assumed to be constant over the height H

$$v = \sqrt{2gH} \ . \qquad\qquad (8.48b)$$

The formula is simply generalized to include the overpressure in a pressurized tank. Considering the pressure $p_1 > p_0$ at the interface 1 of the fluid to the overlying pressurizing gas in Eq. (8.48a) gives

$$v = \sqrt{2g \left(\frac{p_1 - p_0}{\rho g} + H \right)} \ . \qquad\qquad (8.48c)$$

If the height H in the open vessel varies slowly with time, the flow becomes nonstationary, but the time of efflux to level $H_1 < H$ can still be determined by a quasistationary consideration (again see Fig. 8.5)

$$A_2 v \, dt = -A_1 \, dz_1 \ , \quad dt = -\frac{A_1}{A_2} \frac{dz_1}{v} \ .$$

Substituting the approximation

$$v \approx \sqrt{2g \, z_1(t)} \ ,$$

and integrating yield the time elapsed for lowering the level

$$t_e = \frac{\sqrt{2} \, A_1}{g A_2} \left(\sqrt{g H} - \sqrt{g H_1} \right) \ . \qquad\qquad (8.49)$$

8.5.3. Stationary Flow Round an Immersed Rigid Body at Rest

In the farfield, the flow is assumed to be homogeneous and parallel with the common speed v_∞ ; see Fig. 8.6. Such an inviscid flow remains irrotational also in the vicinity of an immersed rigid body. It is easily recognized that the constant in the *Bernoulli* equation

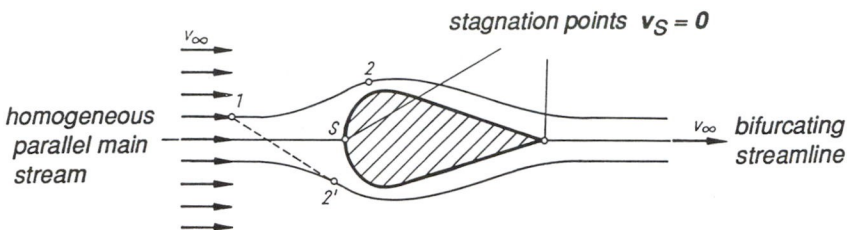

Fig. 8.6. Stationary, inviscid, and incompressible flow round an immersed rigid body at rest. Stagnation points are indicated, as well as the branching of the streamline passing through these points

does not depend on the individual streamline to be selected for application

$$\frac{v^2}{2g} + \frac{p}{\rho g} + z = \frac{v_\infty^2}{2g} + \frac{p_\infty}{\rho g} + z_\infty = \text{Const} .$$

(8.50)

Since $v_\infty = const$ and the hydrostatic pressure due to gravity is linearly distributed, $p_\infty + \rho g z_\infty = const$, the *Bernoulli* constant in Eq. (8.50) is one and the same for all the streamlines. The conservation of mechanical energy per unit of mass applies to the whole fluid body, eg at points 1 and 2′. If there is no surface with assigned pressure, gravity can be eliminated and, thus, does not influence the flow

$$\frac{v^2}{2g} + \frac{p_d}{\rho g} = \text{Const}' .$$

(8.51)

Also, *Const*′ is universal.

8.5.4. Inviscid Flow Along a Rigid Wall

When viscosity is neglected, the streamlines in the nearfield follow a given smooth rigid wall at rest. The difference in the pressure head at two points of a streamline is measured by a manometer; see Fig. 8.7. The static pressure on the left-hand side at the bottom of the fluid column is $p_l = p_1 + \rho g (H + z_1)$. On the other side of the manometer, the pressure is $p_r = p_2 + \rho g (H + z_2)$, and the difference gives $p_l - p_r = (p_1 + \rho g z_1) - (p_2 + \rho g z_2)$. *Bernoulli's* equation renders for the stationary and incompressible flow from point 1 to point 2

$$\frac{v_1^2}{2g} + \frac{p_1}{\rho g} + H + z_1 = \frac{v_2^2}{2g} + \frac{p_2}{\rho g} + H + z_2 .$$

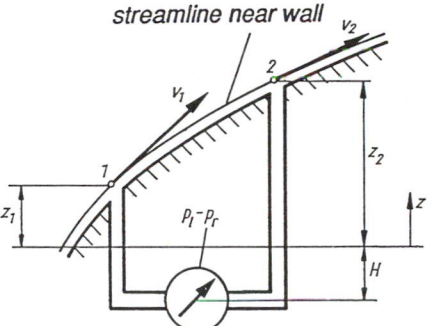

Fig. 8.7. The manometer measures the difference in the velocity head at points 1 and 2

Thus, the manometer measures the difference in the velocity head and not, as may have been expected, the pressure in the flow

$$p_l - p_r = \frac{\rho}{2} \left(v_2^2 - v_1^2 \right) .$$

8.5.5. Pressure in a Pipe Measured by a Gully

The stationary and incompressible flow through a pipe of variable cross-section is considered, and the length of the fluid columns in gullies attached to the pipe at points 1 and 2 is measured. See Fig. 8.8. The hydrostatic condition in the fluid column at rest gives the pressures $p_1 = p_0 + \rho g\, z_1$ and $p_2 = p_0 + \rho g\, (H + z_2)$, respectively, and the difference is

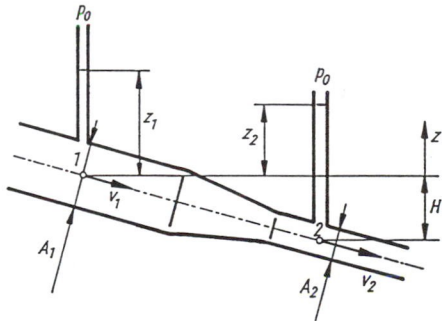

Fig. 8.8. Stationary and incompressible flow through a pipe of variable cross-section. Measurement of the pressure difference by two gullies exposed to the atmospheric pressure p_0

$$z_1 - z_2 = H + \frac{(p_1 - p_2)}{\rho g} \ .$$

When we neglect viscosity, *Bernoulli's* equation applies to the streamline from point 1 to point 2

$$\frac{v_1^2}{2g} + \frac{p_1}{\rho g} + H = \frac{v_2^2}{2g} + \frac{p_2}{\rho g} \ .$$

The pressure difference results

$$\frac{p_1 - p_2}{\rho g} = \frac{v_2^2 - v_1^2}{2g} - H \ ,$$

and renders the measurement to be proportional to the velocity head,

$$z_1 - z_2 = \frac{(v_2^2 - v_1^2)}{2g} \ .$$

Speed can be eliminated from the balance of the mass flow rate, and the measurement of the two pressure heads enters the formula

$$\dot m = \rho A_1 v_1 = \rho A_2 v_2 = \rho \sqrt{2} \ A_1 A_2 \sqrt{g \, (z_1 - z_2)/(A_1^2 - A_2^2)} \ .$$

8.5.6. Prandtl's Tube and Pitot's Tube

The measurement of the velocity profile at a cross-section of a flow can be performed by means of tubes and by the registration of pressure in a "no-flow-through" manometer. The *Prandtl* tube, eg has two openings as shown in Fig. 8.9(a), and the difference of pressure at the stagnation point S and of the slightly disturbed parallel flow at A′ is measured by the manometer. *Bernoulli's*

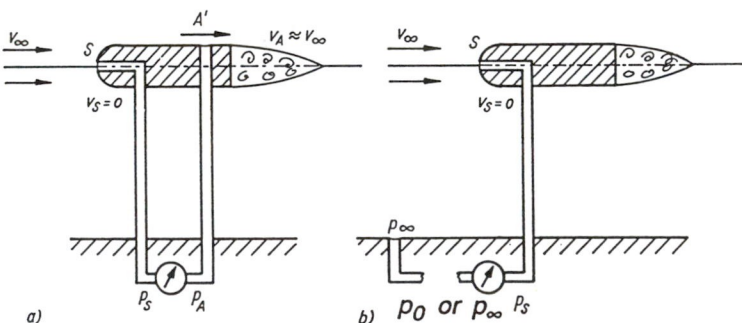

Fig. 8.9. Measuring the local speed v_∞ in a smooth flow by
(a) the *Prandtl* tube and (b) the *Pitot* tube

equation of stationary ideal flow renders, when applied to the streamline through the stagnation point,

$$\frac{\rho}{2} v_S^2 + p_S = \frac{\rho}{2} v_A^2 + p_A \ .$$

Since $v_S = 0$ and $v_A \approx v_\infty$, the tube is assumed to be sufficiently slim, and the pressure difference measured is

$$p_S - p_A \cong \frac{\rho}{2} v_\infty^2 \ . \tag{8.54}$$

A simpler device is a tube with only one opening at the stagnation point; see Fig. 8.9(b). The manometer measures the pressure difference with respect to the atmospheric pressure p_0 , or from a hole at the wall, the pressure p_∞ is considered for reference. *Bernoulli´s* equation gives in that case (any difference in geodesic height is neglected)

$$p_S - p_\infty \cong \frac{\rho}{2} v_\infty^2 \ . \tag{8.55}$$

8.5.7. Transient Flow in a Drain Pipe Controlled by a Cock

Figure 7.8 is reconsidered, but the incompressible flow is assumed to become a free jet of cross-section A_2 under atmospheric conditions after passing through the cock. The nonstationary (inviscid) flow along a streamline of length L may be generalized to include a variability of the cross-section of the drain pipe, ie to consider the speed as dependent on the arc-length s

$$v(s, t) = \frac{A_2}{A(s)} v_2 \equiv a(s) v_2 \ .$$

The generalized *Bernoulli* equation (8.35) is applied at some time instant after suddenly opening the cock and yields (the axis of the pipe is assumed to be horizontal)

$$\dot{v}_2 \int_0^L a(s) \, ds + \frac{v_2^2}{2} + \frac{p_2}{\rho} = \frac{v_1^2}{2} + \frac{p_1}{\rho} \ , \quad p_2 = p_0 \ .$$

The fluid in the large and open container is in a quasistationary flow and the free surface under pressure p_0 is kept at constant height H ; see Fig. 8.5. Thus, relating a point at the free surface with a streamline to the inlet of the drain pipe and applying *Bernoulli´s* equation give approximately

$$0 + \frac{p_0}{\rho} + gH = \frac{v_1^2}{2} + \frac{p_1}{\rho} \quad .$$

Combining the two equations and substituting an effective length L_{eff} for the integral of the (smooth) cross-sectional profile yield the first-order differential equation

$$\dot{v}_2 = \frac{v_\infty^2 - v_2^2}{2 \, L_{eff}} \quad , \quad gH = \frac{v_\infty^2}{2} \quad .$$

It is easily solved by the separation of variables to render explicitly the speed of the jet, which asymptotically approaches the *Torricelli* velocity v_∞ of Eq. (8.48)

$$v_2(t) = v_\infty \tanh \frac{v_\infty t}{2 \, L_{eff}} \quad . \tag{8.56}$$

8.5.8. Free Vibrations of a Fluid in an Open U-Shaped Pipe

The free surfaces of the fluid at rest are at equal geodesic height when exposed to the same atmospheric pressure p_0 (see Fig. 8.10). The pipe is assumed to have a constant cross-section A and the straight parts are inclined under the angles α and β , respectively. A pressure disturbance at level 1 renders a nonstationary flow that is assumed inviscid and incompressible. A streamline still has the original length L of the fluid body, but it is to be considered in an instant configuration and the nonstationary *Bernoulli* equation (8.35) must be applied [$x(t)$ is the coordinate of the free surface 1]

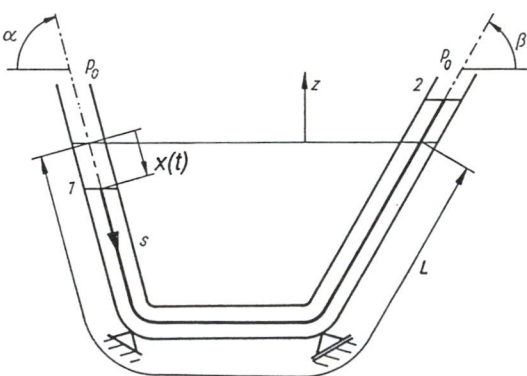

Fig. 8.10. Natural oscillations of an inviscid fluid in a pipe of constant cross-section A

$$\int\limits_{1}^{2} \frac{\partial v}{\partial t}\, ds \;+\; \frac{v_2^2}{2} \;+\; \frac{p_2}{\rho} \;+\; gz_2 \;=\; \frac{v_1^2}{2} \;+\; \frac{p_1}{\rho} \;+\; gz_1 \;,$$

$$v_1 = v_2 = v(t) = \dot{x}\;,\quad \frac{\partial v}{\partial t} = \ddot{x}\;.$$

Since $p_1 = p_2 = p_0$ and the geodesic heights are related to the coordinate x by $z_1 = -\,x\sin\alpha$ and $z_2 = x\sin\beta$, the linear second-order differential equation with constant coefficients results

$$L\,\ddot{x} + g\left(\sin\alpha + \sin\beta\right) x = 0 \;.$$

Hence, due to the constant cross-section, the fluid body oscillates linearly and the normalized equation of motion is

$$\ddot{x} + \omega_0^2\, x = 0 \;,\quad \omega_0 = \sqrt{\frac{g}{L}\left(\sin\alpha + \sin\beta\right)}\;,\quad \alpha, \beta > 0 \;. \tag{8.57}$$

The natural period is $T = 2\pi/\omega_0$. For vertical arms, $\alpha = \beta = \pi/2$, it takes on the value $T = 2\pi\,\sqrt{(L/2g)}$ that may be compared to the linear period of the mathematical pendulum in Eq. (7.116).

To avoid repeated water hammer loadings, the controlled drain pipe of a reservoir in the upper part is connected to a water tank. If the cock is closed downstream, the fluid body may oscillate in a piping system with variable cross-section. Such a system is shown in Fig. 8.11 with the streamline to be considered at some time instant. The nonstationary fluid motion is strongly idealized by the assumption of a lossless flow (despite, eg the fact that *Carnot's* pressure loss may become applicable at position 4 with the flow

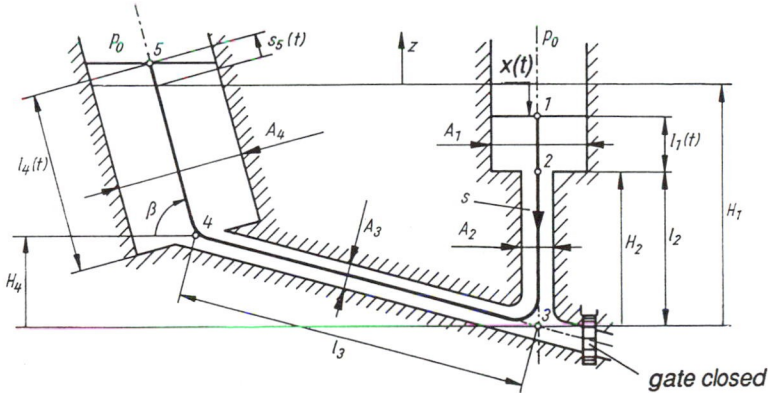

Fig. 8.11. Nonlinear vibrations of a fluid body in
a piping system of variable cross-section

direction indicated). The equation of motion expressed in terms of the coordinate $x(t)$ becomes nonlinear. For sectionally constant cross-sections, the local accelerations are (see Fig. 8.11)

$$1 \to 2: \quad v = \dot{x}(t) \,, \qquad \frac{\partial v}{\partial t} = \ddot{x}(t) \,, \quad \text{arclength } l_1(x)$$

$$2 \to 3: \quad v = \frac{A_1}{A_2} \dot{x}(t) \,, \qquad \frac{\partial v}{\partial t} = \ddot{x} \frac{A_1}{A_2} \,, \quad \text{arclength } l_2$$

$$3 \to 4: \quad v = \frac{A_1}{A_3} \dot{x}(t) \,, \qquad \frac{\partial v}{\partial t} = \ddot{x} \frac{A_1}{A_3} \,, \quad \text{arclength } l_3$$

$$4 \to 5: \quad v = \frac{A_1}{A_4} \dot{x}(t) \,, \qquad \frac{\partial v}{\partial t} = \ddot{x} \frac{A_1}{A_4} \,, \quad \text{arclength } l_4(x)$$

The line integral in the nonstationary *Bernoulli* equation section-wise is easily evaluated

$$\int_1^5 \frac{\partial v}{\partial t} \, ds + \frac{v_5^2}{2} + \frac{p_5}{\rho} + gz_5 = \frac{v_1^2}{2} + \frac{p_1}{\rho} + gz_1 \,, \quad v_1 = \dot{x} \,, \quad v_5 = \frac{A_1}{A_4} \dot{x} \,,$$

$$s_5 = \frac{A_1}{A_4} s_1 \,, \quad p_1 = p_5 = p_0 \,.$$

Putting $z_5 = s_5 \sin\beta$ and $z_1 = -x$ yields the nonlinear oscillator equation

$$f(x) \ddot{x} - \frac{b}{2} \dot{x}^2 + cx = 0 \,, \tag{8.58}$$

where

$$f(x) = l_1 + \frac{A_1}{A_2} l_2 + \frac{A_1}{A_3} l_3 + \frac{A_1}{A_4} l_4 \,, \quad b = \left[1 - \left(\frac{A_1}{A_4} \right)^2 \right] \,,$$

$$l_1(x) = H_1 - H_2 - x \,, \quad l_2 = H_2 \,, \quad l_3 \,, \quad c = g \left(\frac{A_1}{A_4} \sin\beta + 1 \right) \,,$$

$$l_4(x) = \left(H_1 - H_4 + x \frac{A_1}{A_4} \sin\beta \right) / \sin\beta \,.$$

The nonlinear terms are of pure and of mixed quadratic order. The superposition of solutions is no longer possible. Damping terms must be added for practical applications of Eq. (8.58). Only solutions by numerical routines are visible.

8.5.9. Lossless Flow Through a Diffusor

In a subsonic and stationary flow through a pipe of increasing cross-section, kinetic energy is transformed into pressure head. When we consider the idealized diffusor and a lossless and incompressible flow from a cross-section A_1 to $A_2 > A_1$, *Bernoulli´s* equation renders

$$\frac{v_1^2}{2g} + \frac{p_1}{\rho g} = \frac{v_2^2}{2g} + \frac{\bar{p}_2}{\rho g} \quad , \quad \rho A_1 v_1 = \rho A_2 v_2 \quad .$$

Hence, the maximal increase of pressure under these idealized conditions becomes

$$\bar{p}_2 - p_1 = \frac{\rho}{2} (v_1^2 - v_2^2) = \frac{\rho v_1^2}{2} \left(1 - \frac{A_1^2}{A_2^2}\right) \quad . \tag{8.59}$$

A sudden increase in the cross-sectional area, however, has been considered in Sec. 7.3.5 with *Carnot´s* loss of pressure head according to Eq. (7.49). Thus, if we consider the pressure p_1 to be common to both configurations, the loss in the pressure head and the loss factor are respectively given by

$$\bar{p}_2 - p_2 = \frac{\rho v_1^2}{2} \left(1 - \frac{A_1}{A_2}\right)^2 \quad , \quad \frac{p_2 - p_1}{\bar{p}_2 - p_1} = \frac{2 A_1 / A_2}{1 + A_1 / A_2} \quad , \quad A_2 > A_1 \quad . \tag{8.60}$$

8.5.10. A Bernoulli-Type Equation in a Rotating Reference System

The application of *Bernoulli´s* equation to the flow through rotating machines is greatly simplified if the streamlines of the relative motion are considered, ie with respect to the frame that is fixed to the rotating machine. In the case of steady rotations, the relative flow becomes stationary. The altered form of Eq. (8.31) is determined by the integration of the *Euler* equation of motion (8.30) of the inviscid flow along the relative streamline. The kinematics of relative motion renders the absolute acceleration **a** by the superposition of three vectors: acceleration of the material point when assumed to be fixed to the moving reference frame (a rigid-body motion described by the moving origin A´ and the instant rotation with the angular velocity $\mathbf{\Omega}$), plus *Coriolis* acceleration, plus the acceleration relative to the rigid reference frame. The expressions are straightforwardly determined by differentiating the point vector measured from the origin 0 of the inertial system twice with respect to time. In successive order,

$$\mathbf{r} = \mathbf{r}_A + \mathbf{r}' \quad \rightarrow \quad \mathbf{v} = \frac{d\mathbf{r}}{dt} = \mathbf{v}_g + \mathbf{v}' \quad , \quad \mathbf{v}' = \frac{d'\mathbf{r}'}{dt} \quad , \tag{8.61}$$

where a prime denotes the relative part of the vector and d'/dt is the time derivative with respect to the moving frame; see Eq. (7.58). Thus, the guiding velocity $\mathbf{v}_g$ is determined by considering the relative point vector $\mathbf{r}'$ to be fixed to the moving frame that is in rigid-body motion

$$\mathbf{v}_g = \mathbf{v}_A + \boldsymbol{\Omega} \times \mathbf{r}' \quad , \quad \mathbf{v}_A = \frac{d\mathbf{r}_A}{dt} \quad .$$

A second time derivative is performed termwise and gives the acceleration in the required form

$$\mathbf{a} = \frac{d\mathbf{v}}{dt} = \mathbf{a}_g + \mathbf{a}_c + \mathbf{a}' \quad , \quad \mathbf{a}' = \frac{d'\mathbf{v}'}{dt} \quad , \quad \mathbf{a}_c = 2\left(\boldsymbol{\Omega} \times \mathbf{v}'\right) \quad . \tag{8.62}$$

The guiding acceleration corresponds to the rigid-body acceleration of the material point when fixed to the reference frame [see Eqs. (1.9) and (1.10)]

$$\mathbf{a}_g = \mathbf{a}_A + \frac{d\boldsymbol{\Omega}}{dt} \times \mathbf{r}' + \boldsymbol{\Omega} \times \left(\boldsymbol{\Omega} \times \mathbf{r}'\right) \quad , \quad \mathbf{a}_A = \frac{d\mathbf{v}_A}{dt} \quad .$$

The *Coriolis* acceleration $\mathbf{a}_c$ is always pointing in the direction normal to the relative velocity $\mathbf{v}'$ (see also Exercise A 7.1), and thus, it has no component in the direction tangential to the relative streamline. The latter is determined by the vector lines of the field of the relative velocities. The relative acceleration $\mathbf{a}'$ is the relative rate of the relative velocity with respect to the moving frame. Analogous to Eq. (8.31), the projection is now performed in the direction of $\mathbf{v}' = v'(t, s')\, \mathbf{e}_t'$, and the integration over the arc-length s' of the relative streamline at fixed time yields

$$\int_{s_1'}^{s_2'} \mathbf{a} \cdot \mathbf{e}_t'\, ds' = \int_{s_1'}^{s_2'} \frac{1}{\rho}\, k_{t'}\, ds' - \int_{s_1'}^{s_2'} \frac{1}{\rho} \frac{\partial p}{\partial s'}\, ds' \quad . \tag{8.63}$$

The projection of the acceleration gives

$$\mathbf{a} \cdot \mathbf{e}_t' = \mathbf{a}_g \cdot \mathbf{e}_t' + \mathbf{a}' \cdot \mathbf{e}_t' \quad , \tag{8.64}$$

and the tangential component of the relative acceleration in *eulerian* decomposition is in general determined by

$$\mathbf{a}' \cdot \mathbf{e}_t' = \frac{\partial' v'}{\partial t} + \frac{\partial'}{\partial s'}\left(\frac{v'^2}{2}\right) \quad .$$

A stationary relative flow will be assumed further on [Eq. (8.63) takes on its stationary form],

$$\frac{1}{2}\left(v_2'^2 - v_1'^2\right) = \int_{s_1'}^{s_2'} \frac{1}{\rho} k_{t'}\, ds' - \frac{1}{\rho}\left(p_2 - p_1\right) - \int_{s_1'}^{s_2'} \left(\mathbf{a}_g \cdot \mathbf{e}_t'\right) ds' ,$$

$$\frac{\partial' v'}{\partial t} \equiv 0 .$$

$$(8.65)$$

It is formally equal in appearance to the classical stationary *Bernoulli* equation, except for an additional term on the right-hand side, which is given by the line integral of the tangential component of the guide acceleration along the relative streamline. The relative velocities have to be substituted in the expressions of kinetic energy per unit of mass. Commonly, the point of reference A' can be assumed to be at rest, $A' = 0$, and $a_A = 0$.

Equation (8.65) when applied to a relative streamline in a horizontal plane, eg in the case of a water turbine with a vertical axis and in constant rotation, $\Omega = const$, is greatly simplified due to $(\mathbf{e}_r' \cdot \mathbf{e}_t')\, ds' = dr'$, which is just the differential of the radial coordinate

$$\frac{1}{2}\left(w_2^2 - w_1^2\right) = -\frac{1}{\rho}\left(p_2 - p_1\right) + \frac{\Omega^2}{2}\left(r_2^2 - r_1^2\right) , \quad \mathbf{v}' \equiv \mathbf{w} .$$

$$(8.66)$$

Hence, only radial projections of the relative streamline contribute to the additional term. It is common practice to rename the velocities in mechanical engineering applications. Thus the relative velocity is used to be $\mathbf{w} = \mathbf{v} - \mathbf{v}_g$, the absolute velocity is denoted as $\mathbf{v} \equiv \mathbf{c}$, and the guide velocity becomes $\mathbf{v}_g \equiv \mathbf{u}$ where $\mathbf{u}$ is the vector of the circumferential speed; see Sec. 7.3.3. Usually, the pressure p_2 at the outlet $r = r_2$ is determined from Eq. (8.66). It should be well above the evaporation pressure of the fluid to avoid cavitation.

(§)   Example: *Segner's* Water Wheel. A simple but not very efficient type of reaction turbine is shown in Fig. 8.12 in the form of a rotating pipe. If we assume that the action of a proper external moment will keep the rotational speed constant, the relative flow through the rotating pipe becomes stationary and Eq. (8.66) applies. When we put $r_1 = 0$ and, thus, $w_1 = v_1$, and consider the pressure in the free jet $p_2 = p_0$, which is the atmospheric pressure, the relative velocity at the outlet becomes at once

$$w_2 = \left[v_1^2 - \frac{2}{\rho}\left(p_0 - p_1\right) + \left(r_2\,\Omega\right)^2\right]^{1/2} .$$

$$(8.67)$$

If the water is supplied from an open container with the water level at $H = const$ above the outlet and with the same atmospheric

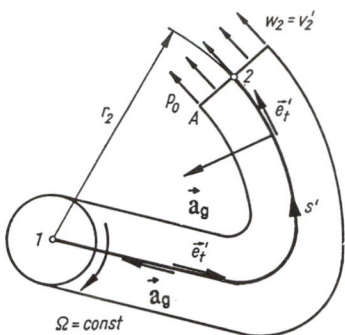

Fig. 8.12. *Segner's* water wheel, an illustration of a reaction turbine

pressure acting on the free surface, *Bernoulli's* equation along a streamline to the inlet at point 1 gives

$$\frac{v_1^2}{2g} + \frac{p_1}{\rho g} = \frac{p_0}{\rho g} + H \ ,$$

and substitution into Eq. (8.67) eliminates the state at the inlet to render

$$w_2 = \sqrt{2gH + (r_2\,\Omega)^2} \ . \tag{8.68a}$$

A comparison of the expression to *Torricelli's* formula (8.48) is quite illustrative. The circular part of the revolving arm does not contribute to the correction term in Eq. (8.67) since a_g in that part is normal to the tangent $e_t{}'$. The (constant) mass flow rate is given by $\rho A w_2$. The moment M_w is easily calculated from the conservation of angular momentum; the control volume may rotate with the arm of the water wheel. The ideal power produced by the turbine is (without any losses), $P < 0$,

$$P = -M_w\,\Omega = \rho A\,v_\infty^3\ \xi\left(\xi - \sqrt{1+\xi^2}\right)\sqrt{1+\xi^2} \ , \ \xi = r_2\Omega/v_\infty \ , \ v_\infty = \sqrt{2gH}\ . \tag{8.68b}$$

The theoretical efficiency is thus given by relating this ideal power to the theoretically available power P_0, $\eta_{th} = P/P_0 < 1$. The latter theoretical power is just the potential energy of the mass flow rate at the utmost elevation H

$$P_0 = -\dot{m}gH = -\frac{1}{2}\,\rho A\,v_\infty^3\ \sqrt{1+\xi^2} \ . \tag{8.68c}$$

8.6. Remarks on the First Law of Thermodynamics (Conservation of Energy)

If friction is to be considered in a motion including material damping during deformation (eg of visco-plastic materials), the field of (stationary) forces is always nonconservative: The conservation of mechanical energy expressed by Eq. (8.8) is no longer applicable. The work of the dissipating forces cannot be fully recovered in the form of mechanical energy, since the heat that is thereby produced is subject to the irreversible process of heat conduction. Due to the dissipation of energy during deformation, the nonconservative field of forces is more precisely termed dissipative. The rate form of Eq. (8.8) is easily extended to account for heat production and heat flux in a material volume. Taking the time derivative of $T + V = E = const$ with $V = U + W$ (U is the elastic potential) gives

$$\frac{d}{dt}(T + U) = -\frac{dW}{dt} = P^{(e)} .$$
(8.69)

The rate of increasing the kinetic energy and the elastic potential equals the power of the external forces in lossless motion. Hence, Eq. (8.69) is to be generalized by adding the heat production per unit of time (which is due to external sources acting in the body) and by considering the net influx of heat through the material surface per unit of time (the outward normal is considered). On the left-hand side, the density of the elastic potential is to be generalized to the internal energy u per unit of mass: It accounts for the kinetic and potential energy of the molecules of the material and is a thermodynamic state variable (it is completely distinct from the mechanical energy per unit of mass $v^2/2$ associated with the motion of the continuum)

$$\frac{d}{dt}\int_{V(t)}\left(\frac{\rho v^2}{2} + \rho u\right)dV = P^{(e)} + P^{(q)} - \oint_{\partial V(t)}q_n \, dS .$$
(8.70)

Integration is performed over the material volume and the material surface (with no mass flow). After that generalization, Eq. (8.70) states the following: The rate of total energy of the body equals the power of the external forces, as with Eq. (8.69), plus the power supplied by external sources of heat in the body and the power supplied by the net influx of heat per unit of time through the noninsulated surface. Other forms of electromagnetic energy are not considered here. In solids, the radiation of heat may be negligible and the heat flux vector is proportional to the temperature gradient according to *Fourier's* law of heat conduction, Eq. (6.353). Substituting Eq. (8.5) above eliminates the rate of kinetic energy and the power of the external forces as well, and Eq. (8.70) takes on the commonly encountered form of the first law of thermodynamics

$$\frac{d}{dt} \int_{V(t)} \rho u \, dV = -P^{(i)} + P^{(q)} - \oint_{\partial V(t)} q_n \, dS \ .$$

(8.71)

<<The rate of internal energy equals the heat supplied per unit of time less the power of the internal forces.>> If the heat production terms balance

$$P^{(q)} - \oint_{\partial V(t)} q_n \, dS = 0 \ ,$$

the specific internal energy ρu becomes the strain energy density U' of Eq. (3.30). Changing the surface integral in Eq. (8.71) into a material volume integral by the *Gauss* integral theorem and considering further that the equation must hold for any part of the material volume render the local form of conservation of energy (a differential relation) just by omitting the common integration. By means of the second law of thermodynamics, frequently used in the form of the *Clausius-Duhem* inequality of entropy production, the irreversibility of any deformation process can be proved. Also, the consistency of constitutive relations has to be checked by such thermodynamic considerations, see, eg *H. Parkus: Thermoelasticity. Springer-Verlag, Wien-New York, 1976, 2nd ed., Chap.5.*
    Equation (8.70) can be transformed by means of the *Reynolds* transport theorem, Eq. (1.72), to hold for a control volume V fixed in space with mass flow through the control surface ∂V (note the production terms of energy on the right-hand side)

$$\int_V \frac{\partial}{\partial t} \left(\frac{\rho v^2}{2} + \rho u \right) dV + \oint_{\partial V} \mu \left(\frac{v^2}{2} + u \right) dS = P^{(e)} + P^{(q)} - \oint_{\partial V} q_n \, dS \ ,$$

$$\mu = \rho \, (\mathbf{v} \cdot \mathbf{e}_n) \ .$$

(8.72)

The volume integral represents the nonstationary rate of total energy, the surface integral determines the loss of total energy through a net flow of mass and, thus, of energy outward of the control volume. The sum is equal to the power of the external forces acting on the mass within the control volume, plus the heat production per unit of time. Equation (8.72) completes the set of conservation laws for mass, momentum, and angular momentum by a consideration of energy (which is a scalar quantity). By putting the partial time derivative outside the volume integral and considering the specific relative mass flow rate by the scalar product, $\mu = \rho \, (\mathbf{v} - \mathbf{w}) \cdot \mathbf{e}_n$, the conservation of energy of mass is extended to a moving control volume; see Eq. (1.82).

8.7. Exercises A 8.1 to A 8.5 and Solutions

A 8.1: A vehicle with rigid-body mass m_1 on wheels of diameter $2a$ is pulled by a cable running over a drum of diameter a that is attached to the front wheel. Front and rear wheels are connected by a rigid rod of mass m_4 ; see Fig. A 8.1. Determine the equation of motion in terms of $\varphi(t)$, assuming pure rolling contact, by means of Eq. (8.5). Also determine the pulling force $F(t)$ for a stationary motion $d\varphi/dt = \omega = const$.

Solution: The mass centers of the partial masses m_1 , m_2 , and m_3 have the same velocity at some instant of time, $v = v_1 = v_2 = v_3 = a \, d\varphi/dt = a \, \omega$. The angular speeds are $\omega_1 = 0, \, \omega_2 = \omega_3 = \omega$; the connecting rod is in translational motion, $\omega_4 = 0$; the velocity is $\mathbf{v_4} = (v + a \, \omega \cos \varphi) \, \mathbf{e_x} - a \, \omega \sin \varphi \, \mathbf{e_y}$; and the absolute value is $|\mathbf{v_4}| = 2 \, a \, \omega \cos (\varphi/2)$. The kinetic energy becomes by superposition

$$T(t) = \frac{1}{2} \left(m_1 v_1^2 + m_2 v_2^2 + m_3 v_3^2 + m_4 v_4^2 + I_2 \omega_2^2 + I_3 \omega_3^2 \right) \; .$$

The power of the external forces is simply given by

$$P^{(e)}(t) = F \, v_5 + (m_1 + m_2 + m_3) \, gv \cos\left(\alpha + \frac{\pi}{2}\right) + m_4 \, gv_4 \cos\left(\alpha + \frac{\pi}{2} - \frac{\varphi}{2}\right) ,$$

$$v_5 = \frac{3}{2} \, v = \frac{3}{2} \, a\omega \; ,$$

the internal forces do not contribute to the total power. $dT/dt = P$ yields, after we cancel $\omega \neq 0$, the second-order differential equation of motion

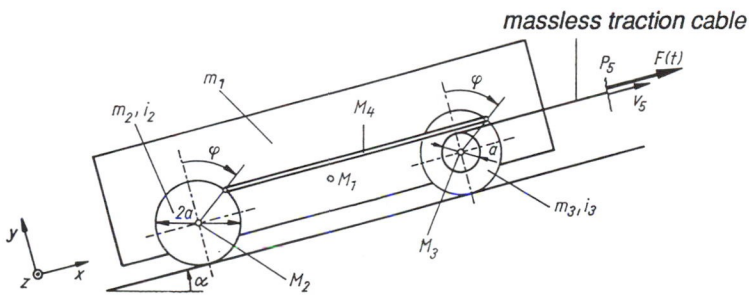

Fig. A 8.1. SDOF-dynamic system

$$\left[m_1 + m_2\left(1 + \frac{i_2^2}{a^2}\right) + m_3\left(1 + \frac{i_3^2}{a^2}\right) + 2\,m_4\,(1 + \cos\varphi)\right] a\,\ddot{\varphi} - m_4\,a\,\dot{\varphi}^2 \sin\varphi$$

$$+ g\left[(m_1 + m_2 + m_3)\sin\alpha + 2\,m_4\sin\left(\alpha - \frac{\varphi}{2}\right)\cos\frac{\varphi}{2}\right] = \frac{3}{2}\,F(t) \ .$$

The pulling force must be timevariant to achieve constant speed; and by substituting $\omega = const$, $d\omega/dt = 0$, and $\varphi = \omega t$, the required force $F(t)$ is determined.

A 8.2: Determine the relationship between the angle φ of the tails of the blades of a radial water turbine and the power P for an assigned height H_1 of the water level; such a system is shown in Fig. A 8.2. The size of the rotor is given by internal radius r_1, external radius r_2, and the channel width B. The tails of the guiding blades at rest direct the flow, so that it approaches under the angle α_1; steady rotation of the rotor is assumed, $\Omega = const$.

Solution: Both the free surfaces are assumed to be under the same atmospheric pressure and at constant levels, separated by H. *Bernoulli's* generalized equation (8.45) renders at once the approximation

$$0 = gH - P/\dot{m} \ , \quad P > 0 \ .$$

Euler's turbine equation (7.46) may be solved for the tangential component of the absolute velocity at the outlet of the rotor

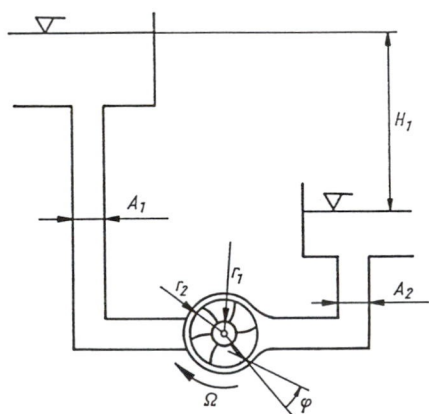

Fig. A 8.2. Power plant with radial turbine

$$c_{2u} = \frac{1}{r_2 \, \Omega} \left(g \, H_1 - \frac{\dot{m} \, \Omega}{2 \pi \, \rho B} \tan \alpha_1 \right) \, .$$

The relative flow is assumed to be tangential to the blade at the tail when leaving the rotor; hence, the wanted angle becomes

$$\varphi = \tan^{-1} \left(\frac{r_2 \, \Omega - c_{2u}}{c_{2r}} \right) \quad , \quad c_{2r} = \frac{\dot{m}}{2 \pi \, r_2 \, B \, \rho} \quad .$$

A 8.3: A spherical shell of aperture 2α and weight G is riding on a vertical free jet as shown in Fig. A 8.3. Determine the equilibrium configuration when the draining pipe is connected to a pressurized container, $p_1 - p_0 = const$, $H = const$. The stationary flow is assumed to be inviscid and incompressible.

Solution: The speed of the free jet at the outlet of the nozzle as well as the velocity profile of the upward motion are determined by *Bernoulli's* equation

$$v_0 = \left[2 \, g \, H + 2 \, \frac{p_1 - p_0}{\rho} \right]^{1/2} \quad \text{and} \quad v(z) = \sqrt{v_0{}^2 - 2 \, g \, z} \quad .$$

Since the mass flow rate $\rho A(z) \, v(z) = \rho A_0 \, v_0 = const$, the cross-section of the jet changes nonlinearly: $A(z) = A_0 \, v_0 / \sqrt{(v_0{}^2 - 2gz)}$. If we neglect the changes of geodesic height in the contact region of the impinging jet, the conservation of momentum of the control volume shown by a dashed line in Fig. A 8.3 yields (p_0 is the atmospheric reference pressure)

$$- \rho v^2(z_S) \, A(z_S) \, (1 + \cos \alpha) = - F_z \quad .$$

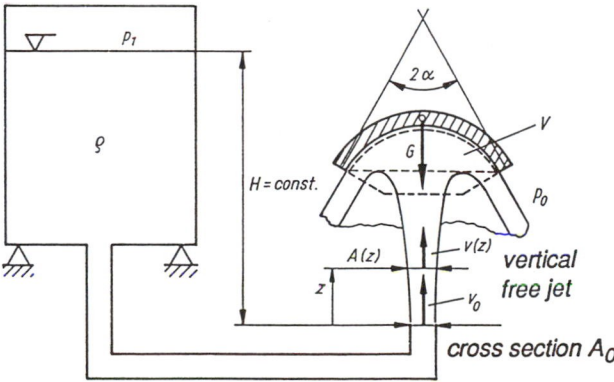

Fig. A 8.3.

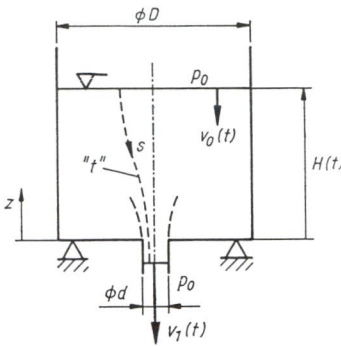

Fig. A 8.4. Nonstationary flow

By means of the equilibrium condition $F_z - G = 0$, the coordinate becomes

$$z_S = \frac{v_0^2}{2g} - \frac{1}{2g} \left[\frac{G}{\dot{m} \, (1 + \cos \alpha)} \right]^2 .$$

A 8.4: An open container of circular cross-section and diameter D is drained through a circular opening in the floor of diameter d (see Fig. A 8.4). Assuming the nonstationary flow is incompressible, determine the time dependence of the height $H(t)$ of the free surface (a proper differential equation suffices).

Solution: Bernoulli's equation (8.35) in its nonstationary form when applied to a streamline from a point at the free surface to the free jet (the pressure there is p_0) and the condition on the mass flow rate $\rho v_1 d^2 = \rho v_0 D^2$, together with the differential relation $v_0 \, dt = (-dH)$, yields the nonlinear differential equation

$$\ddot{H} H - \frac{1}{2} \left(\frac{D^4}{d^4} - 1 \right) \dot{H}^2 + gH = 0 .$$

The quasistationary approximation

$$\dot{H} \approx \frac{d^2}{D^2} \sqrt{2gH} \quad , \quad \ddot{H} \to 0 \ ,$$

which was derived earlier, is a solution in the limit $D \gg d$; see Eq. (8.49).

A 8.5: A piston of mass m is supported by a linear spring of stiffness c and loaded by the fluid pressure as shown in Fig. A 8.5.

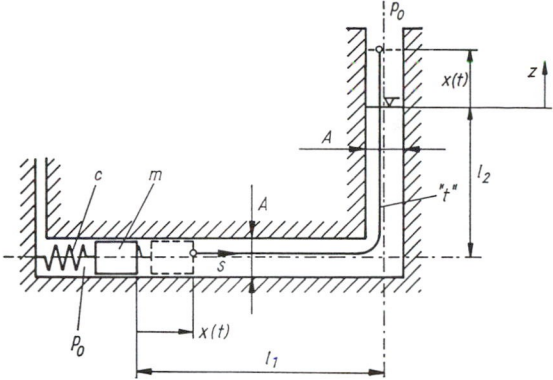

Fig. A 8.5. Free vibrations of a coupled system

Determine the equation of motion of natural vibrations about the equilibrium position under the assumption of an incompressible and inviscid nonstationary flow of the fluid.

Solution: The L-shaped pipe has a constant cross-section and Eq. (8.35) when applied to an instant streamline renders (in combination with the condition of continuity)

$$\int_0^{l_1 + l_2} \frac{\partial v}{\partial t} \, ds = \ddot{x}(l_1 + l_2) = -\frac{1}{\rho}(p_0 - p_1) - gx \ .$$

A second equation results from *Newton's* law when applied to the piston in its free-body diagram

$$m\ddot{x} = -cx + (p_0 - p_1)A \ .$$

The elimination of the pressure p_1 at the piston yields the linear oscillator equation

$$\ddot{x}\left[m + \rho A(l_1 + l_2)\right] + (c + \rho gA)x = 0 \ .$$

The spring force fluctuates about the mean value $\rho g l_2 A$ (the static value in the equilibrium configuration).

9
Stability Problems

The stability of the equilibrium configuration of floating bodies has already been considered by simple static means, and proper conditions were found in the form of inequalities of the type (2.115). In this section, the stability of conservative systems at rest is observed by analyzing the motion that follows any perturbation of the equilibrium state (ie by the dynamic method of small perturbations) and, equivalently, by the static *Dirichlet* stability criterion. Simple applications are given, including the balance problem of rigid heavy bodies in contact, the structural problem of the buckling of slender columns and thin plates (a bifurcation problem), as well as the snapping of shallow arches when the lateral load becomes critical. The extension of the dynamic method of small perturbations to include the stability of a given (main) motion is illustrated through the analyses of a mechanical control device, the centrifugal governor, and the gyroscope without moment. In a third subsection, the stability of elastic-plastic structures is considered statically by limit load analysis and, quasistatically, for alternating loadings by the shake-down theorems of *Melan* and *Koiter* . Furthermore, the hydrodynamic stability of incompressible flow in open channels is discussed and the loss of energy in the hydraulic jump where the rapid flow changes "abruptly" to tranquil streaming is given. Finally, flutter instability is described well by the self-excited oscillations of a simplified model of an airfoil that occur above a critical speed of flight.

9.1. Stability of an Equilibrium Configuration

Equilibrium configurations of a system may be qualitatively quite different when it comes to the consequences of any perturbations of the state at rest. The consideration of a planar pendulum with two positions of equilibrium illustrates drastically these differences. If the center of gravity falls below the axis of support, the position is obviously stable. If, however, the center lies above the support, the reaction force is still equilibrating the weight, but even the smallest perturbation yields a rotational motion of the rigid body: The upright position exhibits instability. A definition of stability by means of the motion of the perturbed system can be given as follows. <<The equilibrium configuration is stable if the motion following a sufficiently small perturbation is restricted to the neighborhood of the configuration at rest and, further, the motion is supposed to decrease with respect to any reduction in the severity of the disturbance.>> Mathematically, the definition can be expressed by considering the state vector of the motion $x(t)$ according to Eq. (7.109): For any $\varepsilon^* > 0$, a number $\eta^* > 0$ exists such that for $|x^*(0)| < \eta^*$, for all subsequent times $|x^*(t)| < \varepsilon*$ holds. From Eq. (9.4), it is concluded that $|\varepsilon^*(t)| < \varepsilon^* = \eta^*$ if $\alpha < \eta^* \ll 1$.

The hanging pendulum is at rest in the vertical position $\varphi_0 = 0$; the upright configuration is given by $\varphi_1 = \pi$. Conservation of angular momentum renders the nonlinear equation of motion of the conservative system [see Eq. (7.48a)]

$$\ddot{\varphi} + \omega_0^2 \sin \varphi = 0 \quad , \quad \omega_0^2 = gs/i^2 \quad . \tag{9.1}$$

Since the perturbations are assumed to be small, the equation is linearized. It is hoped that linearization does not alter the conditions that indicate stability. A linear approximation gives for small amplitudes ε of the perturbing motion

$$\ddot{\varepsilon} + \omega_0^2 \varepsilon \cos \varphi_0 = 0 \ , \ \varphi = \varphi_0 + \varepsilon \ , \ \sin (\varphi + \varepsilon) = \sin \varphi_0 + \varepsilon \cos \varphi_0 \ , \ \varepsilon \ll 1. \tag{9.2}$$

(§)   Hanging Pendulum. Since $\varphi_0 = 0$ in that case, $\cos \varphi_0 = 1$, and

$$\ddot{\varepsilon} + \omega_0^2 \varepsilon = 0 \quad , \tag{9.3}$$

is the linearized equation of the undamped oscillator with the general solution

$$\varepsilon(t) = \alpha \cos (\omega_0 t - \eta) \quad . \tag{9.4}$$

With $\alpha \ll 1$, the motion is restricted to the neighborhood of the equilibrium configuration $\varphi_0 = 0$, $|\varepsilon| \ll 1$, and furthermore, α decreases with any reduction of the initial perturbation. The addition of damping renders the linearized system asymptotically stable, and the motion of the perturbed system dies out.

(§)   Upright Position of the Pendulum. For a pendulum in an upright position, $\varphi_0 = \pi$ and $\cos \varphi_0 = -1$ give

$$\ddot{\varepsilon} - \omega_0^2 \varepsilon = 0 \quad , \tag{9.5}$$

and the solution grows beyond any limits with time

$$\varepsilon = a \exp(\omega_0 t) + b \exp(-\omega_0 t) . \tag{9.6}$$

The assumption of linearization $|\varepsilon| \ll 1$ does not hold, but the linearized equation indicates the instability of the equilibrium configuration qualitatively by the exponential growth of the solution. Quantitatively, the motion is determined by the nonlinear equation (9.1). See also Sec. 10.5.

Hence, by definition, the stability of an equilibrium configuration is to be determined by considering the perturbed system in motion. The equations of motion are to be set up a priori and their solution with the given initial conditions must be found by integration. The time behavior of the motion indicates stability or renders instability. In many cases of technical importance, linearization of the equations of motion is possible without altering the qualitative properties with respect to stability: The theorem of *A. M. Ljapunow* gives the proper limitations, stating that asymptotically stable linearized systems render stability also for the nonlinear case if the nonlinear restoring forces are restricted in some sense. Thus, it suffices in those cases to solve the linear eigenvalue problem that then should have all complex eigenvalues with a negative real part. See, *I. G. Malkin: Theory of Stability of Motions. Munich, 1959.*

The analysis would be greatly simplified if a class of systems existed in which a static consideration sufficed to determine whether the equilibrium was stable or not. Conservative systems are tested for such a simple criterion. Conservation of mechanical energy of the motion following a perturbation applies in that case. Hence, if such a criterion is applicable, the bounds of the kinetic as well as those of the potential energy must be determined separately. The equilibrium state is perturbed by nonhomogeneous initial conditions with total energy $E_0 = T_0 + V_0$ at $t = 0$. Without a loss of generality, the potential is assumed to vanish in the configuration at rest. Subsequently, the conservation of energy of the motion gives [see Eq. (8.8)]

$$T(t) + V(t) = E_0 . \tag{9.7}$$

Since the kinetic energy is positive-semidefinite, $T \geq 0$, an inequality results by subtracting T from Eq. (9.7)

$$V(t) = E_0 - T(t) \leq E_0 . \tag{9.8}$$

The condition of equilibrium is given by requiring that $V = 0$ be a stationary value that does not change in a small neighborhood, $\delta V =$

0, and furthermore, Eq. (9.8) limits $V(t) \leq E_0$ from above. If by an additional assumption the condition is enforced so that $V(t)$ takes on a minimal value of $V = 0$, the potential also becomes a positive-semidefinite function, $V(t) \geq 0$. Consequently, the energy measure of the perturbation becomes a positive value, $E_0 > 0$. Two bounds apply in that case to the potential energy

$$0 \leq V(t) \leq E_0 . \tag{9.9}$$

Substitution into Eq. (9.7) limits also the kinetic energy from above

$$0 \leq T(t) \leq E_0 . \tag{9.10}$$

Considering the motion to be limited in terms of the energy ncrm renders the equilibrium state stable by definition of the dynamic stability criterion. <<Since the potential energy and kinetic energy are bound, the motion is restricted in an integral sense to some neighborhood of the equilibrium configuration and it decreases with diminishing intensity E_0 of the initial perturbation.>> The stability criterion formulated in the energy norm holds for most cases of practical interest, except some pathological continuous systems. For MDOF-systems, the finite number of degrees of freedom guarantees that all stability norms are equivalent. Unfortunately, this is not true for a continuum. Thus, in some rare cases, discretization may change the character of the equilibrium.

Consideration of stability of conservative systems in the energy norm allows the formulation of a simple static criterion, the *Dirichlet stability criterion.* <<A conservative system is in a stable equilibrium configuration if the potential energy $V = W_P + U$ takes on a local minimum. Otherwise, it is unstable.>> Thus, knowing the potential energy of a conservative system renders the conditions of equilibrium from the vanishing first variation, $\delta V = 0$; see Eq. (5.15). If the second variation in the equilibrium configuration $\delta^2 V > 0$, the latter is stable (*Dirichlet's* stability criterion applies). Thus, the dynamic criterion of stability is also met without considering the motion by

$$\delta V = 0 , \quad \delta^2 V > 0 . \tag{9.11}$$

In case it happens that also $\delta^2 V = 0$, then it is necessary that $\delta^3 V = 0$ and the fourth-order variation decides about stability, $\delta^4 V > 0$.

The potential energy of a MDOF conservative system is merely a function of the independent generalized coordinates $V = V(q_1, q_2, ..., q_n)$. The first variation must vanish and renders n conditions of equilibrium

$$\delta V = \sum_{k=1}^{n} \frac{\partial V}{\partial q_k} \, \delta q_k = 0 \quad, \quad \frac{\partial V}{\partial q_i} = 0 \quad, \quad i = 1, 2, ..., n .$$

(9.12)

Requiring the second-order variation to be positive in the equilibrium configuration

$$\delta^2 V > 0 \quad,$$

(9.13)

is the necessary and sufficient condition of stability and may be expressed by considering all the determinants as positive when substituting the coordinates of the state at rest

$$\begin{vmatrix} \dfrac{\partial^2 V}{\partial q_1{}^2} & \dfrac{\partial^2 V}{\partial q_1 \partial q_2} & \cdots & \dfrac{\partial^2 V}{\partial q_1 \partial q_k} \\[2ex] & \dfrac{\partial^2 V}{\partial q_2{}^2} & \cdots & \dfrac{\partial^2 V}{\partial q_2 \partial q_k} \\[2ex] & & \cdots & \cdot \\[2ex] \underline{\text{symm}} & & \cdot & \dfrac{\partial^2 V}{\partial q_k{}^2} \end{vmatrix} > 0 \quad, \quad k = 1, 2, ..., n \quad.$$

(9.14)

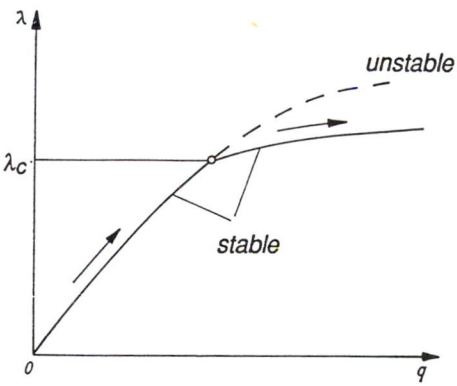

Fig. 9.1. Divergence of equilibrium, bifurcation

All the second-order derivatives in the main diagonal must necessarily be positive.

It is important to know the limits of stability of an equilibrium configuration with respect to a variation of the system parameters. Only the variation of the external loading by a common load factor $\lambda > 0$ is considered here; see also Chap. 6. In that case of uniformly increasing the loading, the potential of the external forces is λW_P , and at the limit of stability where

$$\det \left| \frac{\partial^2 V}{\partial q_k \partial q_n} \right| = 0 \; , \tag{9.15}$$

a nonlinear (characteristic) equation is derived for the determination of the critical load factor λ_c . The stability of the equilibrium under the assigned loading configuration thus requires the load factor $0 < \lambda < \lambda_c$. If there are further stable equilibrium states to be determined by continuously increasing the loading beyond its critical value, $\lambda > \lambda_c$, *bifurcation* occurs as depicted in Fig. 9.1. An example is the buckling of a slender column (*Euler* buckling).

Another, more dangerous way of losing stability in an equilibrium configuration is *snap-through* buckling as depicted in Fig. 9.2. For $\lambda > \lambda_c$, no smooth transition to a stable equilibrium

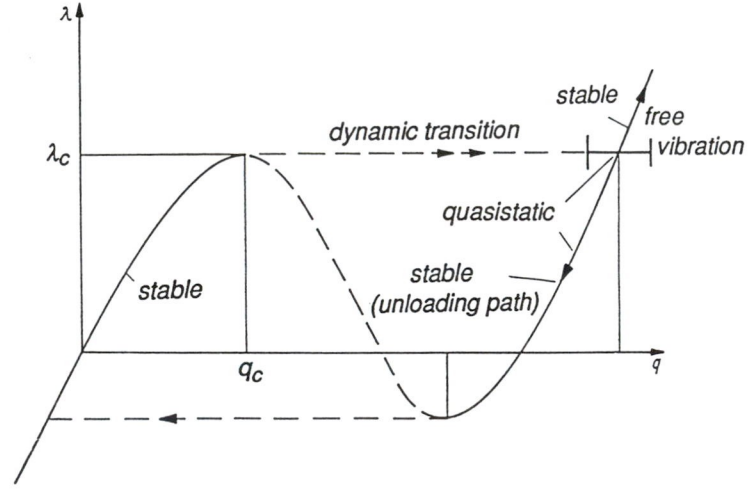

Fig. 9.2. Snap-through buckling. The dynamic transition to a new, possibly stable configuration is indicated

configuration exists. Possibly, in dynamic motion after snapping, the system reaches a different shape where it comes to rest. Loss of stability in a critical configuration is experienced by laterally loaded shallow structures and snap-through is also the common loss of stability of all imperfection-sensitive structures when predominantly loaded in compression. Examples are the buckling of shells and similar structures with particular nonlinearities; see, eg Fig. 9.7b. The limit load of elastic-plastic structures is of a similar nature and is discussed in Sec. 9.3.

9.1.1. Example: The Balancing Problem of Heavy Rigid Cylinders

In Fig. 9.3, the equilibrium position is given if the action lines of the weight mg and of the opposite normal supporting force $N = mg$ coincide. The stability of the balancing (upper) cylinder should be considered under the assumption of pure rolling contact for any perturbing motion. Thus, the elementary condition $r\,d\varphi = R\,d\Phi$ holds (Fig. 9.3). Pure rolling in a constant field of vertical gravity is conservative and *Dirichlet's* stability criterion (9.13) becomes applicable. At rest, the angles $\varphi = \Phi = 0$ and the height of the center of gravity is just $z_S = (R + h)$. r^{-1} and R^{-1} are the curvatures of the cylinders in contact. The height changes in the neighborhood of the equilibrium position to $z_S = (R + r) \cos \Phi - (r - h) \cos (\varphi + \Phi)$ and the gravity potential is increasing with z_S

$$E_p = W_p = mg\,z_S \ . \tag{9.16}$$

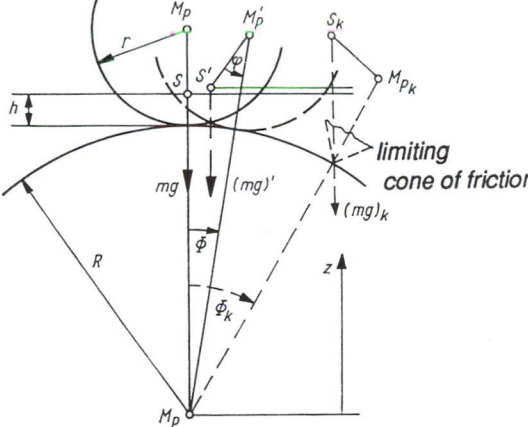

Fig. 9.3. Balancing cylinder. Perturbing motion is assumed with pure rolling contact

The configurational coordinate of the SDOF-system may be the angle φ. The condition of equilibrium is then the derivative

$$\frac{dE_p}{d\varphi} = \frac{\partial E_p}{\partial \varphi} + \frac{\partial E_p}{\partial \Phi} \frac{d\Phi}{d\varphi} = \frac{\partial E_p}{\partial \varphi} + \frac{r}{R} \frac{\partial E_p}{\partial \Phi}$$

$$= mg \left[(r - h) \sin\left(\varphi + \Phi\right) - r \sin \Phi\right] \left(1 + \frac{r}{R}\right) = 0 \ ,$$

with the trivial solution $\varphi = \Phi = 0$. The second derivative becomes by the application of the above operator

$$\frac{d^2 E_p}{d\varphi^2} = mg \left\{(r - h) \cos\left(\varphi + \Phi\right) + \frac{r}{R}\left[(r - h)\cos\left(\varphi + \Phi\right) - r\cos\Phi\right]\right\} \left(1 + \frac{r}{R}\right) . \tag{9.17}$$

Thus, the condition of stability of the upright position $\varphi = \Phi = 0$ is the inequality

$$\left(1 + \frac{r}{R}\right) \left[r - h\left(1 + \frac{r}{R}\right)\right] > 0 \ .$$

Solving for the distance h of the center of gravity renders the stability condition

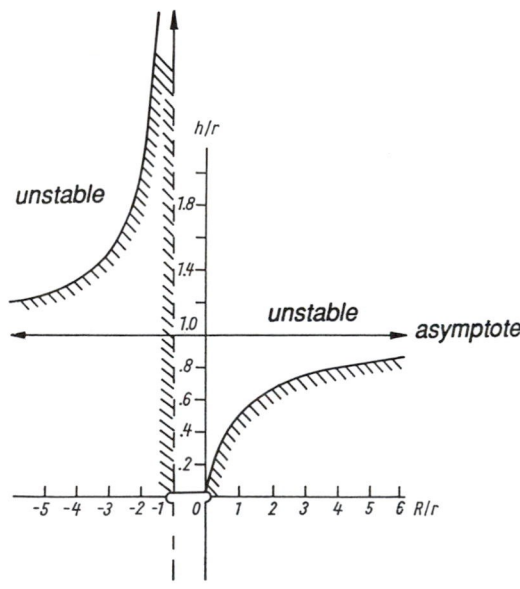

Fig. 9.4. Stability map of the balancing cylinder

$$h < \frac{r}{[1 + (r/R)]} \; . \tag{9.18}$$

Analogous to the simple approach applied to a floating body, the restoring moment of the balancing cylinder can be considered. After a small perturbation of the equilibrium configuration $\varphi = \Phi = 0$, expressed by $d\varphi$ and $d\Phi = r \, d\varphi/R$, it is given by

$$-\frac{dE_p}{d(\varphi + \Phi)} = mg \left\{ R \sin \Phi - [(R + r) \sin \Phi - (r - h) \sin (\varphi + \Phi)] \right\} > 0 \; ,$$

it yields the condition (9.18) as the positive factor of any small angle of disturbance $d\varphi$.

In the limit $R \to \infty$, the condition of stability of a cylinder resting on a rigid plane is derived: $h < r$. If the cylinder is sitting at the bottom within a hollow cylinder where $R^{-1} < 0$, and $|r/R| < 1$, the region of stability is as shown in Fig. 9.4. The mobility of rocking stones found in nature may be interpreted by the height h to be close to the limit of stability. The danger of losing contact by sliding, is not considered above, but the limiting cone of dry friction is indicated in Fig. 9.3.

9.1.2. Example: A Simple Model of Buckling

A rigid column of length l carries a centrally applied dead load F , and at the simple support, a rotational (linear or nonlinear) spring is activated by any perturbation of the upright position $\varphi = 0$ (Fig. 9.5). The stability of this configuration and its limit are to be

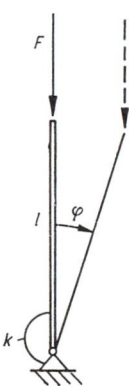

Fig. 9.5. Model for buckling with linear and nonlinear restoring springs

determined by slowly increasing the load and, after bifurcation, the postbuckling behavior is to be studied. Any motion following a disturbance is conservative and *Dirichlet's* criterion (9.13) is applicable. The potential energy varies according to the motion of the tip, the pendulum is assumed weightless, W_p is the potential of the dead weight F, and U is considered according to the constitutive law, $M(\varphi)$, of the restoring spring

$$E_p = W_p + U \quad , \quad W_p = -Fl\left(1 - \cos\varphi\right) ,$$

$$M = k\varphi \quad : \quad U = k\varphi^2/2 , \tag{9.19}$$

$$M = k\varphi\left(1 - a\varphi + b\varphi^2\right) \quad : \quad U = k\varphi^2/2 - ka\varphi^3/3 + kb\varphi^4/4 .$$

$a = 0$ renders the (physically) nonlinear spring symmetric, and $a \neq 0$ gives a restoring moment that depends on the direction of the motion. Such a polynomial approximation of the elastic energy is often encountered in practical applications and, hence, it suffices to approximate also the potential of the external force F by a polynomial of fourth order. In the neighborhood of $\varphi = 0$, $\cos\varphi \approx 1 - \varphi^2/2 + \varphi^4/24$, and the derivatives up to the fourth order are approximated by

$$\frac{dW}{d\varphi} = -Fl\sin\varphi \approx -Fl\left(\varphi - \varphi^3/6\right) , \quad \frac{d^2W}{d\varphi^2} = -Fl\cos\varphi \approx -Fl\left(1 - \varphi^2/2\right) ,$$

$$\frac{d^3W}{d\varphi^3} = Fl\sin\varphi \approx Fl\varphi , \quad \frac{d^4W}{d\varphi^4} = Fl\cos\varphi \approx Fl ,$$

$$\frac{dU}{d\varphi} = k\varphi\left(1 - a\varphi + b\varphi^2\right) , \quad \frac{d^2U}{d\varphi^2} = k\left(1 - 2a\varphi + 3b\varphi^2\right) ,$$

$$\frac{d^3U}{d\varphi^3} = k\left(-2a + 6b\varphi\right) , \quad \frac{d^4U}{d\varphi^4} = 6kb .$$

The condition of equilibrium,

$$\frac{dE_p}{d\varphi} = \frac{d}{d\varphi}\left(W_p + U\right) = 0 \quad \rightarrow$$

$$\varphi\left[-Fl\left(1 - \varphi^2/6\right) + k\left(1 - a\varphi + b\varphi^2\right)\right] = 0 , \tag{9.20}$$

has the trivial root $\varphi = 0$ which is independent of the value of F, and possibly two more configurations are given by the real solution of the vanishing second factor

$$\varphi^2 (kb + Fl/6) - ka\, \varphi - (Fl - k) = 0 \quad \rightarrow \tag{9.21}$$

$$\varphi_{1,2} = \frac{1}{2b + Fl/3k} \left\{ a \pm \sqrt{a^2 + 2\,[(Fl/k) - 1]\,(2b + Fl/3k)} \right\} . \tag{9.22}$$

The condition of stability at $\varphi = 0$ renders the inequality independent of any nonlinearity of the restoring spring

$$\frac{d^2E_p}{d\varphi^2} > 0 \quad \rightarrow \quad -Fl + k > 0 . \tag{9.24}$$

The critical load is derived by considering the equality sign; it turns out to be a system parameter

$$F_c = k/l . \tag{9.24}$$

Thus, for all centrally applied loadings F below that critical value, the configuration $\varphi = 0$ is stable. For $F = F_c$, the second-order derivative of the potential energy vanishes, and

$$\left. \frac{d^3E_p}{d\varphi^3} \right|_{\varphi=0} = -2ak , \tag{9.25}$$

is nonzero for the unsymmetric nonlinear spring, indicating that the point of bifurcation is unstable in that case. For $a = 0$ in the constitutive relation, the third derivative vanishes and the sign of the fourth-order derivative decides on the stability of bifurcation in the case of the symmetric restoring moment. Hence,

$$\left. \frac{d^4E_p}{d\varphi^4} \right|_{\varphi=0,\,F=k/l} = k(1 + 6b) > 0 , \tag{9.26}$$

renders a stable point and $b \geq 0$ (a linear or hardening spring) turns out to be a sufficient condition. Softening springs with $b < -1/6$ render an unstable point of bifurcation.

Stability in the postbuckling range $F > F_c$ [see Eq. (9.22)] requires the load factor $\lambda = F/F_c > 1$

$$\varphi_{1,2}{}^2 (\lambda + 6b) - 4a\,\varphi_{1,2} - 2(\lambda - 1) > 0 . \tag{9.27}$$

The inequality may be further reduced by means of Eq. (9.21) to the form

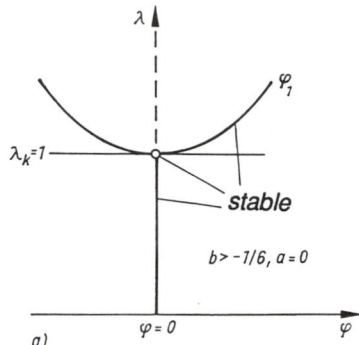

Fig. 9.6 (a). *Safe* bifurcation at $\lambda_c = 1$. Structure is insensitive to imperfections

$$a\, \varphi_{1,2} + 2(\lambda - 1) > 0 \ . \tag{9.28}$$

Various cases of parameter combinations are depicted in Figs. 9.6(a) and 9.6(b) in the load deflection plane. An unstable branch of equilibrium configurations emanating from the (unstable) point of bifurcation into the postbuckling range indicates the *imperfection sensitivity* of the structure: The imperfect structure, eg when loaded eccentrically, becomes unstable (by snapping) far below the (theoretical) critical load F_c .

   If such a qualitative analysis is not sufficient and the actual value of the critical load is to be determined, the imperfect structure to be considered might be quite cumbersome. Selection of the worst imperfection is crucial. For example, a value for the eccentricity *e* of application of the force *F* can be prescribed for a

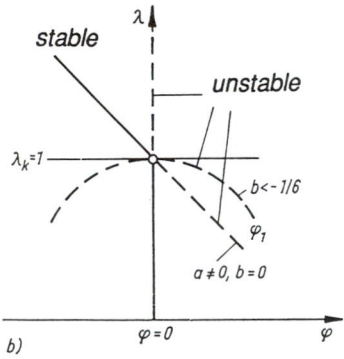

Fig. 9.6 (b). *Unsafe* bifurcation at $\lambda_c = 1$. Imperfection-sensitive structure

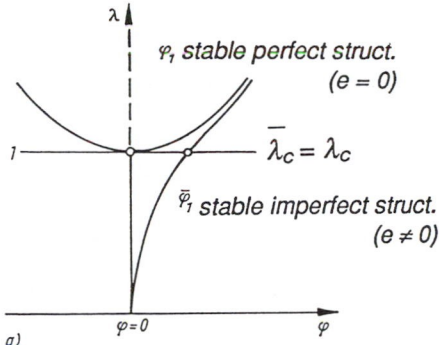

Fig. 9.7 (a). Load deflection path of a structure that is insensitive to imperfection: The critical load of the perfect structure is reached,
but the actual deformations are increased

certain design class of columns in relation to their length l; $\varphi = 0$ is no longer an equilibrium configuration of the imperfect structure. The potential is changed, $\varepsilon = e/l \ll 1$, and may be approximated for sufficiently small angles by

$$W_p = -Fl\left(1 - \cos\varphi\right) - Fe\sin\varphi \approx -Fl\left(\varepsilon\varphi + \frac{\varphi^2}{2} - \varepsilon\frac{\varphi^3}{6} + \frac{\varphi^4}{24}\right). \quad (9.29)$$

Investigation of the nontrivial equilibrium configurations of the imperfect structure renders stable and unstable load deflection branches as shown in Figs. 9.7(a) and 9.7(b). Also illustrated are the

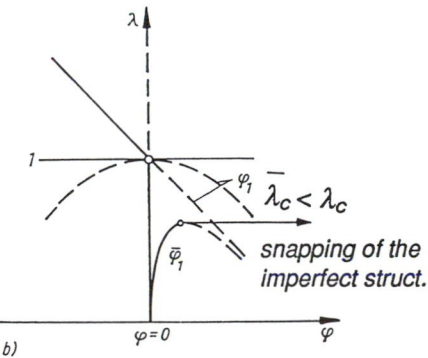

Fig. 9.7 (b). Snapping of an imperfection-sensitive structure at a much lower actual critical load

corresponding curves of the perfect structure. The snapping indicated in Fig. 9.7 (b) yields dynamic motion and may or may not lead to another stable equilibrium in the farfield.

9.1.3. Example: Stability of a Shallow Structure Under Lateral Load

A simple structure that may be subject to snap-through buckling is a shallow *von Mises* truss with linear elastic members when laterally loaded; for a more general three-hinged arch, see Sec. 2.5.1.4. The potential energy of the conservative system as illustrated in Fig. 9.8 is given by the sum of the potential of the dead load F and the elastic potential of the two columns of stiffness c (the compressive members are assumed to be straight and, for the convenience of analysis, weightless)

$$E_p = W_p + U = F(z - b_0) + c(l - l_0)^2 . \qquad (9.30)$$

The subscript zero refers to the undeformed configuration, $l_0^2 = a^2 + b_0^2$. The substitution of $l^2 = a^2 + z^2$ and selection of the nondimensional variable $\zeta = z/a = \tan \alpha$ yield

$$E_p(\zeta) = a F (\zeta - b_0/a) + c a^2 \left(\sqrt{1 + \zeta^2} - l_0/a\right)^2 . \qquad (9.31)$$

For a sufficiently shallow arch, $|\zeta| << 1$, and the condition of equilibrium becomes a cubic equation, since $(1 + \zeta^2)^{-1/2}$ in that case is approximated by $(1 - \zeta^2/2)$

$$\frac{dE_p(\zeta)}{d\zeta} = a F + 2c\, a^2\, \zeta \left(1 - l_0/a\sqrt{1 + \zeta^2}\right) = 0 . \qquad (9.32)$$

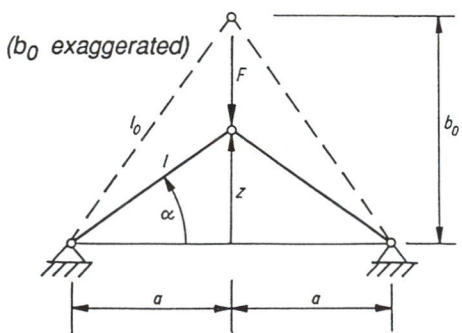

Fig. 9.8. Shallow linear elastic three-hinged arch. The figure shows exaggerated heights

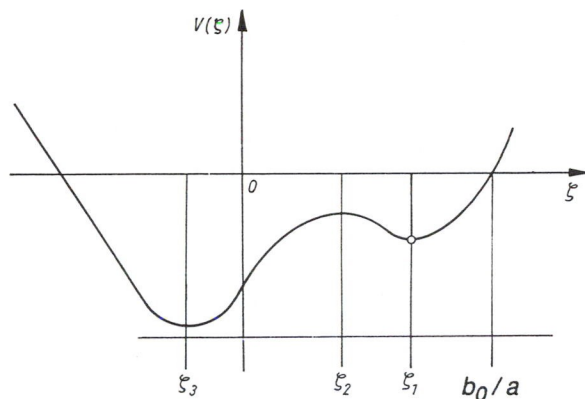

Fig. 9.9. Potential energy with three stationary points

There is always a single real solution, but changing the load gives
three real roots as shown in Fig. 9.9.

The stability of the upper equilibrium configuration $\zeta_1 > 0$ is
considered by

$$\frac{d^2 E_p(\zeta)}{d\zeta^2} = 2c\, a^2 \left[1 - \frac{l_0}{a}\left(1/\sqrt{1+\zeta^2} - \zeta^2(1+\zeta^2)^{-3/2}\right)\right]\Bigg|_{\zeta=\zeta_1} > 0 \; .$$

(9.33)

The stability limit is reached when the second derivative vanishes.
The load takes on its critical value, $F = F_c$, and the configuration,
just before snapping occurs, is determined by the critical angle, $\zeta_1 = \zeta_c$,

$$\left(1 + \zeta^2\right)^{3/2} - \frac{l_0}{a} = 0 \quad \rightarrow \quad \zeta_c = \left[(l_0/a)^{2/3} - 1\right]^{1/2} \; .$$

(9.34)

Equation (9.32) renders upon the substitution of the critical angle
the critical lateral load that turns out to be a system parameter
(see also Exercise A 10.1)

$$F_c = 2c\, l_0 \left[1 - (a/l_0)^{2/3}\right]^{3/2} \; .$$

(9.35)

If the load reaches that critical value, dynamic snap-through occurs
and the arch eventually comes to a rest in the lowered configuration.
The latter is indicated in Fig. 9.9 by $\zeta_3 < 0$. Snapping is accompanied
by high compressive stresses and, thus, by yielding of ductile
members, or fracture may even occur. The structural safety of
shallow arches requires $F < F_c$. Furthermore, a second stability

problem arises for slender bars that may buckle before the onset of snapping; see Sec. 9.1.4 for a discussion of buckling load. The normal force

$$N = -F/(2 \sin \alpha) \quad , \tag{9.36}$$

takes on its extreme static value at the onset of snap-through,

$$N = -cl_0 \left[1 - (a/l_0)^{2/3}\right] \quad , \quad \zeta = \zeta_c \; . \tag{9.37}$$

Shallow structures may creep under lateral loadings far below the critical one, $F < F_c$, with the effect of decreasing the height of the apex z with time. Analogous to Eq. (4.119) a *lifetime* of the viscous shallow structure can be determined before it reaches the critical configuration, Eq. (9.34), and, subsequently fails by (elastic) snap-through.

9.1.4. Example: Buckling of Slender Elastic Columns (Euler Buckling)

The stability of a simply supported straight linear elastic rod of constant (full) cross-section when centrally loaded in compression is considered in Fig. 9.10; such a case was investigated by *L. Euler*. With the assumption of a dead-weight load F and a bar, obeying *Hooke's* law, the perturbing flexural vibration is known to be a conservative motion. The potential in the neighborhood of the straight configuration is the sum of the potential energy of the external force F and the strain energy of both deformations, compression and principal bending about the y axis

$$E_p = W_p + U \; . \tag{9.38}$$

$E_p = 0$, when the axis is straight, and

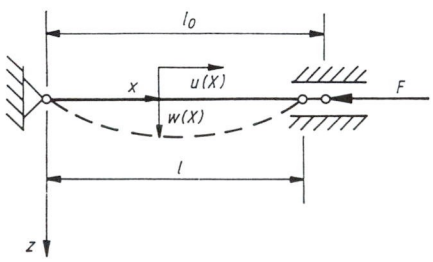

Fig. 9.10. Buckling of the *Euler* column (a slender linear elastic bar)

$$W_p = F \int_0^{l_0} \frac{du}{dX} \, dX \ .$$

(9.39)

Substituting the uniaxial *Hooke's* law, $\sigma_{xx} = E \, \varepsilon_{xx}$, into the strain energy density, $U' = (1/2) \, \sigma_{xx} \, \varepsilon_{xx}$, and performing the volume integration give

$$U = \frac{1}{2} \int_0^{l_0} \int_A \sigma_{xx} \varepsilon_{xx} \, dA \, dX = \int_0^{l_0} \frac{E}{2} \int_A \varepsilon_{xx}^2 \, dA \, dX \ .$$

(9.40)

The linearly distributed strains obey a reduced nonlinear geometric relation (1.20): $\varepsilon_{xx} = du/dX + (1/2) \, (dw/dX)^2 - Z \, d^2w/dX^2$, since $(du/dX)^2 \ll |du/dX|$. Performing the cross-sectional integration yields two terms (the y axis passes through the centroid) that are easily recognized to be the strain energy of axial deformation and of bending, respectively [compare these with Eqs. (5.26) and (5.27)]

$$U = U_S + U_B \ , \ U_S = \int_0^{l_0} \frac{EA}{2} \left[\frac{du}{dX} + \frac{1}{2} \left(\frac{dw}{dX} \right)^2 \right]^2 dX \ , \ U_B = \int_0^{l_0} \frac{EJ}{2} \left(\frac{d^2w}{dX^2} \right)^2 dX \ .$$

(9.41)

The axis of the beam is assumed to be inextensible during deflection; hence,

$$U_S = 0 \ , \ \varepsilon_{xx}^{(0)} = \frac{du}{dX} + \frac{1}{2} \left(\frac{dw}{dX} \right)^2 = 0 \ \rightarrow \ \frac{du}{dX} = -\frac{1}{2} \left(\frac{dw}{dX} \right)^2 \ .$$

Substitution into Eq. (9.39) gives finally the potential of the deflected beam composed of two terms,

$$E_p = \int_0^{l_0} \frac{EJ}{2} \left(\frac{d^2w}{dX^2} \right)^2 dX - \frac{F}{2} \int_0^{l_0} \left(\frac{dw}{dX} \right)^2 dX \ .$$

(9.42)

Its first variation must vanish [see Eq. (5.15)]

$$\delta E_p = 0 \ .$$

Euler's equation of that variational problem (the system has an infinite number of degrees of freedom) is the linear and homogeneous (no lateral loads) differential equation of the deflection according to the second-order theory of elasticity [see Eq. (5.22)]

$$\frac{d^2}{dX^2}\left(EJ \frac{d^2w}{dX^2}\right) + F \frac{d^2w}{dX^2} = 0 \ .$$

(9.43)

Integrating twice and considering the hinged-hinged boundary conditions at $X = 0, l_0$ [see Eq. (5.22a)] give the moment curvature relation

$$EJ \frac{d^2w}{dX^2} = -M = -Fw \ .$$

(9.44)

That relation also follows directly by applying the condition of equilibrium to a deflected finite element of the beam, say, of length x . The proper solution to this oscillator type of equation is given by $[w(0) = w(l_0) = 0]$

$$w(x) = a \sin \alpha_n X \quad \text{(with the eigenvalues)} \quad \alpha_n = n\pi/l_0 = \sqrt{F/EJ} \ .$$

(9.45)

In addition to the trivial solution, $w \equiv 0$, the axis remains straight, deflected equilibrium configurations exist when the load exceeds the first eigenvalue, the critical load or *Euler* buckling load, $F > F_1 = F_c = \pi^2 EJ/l_0^2$. At $F = F_c$, the first bifurcation occurs and the column buckles according to the first mode, $w_1 = a \sin \alpha_1 X$. The buckling modes are affined to the modes of free flexural vibrations; see Sec. 11.3.1 and Eq. (7.152). It can be shown by a nonlinear theory that is valid in the postbuckling range, that for $F > F_c$, the deflected configuration is stable and the equilibrium of the straight column unstable. The second variation of the potential energy is considered here to explore the range of stability of the straight column. Putting $w = w_1 + \delta w$, with the virtual deflection $\delta w = \varepsilon w_1$ assumed to be affined to the first mode, $\varepsilon \ll 1$, yields upon the integration of Eq. (9.42)

$$E_p(\varepsilon) = \frac{l a^2}{4} \alpha_1^2 \left(EJ \alpha_1^2 - F\right) \left(1 + 2\varepsilon + \varepsilon^2\right) \ .$$

(9.46)

Putting the first variation to zero renders the critical load and at this point of bifurcation the corresponding equilibrium configurations to be $w = a = 0 , w = a \sin \alpha_1 X$

$$\delta E_p = \left.\frac{\partial E_p}{\partial \varepsilon}\right|_{\varepsilon = 0} d\varepsilon = 0 \ \rightarrow \ F = F_c = EJ \alpha_1^2 \ .$$

(9.47)

The second variation yields the stability condition for the straight beam only (due to the linearized theory), $w = 0 : F < F_c$, since

$$\delta^2 E_p > 0 \quad \rightarrow \quad \left.\frac{\partial^2 E_p}{\partial \varepsilon^2}\right|_{\varepsilon=0} = \frac{1}{2}\,\alpha_1{}^2\,(F_c - F) > 0 \ .$$

$$(9.48)$$

At the point of bifurcation and also in the deflected configuration in the neighborhood of the straight axis, the load is constant and equals the critical load, $F = F_c$. Thus, the second variation vanishes. It can be concluded that a nonlinear theory of postbuckling is required to show that the point of bifurcation (as well as the deflected branch of equilibrium) is stable since the variation of the fourth order of the associated potential energy is to be considered. Hence, in such a case, it becomes obvious that the linear elastic *Euler* column is insensitive to small geometric imperfections. That is, the load-carrying capacity is not affected by a small eccentricity of the loading or a slight initial deflection of the beam axis.

Generally, it suffices within the elastic range, to consider the perfectly straight and centrally loaded slender column and its critical load also from a practical engineering point of view.

Equilibrium in the deformed configuration of a slender clamped-hinged beam, as shown in Fig. 9.11, renders the bending moment $M(x) = Fw(x) - H(l - x)$ and the edge moment $M_e = -Hl$. Substitution into the linearized differential equation (5.22a) gives (shear deformations are neglected)

$$\frac{d^2w}{dx^2} = -\frac{M}{EJ} \quad \rightarrow \quad \frac{d^2w}{dx^2} + \frac{F}{EJ}\,w = \frac{H}{EJ}\,(l - x) \ .$$

$$(9.49)$$

The general solution is found by superposition [see also the oscillator equation (7.81)] to be

$$w(x) = C_1 \cos \alpha x + C_2 \sin \alpha x + \frac{H}{EJ\,\alpha^2}\,(l - x) \ .$$

$$(9.50)$$

The boundary conditions according to the supports shown in Fig. 9.11 are

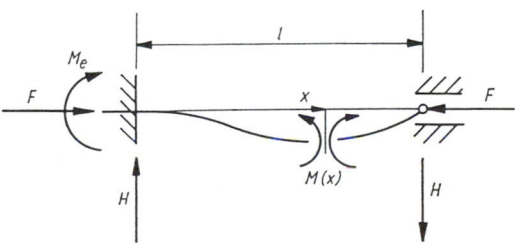

Fig. 9.11. Buckling of a redundant slender column

$$x = 0: \ w = 0 \ , \ \frac{dw}{dx} = 0 \ \text{(geometric conditions)},$$

$$x = l: \ w = 0 \ , \ \frac{d^2w}{dx^2} = 0 \quad \text{(a geometric and a dynamic condition)} .$$
$$(9.51)$$

Thus, the proper buckling mode follows by considering the three homogeneous linear equations

$$C_1 + \frac{l}{EJ\alpha^2} H = 0, \ \alpha C_2 - \frac{1}{EJ\alpha^2} H = 0, \ C_1 \cos \alpha l + C_2 \sin \alpha l = 0 .$$
$$(9.52)$$

A nontrivial solution indicates that a deflected equilibrium configuration exists. Hence, the characteristic equation determines the eigenvalues α ; the lowest one is the critical value α_1 . The buckling determinant must vanish, and the coefficients are determined by inspection of Eq. (9.52)

$$\begin{vmatrix} 1 & 0 & l/EJ\alpha^2 \\ 0 & \alpha & -1/EJ\alpha^2 \\ \cos \alpha l & \sin \alpha l & 0 \end{vmatrix} = 0 \ , \ \text{or} , \ \alpha l = \tan \alpha l .$$
$$(9.53)$$

The roots of this important transcendental equation are tabulated, eg in: *Handbook of Mathematical Functions (eds. M. Abramowitz and I. A. Stegun) Dover, New York, 1965, p. 224.*
The smallest eigenvalue is $\alpha_1 l = 4.49$ and the critical load becomes

$$F_c = EJ \, \alpha_1^2 = \frac{\pi^2 \, EJ}{(0.7 \, l)^2} .$$
$$(9.54)$$

A comparison with the buckling load (9.47) of the *Euler* column of the same span l shows the stiffening effect of the clamped edge since the critical load is increased by a *buckling factor* of *2.04* . The span of a fictitious hinged-hinged bar with the same buckling load as given by Eq. (9.54), the so-called *buckling length* , is thus given by

$$l_c = 0.7 \ l .$$
$$(9.55)$$

The buckling lengths or squared reciprocals of the coefficients, the buckling factors, are tabulated for several cases of simple supports, eg.in *A. Pflüger: Stability Problems of Elastostatics. Springer-Verlag, Berlin, 1975, p. 340 (in German).*

The five different versions of the single-span compressive bar are illustrated in Fig. A 9.3 that is similar to a figure in *H. Ziegler: Principles of Structural Stability. Blaisdell, Waltham, Mass., 1968.* For example, the cantilever of span l has an associated buckling length of $l_c = 2 l$, but the clamped-clamped column with laterally immovable supports has $l_c = l/2$. See Exercise A 9.3. The latter bar has a buckling factor of *1* in the case of a free sidesway of one clamped support.

The *buckling stress* , ie the critical compressive stress at the onset of buckling, is determined by dividing the buckling load by the cross-sectional area A . By means of the radius of inertia $i = (J/A)^{1/2}$ (J is the principal moment of inertia about the buckling axis, commonly the smallest of the two) and the so-called slenderness Λ , Eq. (9.54) takes on the general form

$$\sigma_c = \pi^2 E / \Lambda^2 < \sigma_p \ , \ \Lambda = l_c / i \ . \tag{9.56}$$

This formula is valid in the range of linear elastic material behavior. For example, in the case of a mild steel bar made of *St 360,* the limit stress of proportionality is σ_p = *192 N/mm²* and *Young's* modulus E = *2.06 x 10⁵ N/mm²* . Thus, the minimal value of the slenderness for *Euler's* buckling formula to remain valid is about 104. The buckling load of more compact bars is overestimated by Eq. (9.56) for smaller slenderness ratios! Approximate theories have been developed by *Engesser, von Karman,* and *Shanley.* In a first step, *Young's* modulus is to be replaced by the tangent modulus to extend the range of validity of Eq. (9.56). If these compact bars are made of elastic-plastic materials, limit load analysis is more appropriate; see Sec. 9.3. Elastic buckling is no longer possible since plastic deformations develop far below the critical load.

9.1.5. The Eccentrically Loaded Linear Elastic Column

The *Euler* column of Fig. 9.10 is reconsidered with an eccentrically applied compressive double force F without moment; see Fig. 9.12. The constant moment $e F$ of statically equivalent reduction produces an initial curvature. Equilibrium is considered in the deflected configuration according to the second-order theory and renders the bending moment variable within the span, $M(x) = F[e + w(x)]$. Substitution into the linearized differential equation of deflection (y is a principal axis of the cross-section, and the curvatures must remain sufficiently small) renders Eq. (9.44) inhomogeneous

$$\frac{d^2w}{dx^2} + \alpha^2 w = -\alpha^2 e \ , \ \alpha = \sqrt{F/EJ} \ . \tag{9.57}$$

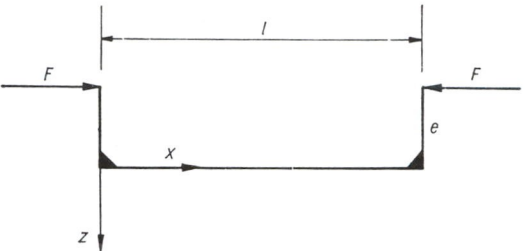

Fig. 9.12. The imperfect *Euler* column

With $e \neq 0$, Eq. (9.57) constitutes a stress rather than an eigen-value problem. Its general solution by superposition is simply

$$w(x) = C_1 \cos \alpha x + C_2 \sin \alpha x - e .\qquad (9.58)$$

The constants of integration are determined by the geometric boundary conditions of vanishing deflections at $x = 0, l$ to be $C_1 = e$ and $C_2 = e$ $(1 - cos\ \alpha l)/sin\ \alpha l$. The deflections grow beyond any bounds when the force F reaches the *Euler* buckling load $\pi^2 EJ/l^2$; this is, of course, a deficiency of the linearized theory. The bending moment

$$M(x) = F\,(e + w) = \frac{e\,F}{\sin \alpha\, l}\left[\sin \alpha x + \sin \alpha\,(l - x)\right] ,\qquad (9.59)$$

the bending stresses $\sigma_{xx} = z\,M(x)/J$, as well as the deflection $w(x)$, are all nonlinearly related to the dead load F (or α). The superposition principle is no longer valid with respect to combined axial loadings. For elastic-plastic (ie ductile) materials, limit load analysis is appropriate; see Sec. 9.3. For a load regime $v\,F$ and a proper design eccentricity e , the load factor v_y may be determined such that the maximal stress just reaches the yield stress, say, in a steel structure.

Stress problems in certain cases may show bifurcations of equilibrium at critical loads below the limit load. A criterion that such stability problems meet, which is also applicable to more complex structures, is due to *K. Klöppel* and *K. Lee* and may be found illustrated by several examples of engineering importance in *C. F. Kollbrunner and M. Meister: Knicken, Biegedrillknicken, Kippen. Springer-Verlag, Berlin, 1961.* A simple case, buckling in tension, is considered in Exercise A 9.4.

9.1.6. Buckling of Thin Plates

Analogous to the buckling of slender columns, the critical in-plane loading of linear elastic and thin plates is determined by the equilibrium method of second order. Hence, Eq. (5.30) is considered, with assigned membrane forces n_{ij} but without any lateral pressure, to render the proper homogeneous fourth-order partial differential equation

$$K\nabla^2\nabla^2 w - n_x \frac{\partial^2 w}{\partial x^2} - n_y \frac{\partial^2 w}{\partial y^2} - 2n_{xy} \frac{\partial^2 w}{\partial x \partial y} = 0 \ , \quad K = Eh^3/12\left(1 - \nu^2\right) \ . \tag{9.60}$$

Contrary to Eq. (5.29), the mid-plane area is assumed to be inextensible; U_M of Eq. (5.26) is supposed to vanish. To Eq. (5.28) the work of the externally assigned membrane forces must be added instead; see Eq. (9.39). Equation (9.60) always has the trivial solution of the undeflected plate. Bifurcation occurs for a given distribution of membrane stresses λn_{ij} when the load factor becomes critical λ_c , ie when it takes on the lowest absolute eigenvalue. In that case, a neighboring deflected equilibrium configuration becomes possible. The eigenvalues of partial differential equations with non-constant coefficients, a problem that is encountered here in general, are approximately determined by means of the *Ritz-Galerkin* procedure. See Chap. 11.

(§)   Rectangular Simply Supported Plate. This special case is further simplified by the assumption of a natural number for the ratio of the spans a/b. Loading, say, in the x direction, is considered here. Such a membrane load can be expanded in the cosine *Fourier* series

$$n_x(y) = -\frac{\pi^2 K}{b^2} \sum_{m=0}^{\infty} p_m \cos \frac{m\pi y}{b} \ , \quad n_y = n_{xy} = 0 \ , \tag{9.61}$$

and its critical intensity $\lambda_c n_x$ is easily determined. Substitution of the deflection with unknown coefficients a_k and with (termwise) built-in boundary conditions at the edges $x = 0, a$ and $y = 0, b$,

$$w(x, y) = \sin \frac{n\pi x}{b} \sum_{k=1}^{\infty} a_k \sin \frac{k\pi y}{b} \ , \tag{9.62}$$

and comparing coefficients of $\sin l\pi y/b$, $l = 1, 2, \ldots$ in Eq. (9.60) yield an (infinite) linear and homogeneous system of equations for the *Fourier* coefficients a_k . The (symmetric) buckling determinant must vanish and the smallest root renders the critical load factor. With

$$\alpha_k = \left(\frac{n^2 + k^2}{n}\right)^2 ,$$

$$\begin{vmatrix} 2(p_0 - \alpha_1/\lambda) - p_2 & p_1 - p_3 & p_2 - p_4 & \cdot\cdot \\ p_1 - p_3 & 2(p_0 - \alpha_2/\lambda) - p_4 & p_1 - p_5 & \cdot\cdot \\ p_2 - p_4 & p_1 - p_5 & 2(p_0 - \alpha_3/\lambda) - p_6 & \cdot\cdot \end{vmatrix} = 0 . \qquad (9.63)$$

In the general case of the infinite determinant, only an approximate solution is possible. For constant prestressing, $n_x = -\pi^2 K/b^2 = const$, $p_0 = 1$, $p_m = 0$, $m > 0$, the (infinite) product of the terms in the main diagonal remains to vanish and the critical load factor is given by the minimum of

$$\lambda_c = \min_{n = sb/a} [(n^2 + 1)/n]^2 , \quad (n_x)_c = -(n^2 + 1)^2 \pi^2 K / n^2 b^2 . \qquad (9.64)$$

The buckling mode shows a single half-wave lateral to the prestressing, but may exhibit several half-waves in the direction of the compressive loading in relationship to both the span ratio a/b and the natural number. The latter is an output of the above minimum problem.

Critical load factors are tabulated for many loadings and planforms of practical interest. See also Exercises A 11.7 through A 11.9, and for reference, A. Pflüger: Stability Problems of Elastostatics. Springer-Verlag, Berlin, 1975, p. 340, (in German).

9.2. Stability of Motion

Stability of motion is considered within the limits of the dynamic method of small perturbations. Analogous to the considerations of stability of equilibrium (Sec. 9.1), a small perturbation is applied to the given motion and the time behavior of the resulting deviations determined. In many cases, it suffices to consider the linearized equations of motion of the perturbing deviations. <<The given motion is stable if the perturbations remain small and decrease with the reduction of the initial disturbance.>> A general procedure that renders stability conditions also for large perturbations is the direct or second method of Ljapunow . Due to its mathematical difficulties, this method is beyond the scope of this textbook. Solutions of problems of engineering interest may be found in the monograph (in German), W. Hahn: Theory and Application of the Direct Method of Ljapunow. Springer-Verlag, Berlin,1959. See also, P. C. Müller: Stability and Matrices. Springer-Verlag, Berlin, 1977, (in

German), and, I. G. Malkin: Theory of the Stability of Motion, Munich, 1959, (in German).

9.2.1. Example: The Centrifugal Governor

The method of small perturbations is illustrated by considering the stability of the stationary rotation of the centrifugal governor sketched in Fig. 9.13. All guiding members are assumed mass-less; thus, the angular momentum about the vertical axis includes that of the two point masses m only. $D = [M\,i^2 + 2m\,(e + l\sin\varphi)^2]\,\omega$ and as Eq. (7.22) gives for the nonstationary motion

$$[M\,i^2 + 2\,m\,(e + l\sin\varphi)^2]\,\dot{\omega} + 4\,m\,(e + l\sin\varphi)\,\omega l\,\dot{\varphi}\cos\varphi = M_R, \quad (9.65)$$

where M_R is the externally applied perturbing moment. It is supplied by the engine under speed control. Conservation of momentum (7.7) of the socket joint with mass M (the free-body diagram in axial sliding motion is to be considered) yields (S_1 is the force in one of the two lower connecting members)

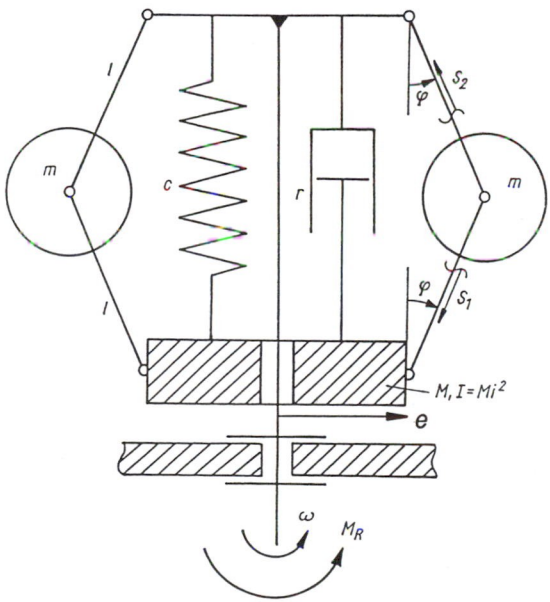

Fig. 9.13. Basic system of a centrifugal governor.
Instantaneous configuration shown when $\omega(t)$ depends on time

$$M \frac{d^2}{dt^2}(2l \cos \varphi) = Mg + c\left[s_0 + 2l(1 - \cos \varphi)\right] - r\frac{d}{dt}(2l \cos \varphi) - 2S_1 \cos \varphi.$$
(9.66)

The conservation of the components of momentum of the single point mass m in the radial and vertical direction completes the set of equations. With S_2 , the force in one of the two upper connecting members, the free-body diagram renders by inspection

$$m \frac{d^2}{dt^2}(e + l \sin \varphi) - m(e + l \sin \varphi)\omega^2 = -(S_1 + S_2) \sin \varphi ,$$

$$m \frac{d^2}{dt^2}(l \cos \varphi) = mg + (S_1 - S_2) \cos \varphi .$$
(9.67)

Eliminating S_2 gives

$$m l \ddot{\varphi} - m(e + l \sin \varphi)\omega^2 \cos \varphi = - mg \sin \varphi - 2S_1 \cos \varphi \sin \varphi .$$
(9.68)

The substitution of S_1 from Eq. (9.66) eliminates the unwanted member forces altogether

$$\left(m + 2 M \sin^2 \varphi\right) l \ddot{\varphi} + 2 M l \dot{\varphi}^2 \sin \varphi \cos \varphi - m(e + l \sin \varphi)\omega^2 \cos \varphi$$

$$= -(m + M) g \sin \varphi - c\left[s_0 + 2l(1 - \cos \varphi)\right] \sin \varphi - 2 r l \dot{\varphi} \sin^2 \varphi .$$
(9.69)

Equations (9.65) and (9.69) are the resulting set of highly nonlinear control equations of the nonstationary rotation. The stationary motion where $M_R = 0$ and $\omega = const$, $\varphi = const$, corresponds to a governor configuration given by the so-called control equation (all time derivatives vanish above), $\varphi_0 = \varphi(\omega_0)$,

$$m(e + l \sin \varphi_0) \omega_0^2 \cos \varphi_0 - (m + M) g \sin \varphi_0 - c\left[s_0 + 2l(1 - \cos \varphi_0)\right] \sin \varphi_0 = 0 .$$
(9.70)

      The stability of that steady state may be considered within the limits of the method of small perturbations if we assume the deviations are small, $|\varepsilon| \ll \omega_0$, $|\eta| \ll 1$ and $\omega(t) = \omega_0 + \varepsilon$, $\varphi(t) = \varphi_0 + \eta$, for all times t . The moment at the control shaft is the function $M_R(\varphi)$ with $M_R(\varphi_0) = 0$. Thus, linearization of the remaining Eqs. (9.65) and (9.69) in the variables of the perturbed motion with respect to the stationary state can be performed by dropping all nonlinear terms, also the mixed quadratic ones after substituting the linear approximations. Some are given below:

$$\sin(\varphi_0 + \eta) \approx \sin \varphi_0 + \eta \cos \varphi_0 , \quad M_R(\varphi_0 + \eta) = 0 + \eta M'(\varphi_0) , \quad M' = \frac{dM_R}{d\eta} , \text{ etc.}$$
(9.71)

Observing the control equation, we find that the set of linearized equations of motion becomes

$$p_2 \ddot{\eta} + p_1 \dot{\eta} + p_0 \eta - p\varepsilon = 0 ,$$

$$q_1 \dot{\varepsilon} + q \dot{\eta} - M'\eta = 0 . \tag{9.72}$$

The constant coefficients are given through the steady-state parameters

$$p_0 = m\omega_0^2 \left[e \sin \varphi_0 + I \left(2 \sin^2 \varphi_0 - 1 \right) \right] + 2 \, cI \left(\cos \varphi_0 + 2 \sin^2 \varphi_0 - 1 \right)$$

$$+ (m + M) \, g \cos \varphi_0 ,$$

$$p_1 = 2 \, rI \sin^2 \varphi_0 , \quad p_2 = mI + 2 \, MI \sin^2 \varphi_0 , \quad p = 2 \, m \left(e + I \sin \varphi_0 \right) \omega_0 \cos \varphi_0 ,$$

$$q_1 = 2m \left(e + I \sin \varphi_0 \right)^2 + Mi^2 , \quad q = 2 \, pI .$$

The solution of the homogeneous linearized set of coupled differential equations is of the exponential type, $(\varepsilon, \eta) = (A, B) \, [\exp (\alpha t)]$. When substituted, the linear homogeneous set of equations follows at once

$$-p A + \left(p_2 \, \alpha^2 + p_1 \, \alpha + p_0 \right) B = 0 ,$$

$$\alpha \, q_1 \, A + \left(\alpha \, q - M' \right) B = 0 . \tag{9.73}$$

They must be linearly dependent to yield a nontrivial solution. Thus, their determinant of coefficients vanishes and gives the characteristic equation

$$p_2 \, q_1 \, \alpha^3 + p_1 \, q_1 \, \alpha^2 + (p_0 \, q_1 + p \, q) \, \alpha - M' p = 0 . \tag{9.74}$$

The stationary motion of the centrifugal governor is stable if the linearized perturbing motion is asymptotically stable, ie if it decays with time such that the original configuration is reached again. Hence, the eigenvalues must have a negative real part, $\mathrm{Re}(\alpha) < 0$. Without solving the polynomial equation (9.74), the conditions can be given by the theorem of *Hurwitz* . <<The algebraic equation of degree n with coefficient $a_0 > 0$,

$$\sum_{k=0}^{n} a_k \, \alpha^{n-k} = 0 ,$$

has only roots with negative real parts, iff all the determinants, $r = 1, 2, ..., n$, of the form given by Eq. (9.75) are positive.>> The condition is not satisfied if any one of the coefficients $a_i < 0$.

$$D_r = \begin{vmatrix} a_1 & a_3 & a_5 & . & . & a_{2r-1} \\ a_0 & a_2 & a_4 & . & . & a_{2r-2} \\ 0 & a_1 & a_3 & . & . & a_{2r-3} \\ 0 & a_0 & a_2 & . & . & a_{2r-4} \\ . & . & . & . & . & . \\ 0 & 0 & 0 & 0 & . & a_r \end{vmatrix} > 0 . \tag{9.75}$$

The stability conditions according to the *Hurwitz* criterion when applied to the third-order characteristic equation (9.74) are subsequently listed. They provide limits on the design parameters as follows:

1. $p_2 q_1 > 0$, p_2 and q_1 are individually and essentially positive.

2. $D_1 = p_1 q_1 > 0$, $q_1 > 0$, essentially positive per definition. Hence, $p_1 > 0$ requires $r > 0$, ie damping of the governor is necessary.

3. $D_2 = \begin{vmatrix} p_1 q_1 & -M' \\ p_2 q_1 & p_0 q_1 + pq \end{vmatrix} = q_1 [p_1 (p_0 q_1 + p q) + p_2 M'] > 0$,

or equivalently, $(p_1/p_2)(p_0 q_1 + p q) > - M'$. Thus, the minimal value of the damping coefficient r follows from $(p_0 q_1 + p q) > 0$.

4. $D_3 = \begin{vmatrix} p_1 q_1 & -p M' & 0 \\ p_2 q_1 & (p_0 q_1 + p q) & 0 \\ 0 & p_1 q_1 & -p M' \end{vmatrix} = -p M' > 0$.

Since $p > 0$, the condition requires that the slope of the external moment, $M' = dM_R/d\varphi < 0$, be negative. The governor controls the engine such that the moment M_R decreases with the increasing speed of revolution ω .

A theorem of *Ljapunow* states that the motion of the nonlinear system is stable if the linearized perturbing motion is asymptotically stable and the absolute values of the nonlinear terms in the equations of motion as well as the absolute measure of the initial disturbance are bounded within certain limits. The latter are to be determined in each individual case. See, eg *I. G. Malkin: Theory of the Stability of Motion. Munich, 1959, (in German).*

9.2.2. Stability of the Steady State of the Spinning Unsymmetric Gyroscope

Self-equilibrating external forces are assumed that are equivalent to a rigid body with a frictionless point support at its center of gravity. *Euler's* equations without moments [see Eq. (7.61)]

$$I_1 \dot\omega_1 - (I_2 - I_3)\, \omega_2\, \omega_3 = 0\,,$$

$$I_2 \dot\omega_2 - (I_3 - I_1)\, \omega_3\, \omega_1 = 0\,, \tag{9.76}$$

$$I_3 \dot\omega_3 - (I_1 - I_2)\, \omega_1\, \omega_2 = 0\,,$$

have special solutions that correspond to steady-state rotations about the principal axes of inertia, eg $\omega_1 = const,\ \omega_2 = \omega_3 = 0$. The stability of such motions is checked by the method of small perturbations. Spinning about the axis 1 of the maximal moment of inertia is perturbed. Substitution of the resulting time-dependent angular velocities,

$$\omega_1 = \omega + \varepsilon\,,\ \ \omega_2 = \lambda\,,\ \ \omega_3 = \mu \quad |\varepsilon| \ll \omega\,,\ |\lambda| \ll \omega\,,\ |\mu| \ll \omega\,, \tag{9.77}$$

into Eqs. (9.76) and subsequent linearization render the set of differential equations with constant coefficients of the perturbing motion

$$I_1 \dot\varepsilon = 0\,,\ \ I_2 \dot\lambda - \omega(I_3 - I_1)\,\mu = 0\,,\ \ I_3 \dot\mu - \omega(I_1 - I_2)\,\lambda = 0\,, \tag{9.78}$$

they are easily integrated. The solutions are: the angular speed $\varepsilon = const$ and the remaining speeds are of the exponential type, $(\lambda,\ \mu) = (A,\ B)\ exp\ (vt)$, where

$$A\,v\,I_2 - B\,\omega(I_3 - I_1) = 0\,,$$
$$-A\,\omega(I_1 - I_2) - B\,v\,I_3 = 0\,. \tag{9.79}$$

Nonvanishing roots are guaranteed if the characteristic equation holds

$$I_2\,I_3\,v^2 - \omega^2(I_1 - I_2)\,(I_3 - I_1) = 0\,. \tag{9.80}$$

The eigenvalues are

$$v = \pm\, i\omega \sqrt{(I_1 - I_2)\,(I_1 - I_3)/I_2 I_3}\,. \tag{9.81}$$

The rotations about the axes 2 and 3 in the body-fixed frame are bounded only if the eigenvalues are purely imaginary. Hence, I_1 must be the largest of the principal moments of inertia. The same condition holds if I_1 is considered the smallest moment of inertia: Axis 3 is just renumbered as 1. Spinning about the axis 2 of the intermediate inertia moment is unstable. The linearized perturbed motion is not asymptotically stable; thus, it cannot be concluded that the corresponding nonlinear motions are stable at all. Considering light viscous damping, however, makes the proper linearized problems asymptotically stable and spinning about the axes 1 and 3 (of the largest and smallest inertia moment, respectively) are stable at large. See also Sec. 7.4.5 on the dynamic balancing of rotors. And, for deeper insight, consult *J. La Salle and S. Lefschetz: The Stability Theory of Ljapunow. Bl-Hochschultaschenbuch 194, Mannheim, 1967, (in German)*.

Note that damping does not always have a stabilizing effect.

9.3. Bounds of Stability of Equilibrium of Elastic-Plastic Structures: Limit Load Analysis

The equilibrium configuration of an elastic-plastic system may become unstable due to plastic deformations when the loading grows beyond the yield limit. A critical loading intensity may be reached where the resistance of the structure is exhausted. At this stage, the loading factor takes on its critical value λ_c , corresponding to the limit load, the structure fails, and possibly large deformations occur. See Fig. 9.14.

According to Sec. 4.3.2, ideal plasticity is assumed by neglecting any hardening of the material in the illustrative example of a single-span, simply supported beam, symmetrically loaded beyond the yield limit by the single force F and with a rectangular full cross-section of area $A = B H$ (see Fig. 9.15). *Bernoulli´s* hypothesis of plane cross-sections remaining plane after deflection of the beam is assumed to hold also in the yielding zone that spreads

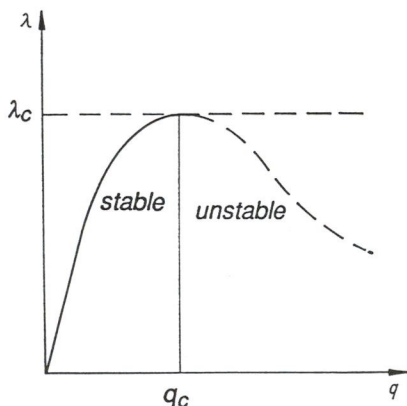

Fig. 9.14. Load factor deformation path of an elastic-plastic structure exhibiting limit load instability

from the midspan area outward, Eq. (6.21). The linear distribution of strain is proportional to the linearized (sufficiently small) curvature

$$\varepsilon_{xx} = -z \frac{d^2w}{dx^2} \; . \tag{9.82}$$

Neglecting the shear stresses simplifies the yield condition considerably

$$\sigma_{xx}^2 - \sigma_Y^2 = 0 \; , \tag{9.83}$$

where σ_Y denotes the yield limit in a uniaxial tension test. The linear distribution of the bending stresses in the elastic regime, Eq. (4.31), holds until the outer fibers at $z = \pm H/2$ start to yield (one and the same yield limit is assumed in tension and in compression)

$$(\sigma_{xx})_{el} = \frac{M}{J} z \; , \quad J = BH^3/12 \; .$$

At this stage of loading, at the onset of yielding, the midspan bending moment takes on the value of the yield moment $M_Y = \sigma_Y BH^2/6$. Increasing the force F further renders a spreading of the yielding zone, where the stress is constant, $\sigma_{xx} = \sigma_Y = const$, within the cross-section toward the center, and from midspan symmetrically toward the supports. The bending moment in the left

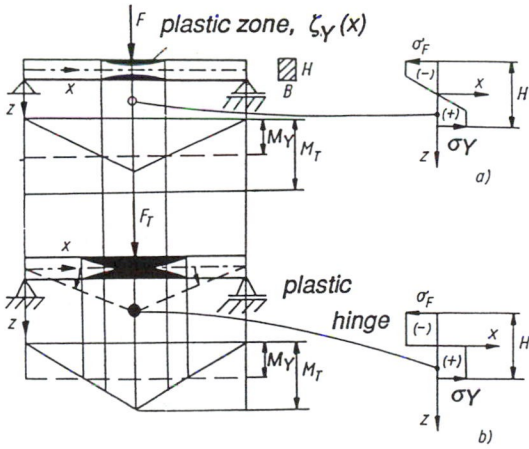

Fig. 9.15. Elastic-plastic single-span beam loaded beyond the yield limit. (a) Nonlinear stress distribution in the yielding zone. (b) Stresses in the plastic hinge. The limit load

half of the span varies linearly with x and is equivalent to the resulting moment of the stress distribution in a cross-section in the plastic zone. Integration thus renders

$$M(x) = x\, F/2 = 2\left[\int_{0}^{\zeta_Y(x)} z\left(\sigma_Y \frac{z}{\zeta_Y}\right) dA + \int_{\zeta_Y(x)}^{H/2} z\,\sigma_Y\,dA\right] = \frac{BH^2}{4}\,\sigma_Y\left[1 - \frac{1}{3}(2\,\zeta_Y/H)^2\right].$$

(9.84)

The equation may be solved for the contour of the plastic zone

$$\zeta_Y(x) = \pm\frac{H}{2}\,\sqrt{3\left(1 - \frac{IF}{HA\,\sigma_Y}\frac{2\,x}{I}\right)}\,,\qquad \frac{HA\,\sigma_Y}{3IF} \le x \le \frac{I}{2}\,.$$

(9.85)

The elastic core has the lateral extension $-\zeta_Y < z < \zeta_Y$. Due to the statically determinate support, the limit load is reached when the midspan cross-section becomes fully plastic, ie when a so-called plastic hinge is formed, $\zeta_Y\,(x = l/2) = 0$. The maximal moment at the limit load thus becomes

$$M_L = \sigma_Y\frac{BH^2}{4} = \frac{I\,F_c}{4}\,.$$

(9.86)

The limit load results explicitly since the plastic reserve of the structure and the plastic reserve of the cross-section are identical, and, hence, are exhausted at the same time

$$F_c = F_L = \sigma_Y\frac{BH^2}{I}\,.$$

(9.87)

If the force F is increased to its critical value, the limit load F_L, the loading factor $\lambda = F/F_L$ reaches the critical value $\lambda_c = 1$, and a plastic hinge is formed at midspan. The structure becomes a mechanism and the equilibrium configuration loses its stability; see the dashed lines in Fig. 9.15(b).

The difference, $M_L - M_Y$, is a measure of the overload-carrying capacity in the plastic zone, the so-called plastic cross-sectional reserve. In addition to the rectangular cross-section considered above, the values for a full circular cross-section of radius r are easily determined to be $M_Y = \sigma_Y\,\pi\,r^3/4$ and $M_L = \sigma_Y\,4\,r^3/3$. For a thin-walled $\bot$-profile with lower girder $B \times t_1$ and with the slab dimensions $H \times t_2$, the ratio of the areas is denoted $\alpha = Ht_2/Bt_1$. The values of these important design parameters become $M_Y = \sigma_Y\,BH\,t_1\,\alpha(4 + \alpha)/6(2 + \alpha)$ and $M_L = \sigma_Y\,BH\,t_1\,\alpha(2 + \alpha)^2/(1 + \alpha)^2$.

Using a portion of the plastic cross-sectional reserve makes it necessary to control the deflections. In the elastic region $0 \le x <$

$l/3\lambda$, where $\lambda = F/F_L$, the linearized Eq. (6.72) for the beam with a rectangular cross-section,

$$\frac{d^2 w_1}{dx^2} = -\frac{M}{EJ} = -\frac{6\,F}{E\,B\,H^3}\,x\;,$$

(9.88)

is easily integrated when taking into account the boundary condition $w_1\,(x = 0) = 0$

$$w_1(x) = -\lambda\,\frac{\sigma_Y}{E}\,\frac{l^2}{H}\,\frac{x}{l}\left(\frac{x^2}{l^2} + C_1\right)\;.$$

(9.89)

The strain at the contour of the yielding zone where $z = \zeta_Y$ within $l/3\lambda \le x \le l/2$ is proportional to the linearized curvature, $\varepsilon_{xx} = -\zeta_Y\,d^2w/dx^2 = \sigma_Y/E$. Hence,

$$\frac{d^2 w_2}{dx^2} = -\frac{2\,\sigma_Y}{EH\,\sqrt{3}\,(1 - 2\,\lambda\,x/l)}\;.$$

(9.90)

Taking into account the condition of symmetry at midspan, $dw_2\,/dx = 0|_{x = l/2}$, gives upon integration

$$w_2(x) = -\frac{2\,l^2\,\sigma_Y}{EH\,\lambda\,\sqrt{3}}\left[\frac{1}{3\lambda}\,(1 - 2\,\lambda\,x/l)^{3/2} + (1 - \lambda)^{1/2}\,x/l\right] + C_2\;.$$

(9.91)

At $x = l/3\lambda$, the conditions of continuous deflection $w_1 = w_2$, and of continuous slope $dw_1\,/dx = dw_2\,/dx$, render two equations and the remaining constants of integration are determined to be

$$C_1 = \frac{1}{\lambda^2}\left[\frac{2}{\sqrt{3}}\,(1 - \lambda)^{1/2} - 1\right]\;,\quad C_2 = 10\,\sigma_Y\,(l/\lambda)^2/27\;EH\;.$$

(9.92)

The maximal value of the deflection becomes

$$w_2(x = l/2) = \frac{\sigma_Y}{27\,E}\,\frac{(l/\lambda)^2}{H}\left[10 - 3\,(2 + \lambda)\sqrt{3\,(1 - \lambda)}\right]\;.$$

(9.93)

Hence, at the onset of yielding and in that moment of the plastic hinge formation, the midspan deflection becomes, respectively,

$$\lambda = 2/3:\quad \max\,(w) = \frac{\sigma_Y}{6\,E}\,\frac{l^2}{H}\;,$$

(9.94)

$$\lambda = \lambda_c = 1:\quad \max\,(w) = \frac{20}{9}\,\frac{\sigma_Y}{6\,E}\,\frac{l^2}{H}\;.$$

(9.95)

A plastic system reserve may still be available in statically indeterminate structures, despite the formation of plastic hinges (where the cross-sectional reserve is exhausted), and loading may thus be further increased until the limit load is reached. In engineering practice, it is often sufficient to determine the bounds of the limit load. *E. Melan* , late Professor of structural mechanics at the Technical University of Vienna, developed shake-down theorems. Independently, *W. T. Koiter* and *W. Prager* found methods to determine the upper and lower bounds of the limit load that are special cases of *Melan´s* theorem. The stability limit may be determined by considering the rigid plastic material of Sec. 4.3.1, since the limit load does not depend on elastic properties. Without proof, the essence of the theorem is the following:

1.   In a case where it is possible to find a statically admissible stress state (ie an equilibrium state) to a given loading, eg the elastic one, which has a maximal intensity below the yield limit or just reaches the limit, it can be concluded that the loading is below the limit load or just equals it. The load factor is thus $\lambda \leq \lambda_c$.

2.   In a case where it is possible to find a kinematically admissible plastic strain state to a given loading, such that the work done by the external forces is larger than or equals the dissipated work [equilibrium requires the total work to vanish, $W^{(e)} + W^{(i)} = 0$], it can be concluded that the loading is above the limit load or equals it. The load factor is thus $\lambda \geq \lambda_c$.

3.   Supplementary to 1 and 2, it is noted that associated with the limit load is a deformation mechanism that is kinematically and statically admissible.

4.   The limit load does not decrease by adding material at the traction-free surface of the structure. Hence, by virtually removing material, the structure may become simpler to analyze and its limit load decreases.

5.   The limit load cannot become larger if material is removed from the structure.

As an example, 1 and 2 are applied at first to the single-span elastic-plastic beam of constant cross-section and with the redundant supports of Fig. 9.16 when uniformly loaded. The kinematically admissible deformation is selected by assuming the beam to be rigid until yielding occurs with the formation of two plastic hinges. Furthermore, the positions of the hinges are "freely" chosen at the clamped edge and at midspan. A small rotation through the angle $\alpha \ll 1$ renders the dissipated work according to Eq. (3.7)

$$W_{pl} = \alpha\, M_T + 2\,\alpha\, M_T = 3\,\alpha\, M_T \ , \qquad (9.96)$$

where M_T denotes the ultimate bending moment of the fully plastic cross-section (when the cross-sectional reserve is exhausted). See, eg Eq. (9.86) for a rectangular cross-section. The work of the external forces is due to the uniform loading q_0 and through consideration of symmetry it is given by

$$W^{(e)} = 2\alpha \frac{q_0 l}{2} \frac{l}{4} = \alpha \frac{q_0 l^2}{4} \ .$$
(9.97)

According to 2, the condition

$$\alpha \frac{q_0 l^2}{4} \geq 3\alpha \, M_T \ ,$$
(9.98)

renders an upper bound of the limit load q_L ,

$$q_L < q_0 = 12 \, M_T \, /l^2 \ .$$
(9.99)

A statically admissible stress state is given according to Fig. 9.16 by assuming a single plastic hinge at the clamped edge. In that case, the edge moment is $M = - M_T$ and the field moments are given by

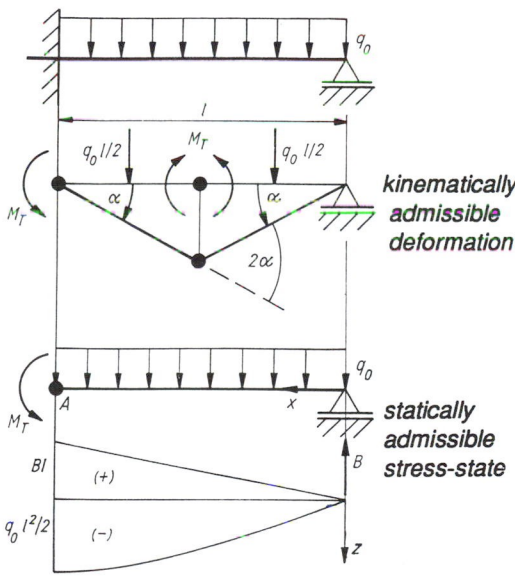

Fig. 9.16. Single-span, elastic-plastic beam with redundant supports having therefore a plastic cross-sectional reserve and a still higher plastic system reserve

$$M(x) = xB - \frac{q_0 x^2}{2} \quad , \quad B = \frac{q_0 l}{2} - \frac{M_T}{l} \; . \tag{9.100}$$

The largest bending moment is the extreme of (9.100)

$$\frac{dM(x)}{dx} = 0 = B - q_0 x \quad , \quad x_m = B/q_0 \quad , \quad M_{max} = M(x_m) = B^2/2q_0 \; . \tag{9.101}$$

Putting $M_{max} = M_T$ renders the lower bound of the limit load according to 1 maximum

$$q_L \geq q_0 = 2(3 + 2\sqrt{2}) \, M_T \, /l^2 = 11.657 \, M_T \, /l^2 \; . \tag{9.102}$$

Repeating the analysis with the second plastic hinge to be placed at position x_m renders according to 3 a both statically and kinematically admissible deformation, and hence, the limit load is equal to the lower bound given in Eq. (9.102)

$$q_L = 11.657 \, M_T \, /l^2 \quad . \tag{9.103}$$

If the uniform loading takes on this value, the stability limit is reached and the plastic structural reserve exhausted.

9.3.1. The Limit Load of the Single-Story Elastic-Plastic Frame

The quadratic planar frame with redundant supports is loaded according to Fig. 9.17. The utmost bending moment in the cross-bar is given by M_T, and a fully plastic cross-section in the (stronger) columns is assumed to take a bending moment of $2 \, M_T$. See Eq. (9.86) for a rectangular cross-section. According to theorem 2 the following admissible mechanisms are investigated: the beam mechanism of Fig. 9.17(a) and the frame mechanism of Fig. 9.17(b), where the plastic hinges are indicated by black spots. The dissipated work is given by $4 \, \alpha_1 \, M_T$ and $6 \, \alpha_2 \, M_T$, respectively. The work of the external forces becomes in the two cases considered $2 \, \alpha_1 \, F_1 \, l/4$ and $\alpha_2 \, F_2 \, l$. Hence, 2 renders with the condition

$$2 \alpha_1 F_1 \, l/4 \geq 4 \alpha_1 M_T \quad \text{and} \quad \alpha_2 F_2 \, l \geq 6 \alpha_2 M_T \; , \tag{9.104}$$

the upper bounds of the limit load F_L

$$F_L \leq F_1 = 8 \, M_T \, /l \quad \text{and} \quad F_L \leq F_2 = 6 \, M_T \, /l \; . \tag{9.105}$$

The lower value of the upper bound is relevant. The moment distribution in the cross-bar associated with this mechanism,

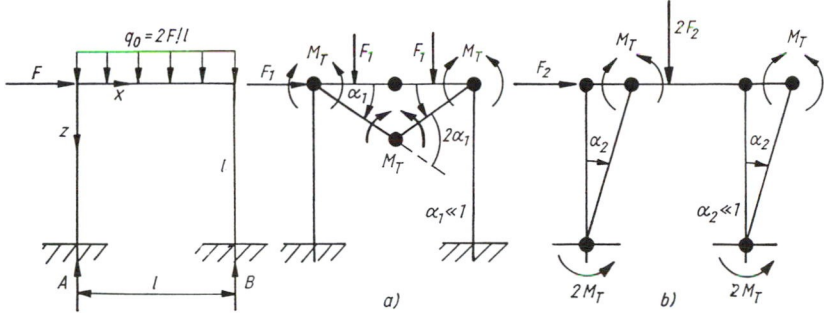

Fig. 9.17. Redundant elastic-plastic frame with given distribution of the cross-sectional reserve and with a still higher plastic system reserve.
(a) Beam mechanism and (b) frame mechanism

$$M(x) = x A + M_T - \frac{12\,M_T}{l^2}\frac{x^2}{2} \quad , \quad M(l) = -M_T \;\rightarrow\; A = 4\,M_T/l \; , \tag{9.106}$$

exhibits an extreme at $x = l/3$ that is larger than the assigned ultimate moment, $M_{max} = (5/3)\,M_T > M_T$. Hence, the stress state is statically inadmissible. The selection of the loading as $(3/5)\,F_2$ renders the stress state statically admissible, but now the deformation is kinematically inadmissible. Theorem 1 in that case gives a lower bound of the limit load and

$$F_2 = \frac{18}{5}\frac{M_T}{l} = 3.6\,\frac{M_T}{l} < F_L < 6\,\frac{M_T}{l} \; . \tag{9.107}$$

The bounds should be narrowed by selecting an improved mechanism, eg the mixed type of Fig. 9.18 is considered next, with the position of the plastic hinge in the cross-beam left open as a parameter, $\xi = x_Y/l$. Geometric conditions render $\varphi = \alpha_3 + \alpha_4$ and $\xi\,\alpha_3 = (1 - \xi)\,\alpha_4$. In that case, the dissipated work is (see Fig. 9.18)

$$W_{pl} = 2\,\alpha_3\,M_T + \varphi\,M_T + \varphi\,M_T + 2\,\alpha_3\,M_T = 2\alpha_3\,M_T(3 - 2\,\xi)/(1 - \xi). \tag{9.108}$$

The work of the external forces becomes

$$W^{(e)} = \alpha_3\,F_3\,l + 2\,\alpha_3\,F_3\,\xi\frac{\xi l}{2} + 2\,\alpha_4\,F_3\,(1 - \xi)\,\frac{(1 - \xi)\,l}{2} = (1 + \xi)\,\alpha_3\,F_3\,l. \tag{9.109}$$

Theorem 2 yields by means of the condition

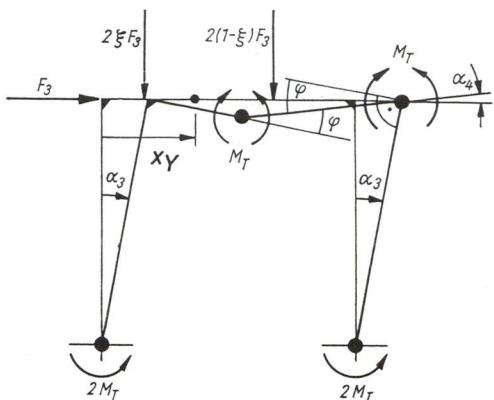

Fig. 9.18. Improved mechanism of mixed type. Parameter x_Y

$$(1 + \xi)\, \alpha_3\, F_3\, l \geq 2\, \alpha_3\, M_T\, (3 - 2\, \xi)/(1 - \xi) \, ,$$

$$l\, F_3 = 2\, M_T\, (3 - 2\, \xi)/(1 - \xi^2) \, ,$$

(9.110)

and by considering the value of $F_3\, (\xi)$ to take on a minimum,

$$\frac{dF_3}{d\xi} = 0 \; \rightarrow \; \xi = (3 - \sqrt{5})/2 \, ,$$

(9.111)

the smallest upper bound of the limit load within the selected mixed mode,

$$F_L \leq \min (F_3) = \frac{4}{3 - \sqrt{5}}\, \frac{M_T}{l} = 5.236\, M_T\, /l \; .$$

(9.112)

Since the stress distribution according to the loading *min (F₃)* is statically admissible, all conditions of theorem 3 are satisfied and the equality sign applies in Eq. (9.112). The critical load at the stability limit is found to be F_L = *5.236 M_T /l* , and the plastic structural reserve of the frame is exhausted when the load reaches that intensity.

The state of the art of limit load analysis even within a second-order theory (which is of special importance for frames) may be found in *H. Rubin and U. Vogel: "Baustatik ebener Stabwerke,"* in: *Handbook of Steel Constructions, Vol. 1. Stahlbau Verlag, Köln, 1982, (in German).*

9.4. Stability of Motion of Elastic-Plastic Bodies (Cyclic Plasticity)

Contrary to Sect. 9.3, where the load is assumed to increase monotonically until the limit load is reached, load cycles are considered. Within elastic limits, the possibility of material fatigue becomes important and the admissible stress amplitude decreases with the number of load cycles according to *Wöhler's* curve. Further information is given in *E. Chwalla: Introduction to Structural Mechanics. Stahlbau Verlag, Köln, 1954, p. 265, (in German).*

In the case of more severe loading, plastic deformation develops and, with the assumption of localized plastic zones, three possibilities exist:

1.   Alternating plasticity occurs, also unloading becomes inelastic. After a few cycles, fracture is encountered (ductile, low-cycle fatigue). The motion is unstable in the sense of the discussion of Sec. 9.2.

2.   Progressive yielding during the loading phase (unloading is elastic) results also finally in a fracture: The motion is unstable.

3.   *Shake-down* is a motion, where, in the course of time of cyclic loading, the size of the plastic zones becomes smaller and smaller. Such promising behavior of a structure requires the development of a favorable elastic state of self-equilibrating stresses (including the associated support reactions but in the absence of the external forcing) during the unloading phase, such that during subsequent loadings, the total stress remains finally below the yield limit. Such a motion is considered stable. Shake-down, eg is not possible in a single rod under cyclic axial loading.

Conditions of shake-down are illustrated by considering a *hollow sphere*, $R_i \le R \le R_e$, under the action of a sufficiently slow variation of the internal pressure p (inertia effects are neglected). That case of practical importance corresponds to an idealized thick-walled pressure vessel under lifeloads. The quasistatic elastic solution can be taken from Eqs. (6.14) and (6.17) by substituting the current value of the pressure. The *von Mises* equivalent stress becomes [see Eqs. (4.110) and (4.121)]

$$\sigma_e = \left\{ \frac{1}{2} \left[\left(\sigma_{rr} - \sigma_{\varphi\varphi} \right)^2 + \left(\sigma_{\varphi\varphi} - \sigma_{rr} \right)^2 + \left(\sigma_{\varphi\varphi} - \sigma_{\varphi\varphi} \right)^2 \right] \right\}^{1/2} = \left| \sigma_{rr} - \sigma_{\varphi\varphi} \right|$$

$$= \frac{3}{2} \frac{q}{R} \left(\frac{R_e}{R} \right)^3 \ , \quad q = p / \left[(R_e / R_i)^3 - 1 \right] \ , \quad \sigma_e \le \sigma_Y \ . \tag{9.113}$$

Putting $\sigma_e (R = R_i) = \sigma_Y$ renders the pressure at the onset of yielding to be proportional to

$$q_Y = \frac{2 \, \sigma_Y}{3} \, (R_i / R_e)^3 \ . \tag{9.114}$$

A plastic zone develops if the pressure is further increased, $q > q_Y$, and, thus, in the shell $R_i \le R \le R_p$ the yield condition $\sigma_{\varphi\varphi} - \sigma_{rr} = \sigma_Y$ holds. The application of the local condition of equilibrium (2.19) in spherical coordinates gives $d\sigma_{rr} / dR = 2 \, \sigma_Y / R$. Integration determines, with the boundary condition $\sigma_{rr} = -p$ at $R = R_i$ taken into account, the statically determinate stress state within the plastic zone

$$\sigma_{rr} = 2 \, \sigma_Y \ln \frac{R}{R_i} - p \ , \ \sigma_{\varphi\varphi} = \sigma_{rr} + \sigma_Y \ , \ R \le R_p \ . \tag{9.115}$$

At $R = R_p$, a steady transition to the elastic solution (to be valid in $R > R_p$) with respect to both the radial stress component and the displacement u [see Eq. (6.15)] is required. The evaluation of the plastic deformation by means of the *Prandtl-Reuss* equations. (4.136) is complex. *Hencky's deformation method* is applicable to the above given problem, assuming the unlimited elastic compressibility of modulus K under hydrostatic compression of the material, $3p^* = \sigma_{rr} + 2 \, \sigma_{\varphi\varphi}$,

$$\varepsilon_{rr} = \psi \, \sigma_{rr}' + K^{-1} p^* \ , \ \varepsilon_{\varphi\varphi} = \psi \, \sigma_{\varphi\varphi}' + K^{-1} p^* , \tag{9.116}$$

[see Eqs. (4.12) and (4.14) and substitute $\psi = \psi(R)$ for $1/2G$]. Combining Eq. (9.116) and the condition of compatibility of the strain field (1.22) [the relevant condition is rewritten for convenience]

$$\frac{d\varepsilon_{\varphi\varphi}}{dR} + \frac{\varepsilon_{\varphi\varphi} - \varepsilon_{rr}}{R} = 0 \ , \tag{9.117}$$

yields the first-order differential equation that is easily integrated

$$\frac{d\psi}{dR} + \frac{3\psi}{R} = 0 \ \rightarrow \psi(R) = -2 \, K^{-1} + C/R^3 \ . \tag{9.118}$$

Since $\psi(R = R_p) = 1/2G$, the constant C is determined, and the solution in the plastic zone becomes

$$\psi(R) = -2 \, K^{-1} + \left(2 \, K^{-1} + 1/2G\right) \left(R_p / R\right)^3 \ , \ R \le R_p . \tag{9.119}$$

The radial displacement $u = R \, \varepsilon_{\varphi\varphi}$ (the geometric relation holds under point symmetry) and its steady transition at $R = R_p$ render a nonlinear equation that must be solved by numerical methods, to determine the extension of the plastic zone R_p

$$(1/3)\,(R_p/R_e)^3 - \ln\,(R_p/R_i) = (1/3) - p/2\,\sigma_Y \; . \tag{9.120}$$

The critical pressure, the limit load p_L, follows under the condition of a fully plastic sphere, $R_p = R_e$, explicitly in terms of the yield stress and a size factor

$$p_L = 2\,\sigma_Y \ln\,(R_e/R_i) \; . \tag{9.121}$$

If the pressure is monotonically increased to its critical value, the sphere loses its plastic structural reserve and the equilibrium becomes unstable. By assuming further that $p < p_L$ and initiating an unloading process such that the internal pressure becomes zero, together with the additional assumption that the residual stresses do not initiate yielding again, the latter are determined through Eqs. (6.14) and (6.17)

$$
\left.
\begin{aligned}
\sigma_{rr}^{(0)} &= 2\,\sigma_Y \ln\,(R/R_i) - p - q\left[1 - (R_e/R)^3\right] \\
\sigma_{\varphi\varphi}^{(0)} &= \sigma_{rr}^{(0)} + \sigma_Y - (3q\,/2)\,(R_e/R)^3
\end{aligned}
\right\} \; R_i \le R \le R_p,
$$

$$
\left.
\begin{aligned}
\sigma_{rr}^{(0)} &= -(q + q_p)\left[1 - (R_e/R)^3\right] \\
\sigma_{\varphi\varphi}^{(0)} &= -(q + q_p)\left[1 + (1/2)(R_e/R)^3\right]
\end{aligned}
\right\} \; R_p \le R \le R_e,
$$

$$\tag{9.122}$$

where

$$q_p = p\left[\frac{2\,\sigma_Y}{p} \ln\left(\frac{R_p}{R_i}\right) - 1\right] \frac{R_p{}^3}{R_e{}^3 - R_p{}^3} \; . \tag{9.123}$$

The equivalence stress that is associated with the residual stress state, $\sigma_e^{(0)} = \sigma_{\varphi\varphi}^{(0)} - \sigma_{rr}^{(0)}$, has an absolute maximum at the traction-free inner boundary $R = R_i$, and there, it is equal to the compressive hoop stress, $q > q_Y$ [see Eq. (9.114)]

$$\sigma_e^{(0)} = \sigma_{\varphi\varphi}^{(0)}(R = R_i) = \frac{3}{2}\,(q_Y - q)\,(R_e/R_i)^3 < 0 \; . \tag{9.124}$$

For shake-down, its absolute value must be smaller than the yield stress in compression. For an elastic-plastic material with symmetry in tension and compression, the inequality must hold: $\sigma_e^{(0)} > -\sigma_Y$. And the latter is given by Eq (9.114) in terms of $-q_Y$, and the condition of shake-down, which is the stability condition of the motion, assumes the simple form, by means of Eq. (9.124),

$$q \leq 2\,q_Y \ . \tag{9.125}$$

Hence, the equivalence stress in an auxiliary problem where the sphere has unlimited elastic properties may take on a maximal value of twice the yield stress. From that result, a sufficient condition for instability of the motion, namely, for plastic failure during the cyclic loading, can be deduced, which is assumed to hold also for more general structural configurations. The shake-down state will not be reached if the equivalence stress in the associated unlimited elastic body varies within an interval that is wider than twice the yield stress. Without proof *Melan's* first shake-down theorem is stated as follows: <<Given the time-dependent stress field $\sigma_{ij}^{(el)}(t)$ of the auxiliary problem where the structure has assigned unlimited elasticity, shake-down is to be expected , if a time-independent and self-equilibrating stress field $\sigma_{ij}^{(0)}$ can be constructed such that the equivalence values of the fictitious resulting stresses $/\,\sigma_{ij}^{(el)}(t) + \sigma_{ij}^{(0)}/$ are permanently below the yield limit.>> Shake-down state will not be reached if such a constant field of self-stresses $\sigma_{ij}^{(0)}$ with the above property does not exist; the motion in that case is unstable.

9.5. Stability of the Flow in Dipping Open Channels, the Hydraulic Jump

The inviscid flow of an incompressible fluid in an open channel of constant width *B* but with a dipping bottom is considered; see Fig. 9.19. *Bernoulli's* equation (8.37), when taken along a streamline of the stationary flow that is assumed to be a straight line as shown in the figure, gives the relation (the absolute pressure applies)

$$\frac{v_2^2}{2g} + \frac{p_2}{\rho g} + z_2 = \frac{v_1^2}{2g} + \frac{p_1}{\rho g} + z_1 = D + \overline{H}_E \ . \tag{9.126}$$

Since friction at the walls is always present, Eq. (9.126) is only good for rather small distances between points 1 and 2. The component of *Euler's* equation (8.30) lateral to a straight streamline yields just the hydrostatic pressure distribution and, especially for small angles α ,

$$p_1(z_1) = p_0 + \rho g\,(D + d_1 - z_1) \ ,$$

$$p_2(z_2) = p_0 + \rho g\,(D - \alpha\,x + d_2 - z_2) \ , \quad \alpha \ll 1 \ . \tag{9.127}$$

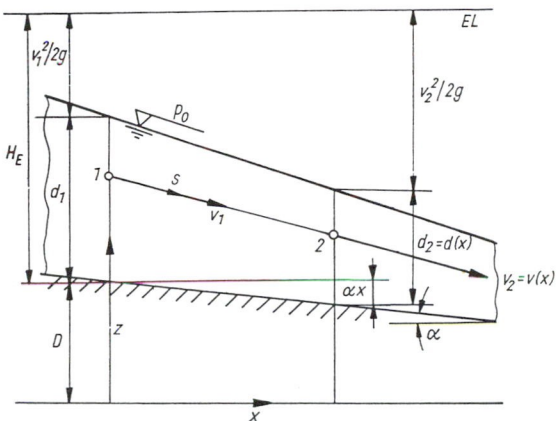

Fig. 9.19. Stationary flow in a dipping open channel with constant width B

When substituted above, the pressure p and atmospheric pressure p_0 are eliminated, and with the energy head H_E of the ideal flow shown in Fig. 9.19, *Bernoulli's* equation takes on a suitable form

$$\frac{v_2^2}{2g} + d_2 - \alpha x = \frac{v_1^2}{2g} + d_1 = \overline{H}_E - \frac{p_0}{\rho g} = H_E \,.$$

(9.128)

By means of the continuity equation

$$\dot{m} = \rho \, B d_1 \, v_1 = \rho \, B d_2 \, v_2 \,, \quad v_2 = v(x) \,,$$

(9.129)

the (averaged) velocities can be eliminated, and by putting $H(x) = H_E + \alpha x$, which is commonly called the specific energy head and which refers to points of the lowermost streamline, a cubic equation for the depth $d_2 = d(x)$ results [indicating an instability like Fig. 9.20(a)]

$$d^3 - H \, d^2 + k = 0 \,, \quad k = \dot{m}^2 / 2g \, (\rho B)^2 \,.$$

(9.130)

Solving for $H(d)$ (its nondimensional form is shown in Fig. 9.21)

$$H(d) = d + k/d^2 \,,$$

(9.131)

circumvents any difficulties that may arise with respect to the nonlinear equation. Since the specific energy has a minimum at $d_c = (2k)^{1/3}$ at a given rate of discharge, k, such a value is considered critical: Two of the three real roots of the cubic equation (9.130)

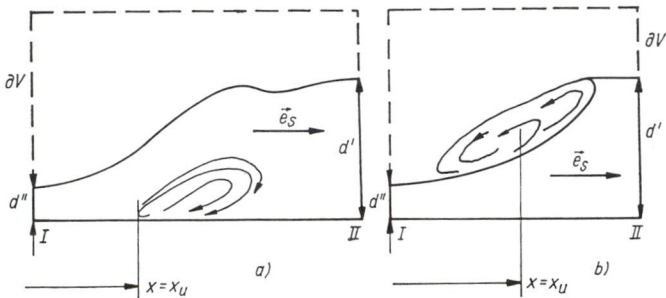

Fig. 9.20. Ground (a) or surface roller (b) (the latter is a standing surge and not very likely to occur) due to a hydraulic jump from a rapid to a tranquil flow in an open channel; see, eg *J. Kozeny: Hydraulics. Springer-Verlag, Wien, 1953, (in German), and H. Rouse: Fluid Mechanics for Hydraulic Engineers. Dover, New York, 1961*

coalesce in that case. For $H > H_c$, where the critical head at that given rate of discharge takes on the value

$$H_c = \frac{3}{2} (2k)^{1/3} ,$$

(9.132)

two of the roots are positive, $d' > d''$. That solution corresponds to two different states of flow. The first slow one, ie that of larger depth, is called tranquil flow, the second the rapid flow. The latter is especially not stable if the disturbance is sufficiently strong, and thus, it may "suddenly" change to tranquil flow. In the region where the flow turns into another mode, viscosity effects become dominant due to the large velocity gradients, and a roller wave forms near the ground, Fig. 9.20(a) [or at the surface, Fig. 9.20(b)]. By keeping the cross-section at x fixed, the energy head is changed by $\Delta H > 0$ according to the two flow conditions. A dimensionless scaled graph illustrates the hydraulic jump in Fig. 9.21.

The energy loss of that hydraulic jump can be evaluated after the application of conservation of momentum (7.13) under stationary conditions, analogous to *Carnot's* loss (see Sec. 7.3.5). Friction at the boundary is a secondary effect over such a short distance and is completely ignored. According to Fig. 9.20, a control surface is selected that encloses the roller wave and the net flux of momentum must equal the resultant external force. The bottom is either considered horizontal over that short distance, or approximately in the case of a small dipping angle (the component of gravity in the direction of flow is considered small and its effect is neglected)

$$\oint_{\partial V} \mu \mathbf{v} \, dS = \mathbf{R} \quad \rightarrow \quad \mu_I = -\rho v_I , \quad \mu_{II} = \rho v_{II} ,$$

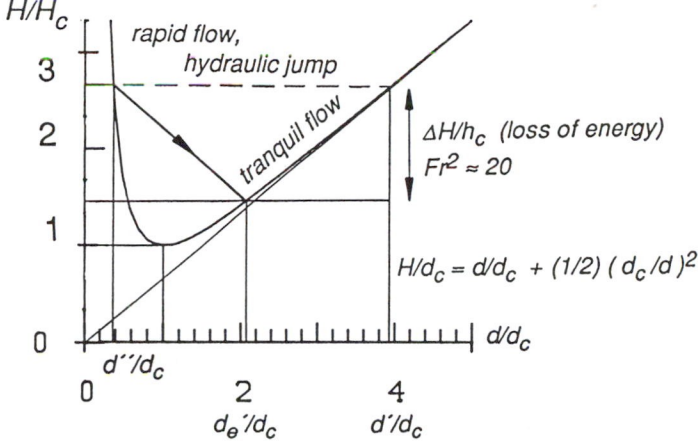

Fig. 9.21. Specific energy diagram: Its head as a function of depth d without loss of energy (constant, dashed line) and with viscosity taken into account during the change of the flow: Rapid flow depth d'', tranquil flow depth increased to d_e'. $H_c = 3\,d_c/2$

$$e_s: \quad \to \quad -\rho\,v_I^2\,B\,d'' + \rho\,v_{II}^2\,B\,d_e' = g\rho\,\frac{d''}{2}\,B\,d'' - g\rho\,\frac{d_e'}{2}\,B\,d_e' \;. \tag{9.133}$$

The static pressure distribution over the depth is already considered above. Elimination of $v_{II} = v_I\,d''/d_e'$ gives a cubic equation for the depth d_e' at the exit of the control surface, ie downstream of the roller,

$$\left(d_e' - d''\right)\left[d_e'\left(d_e' + d''\right) - \frac{2v_I^2}{g}\,d''\right] = 0 \;. \tag{9.134}$$

The trivial solution of keeping the rapid flow depth $d'' = const$ is abandoned, and the remaining quadratic equation is easily solved to render the root, $d_e' > d''$, that proves the direction of the hydraulic jump

$$\frac{d_e'}{d''} = \frac{1}{2}\left(\sqrt{1 + 8\,F_I^2} - 1\right) \;, \quad F_I = v_I/\sqrt{gd''} = \left(d_c/d''\right)^{3/2} \;. \tag{9.135}$$

The characteristic coefficient $F_I \geq 1$, which is evaluated at the cross-section I, is known as the *Froude* number, which is of importance in all gravity-controlled flows; see Eq. (13.26). $\sqrt{(gd'')}$ is of dimension velocity and represents the characteristic wave speed of gravity waves in shallow water, the so-called celerity. The limiting value $F_I^* = 1$ corresponds to the undisturbed flow $d_e' = d_g''$

$= v_I^2/g$. Only the rapid flow may become unstable and a roller wave may be formed if $F_I > 1$; hence, $d'' < d_g''$. The loss is given by the difference of the energy heads and must be positive

$$H_I = d'' + v_I^2/2g \ , \ H_{II} = d_e' + v_{II}^2/2g \quad \rightarrow \quad \Delta H = H_I - H_{II} = \zeta \, v_{II}^2/2g \ ,$$

$$\zeta = \frac{1}{32 \, F_I^2} \left(1 - \sqrt{8 \, F_I^2 + 1}\right) \left(3 - \sqrt{8 \, F_I^2 + 1}\right)^3 .$$

$$(9.136)$$

In terms of the velocity head, the loss factor is given by ζ ; see also H. Press and R. Schröder: Hydromechanik im Wasserbau. Ernst, Berlin, 1966. In terms of the standing surge interpretation of the hydraulic jump, the region $1 \leq F^2 \leq 3$ is the undular zone and $F^2 > 3$ the shock zone. The boundary between jumps of the undular and shock types is thus given by $F^2 = 3$ when $d_e'/d'' = 2$.

Analogously to the specific energy diagram, a dimensionless discharge diagram may be constructed from

$$\frac{1}{2}\left(\frac{\dot{m}}{\dot{m}_c}\right)^2 = \frac{3}{2}\left(\frac{d}{d_c}\right)^2 - \left(\frac{d}{d_c}\right)^3 \ , \quad \dot{m}_c = \rho B \sqrt{g} \, d_c^{3/2} .$$

Illustrative applications of that relation to the flow initiation at a channel inlet at the side of a large reservoir, as well as to the flow through a sluice gate, are left to the reader, who is more specifically interested in hydraulic engineering. The specific energy diagram (Fig. 9.21), however, is useful when it comes to the determination of the inviscid flow in a channel with an elevation of the bottom. It may be assumed that the energy line remains horizontal, the change in H resulting from the fact that the geodesic height rises according to the contour of the lower boundary. Dividing Eq. (9.131) through the depth d renders a dimensionless relation that is known as the energy-discharge diagram. It is, as such, a combination of the other two, and, eg it may be conveniently applied to the overfall.

9.6.   Flutter Instability

Flutter is a special form of a self-excited vibration of an elastic system loaded by follower forces. Historically, the flutter of thin airfoils was encountered in airplanes as soon as their speed was increased and thereby reached a critical value. It was typically regarded in such instances as kinetic instability. The following illustrative example is reduced as much as possible in complexity but still exhibits this important phenomenon: The airfoil in a stationary parallel flow of constant speed v is modeled by a

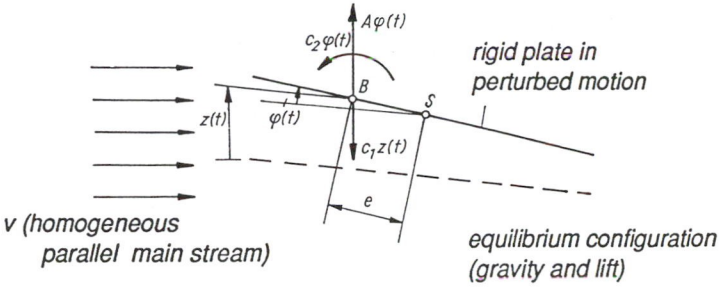

Fig. 9.22. Perturbed configuration of a rigid panel with two degrees of freedom in a parallel flow of fluid. Excitation by the additional lift $A\,\varphi$. Linearized restoring force $z\,c_1$ and moment $\varphi\,c_2$

rectangular rigid panel on a linear elastic foundation with two degrees of freedom. The perturbation of the stationary flight configuration (the latter is indicated by the dashed line in Fig. 9.22) renders with the translatoric foundation modulus c_1 a linearized restoring force zc_1 of a *Winkler* foundation and with the coefficient of angular stiffness c_2 the restoring moment φc_2. The small angle φ of disturbance is considered positive when the stationary angle of attack α is increased. When S denotes the centroid of this section of unit length, the line of attack of the additional lift force $A\varphi$ passes through the point B, which is different from S; see Sec. 13.2.1. The frame of coordinates z, φ is fixed to the airfoil when it moves translational with the constant speed v in the steady state of flight; thus, it is still an inertial system. Given the mass m and central radius of gyration i_S and if we assume further that the eccentricity e is known, the equations of the perturbing motion are (see also Sec. 10.1)

$$m\left(\ddot{z} - e\,\ddot{\varphi}\right) = -c_1\,z + A\,\varphi \ ,$$

$$m\,i_S^2\,\ddot{\varphi} = \left(A\,\varphi - c_1\,z\right)e - c_2\,\varphi \ . \tag{9.137}$$

Simple manipulations render the coupled system of linearized equations in the form

$$\ddot{z} - e\,\ddot{\varphi} + \Omega_z^2\,z - \alpha\,\varphi = 0 \ , \quad \Omega_z^2 = c_1/m \ , \quad \alpha = A/m \approx v^2 \ ,$$

$$-e\,\ddot{z} + i_B^2\,\ddot{\varphi} + 0 + i_B^2\,\Omega_\varphi^2\,\varphi = 0 \ , \quad \Omega_\varphi^2 = c_2/mi_B^2 \ , \quad i_B^2 = i_S^2 + e^2 \ . \tag{9.138}$$

where the mass matrix is easily recognized to be symmetric, but the stiffness matrix turns out to be asymmetric. Such mechanical systems are called *circulatory* ; another example of that kind is buckling of a column when loaded by a tangential follower force. The characteristic equation [see also Eq. (7.139)], which is the frequency equation of the perturbing motion, is found by putting the determinant of the coefficient matrix to zero

$$i_B^2\left(-\omega^2 + \Omega_z^2\right)\left(-\omega^2 + \Omega_\varphi^2\right) + e\omega^2\left(-e\omega^2 + \alpha\right) = 0 \ . \tag{9.139}$$

Or, taking its polynomial form,

$$p_0\,\omega^4 - p_2\,\omega^2 + p_4 = 0 \ , \quad p_0 = \left(i_S\,/i_B\right)^2 > 0 \ , \quad p_2(\alpha) = \Omega_z^2 + \Omega_\varphi^2 - e\,\alpha/i_B^2 \ ,$$

$$p_4 = \Omega_z^2\,\Omega_\varphi^2 > 0 \ .$$

Real and positive solutions are found in the cases, where

$$\omega^2 = \frac{1}{2\,p_0}\left[p_2 \pm \sqrt{p_2^2 - 4\,p_0\,p_4}\right] \ , \ \Delta = p_2^2 - 4\,p_0\,p_4 > 0 \ , \ p_2(\alpha) > 0 \ . \tag{9.140}$$

The critical values of the excitation amplitude are the roots of the

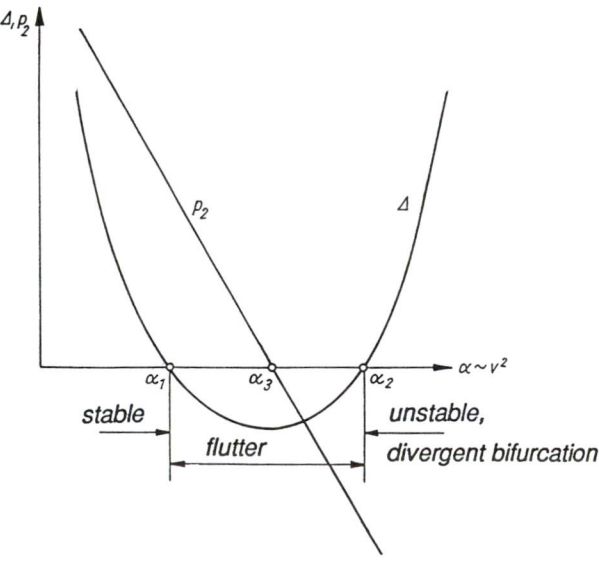

Fig. 9.23. Stable and unstable regions of the flight speed *v*.
Flutter instability is observed for $\alpha_1 \le \alpha \le \alpha_2$, $\alpha = A/m$

discriminating expression Δ (see Fig. 9.23)

$$\alpha = \alpha_{2,1} = i_B^2 \left[\left(\Omega_z^2 + \Omega_\varphi^2 \right) \pm 2 \frac{i_S}{i_B} \Omega_z \Omega_\varphi \right] / e \, .$$ (9.141)

In the range $\alpha_1 < \alpha < \alpha_2$, the discriminate is negative, $\Delta < 0$, for $\alpha > \alpha_3 = i_B^2 (\Omega_z^2 + \Omega_\varphi^2)/e > \alpha_1$; the coefficient $p_2 < 0$. It is easily proven that according to Fig. 9.23 the lower range of flight speeds, ie when $\alpha < \alpha_1$, is stable, where α_1 gives the critical flight speed v_c . For $\alpha > \alpha_1$, the equilibrium configuration of the steady-state motion is unstable: In the range $\alpha_1 < \alpha < \alpha_2$, the eigenvalues ω are conjugate complex, and hence, the linearized perturbing motion grows exponentially. The latter case is referred to as flutter instability. For still higher flight speeds, $\alpha > \alpha_2$, the discriminating expression is positive and the equilibrium configuration becomes unstable by divergent bifurcation.

Similar effects are observed when wide-span bridges are designed with a low torsional rigidity: In 1941, self-excited flutter vibrations became so severe that the *Tacoma Narrows Bridge* collapsed at rather low speeds of lateral wind after vibrating excessively for several hours. A movie was recorded on site that shows the consecutive flow of energy from the bending to the torsional mode of vibration and vice versa in the final stages until failure occurs. It clearly indicates the danger of flutter. Also when considering buckling of slender elastic structures (eg *Beck´s column*) under various types of follower forces, flutter may be observed. Damping may have a destabilizing effect. Flow-induced vibrations of tubes may lead to flutter bifurcation of the equilibrium. See, *H. Leipholz: Theory of Stability. Teubner, Stuttgart, 1958, (in German); H. W. Försching: Grundlagen der Aeroelastik. Springer-Verlag, Berlin, 1974; R. L. Bisplinghoff and H. Ashley: Principles of Aeroelasticity. Dover, New York, 1975; H. Ziegler: Principles of Structural Stability. Blaisdell, Waltham, Mass.,1968.*

9.7. Exercises A 9.1 to A 9.7 and Solutions

A 9.1: Investigate the stability of the equilibrium positions of a heavy point mass m supported by a frictionless circular rigid wire of radius R and attached to a linear elastic spring of stiffness c (see Fig. A 9.1). The gravity field is assumed to be parallel and homogeneous and the supporting ring is fixed in space.

Solution: The system is conservative and *Dirichlet´s* criterion applies. The potential $V = W_P + U$, where $W_P = -2 mg R \cos^2 \varphi$ and $U = cs^2/2$, $s = -l_0 + 2 R \cos \varphi$. The condition of equilibrium

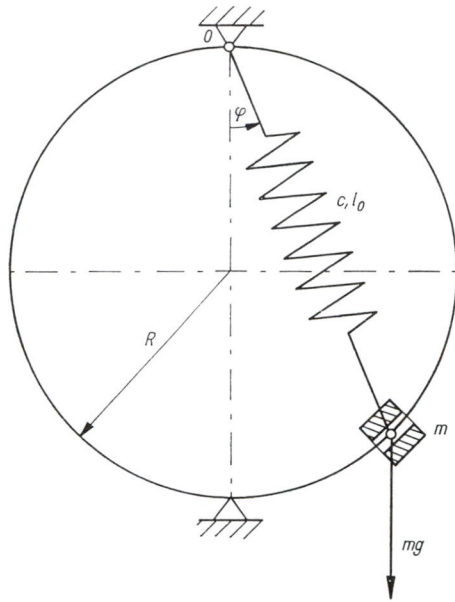

Fig. A 9.1.

$$\frac{dV}{d\phi} = 0 = 2\,R\,\sin\phi\left[c\,l_0 + 2\,(mg - c\,R)\cos\phi\right]\,,$$

always has the trivial solution $\phi_1 = 0$. Other (symmetric) configurations exist if the equation $\cos\phi = c l_0 / 2\,(c\,R - mg) > 0$ has real roots, $\phi_1 \neq 0$, $mg < c\,(R - l_0/2) > 0$. The stability condition

$$\left.\frac{d^2V}{d\phi^2}\right|_{\phi = \phi_1} = 2\,R\left[c\,l_0\cos\phi_1 + 2\,(mg - c\,R)\cos 2\phi_1\right] > 0\,.$$

At $\phi_1 = 0$, the condition reduces to $mg > c\,(R - l_0/2)$. The configurations $\phi_1 \neq 0$ are stable if $mg < 2\,c\,(R - l_0) > 0$. It is left to the reader to apply the dynamic method of small perturbations and to verify the results.

A 9.2:   Determine the buckling load of the slender linear elastic column that is supported and loaded according to Fig. A 9.2. Principal bending stiffness $B = EJ_y = const$.

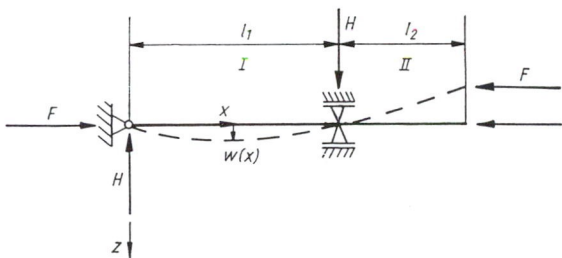

Fig. A 9.2.

Solution: The bending moment in the deflected configuration is given section-wise by: $M_I(x) = F \, w(x) + H \, x$, $0 \le x \le l_1$ and by $M_{II}(x) = F$ $w(x) + H \, l_1$ in the remaining interval $l_1 \le x \le (l_1 + l_2) = l$. The linearized curvature is proportional to the moment and the resulting inhomogeneous differential equation has the general solution that is valid within the supports, $w_I(x) = C_1 \cos \alpha x + D_1 \sin \alpha x - (H/F) \, x$, and outside, $w_{II}(x) = C_2 \cos \alpha x + D_2 \sin \alpha x - (H/F) \, l_1$. A system of five homogeneous linear equations results by considering the boundary conditions and the condition of continuity of the slope at $x = l_1$: $0 = C_1$, $0 = D_1 \sin \alpha l_1 - (H/F) \, l_1$, $0 = C_2 \cos \alpha l_1 + D_2 \sin \alpha l_1 - (H/F) l_1$, $0 = - C_2 \, \alpha^2 \cos \alpha l - D_2 \, \alpha^2 \sin \alpha l$ and $D_1 \, \alpha \cos \alpha l_1 + C_2 \, \alpha \sin \alpha l_1 - D_2 \, \alpha \cos \alpha l_1 - (H/F) = 0$. Putting the buckling determinant to zero gives the transcendental equation

$$\alpha l_1 \sin \alpha l - \sin \alpha l_1 \sin \alpha l_2 = 0 .$$

The smallest root that must be determined by a numerical routine renders the critical load F_c or, alternatively, the buckling length l_c . In the case of $l_1 = l_2 = l/2$, the equation is factorized, $[\alpha l \cos (\alpha l/2) - \sin (\alpha l/2)] \sin (\alpha l/2) = 0$, and the smallest eigenvalue yields approximately $l_c = 1.35 \, l$.

A 9.3: A slender linear elastic column is classically supported according to Fig. A 9.3. Determine the buckling lengths l_c in the *Euler* formula of the critical load, $F_c = \pi^2 EJ/l_c^2$, given the span l .

Solution: Fig. A 9.3.

A 9.4: Even the tensile force F may become critical if it pulls at the end of a (rigid) cantilever, whose other end is welded to the slender linear elastic rod of constant bending rigidity $B = EJ$; such a configuration is shown in Fig. A 9.4. Determine the condition of buckling in tension.

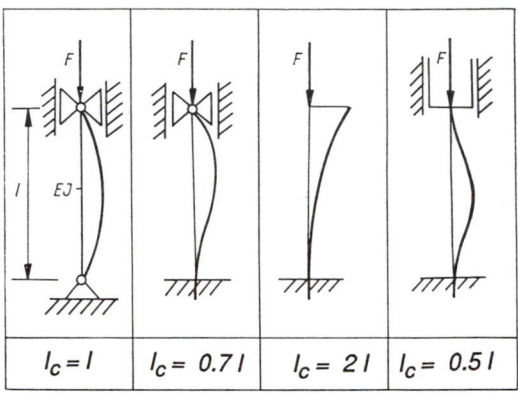

| $l_c = l$ | $l_c = 0.7 l$ | $l_c = 2 l$ | $l_c = 0.5 l$ |

Fig. A 9.3. *Euler's* column $l_c = l$

Solution: The bending moment in the deflected configuration is $M(x)$ $= H x - F w(x)$, where $H l + F e (\partial w/\partial x)|_{x = l} = 0$. The reaction force H can be eliminated and the linearized differential equation of deflection becomes ($\alpha^2 = F/EJ$)

$$\frac{\partial^2 w}{\partial x^2} - \alpha^2 w = \alpha^2 \frac{e x}{l} \left.\frac{\partial w}{\partial x}\right|_{x = l} .$$

The general solution by superposition is easily cast in the form

$$w(x) = C_1 \cosh \alpha x + C_2 \sinh \alpha x - \frac{e x}{l} \left.\frac{\partial w}{\partial x}\right|_{x = l} .$$

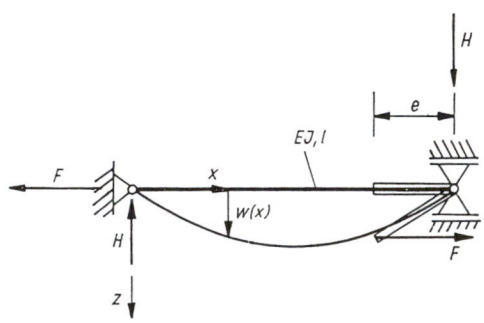

Fig. A 9.4. Buckling in tension

It is subject to the boundary conditions $w(0) = w(l)$ and $w'(l) = (\partial w/\partial x)|_{x=l}$. Hence, $C_1 = 0$, and the remaining system of two linear and homogeneous equations for the two unknowns C_2 and $w'(l)$ has a nontrivial solution only if the buckling determinant vanishes

$$\tanh \alpha l = \frac{\alpha\, e}{(1 + e/l)} \; .$$

The single root of that characteristic equation renders the critical tension F_c. In the case of a very short cantilever $e \ll l$, the buckling load approaches the value $F_c \approx EJ/e^2$. Further problems of this type have been discussed by R. Grammel, Ingenieur Archiv, **17** (1949), p. 107, (in German); see also G. Herrmann and A. E. Armenakas, ASME Journal Applied Mechanics, **31** (1960), p. 455.

A 9.5: Determine the reduction of the critical load of the *Euler* column (Fig. 9.10) that follows from the consideration of the shear deformation.

Solution: Putting $q_z = -d^2M_y/dx^2$ in Eq. (6.70) and considering the moment in the deflected configuration as $M_y(x) = F\, w(x)$ render the linearized differential equation (the constant coefficient is altered with respect to that of the beam rigid in shear)

$$\frac{d^2w}{dx^2} + \alpha_S^2\, w = 0 \; , \quad \alpha_S^2 = F/\left(1 - \kappa_z\, F/GA\right)/EJ_y \; .$$

Since the type of the equation is the same as Eq. (9.44), the critical load can be given at once for hinged-hinged boundary conditions

$$F_c = \frac{F_c^{(\text{Euler})}}{\left(1 + \kappa_z\, F_c^{(\text{Euler})}/GA\right)} \; .$$

Its value is smaller than $F_c^{(Euler)} = \pi^2\, EJ_y/l^2$, since the denominator is larger than 1. In the case of a sandwich beam, a second problem of stability arises, namely, the possibility of wrinkling of the thin sheets of the girders (a local buckling phenomenon) may become important. By putting the stiffness of the foundation k equal to that of the core material, the results of Sec. 11.3.2 can be directly applied for this investigation.

A 9.6: A thin rectangular plate *(a, b)* with two parallel edges $x = 0$, a simply supported is subject to constant in-plane loading, $n_x = -n$, $n_y = n_{xy} = 0$. The buckling mode should be chosen according to $w(x, y) = f(y) \sin(m\pi x/a)$, $m = 1, 2, ...$ to account for the given boundary conditions. Determine the critical load factors by the selection of the boundary conditions of the remaining parallel edges $y = 0, b$ to be either both hinged or both rigidly clamped.

Solution: The linearized *von Karman* equation reduces to

$$K \Delta\Delta w + \lambda n \frac{\partial^2 w}{\partial x^2} = 0 .$$

The substitution of the buckling mode gives the ordinary differential equation of the fourth order (like that of the prestressed beam)

$$f^{IV} - 2 \left(\frac{m\pi}{a}\right)^2 f'' + \left[\left(\frac{m\pi}{a}\right)^4 - \lambda \frac{n}{K} \left(\frac{m\pi}{a}\right)^2\right] f = 0 ,$$

with the general solution, namely, the buckling modes of a slender beam of span b ,

$$f(y) = C_1 \cosh \alpha_1 y + C_2 \sinh \alpha_1 y + C_3 \sin \alpha_2 y + C_4 \cos \alpha_2 y ,$$

$$\alpha_{1,2} = \left\{\left[\lambda \frac{n}{K} \left(\frac{m\pi}{a}\right)^2\right]^{1/2} \pm \left(\frac{m\pi}{a}\right)^2\right\}^{1/2} .$$

Three of the constants C_i and the critical value of the load factor are all to be determined by the boundary conditions at $y = 0, b$. For example, the plate with all four edges simply supported has the eigenvalues

$$\lambda_m = \pi^2 K \left[2 + \left(\frac{a}{mb}\right)^2 + \left(\frac{mb}{a}\right)^2\right] / nb^2 .$$

In the case where the remaining pair of parallel edges is rigidly clamped, the characteristic values become (the plate is stiffened when compared to that having hinged supports)

$$\lambda_m = \pi^2 K \left[\frac{5}{2} + 5\left(\frac{a}{mb}\right)^2 + \left(\frac{mb}{a}\right)^2\right] / nb^2 .$$

The critical value is the minimum $\lambda_c = min \ \lambda_m$. Further reading is suggested; see, eg *C. F. Kollbrunner and M. Meister: Ausbeulen. Springer-Verlag, Berlin, 1958, p. 94.*

A 9.7: Relations are to be established between the natural vibrations of membranes, uniformly and isotropically stressed by S and having a polygonal planform, and that of thin plates of the same planform, all edges simply supported, under hydrostatic in-plane prestressing.

Solution: The static Eq. (6.191) may be used with inertial loading of the membrane, $p = -\rho^* h^* (d^2 u/dt^2)$ added. For time-harmonic vibrations, $u = U \exp(i\omega^* t)$ and the *Helmholtz* partial differential equation results that is reduced in time

$$\Delta U + \rho^* h^* \omega^{*2} U/S = 0 \ .$$

At the boundary, $U = 0$ and hence, $\Delta U = 0$ are prescribed .

A biharmonic operator appears in the plate equation. Thus, the *Laplace* operator is applied to the *Helmholtz* equation to increase the order properly. The result is

$$\Delta\Delta U + \rho^* h^* \omega^{*2} \Delta U/S = 0 \ .$$

Subtracting the two equations renders a biharmonic equation of such a type that follows also after reduction in time from the first of the (linearized) *von Karman* equations (5.30a) under the assumption of hydrostatic prestress $n_x = n_y = n = const$, $n_{xy} = 0$ and by substituting $p = -\rho h (d^2 w/dt^2)$. Deflection $w(x, y, t) = W(x, y) \exp(i\omega t)$ gives

$$K \Delta\Delta W - n \Delta W - \rho h \omega^2 W = 0 \ .$$

At straight and simply supported edges, the boundary conditions are the same as those given above: $W = 0$, $\Delta W = 0$. See Sec. 6.6.3. By using the affinity $\Delta W + \lambda^2 W = 0$ and considering the discrete spectrum of the membrane $\lambda_j^{*2} = \rho^* h^* \omega_j^{*2}/S$, the natural frequencies of the prestressed plate follow at once

$$\omega_j^2 = K \lambda_j^4 (1 - n/n_j)/\rho h \ , \quad n_j = -K \lambda_j^2 \ , \quad j = 1, 2, 3, ..$$

in terms of the discrete spectrum λ_j of the associated *Dirichlet* boundary value problem of the second-order *Helmholtz* equation. The in-plane forces n_j are the buckling loads where $\omega_j = 0$. Their absolute smallest value is the critical load of the plate, n_c . The eigenvalues λ_j of the membrane are approximately determined by means of the *Ritz-Galerkin* procedure; see Chap. 11. For the rectangular membrane *(a, b)* , they are easily found to be

$$\lambda_{jk} = \pi \sqrt{(j/a)^2 + (k/b)^2} \ , \quad j, k = 1, 2, 3, ..$$

Solutions of the natural and forced vibrations of membranes of general planform are efficiently derived by the boundary element method.

10
D´Alembert´s Principle and Lagrange Equations of Motion

The principle of virtual work, Eq. (5.3), can be generalized to include the inertia forces of dynamics. In statics, the equilibrium configuration of a system at rest has to be considered; in dynamics, the instant configuration of a moving body at some time t is to be observed. In analogy to the virtual variation of the equilibrium configuration, virtual displacements are applied to the instant configuration, keeping the time t constant. Scalar multiplication of the vector *Euler-Cauchy* equations of motion (7.1) with the virtual displacement $\delta \mathbf{r}$ of the material point causes the specific generalized virtual work to vanish

$$(\mathbf{f} - \rho \mathbf{a}) \cdot \delta \mathbf{r} = 0 . \tag{10.1}$$

Integration over the material volume $V(t)$ of the moving body renders the total virtual work of the external and internal forces [see Eq. (5.3)]

$$\int_{V(t)} \mathbf{f} \cdot \delta \mathbf{r} \, dV = \delta W = \delta W^{(e)} + \delta W^{(i)} . \tag{10.2}$$

Hence, the general form of *D´Alembert´s* principle is derived

$$\delta W - \int_m (\mathbf{a} \cdot \delta \mathbf{r}) \, dm = 0 . \tag{10.3}$$

<<The sum of the total virtual work and of the generalized virtual work due to the inertia forces, $- \mathbf{a} \, dm$, vanishes for all admissible virtual variations of the instant configuration.>>

Equation (10.3) is a scalar relation and holds for the arbitrary motion of deformable bodies. It will be used for the mechanical interpretation of *Galerkin´s* procedure (Sec. 11.1.2) when applied to dynamic problems. Within this chapter, the principle is specified for the dynamics of (discretized) multiple-degree-of-freedom systems (MDOF-S). In the case of a finite number n of degrees of freedom, a finite number of independent scalar coordinates $q_k(t)$, $k = 1, 2, .., n$ determines the instant configuration, and any position vector is a function of these coordinates and possibly depends on time also explicitly

$$r = r(q_1, q_2, .., q_n; t) .$$

(10.4)

Especially, the radius vector of the point of application of a single force F is such a vector function and the virtual work $F . \delta r$ (time t is fixed) is related to the variation of the generalized coordinates δq_k , $k = 1, 2, ..., n$ through

$$\delta r = \sum_{k=1}^{n} \frac{\partial r(q_1, q_2, .., q_n; t)}{\partial q_k} \delta q_k .$$

(10.5)

The virtual displacement is analogous to the total derivative of the radius vector keeping the time constant. The gross virtual work of a number K of internal and external single forces is just the sum of the individual contributions

$$\delta W = \sum_{l=1}^{K} (F_l . \delta r_l) = \sum_{l=1}^{K} F_l . \sum_{k=1}^{n} \frac{\partial r_l}{\partial q_k} \delta q_k = \sum_{k=1}^{n} Q_k \delta q_k .$$

(10.6)

Comparing coefficients renders the abbreviations

$$Q_k = \sum_{l=1}^{K} \left(F_l . \frac{\partial r_l}{\partial q_k} \right) ,$$

(10.7)

which are n generalized forces. They give, when multiplied by the virtual variation of the associated generalized coordinate, say, δq_k , and after summation of these products, the proper virtual work. Hence, these generalized forces replace the external and internal forces in the system equivalently in the sense that they produce the same virtual work. If the forces that contribute to the virtual work are members of irrotational vector fields, their potential is a function $V = V(q_1, q_2, ..., q_n; t)$. Hence, the virtual work can be related to the variation of the potential energy δV , with fixed time,

$$\delta W = -\delta V = -\sum_{k=1}^{n} \frac{\partial V}{\partial q_k} \delta q_k = \sum_{k=1}^{n} Q_k \delta q_k .$$

(10.8)

Comparing coefficients determines the generalized forces in terms of the negative gradient of the potential function with respect to the generalized coordinates

$$Q_k = -\frac{\partial V}{\partial q_k} \quad .$$

(10.9)

The above relations are equally useful in statics. What remains to be done is a proper transformation of the virtual work of the inertia forces to a form, which factors the variations of the generalized coordinates. By putting through the definition $a = dv/dt$ and interchanging the derivatives $(d/dt)(\partial r/\partial q_k) = \partial v/\partial q_k$, the scalar product is delineated to

$$(\mathbf{a} \cdot \delta \mathbf{r}) = \frac{d\mathbf{v}}{dt} \cdot \sum_{k=1}^{n} \frac{\partial \mathbf{r}}{\partial q_k} \delta q_k = \sum_{k=1}^{n} \left[\frac{d}{dt} \left(\mathbf{v} \cdot \frac{\partial \mathbf{r}}{\partial q_k} \right) - \left(\mathbf{v} \cdot \frac{\partial \mathbf{v}}{\partial q_k} \right) \right] \delta q_k \quad .$$

Since, also through the definition, the velocity vector is the time derivative

$$\mathbf{v} = \frac{d\mathbf{r}}{dt} = \frac{\partial \mathbf{r}}{\partial t} + \sum_{k=1}^{n} \frac{\partial \mathbf{r}}{\partial q_k} \frac{dq_k}{dt} \quad , \quad \frac{dq_k}{dt} \equiv \dot{q}_k \quad ,$$

its partial derivative with respect to one of the generalized velocities becomes

$$\frac{\partial \mathbf{v}}{\partial \dot{q}_k} = \frac{\partial \mathbf{r}}{\partial q_k} \quad .$$

(10.10)

Substitution of Eq. (10.10) into the above bracketed term allows the application of the identical transformation twice, $(\mathbf{v} \cdot \partial v/\partial q_k) = \partial(v^2/2)/\partial q_k$ and the analogous substitution of the generalized velocity for the coordinate. Hence, the result is the finite sum

$$(\mathbf{a} \cdot \delta \mathbf{r}) = \sum_{k=1}^{n} \left[\frac{d}{dt} \left(\frac{1}{2} \frac{\partial v^2}{\partial \dot{q}_k} \right) - \frac{1}{2} \frac{\partial v^2}{\partial q_k} \right] \delta q_k \quad .$$

(10.11)

Since the mass of the MDOF-system is assumed to be constant, integration in Eq. (10.3) and parametric differentiation can be interchanged when Eq. (10.11) is substituted, and with the kinetic energy,

$$T = \frac{1}{2} \int_m v^2 \, dm \quad ,$$

(10.12)

D´Alembert´s principle takes on the form appropriate for dynamic MDOF-systems

$$\sum_{k=1}^{n} \left[Q_k - \frac{d}{dt} \left(\frac{\partial T}{\partial \dot{q}_k} \right) + \frac{\partial T}{\partial q_k} \right] \delta q_k = 0 \quad .$$

(10.13)

To perform the derivatives in Eq. (10.13) properly, the kinetic energy is to be expressed as the function of the generalized velocities, the generalized coordinates, and possibly it may also explicitly depend on time t

$$T = T(\dot{q}_1, \dot{q}_2, .., \dot{q}_n; q_1, q_2, .., q_n; t) .$$

Note that in a system of cartesian coordinates, the partial derivative of T with respect to q_k vanishes. Such terms arise from the curvature of the generalized coordinates, eg in the case where a polar angle is selected in cylindrical coordinates.

The independence of the generalized coordinates of dynamic systems does not automatically imply the independence of their virtual variations. There are systems under guiding conditions that render so-called *nonholonomic* constraints and the number of degrees of freedom N of infinitesimal motions becomes smaller than n, which is the number of degrees of freedom at large. There is a class of unilateral constraints that involves inequalities and there are nonintegrable differential constraints, eg of the type

$$\sum_{i=1}^{n} f_i(q_1, q_2, .., q_n) \dot{q}_i + g(q_1, q_2, .., q_n) = 0 \quad , \quad \frac{\partial f_i}{\partial q_k} \neq \frac{\partial f_k}{\partial q_i} \quad , \quad i \neq k .$$

Illustrations of the latter type of nonholonomic constraints are the pure rolling contact of a rigid sphere or a single rigid disk (the motion is constrained such that the disk´s inclination is constant) on a rigid plane, ot the ideal motion of a sharp runner say on ice, two wheels mounted at the ends of a common axle when rotating independently if the whole combination rolls without slipping on a rigid plane, and so on. When a particle moves in the plane under the constraint that its velocity vector is permanently directed toward a point that moves along a given straight line with the position given by a time differentiable function, the constraint is also nonholonomic. For these systems, the method of *Lagrangean* multipliers is to be applied.

There is, however, a large class of mechanical systems of engineering importance in which the constraints are *holonomic* , ie the number of degrees of freedom at large and of differential motions is the same. The virtual variations of the generalized coordinates are independent and Eq. (10.13) can be satisfied in general for all admissible variations only when the coefficients in the brackets vanish individually, ie

$$\frac{d}{dt}\left(\frac{\partial T}{\partial \dot{q}_k}\right) - \frac{\partial T}{\partial q_k} = Q_k \left(= -\frac{\partial V}{\partial q_k}\right) \quad , \quad k = 1, 2, .., n .$$

(10.14)

This set of *Lagrange* equations of motion is equivalent to *D´Alembert´s* principle when applied to holonomic MDOF-systems. The number of equations equals the number of degrees of freedom. Application of the differential operator requires knowledge of the system´s kinetic energy in the functional form of Eq. (10.12) and the determination of the generalized forces Q_k . In the general case, the virtual work of Eq. (10.6) renders the generalized forces by comparing coefficients. All those internal and external forces having a potential function that depends on the generalized coordinates and, possibly, on time explicitly, $V = V(q_1 , q_2 , .. , q_n ; t)$, contribute according to the gradient, Eq. (10.9). The remaining forces must be considered equivalent to generalized forces by their virtual work. An important example of the latter type is the damping force. If all the forces that contribute to the virtual work have a potential (necessarily of an undamped system), it is convenient to introduce the *Lagrangean* function $L = T - V$. *Lagrange* equations of motion in that case take on a simpler form,

$$\frac{d}{dt}\left(\frac{\partial L}{\partial \dot{q}_k}\right) - \frac{\partial L}{\partial q_k} = 0 \ , \ k = 1, 2, .., n \ , \ L = T - V \ .$$

(10.14a)

However, in that form, they can be generalized, eg to applications in electrodynamics. The *Lorentz* force is derived from a generalized potential function that depends on the velocity of motion, but the operator remains the same. Symmetry is to be noted when deriving the generalized forces by means of the newly defined operator, $k = 1, 2, ..., n$,

$$Q_k = -\frac{\partial V}{\partial q_k} + \frac{d}{dt}\left(\frac{\partial V}{\partial \dot{q}_k}\right) \ , \ V = V^*(q_1, q_2, .., q_n; \dot{q}_1, \dot{q}_2,.., \dot{q}_n; t) \ .$$

(10.14b)

(§) A Point Mass m Carrying a Charge q , the *Lorentz* Force. The charged particle moves with velocity v in an electromagnetic field. In such a case, the force is given in *Gaussian* units by (c is the speed of light)

$$F = q\left[E + \frac{1}{c}\left(v \times B\right)\right] \ , \ \nabla.B = 0 \ \rightarrow B = \nabla \times A \ .$$

(10.15)

A is the magnetic vector potential. By using the *Maxwell* equations (they are not discussed here, but their notation is standard), it can be shown that the *Lagrange* function takes on the form, $T = m \, v^2/2$,

$$L = T - q \phi + \frac{q}{c} \left(A . v\right) \ .$$

(10.16)

The scalar potential refers to the irrotational vector field $E + (1/c)(\partial A/\partial t) = -$ grad ϕ . The *Lorentz* force is given by applying the

newly defined operator of Eq. (10.14b) to the generalized potential
that explicitly depends on the velocity

$$V = q\left[\phi - \frac{1}{c} (\mathbf{A} \cdot \mathbf{v})\right] .$$

(10.17)

Subsequently, simple examples illustrate the application of
the *Lagrangean* formalism to purely mechanical systems. Further
results on free and forced vibrations are derived and the natural
modes are discussed. Further use of *D´Alembert´s* principle is made
when dealing with the *Ritz* approximation (see Sec. 11.1.1).

10.1. Natural Vibrations of an Elastically Supported Foundation

The stiffness of the supporting linear elastic springs is selected
such that the rigid "beam" of weight mg is at rest in a horizontal
position. Natural vibrations of small amplitude are investigated
around that equilibrium configuration. Two degrees of freedom are
considered and the generalized coordinates selected are $q_1 = z_S(t)$,
the vertical displacement of the center of gravity, and the skewness
angle $q_2 = \varphi(t)$ (see Fig.10.1). If we consider the springs mass-less,
the kinetic energy is given for the rigid foundation by the sum of the
translational and rotatory energy (the centroid is the reference
point)

$$T = \frac{m}{2}\left(\dot{z}_S^2 + i_S^2\, \dot{\varphi}^2\right) , \quad I_S = m i_S^2 .$$

(10.18a)

Since damping is neglected, only the gravity and forces in the
springs contribute to the virtual work if the instant configuration is
varied virtually by δz_S and independently by $\delta\varphi$. The motion is

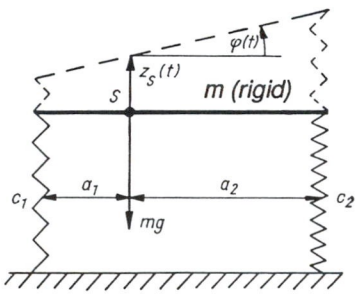

Fig. 10.1. Natural vibrations of a rigid foundation

conservative and all three forces can be considered by the potential energy, $V = W_P + U$. Since the weight does not influence the free vibrations, the relative potential of the springs enters the picture. The potential energy increases with respect to the equilibrium state by [refer to Eq. (3.40)]

$$V = U = \frac{c_1 s_1^2}{2} + \frac{c_2 s_2^2}{2} \quad , \quad s_1 = z_S - a_1 \sin \varphi \quad , \quad s_2 = z_S + a_2 \sin \varphi . \tag{10.18b}$$

The derivatives according to the *Lagrange* equations (10.14) are listed below:

$$\frac{\partial V}{\partial z_S} = c_1 s_1 + c_2 s_2 \quad , \quad \frac{\partial V}{\partial \varphi} = c_1 s_1 \left(-a_1 \cos \varphi\right) + c_2 s_2 a_2 \cos \varphi ,$$

$$\frac{\partial T}{\partial \dot{z}_S} = m \dot{z}_S \quad , \quad \frac{\partial T}{\partial \dot{\varphi}} = m i_S^2 \dot{\varphi} \quad , \quad \frac{d}{dt}\left(\frac{\partial T}{\partial \dot{z}_S}\right) = m \ddot{z}_S \quad , \quad \frac{d}{dt}\left(\frac{\partial T}{\partial \dot{\varphi}}\right) = m i_S^2 \ddot{\varphi} .$$

The kinetic energy does not explicitly depend on the generalized coordinates, and thus, the partial derivatives must vanish. Since $|\varphi| \ll 1$ is assumed, the terms may be linearized after performing the differentiation, and the set of coupled linear equations results (it holds approximately for small vibrations)

$$\ddot{z}_S + \Omega_z^2 z_S + k_1^2 \varphi = 0 \quad , \quad \Omega_z^2 = (c_1 + c_2)/m \quad , \quad k_1^2 = (c_2 a_2 - c_1 a_1)/m > 0 , \tag{10.18c}$$

$$\ddot{\varphi} + \Omega_\varphi^2 \varphi + k_2^2 z_S = 0 \quad , \quad \Omega_\varphi^2 = (c_1 a_1^2 + c_2 a_2^2)/m i_S^2 \quad , \quad k_2^2 = k_1^2 / i_S^2 .$$

Time-harmonic solutions are expected and it suffices to substitute

$$z_S = A \cos \omega t \quad , \quad \varphi = B \cos \omega t \tag{10.19}$$

into the linearized equations of motion to derive the linear system of homogeneous equations for the amplitudes A and B. A nontrivial solution is possible only for discrete values of the circular frequency ω, the two natural frequencies of the linearized system. They are the roots of the frequency equation; the determinant of the coefficients must vanish (see also Sec. 7.4.9)

$$D(\omega) = \omega^4 - \left(\Omega_z^2 + \Omega_\varphi^2\right) \omega^2 + \Omega_z^2 \Omega_\varphi^2 - k_1^2 k_2^2 = 0 .$$

The system has two degrees of freedom and, in general, two separate roots render the basic and first higher order natural frequency, $\omega_1 < \omega_2$,

$$\omega_{2,1}^2 = \frac{1}{2}\left(\Omega_z^2 + \Omega_\varphi^2\right) \pm \sqrt{\frac{1}{4}\left(\Omega_z^2 - \Omega_\varphi^2\right)^2 + k_1^2 k_2^2} \quad , \quad \omega_1 \leq \left(\Omega_z, \Omega_\varphi\right) \leq \omega_2 \,.$$

$$(10.20)$$

Substituting these eigenvalues subsequently into one of the linear equations,

$$\left(\Omega_z^2 - \omega^2\right)A + k_1^2 B = 0 \,, \tag{10.21}$$

relates the constants of integration $B_{1,2} = \kappa_{1,2} A_{1,2}$, where the coefficients of proportionality are system parameters and are given by $\kappa_{1,2} = (\omega_{1,2}^2 - \Omega_z^2) / k_1^2$. The general solution describing the coupled vibrations becomes

$$z_S = A_1 \cos(\omega_1 t - \varepsilon_1) + A_2 \cos(\omega_2 t - \varepsilon_2),$$

$$(10.22)$$

$$\varphi = \kappa_1 A_1 \cos(\omega_1 t - \varepsilon_1) + \kappa_2 A_2 \cos(\omega_2 t - \varepsilon_2).$$

In addition to the amplitudes $A_{1,2}$, the phase angles $\varepsilon_{1,2}$ are to be determined through the initial conditions. For example, $z_S(t = 0) = z_0, dz_S/dt = \varphi = d\varphi/dt = 0$ at $t = 0$ render the phase as zero, and the remaining constants of integration are proportional to the translational disturbance of the equilibrium

$$A_1 = \frac{\omega_2^2 - \Omega_z^2}{\omega_2^2 - \omega_1^2} z_0 \,, \quad A_2 = \frac{\Omega_z^2 - \omega_1^2}{\omega_2^2 - \omega_1^2} z_0 \,, \quad \varepsilon_1 = \varepsilon_2 = 0 \,.$$

$$(10.23a)$$

A phenomenon of considerable interest is observed when the natural frequencies become closely spaced. This happens to exist if, eg the system parameters are such that $\Omega_z^2 = \Omega_\varphi^2 = \Omega^2$ and $k_1 k_2 = \eta^2 \ll \Omega^2$. By means of an addition theorem, Eq. (10.22) and the special solution (10.23a), in that case, take on the form

$$z_S = \frac{z_0}{2}(\cos\omega_1 t + \cos\omega_2 t) = z_0 \cos\left(\frac{\eta^2}{\omega^*}t\right)\cos(\omega^* t) \,, \quad \omega^* = (\omega_1 + \omega_2)/2 \approx \Omega \,,$$

$$(10.23b)$$

$$\text{is } \varphi = \frac{z_0}{2}(-\cos\omega_1 t + \cos\omega_2 t) = z_0 \sin\left(\frac{\eta^2}{\omega^*}t\right)\cos(\omega^* t + \pi/2).$$

The fast vibration with the period $2\pi/\omega^*$ is modulated since the amplitude is considered time-dependent and varies slowly also in a time-harmonic fashion. Such a *beat* motion shows a phase shift of $\pi/2$, and thus, the energy flows between the modes of vibration, alternatively exciting the translational and rotational motion. The amplitudes take on the maximal possible values, z_0 in Eq. (10.23b). Structures sensitive to torsion may fail in such a beat motion. In

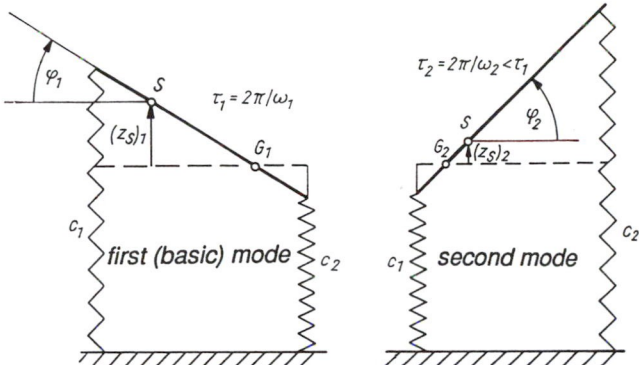

Fig. 10.2. Natural modes of (uncoupled) vibrations of the system sketched in Fig. 10.1.

general terms of control theory, $z_S(t)$ is a modulated vibration with carrier frequency ω^* and modulating frequency η^2/ω^*.

The natural vibration, Eq. (10.22), splits into two independent modes, which are individual solutions of uncoupled oscillator equations,

$$\xi_1 = A_1 \cos(\omega_1 t - \varepsilon_1) \; , \; \ddot{\xi}_1 + \omega_1^2 \xi_1 = 0 \; ,$$

$$\xi_2 = A_2 \cos(\omega_2 t - \varepsilon_2) \; , \; \ddot{\xi}_2 + \omega_2^2 \xi_1 = 0 \; . \tag{10.24}$$

$\xi_{1,\,2}$ are the principal generalized coordinates that uncouple the vibrational motion. The geometry of the basic mode is easily recognized after eliminating the coordinate ξ_1

$$(z_S)_1 = \xi_1 \; , \; \varphi_1 = \kappa_1 \xi_1 \;\rightarrow\; (z_S)_1 = \frac{1}{\kappa_1} \varphi_1 \; . \tag{10.25}$$

The higher, second-order (fast) mode that vibrates with the period $2\pi/\omega_2$ is found analogously by eliminating the coordinate ξ_2

$$(z_S)_2 = \xi_2 \; , \; \varphi_2 = \kappa_2 \xi_2 \;\rightarrow\; (z_S)_2 = \frac{1}{\kappa_2} \varphi_2 \; . \tag{10.26}$$

Figure 10.2 shows the modes for the parameters $\kappa_1 < 0$ and $\kappa_2 > 0$. Also marked are the velocity poles G_i, $i = 1, 2$, and the clockwise and counterclockwise rotations of mode 1 and 2, respectively. The vector form of natural modes is discussed in Sec. 10.4.

10.2. Pendulum with Moving Support

The plane motion of rigid bodies is considered when the supporting axle moves in a straight guide restrained by a restoring linear elastic spring.

10.2.1. Horizontal Motion of the Support

The system of two rigid masses (m, i_S) of the pendulum, and m_1 of the sliding support, according to Fig.10.3 has two degrees of freedom. The generalized coordinates are selected as $q_1 = x_1$ and $q_2 = \varphi$. The cartesian coordinates of the centroid S of the pendulum are geometrically related to the generalized coordinates by $x_S = x_1 + (s \sin \varphi)$, $z_S = s \cos \varphi$. The kinetic energy of the system becomes by superposition [S must be the point of reference of the pendulum; see Eq. (8.10)]

$$T = \frac{1}{2} m_1 \dot{x}_1^2 + \left(\frac{1}{2} m v_S^2 + \frac{1}{2} m i_S^2 \dot{\varphi}^2 \right) , \quad i_S^2 = I_S / m ,$$

$$(10.27)$$

$$v_S^2 = \dot{x}_S^2 + \dot{z}_S^2 = \dot{x}_1^2 + \left(s \dot{\varphi} \right)^2 + 2 s \dot{x}_1 \dot{\varphi} \cos \varphi .$$

Hence, the partial derivatives with respect to the generalized velocities and coordinates are readily found and the total time derivative is considered subsequently

$$\frac{\partial T}{\partial \dot{x}_1} = (m + m_1) \dot{x}_1 + ms \dot{\varphi} \cos \varphi , \quad \frac{\partial T}{\partial \dot{\varphi}} = m \left(s^2 + i_S^2 \right) \dot{\varphi} + ms \dot{x}_1 \cos \varphi ,$$

$$\frac{\partial T}{\partial x_1} = 0 , \quad \frac{\partial T}{\partial \varphi} = - ms \dot{x}_1 \dot{\varphi} \sin \varphi ,$$

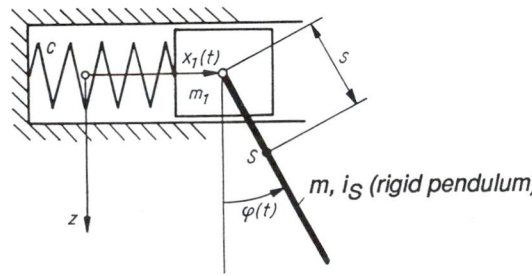

Fig. 10.3. Plane pendulum with horizontally mobile support. Model of a bell tower

$$\frac{d}{dt}\left(\frac{\partial T}{\partial \dot{x}_1}\right) = (m + m_1)\,\ddot{x}_1 + m\left(s\ddot{\varphi}\cos\varphi - s\dot{\varphi}^2\sin\varphi\right),$$

$$\frac{d}{dt}\left(\frac{\partial T}{\partial \dot{\varphi}}\right) = m\left(s^2 + is^2\right)\ddot{\varphi} + ms\left(\ddot{x}_1\cos\varphi - \dot{x}_1\,\dot{\varphi}\sin\varphi\right).$$

Virtual work during a variation of any instant configuration is performed by the weight of the pendulum and by the internal force of the spring of stiffness c . Hence, the potential energy and its derivatives are to be considered (since the system is conservative, any friction or damping is neglected)

$$V = -\,mg\,s\cos\varphi + \frac{c}{2}\,x_1^2 \;\to\; \frac{\partial V}{\partial x_1} = cx_1 = -Q_x\,,\quad \frac{\partial V}{\partial \varphi} = mg\,s\sin\varphi = -Q_\varphi\,.$$

$$(10.28)$$

Combining the terms according to the *Lagrange* equations of motion and dividing through the coefficients of the highest order derivatives render the nonlinear set of two second-order differential equations

$$\ddot{x}_1 + \Omega_x^{\,2}\,x_1 + k_1\,\ddot{\varphi}\cos\varphi - k_1\,\dot{\varphi}^2\sin\varphi = 0\;,\quad \Omega_x^{\,2} = c/(m + m_1),$$

$$(10.29)$$

$$\ddot{\varphi} + \Omega_\varphi^{\,2}\sin\varphi + \frac{\Omega_\varphi^{\,2}}{g}\,\ddot{x}_1\cos\varphi = 0\;,\quad \Omega_\varphi^{\,2} = gs/(s^2 + is^2)\;,\quad k_1 = ms/(m + m_1)\;.$$

Linearization in the case of small vibrations about the equilibrium configuration, $x_1 = \varphi = 0$, keeps the leading inertia coupling terms

$$\ddot{x}_1 + \Omega_x^{\,2}\,x_1 + k_1\,\ddot{\varphi} = 0\;,\quad \ddot{\varphi} + \Omega_\varphi^{\,2}\,\varphi + \frac{\Omega_\varphi^{\,2}}{g}\,\ddot{x}_1 = 0\;.$$

$$(10.30)$$

Substituting the time-harmonic terms according to Eq. (10.19), where x_1 replaces z_S , and solving the frequency equation, which is again given in the form of the vanishing determinant, yield the (squared) linear natural frequencies

$$\omega_{2,1}^2 = \frac{1}{2}\,\frac{\Omega_x^{\,2} + \Omega_\varphi^{\,2}}{1 - k_1\,\Omega_\varphi^{\,2}/g}\left\{1 \pm \sqrt{\left(\frac{\Omega_x^{\,2} - \Omega_\varphi^{\,2}}{\Omega_x^{\,2} + \Omega_\varphi^{\,2}}\right)^2 + k_1\,\frac{\Omega_\varphi^{\,2}}{g}\,\frac{4\,\Omega_x^{\,2}\Omega_\varphi^{\,2}}{\left(\Omega_x^{\,2} + \Omega_\varphi^{\,2}\right)^2}}\right\}.$$

$$(10.31)$$

The system considered in this section is an oversimplified model of the natural vibrations of a bell when supported in a slender elastic tower.

10.2.2. Vertical Motion of the Support

According to Fig. 10.4, the generalized coordinates are z_1 and the angle φ. The cartesian coordinates of the center of mass of the pendulum are thus $x_S = s \sin \varphi$ and $z_S = z_1 + s \cos \varphi$. The kinetic energy, Eq. (8.10), by superposition becomes

$$T = \frac{m_1}{2} \dot{z}_1^2 + \frac{m}{2} \left(v_S^2 + i_S^2 \dot{\varphi}^2 \right) \ , \ v_S^2 = \dot{x}_S^2 + \dot{z}_S^2 \tag{10.32}$$

$$= \dot{z}_1^2 + s^2 \dot{\varphi}^2 - 2 s \dot{z}_1 \dot{\varphi} \sin \varphi \ .$$

The derivatives are subsequently listed

$$\frac{\partial T}{\partial \dot{z}_1} = (m + m_1) \dot{z}_1 - ms\dot{\varphi} \sin \varphi \ , \ \frac{\partial T}{\partial \dot{\varphi}} = m \left(s^2 + i_S^2 \right) \dot{\varphi} - ms \dot{z}_1 \sin \varphi \ ,$$

$$\frac{\partial T}{\partial z_1} = 0 \ , \ \frac{\partial T}{\partial \varphi} = - m s \dot{z}_1 \dot{\varphi} \cos \varphi \ .$$

$$\frac{d}{dt} \left(\frac{\partial T}{\partial \dot{z}_1} \right) = (m + m_1) \ddot{z}_1 - ms \left(\ddot{\varphi} \sin \varphi + \dot{\varphi}^2 \cos \varphi \right) \ ,$$

$$\frac{d}{dt} \left(\frac{\partial T}{\partial \dot{\varphi}} \right) = m \left(s^2 + i_S^2 \right) \ddot{\varphi} - ms \left(\ddot{z}_1 \sin \varphi + \dot{z}_1 \dot{\varphi} \cos \varphi \right) \ .$$

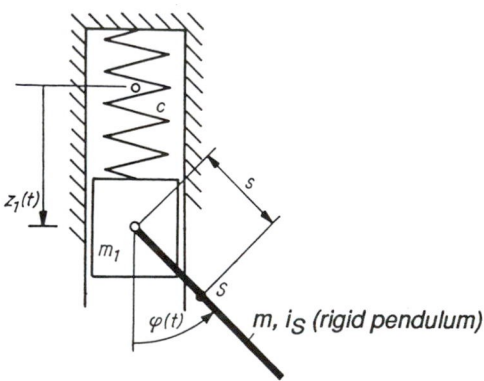

Fig. 10.4. Plane pendulum with vertically moving support.
Illustration of parametric excitation

The potential energy due to gravity and that part stored in the linear elastic spring of stiffness c is given by superposition (again see Fig. 10.4)

$$V = - mg \left(z_1 + s \cos \varphi\right) - m_1 g \, z_1 + \frac{c}{2} \, z_1^2 .$$

(10.33)

The partial derivatives render the generalized forces

$$\frac{\partial V}{\partial z_1} = - mg - m_1 g + c z_1 = - Q_z \quad , \quad \frac{\partial V}{\partial \varphi} = mg \, s \sin \varphi = - Q_\varphi .$$

(10.34)

According to Eq. (10.14), the nonlinear equations of motion result

$$\ddot{z}_1 + \Omega_z^2 z_1 - k_1 \ddot{\varphi} \sin \varphi - k_1 \dot{\varphi}^2 \cos \varphi - g = 0 \quad , \quad \Omega_z^2 = c / (m + m_1) ,$$

(10.35)

$$\ddot{\varphi} + \Omega_\varphi^2 \sin \varphi - \frac{\Omega_\varphi^2}{g} \ddot{z}_1 \sin \varphi = 0 \quad , \quad \Omega_\varphi^2 = gs / \left(s^2 + i s^2\right) \quad , \quad k_1 = ms / (m + m_1) .$$

The direct linearization technique commonly applied to account for small vibrations fails in that case, since the linearized equations decouple. Putting $\varphi = d\varphi/dt = 0$ in the first equation yields the linear oscillator equation of translational motion

$$\ddot{z}_1^* + \Omega_z^2 z_1^* = 0 \quad , \quad z_1^* = z_1 - z_{1S} .$$

(10.36)

Its solution may be called the main vibration

$$z_1^* = A \cos \left(\Omega_z t - \varepsilon\right) \quad , \quad \varphi \equiv 0,$$

(10.37)

and the second Eq. (10.35) is identically zero. The partial linearization of that equation with respect to the angular coordinate, $|\varphi*| \ll 1$, and substitution of the main vibration for the translational acceleration term yield the proper form

$$\ddot{\varphi}^* + \Omega_\varphi^2 \left[1 + \frac{A \, \Omega_z^2}{g} \cos \left(\Omega_z t - \varepsilon\right)\right] \varphi^* = 0 .$$

(10.38)

A differential equation with a periodic coefficient results. That is, initiating the main vibration in the vertical direction forces the rotational motion through a parametric excitation. The first of Eqs. (10.35) still holds in the linear approximation. Equation (10.38) is the *Mathieu* differential equation, a special case of the more general *Hill´s* equation. See, eg K. Klotter: *Technische Schwingungslehre. Vol. 1, Part A.* Springer-Verlag, Berlin, 1978, 3rd ed.

Instabilities of motion can be investigated by Eq. (10.38) and may be found in the neighborhood of the parameters $\Omega_z = 2\,\Omega_\varphi /n$, $n =$ 1, 2, ... The system is conservative in nature and the transition from a purely translational mode to rotational motion must have the characteristic of coupled vibration. Therefore, the classical and direct linearization of Eq. (10.35) fails to give reliable results. Parametric resonance is an important, yet unwanted phenomenon in many engineering structures and machines.

10.3. MDOF-Vibrational System of Point Masses Supported by a Mass-Less String

To illustrate such a system, three equal point masses *m* are considered supported by a string under the tension *S* . According to Fig. 10.5, restoring forces are initiated by the deflection of the (mass-less) string in lateral vibrations. During small amplitude motions, the normal force is assumed to be approximately constant and equal to the prestress, *N = S* . The generalized coordinates are selected as the lateral coordinates of the mass points, $q_k = w_k$, $k =$ 1, 2, 3.   The kinetic energy is the sum of the translational portions. The generalized forces are determined by the equivalence of virtual work, as expressed by Eq. (10.6),

$$\delta W = \sum_{k=1}^{3} Q_k\,\delta q_k = -\,S \sum_{j=1}^{4} \delta s_j \quad , \quad s_j = \sqrt{a^2 + \left(w_j - w_{j-1}\right)^2}$$

(10.39)

$$\to \; \delta s_j = \frac{w_j - w_{j-1}}{s_j}\left(\delta w_j - \delta w_{j-1}\right) .$$

Comparing coefficients, $\delta q_k = \delta w_k$, *k = 1, 2, 3*, renders

$$Q_1 = -\,S\left(\frac{w_1}{s_1} - \frac{w_2 - w_1}{s_2}\right) \quad , \quad Q_2 = -\,S\left(\frac{w_2 - w_1}{s_2} - \frac{w_3 - w_2}{s_3}\right) ,$$

(10.40)

$$Q_3 = -\,S\left(\frac{w_3 - w_2}{s_3} - \frac{w_3}{s_4}\right) .$$

The *Lagrange* equations of motion (10.14) yield the nonlinear and coupled system (the kinetic energy is easily differentiated according to the rules)

$$T = \frac{m}{2} \sum_{k=1}^{3} \dot{w}_k^2 \; \to$$

$$\ddot{w}_1 + \alpha^2 \left[\left(\frac{a}{s_1} + \frac{a}{s_2}\right) w_1 - \frac{a}{s_2}\, w_2\right] = 0 \; , \; \alpha^2 = \frac{S}{ma} \; ,$$

$$\dot{w}_2 + \alpha^2 \left[\left(\frac{a}{s_2} + \frac{a}{s_3}\right) w_2 - \frac{a}{s_2}\, w_1 - \frac{a}{s_3}\, w_3\right] = 0 \ ,$$

$$\dot{w}_3 + \alpha^2 \left[\left(\frac{a}{s_3} + \frac{a}{s_4}\right) w_3 - \frac{a}{s_3}\, w_2\right] = 0 \ . \tag{10.41}$$

The restoring forces may be linearized by means of the geometric approximations $(a/s_j) \approx 1 - (w_j - w_{j-1})^2/2a^2$. The linearized system of equations of motion with constant coefficients, approximately valid for small natural vibrations, results

$$\dot{w}_1 + \alpha^2\, [2\, w_1 - w_2] = 0 \ , \ \ \dot{w}_2 + \alpha^2\, [2\, w_2 - w_1 - w_3] = 0 \ ,$$

$$\dot{w}_3 + \alpha^2\, [2\, w_3 - w_2] = 0 . \tag{10.42}$$

A cubic equation is characteristic of the linear eigenvalue problem of the (squared) linear natural frequencies that has three real and positive roots, in agreement with the general rules. The eigenfrequencies given below are ordered by their increasing value

$$\omega_1 = \alpha\sqrt{2 - \sqrt{2}} \ \ , \ \ \omega_2 = \alpha\sqrt{2} \ \ , \ \ \omega_3 = \alpha\sqrt{2 + \sqrt{2}} \ . \tag{10.43}$$

The associated linear mode shapes are given by the following amplitude ratios:
1.   The symmetric basic mode $(1/\sqrt{2}) : 1 : (1/\sqrt{2})$. Its period is $2\pi/\omega_1$.
2.   The skew-symmetric second mode $1 : 0 : (-1)$ (with one nodal point). Its period is $2\pi/\omega_2$.

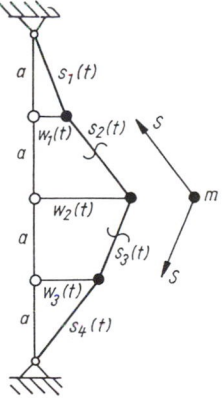

Fig. 10.5. Free lateral vibrations of three point masses supported by a mass-less string prestressed by tension S

3.   The symmetric third mode $(1/\sqrt{2}) : (- 1) : (1/\sqrt{2})$ (with two nodal points in the field). Its period is $2\pi/\omega_3$.
Placing point supports at the proper nodes does not change the associated single mode of natural vibrations.

10.4. MDOF-Vibrational System of Point Masses Supported by a Mass-Less Beam

A simply supported single-span beam with constant bending rigidity, $B = EJ$, is considered, but without inertia, in flexural vibrations. It carries two point masses m and M in the distance c and d from the left support, the span is l. Free undamped vibrations, as well as forced vibrations by the given single forces $F_j (t)$, $j = 1, 2$, are to be considered by means of the *Lagrangean* form of the equations of motion. The system of Fig. 10.6 exhibits two degrees of freedom and the generalized coordinates selected are $q_1 = w_1$ and $q_2 = w_2$. Deflections are measured from the equilibrium configuration of the system under its own weight. The kinetic energy is simply given by superposition (rotational inertia and hence, energy is neglected)

$$T = \frac{1}{2} \left(m \dot{w}_1{}^2 + M \dot{w}_2{}^2\right) .$$

(10.44)

The elastic potential of the linear elastic beam must be expressed in terms of the discrete deflections $w_{1, 2}$, $U = U(w_1, w_2)$. Since the supports of the beam are statically determinate, it is simpler to calculate the influence coefficients α_{ij} , $i, j = 1, 2$ by considering the static unit forces $F_{i, j}{}^* = 1$ to be applied at the location of the point masses. With Eq. (3.46), the static deflections are given by superposition

$$w_1{}^* = F_1{}^* \alpha_{11} + F_2{}^* \alpha_{12} , \quad w_2{}^* = F_1{}^* \alpha_{21} + F_2{}^* \alpha_{22} , \quad \alpha_{12} = \alpha_{21} . \quad (10.45)$$

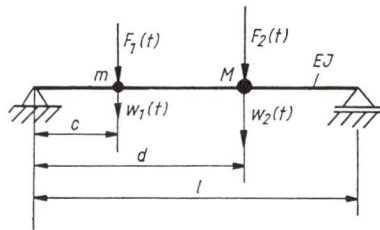

Fig. 10.6. Elastic beam with two concentrated masses. Solution by modal expansion

Solving for the forces renders (thereby the flexibility matrix is inverted)

$$F_1{}^* = k_{11}w_1{}^* + k_{12}\,w_2{}^* \quad , \quad F_2{}^* = k_{21}w_1{}^* + k_{22}\,w_2{}^* \quad , \quad k_{12} = k_{21} \cdot \quad (10.46)$$

The stiffness coefficients k_{ij} (the elements of the symmetric stiffness matrix) are expressed in terms of the influence coefficients by

$$k_{11} = \alpha_{22} / \left(\alpha_{11}\,\alpha_{22} - \alpha_{12}{}^2\right) \quad , \quad k_{12} = k_{21} = -\alpha_{12} / \left(\alpha_{11}\,\alpha_{22} - \alpha_{12}{}^2\right) \quad ,$$

$$k_{22} = \alpha_{11} / \left(\alpha_{11}\,\alpha_{22} - \alpha_{12}{}^2\right) \quad . \quad\quad (10.47)$$

Equation (3.45) gives the solution by expanding the scalar matrix product, $U = (1/2)\,\mathbf{w}^T\,\mathbf{k}\,\mathbf{w} = (1/2)\,(k_{11}\,w_1{}^2 + k_{22}\,w_2{}^2 + 2\,k_{12}\,w_1\,w_2)$. By adding the (time-dependent) potential of the external forces $W_P = - w_1\,F_1 - w_2\,F_2$, the partial derivatives of the total potential V determine the generalized forces

$$Q_1 = -\frac{\partial V}{\partial w_1} = F_1 - (k_{11}\,w_1 + k_{12}\,w_2) \quad , \quad Q_2 = -\frac{\partial V}{\partial w_2} = -F_2 - (k_{22}\,w_2 + k_{12}\,w_1) \, .$$

$$(10.48)$$

The influence coefficients of the single-span beam of Fig. 10.6 are easily determined by static methods and are listed below for convenience:

$$\alpha_{11} = \frac{1}{6\,B}\left(\frac{2\,c^4}{I} + 2\,I\,c^2 - 4\,c^3\right) \, , \quad \alpha_{12} = \frac{1}{6\,B}\left(\frac{d\,c^3 + c\,d^3}{I} + 2\,I\,c\,d - 3\,c\,d^2 - c^3\right) \, ,$$

$$\alpha_{22} = \frac{1}{6\,B}\left(\frac{2\,d^4}{I} + 2\,I\,d^2 - 4\,d^3\right) \, . \quad\quad (10.49)$$

The coupled equations of motion according to Eq. (10.14) take on matrix form

$$\mathbf{m}\,\dot{\mathbf{w}} + \mathbf{k}\,\mathbf{w} = \mathbf{F}(t) \quad , \quad \mathbf{w}^T(t) = (w_1(t),\,w_2(t)) \quad , \quad \mathbf{F}^T(t) = (F_1(t),\,F_2(t)) \, ,$$

$$\mathbf{m} = \begin{pmatrix} m & 0 \\ 0 & M \end{pmatrix} \quad , \quad \mathbf{k} = \begin{pmatrix} k_{11} & k_{12} \\ k_{21} & k_{22} \end{pmatrix} \, . \quad\quad (10.50)$$

The mass matrix is diagonal and the stiffness matrix symmetric. If we put the forcing vector to zero, the frequency equation may be set up, say, in the form of the vanishing determinant, and solved for the natural (squared) frequencies

$$\omega_{2,\,1}{}^2 = \frac{1}{2\,m\,M}\left[(M\,k_{11} + m\,k_{22}) \pm \sqrt{(M\,k_{11} - m\,k_{22})^2 + 4\,m\,M\,k_{12}{}^2}\right]. \quad (10.50a)$$

The mode shapes are found by solving the linearly dependent homogeneous equations of the eigenvalue problem

$$(k - \omega_k^2 m)\, \Phi_k = 0 \;\rightarrow\; \Phi_k^T = \left(1, \; (k_{11} - m\omega_k^2)\,/\,(-k_{12})\right) \;, \quad k = 1, 2 .\tag{10.50b}$$

The expression in parentheses is called the dynamic matrix. The eigenvectors are mutually orthogonal, and thus, any solution of Eq. (10.50) has a modal expansion, $\omega_1 \neq \omega_2$ (the natural frequencies are assumed to be well separated)

$$w = \sum_{k=1}^{2} Y_k(t)\, \Phi_k \;,\tag{10.50c}$$

where $Y_k(t)$ are the principal (modal) coordinates. By substituting Eq. (10.50c) into (10.50) and after scalar multiplication from the left with any one of the transposed eigenvectors, the uncoupled oscillator equations are derived, noting the orthogonality condition that holds with respect to the mass as well as the stiffness matrix,

$$\ddot{Y}_k + \omega_k^2\, Y_k = \frac{1}{m_k}\, \Phi_k^T . F \;, \quad k = 1, 2 \;, \quad m_k = \Phi_k^T . m . \Phi_k .\tag{10.50d}$$

The effective excitation of each of these modal oscillators is given by the projection of the forcing vector onto the proper eigenvector. Hence, the actual forces F_1 and F_2, in general, are multiplied by different participation factors. The orthonormalization of Φ_k is derived by premultiplication with the factor $1/\sqrt{m_k}$. Thus, by considering the modal masses $m_k = 1$, those eigenvectors become a base of unit vectors. It is common practice to consider light material damping at this stage by adding a proper amount of viscous damping $2\,\zeta_k\,\omega_k\,(dY_k/dt)$ to the modal equations [see Eq. (7.81) and also the discussion on efficient solution techniques in the time and in the frequency domain].

10.5. Planar Framed System with External Viscous Damping

The SDOF-system in Fig. 10.7 is considered that consists of several rigid and homogeneous members connected by ideal hinges. When we select the angular coordinate $q = \varphi$, the application of the *Lagrange* equation (10.14) is quite convenient since it saves any individual free-body diagrams. The motion of bodies 1 and 3 is purely rotational [see Eq. (8.11)], the cross bar 2 moves in a translational sense. Thus, the kinetic energy is the sum of the individual contributions

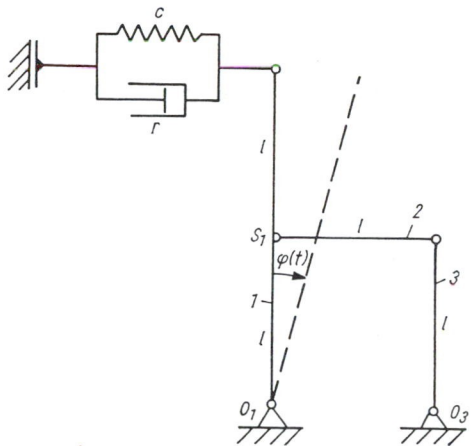

Fig. 10.7. Stability of a heavy framed system by means of the dynamic method

$$T = \frac{1}{2}\left(m_1\, i_1^2\, \dot{\varphi}^2 + m_3\, i_3^2\, \dot{\varphi}^2 + m_2\, v_2^2\right) \quad, \quad v_2 = l\,\dot{\varphi} \ . \tag{10.51}$$

The moments of inertia must be taken with respect to the points 0_1 and 0_3, respectively. The motion is in a homogeneous gravity field and the spring is considered linear elastic with stiffness c . The potential energy is

$$V^* = W_P + U = -m_1\, g\, l\left(1 - \cos\varphi\right) - m_3\, g\, \frac{l}{2}\left(1 - \cos\varphi\right) - m_2\, g\, l\left(1 - \cos\varphi\right)$$

$$+ \frac{c}{2}\left(2l \sin\varphi\right)^2 . \tag{10.52}$$

The contribution of the external viscous force of the dashpot to the generalized force in the *Lagrange* equation of motion is determined by equating its virtual work and comparing coefficients

$$Q_\varphi^{**}\, \delta\varphi = -r\, \frac{d\left(2\,l\,\sin\varphi\right)}{dt}\, \delta\left(2\,l\,\sin\varphi\right) = -r\left(2\,l\,\dot{\varphi}\cos\varphi\right)\left(2\,l\,\cos\varphi\,\delta\varphi\right)$$

$$\rightarrow Q_\varphi^{**} = -4\, r\, l^2\, \dot{\varphi}\, \cos^2\varphi \ . \tag{10.53}$$

By putting specifically $m_2 = m_3 = m$ and $m_1 = 2\, m$, the nonlinear equation of motion results

$$\ddot{\varphi} - \frac{7}{8}\frac{g}{l}\sin\varphi + \frac{c}{2m}\sin 2\varphi = -\frac{r}{m}\dot{\varphi}\cos^2\varphi.$$ (10.54)

Small and damped angular vibrations about the equilibrium configuration $\varphi = 0$ are approximately determined by means of the linearized equation

$$\ddot{\varphi} + 2\zeta\omega_0\dot{\varphi} + \omega_0^2\varphi = 0 , \; \zeta = r/2m\omega_0 \; \text{and} \; \omega_0^2 = \left(\frac{c}{m} - \frac{7}{8}\frac{g}{l}\right) > 0.$$ (10.55)

The asymptotic stability of the state at rest is indicated by the condition of a real-valued eigenfrequency. The linear period of the damped vibration is given by $\tau = 2\pi/\omega_0 \sqrt{(1 - \zeta^2)} , \zeta < 1$; see also Sec. 9.1.2.

10.6. Vibrational Testing by an Unbalanced Rotor

In situ dynamic testing of all kinds of elastic structures requires a source of time-harmonic excitation with a sweeping of the forcing frequency by easy means. A simple mechanical device is given by an unbalanced rotor that is driven by an electric motor with speed control. Uniaxial excitation is achieved by using a pair of counter rotating wheels when splitting the unbalanced mass m_u properly; a sketch is given in Fig. 10.8(b). The testing of a single-span (mass-less) beam is illustrated in Fig. 10.8(a), in which the unbalanced rotor is mounted at midspan where also a single mass m is assumed to be concentrated. The material damping of the linear elastic beam is symbolized by the viscous damper. The *Lagrange* equation of motion of the SDOF combined system is to be set up in terms of the midspan deflection $q(t) = w(t)$. Masses M and m_u , the adjustable parameters eccentricity e and the revolutions per minute $60v/2\pi$ (rpm) are given. The relevant part of the kinetic energy is the sum

$$T = \frac{(m + M)}{2}\dot{w}^2 + \frac{1}{2}\frac{m_u}{2}\left(v_1^2 + v_2^2\right) ,$$ (10.56)

where $v_{1,2}$ are the absolute speeds of the unbalanced point masses. They are easily determined by taking the time derivative of the cartesian coordinates $x_1 = (l/2) - e - e\sin\varphi , z_1 = - H - e\cos\varphi + w ,$ $x_2 = (l/2) + e + e\sin\varphi , z_2 = z_1$ (and considering the stationary rotations and the angle $\varphi = vt$)

$$\dot{x}_1 = - ev\cos\varphi , \; \dot{z}_1 = ev\sin\varphi + \dot{w} , \; \dot{x}_2 = ev\cos\varphi , \; \dot{z}_1 = \dot{z}_2 ,$$ (10.57)

$$v_1^2 = \dot{x}_1^2 + \dot{z}_1^2 = (ev)^2 + 2ev\dot{w}\sin\varphi + \dot{w}^2 = v_2^2 .$$

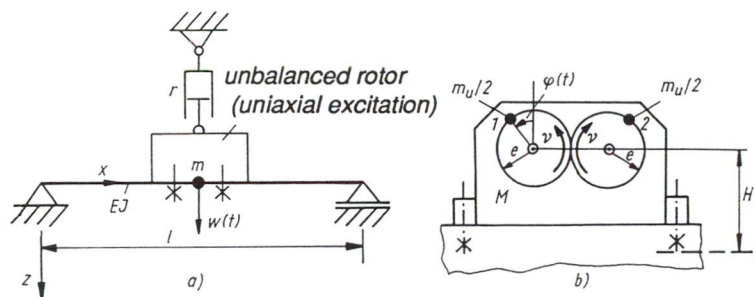

Fig. 10.8. (a) Resonance testing of a beam with concentrated mass m.
(b) Sketch of an unbalanced rotor to achieve a one-dimensional
time-harmonic excitation in the vertical direction

Since the rotation is assumed to be stationary, the virtual work is simply given by

$$Q_w \, \delta w = -\frac{\partial U_B}{\partial w} \, \delta w - r \dot{w} \, \delta w \, , \quad U_B = \frac{c}{2} \, w^2 \, , \quad c = 48 \, EJ \, / \, l^3 \, .$$
(10.58)

Bending rigidity is assumed constant over the whole span. Taking the proper derivatives of the kinetic energy renders the equation of motion [the forcing term is proportional to the squared angular speed, Eq. (7.135), with period $\tau_e = 2\pi/v$]

$$(m + M + m_u) \, \ddot{w} + r \dot{w} + c \, w = -m_u \, e \, v^2 \cos vt \, .$$
(10.59)

Dynamic deflection *w(t)* is measured against the equilibrium configuration where the system is at rest under its own weight. Damping is usually small and care has to be taken when driving the system into resonance; $e \, m_u$ is adjustable. Most important, the damping coefficient is identified by considering the resonance curve, Eq. (7.94). In addition to the maximal amplitude a_c , the forcing frequencies v_1 and v_2 are recorded, where the amplitude takes on the value $a_c \, /\sqrt{2}$, by the so-called bandwidth method. It is easy to show that the critical damping coefficient is approximately given by the ratio of the difference to the sum of these measured frequencies

$$0 < \zeta \cong \frac{v_2 - v_1}{v_2 + v_1} \ll 1 \, .$$
(10.60)

Equation (10.60) is quite insensitive to a small shift on the frequency axis caused by any error in measuring the amplitude at resonance; see Fig. 7.17(a).

10.7. Exercises A 10.1 to A 10.3 and Solutions

A 10.1:   The roof structure of Fig. A 10.1 consists of two homogeneous and rather rigid member rods of length l and with a given mass supported by a linear elastic tendon spring of stiffness c. Only vertical reaction forces act at the supports when the structure is statically loaded by its own weight and the single force F, as well as when it vibrates after any disturbance of the state at rest. The closed-loop triangle is considered the basic element of a truss. Set up the linearized equation of motion for small natural vibrations about the equilibrium configuration in the form of the *Lagrange* equation; damping may be neglected in this exercise.

Solution: The triangle remains isosceles during deformation and the single (angular) coordinate $\alpha(t)$ determines the configuration. The kinetic energy of the two member bars is (the tendon is assumed to be mass-less)

$$T = T_1 + T_2 = \frac{m}{2} v_{S1}{}^2 + \frac{I_{S1}}{2} \dot{\alpha}^2 + \frac{I_A}{2} \dot{\alpha}^2 , \quad I_{S1} = \frac{m\,l^2}{12} , \quad I_A = \frac{m\,l^2}{3} ,$$

$$v_{S1}{}^2 = \dot{x}_{S1}{}^2 + \dot{y}_{S1}{}^2 = \frac{l^2}{4} \dot{\alpha}^2 \left(1 + 8 \sin^2 \alpha\right) .$$

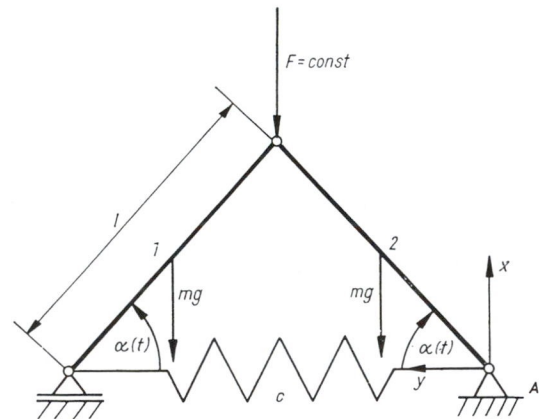

Fig. A 10.1. An isolated portion of a roof structure

The potential energy $V = W_P + U$ measured with respect to the unstressed configuration is $W_P = (mg + F) \, l \sin \alpha$ and $U = 2cl^2 (\cos \alpha - \cos \alpha_0)^2$. It is a function of $\alpha(t)$. The nonlinear *Lagrange* equation of motion thus contains the gravity

$$\frac{2m}{3} \, l \ddot{\alpha} \left(1 + 3 \sin^2 \alpha \right) + m \, l \dot{\alpha}^2 \sin 2\alpha - 4 \, c \, l \left(\cos \alpha - \cos \alpha_0\right) \sin \alpha$$

$$+ \left(F + mg\right) \cos \alpha = 0.$$

The equilibrium configuration $\alpha = \alpha_s$ is determined by solving the static part of the dynamic equation

$$- 4c \, l \left(\cos \alpha - \cos \alpha_0\right) \sin \alpha + \left(F + mg\right) \cos \alpha = 0 \; .$$

Small vibrations are considered by putting $\alpha = \alpha_s + \varepsilon$ and requiring $|\varepsilon| \ll 1$. Linearization with respect to ε renders

$$\ddot{\varepsilon} + \omega_0^2 \, \varepsilon = 0 \; , \quad \omega_0^2 = \gamma c / 2m \; ,$$

$$\gamma = 12 \left\{ \sin^2 \alpha_s - \left(1 - \frac{\cos \alpha_0}{\cos \alpha_s}\right) \cos 2\alpha_s \right\} / \left(1 + 3 \sin^2 \alpha_s\right).$$

Note that $\omega_0 = 0$ determines the critical configuration $\alpha_s = \alpha_c$ at the onset of snap-through and the condition of equilibrium for that critical value of α determines the critical load (see Sec. 9.1.3).

A 10.2: In Fig. A 10.2, a usually hand-held vibrator used for densifying soil is sketched when resting on the simplest possible linear dynamic model of the ground. An internal periodic force $F(t)$ is applied between the piston m_1 and the cylinder m_2. Derive the equation of motion under the condition of no uplift.

Solution: By taking proper derivatives of the kinetic energy,

$$T = \frac{m_1}{2} \, \dot{x}_1^2 + \frac{m_2 + m_3}{2} \, \dot{x}_2^2 \; ,$$

and by the determination of the generalized forces from virtual work,

$$Q_1 \, \delta x_1 + Q_2 \, \delta x_2 = F \, \delta x_1 - F \, \delta x_2 - r_1 \left(\dot{x}_1 - \dot{x}_2\right)\left(\delta x_1 - \delta x_2\right) - c_1 \left(x_1 - x_2\right)\left(\delta x_1 - \delta x_2\right)$$

$$- r_2 \dot{x}_2 \, \delta x_2 - c_2 x_2 \, \delta x_2 + m_1 g \, \delta x_1 + \left(m_2 + m_3\right) g \, \delta x_2 \; ,$$

the *Lagrange* equations of motion of the two-degree-of-freedom system are set up in the coupled form

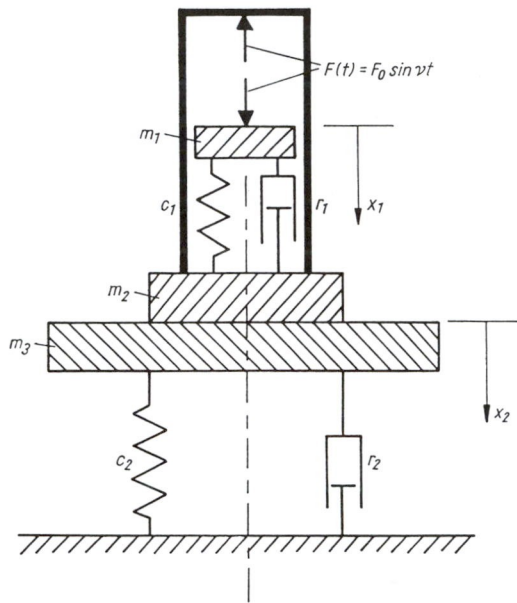

Fig. A 10.2. Model of a soil compactor. No uplift is considered

$$m_1 \ddot{x}_1 + r_1 (\dot{x}_1 - \dot{x}_2) + c_1 (x_1 - x_2) = m_1 \, g + F(t) \, ,$$

$$(m_2 + m_3) \, \ddot{x}_2 - r_1 \, \dot{x}_1 + (r_1 + r_2) \, \dot{x}_2 - c_1 \, x_1 + (c_1 + c_2) \, x_2 = (m_2 + m_3) \, g - F(t) \, .$$

A 10.3: Determine the equation of motion of a plane rigid pendulum of mass m_2 with a nonuniformly rotating point support A, as shown in Fig. A 10.3. Calculate the linear natural frequency of small vibrations under the condition of stationary rotation of the carrier disk: $d\varphi/dt = \Omega = const$. The pendulum can be designed as an dynamic absorber in the case of unwanted torsional vibrations of the carrier disk of mass m_1 , which are superposed on the stationary rotation of a machine. The external moment in the free-body diagram in that event becomes $M(t) = -k \, \theta + M'(t)$, where $\theta = \varphi - \Omega t$ is the relative angle of rotation (see Fig. A 10.3) and $M'(t) = M_0 \cos vt$, the forcing moment, is assumed to be time-harmonic. Determine the tuning of such a *Sarazin* pendulum in the absence of (dissipating) damping.

Solution: The system is considered to have two degrees of freedom and the generalized coordinates selected are the angles φ and ψ. The kinetic energy is the sum

$$T = \frac{m_1 \, i_1^2}{2} \, \dot{\varphi}^2 + \frac{m_2}{2} \, v_2^2 + \frac{m_2 \, i_2^2}{2} \left(\dot{\varphi} + \dot{\psi} \right)^2 \, , \quad v_2^2 = \dot{x}_2^2 + \dot{y}_2^2 \, ,$$

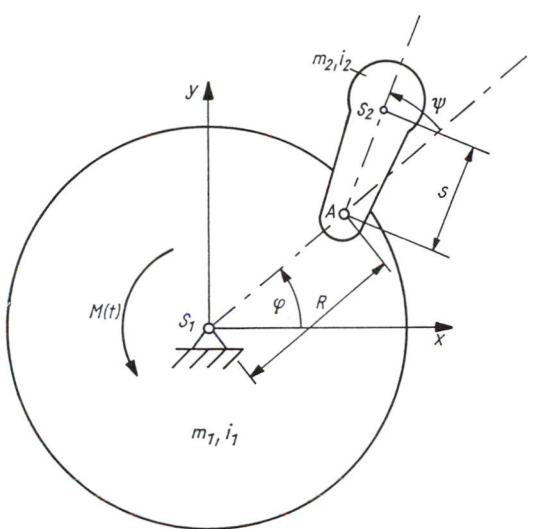

Fig. A 10.3. *Sarazin* pendulum with vertical axis: Absorber of torsional vibrations

where

$$x_2 = R \cos \varphi + s \cos(\varphi + \psi) \ , \quad y_2 = R \sin \varphi + s \sin(\varphi + \psi) \ ,$$

$$\dot{x}_2 = -R \dot{\varphi} \sin \varphi - s(\dot{\varphi} + \dot{\psi}) \sin(\varphi + \psi) \ , \quad \dot{y}_2 = R \dot{\varphi} \cos \varphi + s(\dot{\varphi} + \dot{\psi}) \cos(\varphi + \psi).$$

Considering the virtual work $\delta W = M \, d\varphi = Q_\varphi \, \delta\varphi + Q_\psi \, \delta\psi$ and comparing coefficients give $Q_\varphi = M$ and $Q_\psi = 0$. Nonlinear *Lagrange* equations of motion result

$$\left(m_1 i_1{}^2 + m_2 i_0{}^2 + 2m_2 \, Rs \cos \psi\right) \ddot{\varphi} + m_2 \left(i_A{}^2 + Rs \cos \psi\right) \ddot{\psi}$$

$$- m_2 \, Rs \left(2\dot{\varphi}\dot{\psi} + \dot{\psi}^2\right) \sin \psi = M(t) \ ,$$

$$i_A{}^2 \ddot{\psi} + \left(i_A{}^2 + Rs \cos \psi\right) \ddot{\varphi} + Rs \, \dot{\varphi}^2 \sin \psi = 0 \ , \quad i_A{}^2 = i_2{}^2 + s^2 \ ,$$

$$i_0{}^2 = i_A{}^2 + R^2 \ .$$

Putting $d\varphi/dt = \Omega = const$ and linearizing under the condition $|\psi| \ll 1$ give an oscillator equation and render the linear natural frequency of small vibrations about the radially straight configuration $\psi = 0$

$$\ddot{\psi} + \omega^2 \psi = 0 \ , \quad \omega = \Omega \sqrt{Rs} / i_A \ .$$

The condition $\Omega = const$ requires a moment of approximately $M \approx [-m_2 (i_A{}^2 + Rs) \omega^2 \psi]$. Linearization in the case of small torsional vibrations $|\theta| \ll 1$ yields, however,

$$\left(m_1 i_1{}^2 + m_2 i_0{}^2 + 2m_2 Rs\right)\ddot{\theta} + m_2 \left(i_A{}^2 + Rs\right)\ddot{\psi} = -k\theta + M'(t) \ ,$$

$$\ddot{\psi} i_A{}^2 + \left(i_A{}^2 + Rs\right)\ddot{\theta} + Rs\, \Omega^2 \psi = 0 \ .$$

In the case of a time-harmonic forcing moment $M'(t)$, particular solutions are $\theta = A \cos vt$ and $\psi = B \cos vt$, and substitution renders the linear but inhomogeneous set of equations

$$\left\{k - v^2 \left(m_1 i_1{}^2 + m_2 i_0{}^2 + 2m_2 Rs\right)\right\} A - v^2 m_2 \left(i_A{}^2 + Rs\right) B = M_0 \ ,$$

$$- v^2 \left(i_A{}^2 + Rs\right) A + \left(Rs\, \Omega^2 - v^2 i_A{}^2\right) B = 0 \ .$$

Elimination of the torsional vibration, $\theta \equiv 0$, for a discrete and assigned value of the forcing frequency v requires that the amplitude be zero, $A = 0$. Since $B \neq 0$, its coefficient must vanish: $i_A{}^2 = Rs\, \Omega^2 / v^2$. Tuning of the dynamic absorber thus requires that the linear natural frequency be equal to the discrete value of the forcing frequency $\omega = v$. See also Fig. 7.31. The pendulum's motion is antiphase with the forcing moment with the linear period $2\pi/v$ and amplitude $B = - M_0 / m_2 Rs\, v^2 [1 + (\Omega/v)^2]$.

1 1
Some Approximation Methods of Dynamics and Statics

The deformed configuration of a body, in general, is determined by the field of displacement vectors $\boldsymbol{u} = \boldsymbol{u}(x, y, z, t; X, Y, Z)$, ie a continuum possesses an infinite number of degrees of freedom. The basic partial differential equations of such distributed parameter systems, with associated boundary and initial conditions, even in the case of linear elastic solids, but with non-simple geometry, can hardly be solved in an exact manner. Two classes of approximation techniques are commonly used to overcome these difficulties: (1) The essential boundary conditions are built into the approximation that is not a solution of the basic differential equations. The *Rayleigh-Ritz-Galerkin* method based on such a set of admissible functions is discussed below, and examples of discretization by the finite-element method (FEM) are given. (2) A solution of the basic equations is known that takes on the prescribed boundary values at a number of discrete points only. Such an approximation is the output of the class of collocation methods. The boundary element method (BEM) should be mentioned here. Collocation methods are not discussed in this text any further.

The motion of the deformable body is projected onto a finite functional space such that the resulting MDOF-system reflects the characteristic dynamic properties. In the case of linear problems, such a MDOF equivalence must have closely approximated eigenvalues, as the first few natural periods of a vibrational system must be nearly the same, or the critical buckling load of a bifurcating static problem must nearly be coincident. The fields of deformation and stress in space have to be approximated in the mean square sense with (preferred) or without convergence. Convergence is considered with respect to the increase of the number of degrees of freedom of the equivalent system

$$\lim_{n \to \infty} \int_V (w - w^*)^2 \, dV \to 0 \, .$$

The application of the *Galerkin* procedure to the *Lagrange* equations of motion renders an approximate set of (nonlinear) equations for the amplitudes, which, in special cases, is more economically solved by numerical methods of algebraic problems than the original system of differential equations. Finally, the solution technique applicable to vibrational systems, called the harmonic balance, is discussed. The equations of motion that result from a FEM approximation of a dynamic problem are often solved numerically by means of the *Wilson* θ-method (method of linear acceleration) in an appropriate time-stepping manner. Therefore, a short account is given in Sec. 11.6 for a SDOF-system.

11.1. The *Rayleigh-Ritz-Galerkin* Approximation Method

The basic idea is to approximate the mechanical field, eg the displacements, say, $w(x, y, z, t)$, of a deformable body by a finite series of functions separable in space and time, the *Ritz* approximation

$$w^*(x, y, z, t) = \sum_{k=1}^{n} q_k(t)\, \varphi_k(x, y, z) \quad ,$$

(11.1)

where $q_k(t)$ are n generalized coordinates of the MDOF equivalent system of the continuum. The finite set of functions $\varphi_k\,(x, y, z)$, $k = 1, 2, ..., n$, is properly selected if the essential boundary conditions are built in. Each of the functions must necessarily comply with the geometric boundary conditions and should as far as possible take into account any dynamic boundary conditions as well. Specifically, Eq. (11.1) may be the approximation of the deflection of a beam or a plate. In that case, dependence on the lateral coordinate z is to be suppressed, and for a beam, it suffices to consider dependence on the axial coordinate x only. The generalized coordinates that are constants in static problems have to be determined in a sense of best fit. There are at least two ways of determining the equations of motion of the "optimal" equivalent MDOF-system after properly selecting the spatial functions. For *Rayleigh´s quotient* see Sec. 8.4.2.

11.1.1. The Rayleigh-Ritz Method and the Lagrange Equations of the Equivalent MDOF-System

The kinetic energy of the body under consideration is given by the definition (8.3). By substituting the *Ritz* approximation of the type of Eq. (11.1), it becomes a function of the generalized velocities and possibly also of the generalized coordinates and time. In the case of irrotational internal and external forces, potential energy exists, $V = U + W_P$, which becomes a function of the generalized coordinates (and possibly of time) after substitution of the *Ritz* approximation. Hence, the *Lagrange* equations of motion of a MDOF-system, Eq. (10.14), render the set of ordinary differential equations

$$\frac{d}{dt}\left(\frac{\partial T}{\partial \dot{q}_k}\right) - \frac{\partial T}{\partial q_k} + \frac{\partial V}{\partial q_k} = 0 \ , \ k = 1, 2, ..., n \ ,$$

(11.2)

$$T = T(\dot{q}_1, \dot{q}_2, .., \dot{q}_n; q_1, q_2, .., q_n; t) \ .$$

Forces that are not derived from a potential, but do contribute to the virtual work through the virtual displacements of their points of application are considered by generalized forces and, thus, by equivalence of that virtual work. For a single force, see Eq. (10.6). With Eq. (11.1), the field of virtual variations is expressed by

$$\delta w^* = \sum_{k=1}^{n} \delta q_k \ \varphi_k(x, y, z) \ , \ \ t = const \ .$$

In the case of static deformations, $T \equiv 0$, and the finite set of equilibrium conditions results [see Eqs. (5.9) and (5.15)]

$$\frac{\partial V}{\partial q_k} = 0 \ , \ or, more general, \ \ Q_k = 0 \ , \ k = 1, 2, .., n \ .$$

(11.3)

The projection of the distributed parameter system renders an approximate MDOF-system of the best possible fit within the a priori selected *Ritz* approximation (11.1). The statement is verified since in the case of a dynamic problem, *D´Alembert´s* principle when applied to the equivalent system gives the *Lagrange* equations of motion, Eq. (11.2), and in the case of a static problem, the principle of virtual work renders the conditions of equilibrium, Eq. (11.3). Personal experience must be gained by solving simple problems first. Examples are given below.

11.1.2. The Galerkin Procedure

It is assumed that the (partial) differential equations of the distributed parameter system are given

$$D\{w\} = 0 \ ,$$

(11.4)

where D in symbolic notation is the (nonlinear) partial differential operator according to an infinite number of degrees of freedom. Any external forcing functions are shifted to the left-hand side and included in the operational form of Eq. (11.4). Analogous to *D´Alembert´s* principle (10.3), Eq. (11.4) is multiplied by the virtual variation δw of the instant configuration and integrated over the domain of definition of the mechanical field w (in general, the body volume B)

$$\int_B D\{w\}\, \delta w\, dB = 0 \quad .$$

(11.5)

Substituting the *Ritz* approximation (11.1) for the field w in Eq. (11.4) yields an error p^*, which in the case of a proper mechanical equation (11.4) can be interpreted as fictitious loading of the given body,

$$D\{w^*\} = p^* \neq 0 \quad .$$

(11.6)

In rare cases, where all the boundary conditions are built into the *Ritz* approximation, it suffices that those fictitious loadings are self-equilibrating in the course of time. The principle of virtual work (5.9) thus requires

$$\int_B p^*\, \delta w^*\, dB = 0 \quad .$$

(11.7)

If, however, in the *Ritz* approximation, the dynamic boundary conditions are not properly taken care of, in addition to the error p^*, fictitious surface tractions Q_w^* are to be considered. In that case, it must be required that the whole system of fictitious forces be self-equilibrating and Eq. (11.7) is generalized to include the additional virtual work of the fictitious surface tractions

$$\int_B p^*\, \delta w^*\, dB + \oint_{\partial B} Q_w^*\, \delta w^*\, dS = 0 \quad .$$

(11.8)

The variation of the *Ritz* approximation

$$\delta w^* = \sum_{k=1}^n \varphi_k(x, y, z)\, \delta q_k \quad , \quad t = \text{const} \quad ,$$

(11.9)

is substituted and the virtual variations of the generalized coordinates are assumed to be independent (the equivalent system is holonomic). Hence, each of the coefficients of δq_k must vanish individually and *Galerkin's* rule in its general form

$$\int_B p^*\, \varphi_k\, dB + \oint_{\partial B} Q_w^*\, \varphi_k\, dS = 0 \quad , \quad k = 1, 2, .. , n \quad ,$$

(11.10)

yields n equations of motion expressed in the generalized coordinates. The classical form applies if all boundary conditions are considered in the *Ritz* approximation. In that case, $Q_w{}^* \equiv 0$, and *Galerkin's* rule is reduced to

$$\int_B p^* \, \varphi_k \, dB = 0 \quad , \quad k = 1, 2, .. ,n \quad .$$

(11.11)

Equation (11.11) determines the scalar product of two functions to be zero: <<The error made by substituting the *Ritz* approximation in the basic partial differential equation, the function p^*, is thus orthogonal to each of the functional elements φ_k in the domain B.>> That is, the error is minimized with respect to the given set of spatial functions in the *Ritz* approximation. The projection according to the *Galerkin* rule renders the *Lagrange* equations of motion (11.2) if the operator D is self-adjoint and if the *Ritz* approximation is common to the procedures as discussed in Sec. 11.1.1 and 11.1.2. Given the basic partial differential equations, illustrated by Eq. (11.4), *Galerkin's* rule may be judged more convenient in applications when compared to the original *Ritz* procedure for the determination of the set of (nonlinear) equations of motion. Moreover, it can be applied to this system of time differential equations as well to reduce it to an algebraic problem.

11.1.3. Complete Algebraization of the Lagrange Equations of Motion

Equations (10.14) are, in general, nonlinear time differential equations. Their operational form is

$$L_k (q_1, q_2, .. , q_n; t) = 0 \quad , \quad k = 1, 2, ... , n \quad .$$

(11.12)

In some cases, it is possible to find sets of proper functions $\psi_{kj} (t)$ such that the generalized coordinate $q_k (t)$ has assigned a generalized *Ritz* approximation,

$$q_k{}^*(t) = \sum_{j=1}^{m_k} a_{kj} \, \psi_{kj}(t) \quad , \quad k = 1, 2, ... , n \quad .$$

(11.13)

Initial conditions and also terminal conditions at some time instant $t = \tau$, must be the same for the actual motion and its approximation. Under these quite restrictive conditions, *Galerkin's* rule (11.11) is generalized to

$$\int_0^\tau L_k (q_1{}^*, q_2{}^*, .. ,q_n{}^*; t) \, \psi_{kj}(t) \, dt = 0 \quad , \quad j = 1, 2, ... , m_k \quad , \quad k = 1, 2, ..., n \quad .$$

(11.14)

The resulting (nonlinear) algebraic equations of the unknown constants a_{kj} are numerically solved. The application of this approximation strategy is more or less successful only for vibrational problems with period τ given in advance. In the case of weak nonlinearities and free vibrations, the period may be approximated by the constant linear natural period: The approximation is then derived within the limits of asymptotic methods. Since the difficulties of solving the nonlinear equations remain, the method of harmonic balance (see Sec. 11.5) is preferable. In engineering practice, incremental numerical methods prevail, an example is given in Sec. 11.6.

11.1.4. Forced Vibrations of a Nonlinear Oscillator

The nonlinear restoring force is given in symmetric form by a cubic polynomial, $F = - (cx + bx^3)$. A mass m attached to the spring is forced time-harmonically by the external force $S \cos vt$. If we neglect any damping, the equation of motion becomes

$$m \ddot{x} = -(cx + bx^3) + S \cos vt \ .$$

(11.15)

That *Duffing* equation is rewritten in standard form

$$D\{x\} = \ddot{x} + \omega^2 x + \beta x^3 - \frac{S}{m} \cos vt = 0 \ , \quad \omega^2 = c/m \ , \quad \beta = b/m \ .$$

(11.16)

A stationary periodic solution of period $2\pi/v$ is expected and a proper *Ritz* approximation is, eg a harmonic function of time,

$$x^*(t) = a \ \psi(t) \ , \quad \psi(t) = \cos vt \ .$$

(11.17)

Subharmonic vibrations, eg with the frequency $v/3$, which are possible for discrete values of the forcing frequency are not considered here. *Galerkin's* rule (11.14),

$$\int_0^{\tau = 2\pi/v} D\{x^*(t)\} \ \psi(t) \ dt = 0 \ , \quad \psi(t) = \cos vt \ ,$$

(11.18)

renders upon (analytic) integration the following cubic equation:

$$\frac{3\beta}{4} a^3 + (\omega^2 - v^2) a - \frac{S}{m} = 0 \ .$$

(11.19)

Putting $S = 0$ gives an approximation of the natural vibrations and Eq. (11.19) determines the backbone curve that is the relation between the eigenfrequency and the amplitude, $v_e = v_e (a)$. Hence, within this approximation

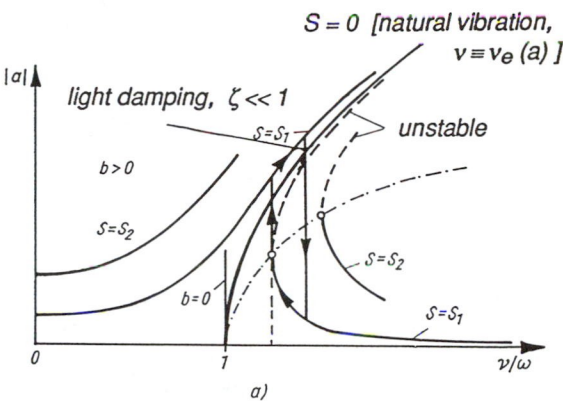

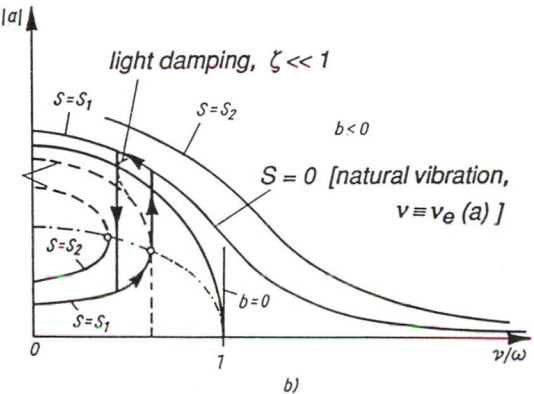

Fig. 11.1. Resonance curves of an oscillator with hardening (a) and with softening (b) spring. The jump phenomenon is indicated by vertical lines with arrows

$$v_e = \sqrt{\omega^2 + \frac{3\,\beta}{4}\,a^2} \;.$$
(11.20)

In the case of a hardening spring, the eigenfrequency is increased with respect to the linear natural frequency ω , and a softening spring renders a system where the eigenfrequency is progressively decreased.

The approximating resonance curve is more easily determined by solving Eq. (11.19) inversely

$$\left(\frac{v}{\omega}\right)^2 = 1 - \frac{S}{c\,a} + \frac{3\,b}{4\,c}\,a^2 ,$$
(11.21)

sketches are given in Fig. 11.1. The jump phenomenon that is encountered by sweeping the forcing frequency is indicated . Actually, a violent transient vibration is observed when the amplitudes of the steady states are abruptly changing.

11.2. Illustrations of Linearized Elastic Systems with Heavy Mass and Soft Spring, SDOF Equivalent System

Those structures that vibrate mainly in their basic mode when periodically forced may be projected onto a SDOF equivalent oscillator by selecting a proper one-term *Ritz* approximation. The output of the *Ritz-Galerkin* procedure is a single approximating equation of motion. Thus, it suffices to determine a proper equivalent mass (usually the given heavy mass plus a portion of the mass of the elastic spring) and an associated effective stiffness (in the case of a linear problem, a constant). If a single force acts on the concentrated mass, it directly enters (unchanged) the equivalent system.

11.2.1. Longitudinal Vibrations

At the end of a linear elastic rod of length l_0 , a heavy mass m is attached. If we consider the mass per unit of length ρA , as well as the tensile stiffness EA of the spring, constant (for convenience of integration), an approximation of the equation of motion of the longitudinal vibration in the first natural mode can be determined. According to the support shown in Fig. 11.2, the displacement field $u(x, t)$ increases in that mode monotonically from the fixed end over x toward the sliding mass. An admissible *Ritz* approximation can always be given affined to a proper static deformation. Hence, it suffices to choose a linear function in x

$$u^*(x, t) = q(t)\,\varphi(x) \quad, \quad \varphi(x) = x/l_0 \quad, \quad \varphi(x = l_0) = 1 \quad. \tag{11.22}$$

With that normalization, the generalized coordinate $q(t)$ is well illustrated as the measure of the translational motion of the heavy mass, and the kinetic energy becomes simply

$$T = \frac{m\dot{q}^2}{2} + \frac{1}{2}\int_0^{l_0} \dot{u}^{*2}(x, t)\,\rho A\,dx = \frac{1}{2}\left(m + \frac{m_S}{3}\right)\dot{q}^2 ,$$

$$\dot{u}^* = \dot{q}\,\varphi(x) \quad, \quad m_S = \rho A\,l_0 \quad. \tag{11.23}$$

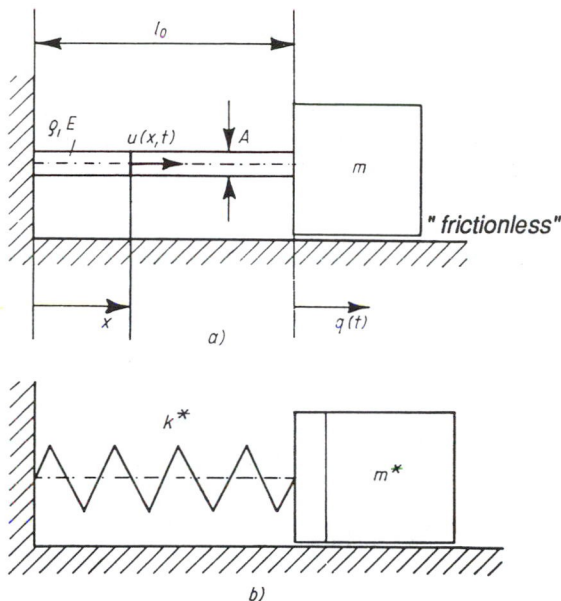

Fig. 11.2. Longitudinal vibrations: (a) Distributed parameter system.
(b) Equivalent system, $m^* = m + m_S /3$, $k^* = EA/l_0$

Thus, the kinetic energy is equivalent to that of a single equivalent mass $m^* = m + m_S /3$, moving with the speed of the heavy mass m . One-third of the homogeneously distributed mass of the spring is to be added. The potential energy is approximated by the strain energy of the spring, deforming according to Eq. (11.22),

$$V = U = \frac{1}{2} \int_0^{l_0} EA \left(\frac{\partial u^*}{\partial x}\right)^2 dx = \frac{1}{2} \frac{EA}{l_0} q^2 \quad , \quad \frac{\partial u^*}{\partial x} = q(t) \frac{\partial \varphi}{\partial x} \quad . \tag{11.24}$$

The effective stiffness $k^* = EA/l_0$ of a mass-less equivalent spring equals the static stiffness, Eq. (3.40), due to the linear *Ritz* approximation. The *Lagrange* equation of motion is that of the linear oscillator shown in Fig. 11.2(b),

$$\ddot{q} + \omega_0^2 q = 0 \quad , \quad \omega_0 = \sqrt{k^*/m^*} \quad . \tag{11.25}$$

The basic natural period of the actual spring mass system is approximated by $2\pi/\omega_0$. The eigenfrequency is approximated from above; the system with an infinite number of degrees of freedom becomes stiffer when projected onto a SDOF-system.

11.2.2. Bending Vibrations

A simply supported single-span and linear elastic beam is considered. The span is denoted l and a homogeneous distribution of mass ρA and bending stiffness $B = EJ$ (for convenience of integration) is assumed. A heavy mass m is supported at midspan and the basic mode of vibration must be convex as shown in Fig. 11.3. A sine half-wave is a proper choice in the *Ritz* approximation, also in the case of a variable cross-section,

$$w^*(x, t) = q(t)\, \varphi(x) \quad , \quad \varphi(x) = \sin\frac{\pi x}{l} \quad , \quad \varphi(x = l/2) = 1 \quad . \tag{11.26}$$

The kinematic as well as dynamic boundary conditions are considered in the *Ritz* approximation, $w = \partial^2 w/\partial x^2 = 0$ at $x = 0, l$. For a slender beam, the rotational energy of a "cross-section" is negligible and the kinetic energy is approximated by

$$T = \frac{m\,\dot{q}^2}{2} + \frac{1}{2}\int_0^l \dot{w}^{*2}(x, t)\, \rho\, A\, dx = \frac{1}{2}\left(m + \frac{m_S}{2}\right)\dot{q}^2(t)\,, \tag{11.27}$$

$$\dot{w}^* = \dot{q}\,\varphi(x)\,, \; m_S = \rho\, A\, l\,.$$

When we consider small amplitude vibrations about the equilibrium configuration and neglect the shear deformations in the potential energy, Eq. (5.17) renders the approximation

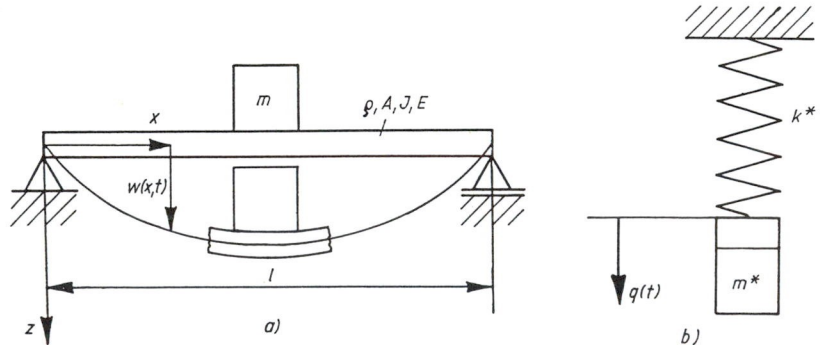

Fig. 11.3. Bending vibrations: (a) Basic mode of the distributed parameter system. (b) Equivalent SDOF-oscillator

$$V = U = \frac{1}{2} \int_0^l EJ \left(\frac{\partial^2 w^*}{\partial x^2} \right)^2 dx = \frac{1}{2} \frac{\pi^4 \, EJ}{2 \, l^3} \, q^2(t) \; ,$$

(11.28)

$$\frac{\partial^2 w^*}{\partial x^2} = q(t) \frac{\partial^2 \varphi}{\partial x^2} \; , \; k^* = \frac{\pi^4 \, EJ}{2 \, l^3} \; .$$

It can be concluded that the equivalent system of Fig. 11.3(b) has an effective mass $m^* = m + m_S/2$ (half of the mass of the beam must be added consistently to the heavy mass m) and effective stiffness of the mass-less spring of k^*. Note that in such a case, the static stiffness $c = F/w_0 = 48 \, EJ/l^3 < k^*$. The *Lagrange* equation of motion in terms of the midspan deflection is approximated by

$$\ddot{q} + \omega_0^2 \, q = 0 \; , \; \omega_0 = \sqrt{k^*/m^*} \; .$$

(11.29)

Affinity between the *Ritz* approximation and the basic mode of free flexural vibrations of the beam without the additional mass m is recognized. Hence, $lim_{m \to 0} \, \omega_0 = \omega_{0exact} \vert_{m = 0}$, but, $lim_{m_S \to 0} \, \omega_0 > \sqrt{(c/m)}$, $c = 48 \, EJ/l^3$; the error is rather small. Alternatively, a polynomial *Ritz* approximation is selected that is affined to the static deflection under a single force loading, $\varphi(x) = (3x/l) \, (1 - 4x^2/3l^2)$, $\varphi(x = l/2) = 1$, $0 \le x \le l/2$ (symmetric). Consistent mass $m^* = m + 17 \, m_S/35$ and $k^* = c$. In that case, $lim_{m_S \to 0} \, \omega_0 = \sqrt{(c/m)}$, but, $lim_{m \to 0} \, \omega_0 > \omega_{0exact}\vert_{m = 0} = \pi^2 \, \sqrt{(EJ/m_S \, l^3)}$. Again, the error is small. It may be concluded from the above considerations that, in general, the natural frequency is insensitive against any variation in the shape of the convex *Ritz* approximation.

11.2.3. Torsional Vibrations

A linear elastic shaft of length l is rigidly clamped at one end and a rigid disk with mass properties $m, \, I_S = m i_S^2$, is attached to the other end; see Fig. 11.4(a). The free torsional vibrations are considered under the assumption that the circular cylindrical rod is homogeneous (for convenience of integration). The affinity of the basic natural mode to the static deformation under a torque applied to the disk renders an admissible *Ritz* approximation

$$\chi^*(x, t) = q(t) \, \varphi(x) \; , \; \varphi(x) = x/l \; , \; \varphi(x = l) = 1 \; .$$

(11.30)

The generalized coordinate $q(t)$ equals the angle of rotation of the disk, $\alpha(t)$. The kinetic energy is especially evaluated for the circular cylindrical shaft, where $A = \pi R^2$, $dI_x/dx = \rho A \, R^2/2$,

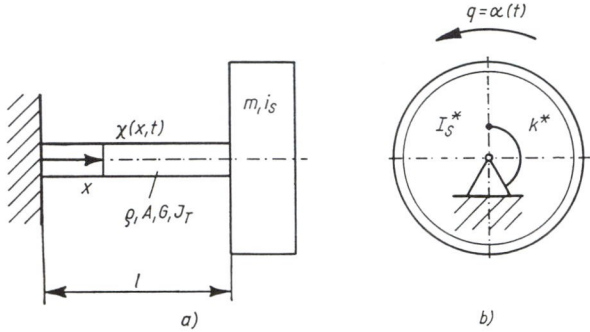

Fig. 11.4. Torsional vibrations: (a) The distributed parameter system. (b) The equivalent torsional oscillator with a mass-less restoring spring of angular stiffness k^*

$$T = \frac{m\,i_S^2\,\dot{\alpha}^2}{2} + \frac{1}{2} \int_0^l \dot{\chi}^{*2}(x,t)\frac{dI_x}{dx}\,dx = \frac{1}{2}\left(m\,i_S^2 + \frac{1}{3}\,m_S\,\frac{R^2}{2}\right)\dot{\alpha}^2(t)\,,$$

$$\tag{11.31}$$

$$\dot{\chi}^* = \dot{\alpha}\,\varphi(x)\,,\quad I^* = \left(m\,i_S^2 + \frac{1}{3}\,m_S\,\frac{R^2}{2}\right).$$

The potential energy equals the strain energy of *Saint Venant* torsion [see Eqs. (5.51) and (6.104)]

$$V = U = \frac{1}{2}\int_0^l GJ_T\left(\frac{\partial\chi^*}{\partial x}\right)^2 dx = \frac{1}{2}\frac{GJ_T}{l}\,\alpha^2,\ J_T = J_p = A\,R^2/2\,,\ k^* = GJ_T/l\,.$$

$$\tag{11.32}$$

The equivalent system is a SDOF torsional oscillator: Its rigid disk has an apparent moment of inertia I^* and a mass-less restoring angular spring of stiffness k^*. The *Lagrange* equation of motion in the angular coordinate becomes simply

$$\ddot{\alpha} + \omega_0^2\,\alpha = 0\,,\quad \omega_0 = \sqrt{k^*/I^*}\,,\tag{11.33}$$

where the angular stiffness equals the static one, $k^* = c = M/\chi_0 = GJ_T/l$.

11.2.4. Single-Story Frame

The beam (cross bar) of heavy mass m is assumed to be rigid (to keep the dynamic system as simple as possible), and it is supported by two clamped-clamped, homogeneous, and slender linear elastic

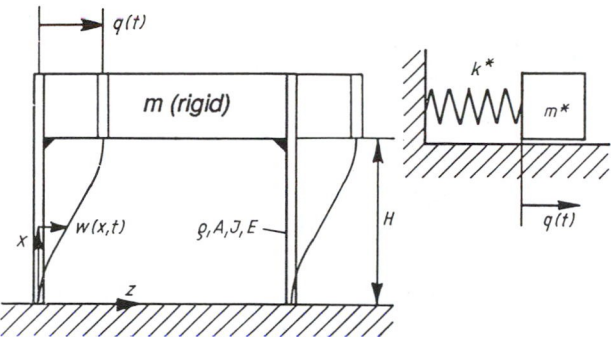

Fig. 11.5. Basic natural vibration of a framed structure. Distributed parameter system and its equivalent oscillator

columns. See Fig. 11.5. The redundant frame vibrates in its plane, and the basic mode may be approximated by the *Ritz* approximation

$$w^*(x, t) = q(t)\, \varphi(x) \ , \quad \varphi(x) = \frac{1}{2}\left(1 - \cos\frac{\pi x}{H}\right) \ , \quad \varphi(\,x = H) = 1 \ . \tag{11.34}$$

The generalized coordinate is very visible, since it measures the horizontal translation of the rigid mass. The kinetic energy is approximated by

$$T = \frac{m\dot{q}^2}{2} + 2\,\frac{1}{2}\int_0^H \dot{w}^{*2}(x, t)\, \rho\, A\, dx = m^*\,\frac{\dot{q}^2}{2} \ , \quad m^* = (m + 2\, m_1^*) \ . \tag{11.35}$$

That is, a portion of the mass of each of the columns is added to the beam,

$$m_1^* = \int_0^H \rho A\, \varphi^2(x)\, dx = \frac{3\, m_S}{8} \ , \quad m_S = \rho\, A\, H \ . \tag{11.36}$$

The columns are statically prestressed in compression by the large weight *mg* that reduces their bending rigidity. Thus, the strain energy is considered according to second-order theory [see Eq. (9.42)]; the gravity of the columns is negligible, and $N = -\ mg/2 = const$,

$$V = 2\,\frac{1}{2}\int_0^H EJ\left(\frac{\partial^2 w^*}{\partial x^2}\right)^2 dx - 2\,\frac{m\,g}{2}\,\frac{1}{2}\int_0^H\left(\frac{\partial w^*}{\partial x}\right)^2 dx = \frac{k^*\, q^2}{2} \ . \tag{11.37}$$

The effective stiffness with a constant geometric correction becomes, in the case of a constant bending rigidity,

$$k^* = 2 \int_0^H EJ \left(\frac{\partial^2 \varphi}{\partial x^2}\right)^2 dx - mg \int_0^H \left(\frac{\partial \varphi}{\partial x}\right)^2 dx = 2 \frac{\pi^4 EJ}{8 H^3} \left(1 - \frac{mg}{2} \frac{H^2}{\pi^2 EJ}\right).$$

(11.38)

The *Lagrange* equation of motion is

$$\ddot{q} + \omega_0^2 q = 0 , \qquad (11.39)$$

with the approximate value of the basic natural frequency of the single-story frame (see again Fig. 11.5)

$$\omega_0 = \sqrt{k^*/m^*} . \qquad (11.40)$$

Light material damping could be added to the modal form of Eq. (11.39). The flexibility of the beam is considered in a rough approximation by placing in a nonconsistent manner $m/2$ at the frame's corners and adding the strain energy of that cross-bar to Eq. (11.37).

11.2.5. Thin Elastic Circular Plate with a Centrally Attached Heavy Mass

Assuming the boundary $r = R$ to be rigidly clamped renders the basic mode axisymmetric. An admissible *Ritz* approximation of the deflection is, eg the polynomial function

$$w^*(r, t) = q(t)\, \varphi(r) , \quad \varphi(r) = \left(1 - \frac{r^2}{R^2}\right)^2 , \quad \varphi(r = R) = 0 , \; \varphi'(r = R) = 0 , \; \varphi(0) = 1 .$$

(11.41)

The kinetic energy of the central mass m and of the plate of constant thickness h and density ρ is approximated by substituting Eq. (11.41)

$$T = \frac{m \dot{q}^2}{2} + \frac{1}{2} \int_m \dot{w}^{*2}(r, t)\, dm = \frac{m \dot{q}^2}{2} + \frac{1}{2} \int_0^R 2\pi\, \rho h\, \dot{w}^{*2}(r, t)\, r\, dr = m^* \frac{\dot{q}^2}{2} ,$$

(11.42)

$$m^* = \left(m + \frac{m_S}{5}\right) , \quad m_S = \rho\, \pi\, R^2\, h .$$

The strain energy of the dynamic deflection determines the potential energy

$$V = \frac{1}{2} \int_0^R 2\pi\, K \left(\frac{\partial^2 w^*}{\partial r^2} + \frac{1}{r} \frac{\partial w^*}{\partial r}\right)^2 r\, dr = k^* \frac{q^2}{2} ,$$

(11.43)

$$k^* = \frac{64 \pi K}{3 R^2} \quad , \quad K = Eh^3 / 12 (1 - \nu^2) .$$

The equivalent oscillator is given by m^* and k^*. The basic natural frequency in the *Lagrange* equation of motion (11.39) is determined by substituting the effective parameters in Eq. (11.40). For a *Ritz* approximation of the simply supported circular plate see Exercise A 11.8.

11.3. Examples of Elastic Structures with Abstract Equivalent Systems

The *Ritz-Galerkin* approximation is illustrated below when the equivalent system has an MDOF or is of a completely abstract nature.

11.3.1. Free Flexural Vibrations of a Prestressed Slender Beam

The simply supported beam of Fig. 11.6 of span l under constant tensile prestress S is subject to bending vibrations. A *Ritz* approximation with admissible functions with respect to hinged boundaries is given by the finite sum

$$w^*(x, t) = \sum_{k=1}^{n} q_k(t) \, \varphi_k(x) \quad , \quad \varphi_k(x) = \sin \frac{k \pi x}{l} .$$
(11.44)

The basic linear partial differential equation of the *Bernoulli-Euler* bending theory when considering the constant normal force $N = S$ (in a second-order theory of elasticity approximation) is of the fourth order [see Eqs. (9.60) and (7.147)]. Inertia forces are substituted for the lateral load, and the equation is homogeneous

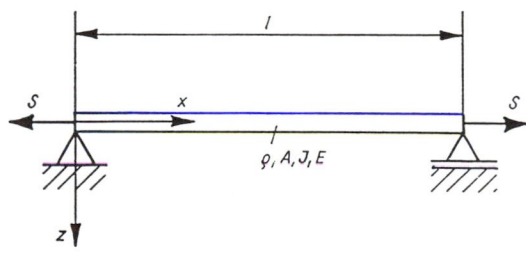

Fig. 11.6. Flexural vibrations of a prestressed beam (or string)

$$\frac{\partial^2}{\partial x^2}\left(EJ\,\frac{\partial^2 w}{\partial x^2}\right) - N\,\frac{\partial^2 w}{\partial x^2} + \rho A\,\frac{\partial^2 w}{\partial t^2} = 0 \ .$$
(11.45)

The substitution of the *Ritz* approximation renders an error that corresponds to a fictitious lateral load. Assuming the bending rigidity *EJ* to be constant (for convenience of integration) gives

$$p^*(x, t) = D\{w^*\} = \sum_{k=1}^{n}\left\{\rho A\,\ddot{q}_k + \left[\left(\frac{k\pi}{l}\right)^4 EJ + N\left(\frac{k\pi}{l}\right)^2\right]q_k\right\}\sin k\pi x \ .$$
(11.46)

In that special case, the error vanishes in the limit $n \to \infty$, since the admissible functions are the set of (orthogonal) eigenfunctions of Eq. (11.45), and Eq. (11.44) is thus just an eigenfunction expansion. *Galerkin´s* rule (11.11) requires the orthogonality of the error to each of the admissible functions. The dynamic boundary conditions are satisfied

$$\int_0^l p^*\,\varphi(x)\,dx = 0 \ , \quad k = 1, 2, ..., n \ .$$
(11.47)

Due to constant bending rigidity, integration is performed analytically considering the orthogonality conditions within the span *l*

$$\int_0^l \varphi_k(x)\,\varphi_j(x)\,dx = \frac{1}{2}\,\delta_{jk} \ , \quad \delta_{jk} = \begin{cases} 1, \ j=k \\ 0, \ j \neq k \end{cases} \ .$$
(11.48)

The resulting equations of motion determine uncoupled, so-called modal oscillators, and the linear eigenfrequencies are properly approximated from above

$$\ddot{q}_k + \omega_k^2\,q_k = 0 \ , \quad k = 1, 2, ..., n \ , \quad \omega_k = \frac{k\pi}{l}\,\sqrt{\left(\frac{k\pi}{l}\right)^2 \frac{EJ}{\rho A} + \frac{S}{\rho A}} \ .$$
(11.49)

For the initially unstressed beam, $N = S = 0$, the discrete spectrum given turns out to be exact within *Bernoulli-Euler* theory. Putting the bending stiffness $EJ = 0$ and keeping the prestress *S* fixed render a vibrating homogeneous string. Its eigenfrequencies ω_{0k} are again exactly given by the formula (11.49) with that substitution, since the admissible functions φ_k are members of the infinite set of orthogonal eigenfunctions in both of these limiting cases. Especially, the effect of a small bending stiffness *EJ* , present in any string, on its spectrum of natural frequencies is described by rearranging terms in Eq. (11.49)

$$\omega_k = \omega_{0k} \sqrt{1 + \left(\frac{k\pi}{l}\right)^2 \frac{EJ}{S}} \quad , \quad \omega_{0k} = \frac{k\pi}{l} \sqrt{\frac{l \, S}{m}} \quad , \quad m = \rho A \, l \; . \tag{11.50}$$

The eigenvalues remain valid for initial compression by an axial force $S = -F$. Substitution into Eq. (11.49) approximates the eigenfrequencies that are decreased with respect to the initially unstressed beam with constant bending rigidity

$$\omega_k = \frac{k\pi}{l} \sqrt{\frac{l}{m}} \sqrt{\left(\frac{k\pi}{l}\right)^2 EJ - F} \; . \tag{11.51}$$

When compression is increased, the basic natural frequency ω_1 is supposed to vanish at first. Free vibrations about the straight equilibrium configuration for higher prestress are no longer possible and bifurcation occurs at the critical *Euler* buckling load

$$\omega_1 \to 0: \quad F_c = \frac{\pi^2 \, EJ}{l^2} \; . \tag{11.52}$$

The buckling mode and basic mode of free vibrations exhibit affinity; see Sec. 9.1.4.

11.3.2. Buckling Load of the Euler Column on an Elastic Foundation

A linear *Winkler* foundation is considered in Fig. 11.7; it exerts a restitution per unit of length proportional to the deflection in the postbuckling range, $q_z = -kw$. A two-term *Ritz* approximation

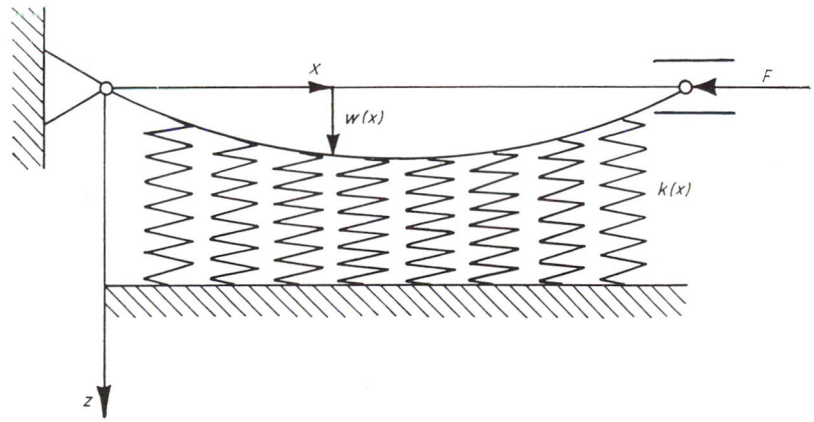

Fig. 11.7. *Euler* column on a *Winkler* foundation. Mode jumping

suffices to illustrate the phenomenon of mode jumping when the axial compression F becomes critical

$$w^*(x) = q_1 \sin \frac{\pi x}{l} + q_2 \sin \frac{2\pi x}{l} \ .$$

(11.53a)

The generalized coordinates are the constant amplitudes of the buckling modes. The potential energy includes the elastic energy of the foundation in addition to Eq. (9.42) (see Fig. 11.7)

$$V = \frac{1}{2} \int_0^l EJ \left(\frac{\partial^2 w^*}{\partial x^2} \right)^2 dx \ - \ \frac{F}{2} \int_0^l \left(\frac{\partial w^*}{\partial x} \right)^2 dx \ + \ \frac{1}{2} \int_0^l k \, w^{*2} \, dx \ .$$

(11.53b)

The orthogonality relations hold for the approximating functions that are the eigenfunctions in the case of a constant bending rigidity, EJ , and the mixed term vanishes upon integration

$$V = \frac{k_1^* \, q_1^2}{2} + \frac{k_2^* \, q_2^2}{2} \ , \quad
\begin{cases}
k_1^* = \dfrac{\pi^4 \, EJ}{2 \, l^3} - F \, \dfrac{\pi^2}{2 \, l} + \dfrac{k \, l}{2} \\[2mm]
k_2^* = \dfrac{8\pi^4 \, EJ}{l^3} - F \, \dfrac{2\pi^2}{l} + \dfrac{k \, l}{2}
\end{cases} \ .$$

(11.53c)

The conditions of equilibrium are independently derived from $\delta V = 0$

$$\frac{\partial V}{\partial q_1} = k_1^* \, q_1 = 0 \ , \quad \frac{\partial V}{\partial q_2} = k_2^* \, q_2 = 0 \ .$$

The straight column, $q_1 = q_2 = 0$, is stable according to the *Dirichlet* criterion, as long as the potential energy $V(q_1 = q_2 = 0)$ takes on a minimal value [see Sec. 9.1]

$$\frac{\partial^2 V}{\partial q_1^2} \bigg|_{q_1 = q_2 = 0} = k_1^* > 0 \ \text{ and } \ \frac{\partial^2 V}{\partial q_2^2} \bigg|_{q_1 = q_2 = 0} = k_2^* > 0 \ .$$

(11.54a)

The stability limit is reached if $k_1^* = 0$ or $k_2^* = 0$. The critical load is the minimum of the buckling loads, which are expressed in terms of the *Euler* load,

$$F_1 = F_E \left(1 + \frac{k \, l^4}{\pi^4 \, EJ} \right) \ , \text{ and } \ F_2 = F_E \left(4 + \frac{k \, l^4}{4 \, \pi^4 \, EJ} \right) \ , \quad F_E = \pi^2 \, EJ / l^2 \ ,$$

$$\rightarrow \ F_c = \min \{ F_1, F_2, \dots \} \ .$$

(11.54b)

By increasing the stiffness of the foundation, the buckling mode that is associated with the critical load turns to higher waviness. The transition through the double root $F_c = F_1 = F_2$ is not smooth and is called mode jumping. Imperfections determine the form of deflections in some neighborhood of that critical load.

11.3.3. Torsional Rigidity of an Elastic Rod with a Rectangular Cross-Section

The cross-section, area $A = 2B \times 2H$, is simply connected and torsional rigidity according to the first part of Eq. (6.175) is proportional to the integral

$$J_T = 4 \int_A \psi \, dA \ .$$

(11.55)

Poisson's equation (6.170) with the boundary condition $\psi = 0$ is considered, together with an admissible one-term *Ritz* approximation (q is a constant subject to best fit)

$$D \{\psi\} = \frac{\partial^2 \psi}{\partial x^2} + \frac{\partial^2 \psi}{\partial x^2} + 1 = 0 ,$$

(11.56)

$$\psi^*(y, z) = q \left(B^2 - y^2\right)\left(H^2 - z^2\right) , \quad \psi = 0 \ \text{at} \ y = \pm B , z = \pm H .$$

(11.57)

Galerkin's rule (11.11) minimizes the error $D \{\psi^*\}$, which results when the approximation is substituted in Eq. (11.56). The integrand is doubly symmetric; hence,

$$4 \int_0^B \int_0^H D\{\psi^*(y, z)\} \left(B^2 - y^2\right)\left(H^2 - z^2\right) dy \, dz = 0 \ .$$

(11.58)

The integration is easily performed and renders the constant of best fit within the admissible functional approximation

$$q = 5/8\left(B^2 + H^2\right) \ .$$

(11.59)

Integration according to Eq. (11.55) is somewhat smoothing the functional error and the torsional rigidity should be well approximated

$$J_T^* = 4 \int_A \psi^* \, dA \ = 40 \, (H \, B)^3/9 \left(B^2 + H^2\right) \ .$$

(11.60)

By comparing this with a slowly convergent exact series solution for a quadratic cross-section, $2B = 2H$, where $J_T = 2.24 \, B^4$, the approximation turns out to slightly underestimate torsional rigidity by $- 0.8 \%$. The maximum of the shear stress at $y = B$, $z = 0$, however, is determined by taking the partial derivative and is approximated by

$$\tau_{max}^* = -2 \frac{M_T}{J_T^*} \frac{\partial \psi^*}{\partial y} = 9 M_T / 16 \, H \, B^2 .$$

(11.61)

In the case of the quadratic cross-section, the exact value is $\tau_{max} = 0.601 \times M_T / B^3$, and the error is exaggerated due to the differentiation of the *Ritz* approximation to a full $- 6.5 \%$.

11.4.   The Finite-Element Method (FEM)

Only a brief outline of the underlying concept within the *Ritz-Galerkin* approximation can be given in this textbook. Above, the *Ritz* approximation (11.1) is based on a set of (smooth) admissible functions defined in the whole domain (volume) of the body with built-in boundary conditions. In FEM, the body is subdivided into finite elements where neighboring elements are simply connected in joints. The displacement vector is estimated by a *Ritz* approximation within each element. The generalized coordinates of such an approximation are selected as the generalized displacements of these nodes. In that sense, the finite element of some a priori selected shape is considered analogous to the above procedure for the whole body: The equations of motion of the small (possibly low-order MDOF) equivalent system are set up. In a second step, a transformation to nonlocal (gross) coordinates gives the possibility of "superposition" of the equivalent systems. Matrix analysis is available in modern computers, and after the stiffness matrices of all the elements are determined, the gross stiffness matrix of the body or structure is put together. Analogously, the mass matrix is set up in dynamic problems. Subsequently, the procedure is illustrated by considering a vibrating beam element and the triangular element of an in-plane, statically loaded thermoelastic plate.

11.4.1. A Beam Element

A linear elastic beam can be discretized according to the FEM. Within the *Bernoulli-Euler* theory of flexural vibrations, Eq. (7.147) applies and becomes homogeneous for free vibrations

$$\frac{\partial^2}{\partial x^2} \left(EJ \frac{\partial^2 w}{\partial x^2} \right) + \rho A \frac{\partial^2 w}{\partial t^2} = 0 .$$

(11.62)

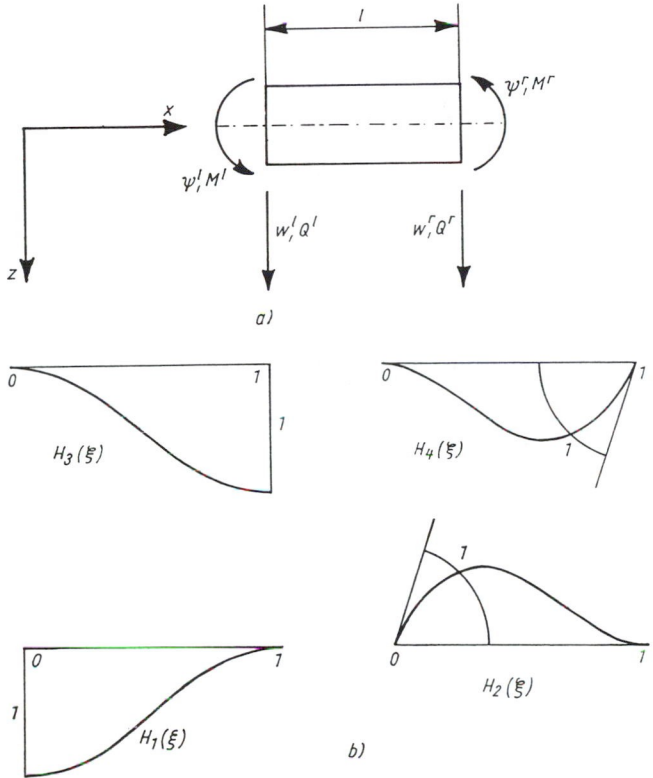

Fig. 11.8. (a) A simple beam element. (b) Low order *Hermite* polynomials

The *Ritz* approximation of the deflection within an element of length l is selected such that the generalized coordinates are the displacements and rotations at the nodal joints (the centroids of the cross-sections at $x = 0, l$ of the local coordinate) shown in Fig. 11.8(a).

The set of lowest-order polynomials that are the influence functions of a beam of constant cross-section built-in at one end and with the boundary conditions shown in Fig. 11.8(b) to be prescribed at the other end are admissible

$$w^*(x, t) = \sum_{k=1}^{4} q_k(t)\, H_k(\xi) \quad , \quad \mathbf{q} = \{w^l(t),\, \psi^l(t),\, w^r(t),\, \psi^r(t)\} \quad , \quad \xi = x/l \; .$$
(11.63)

The superscript l refers to the left end, the letter r to the right end. Solving the simple static problems of unit displacement or the unit angle of rotation renders the *Hermite* polynomials in $0 \leq \xi \leq 1$

$$H_1 = 2\xi^3 - 3\xi^2 + 1 \; , \; H_2 = -l\left(\xi^3 - 2\xi^2 + \xi\right) \; ,$$

$$H_3 = -2\xi^3 + 3\xi^2 \; , \; H_4 = -l\left(\xi^3 - \xi^2\right) \; .$$

(11.64)

A fictitious lateral load results from this approximation in Eq. (11.62) and *Galerkin's* rule (11.7) requires its virtual work to vanish

$$\int_0^l q^* \, \delta w^* \, dx = 0 \; .$$

(11.65)

Integrating the first term partially twice and considering the virtual work of the cross-sectional resultants [see Fig. 11.8(a) for their positive action] render

$$\int_0^1 \frac{EJ}{l^3} \frac{\partial^2 w^*}{\partial \xi^2} \, \delta\!\left(\frac{\partial^2 w^*}{\partial \xi^2}\right) d\xi + \int_0^1 \rho A \, l \, \ddot{w}^* \, \delta w^* \, d\xi - \left(Q^l \, \delta w^l + M^l \, \delta \psi^l\right)$$

$$+ \, Q^r \, \delta w^r + M^r \, \delta \psi^r\,\big) = 0 \; .$$

(11.66)

Integration and comparing coefficients give four differential equations of motion that in matrix notation are

$$\left(k + m \frac{\partial^2}{\partial t^2}\right) D = F \; , \; F^T = \left(Q^l, M^l, Q^r, M^r\right) \; , \; D^T = \left(w^l, \psi^l, w^r, \psi^r\right) \; .$$

(11.67)

The elements of the (4 x 4) symmetric stiffness matrix k are determined by the integrals

$$k_{lm} = \int_0^1 \frac{EJ}{l^3} H_l'' \, H_m'' \, d\xi \; , \; H'' = \frac{d^2 H}{d\xi^2} \; ,$$

(11.68)

and those of the (4 x 4) symmetric mass matrix m are (analogously) given by

$$m_{kl} = \int_0^1 \rho A \, l \, H_k \, H_l \, d\xi \; .$$

(11.69)

These integrals are easily evaluated for the homogeneous element where the bending rigidity and mass per unit of length are given constants

$$k = \frac{EJ}{l^3} \begin{pmatrix} 12 & -6\,l & -12 & -6\,l \\ . & 4\,l^2 & 6\,l & 2\,l^2 \\ . & . & 12 & 6\,l \\ . & . & . & 4\,l^2 \end{pmatrix} ,$$

(11.70)

$$m = \frac{m_S}{420} \begin{pmatrix} 156 & -22\,l & 54 & 13\,l \\ . & 4\,l^2 & -13\,l & -3\,l^2 \\ . & . & 156 & 22\,l \\ . & . & . & 4\,l^2 \end{pmatrix} , \quad m_S = \rho A\,l .$$

(11.71)

Conservation of momentum and of angular momentum of any motion without deformation is subject to superposition. The rigid-body motions of the finite element result in (eg in the case of a translation, where $w^l = w^r$, $\psi = 0$)

$$Q^l + Q^r = \frac{1}{2}\, m_S \left(\frac{\partial^2 w^l}{\partial t^2} + \frac{\partial^2 w^r}{\partial t^2} \right) ,$$

(11.72)

and for a small rotation, where $\psi^l = \psi^r$, $w^l = 0$, $w^r = -l\,\psi^r$,

$$M^l + M^r - l\,Q^r = \frac{l^2}{3}\, m_S \frac{\partial^2 \psi^r}{\partial t^2} .$$

(11.73)

Since there are two degrees of freedom of such rigid-body motions to be considered here, the stiffness matrix is singular of the order two. The inverse, the flexibility matrix, does not exist.

Superposition of the matrices is illustrated by joining two neighboring elements. At the common node 0, the compatibility conditions are considered

$$w_0 = w_1^r = w_2^l , \quad \psi_0 = \psi_1^r = \psi_2^l ,$$

(11.74)

the slope must be steady. In addition to the distributed mass, a concentrated mass m_0 may be attached at the node with the moment of inertia given by $m_0\, i_0^2$. Conservation of momentum and of angular momentum renders in that case two equations (the action of an external single force loading F_0 and of an external moment M_0 at the node is assumed)

$$m_0\, \ddot{w}_0 = F_0 - \left(Q_1^r + Q_2^l \right) , \quad m_0 i_0^2\, \ddot{\psi}_0 = M_0 - \left(M_1^r + M_2^l \right) .$$

(11.75)

Equation (11.67) renders the sum of the shear forces and that of the moments as well, and by superposing the nodal stiffness, the following set of equations results:

$$Q_1^r + Q_2^l = F_0 - m_0 \ddot{w}_0$$

$$= \frac{EJ}{l^3} \left[-12 w_1^l + 6 l\psi_1^l + 12 w_0 + 6 l\psi_0 + 12 w_0 - 6 l\psi_0 - 12 w_2^r - 6 l\psi_2^r \right]$$

$$+ \frac{ms}{420} \left[54 \ddot{w}_1^l - 13 l\ddot{\psi}_1^l + 156 \ddot{w}_0 + 22 l\ddot{\psi}_0 + 156 \ddot{w}_0 - 22 l\ddot{\psi}_0 + 54 \ddot{w}_2^r + 13 l\ddot{\psi}_2^r \right],$$

$$M_1^r + M_2^l = M_0 - m_0 i_0^2 \ddot{\psi}_0$$

$$= \frac{EJ}{l^3} \left[-6 l w_1^l + 2 l^2 \psi_1^l + 6 l w_0 + 4 l^2 \psi_0 - 6 l w_0 + 4 l^2 \psi_0 + 6 l w_2^r + 2 l^2 \psi_2^r \right]$$

$$+ \frac{ms}{420} \left[13 l\ddot{w}_1^l - 3 l^2 \ddot{\psi}_1^l + 22 l\ddot{w}_0 + 4 l^2 \ddot{\psi}_0 - 22 l\ddot{w}_0 + 4 l^2 \ddot{\psi}_0 - 13 l\ddot{w}_2^r - 3 l^2 \ddot{\psi}_2^r \right].$$

If we enlarge the size of the matrices properly, the deformation vector, eg is now a (1 x 6) column matrix, $D_{1,2}^T = (w_1^l, \psi_1^l, w_0, \psi_0, w_2^r, \psi_2^r)$, and arranging the elements of the generalized forcing vector accordingly give the equations of motion of the beam with two elements in matrix form

$$F_{1,2}^T = \left(Q_1^l, M_1^l, [F_0 - m_0 \ddot{w}_0], [M_0 - m_0 i_0^2 \ddot{\psi}_0], Q_2^r, M_2^r \right),$$

$$m_{1,2} \dot{D}_{1,2} + k_{1,2} D_{1,2} = F_{1,2}, \tag{11.76}$$

the inertia forces of the single mass m_0 are left in the forcing vector for the convenience of writing down the gross mass matrix. The resulting (6 x 6) stiffness and mass matrix for the homogeneous beam of two elements considered above becomes

$$k_{1,2} = \frac{EJ}{l^3} \begin{pmatrix} 12 & -6l & -12 & -6l & 0 & 0 \\ -6l & 4l^2 & 6l & 2l^2 & 0 & 0 \\ -12 & 6l & \underline{24} & \underline{0} & -12 & -6l \\ -6l & 2l^2 & \underline{0} & \underline{8l^2} & 6l & 2l^2 \\ 0 & 0 & -12 & 6l & 12 & 6l \\ 0 & 0 & -6l & 12l^2 & 6l & 4l^2 \end{pmatrix}. \tag{11.77}$$

Superposed elements are underlined and, below the terms of the concentrated mass are still not shifted to the mass matrix

$$m_{1,2} = \frac{m_S}{420} \begin{pmatrix} 156 & -22\,l & 54 & 13\,l & 0 & 0 \\ -22\,l & 4\,l^2 & -13\,l & -3\,l^2 & 0 & 0 \\ 54 & -13\,l & \underline{312} & \underline{0} & 54 & 13\,l \\ 13\,l & -3\,l^2 & \underline{0} & \underline{8\,l^2} & -13\,l & -3\,l^2 \\ 0 & 0 & 54 & -13\,l & 156 & 22\,l \\ 0 & 0 & 13\,l & -3\,l^2 & 22\,l & 4\,l^2 \end{pmatrix}. \tag{11.78}$$

The procedure applied above is the first step in the direct stiffness method and is analogously extended to the successive superposition of further beam elements. At the left end of the first element and at the right end of the last one, the boundary conditions are considered. Use of the tools of computer science, like pre and postprocessors are crucial for practical applications. The support reactions remain unknown and are determined subsequently to vibration analysis. By keeping the time constant, the beam in that case is assumed to be loaded by the given load and, in addition, by the inertia forces. Thus, static methods become applicable.

Mass and stiffness matrices may be transformed into a rotated gross coordinate system of reference. By means of such transformations, frames and other beam structures with branches are easily discretized. In general, the normal forces and axial vibrations of the member beams must be taken into account. The following references provide exhaustive information: *K. J. Bathe: Finite Element Procedures in Engineering Analysis. Prentice-Hall, Englewood Cliffs, N. J., 1982* , and *O. C. Zienkiewicz: The Finite Element Method. McGraw-Hill, New York, 1977.*

11.4.2. The Planar Triangular Plate Element

A thin plate is considered under the action of in-plane forces under plane stress conditions. Triangular elements are commonly selected, and by choosing the minimal number of degrees of freedom, the nodes are situated at the corners (see Fig. 11.9). The *Ritz* approximations of the displacements within each element in that case are linear functions of the coordinates x and y such that the nodal displacements become the generalized coordinates and the element thus has six degrees of freedom assigned

$$u(x, y) = \sum_{n=1}^{3} u_n \, \varphi_n(x, y) \ , \ v(x, y) = \sum_{n=1}^{3} v_n \, \varphi_n(x, y) \ ,$$

$$\varphi_n(x, y) = a_n + b_n \, x + c_n \, y \ . \tag{11.79}$$

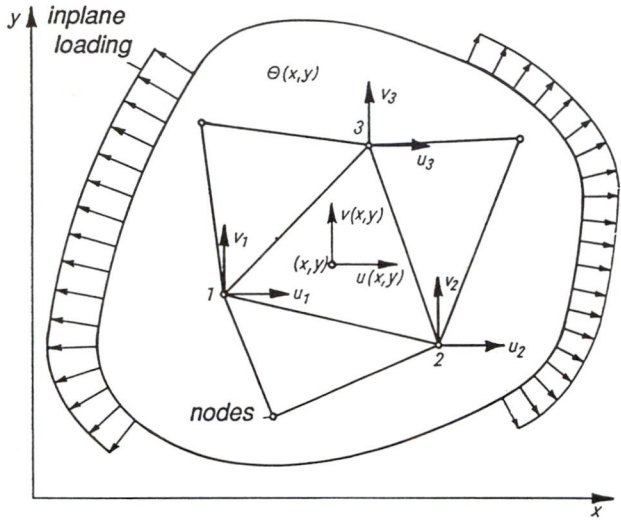

Fig. 11.9. Plate in equilibrium with enlarged triangular elements. In-plane tractions
prescribed at some part of the surface contour and a
(mean) temperature field are considered to be given

The nine coefficients $(a_n , b_n , c_n , n = 1, 2, 3)$ are easily determined
from the (3×3) linear conditions

$$\varphi_n(x_1, y_1) = \delta_{n1} = a_n + b_n x_1 + c_n y_1 ,$$
$$\varphi_n(x_2, y_2) = \delta_{n2} = a_n + b_n x_2 + c_n y_2 , \quad n = 1, 2, 3 , \quad \delta_{nm} = \begin{cases} 1, & n = m \\ 0, & n \neq m \end{cases} .$$
$$\varphi_n(x_3, y_3) = \delta_{n3} = a_n + b_n x_3 + c_n y_3 , \qquad\qquad\qquad (11.80a)$$

By combining the x and y coordinates of the nodes in the column
matrices $\mathbf{x}^T = (x_1 , x_2 , x_3)$ and $\mathbf{y}^T = (y_1 , y_2 , y_3)$ and denoting their
differences by $x_{ij} = x_i - x_j$ and $y_{ij} = y_i - y_j$, the roots are expressed in
economical fashion by

$$\mathbf{a} = \frac{1}{2A} (\mathbf{x} \times \mathbf{y}) , \mathbf{a}^T = (a_1, a_2, a_3) , \quad \begin{aligned} \mathbf{b}^T &= \frac{1}{2A} (y_{23}, y_{31}, y_{12}) \\ \mathbf{c}^T &= \frac{1}{2A} (x_{32}, x_{13}, x_{21}) \end{aligned} , 2A = (\mathbf{b}^T . \mathbf{x}).$$

$$(11.80b)$$

The strains are constant within each element and are given in terms
of the node displacements through the (linearized) geometric
relations

$$\varepsilon_{xx} = \frac{\partial u}{\partial x} = \sum_{n=1}^{3} \frac{\partial \varphi_n}{\partial x} u_n , \; \varepsilon_{yy} = \frac{\partial v}{\partial y} = \sum_{n=1}^{3} \frac{\partial \varphi_n}{\partial y} v_n ,$$

$$2\varepsilon_{xy} = \frac{\partial v}{\partial x} + \frac{\partial u}{\partial y} = \sum_{n=1}^{3} \left(\frac{\partial \varphi_n}{\partial x} v_n + \frac{\partial \varphi_n}{\partial y} u_n \right), \quad \frac{\partial \varphi_n}{\partial x} = b_n, \quad \frac{\partial \varphi_n}{\partial y} = c_n. \tag{11.81}$$

Hooke's law, Eqs. (4.20) and (6.29), takes on matrix form with the column vector of the average stress S and the displacement row vector u, and in generalized form, with the column vector θ, whose components are the values of the temperature field at the nodes

$$S = S\, u - \Psi\, \theta \;,\; S^T = (\sigma_{xx}, \sigma_{yy}, \sigma_{xy}) \;,\; u = (u_1, v_1, u_2, v_2, u_3, v_3)\;,$$

$$\theta(x, y) = \sum_{n=1}^{3} \theta_n\, \varphi_n(x, y) \;,\; \theta^T = (\theta_1, \theta_2, \theta_3)\;,$$

$$S = \frac{G}{(1-v)\,A} \begin{pmatrix} y_{23} & vx_{32} & y_{31} & vx_{13} & y_{12} & vx_{21} \\ vy_{23} & x_{32} & vy_{31} & x_{13} & vy_{12} & x_{21} \\ \lambda x_{32} & \lambda y_{23} & \lambda x_{13} & \lambda y_{31} & \lambda x_{21} & \lambda y_{12} \end{pmatrix} , \quad \Psi = \frac{E\alpha}{1-v} \begin{pmatrix} \varphi_1 & \varphi_2 & \varphi_3 \\ \varphi_1 & \varphi_2 & \varphi_3 \\ 0 & 0 & 0 \end{pmatrix},$$

$$\lambda = (1-v)/2\;. \tag{11.82}$$

The generalized internal forces are applied at the nodes and determined by the gradient of the strain energy. If we assume the thickness $h = const$ and consider Eqs. (3.41) and (3.35), the partial derivatives are

$$\frac{\partial U}{\partial u_n} = h \int_A \left(\sigma_{xx} \frac{\partial \varphi_n}{\partial x} + \sigma_{xy} \frac{\partial \varphi_n}{\partial y} \right) dA \;,\; \frac{\partial U}{\partial v_n} = h \int_A \left(\sigma_{yx} \frac{\partial \varphi_n}{\partial x} + \sigma_{yy} \frac{\partial \varphi_n}{\partial y} \right) dA\;.$$

Elimination of the stresses by *Hooke's* law renders, in matrix notation,

$$\begin{pmatrix} \frac{\partial U}{\partial u_n} \\ \frac{\partial U}{\partial v_n} \end{pmatrix} = hA \begin{pmatrix} b_n & 0 & c_n \\ 0 & c_n & b_n \end{pmatrix} \cdot (S\, u - \Phi\, \theta)\;.$$

Φ is determined by the domain integral,

$$\Phi = \frac{1}{A} \int_A \Psi\, dA = \frac{E\,\alpha}{3\,(1-v)} \begin{pmatrix} 1 & 1 & 1 \\ 1 & 1 & 1 \\ 0 & 0 & 0 \end{pmatrix}\;. \tag{11.83}$$

Properly enlarging the dimension of the matrices yields finally

$$\left(\frac{\partial U}{\partial u_1}, \frac{\partial U}{\partial v_1}, \frac{\partial U}{\partial u_2}, \frac{\partial U}{\partial v_2}, \frac{\partial U}{\partial u_3}, \frac{\partial U}{\partial v_3}\right)^T = \left(K \, u - T \Phi \, \theta\right), \quad K = T S, \tag{11.84}$$

where

$$T = \frac{h}{2} \begin{pmatrix} y_{23} & 0 & x_{32} \\ 0 & x_{32} & y_{23} \\ y_{31} & 0 & x_{13} \\ 0 & x_{13} & y_{31} \\ y_{12} & 0 & x_{21} \\ 0 & x_{21} & y_{12} \end{pmatrix}. \tag{11.85a}$$

K is the symmetric (6 x 6) stiffness matrix of the triangular element and thermal effects are considered by the column matrix

$$\Lambda^T = \left(T \Phi \, \theta\right)^T = \frac{E h \, \alpha \, \bar{\theta}}{2 \, (1 - \nu)} \, (y_{23}, x_{32}, y_{31}, x_{13}, y_{12}, x_{21}), \quad \bar{\theta} = \sum_{i=1}^{3} \theta_i / 3. \tag{11.85b}$$

The six conditions of equilibrium in vector form combine Eq. (11.84) and the external nodal forces that are statically equivalent to any given external loads distributed over the midplane of the element

$$F = K \, u - \Lambda, \quad F^T = (X_1, Y_1, X_2, Y_2, X_3, Y_3). \tag{11.86}$$

The finite elements are combined by adding their equilibrium conditions. The numbering of the nodes should guarantee a banded structure for the resulting large system of equations. The matrices are properly enlarged by null elements and the nodal displacements are found by solving the system of linear equations

$$\left(K_I + K_{II} + K_{III} + \ldots\right) u_r = F_r + \Lambda_r, \quad u_r^T = \left(u_I^T, u_{II}^T, u_{III}^T, \ldots\right). \tag{11.87}$$

Geometric boundary conditions are considered in u_r. The solution is efficiently determined by *Choleski's* method of splitting the coefficient matrix into the product of two triangular ones.

11.5. Linearization of Nonlinear Equations of Motion

A MDOF-system with n degrees of freedom is given by $2n$ first-order differential equations, the state equations [see Eq. (7.109)] $x_i = q_i$, $i = 1, \ldots, n$, $x_i = dq_i / dt$, $i = (n + 1), \ldots, 2n$

$$\dot{x}_i = \sum_{k=1}^{2n} a_{ik} x_k + \mu_i \, f_i \, (x_1, x_2, \ldots, x_{2n}). \tag{11.88}$$

Subsequently, only vibrational systems are considered with at least one periodic solution in free motion. Hence, in a time-harmonic approximation

$$x_k = A_k \sin(\omega t + \varepsilon_k) \equiv A \kappa_k \sin \phi_k \quad , \quad \phi_k = \omega t + \varepsilon_k \quad , \quad \kappa_k = A_k/A \quad . \quad (11.89)$$

Substitution into the nonlinear functions f_i makes these functions periodic with the period $\tau = 2\pi/\omega$, and, following the rules of the harmonic balance method, a *Fourier* series expansion is performed. Restriction to the basic harmonics renders

$$f_i \approx a_i \cos \phi_1 + b_i \sin \phi_1 \quad . \quad\quad\quad\quad (11.90)$$

The *Fourier* coefficients, by definition, are the projections of these functions onto the basic harmonics, which have been selected as the components of the vibrational motion x_1 ,

$$a_i = \frac{1}{\pi} \int_0^{2\pi} f_i \left(A\kappa_1 \sin \phi_1, \, ..., \, A\kappa_{2n} \sin \phi_{2n} \right) \cos \phi_1 \, d\phi_1 \, ,$$

$$b_i = \frac{1}{\pi} \int_0^{2\pi} f_i \left(A\kappa_1 \sin \phi_1, \, ..., \, A\kappa_{2n} \sin \phi_{2n} \right) \sin \phi_1 \, d\phi_1 \, .$$

$$(11.91)$$

In any practical application, the state vector must be rearranged such that the motion x_1 refers to the "leading degree of freedom."

Substitution into the nonlinear state equations (11.88) and comparing coefficients with the linear equivalence,

$$\dot{x}_i = \sum_{k=1}^{2n} \left(a_{ik}{}^* x_k + \overline{a}_{ik} x_k \right) \quad , \quad\quad\quad (11.92)$$

render

$$\overline{a}_{ik} = a_{ik} \, , \quad a_{ik}{}^* = 0 \, , \quad \text{if } k \neq 1 \, ,$$

$$\overline{a}_{i1} = a_{i1} + \frac{\mu_i}{A\kappa_1} \, b_i \, , \quad a_{i1}{}^* = \frac{\mu_i}{\omega A\kappa_1} \, a_i \, . \quad\quad (11.93)$$

(§)   Linearization by Harmonic Balance. The above becomes quite simple for a nonlinear SDOF-oscillator. The free motion in general form is determined by a single equation,

$$\ddot{x} + f(x, \dot{x}) = 0 \, . \quad\quad\quad\quad (11.94)$$

The harmonic approximation, $x = A\cos\omega t$, and its derivative, $dx/dt = -A\omega\sin\omega t$, are substituted, and the *Fourier* coefficients determined,

$$a = \frac{1}{\pi A}\int_0^{2\pi} f(A\cos\omega t, -A\omega\sin\omega t)\cos\omega t \, d(\omega t),$$

$$b = -\frac{1}{\pi A\omega}\int_0^{2\pi} f(A\cos\omega t, -A\omega\sin\omega t)\sin\omega t \, d(\omega t).$$

$$(11.95)$$

The equivalence is defined by the linearized differential equation

$$\ddot{x} + b\dot{x} + ax = 0. \tag{11.96}$$

The coefficients *a* and *b* are functions of the amplitude *A* , and thus, the linearized system still reflects the characteristic property of the nonlinear vibration that the natural period depends on amplitude.

(§)   Transient Vibrations of Nonlinear Systems. The solution can be approximated by the method of *Krylow* and *Bogoljubow* that assumes a slowly varying amplitude and phase. The linear term of the restoring force is separated and the procedure illustrated by considering Eq. (11.94)

$$\ddot{x} + \omega^2 x + f(x, \dot{x}) = 0. \tag{11.97}$$

Amplitude and phase are assumed to be time-dependent in the approximation of the solution

$$x(t) = A(t)\cos\varphi(t) \quad, \quad \varphi(t) = \omega t + \varepsilon(t) \quad. \tag{11.98}$$

The velocity is given by the time derivative

$$\dot{x}(t) = \dot{A}\cos\varphi - (\omega + \dot{\varepsilon})A\sin\varphi, \tag{11.99}$$

with the crucial assumption, with respect to the transformation of the state variables to new variables *A* and *ε* , that it takes on a form similar to that of a linear system

$$\dot{x}(t) = -\omega A\sin\varphi. \tag{11.100}$$

Putting the two expressions equal renders a first relation between the amplitude and the phase

$$\dot{A} \cos \varphi - \dot{\varepsilon} A \sin \varphi = 0 . \tag{11.101}$$

A second relation results after the substitution of the approximations (11.98) and (11.100) in the equation of motion (11.97). Note that the acceleration is approximated by the time derivative of Eq. (11.100)

$$- \omega \dot{A} \sin \varphi - \dot{\varepsilon} \omega A \cos \varphi + f(A \cos \varphi, - \omega A \sin \varphi) = 0 . \tag{11.102}$$

The differential equations can be decoupled in the rates and are finally cast in a form analogous to that of the first-order state equations

$$\dot{A} = \frac{1}{\omega} f(A \cos \varphi, - \omega A \sin \varphi) \sin \varphi ,$$

$$\dot{\varepsilon} = \frac{1}{\omega A} f(A \cos \varphi, - \omega A \sin \varphi) \cos \varphi . \tag{11.103}$$

By considering the variations of the amplitude A and the phase ε to be slow, ie their rates may be assumed to be constant over a single period, integration over that period gives two differential equations that are still nonlinear, but the first one for the averaged amplitude $A(t)$ decouples

$$\dot{A} = \frac{1}{2\pi \omega} \int_0^{2\pi} f(A \cos \varphi, - \omega A \sin \varphi) \sin \varphi \, d\varphi ,$$

$$\dot{\varepsilon} = \frac{1}{2\pi \omega A} \int_0^{2\pi} f(A \cos \varphi, - \omega A \sin \varphi) \cos \varphi \, d\varphi . \tag{11.104}$$

Integration is illustrated by reconsidering the *Duffing* equation (11.16) generalized by adding the nonlinear "turbulence" damping

$$\ddot{x} + \omega^2 x + \beta x^3 + \alpha \dot{x} |\dot{x}| = 0 . \tag{11.105}$$

Performing the integrations in Eqs. (11.104) gives the averaged set of equations

$$\dot{A} = - \frac{4}{3\pi} \alpha \omega A^2 , \quad \dot{\varepsilon} = \frac{3}{8} \frac{\beta}{\omega} A^2 . \tag{11.106}$$

The separation of variables and subsequent integration yield the time-dependent (decaying) amplitude

$$A(t) = \frac{A_0}{1 + (4\alpha/3\pi)\,A_0\,\omega t} \;\to\; A_0 \exp\left(-\frac{4\,\alpha}{3\,\pi}\,A_0\,\omega t\right).$$

$$\tag{11.107}$$

Substitution into the second differential equation (the exponential function is used) renders the time function of the phase angle

$$\varepsilon(t) = \frac{9\,\pi\,\beta\,A_0}{64\,\alpha\,\omega^2}\left[1 - \exp\left(-\frac{8\,\alpha}{3\,\pi}\,A_0\,\omega t\right)\right].$$

$$\tag{11.108}$$

Considering the damping to be negligible, $\alpha = 0$, yields the same approximation of the eigenfrequency as was derived previously by *Galerkin's* rule, Eq. (11.20).

11.6. Numerical Integration of a Nonlinear Equation of Motion

The *Wilson θ-method* dominates in engineering applications since it is an unconditionally stable linear acceleration method, and it is applicable to a rather large number of equations. Implementation in a computer routine requires only an efficient solver of a system of linear equations, eg the *Choleski* procedure that is usually already available from elasto-static considerations. The time incremental procedure is illustrated for a nonlinear SDOF-system. Equation (11.94) is considered at two instants of time, t and $t + \Delta t$, and its difference renders the equation for the increments

$$m\,\Delta\ddot{x}(t) + r(t)\,\Delta\dot{x}(t) + k(t)\,\Delta x(t) = \Delta F(t) .$$

$$\tag{11.109}$$

The tangential stiffness is determined by the derivative of the restitution-deformation relation

$$k(t) = \left.\frac{dF_S}{dx}\right|_t ,$$

$$\tag{11.110}$$

and the tangential coefficient of damping is analogously given by

$$r(t) = \left.\frac{dF_D}{d\dot{x}}\right|_t .$$

$$\tag{11.111}$$

The damping force is assumed to be given by a nonlinear function of the velocity; see Sec. 4.2.3. Acceleration is linearly approximated in the extended time interval

$$\tau = \theta \, \Delta t \ , \quad \theta > 1.37 \ , \tag{11.112}$$

and the coefficients of Eq. (11.109) are kept constant. By means of

$$\ddot{x}(t + s) = \ddot{x}(t) + \frac{\Delta \ddot{x}}{\tau} \, s \ , \quad 0 \le s \le \tau \ , \tag{11.113}$$

the terminal values of the increments of velocity and deformation become, respectively,

$$\widehat{\Delta \dot{x}}(t + s) = \tau \, \ddot{x}(t) + \frac{\tau}{2} \, \widehat{\Delta \ddot{x}} \ , \quad \widehat{\Delta x}(t) = \tau \, \dot{x}(t) + \frac{\tau^2}{2} \, \ddot{x}(t) + \frac{\tau^2}{6} \, \widehat{\Delta \ddot{x}}(t) \ . \tag{11.114}$$

The increment of deformation is considered to be the independent variable, and Eq. (11.109) becomes a quasistatic relation

$$\widehat{k}(t) \, \widehat{\Delta x}(t) = \widehat{\Delta F}(t) \ , \tag{11.115}$$

where

$$\widehat{k}(t) = k(t) + \frac{6}{\tau^2} \, m + \frac{3}{\tau} \, r(t) \ , \tag{11.116}$$

and

$$\widehat{\Delta F}(t) = \widehat{\Delta F}(t) + m \left[\frac{6}{\tau} \, \dot{x}(t) + 3 \, \ddot{x}(t) \right] + r(t) \left[3 \, \dot{x}(t) + \frac{\tau}{2} \, \ddot{x}(t) \right] \ . \tag{11.117}$$

Solving for the terminal increment

$$\widehat{\Delta x}(t) = \widehat{\Delta F}(t) \, / \, \widehat{k}(t) \ , \tag{11.118}$$

determines the increment of acceleration at once

$$\widehat{\Delta \ddot{x}}(t) = \frac{6}{\tau^2} \, \widehat{\Delta x}(t) - \frac{6}{\tau} \, \dot{x}(t) - 3 \, \ddot{x}(t) \ . \tag{11.119}$$

Linear interpolation renders the increment of acceleration at the true time step Δt

$$\Delta \ddot{x}(t) = \frac{1}{\theta} \, \widehat{\Delta \ddot{x}}(t) \ . \tag{11.120}$$

Accordingly, the true increments of velocity and deformation are determined by means of Eq. (11.120)

$$\Delta \dot{x}(t) = \Delta t \; \ddot{x}(t) + \frac{\Delta t}{2} \; \Delta \ddot{x}(t) \;\;, \;\; \Delta x(t) = \Delta t \; \dot{x}(t) + \frac{(\Delta t)^2}{2} \; \ddot{x}(t) + \frac{\tau^2}{6} \; \Delta \ddot{x}(t) \cdot \tag{11.121}$$

Hence, the initial mechanical state for the next time step is determined with sufficient accuracy by

$$x(t + \Delta t) = x(t) + \Delta x(t) \;\;, \;\; \dot{x}(t + \Delta t) = \dot{x}(t) + \Delta \dot{x}(t) \; . \tag{11.122}$$

Acceleration, however, must be determined from the updated equation of motion

$$\ddot{x}(t + \Delta t) = \frac{1}{m} \left[F(t + \Delta t) - F_D(t + \Delta t) - F_S(t + \Delta t) \right] , \tag{11.123}$$

to avoid any accumulation of errors, eg those introduced by taking the tangent stiffness. See also *R. W. Clough* and *J. Penzien: Dynamics of Structures. McGraw-Hill, New York, 1975.*

11.7. Exercises A 11.1 to A 11.11 and Solutions

A 11.1: The flexural vibrations of a linear elastic and homogeneous beam of length *2l* , mass *m* , and with constant bending rigidity *B* = *EJ* are considered for symmetrically arranged hinged-hinged supports at a distance $2 \lambda l \le 2 l$. Determine the relation between the span ratio λ and the natural basic frequency of the symmetric mode.

Solution: The *Ritz* approximation is quite suitable and it suffices to represent the deflection by a single term, $w^*(x, t) = q(t) \, \varphi(x)$. A proper shape function is, eg $\varphi(x) = 1 - (x/\lambda l)^2$. One of the dynamic boundary conditions, $w''(|x| = l) = 0$, is violated. The kinetic and strain energy are, respectively,

$$T = 2 \frac{1}{2} \int_0^l \rho \, \dot{w}^{*2} A \, dx = \frac{m^* \dot{q}^2}{2} \;\;, \;\; V = U = 2 \frac{1}{2} \int_0^l EJ \left(\frac{\partial^2 w^*}{\partial x^2} \right)^2 dx = \frac{k^* q^2}{2} \;,$$

$$m^* = m \left(1 - \frac{2}{3 \lambda^2} + \frac{1}{5 \lambda^4} \right) \;\;, \;\; k^* = \frac{64 \, B}{\lambda^4 \, (2 \, l)^3} \; .$$

The *Lagrange* equation of motion takes on the form of Eq. (11.29) and the squared natural frequency is approximated by

$$\omega_0^2 = k^*/m^* = \frac{320 \, B/(2 \, l)^3}{m \left(5 \lambda^4 - 10 \lambda^2/3 + 1 \right)} \; .$$

Taking extreme values of the span ratio and comparing these with the exact natural frequencies of the *Bernoulli-Euler* beam theory indicate the quality of the approximation

$$\lambda = 1: \quad \omega_0/\omega_{exact} = \frac{10.97}{\pi^2} > 1 \ , \quad \lambda \to 0: \quad \omega_0/\omega_{exact} = \frac{17.89}{14.06} > 1 \ .$$

A 11.2: The point of application (eg a small wheel) of a dead-weight load *F* moves with constant speed *v* over the span *l* of a simply supported and homogeneous beam of total mass *m* and with bending rigidity *B = EJ* ; see Fig. A 11.2. The problem may be considered the simplest model of a truck moving over a single-span bridge. The nonstationary lateral loading renders forced flexural vibrations. Determine a MDOF-system using the shape functions $\varphi_j = sin \ j\pi \ x/l$, which, in the case of a homogeneous beam are the orthogonal eigenfunctions. Determine the first critical speed where the first mode is excited in resonance. Asymptotically, in the limit $v \to 0$ and $t \to \infty$, a fast convergent *Fourier* series of the static influence line of the deflection [see Eqs. (6.78) and (6.79)] results as a byproduct.

Solution: The *Ritz* approximation by the finite series

$$w^*(x, t) = \sum_{j=1}^{n} q_j(t) \ \varphi_j(x) \ , \quad \varphi_j = sin \ j\pi x/l \ ,$$

is proper with respect to the boundary conditions and the kinetic and elastic energy are expressed, respectively, by the integrals

$$T = \frac{1}{2} \int_0^l \dot{w}^{*2} \rho A \ dx = \frac{1}{2} \sum_{j=1}^{n} m_j^* \dot{q}_j^2 \ , \quad m_j^* = m/2 \ , \quad m = \rho A l \ ,$$

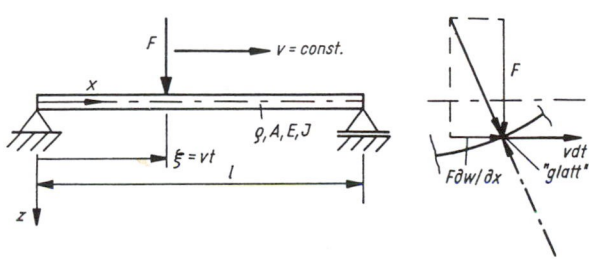

Fig. A 11.2. Moving lateral load *F = const*

$$U = \frac{1}{2} \int_0^1 B \left(\frac{\partial^2 w^*}{\partial x^2} \right)^2 dx = \frac{1}{2} \sum_{j=1}^{n} k_j^* q_j^2 \ , \ k_j^* = j^4 \pi^4 B/2 \ l^3 \ , \ B = EJ \ .$$

The portion of the generalized forces Q_j that is equivalent to the moving force F may be determined by the gradient of the time-dependent potential $W_P = - F w^*(x = \xi = vt, t)$ or, according to the definition, by comparing coefficients in the expression of virtual work

$$\sum_{j=1}^{n} Q_j \, \delta q_j = F \, \delta w^* \ , \ \delta w^* = \sum_{j=1}^{n} \varphi_j(\xi) \, \delta q_j \ \to Q_j = F \, \varphi_j(\xi) = F \sin v_j t \ , \ v_j = j \frac{\pi v}{l} \ .$$

The equivalent system consists of n uncoupled linear oscillators forced effectively for $t > 0$ by time-harmonic excitations. Their *Lagrange* equations of motion are

$$\ddot{q}_j + \omega_j^{*2} q_j = \frac{F}{m_j^*} \sin v_j t \ , \ \omega_j^{*2} = k_j^*/m_j^* \ , \ j = 1, 2, ..., n \ .$$

With quiescent initial conditions, $q_j(0) = dq_j/dt|_{t=0} = 0$, at the moment when the load enters the span at $x = 0$, the transient solutions are given by

$$q_j(t) = \frac{F}{k_j^{(eff)}} \left(\sin v_j t - \frac{v_j}{\omega_j^*} \sin \omega_j^* t \right) , \ k_j^{(eff)} = m_j^* \left(\omega_j^{*2} - v_j^2 \right) = k_j^* \left(1 - \frac{m v^2 l}{j^2 \pi^2 B} \right) .$$

The apparent bending stiffness is that of a beam under the compressive prestress $N = - m v^2/l$; see Eq. (11.51). Refer to Fig. A 11.2 for the resulting force in frictionless contact, ie an additional effect of a moving lateral load is a reduction of the apparent flexural rigidity. The lowest resonance loading is encountered when $v_1 = \omega_1^*$ and the critical speed becomes

$$v_c = \sqrt{\pi B/ml} \ , \ \frac{2\pi}{\omega_1^*} = \frac{2l}{v_c} \ .$$

By means of *l´Hospital´s* rule, the transient resonance response $q_1(t)$ is calculated and its maximum determined. Note that the first mode suffers a maximal deflection when the force just leaves the span, $\xi = l, t_c = l/v$, and the amplitude is

$$q_1(t_c) = \frac{F l^3}{\pi^3 B} \ .$$

The work of the lateral load F within this critical time elapsed must be zero. The energy of the vibrating beam is determined by the work of the negative axial component of the reaction force acting at position ξ [see Fig. A 11.2 for the deflected beam element, F $(\partial w/\partial x)|_{x=\xi}$]

$$U(t) = W(t) = \int_0^{t_c} \left(F \frac{\partial w}{\partial x} \right) v \, dt \quad ,$$

see *E. H. Lee: "On a Paradox in Beam Vibration Theory." Quarterly Applied Mathematics 10, (1952), p. 290* , for further details.

That maximum of the first mode deflection at midspan may be compared to the maximal static deflection. The latter corresponds to the midspan position of the force F and is thus given by $(q_1)_{stat} = F \, l^3/48 \, B$. The ratio determines the dynamic first mode magnification factor to be $\chi_1 = 48/\pi^3 = 1.55$.

Total deflection and inertia loading are approximated by the finite series

$$w^*(x, t) = \sum_{j=1}^{n} q_j(t) \, \varphi_j(x) \quad , \quad \rho A \ddot{w}^*(x, t) = -\rho A \sum_{j=1}^{n} \omega_j^{*2} q_j(t) \, \varphi_j(x) \, .$$

By keeping the time constant, both, shear force and bending moment are commonly determined by static methods taking into account the distributed lateral inertia loading, $p(x, t) = -\rho A \, (\partial^2 w^*/\partial t^2)$.

Considering the limit of w^* with respect to $v \to 0$ and $t \to \infty$ under the condition $\lim_{v \to 0, \, t \to \infty} (\xi = v \, t) \neq 0$ renders the static influence line of deflection [the infinite series $(n \to \infty)$ is convergent; see Eqs. (6.78) and (6.79)]

$$\overline{w}_S(x, \xi) = \sum_{j=1}^{\infty} \frac{F}{k_j^*} \sin \frac{j\pi x}{l} \sin \frac{j\pi \xi}{l} \, .$$

A 11.3: The single-span beam of Fig. A 11.2 is subject to time-dependent lateral loading $F = F(t)$ with $\xi = const$. Determine the dynamic deflection $w(x, \xi, t)$ within the same *Ritz* approximation used in Exercise A 11.2. The solution should be worked out for a time-harmonic loading $F(t) = F_0 \sin vt$. Putting the amplitude to unity, $F_0 = 1$, renders the finite series approximation of the dynamic *Green's* function of the homogeneous beam. Further, it is quite interesting to construct the ratio of the force amplitude over the amplitude of the deformation speed at the point of application of the lateral load, the mechanical impedance.

Solution: The modal masses and modal stiffnesses are those of Exercise A 11.2. Also the generalized forces are formally given by the same expressions, $Q_j = F(t)\,\varphi_j\,(\xi)$, $j = 1, 2, ..., n$. The *Lagrange* equations of motion take on the form

$$\ddot{q}_j + \omega_j^{*2}\,q_j = \frac{F(t)}{m_j^*}\,\sin\frac{j\pi\xi}{l}\ ,\quad \omega_j^{*2} = k_j^*/m_j^*\ ,\quad j = 1, 2, ..., n\ .$$

The coefficient of the forcing function $F(t)$ is called the participation factor. Time-harmonic excitation renders forced vibrations [see Eq. (7.87) and put $\zeta = 0$ in Eq. (7.94)]

$$q_j(t) = a_j\sin vt\ ,\quad a_j = \frac{F_0}{m_j^*\left(\omega_j^{*2} - v^2\right)}\,\sin\frac{j\pi\xi}{l}\ .$$

Some modal vibrations may be in phase with the forcing; the others are in counterphase. The *Green's* function of the deflection follows with $F_0 = 1$, by superposition,

$$\overline{w}^*(x, \xi, t) = a(x, \xi)\sin vt\ ,\quad a(x, \xi) = \sum_{j=1}^{n}\frac{\sin\dfrac{j\pi x}{l}\,\sin\dfrac{j\pi\xi}{l}}{m_j^*\left(\omega_j^{*2} - v^2\right)}\ .$$

At the resonance frequencies $v = \omega_j^*$, the amplitude function exhibits singularities that are cured by adding light modal damping to the denominator. The mechanical impedance of the beam at $x = \xi$ is given by the ratio

$$Z(\xi) = \frac{F_0}{v\,a(\xi, \xi)} = F_0\left\{\sum_{j=1}^{n}\frac{v\left(\sin\dfrac{j\pi\xi}{l}\right)^2}{m_j^*\left(\omega_j^{*2} - v^2\right)}\right\}^{-1}\ .$$

At resonance frequencies, the impedance vanishes, contrary to its reciprocal, which is called the admittance. The latter grows beyond any bounds.

A 11.4: An elastic tower on a rigid base that is simply modeled by the homogeneous cantilever shown in Fig. A 11.4 is subject to ground acceleration, ie the foundation is in an assigned motion $w_g(t)$, eg forced by the horizontal component of an earthquake. Using a one-term *Ritz* approximation, determine the equation of motion within the *Bernoulli-Euler* beam theory.

Solution: The total displacement of a mass element is given by the sum $w_t = w_g\,(t) + w(x, t)$. The absolute velocity enters the kinetic energy of the beam

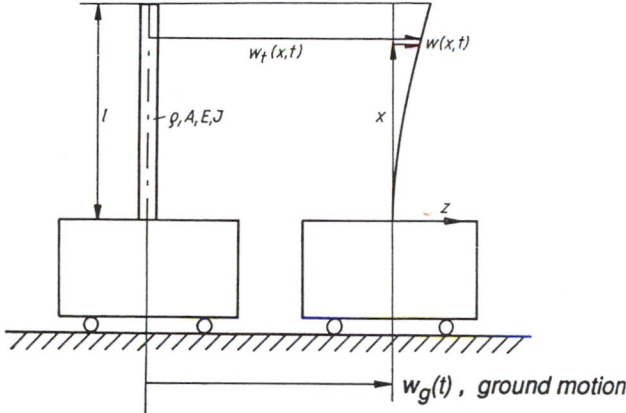

Fig. A 11.4. A towerlike structure in the strong motion phase of an earthquake

$$T = \frac{1}{2} \int_0^l \dot{w}_t^2 \, \rho A \, dx = \frac{1}{2} m \, \dot{w}_g^2 + \dot{w}_g \int_0^l \dot{w} \, \rho A \, dx + \frac{1}{2} \int_0^l \dot{w}^2 \, \rho A \, dx \quad .$$

With the assumption that the excitation of the basic mode dominates, a one-term *Ritz* approximation suffices, $w^*(x, t) = q(t) \, \varphi(x)$, $\varphi(l) = 1$, which considers the kinematic boundary conditions at the clamped edge, $x = 0$: $w = 0$, $\partial w / \partial x = 0$. The approximation is improved if the dynamic boundary conditions at the free end are also built-in, $x = l$: $\partial^2 w / \partial x^2 = 0$, $\partial^3 w / \partial x^3 = 0$. Substitution in kinetic energy gives

$$T = \frac{1}{2} m \, \dot{w}_g^2 + \dot{w}_g \, \dot{q} \int_0^l \varphi(x) \, \rho A \, dx + \frac{1}{2} \dot{q}^2 \int_0^l \varphi^2(x) \, \rho A \, dx \quad .$$

Due to the assigned motion of the foundation, the external forces do not contribute to the virtual work and the remaining elastic potential is expressed by

$$V = U = \frac{1}{2} q^2 \int_0^l EJ \left(\frac{\partial^2 \varphi}{\partial x^2} \right)^2 dx = \frac{k^* \, q^2}{2} \quad .$$

The *Lagrange* equation of motion is derived by proper differentiation

$$\frac{d}{dt} \left(\frac{\partial T}{\partial \dot{q}} \right) = L \, m \, a_g + m^* \, \ddot{q} \quad , \quad a_g = \ddot{w}_g(t) \quad , \quad L = \frac{1}{m} \int_0^l \varphi(x) \, \rho A \, dx \quad ,$$

$$m = \int_0^l \rho A \, dx \ , \quad m^* = \int_0^l \varphi^2(x) \, \rho A \, dx \ ,$$

$$\frac{\partial V}{\partial q} = k^* q \ , \quad k^* = \int_0^l EJ \left(\frac{d^2\varphi}{dx^2}\right)^2 dx \ ,$$

and it has a form corresponding to that of an oscillator driven by the assigned effective motion of the free end of its spring

$$m^* \ddot{q} + k^* q = - L \, m \, a_g \ .$$

The solution is considerably improved by considering the *Ritz* approximation of the dynamic part only, $w_D{}^* = (q - q_S) \, \varphi(x)$, $q_S = - L$ $(m/k^*) \, a_g$, and adding the quasistatic deflection due to the distributed inertia loading of the rigid-body motion with assigned ground acceleration a_g , say, $w_S \, (x, \, t)$,

$$w(x, \, t) = w_S + w_D{}^* \ .$$

$w_S \, (x, \, t)$ may be expressed by means of the static influence function of the cantilevered beam.

Dynamic stress resultants are calculated by considering the distributed lateral loading $p(x, \, t) = - \rho A \, [a_g \, (t) + \partial^2 w(x, \, t)/\partial t^2]$ at constant time by means of static methods.

A proper shape function would be $\varphi(x) = 1 - \cos \pi x/2l$, where only an erroneous shear force remains at the free end of the cantilever. With $\rho A = const$, $m = \rho A l$, and $B = EJ = const$, the modal mass is $m^* = 0.227 \, m$, the modal stiffness $k^* = \pi^4 B/32 \, l^3$, and the participation factor of that first mode becomes $L = 0.3634$. The quasistatic deflection is easily calculated as

$$w_S(x, \, t) = - \frac{m \, a_g(t) \, l^3}{2 \, B} \, \frac{x^2}{2 \, l^2} \left(1 - \frac{2 \, x}{3 \, l} + \frac{x^2}{6 \, l^2}\right).$$

A 11.5: The single-span and simply supported beam of Fig. A 11.5 is loaded by a transient thermal moment $m_\theta \, (x, \, t)$, eg from a thermal shock. Since the boundary conditions are inhomogeneous, the solution is split into a quasistatic and dynamic part. The equations of motion are to be determined within the limits of the *Ritz* approximation of the dynamic portion with the proper shape functions $\varphi_j \, (x) = \sin j\pi x/l$.

Solution: The quasistatic thermal deflection follows by the integration of Eq. (6.36) at a constant time

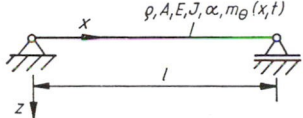

Fig. A 11.5. Thermal shock loading

$$\frac{\partial^2 w_S}{\partial x^2} = - \alpha \, m_\theta \, .$$

Putting the lateral load to zero in Eq. (7.147) gives for the remaining dynamic part the partial differential equation, $B = EJ = const$,

$$B \frac{\partial^4 w_D}{\partial x^4} + \rho A \, (\ddot{w}_D + \ddot{w}_S) = 0 \, ,$$

with homogeneous boundary conditions, $w_D = \partial^2 w_D/\partial x^2 = 0$ at $x = 0, \, l$. *Galerkin's* rule renders with the *Ritz* approximation of n th order (the shape functions are orthogonal)

$$w_D^*(x, t) = \sum_{j=1}^{n} q_j(t) \, \varphi_j(x) \, , \quad \varphi(x) = \sin j\pi x/l \rightarrow \ddot{q}_j + \omega_j^2 \, q_j = - \ddot{u}_j \, , \quad \omega_j^2 = k_j / m_j \, ,$$

$$k_j = B \, j^4 \pi^4/2 \, l^3 \, , \quad m_j = \rho A l / 2 \, , \quad \ddot{u}_j = \frac{2}{l} \int_0^l \ddot{w}_S \sin (j\pi x/l) \, dx \, , \quad j = 1, 2, ..., n \, .$$

The dynamic support reactions and stress resultants are determined by static considerations under the lateral loading $p(x, t) = - \rho A \, \partial^2(w_S + w_D^*)/\partial t^2$. In that rare case, differentiation of the *Ritz* approximation with respect to the axial coordinate x gives the same result.

A 11.6: The blades of an axial turbine or compressor and those of propellers are rods with, in general, strongly winding and variable cross-sections (for hydrodynamic reasons). The torsional vibrations are independent to the first order from the initial angles of rotation $\chi_0 (x)$ of the cross-sections, if the axis is nearly straight. That is, the natural frequencies and linear modes of torsional vibrations are very close to those of a classical rod with no winding of cross-sections. Small amplitude flexural vibrations, however, are strongly influenced by the winding; see Fig. A 11.6. Refer, eg to *W. Traupel: Thermal Turbomachines. Springer-Verlag, Berlin, 1968, Vol. 2, (in German), reference number [26].*

With the simplifying assumption of the coincidence of the cross-sectional centroid S and the torsional center D of the (variable) cross-section $A(x)$, determine the differential equations of the bending vibrations. Vibrations are considered when the rotor is at rest, and in a second-order approximation, the influence of the rotational speed Ω should be taken into account in the steady state of revolutions of the machine. The undeformed axis of the beam with a winding cross-section is assumed to be straight. The lowest natural frequency of the basic flexural mode about the z axis follows from *Galerkin´s* rule with a proper one-term *Ritz* approximation $v^*(x, t) = q(t)\, \varphi(x)$, where the initial winding of the cross-sections is neglected.

Solution: The beam axis is assumed to be inextensible. At each station x , the oblique bending is properly replaced by the two principal (small) components of deflection and *Hooke´s* law provides the linear relation between the linearized principal curvatures and the principal bending moments (see Fig. A 11.6)

$$M_\eta = -\, EJ_\eta\, w''\ ,\quad M_\zeta = EJ_\zeta\, v''\ ,\quad ()' = \frac{\partial}{\partial x}\ .$$

In matrix notation,

$$\begin{pmatrix} M_\eta \\ M_\zeta \end{pmatrix} = E \begin{pmatrix} J_\eta & 0 \\ 0 & J_\zeta \end{pmatrix} \begin{pmatrix} -w'' \\ v'' \end{pmatrix}\ .$$

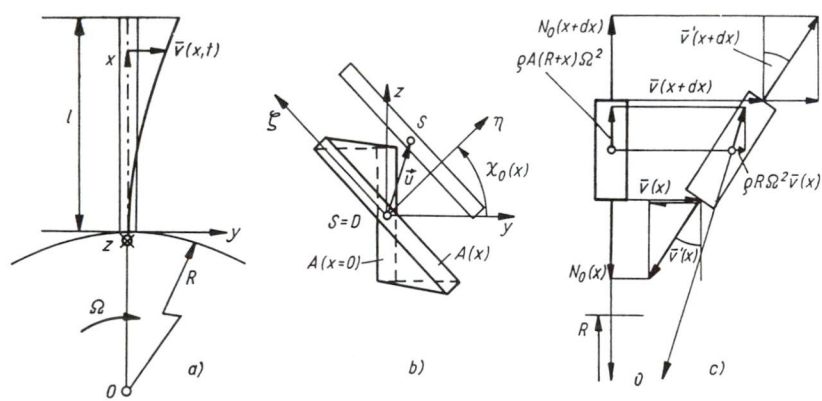

Fig. A 11.6. (a) A blade modeled by a cantilever beam. (b) Winding cross-sections. Principal axes of inertia are denoted η, ζ, the reference system is y, z. (c) Deformed element with axial force $N(x)$ indicated

Any vector may be projected onto the rotated coordinate systems: The point vector in the plane x = const, eg is represented by

$$\begin{pmatrix} \eta \\ \zeta \end{pmatrix} = D \cdot \begin{pmatrix} y \\ z \end{pmatrix} \;, \quad D = \begin{pmatrix} \cos \chi_0 & \sin \chi_0 \\ -\sin \chi_0 & \cos \chi_0 \end{pmatrix} \;.$$

The components of the bending moment vector M are related by the same rotational matrix

$$\begin{pmatrix} M_\eta \\ M_\zeta \end{pmatrix} = D \cdot \begin{pmatrix} M_y \\ M_z \end{pmatrix} \;.$$

Analogously, the linearized components of the curvature vector of the deflected beam axis, de_t/ds (see Sec. 1.1.3), are related according to

$$\begin{pmatrix} -w'' \\ v'' \end{pmatrix} = D \cdot \begin{pmatrix} -\overline{w}'' \\ \overline{v}'' \end{pmatrix} \;.$$

The substitution and multiplication of the principal bending moments by the transposed rotational matrix render with the tensor transformations of the moments of inertia taken into account [see Eq. (2.123) and note the difference in sign of the off-diagonal tensor-element], the proper linear vector transformation of oblique bending, expressed in the global coordinates y and z,

$$\begin{pmatrix} M_y \\ M_z \end{pmatrix} = E\, D^T \cdot \begin{pmatrix} J_\eta & 0 \\ 0 & J_\zeta \end{pmatrix} \cdot D \begin{pmatrix} -\overline{w}'' \\ \overline{v}'' \end{pmatrix} \;, \quad D^T \cdot \begin{pmatrix} J_\eta & 0 \\ 0 & J_\zeta \end{pmatrix} \cdot D = \begin{pmatrix} J_y & J_{yz} \\ J_{yz} & J_z \end{pmatrix} \;.$$

Note the similarity transformation. These linear and coupled relations are of some practical importance

$$M_y = -E J_y\, \overline{w}'' + E J_{yz}\, \overline{v}'' \;, \quad M_z = E J_z\, \overline{v}'' - E J_{yz}\, \overline{w}'' \;. \tag{a}$$

Conservation of momentum and angular momentum of an element yields, after elimination of the shear force [see Eqs. (2.150) and (2.152)]

$$M_y'' = -q_z \;, \quad M_z'' = q_y \;. \tag{b}$$

The lateral force per unit of length is understood to include the inertia loading. The time-reduced equations of free motion of the blade at rest are thus derived by combining Eqs. (a) and (b) [the time factor $exp\,(i\omega t)$ is cancelled]

$$\frac{\partial^2}{\partial x^2}\left(EJ_z\ \overline{v}'' - EJ_{yz}\ \overline{w}''\right) - \rho A\ \omega^2\ \overline{v} = 0\ , \quad \frac{\partial^2}{\partial x^2}\left(EJ_y\ \overline{w}'' - EJ_{yz}\ \overline{v}''\right) - \rho A\ \omega^2\ \overline{w} = 0\ .$$

(c)

Taking into account a stationary speed of revolution renders an axial force $N_0(x)$. According to second-order theory of elasticity, it contributes to the shear force in the y direction as well as z direction

$$dQ_y = N(x + dx)\ \overline{v}'(x + dx) - N(x)\ \overline{v}'(x) = \left[N'(x)\ \overline{v}'(x) + N(x)\ \overline{v}''(x)\right]dx\ ,$$

$$dQ_z = N(x + dx)\ \overline{w}'(x + dx) - N(x)\ \overline{w}'(x) = \left[N'(x)\ \overline{w}'(x) + N(x)\ \overline{w}''(x)\right]dx\ ,$$

where, $N \cong N_0(x)$.

Furthermore, the radial centrifugal force $q_x = \rho A\ (R + x)\ \Omega^2$ contributes to the lateral load q_y (but not to q_z) [see Fig. A 11.6(c)]

$$q_y = \rho A\ \omega^2\ \overline{v} + \rho A\ (R + x)\ \Omega^2\ \frac{\overline{v}}{R + x} = \rho A\ (\omega^2 + \Omega^2)\ \overline{v}\ .$$

Hence, Eq. (b) is to be replaced by

$$M_z'' = \left[q_y + N'\ \overline{v}' + N\ \overline{v}''\right], \quad M_y'' = -\left[q_z + N'\ \overline{w}' + N\ \overline{w}''\right]\ .$$

(d)

Equation (c) is extended and thus becomes with $A = const$, $N_0 = \rho A\ (l - x)\ (2R + l + x)\ \Omega^2/2$,

$$\frac{\partial^2}{\partial x^2}\left(EJ_z\ \overline{v}'' - EJ_{yz}\ \overline{w}''\right) - \rho A\ \frac{\Omega^2}{2}\ (l - x)\ (2R + l + x)\ \frac{\partial^2\overline{v}}{\partial x^2}$$

$$+ \rho A\ \Omega^2(R + x)\ \frac{\partial\overline{v}}{\partial x} - \rho A\ (\omega^2 + \Omega^2)\overline{v} = 0\ ,$$

$$\frac{\partial^2}{\partial x^2}\left(EJ_y\ \overline{w}'' - EJ_{yz}\ \overline{v}''\right) - \rho A\ \frac{\Omega^2}{2}\ (l - x)\ (2R + l + x)\ \frac{\partial^2\overline{w}}{\partial x^2}$$

$$+ \rho A\ \Omega^2(R + x)\ \frac{\partial\overline{w}}{\partial x} - \rho A\ \omega^2\ \overline{w} = 0\ .$$

(e)

The basic natural mode $v_0\ (x)$ is generally determined from the uncoupled equations with sufficient accuracy ($EJ_z = const$ is assumed subsequently)

$$EJ_z\ \frac{\partial^4 v_0}{\partial x^4} - \rho A\ \frac{\Omega^2}{2}\ (l - x)\ (2R + l + x)\ \frac{\partial^2 v_0}{\partial x^2}$$

$$+ \rho A \, \Omega^2 (R + x) \frac{\partial v_0}{\partial x} - \rho A \left(\omega^2 + \Omega^2 \right) v_0 = 0 \, .$$

A proper shape function is selected to be affined to the static deflection under constant lateral load (see Exercise A 11.4)

$$v_0 = a \, \varphi(x) \, , \quad \varphi(x) = \frac{1}{3} \left(\frac{x}{l} \right)^2 \left(6 - 4 \frac{x}{l} + \frac{x^2}{l^2} \right) \, ,$$

Galerkin's rule, ie the orthogonality condition of the error and that shape function with $a \neq 0$, renders the approximation of the natural basic frequency

$$\omega^* = \omega_0^* \left[1 + \frac{\Omega^2}{\omega_0^2} \left(0.173 + 1.556 \frac{R}{l} \right) \right]^{1/2} \, ,$$

where $\omega_0^* = 3.530 \sqrt{EJ_z/m \; l^2}$, "exact" $\omega_0 = 3.516 \sqrt{EJ_z/m \; l^2}$.

ω_0 is the corresponding natural frequency when the rotor is at rest.

A 11.7: A thin rectangular plate with all edges simply supported is subject to pure shear loading (see, eg Exercise A 2.5). The planform is $L \times B$ and the critical buckling load is to be determined by the *Ritz* approximation of the deflection

$$w^*(x, y) = \sum_{m=1}^{M} \sum_{n=1}^{N} A_{mn} \sin \frac{m \pi x}{L} \sin \frac{n \pi y}{B} \, .$$

Solution: The shape functions satisfy all the boundary conditions at the hinged and straight edges $x = 0, L$ and $y = 0, B$. The linearized *von Karman* equation with inplane loading, $n_{xy} = - T = const$, is the starting point and *Galerkin's* rule renders a linear system of equations

$$K \, \Delta\Delta w + 2T \frac{\partial^2 w}{\partial x \partial y} = 0 \, , \quad D \{w^*\} = p^* \; \rightarrow$$

$$\pi^4 K \frac{BL}{4} \left(\frac{m^2}{L^2} + \frac{n^2}{B^2} \right)^2 A_{mn} - 8T \sum_{k=1}^{M} \sum_{j=1}^{N} A_{kj} \frac{m \, n \, k \, j}{\left(k^2 - m^2 \right) \left(j^2 - n^2 \right)} = 0 \, .$$

The subscripts k and j are to be selected such that the expressions $(k + m)$ and $(j + n)$ are odd numbers. The buckling determinant must vanish and the approximate critical shear loading becomes

$$T_{crit} = \begin{cases} \dfrac{\pi^2 K}{B^2}\,(5.34 + 4/\alpha^2) \;, & \alpha \geq 1 \\[2em] \dfrac{\pi^2 K}{B^2}\,(4 + 5.34/\alpha^2) \;, & \alpha \leq 1 \end{cases} \;, \quad \alpha = L/B \;.$$

The critical force of Exercise A 2.5 with respect to buckling of the plate thus is given by $F_{crit} = B\,T_{crit}$. Stable edge beams are understood.

A 11.8: Determine the critical load factor of a simply supported circular plate of radius R and constant bending stiffness K for the point-symmetric in-plane compressive loading $n_r(r) = [- (\pi^2 K/R^2)$ cos $(\pi r/2R)]$. The generalized *Galerkin's* rule must be applied when the *Ritz* approximation of the basic buckling mode $w^*(r) = [q_1$ cos $(\pi r/2R)]$ is selected. An erroneous edge moment $m_r{}^*(R) = \pi\, q_1\, v\, K/2R$ destroys the homogeneous dynamic boundary condition.

Solution: Equation (11.8) is applied and the error $p^*(r)$ is specified in addition to $m_r{}^*(R)$

$$\int_0^R p^* \, \phi \, 2\pi r \, dr - \int_0^{2\pi} m_r{}^* \frac{\partial \phi}{\partial r} R \, d\theta = 0, \; \phi = \cos \pi r/2R, \; p^* = K \, \Delta\Delta w^* - \frac{\lambda}{r} \frac{\partial}{\partial r}\left(r n_r \frac{\partial w^*}{\partial r}\right).$$

Considering properly the difference of infinite integrals and definitions leading to *Euler's* constant C , and the finite value of an infinite integration, gives the critical load factor in the following approximation:

$$\lambda_c = 9\left[2v + \left(\pi^2 - 4\right)/8 + C + \ln \pi - Ci\,(\pi)\right]/4\,(3\pi - 4) \cong 0.8295\, v + 0.9878\,,$$

$$Ci\,(\pi) = -\int_0^\infty \frac{\cos u}{u}\, du\;.$$

A 11.9: A clamped circular plate is considered in the elastic postbuckling range of constant in-plane loading; see Fig. A 11.9. The generalized coordinate q_1 in the *Ritz* approximation of the deflection $w^*(r) = q_1\,(1 - r^2/a^2)^2$ is identical to the approximate deflection at the origin. *Galerkin's* rule must be applied to the nonlinear *von Karman* equations.

    Note: In the absence of any pressure loading and with point symmetry taken into account, the *von Karman* equations take on the following form in polar coordinates:

$$K \frac{d}{dr}(\Delta w) = \frac{1}{r} \frac{dF}{dr} \frac{dw}{dr} \;, \quad \frac{d}{dr}(\Delta F) = -\frac{Eh}{2\,r}\left(\frac{dw}{dr}\right)^2 , \quad \Delta F = \frac{1}{r} \frac{d}{dr}\left(r \frac{dF}{dr}\right).$$

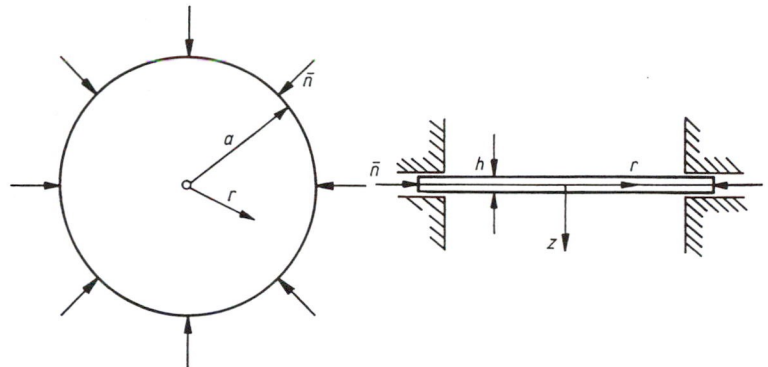

Fig. A 11.9. Undeformed clamped circular plate under constant radial compression

They are of the third order only and, thus, they are more convenient with respect to integration than Eq. (5.30a, b). See, eg *A. S. Wolmir: Biegsame Platten und Schalen. VEB Verlag Bauwesen, Berlin, 1962.*

Solution: The *Ritz* approximation is substituted into the second *von Karman* equation that, when integrated, renders under consideration of the boundary conditions and the point symmetry

$$\frac{dF^*}{dr} = \frac{Eh\,q_1^2}{6\,a}\left(\frac{3\,r}{a} - \frac{6\,r^3}{a^3} + \frac{4\,r^5}{a^5} - \frac{r^7}{a^7}\right) - \bar{n}\,h \; ,$$

$$r = 0: \; \frac{dF}{dr} = 0 \; , \quad r = a: \; n_r = \frac{1}{a}\frac{dF}{dr} = -\bar{n} \; .$$

Both of the approximations are substituted in the first *von Karman* equation. The kinematic boundary conditions at $r = a$, $w^* = dw^*/dr = 0$ are taken care of, and *Galerkin's* rule yields a cubic equation

$$\left[\left(\frac{8\,K}{3} - \frac{\bar{n}\,a^2}{6}\right) + \frac{Eh}{28}\,q_1^2\right]q_1 = 0 \; .$$

The trivial root corresponds to the undeflected configuration $q_1 = 0$. The critical buckling load is determined by taking the limit $q_1 \to 0$ and putting the remaining term in parenthesis to zero, $n_c^* = 16\,K/a^2$; the "exact" value is $n_c = 14.68\,K/a^2 \approx 0.92\,n_c^*$. The amplitude in the narrow postbuckling range follows by putting the bracketed factor to zero (the sign is decided upon any given initial deformation)

$$q_1 = \pm\, 2.16\,a\,\sqrt{(\bar{n} - n_c)/Eh} = 0 \; , \quad \bar{n} > n_c \; .$$

A 11.10:   A shallow cylindrical panel of infinite length with simply supported edges is subject to a uniformly distributed radial pressure $p_r (t)$; see Fig. A 11.10. Consider the forced vibrations by taking into account the essential geometric nonlinearities by means of the one-term *Ritz* approximation of the deflection $w^*(x, t) = [q_1 (t) \sin (\pi y/b)]$. The flexural stiffness K and stretching rigidity $E h$ are given.

    Note: The nonlinear *von Karman* equations can be generalized to include any small initial curvatures imposed on the rectangular plate. A small cylindrical initial deformation of the plate, with curvature $1/R$, gives (see, eg *K. Marguerre: Neuere Festigkeitsprobleme des Ingenieurs. Springer-Verlag, Berlin,1950, p. 234)*

$$K \, \Delta\Delta w = p + n_x \frac{\partial^2 w}{\partial x^2} + n_y \left(\frac{1}{R} + \frac{\partial^2 w}{\partial y^2} \right) + 2n_{xy} \frac{\partial^2 w}{\partial x \partial y} \quad ,$$

$$\Delta\Delta F = Eh \left[\left(\frac{\partial^2 w}{\partial x \partial y} \right)^2 - \frac{\partial^2 w}{\partial x^2} \left(\frac{1}{R} + \frac{\partial^2 w}{\partial y^2} \right) \right] .$$

The curvilinear coordinate y is to be measured along the circular arch of the shallow cylindrical shell and the pressure loading of the plate p is to be substituted by the radial pressure loading p_r .

Solution: The problem is independent of the axial coordinate x , and the second *von Karman* equation is thus homogeneous. Inertia forces, tangential to the surface, are neglected. The membrane force

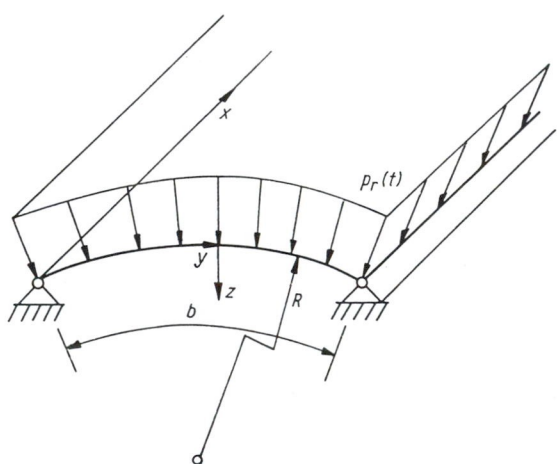

Fig. A 11.10. Forced vibrations of a long shallow panel

n_y in the first equation does not depend on y ; see Eq. (2.98). Its value is determined by the geometric boundary conditions of fixed (immovable, hard) supports, which render the condition at the shell midsurface,

$$\int_0^b \frac{\partial v}{\partial y} \, dy = 0 .$$

Taking into account the nonlinear geometric relation $\partial v/\partial y = - \varepsilon_y - w/R + (1/2) \, (\partial w/\partial x)^2$ and *Hooke's* law, $\varepsilon_y = n_y \, (1 - v^2)/Eh$, $\varepsilon_x \equiv 0$, and substituting the *Ritz* approximation render the membrane force to be approximately given by

$$n_y^* = \frac{Eh}{1 - v^2} \left(\frac{\pi^2}{4} \frac{q_1^2}{b^2} - \frac{2}{\pi} \frac{q_1}{R} \right) .$$

Consequently, *Galerkin's* rule is applied to the first *von Karman* equation where the inertia force $- \rho h \, \partial^2 w/\partial t^2$ is substituted together with $p_r(t)$. The result of integration over arc-length b is a vibrational equation with an *unsymmetric* restoring force

$$\ddot{q}_1 + \omega^2 q_1 + \alpha \, q_1^2 + \beta \, q_1^3 = \frac{4}{\pi \rho h} \, p_r(t) \; , \; \omega^2 = \frac{\pi^4 K}{\rho h b^4} \left(1 + \frac{96 \, b^4}{\pi^6 R^2 h^2} \right) ,$$

$$\alpha = - \frac{3\pi \, E}{\left(1 - v^2\right) \rho R b^2} \; , \; \beta = \frac{\pi^4 E}{4 \left(1 - v^2\right) \rho b^4} .$$

In the static case, a cubic equation results for the central deflection q_1 . It exhibits the snap-through characteristics of the shallow arch; see Sec. 9.1.3. Dynamic stress resultants are determined by means of static methods at constant times considering the distributed inertia loading, $p_D \, (y, t) = [- \rho h \, (d^2 q_1/dt^2) \sin (\pi y/b)]$. Acceleration is expressed by the remaining terms of the above equation of motion. No direct superposition applies to the nonlinear equation of motion, and $q_1 \, (t)$ must be found by means of numerical routines or by some approximation techniques (eg by means of *Galerkin's* rule).

A 11.11: Determine mode shapes and natural frequencies of a given linear dynamic MDOF-system alternatively to Sec. 7.4.9 by matrix iteration, a method originally suggested by A. *Stodola* .

   Note: The eigenvalue problem in matrix form [see Eq. (10.50b)] is rewritten in the form of elastostatics with inertia loading. *a* is the amplitude vector [see Eq. (3.46)]

$$\frac{1}{\omega^2} \mathbf{a} = f m \mathbf{a} \ , \quad f = k^{-1} \ .$$

The product $\mathbf{d} = f m$, the dynamic matrix, is nonsymmetric in general. The basic mode is to be determined iteratively. A start vector is selected, usually all components are set to 1, $\mathbf{a}_1{}^{(0)}$.

Any vector can be expanded with respect to the orthonormal eigenvectors

$$\mathbf{a}_1{}^{(0)} = \sum_{k=1}^{n} c_k{}^{(0)} \ \boldsymbol{\phi}_k \ .$$

Successive multiplication by $\mathbf{d}$ renders a vector that approaches the basic mode shape if the natural frequencies are well separated,

$$\mathbf{a}_1{}^{(s)} = \sum_{k=1}^{n} c_k{}^{(s)} \ \boldsymbol{\phi}_k \to \boldsymbol{\phi}_1 \ , \quad c_k{}^{(s)} = \left(\frac{\omega_1}{\omega_n}\right)^{2s} c_k{}^{(0)} \ .$$

Hence, by removing the component of $\mathbf{a}$ in the direction $\boldsymbol{\phi}_1$ the same iteration scheme renders the next higher mode. Find the sweeping matrix $\mathbf{S}_1$ at first and, by removing all lower order modes, $\mathbf{S}_2$ and so on. Apply the discrete and generalized form of Eq. (8.21) to improve the approximation of the squared natural frequencies.

Solution: Multiplication of the start vector by the dynamic matrix renders a vector $\mathbf{A}_1{}^{(1)}$ which is subject to normalization or, alternatively, the length is reduced by putting the largest component to 1 before starting the iteration again. After a number of s steps the approximation is found to be sufficiently accurate,

$$\mathbf{A}_1{}^{(s)} = \mathbf{d} \ \mathbf{a}_1{}^{(s-1)} \ , \quad \boldsymbol{\phi}_1 = \mathbf{a}_1{}^{(s)} \ , \quad \omega_1^2 \cong \frac{a_{1k}{}^{(s-1)}}{A_{1k}{}^{(s)}} \ \text{ or } \ \omega_1^2 = \frac{\mathbf{A}_1{}^{(s)T} \ m \ \mathbf{a}_1{}^{(s-1)}}{\mathbf{A}_1{}^{(s)T} \ m \ \mathbf{A}_1{}^{(s)}} \ .$$

The component of any start vector $\mathbf{a}_2{}^{(0)}$ to be removed is determined by multiplication of its series expansion by $\boldsymbol{\phi}_1{}^T m$ from the left

$$\boldsymbol{\phi}_1{}^T \ m \ \mathbf{a}_2{}^{(0)} = m_1{}^* \ c_1{}^{(0)} \ , \quad m_1{}^* = \boldsymbol{\phi}_1{}^T \ m \ \boldsymbol{\phi}_1 = 1, \quad \mathbf{a}_{2'}{}^{(0)} = \mathbf{a}_2{}^{(0)} - \boldsymbol{\phi}_1 \ c_1{}^{(0)} = \mathbf{S}_1 \ \mathbf{a}_2{}^{(0)} \to$$

$$\mathbf{S}_1 = \mathbf{I} - \frac{1}{m_1{}^*} \boldsymbol{\phi}_1 \ \boldsymbol{\phi}_1{}^T \ m \ .$$

The sweeping matrix $\mathbf{S}_1$ must be applied in every iteration step to guarantee convergence to the next higher mode which is then the "basic" mode of the system with the dynamic matrix $\mathbf{d} \ \mathbf{S}_1$,

$$\mathbf{A}_{2'}{}^{(s)} = \mathbf{d} \ \mathbf{S}_1 \ \mathbf{a}_2{}^{(s-1)} \ , \quad \boldsymbol{\phi}_2 = \mathbf{a}_{2'}{}^{(s)} \ , \quad \omega_2^2 \cong \frac{a_{2'k}{}^{(s-1)}}{A_{2'k}{}^{(s)}} \ \text{ or } \ \omega_2^2 = \frac{\mathbf{A}_{2'}{}^{(s)T} \ m \ \mathbf{a}_{2'}{}^{(s-1)}}{\mathbf{A}_{2'}{}^{(s)T} \ m \ \mathbf{A}_{2'}{}^{(s)}} \ .$$

The approximation of the second mode is further improved by leaving out a single component in the start vector. That component is subsequently determined by the orthogonality condition

$$\mathbf{a}_{2'}^{(s)\mathsf{T}}\, \mathbf{m}\, \phi_1 = 0 \, .$$

For the next higher mode the sweeping matrix is extended to remove both the component of the start vector in the direction of the basic mode and its component in the direction of the second eigenvector,

$$\mathbf{S}_2 = \mathbf{I} - \frac{1}{m_1^{\,*}}\, \phi_1\, \phi_1^{\mathsf{T}}\, \mathbf{m} - \frac{1}{m_2^{\,*}}\, \phi_2\, \phi_2^{\mathsf{T}}\, \mathbf{m} \, ,$$

$$\mathbf{A}_{3'}^{(s)} = \mathbf{d}\, \mathbf{S}_2\, \mathbf{a}_3^{(s-1)} \, , \quad \phi_3 = \mathbf{a}_{3'}^{(s)} \, , \quad \omega_3^2 \cong \frac{\mathbf{a}_{3'\,k}^{(s-1)}}{\mathbf{A}_{3'\,k}^{(s)}} \quad \text{or} \quad \omega_3^2 = \frac{\mathbf{A}_{3'}^{(s)\mathsf{T}}\, \mathbf{m}\, \mathbf{a}_{3'}^{(s-1)}}{\mathbf{A}_{3'}^{(s)\mathsf{T}}\, \mathbf{m}\, \mathbf{A}_{3'}^{(s)}} \, ,$$

$$\mathbf{a}_{3'}^{(s)}\, \mathbf{m}\, \phi_1 = 0 \, , \quad \mathbf{a}_{3'}^{(s)}\, \mathbf{m}\, \phi_2 = 0 \, .$$

The approximation is improved by leaving out two components of the start vector. Two additional equations result from the above orthogonality relations. The number of modes determined by matrix iteration should be much smaller than the number of degrees of freedom. The eigenvectors may be sampled in the modal matrix and the iteration can be performed at once using a sweeping hypermatrix.

1 2
Impact

Impact is a process of momentum exchange between two colliding bodies within a short time of contact. With respect to a single impacted body or structure, the loading in such a process acts with high intensity during this short period of time. As a result, the initial velocity distribution is rapidly changed (even pressure wave loadings, eg following an explosion, are events of that category). Such rapid loading in the contacting area is a source where waves are emitted that propagate with finite speeds through the body. In the case of sufficiently small amplitudes (in the far field), linear elastic body waves propagate with the speed of sound waves, the distinction between the fast longitudinal P-wave (primary wave) and the slower transverse S-wave (secondary wave) complicates the pattern (see Sec. 7.6 and Exercise A 7.12), and further refractions and diffractions are encountered. An illustration is given in Sec. 12.7 in which an impacted finite rod is considered. A further complication arises in the near field through large deformations and nonlinear material behavior. Visco-plasticity must be considered, and high deformation rates may even render fracturing in statically ductile materials. This short account of the complex processes involved already indicates that a complete analysis of geometrically nonsimple impacted bodies requires numerical methods of high resolution, ie say, in terms of the finite-element method, large systems must be considered when marching in time with extremely fine steps. Such expensive models are restricted to special problems, like the impact following an airplane crash on an atomic power plant and the like. Many "everyday" impact problems of engineering interest, however, can be drastically simplified by idealization. The most important idealizing assumption is that of a sudden jump in the velocity distribution and, thus, of the momentum in the limit of vanishing contacting time. Taking such a limit implies the infinite speed of the stress signal propagation in the body. Furthermore, the impact forces, accelerations, and stresses take on infinite values such that their time integrals remain finite and assume actual values. Idealized impact becomes a sudden mechanical process at some instant of time, and the colliding bodies keep their positions with unchanged configurations. The motion following impact is initiated with the new velocity distributions after completing the sudden exchange of momentum.

A second assumption with respect to the spatial velocity distribution at the time instant after impact is necessary. Such shape functions must be compatible with the boundary conditions and should confirm the integrity of the body. If rigid-body motion is admissible, the linear velocity distribution according to Eq. (1.4) is determined by the velocity vector of a single material point, commonly of the mass center, and by the vector of the angular velocity. It suffices in many cases to assume that an admissible distribution of the velocity of a deformable body is affined to a properly selected static deformation [note the analogy to the *Ritz* approximation, Eq. (11.1)].

A third assumption must be made with respect to the total energy of the colliding bodies: The energy is conserved in the extreme case of an ideal elastic impact; some portion of the energy is dissipated otherwise. In the course of this introduction, only "pure inelastic impacts" are considered in which energy dissipation is extreme.

Within such an idealized theory of the impact process, the rate form of the conservation laws by time integration and through the limit of vanishing duration of the loading is transformed to a set of finite difference relations. Their application is illustrated by typical examples. Considerations of the wave trains in a finite elastic rod and a fluid filled pipe, the so-called water hammer, give some insight into actual problems and, in an informal manner, yield some justification of the oversimplified theory.

12.1. Finite Relations of Momentum and Angular Momentum

Conservation of momentum (7.10) renders together with the assumption of external single force loadings

$$\mathbf{I}(t_2) - \mathbf{I}(t_1) = \int_{t_1}^{t_2} \mathbf{R} \, dt \ , \quad \mathbf{R} = \sum_{i=1}^{n} \mathbf{F}_i \ . \tag{12.1}$$

Finite summation and time integration are interchanged, and taking the limit $(t_1, t_2) \rightarrow t_0$, where t_0 is the time instant of the idealized impact, gives

$$\mathbf{I}' - \mathbf{I} = \sum_{i=1}^{n} \lim_{(t_2 \rightarrow t_{0+}), (t_1 \rightarrow t_{0-})} \int_{t_1}^{t_2} \mathbf{F}_i \, dt = \sum_{j=1}^{m \leq n} \mathbf{S}_j \ . \tag{12.2}$$

During impact, the velocity distribution is suddenly changed and the momentum vector jumps at t_0 by $I' - I = m \, (v_M' - v_M)$. Those forces that are related to the impact and, hence, go to infinity enter Eq. (12.2) by their finite impulse S_i,

$$S_i = \lim_{(t_2 \to t_{0+}), (t_1 \to t_{0-})} \int_{t_1}^{t_2} F_i \, dt \; .$$

$$(12.3)$$

All the external forces that remain finite, like gravity or the restitution force in a mass-less spring, do not contribute, since their time integral goes to zero in the foregoing limiting process. Analogously, the conservation of angular momentum with a special point of reference, Eq. (7.23),

$$H(t_2) - H(t_1) = \int_{t_1}^{t_2} M \, dt = \sum_{i=1}^{n} \int_{t_1}^{t_2} r_i \times F_i \, dt \; ,$$

$$(12.4)$$

gives (in the limit, the material points remain at constant positions during the idealized impact, ie the position vectors do not change, r_i = *const*)

$$H' - H = \sum_{i=1}^{n} r_i \times \lim_{(t_2 \to t_{0+}), (t_1 \to t_{0-})} \int_{t_1}^{t_2} F_i \, dt = \sum_{j=1}^{m \le n} r_j \times S_j \; .$$

$$(12.5)$$

The jump of the angular momentum equals the resulting moment of the externally applied impulses. The finite relations of Eqs. (12.2) and (12.5) are a system of six linear difference equations of the components. The angular momentum immediately after impact is related to the velocity field with an essential assumption with respect to the spatial velocity distribution. Furthermore, the external impulses must be known a priori or must be eliminated by additional and independent relations. If we assume that the impulses are known, the six equations can be solved and the velocity field of a rigid body (with six degrees of freedom) determined. The following two examples illustrate the application of the difference equations.

12.1.1. Example: Impacting a Rigid-Plate Pendulum

A quadratic plate *2a x 2a* is supported at a corner and is free for any spatial rotation (see Fig. 12.1). A lateral impact by a hammer at the corner A supplies the impulse $S = - S e_3$. Prior to impact, the pendulum is at rest. After impact, the velocity is assumed to be linearly distributed according to a rigid-body rotation. The onset of bending vibrations is neglected. Equation (12.5) is applied with point 0 for reference and renders the jump of the angular momentum

$$\mathbf{H_0'} - \mathbf{H_0} = \mathbf{r_A} \times \mathbf{S} \ , \ \mathbf{H_0} = 0 \ , \ \mathbf{H_0'} = I_1 \, \omega_1' \, \mathbf{e_1} + I_2 \, \omega_2' \, \mathbf{e_2} + I_3 \, \omega_3' \, \mathbf{e_3} \ ,$$

$$\mathbf{r_A} = a \sqrt{2} \, (\mathbf{e_1} - \mathbf{e_2}) \ , \ I_1 = \frac{m(2a)^2}{12} + m(2a/\sqrt{2})^2 = \frac{7}{3} \, ma^2 \, , \ I_2 = \frac{m(2a)^2}{12} \ .$$

The finite relation is solved for the three components of angular velocity

$$\omega_1' = \frac{3}{7} \sqrt{2} \, \frac{S}{ma} \ , \ \omega_2' = 3 \sqrt{2} \, \frac{S}{ma} \ , \ \omega_3' = 0 \ .$$

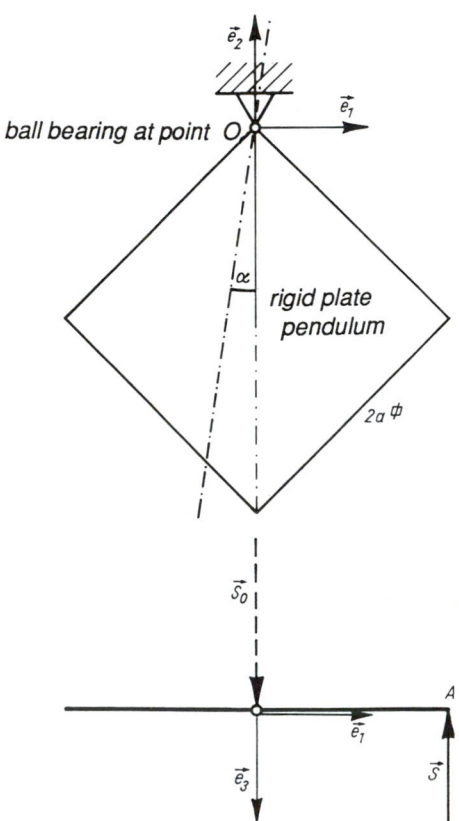

Fig. 12.1. Impacting a rigid-plate pendulum at the corner A

The rigid plate starts to rotate after impact about the instant in-plane axis ω' with an inclination given by $tan\ \alpha = 1/7$. With that information, the impulse of the reaction at the support 0 can be determined from the finite relation for the momentum, Eq. (12.2),

$$I' = S_0 + S\ ,\ v_S' = \omega' \times r_S\ ,\ r_S = -a\sqrt{2}\ e_2\ ,\ I' = m\ v_S' = -\frac{6}{7}\ S e_3\ ,\ S_0 = \frac{S}{7}\ e_3\ .$$

The condition for leaving the support quiet during the impact is discussed in Sec. 12.4.

12.1.2. Example: Axial Impact of a Deformable (Elastic) Column

A hammer hits the free end of a homogeneous column at rest with the impulse S ; the other end is rigidly supported (see Fig. 12.2). Rigid-body motion following impact is impossible. Hence, a plausible assumption of the velocity distribution must be made that renders deformation in the subsequent motion over time. By

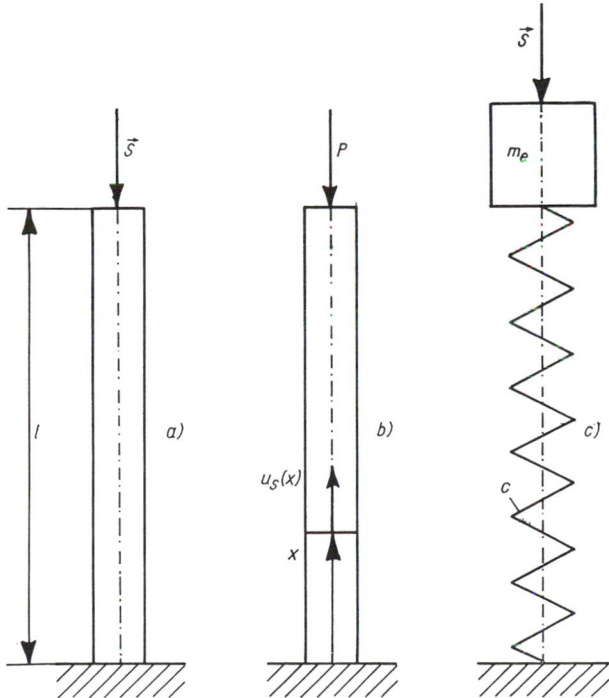

Fig. 12.2. (a) Axial impact (b) Static deformation of an elastic column (c) Simple equivalent system

considering the static axial deformation of a linear elastic column under the action of a dead-weight load P applied at the point of impact and pointing in the same axial direction (again see Fig. 12.2 for that auxiliary problem),

$$u_s(x) = -\frac{P}{EA} x = \frac{l\,P}{EA}\, \varphi(x) \ , \quad \varphi(x) = -x/l \ , \tag{12.6}$$

a compatible velocity distribution can be assumed to be affined and thus has the same linear distribution

$$\dot{u}'(x) = \dot{q}'\, \varphi(x) \ . \tag{12.7}$$

The generalized velocity $(dq/dt)'$ is just the speed of the impacted cross-section. Since momentum is not a positive-definite measure, the kinetic energy, which is stored in the column immediately following impact, must be considered to determine a simple equivalent system

$$T' = \frac{1}{2} \int_m \dot{u}'^2\, dm = \frac{\dot{q}'^2}{2} \int_m \varphi^2\, \rho A\, dx = \frac{1}{2}\, \frac{m}{3}\, \dot{q}'^2 = \frac{m^*}{2}\, \dot{q}'^2 \rightarrow m^* = m/3 \ . \tag{12.8}$$

The single mass m^* when hit by the same impulse takes on the (approximate) kinetic energy of the column and has the speed of the impacted cross-section. According to Fig. 12.2(c), the equivalent system is completed by assuming a mass-less spring of static stiffness $c^* = EA/l$ for support of the mass m^*. The force in the spring remains finite and unchanged during impact and the momentum is easily calculated from

$$I' - I = S, I = 0 \ , \ I' = -\dot{q}'\, m^*\, e_x \ , \ S = -S\, e_x \ , \ m^*\, \dot{q}' = S \ , \ m = \rho A\, l \ ,$$

$$\dot{q}' = \frac{S}{m^*} = \frac{3\,S}{m} \ , \ \dot{u}'(x, t_{0+}) = -\frac{3\,S}{m}\, \frac{x}{l} \ . \tag{12.9}$$

The motion after impact, when time is continuously increasing, is determined by these initial conditions. The linear distribution of the speeds of the cross-sections at t_{0+} suggests the further use of the same linear shape function for the deformation. See Sec. 12.6 for the dynamic magnification factor.

12.2. *Lagrange* Equations of Idealized Impact

The time integration of the *Lagrange* equations of motion of a MDOF-system (10.14), under idealized impact conditions

$$\lim_{(t_2 \to t_{0+}), (t_1 \to t_{0-})} \int_{t_1}^{t_2} \left[\frac{d}{dt}\left(\frac{\partial T}{\partial \dot{q}_i}\right) - \left(\frac{\partial T}{\partial q_i}\right) = Q_i \right] dt \ , \quad i = 1, 2, ..., n \ ,$$

$$(12.10)$$

when performed termwise, renders a set of difference equations $[(\partial T/\partial q_i)$ remains finite during impact]

$$\left(\frac{\partial T}{\partial \dot{q}_i}\right)' - \left(\frac{\partial T}{\partial q_i}\right) = H_i \ , \quad i = 1, 2, ..., n \ .$$

$$(12.11)$$

Note that the number n of degrees of freedom must remain unchanged during impact. The generalized impulse H_i results from the generalized force with infinite contributions, in the time limit of its time integral,

$$H_i = \lim_{(t_2 \to t_{0+}), (t_1 \to t_{0-})} \int_{t_1}^{t_2} Q_i(t) \, dt \ , \quad i = 1, 2, ..., n \ .$$

$$(12.12)$$

By considering the virtual "power" of the actual vector impulses $\boldsymbol{S}_k$, the generalized impulses H_i are determined by comparing coefficients in the relation, analogous to Eq. (10.6),

$$\sum_{k=1}^{m} \left(\boldsymbol{S}_k \cdot \delta \boldsymbol{r}_k\right) = \sum_{i=1}^{n} H_i(t) \, \delta \dot{q}_i \ , \quad \boldsymbol{r}_k = \boldsymbol{r}_k(\dot{q}_1, \dot{q}_2, ..., \dot{q}_n) .$$

$$(12.13)$$

Two examples illustrate the convenience of application of that set of finite difference relations of idealized impact (eg the construction of an equivalent system is no longer necessary).

12.2.1. Example: Impacting a Chain-Type Pendulum (MDOF-S)

A heavy chain of identical rigid rods connected by ideal joints hangs down from a crab of mass M ; a sketch is given in Fig. 12.3. The instantaneous velocity distribution following impact by an impulse S is to be determined. The lower end of the pendulum is free to move in the plane. The MDOF-system has $(n + 1)$ degrees of freedom that do not change during impact. The application of Eqs. (12.11) saves the free-body diagram and, hence, as well the consideration of the impulses transmitted through the joints. The system is assumed to be at rest before impact, $T = 0$, and the jump in kinetic energy is expressed according to Fig. 12.3 by the sum

$$T' = \frac{M}{2} v'^2 + \frac{m}{2} \sum_{i=1}^{n} v_i'^2 + \frac{m \, i_S^2}{2} \sum_{i=1}^{n} \omega_i'^2 \ , \quad i_S^2 = (2 \, l)^2/12 \ .$$

$$(12.14)$$

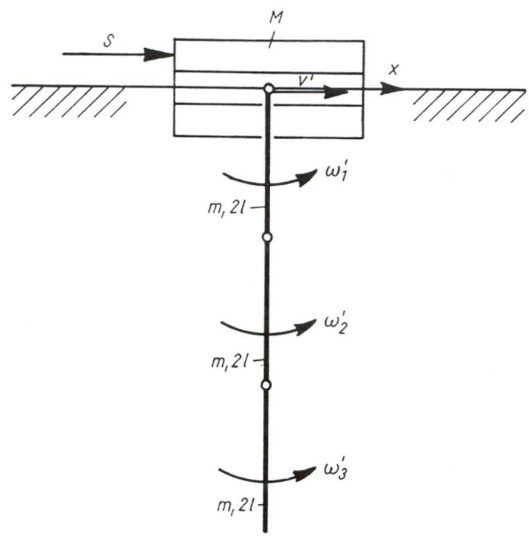

Fig. 12. 3. Impacting the crab with attached heavy chain

The speeds of the individual mass centers of the homogeneous chain links are related by the kinematic relations

$$v_i' = v' + 2l \sum_{k=1}^{i-1} \omega_k' + l\,\omega_i'\,.$$

(12.15)

The generalized impulses are found from the virtual "power" by comparing coefficients

$$S\,\delta v = H_x\,\delta v + \sum_{i=1}^{n} H_i\,\delta\omega_i \;\;\rightarrow\;\; H_x = S \;\;,\;\; H_i = 0 \;\;,\;\; i = 1, 2, ..., n\,.$$

(12.16)

For a chain with three links, Eqs. (12.11) become the 4 x 4 linear system

$$\left(\frac{M}{m} + 3\right)\frac{v'}{l} + 5\,\omega_1' + 3\,\omega_2' + \omega_3' = \frac{S}{m\,l}\,,$$

$$3\frac{v'}{l} + 6\,\omega_1' + 6\,\omega_2' + 4\,\omega_3' = 0 \;\;,\;\; 9\frac{v'}{l} + 18\,\omega_1' + 16\,\omega_2' + 6\,\omega_3' = 0\,,$$

$$15\frac{v'}{l} + 28\,\omega_1' + 18\,\omega_2' + 6\,\omega_3' = 0\,.$$

(12.17a)

Triangulation of the coefficient matrix renders the solution ($\Delta = 15 + 52\ M/m$)

$$v' = \frac{52}{\Delta}\ S/m\ ,\ \omega_1' = -\frac{33}{\Delta\,l}\ S/m\ ,\ \omega_2' = \frac{9}{\Delta\,l}\ S/m\ ,\ \omega_3' = -\frac{3}{\Delta\,l}\ S/m\ . \tag{12.17b}$$

12.2.2. Lateral Impact Loading of a Simply Supported (Elastic) Beam

The homogeneous beam of span l is simply supported and impacted at midspan by the impulse S ; see Fig. 12.4(a). Rigid-body motion is not possible in the case of laterally immobile supports. The velocity distribution of the mass elements $dm = \rho A\ dx$ must be compatible with the boundary conditions and may be assumed affined to a proper elastic deflection, eg to

$$\dot{w}'(x) = \dot{q}'\ \varphi(x)\ ,\ \ \varphi(x) = \cos\frac{\pi x}{l}\ ,\ \ -\frac{l}{2} \le x \le \frac{l}{2}\ . \tag{12.18}$$

An alternative to the above basic natural mode shape of a homogeneous beam is the static deflection under a single force loading $\varphi(x) = 1 - 6\ (x/l)^2 + 4\ (|x/l|)^3$. The generalized velocity is coincident with the speed of the cross-section at $x = 0$ after impact.

The kinetic energy is approximated by substituting the velocity distribution, the cosine function is used, and (for slender beams the translational portion is only considered)

$$T' = \frac{1}{2} \int_m \dot{w}'^2\ dm = \frac{\dot{q}'^2}{2} \int_{-l/2}^{l/2} \varphi^2(x)\ \rho A\ dx = \frac{m_e}{2}\ \dot{q}'^2\ ,\ \ m_e = m/2\ . \tag{12.19}$$

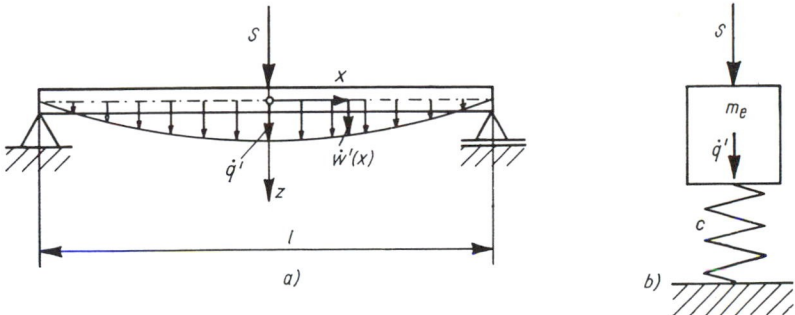

Fig. 12.4. (a) Lateral impact loading of an elastic beam (b) Equivalent system

The foregoing expression corresponds to the equivalent system of Fig. 12.4(b) and the momentum relation analogous to Eq. (12.9) becomes easily applicable. When proceeding along the lines of the integrated *Lagrange* equation, however, the equivalent system is not considered explicitly, and the generalized impulse is determined properly by comparing coefficients (the impulse S is applied at $x = 0$)

$$H \, \delta\dot{q} = S \, \delta w|_{x=0} = S \, \delta\dot{q} \; , \;\; H = S \; . \tag{12.20}$$

Equation (12.11) takes on the form

$$\left(\frac{\partial T}{\partial\dot{q}}\right)' = m_e \, \dot{q}' = S \; , \;\; \dot{q}' = 2S/m \;\rightarrow\; \dot{w}'(x) = \frac{2S}{m} \cos\frac{\pi x}{l} \; . \tag{12.21}$$

The *Lagrange* formulation makes it possible to consider a MDOF equivalent system by properly selecting several admissible functions in the approximation (12.18).

12.3. Idealized Elastic and Inelastic Impact Processes

The impulse of the contact force between two colliding bodies remains undetermined if no assumption is made concerning the physical behavior, ie about the loss of energy during the impact process. Two extreme cases are considered below.

12.3.1. The Idealized Elastic Impact

No mechanical energy is dissipated in this case and the conservation of mechanical energy of the two colliding bodies, Eq. (8.8), holds during impact. Assuming that the bodies are elastic guarantees the existence of the potential of the internal forces, and taking the time limit yields

$$T' + V' = T + V \; , \;\; V = U + W_P \; . \tag{12.22}$$

Since any initial deformations remain frozen during collision, the potential energy is unaltered and the relation reduces to conservation of the gross kinetic energy

$$T' - T = 0 \; , \;\; V' = V. \tag{12.23}$$

See 12.3.3. for a first and simple application and note the derivation of a finite relation which is linear in velocity.

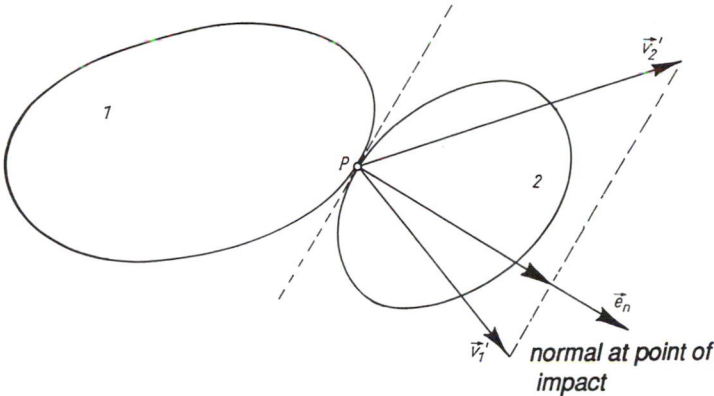

Fig. 12.5. Inelastic impact. The kinematic condition of Eq. (12.24)

12.3.2. The Idealized Inelastic Impact

In the case of utmost dissipation, it is assumed that the colliding bodies do not separate immediately after impact. The surface points of contact (P in Fig. 12.5, where the existence of a tangential plane is indicated), take on a common component of velocity in the direction of the "impact normal" $\boldsymbol{e}_n$, at the end of the collision process,

$$(\mathbf{v_1'} - \mathbf{v_2'}) \cdot \mathbf{e}_n = 0 . \tag{12.24}$$

12.3.3. Example: Collision of Two Point Masses

The collision of two mass points is illustrated by the more appealing problem of central impact (the impact normal passes through both mass centers) of the bodies m_1 and m_2 that are in translational motion, as shown in Fig. 12.6. The free-body diagram

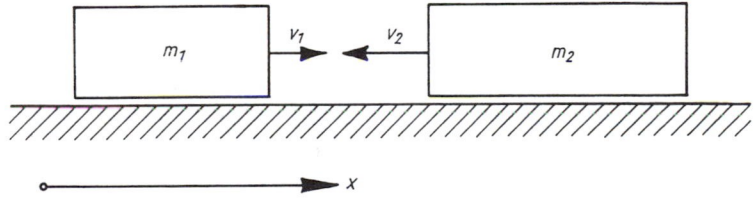

Fig. 12.6. Central impact of two masses.
Idealized elastic and inelastic collision, respectively

at impact is considered with impulse S and $-S$ acting on the mass m_2 and m_1, respectively. The momentum relation (12.2) is applied. The components in the x direction give

$$m_1 \left(v_1' - v_1\right) = -S, \tag{12.25}$$

$$m_2 \left(v_2' - (-v_2)\right) = S. \tag{12.26}$$

Addition eliminates the (internal) impulse and the momentum equation of the gross system becomes

$$m_1 \left(v_1' - v_1\right) + m_2 \left(v_2' + v_2\right) = 0. \tag{12.27}$$

An additional assumption about physical behavior during impact is required to determine the three unknowns v_1', v_2', and S.

(§)   Idealized Elastic Impact. The conservation of energy (12.23) of the gross system renders

$$T' = \frac{m_1 v_1'^2}{2} + \frac{m_2 v_2'^2}{2} = T = \frac{m_1 v_1^2}{2} + \frac{m_2 v_2^2}{2}. \tag{12.28}$$

Expansion of the difference of squares gives

$$m_1 \left(v_1' + v_1\right)\left(v_1' - v_1\right) + m_2 \left(v_2' + v_2\right)\left(v_2' - v_2\right) = 0, \tag{12.29}$$

and by substitution of Eqs. (12.25) and (12.26), the impulse becomes a common factor, $S \neq 0$, and cancels

$$\left(v_1' + v_1\right)\left(-S\right) + \left(v_2' - v_2\right) S = 0. \tag{12.30}$$

A linear equation results and replaces the nonlinear energy relation

$$-v_1' + v_2' - v_1 - v_2 = 0. \tag{12.31}$$

Hence, with Eq. (12.27), the elastic impact terminates with a new velocity distribution, $\mu = m_2/m_1$ is the mass ratio, and the impulse is determined from either one of Eqs. (12.25) and (12.26)

$$v_1' = \left[(1 - \mu) v_1 - 2 \mu v_2\right]/(1 + \mu), \quad v_2' = \left[2 v_1 + (1 - \mu) v_2\right]/(1 + \mu), \tag{12.32}$$

$$\rightarrow S = 2 m_2 \left(v_1 + v_2\right)/(1 + \mu).$$

In the limit $m_1 \rightarrow \infty$, $\mu \rightarrow 0$, the elastic reflection of the mass m_2 at a rigid wall is determined.

(§)   Idealized Inelastic Impact. Since Eq. (12.24) applies, the inelastic collision is completely determined through the third equation

$$v_1' - v_2' = 0 .$$
(12.33)

Equations (12.27) and (12.26) when combined with Eq. (12.33) yield the common velocity after impact, and, hence, the impulse takes on a different value

$$v_1' = v_2' = (v_1 - \mu v_2)(1 + \mu) \quad \rightarrow \quad S = m_2 (v_1 + v_2)(1 + \mu) .$$
(12.34)

The energy dissipation is given by the difference of the gross kinetic energies before and after inelastic impact

$$T' - T = \frac{- m_2}{2 (1 + \mu)} (v_1 + v_2)^2 .$$
(12.35)

In that case, the limit $m_1 \rightarrow \infty$, $\mu \rightarrow 0$, renders the inelastic impact of a mass m_2 at a rigid wall with total dissipation of its energy.

12.4.   The "Ballistic" Pendulum and the Center of Impact

A point mass m_1 , eg a bullet, hits a planar pendulum with a given mass distribution. A rigid-body rotation about the support 0 is possible subsequently following impact. Hence, the linear velocity distribution of the pendulum is compatible with support conditions. With the configuration shown in Fig. 12.7, the equations of momentum and of angular momentum are applied to the free-body diagram of the collision

$$m_1 (v_1' - v_1) = -S , \quad m_2 (v_2' - v_2) = S - S_0 , \quad v_2' = s_2 \omega_2' ,$$

$$m_2 i_0^2 (\omega_2' - \omega_2) = I S , \quad i_0^2 = i_2^2 + s_2^2 , \quad v_2 = \omega_2 = 0 .$$

$$\rightarrow v_1' - v_1 = \frac{- \mu i_0^2}{I} \omega_2' , \quad \mu = m_2 / m_1 .$$
(12.36)

A fourth relation is derived from the physical condition of the impact.

(§)   Conservation of Energy. Such conservation during elastic impact gives

$$T' = T = \frac{1}{2} m_1 v_1'^2 + \frac{1}{2} m_2 i_0^2 \omega_2'^2 = \frac{1}{2} m_1 v_1^2 ,$$
(12.37)

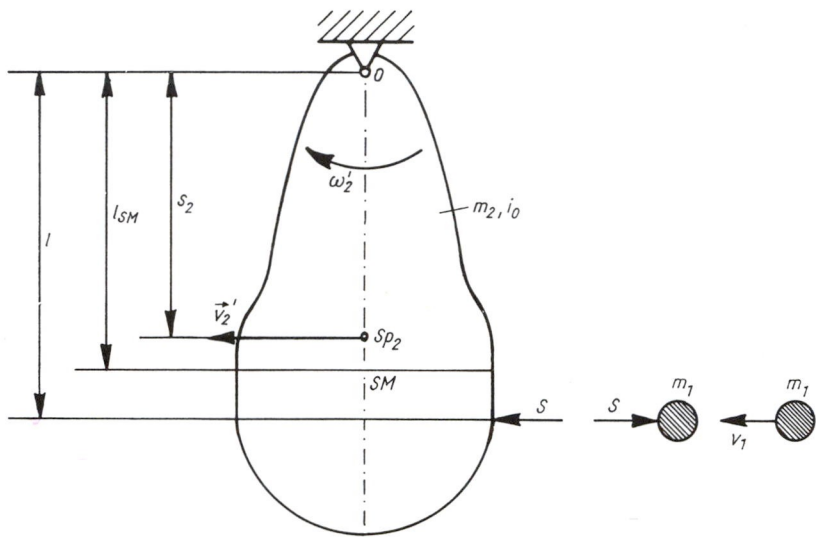

Fig. 12.7. The "ballistic" pendulum. Center of impact SM below the center of gravity Sp_2

or equivalently,

$$\left(v_1' - v_1\right)\left(v_1' + v_1\right) = -\mu i_0^2\, \omega_2'\, \omega_2' . \qquad (12.38)$$

Substituting for $(v_1' - v_1)$ and for ω_2' as well and cancelling $S \neq 0$ render the linear equivalent to Eq. (12.37)

$$v_1' + v_1 = l\, \omega_2' . \qquad (12.39)$$

(§)   Inelastic Impact. Equation (12.24) becomes according to Fig. 12.7, note the common normal,

$$v_1' - l\, \omega_2' = 0 . \qquad (12.40)$$

The solutions of the four linear equations are sampled in Table 12.1 below.

By inspection of the last row of Table 12.1 it is seen that the nominator in the expression for the reaction impulse is one and the same in both idealized cases. That is, the support is kept free from any dynamic reaction if the nominator vanishes

$$i_0^2 - l\, s_2 = 0 . \qquad (12.41)$$

Table 12.1:

Impact	elastic	inelastic
ω_2'	$\dfrac{2l}{l^2 + \mu i_0^2} v_1,$	$\dfrac{l}{l^2 + \mu i_0^2} v_1,$
v_1'	$\dfrac{l^2 - \mu i_0^2}{l^2 + \mu i_0^2} v_1,$	$\dfrac{l^2}{l^2 + \mu i_0^2} v_1,$
S	$\dfrac{2\, m_2 i_0^2}{l^2 + \mu i_0^2} v_1,$	$\dfrac{m_2 i_0^2}{l^2 + \mu i_0^2} v_1,$
S_0	$2 m_2 \dfrac{i_0^2 - l s_2}{l^2 + \mu i_0^2} v_1,$	$m_2 \dfrac{i_0^2 - l s_2}{l^2 + \mu i_0^2} v_1.$

The equation is solved for the distance $l = l_{SM}$, which determines the center of impact

$$l_{SM} = s_2 + i_2^2/s_2 . \qquad (12.42)$$

The radius of inertia i_2 with respect to the center of gravity is used above to explicitly show that $l_{SM} > s_2$. The center of impact is of practical importance for a hammer and similar tools or sporting goods (eg tennis rackets) in which one wants to avoid the dynamic loading of the support 0 during collision.

In the case of the ballistic pendulum, the angle of rotation α to the first turning point of the motion, which subsequently follows impact, is measured. Conservation of energy applies to the pendulum vibration in the gravity field, which follows any kind of impact. Hence, assuming $m_1 \ll m_2$ and that no friction exists renders

$$\frac{m_2\, i_0^2}{2}\, \omega_2'^2 = m_2\, g\, s_2 (1 - \cos \alpha), \qquad (12.43)$$

where the proper initial value ω_2' is to be substituted from Table 12.1.

12.5. Sudden Fixation of an Axis of Rotation

Suddenly fixing an axis of a freely moving rigid body renders inelastic impact. The example of a rolling wheel hitting the edge of a sidewalk illustrates such a sudden fixation in Fig. 12.8. At 0, a

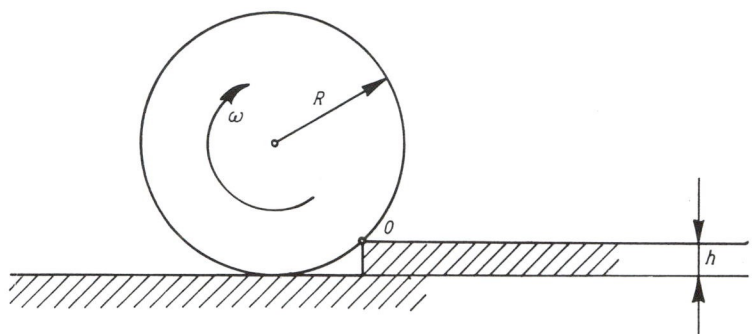

Fig. 12.8. A rolling wheel hitting suddenly the edge of a sidewalk

frictional cone exists, and the impulse of the contact force is assumed to stay within the interior region (slip velocity is assumed to be zero). The kinematic condition of pure rolling contact is understood before impact, and at this point, the contact force remains finite during the fixation. The axis through the point 0 of direction e is suddenly fixed in space, the impulse passes through the axis, and thus, the angular momentum with respect to that axis e remains unchanged [see Eqs. (12.5) and (7.28)] and M is the center of mass

$$\mathbf{H_0'} \cdot \mathbf{e} = \mathbf{H_0} \cdot \mathbf{e} \quad , \quad \mathbf{H_0} = \mathbf{H_M} + \mathbf{r_M} \times m\mathbf{v_M} \ . \tag{12.44}$$

The application of that general result to the plane problem of Fig. 12.8 gives at once

$$H_0' = H_0 = I\omega' + (mR\omega')R = I\omega + (mR\omega)(R-h) \quad \rightarrow \tag{12.45}$$

$$\omega' = \omega \left[1 - \frac{h}{R} \frac{1}{1 + (i_M/R)^2} \right] . \tag{12.46}$$

12.6. Dynamic Magnification Factor of Axial and Lateral Impact

The impulse S was given in Examples 12.1.2 and 12.2.2. Assuming the normal impact by a point mass M with given speed v and applying the equation of momentum to the impacting mass point (in a free-body diagram)

$$M(v' - v) = -S,   \qquad (12.47)$$

eliminate the impulse S, but introduce the new unknown speed v' of reflection. The missing equation is supplied by a further assumption on physical behavior of the collision process. Purely inelastic impact renders, eg

$$v' - \dot{q}' = 0 \;\rightarrow\; S = M(v - \dot{q}') = m_e\,\dot{q}' \;\rightarrow\; \dot{q}' = \frac{v}{1 + \mu_e}, \quad \mu_e = \frac{m_e}{M} = \beta\frac{m}{M}. \qquad (12.48)$$

The speed of the impacted cross-section becomes a fraction of the initial speed of the mass M, and the coefficient β is approximately $1/3$ in the case of axial impact of a rod and $1/2$ for lateral impact at midspan of a beam [see Eqs. (12.8) and (12.19), $m_e = m^*$].

The motion of the elastic rod after impact is conservative, and damping can be neglected since it has little influence on the dynamic amplitude a of the first turning point of the free vibration. Conservation of energy (8.8) yields, the point mass M is assumed to sit on the rod,

$$T' = \frac{1}{2}(M + m_e)\,\dot{q}'^2 = V|_{T=0} = -Mg\,a + \frac{1}{2}c\,a^2, \qquad (12.49)$$

$$c = EA/l \;\text{...axial compression,}$$
$$c = \pi^4\,EJ/2\,l^3 \;\text{...beam loaded at midspan, or } c = 48\,EJ/l^3.$$

Substituting Eq. (12.48) renders the dynamic magnification factor ($a_s = Mg/c$ is the proper portion of static deflection under the additional weight)

$$\chi = a/a_s = 1 + \sqrt{1 + \frac{v^2}{g\,a_s\,(1 + \beta\,m/M)}} \;\geq 2. \qquad (12.50)$$

In the case of an undamped linear vibration, the maximal dynamic stresses can be derived simply from the resultants (see Fig. 12.4)

Elastic column: $\max|N| = \chi\,Mg$, $\qquad (12.51)$

Beam, simply supported: $\max\{M_y(x = 0)\} = \chi\,\dfrac{l\,Mg}{4}$. $\qquad (12.52)$

Note that putting $v = 0$ corresponds to a sudden loading by the weight Mg when multiplied by a *Heaviside* time step function, and the dynamic magnification factor takes on its minimal value: $\chi = 2$. The formula (12.50) is applicable, eg when a crane puts down its load on an elastic structure.

12.7. Axial Impact of a Thin Elastic Rod, Wave Propagation

The rate of axial momentum of a mass element $dm = \rho A\, dx$ is given
by Eq. (7.7), cross-sections are assumed to remain plane, lateral
inertia is neglected, $N(x, t) = A\, \sigma_{xx} (x, t)$ is the axial force, and $u(x,
t)$ is the displacement of the cross-section,

$$\frac{\partial N}{\partial x} = \rho A \frac{\partial^2 u}{\partial t^2} \quad .$$

(12.53)

By means of *Hooke's* law,

$$\sigma = E \frac{\partial u}{\partial x} \;\to\; N = EA \frac{\partial u}{\partial x} \quad ,$$

(12.54)

the normal force may be eliminated to render the (hyperbolic) wave
equation

$$\frac{\partial}{\partial x}\left(EA \frac{\partial u}{\partial x}\right) = \rho A \frac{\partial^2 u}{\partial t^2} \quad .$$

(12.55)

Or alternatively, the displacement is eliminated to yield

$$E \frac{\partial}{\partial x}\left(\frac{1}{\rho A} \frac{\partial (A\,\sigma_{xx})}{\partial x}\right) = \frac{\partial^2 \sigma_{xx}}{\partial t^2} \quad .$$

(12.56)

For a homogeneous rod, the classical wave equation results

$$c^2 \frac{\partial^2 u}{\partial x^2} = \frac{\partial^2 u}{\partial t^2} \quad , \quad c^2 \frac{\partial^2 \sigma_{xx}}{\partial x^2} = \frac{\partial^2 \sigma_{xx}}{\partial t^2} \quad , \quad c = \sqrt{E/\rho} \quad ,$$

(12.57)

where c is the sound velocity of the elastic rod surrounded by a
vacuum, ie the speed of propagation of small disturbances. Instead
of dealing with the two wave equations (12.57), the system of two
equations of the first order in the space coordinate may be used

$$\frac{\partial \sigma_{xx}}{\partial x} = \rho \frac{\partial^2 u}{\partial t^2} \quad , \quad \sigma_{xx} = \rho c^2 \frac{\partial u}{\partial x} \quad .$$

(12.58)

Substituting the deformation rate $v = \partial u/\partial t$ renders the first-order
system of partial differential equations "symmetrized"

$$\frac{\partial \sigma_{xx}}{\partial x} = \rho \frac{\partial v}{\partial t} \quad , \quad \frac{\partial \sigma_{xx}}{\partial t} = \rho c^2 \frac{\partial v}{\partial x} \quad .$$

(12.59)

If the axial force N is inserted, the density is substituted by the mass per unit of length ρA. The general D´Alembert´s solution is the superposition of waves traveling right and left

$$u(x, t) = g\,(t - x/c) + G\,(t + x/c) \;,\; v(x, t) = s\,(t - x/c) + S\,(t + x/c) \;, \quad (12.60)$$

$$v = \frac{\partial u}{\partial t} \;\rightarrow\; s = \dot{g} \;,\; S = \dot{G} \;;\; \sigma_{xx}\,(x, t) = f\,(t - x/c) + F\,(t + x/c) \;,\; \sigma_{xx} = \rho c^2 \frac{\partial u}{\partial x} \;.$$

Thus, the stress functions are further related to

$$f = \rho c^2 \left(-\frac{1}{c}\,\dot{g} \right) = -\rho c\,\dot{g} = -\rho c\,s \;,\; F = \rho c^2 \left(\frac{1}{c}\,\dot{G} \right) = \rho c\,\dot{G} = \rho c\,S \;. \quad (12.61)$$

The stress f and deformation rate s are linearly related and the coefficient of proportionality ρc is the mechanical impedance (in analogy to the electric one).

Subsequently, a stress wave is considered that is initiated at $t = 0$ at the end $x = 0$ by an impact, which propagates in a rod of finite length l,

$$\sigma_0(x, t) = f\,(t - x/c) \;,\; 0 \le t \le l/c \;. \quad (12.62)$$

The wave reaches the free end after traversing the span l in time l/c. The dynamic boundary condition requires that the normal stress vanish

$$\sigma(l, t) = f\,(t - l/c) + \sigma_r\,(l, t) = 0 \;. \quad (12.63)$$

The reflected wave must thus have the same time function, but an alternate sign, and it travels in the opposite direction

$$\sigma_r\,(x, t) = -f\left(t - l/c + \frac{x - l}{c} \right) \;,\; l/c \le t \le 2l/c \;. \quad (12.64)$$

The wave pattern that renders zero stress at $x = l$ is given by superposition

$$\sigma_0 + \sigma_r = f\,(t - x/c) - f\left(t - l/c + \frac{x - l}{c} \right) \;,\; l/c \le t \le 2l/c \;. \quad (12.65)$$

If the cross-section at $x = 0$ is already unloaded at time $t = 2\,l/c$, the second reflection takes place under a zero stress condition, the sign of the stress changes again, and the total field is given by

$$\sigma_0 + \sigma_r + \sigma_{rr} = f\,(t - x/c) - f\left(t - l/c + \frac{x - l}{c} \right) + f\left(t - 2\,l/c - \frac{x}{c} \right) \;,\; 2\,l/c \le t \le 3\,l/c \;. \quad (12.66)$$

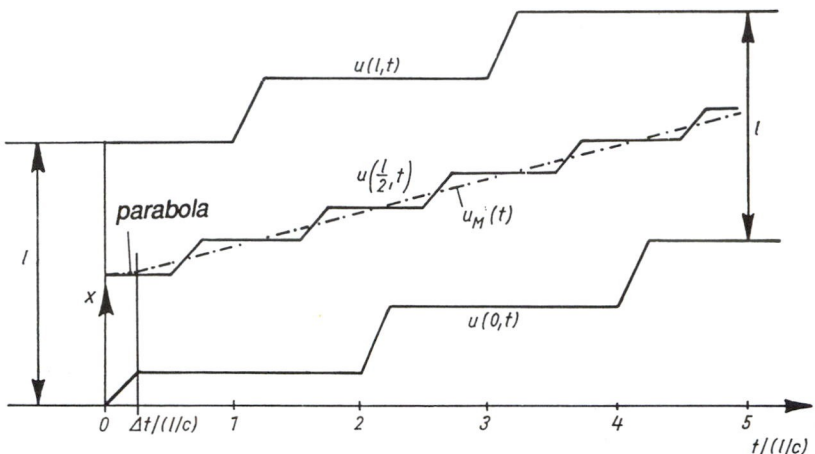

Fig. 12.9. Motion of an elastic rod of length l after impact of short duration Δt

Since material damping is neglected and dispersion of the waves by lateral inertia is not considered, the train of waves keeps its original pulse shape of the initial wave. The motion is greatly simplified in mathematical terms if the impact loading is assumed to be given by a constant force P acting during the time interval Δt. The associated total impulse is $S = P \, \Delta t$ and $\sigma_0 \leq 0$ [$H(t)$ is the *Heaviside* step function]

$$\sigma_0(x, t) = \frac{P}{A} \left[H\left(t - x/c\right) - H\left(t - \frac{x - \delta}{c}\right) \right] , \quad \delta = c \, \Delta t .$$

(12.67)

The compressional wave accelerates the mass elements $\rho A \, \delta$ within the space dimension δ of the box-type pulse when propagating from the left to the right. The velocity becomes

$$v(x, t) = \frac{P}{\rho c A} \left[H\left(t - \frac{x - \delta}{c}\right) - H\left(t - x/c\right) \right] .$$

(12.68a)

At the end of the loading time, however, the mass center takes on the constant speed given by the impulse and Eq. (7.10)

$$\rho A l \, v_M = P \, \Delta t , \quad t > \Delta t .$$

(12.68b)

It is also the mean value of the velocity distribution of the elastically deforming rod. The diagram of Fig. 12.9 shows the common effects of compressive waves propagating to the right and of tensile waves propagating to the left. In the case of sharp loading of a rod of a material sensitive to tension, tensile fracture may occur despite original loading in compression, due to the sign change during reflection. In general, this effect is called "spalling"; see, eg J. D. Achenbach: *Wave Propagation in Elastic Solids. North-Holland, Amsterdam, 1975.*

12.8. Water Hammer, Wave Propagation

The problem of control of the flow through a straight pipe (sketched in Fig. 7.8) is considered taking into account compressibility of the fluid and (linear) elasticity of the pipe wall. Any (small) pressure disturbance propagates with the speed of sound of the combined system. If the pressure of the stationary flow is changed at $x = L$, the pressure head propagates upstream toward the (large) container. At the ends, partial or total reflection with or without a change of sign in the pressure wave takes place; see, eg Eq. (12.63). The basic equations are derived by considering the cylindrical control surface shown in Fig. 12.10. The cross-section is assumed to coincide with the pipe dimensions under stationary flow conditions, $A = \pi a^2$. Due to a nonstationary increase of the pressure, the diameter of the pipe is widened and its length increased. Thus, a radial flow of mass through the control surface must be considered in the mass balance equation (1.72). If we neglect the convective change of the fluid density (the condition $|v\, \partial\rho/\partial x| << |\partial\rho/\partial t|$ is verified later), the nonstationary rate of mass is approximately given by

$$\frac{\partial\rho}{\partial t}\, A\, dx = -A\,(\mu_1 + \mu_2) - \dot{m}_r \ , \ \ \mu_1 = -\rho v \ , \ \ \mu_2 \approx \rho\left(v + \frac{\partial v}{\partial x}\, dx\right). \tag{12.69}$$

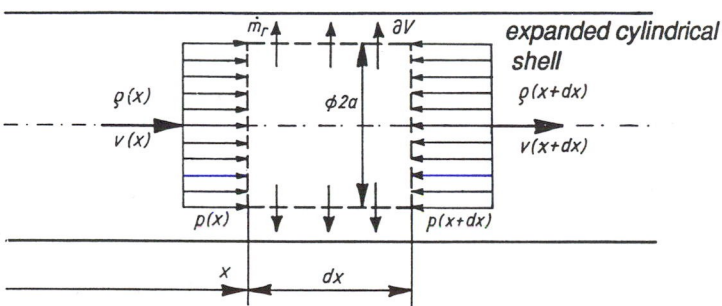

Fig. 12.10. A control volume element in the nonstationary elastically deforming pipe

The radial rate of the flow is approximately determined by considering the quasistatic elastic deformation of the pipe (the inertia of the solid material is neglected) under the additional time-dependent loading by the dynamic pressure $(\partial p/\partial t) \, dt$. The hoop stress in a thin-walled cylindrical shell due to internal pressure is given by Eq. (2.98), $\sigma_{\varphi\varphi} = p \, a/s$, thickness $s \ll a$. The axial stress $\sigma_{xx} = \alpha \, \sigma_{\varphi\varphi}$ depends on the boundary conditions with respect to any support of the pipe. $\alpha = 0$ reflects free expansion in the axial direction, $\alpha = v$ holds under plane strain conditions, $\alpha = 1/2$ if the pipe is closed and not supported externally. *Hooke's* law on isotropic material relates the membrane stresses and principal strains (note the linearized geometric relations in cylindrical coordinates)

$$\varepsilon_{\varphi\varphi} = \frac{u}{a} = \frac{1}{E}\left(\sigma_{\varphi\varphi} - v\,\sigma_{xx}\right) \, , \; \varepsilon_{xx} = \frac{\Delta l}{l} = \frac{1}{E}\left(\sigma_{xx} - v\,\sigma_{\varphi\varphi}\right) . \tag{12.70}$$

The static increase of volume of a pipe of length l under internal pressure p thus becomes

$$\pi\,(a + u)^2\left(l + \Delta l\right) - \pi a^2 \, l \approx \pi a^2 \, l \, \frac{p\,a}{E\,s}\,\beta \, , \; \beta = 2 + \alpha - v\,(1 + 2\alpha) .$$

Hence, the mass leaving the control surface in the radial direction due to the quasistatic pipe deformation when the pressure increases from its value $p(x)$ to $p(x) + (\partial p/\partial t) \, dt$ is approximately given by

$$\dot{m}_r \, dt \approx \rho A \, dx \, \frac{\partial p}{\partial t} \, dt \, \frac{a}{E\,s}\,\beta \; . \tag{12.71}$$

The time increment cancels.

By assuming a linear compressibility of the barotropic fluid, Eq. (2.87),

$$dp = K_F \, \frac{d\rho}{\rho} \; \rightarrow \; \frac{\partial \rho}{\partial t} = \frac{\rho}{K_F}\,\frac{\partial p}{\partial t} \, , \tag{12.72}$$

the fluid density can be eliminated and the continuity equation (12.69) yields the first-order partial differential equation

$$\frac{\partial p}{\partial t} = -\rho c^2 \, \frac{\partial v}{\partial x} \; , \; c^2 = \frac{E_{eff}}{\rho} \, , \; E_{eff} = \left(\frac{1}{K_F} + \frac{\beta \, a}{E\,s}\right)^{-1} . \tag{12.73}$$

The sound velocity c of (steel) life lines is of the order of *1000* m/s . The effective modulus E_{eff} corresponds to springs in a series connection: the compressible fluid and elastic wall of the pipe [see Fig. A 3.1 and Eq. (12.57)].

Conservation of momentum, Eq. (7.13), gives (the local acceleration dominates)

$$\frac{\partial v}{\partial t}\, \rho A\, dx = -A\, \frac{\partial p}{\partial x}\, dx \quad , \quad \left| v\frac{\partial v}{\partial x} \right| \ll \left| \frac{\partial v}{\partial t} \right| ,$$

(12.74)

and hence, a second first-order partial differential equation results

$$\frac{\partial p}{\partial x} = -\rho\, \frac{\partial v}{\partial t} .$$

(12.75)

Alternatively, velocity v or pressure p can be eliminated from Eqs. (12.73) and (12.75) to render the wave equations, respectively,

$$\frac{\partial^2 p}{\partial t^2} = c^2\, \frac{\partial^2 p}{\partial x^2} \quad , \quad \frac{\partial^2 v}{\partial t^2} = c^2\, \frac{\partial^2 v}{\partial x^2} .$$

(12.76)

Their *D´Alembert´s* solutions of right and left traversing waves are

$$p - p_0 = f\left(t - \frac{x}{c}\right) + F\left(t + \frac{x}{c}\right) \,, \quad v - v_0 = \frac{1}{\rho c}\left[f\left(t - \frac{x}{c}\right) - F\left(t + \frac{x}{c}\right) \right].$$

(12.77)

Linearization by omitting the nonlinear convective terms in the basic equations is easily verified to be safe, since, in general, $v \ll c$, and $|v\, \partial v/\partial x| = (v/c)\, |\partial v/\partial t|$, $|v\, \partial p/\partial x| = (v/c)\, |\partial p/\partial x|$, etc. In engineered lifelines $v_0\, /c \le 10^{-2}$.

The rapid closing of a gate in a pipeline conveying a stationary flow yields a jump in pressure

$$p - p_0 = F\left(t + \frac{x-l}{c}\right) \quad , \quad v - v_0 = -\frac{1}{\rho c}\, F\left(t + \frac{x-l}{c}\right) ,$$

(12.78)

whose intensity and shape are determined by the boundary condition at $x = l$, $v(l) = 0$, in the limit of vanishing control time

$$-v_0 = -\frac{1}{\rho c}\, F(t) \,, \quad F(t) = \rho c v_0\, H(t).$$

(12.79)

Hence, the pressure wave that moves in the upstream direction

$$p - p_0 = \rho c v_0\, H\left(t + \frac{x-l}{c}\right) \,, \quad v - v_0 = -v_0\, H\left(t + \frac{x-l}{c}\right) ,$$

(12.80)

has mainly three effects:
1.   The flow behind the wave front vanishes; the fluid is at rest.
2.   The pressure behind the wave front is increased and the pipe

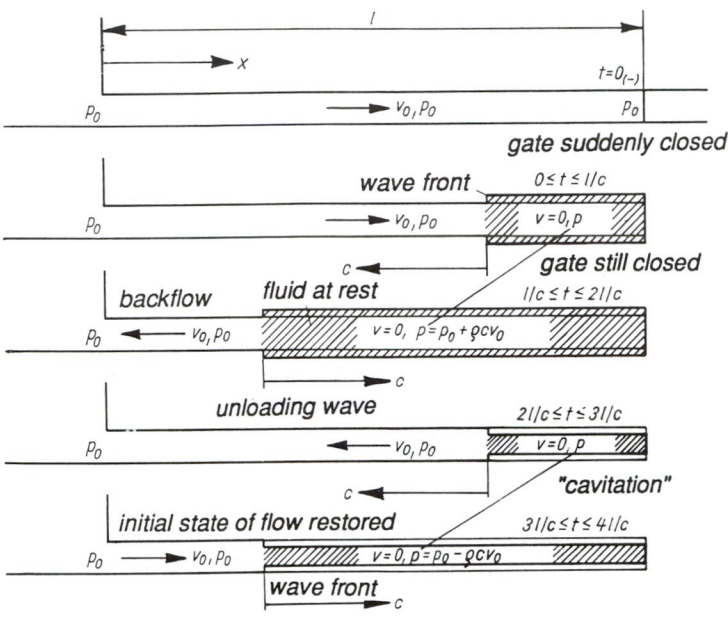

Fig. 12.11. Sudden shut-down of a lifeline. Wave trains without consideration of damping and dispersion (linear elastic quasistatic deformation of the pipe, linear compressible fluid, both taken into account)

is elastically expanded.

3.   The density of the fluid at rest is increased due to (linear) compressibility.

Due to the deformation of the pipe and the increased density, the fluid mass behind the wave front is increased with respect to the amount under stationary flow conditions. At the time of traverse $t = l/c$, the wave arrives at the free end, where at $x = 0$, the pressure of the fluid in a large container is assigned. The increased pressure and elasticity of the pipe thus yield a reversed flow from the pipe into the container. A reflected wave propagates to the right. The idealized boundary condition at $x = 0$ is $p = p_0$, and for $t \geq l/c$, the relation must hold

$$0 = F\left(t - \frac{l}{c}\right) + f\left(t - \frac{l}{c}\right) .$$

(12.81)

Hence, behind the wave front, the flow is reversed and the pressure is again p_0 (see Fig. 12.11)

$$p - p_0 = \rho c v_0 + f\left(t - \frac{l}{c} - \frac{x}{c}\right) = \rho c v_0 - \rho c v_0\, H\left(t - \frac{l}{c} - \frac{x}{c}\right), \quad l/c \leq t \leq 2\,l/c,$$

$$v - v_0 = -v_0 + \frac{1}{\rho c}\, f\left(t - \frac{l}{c} - \frac{x}{c}\right) = -v_0 - v_0\, H\left(t - \frac{l}{c} - \frac{x}{c}\right). \tag{12.82}$$

When the front arrives at the closed gate, at time $t = 2\,l/c$, a total reflection takes place (the boundary condition at $x = l$ is still $v = 0$)

$$0 = f\left(t - \frac{2l}{c}\right) - F\left(t - \frac{2l}{c}\right), \quad t \geq 2\,l/c, \tag{12.83}$$

and a new wave travels toward the container

$$v - v_0 = -2 v_0 - \frac{1}{\rho c}\, F\left(t - \frac{2l}{c} + \frac{x - l}{c}\right) = -2 v_0 + v_0\, H\left(t - \frac{2l}{c} + \frac{x - l}{c}\right),$$

$$p - p_0 = F\left(t - \frac{2l}{c} + \frac{x - l}{c}\right) = -\rho c v_0\, H\left(t - \frac{2l}{c} + \frac{x - l}{c}\right), \quad 2\,l/c \leq t \leq 3\,l/c. \tag{12.84}$$

Behind the wave front, the fluid at reduced density is at rest, and the pressure is decreased to $p = p_0 - \rho c v_0$. There is danger of cavitation near the gate and the thin-walled pipe may even buckle under the external atmospheric pressure when the compressive hoop stress becomes critical. After reflection at the free end at $x = 0$, at time $t = 3\,l/c$, the initial flow conditions are regained: The flow is again directed toward the closed gate. The wave front is sketched in Fig. 12.11. Damping due to viscosity and dispersion of the pressure wave due to lateral inertia weaken the waves after a few reflections. The effects of water hammer in lifeline systems have to be considered in the design stage. Special safety measures are required for elbows. In water plant systems with highly elevated reservoirs, alternating water hammer loadings must be avoided; see, eg Fig. 8.11.

12.9. Exercises A 12.1 to A 12.3 and Solutions

A 12.1: Determine the angle of reflection β of a ball rolling against a rigid wall (Fig. A 12.1). Impact is considered ideal elastic, and a no-slip condition applies at point B of contact.

Solution: The fixation of the axis through B gives the kinematic condition $v'\sin\beta - R\,\omega' = 0$. The angular momentum with respect to B is preserved: $I\,\omega' + m v'\,R\sin\beta - I\,\omega = 0$. The intermediate result is $\omega' = \omega/(1 + R^2/i^2)$ and $v'\sin\beta = v/(1 + R^2/i^2)$. Conservation of energy renders a third condition: $T' = T = (1/2)\,[m\,v'^2 + I\,\omega'^2] = (\omega^2/2)\,(I + m\,R^2)$. The speed of the mass center after impact thus becomes $v' = \omega\,[(R^2 + i^2)^3 - i^6]^{1/2}/(R^2 + i^2)$. The angle of reflection is

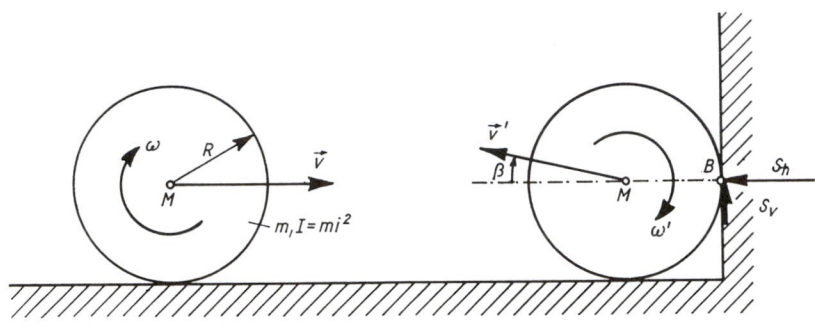

Fig. A 12.1.

determined by the inverse of $\sin \beta = R\,i^2 / [(R^2 + i^2)^3 - i^6]^{1/2}$. For a thin spherical shell, $i^2 = R^2/3$ and $\sin \beta = 1/\sqrt{21}$. The jump in momentum determines the components of the reaction impulse

$$mv' \sin \beta = S_v \quad , \quad mv' \cos \beta + mv = S_h \quad .$$

The reaction force must lie within the friction cone, $S_v/S_h \leq \mu$, and the minimal friction coefficient for the reflection of a thin shell ball takes on the value $\mu_{min} = 1/(4 + 2\sqrt{5})$.

A 12.2: A concentrated mass m_1 hits the tip of an elastic (homogeneous) cantilever (see Fig. A 12.2) with speed v. Assuming an inelastic and normal impact, determine the dynamic magnification factor. The simple admissible shape function of the deflection is $\varphi(x) = (x/l)^2$.

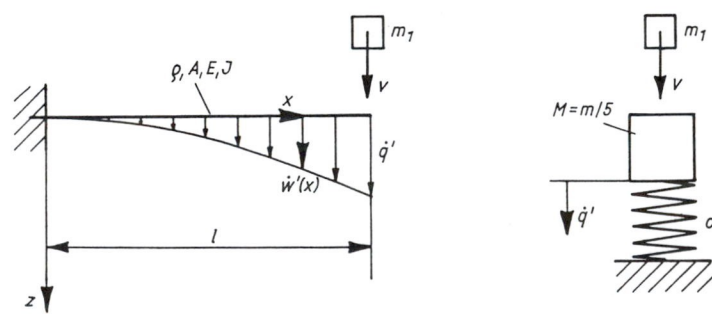

Fig. A 12.2.

Solution: The kinetic energy of the cantilever immediately after impact is given by

$$T' = \frac{1}{2} \int_0^l \dot{w}'^2 \, \rho A \, dx = \frac{M}{2} \dot{q}'^2 \ , \ \dot{w}'(x) = \dot{q}' \, \varphi(x) \ , \ m^* \equiv M = m/5 \ , \ m = \rho A \, l.$$

Consideration of the equivalent system as shown in Fig. A 12.2 or, in a straightforward manner by direct application of the integrated *Lagrange* equation (12.11) to the cantilever, gives, upon elimination of the impulse by means of the jump in momentum of the impacting mass,

$$M \dot{q}' = S \ , \ m(v' - v) = -S \ , \ \dot{q}' = v' \ \rightarrow \ \dot{q}' = v / (1 + M/m_1).$$

If we assume the mass m_1 is permanently attached to the cantilever also during the subsequent undamped natural vibration, the maximal deflection is easily determined from the conservation of energy

$$T' = \frac{(m_1 + M)}{2} \dot{q}'^2 = V(a) = -m_1 \, g \, a + \frac{c^*}{2} a^2 \ , \ a_s = m_1 \, g / c^* \ , \ c^* = 2 \, EJ/l^3.$$

The dynamic magnification factor results by solving the quadratic equation

$$\chi = a/a_s = 1 + \sqrt{1 + 2 \, T'/m_1 g a_s} \ .$$

A 12.3: A rigid and homogeneous pendulum of mass *M* and length *l* falls down from its initially vertical position. As shown in Fig. A

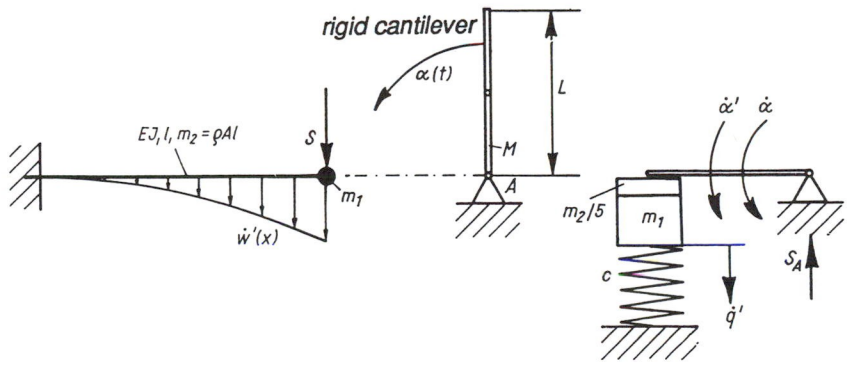

Fig. A 12.3.

12.3, it hits an elastic and homogeneous cantilever of mass m_2, with constant bending rigidity $B = EJ$, at the tip mass m_1. The shape function of Exercise A 12.2 may be used when determining the velocity distribution and internal impulse S after elastic impact.

Solution: Application of the law of conservation of energy to the free fall of the pendulum renders its angular velocity just before impact. The kinetic energy of the cantilever immediately after impact is approximately

$$\frac{m_1}{2} \dot{q}'^2 + \frac{1}{2} \int_0^l \dot{w}'^2 \, \rho A \, dx = \frac{1}{2} \left(m_1 + \frac{m_2}{5} \right) \dot{q}'^2 .$$

The impact equations applied separately to the pendulum and the cantilever yield, together with the assumption of conservation of total energy during elastic impact,

$$\left(m_1 + \frac{m_2}{5} \right) \dot{q}' = S , \quad \frac{ML^2}{3} \left(\dot{\alpha}' - \dot{\alpha} \right) = - SL ,$$

$$\frac{1}{2} \left(m_1 + \frac{m_2}{5} \right) \dot{q}'^2 + \frac{1}{2} \frac{ML^2}{3} \left(\dot{\alpha}'^2 - \dot{\alpha}^2 \right) = 0 .$$

Solving equivalently the linear counterparts of the equations for the unknowns gives the results

$$\dot{\alpha}' = \frac{M/3 - m_1 - m_2/5}{M/3 + m_1 + m_2/5} \dot{\alpha} , \quad \dot{q}' = \frac{2 M L}{M + 3 \, m_1 + 3 \, m_2/5} \dot{\alpha} ,$$

$$S = 2 M \frac{(m_1 + m_2/5) L}{M + 3 \, m_1 + 3 \, m_2/5} \dot{\alpha} .$$

1 3
Elementary Supplements of Fluid Dynamics

Hydrodynamic forces, resultants of pressure distributions, have been discussed in connection with the conservation of momentum of control volume. One ingredient of the theory is still missing: the "circulation." It is basic for the understanding of the most important of these forces, the *lift* . Its connection to the vortex vector, Eq. (1.50), ie to the rotation, is derived below. The effects of viscosity on *Newtonian* fluids (for the constitutive relations, see Sec. 4.2.1) are illustrated and the *Navier-Stokes* equations derived. The essential parameters of similarity solutions and, hence, of model testing in wind tunnels as well as in water channels are discussed. The boundary layer that develops in the flow along a plate is calculated. The irrotational motion of ideal fluids, which enters the solution, eg of the outer flow, is studied by means of potential theory, and a brief account of the singularity method is given. One application is the *von Karman* vortex trail in the wake behind a blunt body moving through a viscous fluid, with the notion of *drag* as a byproduct. Moving walls in contact with a fluid pose a nonstationary boundary value problem of interest in earthquake engineering and, eg they function as wave makers in a swimming pool. The effects of compressibility and *Mach* number are discussed in connection with stationary efflux from a pressure vessel into open air, a problem often encountered in engineering. It is, eg a strong source of noise in connection with the operation of steam power plants. For shock waves formed in supersonic flows, the reader is referred to the special literature on gas dynamics.

13.1. Circulation and the Vortex Vector

Circulation Γ in an ideal flow is of fundamental importance, since it is conserved when taken along a closed material line. It is a scalar kinematic parameter and it is defined as the line integral of the tangential velocity component around any closed curve C to be considered in the flow field and the loop integral is to be evaluated

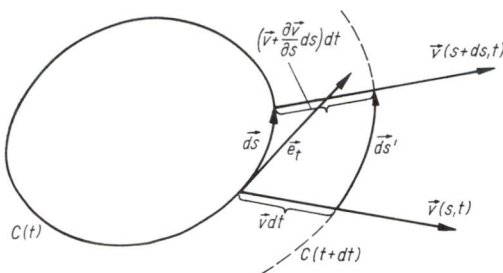

Fig. 13.1. Circulation. Deformation of an element of a fluidic (material) line $C(t)$

keeping the time t fixed. As shown in Fig. 13.1, the tangential component of velocity $\mathbf{v}$ is constructed by taking the scalar product, the arc-length is denoted s, the tangential unit vector is $\mathbf{e}_t$,

$$\Gamma = \oint_C (\mathbf{v} \cdot \mathbf{e}_t)\, ds .$$

(13.1)

Thomson (Lord Kelvin), inspired by the classical paper of *H. Helmholtz*, "On Integrals of the Hydrodynamic Equations Corresponding to Vortex Motions", in 1858, observed the rate of Γ to vanish when the closed curve C is a material line of an inviscid flow [$C(t)$ is then called a fluidic line]

$$\frac{d\Gamma}{dt} = 0 .$$

(13.2)

<<Circulation along a closed material line $C(t)$ remains constant in an inviscid flow if the external forces are irrotational.>>

    Thomson´s theorem on the permanence of circulation is easily proven For that purpose, the time derivative of Eq. (13.1) is taken, considering the closed fluidic line $C(t)$,

$$\frac{d\Gamma}{dt} = \frac{d}{dt} \oint_{C(t)} (\mathbf{v} \cdot \mathbf{ds}) = \oint_{C(t)} \left[\frac{d\mathbf{v}}{dt} \cdot \mathbf{ds} + \mathbf{v} \cdot \frac{d}{dt}(\mathbf{ds}) \right] ,$$

where acceleration $d\mathbf{v}/dt = \mathbf{a} = -(1/\rho)\, grad\, (W'_p + p)$ can be eliminated by substituting the *Euler* equation (8.30) (W'_p is the potential density of the irrotational external forces). The rate of deformation of the material arc-length differential $\mathbf{e}_t\, ds = \mathbf{ds}$ can be determined by inspection from Fig. 13.1, where the motion during time dt is sketched

$$\mathbf{v}\,dt + \mathbf{ds'} = \mathbf{ds} + \left(\mathbf{v} + \frac{\partial \mathbf{v}}{\partial s}\,ds\right)dt \quad \rightarrow \quad \lim_{dt \to 0} \frac{\mathbf{ds'} - \mathbf{ds}}{dt} = \frac{d}{dt}(\mathbf{ds}) = \frac{\partial \mathbf{v}}{\partial s}\,ds \;.$$

Substitution gives, when taking into account the identity $(\mathbf{v} \cdot \partial \mathbf{v}/\partial s)$ = $(\partial/\partial s)(v^2/2)$,

$$\frac{d\Gamma}{dt} = \oint_{C(t)} \frac{\partial}{\partial s}\left(\frac{v^2}{2} - \frac{p}{\rho} - \frac{W'}{\rho}\right)ds = 0 \;.$$

Incompressibility of the flow is assumed for convenience. The loop integral renders the term in parenthesis and vanishes for continuous velocity fields, QED. The theorem holds also in the more general case of a compressible barotropic flow, $\rho = \rho(p)$. An extension of the theorem to include the inhomogeneity of the fluid, which is of importance in meteorology, can be credited to *V. Bjerkness* in 1900. By inspection of Eq. (13.1), it is concluded that the circulation in an irrotational flow must be zero in the absence of singularities: Since $\mathbf{v} = grad\ \phi$ [see Eq. (1.51)], the integrand is the total derivative $d\phi$ and the loop integral vanishes.

From the above definition, it can be concluded that the circulation taken along a closed curve C and the vortex vector (1.50), ie the *curl* of the velocity field, must be related. The boundary of an infinitesimal rectangle in the *(x, y)* plane, Fig. A 1.6, may be considered as a special planar curve C. The infinitesimal circulation in the counterclockwise direction in a linear approximation is then given by the sum

$$d\Gamma = v_x\,dx + \left(v_y + \frac{\partial v_y}{\partial x}\,dx\right)dy - \left(v_x + \frac{\partial v_x}{\partial y}\,dy\right)dx - v_y\,dy$$

$$= \left(\frac{\partial v_y}{\partial x} - \frac{\partial v_x}{\partial y}\right)dx\,dy = 2\,\omega_z\,dA \;, \quad dA = dx\,dy \;, \quad \boldsymbol{\omega} = \frac{1}{2}\,curl\,\mathbf{v} \;. \tag{13.3}$$

The last expression contains the normal component of the vortex vector $\boldsymbol{\omega}$ and the small area dA enclosed by an infinitesimal closed curve C of any shape; thus, it is independent of the original cartesian frame. Even the assumption of a plane curve can be removed, and in the limit, as the area approaches zero, the relation of the infinitesimal circulation and the normal component of the vortex vector hold, in an exact manner, for any doubly curved surface element dS with the normal $\mathbf{e}_n$

$$d\Gamma = 2\,\omega_n\,dS \;, \tag{13.4}$$

where $\omega_n = (\boldsymbol{\omega} \cdot \mathbf{e}_n)$.

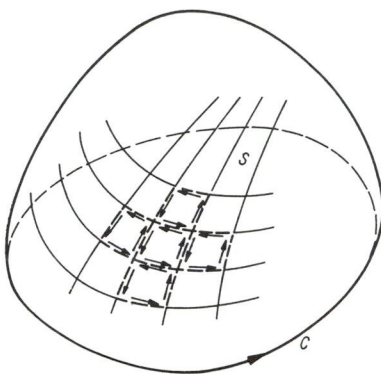

Fig. 13.2. (a) Surface S with edge C in an ideal flow. Definition of circulation by
loop integration and, alternatively, by the flow of rotation through S

Consider a closed three-dimensional curve C as shown in Fig.
13.2(a), as the base of a surface S. The latter is determined by the
net of curved coordinate lines sketched in the figure. Circulation can
be determined by summing the left-hand side of Eq. (13.4) along all
subdivisions (what remains is the integral along the perimeter of C)
and conversely by summing the right-hand side with respect to any
surface S with base C. The result is *Stokes's* integral theorem

$$\Gamma = \int_S d\Gamma = \int_S \text{curl } \mathbf{v} \cdot \mathbf{e}_n \, dS = \oint_C \mathbf{v} \cdot \mathbf{e}_t \, ds \, .$$

(13.5)

The region surrounded by C must be simply connected. The physical
meaning of the surface integral is the "flow" of the rotation through
the surface S, which may be called the intensity of vorticity.
Analogous to the streamlines, the vector lines of ω are considered
the *vortex lines* and the generalized local continuity equation *div ω
= 0* applies. The important hydrodynamic conclusion that, in an
irrotational flow, the circulation along any closed curve C vanishes
also can be immediately drawn from the relation (13.5). The two-
dimensional inviscid flow round an immersed (fixed) rigid
cylindrical body is considered in Fig. 13.2(b): Circulation is assumed
to vanish along any proper closed curve. Hence, such a proper loop
consists of the curves C_1 and C_2 with a slit connection when
considering a simply connected region outside of the rigid body. The
vanishing total circulation is the sum of the circulations along the
closed curves C_1 and C_2; thus, the circulation taken along any loop C
enclosing the rigid body is constant or trivially vanishes

$$\Gamma_1 + \Gamma_2 = 0 \;\rightarrow\; \Gamma_1 = -\Gamma_2 = \Gamma \, .$$

(13.6)

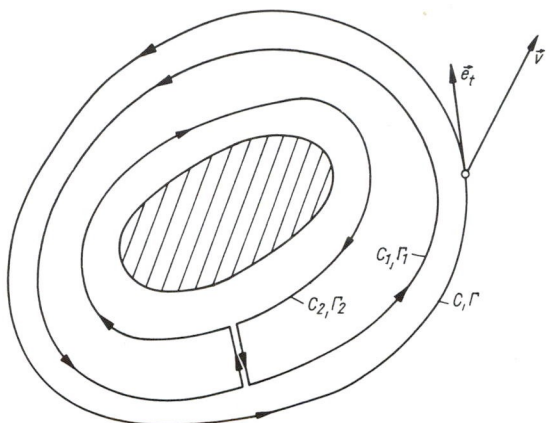

Fig. 13.2. (b) Irrotational and inviscid flow round a rigid cylinder with a
production of circulation Γ possible

The plane ideal flow round an immersed cylinder without any mass-producing sources can be defined by the continuity equation $div\ v = 0$ and, possibly, by the assumption of a constant circulation $\Gamma \neq 0$ along any curve C enclosing the contour of the rigid body.

13.2. The Hydrodynamic Lift Force

Lift, a lateral force, is one of the most important phenomena of flow round immersed slender, circulation-producing bodies. The plane stationary and periodic flow through an airfoil cascade (a grid or lattice) with equidistant spacing of thin cylindrical blades is considered first, as in Fig. 13.3. The net force acting as the resultant of the pressure distribution on a single blade can be calculated. The control surface shown by a dashed line in the figure cuts out a single airfoil, and the outer part is formed by two similar streamlines (streamtube) with conditions of periodicity taken into account. Inlet and outlet have the same cross-section t measured parallel to the cascade. Sufficiently distant from the guiding vanes, the flow is assumed to be homogeneous and parallel with constant velocities v_1 and v_2, respectively. The mass flow rate being constant gives, with the assumption of incompressibility,

$$\dot{m} = \rho\, t\, b\, u_1 = \rho\, t\, b\, u_2 \rightarrow u_1 = u_2 = u .$$

$$(13.7a)$$

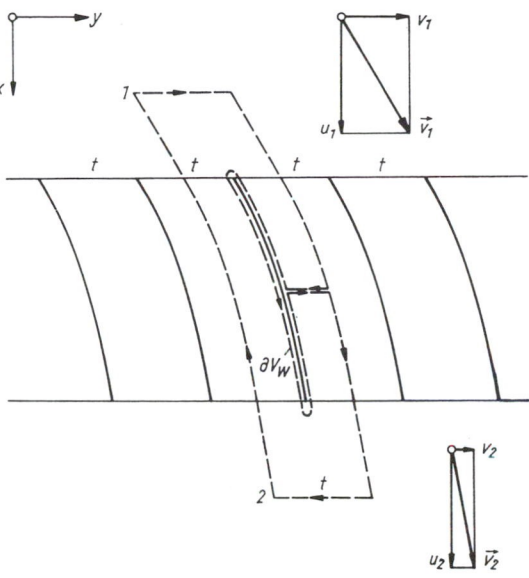

Fig. 13.3. Deflection of a stationary parallel flow by a periodic array of cylindrical guiding vanes. Control surface cutting out a single airfoil and taking into account the periodicity of the channel stream

The velocity component in the x direction, ie orthogonal to the guiding array, remains uneffected and is constant. Conservation of momentum, Eq. (7.13), is applied to the control volume, and the reaction of the resulting force on the "wall" ∂V_w , $-\boldsymbol{F}_w = -F_x \, \boldsymbol{e}_x - F_y \, \boldsymbol{e}_y$, enters (again see Fig. 13.3)

$$\dot{m}\,(u_2 - u_1) = (p_1 - p_2)\,t\,b - F_x = 0 , \quad \dot{m}\,(v_2 - v_1) = -F_y . \qquad (13.7b)$$

Bernoulli´s equation (8.40), when applied to the streamline from upstream point 1 in the advancing parallel flow to point 2 downstream in the deflected parallel stream, renders the pressure difference

$$p_1 + \frac{\rho}{2}\,(u_1^2 + v_1^2) = p_2 + \frac{\rho}{2}\,(u_2^2 + v_2^2) \ \rightarrow$$

$$p_1 - p_2 = \frac{\rho}{2}\,(v_2^2 - v_1^2) = -\rho\,(v_1 - v_2)\,\frac{v_1 + v_2}{2} . \qquad (13.8)$$

It is substituted in Eq. (13.7b). The force components acting on the single blade result in this manner in suitable form

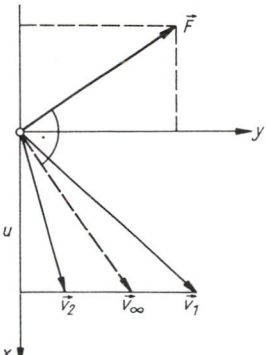

Fig. 13.4. Deflection of a parallel flow with deceleration and the lateral force acting on one of the guiding airfoils

$$F_x = - \rho \, t \, b \, (v_1 - v_2) \, \frac{v_1 + v_2}{2} \; , \quad F_y = \rho \, t \, b \, (v_1 - v_2) \, u \; . \tag{13.9}$$

By means of the vector of the average velocity,

$$\mathbf{v}_\infty = \frac{1}{2} \, (\mathbf{v}_1 + \mathbf{v}_2) = u \, \mathbf{e}_x + \frac{v_1 + v_2}{2} \, \mathbf{e}_y \; , \tag{13.10}$$

and by consideration of the circulation, which is to be calculated along the outer closed curve C of the control surface, the periodicity of the velocity distribution is taken into account

$$\Gamma = \oint_C (\mathbf{v} \cdot \mathbf{e}_t) \, ds = t \, (v_1 - v_2) = \text{const} \; , \tag{13.11}$$

the force components take on their final form (see Fig. 13.4)

$$F_x = - \rho \, v_{\infty y} \, \Gamma \, b \; , \quad v_{\infty x} = u \; , \quad F_y = \rho \, v_{\infty x} \, \Gamma \, b \; , \quad v_{\infty y} = \frac{v_1 + v_2}{2} \; , \tag{13.12}$$

$$\to \; |\mathbf{F}| = \rho \, v_\infty \, \Gamma \, b \; , \quad \mathbf{F} \perp \mathbf{v}_\infty \; .$$

If the cascade spacing t grows beyond any bounds, $t \to \infty$, the parallel stream is no longer deflected ($\mathbf{v}_1 \to \mathbf{v}_2 \to \mathbf{v}_\infty$), but the circulation $\Gamma = \lim_{t \to \infty} t \, (v_1 - v_2)$ due to the indefinite product may still be nonzero. Downstream, the flow may be shifted somewhat downward. In the case of a single, circulation-producing profile immersed in a parallel flow of velocity $\mathbf{v}_\infty$, when measured sufficiently distant ahead and behind the foil, the lateral force is

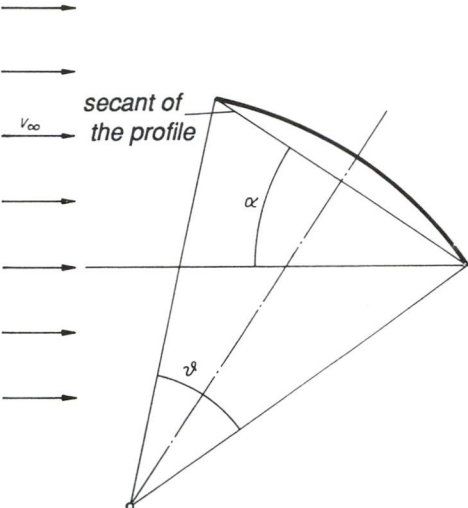

Fig. 13.5. A shallow circular profile in a parallel flow (angle α exaggerated)

known as the *lift* and the limit of Eq. (13.12) is known as the *Kutta-Joukowsky* theorem

$$F_L = \rho\, v_\infty\, \Gamma\, b \;,\quad \mathbf{F}\perp \mathbf{v}_\infty \;. \tag{13.13}$$

The lift force is orthogonal to the main stream, which is parallel in the farfield. In an inviscid flow, there is no force component in the direction of flow (ie the drag is zero). In the case of wings of finite span b and free (open) ends, however, the trailing vortices from the wing tips render varying lift and induced drag even in inviscid flow. The effects of such a horse-shoe vortex line are not discussed here any further. Since the circulation is proportional to the speed v_∞ and to the chord l of the airfoil, and its value depends on the shape of the slender blade as well as the angle of attack α (see Fig. 13.5 and also Exercise A 7.7), it is reasonable to separate these influences, and the engineering formula of the lift results (b is the span)

$$F_L = c_L(\alpha)\, \frac{\rho\, v_\infty^2}{2}\, b\, l \;. \tag{13.14}$$

The dimensionless lift coefficient $c_L(\alpha)$ is generally determined in wind-tunnel testing and the diagrams of so-called NACA profiles are readily available; see, eg Fig. 13.6. The parameters of the actual main flow are summarized in the

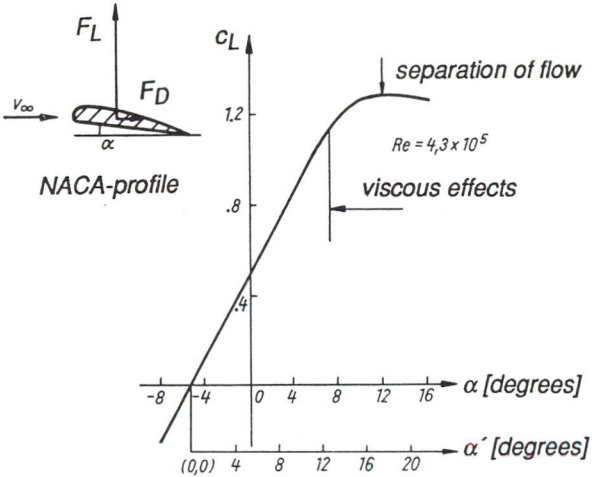

Fig. 13.6. The lift coefficient of a NACA profile. The linear function is extended into the nonlinear range where viscosity effects become dominant

stagnation pressure. For a thin rigid plate and small angles of attack, $\alpha \ll 1$, a complex theoretical investigation of the circulation renders the formula

$$c_L(\alpha) = 2\pi \sin \alpha \;\to\; c_L(\alpha) = 2\pi \alpha \;,\; |\alpha| \ll 1 \;. \tag{13.15}$$

The lift coefficient of a shallow cylindrical thin shell with a given angle of aperture becomes, also by a theoretical investigation (see Fig. 13.5),

$$c_L(\alpha) = 2\pi \sin\left(\alpha + \frac{\vartheta}{4}\right) \;. \tag{13.16}$$

Analogous to the circular profile, a "zero angle of attack" $\alpha = \alpha_0$ can be determined in general, where the circulation and, hence, the lift vanishes. With reference to this configuration, the angle of attack is measured, $\alpha' = \alpha - \alpha_0$, and the linearized relation when expressed by α' then holds for slender profiles of any convex shallow shape (see Fig. 13.6)

$$c_L(\alpha) \cong 2\pi \alpha' \;,\; |\alpha'| \ll 1 \;. \tag{13.17}$$

For larger angles α' , the lift becomes nonlinear and viscosity effects become dominant; the lift coefficient reaches a maximum. A second fluid parameter, the dynamic coefficient of viscosity [see Eq. (4.62)], enters, and the lift coefficient depends on the dimensionless *Reynolds* number as well [see Eqs. (4.68) and (13.24)]

$$Re = \frac{v_\infty l}{v} \quad , \quad v = \eta/\rho \; .$$

$$(13.18)$$

A second force component parallel to v_∞, the *drag* F_D, starts to increase with α and is expressed analogously to the lift by the formula

$$F_D = c_D(\alpha, Re) \; \frac{\rho v_\infty^2}{2} \; bl \; .$$

$$(13.19)$$

Figure 13.7 shows the rather small coefficient of the unwanted drag of an airfoil , $c_D \ll c_L$, $(c_D \approx 10^{-2} \, c_L)$, for not too large angles of attack. The drag coefficient increases rapidly when the upper surface increasingly becomes a region of separated flow. Lift and drag are maximum when the airfoil stalls. Both coefficients are combined in polar diagrams at assigned *Reynolds* numbers. The separation of flow is also observed at stream-lined bodies with a rounded trailing edge. Their drag coefficient is of the order of 0.2 to 1. Wind-tunnel testing, initiated by the aircraft industry, is now common for all kinds of earthbound vehicles and even the suits of ski racers and their posture are tested for low drag.

A sufficiently high *Reynolds* number is understood, since the phenomena change in so-called creeping flows according to *Oseen*.

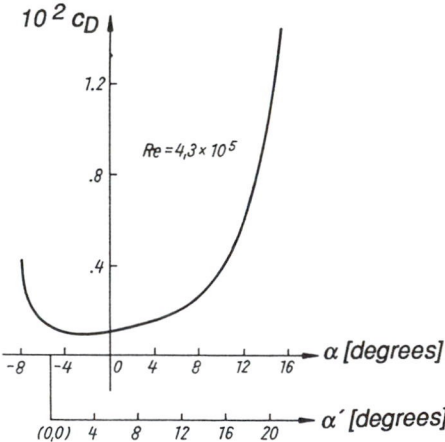

Fig. 13.7. The drag coefficient of a NACA-profile. Note the difference in scale

13.3. The *Navier-Stokes* Equations, Similarity Solutions

Equation (7.1) holds in every material point, irrespective of any constitutive relation. The substitution of *Stokes* material law of a *Newtonian* fluid, Eq. (4.62), eliminates the viscous shear stresses and renders, with the assumption of an incompressible flow, *div* v = 0, and constant dynamic viscosity, η = *const* (any temperature fluctuations must be small), a special form of the *Navier-Stokes* equations

$$\rho\mathbf{a} = \mathbf{k} - \text{grad } p + \eta \, \Delta\mathbf{v} \ , \quad \Delta = \text{Laplace operator.} \tag{13.20}$$

The partial differential equations are of the second order due to the viscosity term, and thus, they are fundamentally different from the *Euler* equations of inviscid flow, Eq. (8.30). Even in the case of small viscosity are the solutions qualitatively different, and a singular perturbation of the inviscid flow is appropriate. An additional boundary condition arises, eg at a wall at rest, where in a viscous flow $v = 0$, ie $(\mathbf{v} \cdot \mathbf{e}_n) = 0$ (as with inviscid flow) and $(\mathbf{v} \cdot \mathbf{e}_t)$ $= 0$ (a no-slip condition, the fluid particles stick to the wall). Equation (13.20) holds for laminar as well as turbulent flows. In the latter case, the velocity may have a stationary mean value, and, in addition, a time-dependent fluctuating component, which is often considered a sample of a stochastic process in space and time,

$$\mathbf{v}(t, \mathbf{r}) = \mathbf{v}_m(\mathbf{r}) + \mathbf{v}'(t, \mathbf{r}) \ , \quad <\mathbf{v}'> = 0 \ . \tag{13.21}$$

Detailed considerations are given in special books on turbulence, where also the important notion of "turbulent shear stress" is introduced. See, *H. Schlichting: Boundary Layer Theory. Braun, Karlsruhe, 1965, (in German)* a classic; also refer to *K. Wieghardt: Theoretische Strömungslehre. Teubner, Stuttgart, 1965* (for *Reynolds* stresses, p. 195).

The similarity of solutions of Eq. (13.20) with respect to both geometry and mechanical properties is found after introducing dimensionless coordinates, eg $x^* = x/L$, L is the characteristic length, and dimensionless time $t^* = t \, u/L$, where u is the characteristic speed. Substituting $\mathbf{v}^* = v/u$ and rearranging terms in Eq. (13.20) render the *Navier-Stokes* equations dimensionless

$$\frac{d\mathbf{v}^*}{dt^*} = -\text{grad}^*\left(\frac{p}{\rho u^2} + \frac{g\,L}{u^2} z^*\right) + \frac{\eta}{\rho u\,L} \, \Delta^*\mathbf{v}^* \ . \tag{13.22}$$

The incompressible flow is assumed in the homogeneous and parallel gravity field; the vertical unit vector e_z points upward. The effects may be analyzed separately.

1.   In a flow with no influence of gravity and a negligible influence of viscosity, similitude is determined by the pressure field $p(t, r)$, with proportionality to the stagnation pressure $\rho u^2/2$, according to the first of the dimensionless terms. An illustration is given by considering a thin plate at rest in a parallel stream, such that its blunt side faces the flow. The drag F_D is the resulting net force of the pressure distributions at the front and at the rear face, and as such, it is delineated in product form [see also Eq. (13.19)]

$$F_D = c_D \frac{\rho u^2}{2} A.$$

$$(13.23)$$

The dimensionless coefficient of drag c_D in an inviscid flow depends solely on the planform of the plate, and it is determined in a wind tunnel or a water channel using a model irrespective of size (see the coefficient of area A); eg for a circular plate, $c_D = 1.17$ is found. Hence, for blunt bodies, the drag coefficient of large *Reynolds* numbers is independent of the fluid parameters and takes on the asymptotic values of Fig. 13.8. Note the difference in scale.

2.   In most cases of flow of engineering interest, similitude is also strongly affected by viscosity and only gravity has no influence. Inspection of the dimensionless coefficient of the highest derivative of Eq. (13.22) identifies the *Reynolds* number as characteristic [see also Eq. (13.18)]

$$Re = u\,L/v \ , \quad v = \eta/\rho.$$

$$(13.24)$$

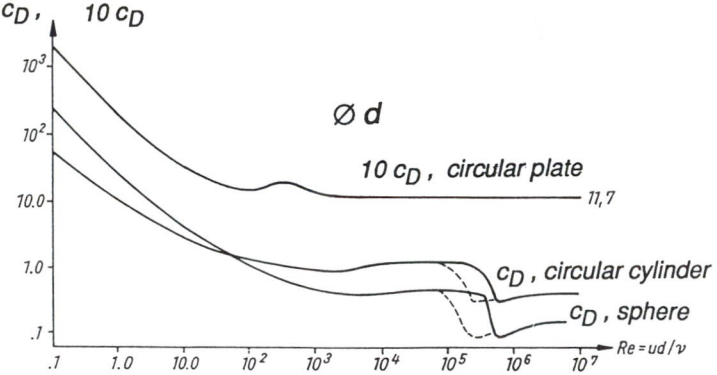

Fig. 13.8. Drag coefficient of bluff and round bodies, measured

All details of a flow, like the location of stagnation points, laminar and turbulent boundary-layer formation, and the drag coefficient, are dependent on the *Reynolds* number (see also Sec. 13.4.4)

$$c_D = c_D(Re) .$$

(13.25)

Figure 13.8 illustrates this function for the circular plate when held normal to a main stream, for the two-dimensional flow around a circular cylinder with a lateral axis and for the axisymmetric flow round a sphere, respectively. Note the difference in the logarithmic scale for the drag of bluff bodies and round bodies. With respect to the testing of models in parallel approaching main flow, air or water may be used if the *Reynolds* number is kept constant.
3.   In the case of flow with a free surface, similitude also requires that the dimensionless coefficient of gravity be constant. The square root of the reciprocal is known as the *Froude* number

$$Fr = \frac{u}{\sqrt{g\,L}} .$$

(13.26)

See also Eq. (9.135). This characteristic number plays an important role in ship model testing and in problems related to shallow water waves.
        Summarizing the results of the analysis of Eq. (13.22), we observe that the physical similitude of flow requires, in addition to the geometric affinity, the observation of characteristic numbers. Most important, the *Reynolds* number must be kept constant. In the case of the strong influence of gravity, the *Froude* number must also be the same.

13.3.1. Viscous Pipe Flow

The stationary and incompressible laminar flow through a pipe with a circular cross-section is axisymmetric. When we consider the *Navier-Stokes* equation (13.20) in cylindrical coordinates, r and x, the radial component reduces to the static condition

$$0 = -\frac{1}{\rho}\frac{\partial p}{\partial r} \quad \rightarrow \quad p = p(x) .$$

(13.27)

By taking into account the continuity relation, $\partial u/\partial x = 0$, the axial component of Eq. (13.20) gives [see Eq. (6.181) for the *Laplace* operator]

$$0 = -\frac{1}{\rho}\frac{dp}{dx} + v\left(\frac{d^2u}{dr^2} + \frac{1}{r}\frac{du}{dr} + 0\right) .$$

(13.28)

Separation of variables and subsequently integrating twice yield the parabolic velocity distribution in a cross-section (see also Fig. 4.6)

$$u(r) = \frac{1}{4\eta} \frac{dp}{dx} r^2 + C_1 \ln r + C_2 \ , \quad \begin{cases} r = 0 : u \text{ finite} \rightarrow C_1 = 0 \ , \\ r = R : u = 0 \ \rightarrow C_2 = -\frac{1}{4\eta} \frac{dp}{dx} R^2 , \end{cases}$$

(13.29)

the fluid particles stick to the wall at $r = R$, the no-slip condition.

Since the right-hand side of Eq. (13.29) must be a function of r and does not depend on x , the pressure gradient is constant and the pressure decreases linearly with the distance downstream

$$\frac{dp}{dx} = -\frac{\Delta p}{L} \ , \quad \Delta p = p(x) - p(x + L) \ .$$

(13.30)

Since the kinetic energy density must be constant, the dissipated energy reduces the potential energy of the internal forces, ie the pressure, which is the potential energy per unit of volume. The rate of dissipation is constant because neither the velocity distribution nor the viscous shear distribution depend on x ; see Eq. (8.45) with $P < 0$. The average speed according to the parabolic velocity distribution (13.29) of the laminar flow is known as the law of *Hagen-Poiseuille*

$$u_m = \frac{\dot{m}}{\rho A} = \frac{1}{A} \int_0^R 2\pi \, r \, u \, dr = \frac{1}{8\eta} \frac{\Delta p}{L} R^2 = \frac{1}{2} u_{max} \ .$$

(13.31)

The loss factor λ of the laminar viscous flow is defined by the dimensionless pressure head difference and is a function of the *Reynolds* number [see also Eq. (4.69)]

$$\frac{\Delta p}{\frac{\rho}{2} u_m^2} = \frac{L}{2R} \lambda(Re) \ , \quad \lambda(Re) = 64/Re \ , \quad Re = \frac{2 R \, u_m}{\nu} \ .$$

(13.32)

The velocity distribution of a laminar flow in a duct of noncircular cross-section is a solution of the *Poisson* equation [see Eq. (6.170)]

$$\Delta u = \frac{1}{\eta} \frac{dp}{dx} = const \ , \quad \Delta = \frac{\partial^2}{\partial y^2} + \frac{\partial^2}{\partial z^2} \ .$$

(13.33)

Using Eq. (13.29) as a particular solution, $r^2 = y^2 + z^2$, reduces the solution by superposition to the boundary value problem of the *Laplace* equation of potential theory

$$\Delta\psi = 0 \; , \; \psi\big|_{\text{at boundary B}} = -\frac{1}{4\eta}\frac{dp}{dx}\left(y_B^2 + z_B^2\right) \; , \; \Delta = \frac{\partial^2}{\partial y^2} + \frac{\partial^2}{\partial z^2} \; ,$$

$$u(y, z) = \frac{1}{4\eta}\frac{dp}{dx}\left(y^2 + z^2\right) + \psi(y, z) \, . \tag{13.34}$$

The planform of the cross-section is assumed to be implicitly given by $f(y_B, z_B) = 0$.

The laminar axisymmetric flow through a tubular pipe with a circular ring cross-section is easily found by the above method of superposition

$$u(r) = \frac{1}{4\eta}\frac{dp}{dx}\left(r^2 - R_i^2 - \frac{R_e^2 - R_i^2}{\ln\left(R_e/R_i\right)}\ln\left(r/R_i\right)\right) \; , \; R_i \le r \le R_e \, . \tag{13.35}$$

A rapid increase of the loss factor is observed due to the friction at the inner surface: The mass flow rate is reduced to a fraction of $2/3$ in such a tubular pipe even for a small ratio, $R_i / R_e = 1/20$, when compared to the undisturbed flow in the circular pipe of the same radius R_e , $R_i = 0$, $u(0) = u_{max}$. The second boundary condition $u = 0$ at $r = R_i$ renders additional lateral velocity gradients and, hence, viscous shear stresses. See Eq. (4.59), where x is to be substituted for y .

The *laminar flow* in the pipe is stable only for rather small values of the *Reynolds* number. If, eg the velocity is increased, the flow becomes *turbulent* (mixing). The critical *Reynolds* number of $Re_c = 2320$ is followed by a transition zone, and for $Re > 5000$, the flow is fully turbulent. Under this condition, the distribution of the

boundary layer (hydraulic smooth wall)

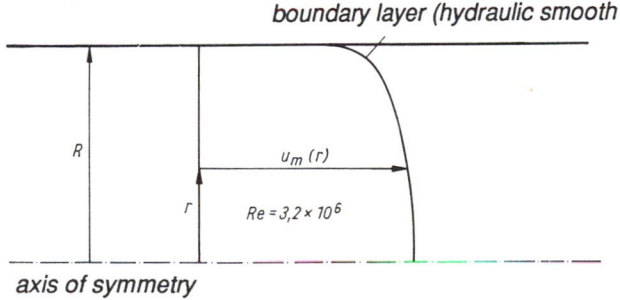

axis of symmetry

Fig. 13.9. Mean axial velocity distribution of the turbulent flow in a circular pipe. Laminar sublayer of the boundary layer not visible. Compare with the parabolic distribution of the laminar flow, Fig. 4.6

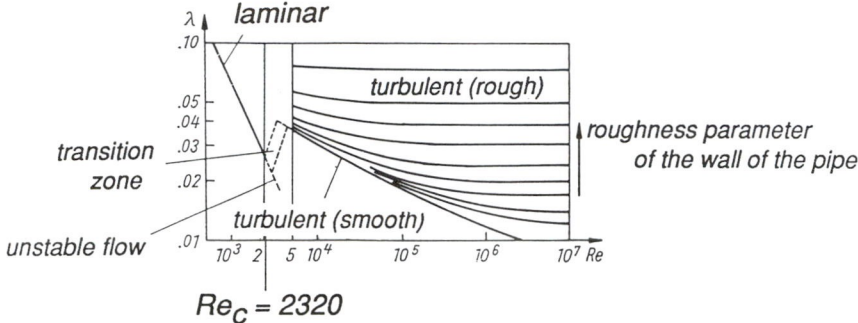

Fig. 13.10. The loss factor of a pipe under laminar and turbulent flow conditions.
Note the critical *Reynolds* number and transition zone

mean axial velocity increases rapidly within a thin boundary layer
that has a laminar sublayer at the wall and is nearly constant
outside; see Fig. 13.9. It is no longer parabolic. The loss factor
nearly shows a "jump" in the transition zone and is increased for the
turbulent motion when compared to the laminar flow. Look at Fig.
13.10 and note the logarithmic scale. Furthermore, the loss factor
decreases in the region of turbulent motion much more slowly with
an increasing *Reynolds* number or the decreasing influence of
viscosity. It is worth noting that turbulence is solely a viscous
effect. The loss factor is no longer a unique function of the *Reynolds*
number. It depends on whether the surface roughness is covered by
the laminar boundary sublayer or not. The notion of hydraulic smooth
pipes or hydraulic rough pipes with all intermediate stages possible
describes this fact in Fig. 13.10. In the first case, the outer flow is
not influenced by the roughness of the surface. The loss factor is
determined by empirical formulas explicitly or implicitly,
respectively,

$$\lambda = 0.316 \, Re^{-1/4} \, , \quad 5000 < Re < 10^5 \quad (Blasius) \, ,$$

$$\frac{1}{\sqrt{\lambda}} = 2 \log\left(Re \, \sqrt{\lambda}\right) - 0.8 \, , \quad Re > 5000 \quad (Prandtl \text{ - } Nikuradse) \, .$$

$$(13.36)$$

See the lower curve in Fig. 13.10. The flow in lifelines is always
turbulent. However, the loss factor may be dramatically reduced by
adding a small amount of a second phase with long chains of
molecules. Energy consumption at the pumping stations of such a
two-phase flow can be reduced considerably. The problem of
discharging the particles needs further investigation.

13.3.2. The Boundary Layer of a Plate

Ludwig Prandtl recognized in 1904 that the flow along a rigid wall
separates into an (inviscid) outer flow and an inner (viscous) flow.
The latter is referred to as a thin boundary layer. If we consider the
boundary-value problem of the *Navier-Stokes* equation (13.20) with
small viscosity η , or Eq. (13.22) with a large *Reynolds* number *Re* ,
such a separation into zones of dominating influence of viscosity
and of negligible influence, respectively, seems quite natural in
light of modern singular perturbation theory. The boundary condition
at the wall at rest requires the velocity to vanish (a no-slip
condition), which is only possible by keeping the viscosity term.
This term dominates the flow in a small surrounding of the wall
where the lateral velocity gradient is necessarily large. At some
distance from the wall, the velocity of the "free" flow according to
the mass flow rate is reached, say, 99 or 99.5%. Its value is
independent of the condition that the fluid particles stick to the
wall. The lateral velocity gradient and, hence, the shear stresses
become negligible outside of this layer, and in the limit $Re \rightarrow \infty$,
Euler´s equation (8.30) applies to the outer flow. Following the
rules of matched asymptotic expansions, the coordinate normal to
the wall is properly stretched within the boundary layer, and the
inner solution (of first order of perturbation analysis) is
subsequently matched at the interface to the outer flow conditions.
The coupling of the two flow fields requires a steady tangential
velocity component: The velocity at the outer surface of the viscous
boundary layer (ie asymptotically, the stretched coordinate goes to
infinity) must be equal to the velocity of the inviscid potential flow
at its inner boundary (ie at the wall at rest, where the normal
coordinate is zero). The wall is a streamline for the outer inviscid
flow. The mathematical formalism in a second iteration is
sometimes used to determine the flow in the boundary layer in a
second-order approximation. This analysis is outlined in *W.
Schneider: Mathematical Methods of Fluid Mechanics. Vieweg,
Braunschweig, 1978, (in German)*.
   In general, in the first step of the computational program of a
flow with a large *Reynolds* number, the inviscid potential flow field
with a streamline at the wall is determined, ie its velocity and
pressure. For the special problem of a (homogeneous) parallel flow
along a plate, to be considered below, $U = const$ and $p = const$ are
the proper solutions of the main stream. The solution is substituted
into the *Navier-Stokes* equation (13.20), say, in the x component,

$$\frac{\partial u}{\partial t} + u \frac{\partial u}{\partial x} + v \frac{\partial u}{\partial y} = \left\{ -\frac{1}{\rho} \frac{\partial p}{\partial x} \right\} + v \frac{\partial^2 u}{\partial y^2} = \left\{ \frac{\partial U}{\partial t} + U \frac{\partial U}{\partial x} \right\} + v \frac{\partial^2 u}{\partial y^2} , \quad (13.37)$$

to eliminate the pressure gradient. Only the leading derivative is kept in the viscosity term, $|\partial^2 u/\partial x^2| << |\partial^2 u/\partial y^2|$. With that approximation, Eq. (13.37) is called a boundary layer equation. For the stationary viscous and laminar flow in the boundary layer of the plate,

$$u\frac{\partial u}{\partial x} + v\frac{\partial u}{\partial y} = v\frac{\partial^2 u}{\partial y^2} \quad , \quad y = 0 : u = v = 0 \quad , \quad y \to \delta : u \to U \ . \tag{13.38}$$

The thickness of a (laminar) boundary layer for analytical purposes must be defined as a parameter of the flow, ie by considering, eg the mass flow rate [thus, the displacement thickness δ^* of a stagnated layer results that has the same integrated velocity deficit $(1 - u/U)$] as the actual boundary layer. Analogously, the momentum thickness θ results by considering the integrated momentum deficit in terms of the actual flow, $(u/U)(1 - u/U)$; see Eq. (13.100). Also an energy thickness δ^{**} may be used by considering the integral of $(u/U)(1 - u^2/U^2)$ over the normal coordinate y . See Sec. 13.5 for the details of derivation. By comparing the order of terms in Eq. (13.38), $u\partial u/\partial x \propto U^2/L$ and $v\,\partial^2 u/\partial x^2 \propto v\,U/\delta^2$, the order of $\delta \propto L/\sqrt{Re}$ is found. Since the plate has no characteristic length L , the similarity of the flow in the boundary layer is to be expected for all values of x

$$\frac{u}{U} = f(Y) \ , \ Y = y/\delta \text{ is the stretched normal coordinate} \ .$$

By selecting $L = x$ as equal to the upstream distance from the sharp leading edge of the semiinfinte plate, the boundary-layer thickness grows according to the square root order

$$\delta \approx \sqrt{v\ x/U} \ . \tag{13.39}$$

The outer potential flow is no longer parallel. The interface to the boundary layer is curved, but it is not a streamline, for some flow crosses it transversely. By means of the so-called stream function ψ [see Eq. (13.48) for a definition], the continuity equation $div\ \mathbf{v} = 0$ is identically satisfied. Integration within the viscous flow renders

$$\psi = \int_0^y u\ dy = U\delta \int_0^Y f(Y)\ dY = \sqrt{vUx}\ F(Y) \ . \tag{13.40}$$

The substitution of the separable function in Eq. (13.38) gives a nonlinear ordinary differential equation of the third order in the stretched coordinate Y

$$F F'' + 2 F''' = 0 \quad , \quad Y = 0 \colon \; F = F' = 0 \quad , \quad \lim_{Y \to \infty} F' = 1 \quad . \tag{13.41}$$

It is numerically integrated with homogeneous geometric boundary conditions at the wall and an asymptotic condition as shown above. At a distance $y \approx 5 \, \delta$, $u/U = 0.99$.

The shear-stress distribution in the laminar boundary layer determines the dimensionless coefficient of the local drag per unit width for one side of the plate, $0 \le x \le L$, to be

$$c_D = \frac{F_D}{\dfrac{\rho\, U^2}{2}\, B\, x} = \frac{1.328}{\sqrt{Re(x)}} \quad . \tag{13.42a}$$

The local skin friction coefficient is simply one-half the above value. The calculated results agree very well with those obtained experimentally. The laminar boundary layer extends a critical distance x_c from the tip, according to a range of critical *Reynolds* numbers of $3 \times 10^5 < Re_c < 5 \times 10^5$. For a parallel stream of air with $U = 20$ m/s , $v = 15 \times 10^{-6}$ m²/s , the critical *Reynolds* number is $Re_c = 4 \times 10^5$ and $x_c = 300$ mm . The 99% laminar boundary-layer thickness grows under those conditions to $\delta_c = 2.4$ mm . It is followed by a transition zone before the turbulent layer is fully developed. Its thickness grows faster by $\propto (x - x_c)^{0.8}$. By means of the momentum integral method (see Sec. 13.5), the local friction coefficient of the turbulent layer is approximated within the *Blasius* relation by [see Eq. (13.36) for the pipe flow]

$$c_f = \frac{0.0592}{Re_x^{1/5}} \quad , \tag{13.42b}$$

and the drag coefficient becomes by integration

$$c_D = \frac{0.074}{Re_x^{1/5}} \quad . \tag{13.42c}$$

An improved formula for the one-sided drag coefficient that holds for the whole range of *Reynolds* numbers of the turbulent layer is based on a logarithmic velocity distribution according to *Prandtl-Schlichting*

$$c_D = \frac{0.455}{(\log Re_x)^{2.58}} \quad . \tag{13.42d}$$

The viscous flow in boundaries, especially along curved surfaces when the pressure is increasing in the flow direction, is subject to separation. At the separation point, the ordered flow stops.

Downstream, the flow is composed of eddies. A wake forms behind an immersed body, and due to the loss of momentum, the drag increases. The separation of a turbulent layer is observed further downstream when compared to a laminar one. Thus, in addition to controlling the boundary layer by means of a proper body shape, wires may be attached to the surface to produce turbulence. A more expensive alternative is to energize the slow part of fluid motion either by "blowing", ie directing fluid of higher speed into the layer, or by suction, ie removing fluid, thereby reducing the thickness of the layer. The latter method finds standard application in wind tunnels.

13.4.   Potential Flow, the Singularity Method

Knowledge of inviscid flow forms the basis of many actual flow problems, like flow round immersed bodies, where the outer flow must be determined, or even the flow in curved ducts is affected. The assumption of irrotational motion greatly simplifies any analysis

$$2 \, \omega = \text{curl } \mathbf{v} \equiv 0 \ . \tag{13.43}$$

The vector field of velocity $\mathbf{v}$ becomes the gradient of the velocity potential Φ [see Eq. (3.12) for an irrotational force field and Eq. (1.51)]

$$\mathbf{v} = \text{grad } \Phi \ , \tag{13.44}$$

and Eq. (13.43) is identically solved. Assuming that the flow is incompressible reduces the continuity equation to $div \ \mathbf{v} = 0$ [see Eq. (1.77)] and the *Laplace* equation results upon the substitution of Eq. (13.44)

$$\text{div grad } \Phi = \Delta\Phi = 0 \ , \ \Delta = \frac{\partial^2}{\partial x^2} + \frac{\partial^2}{\partial y^2} + \frac{\partial^2}{\partial z^2} \ . \tag{13.45}$$

Time dependence, which may result from nonstationary boundary conditions, appears merely in the form of a parameter in the solution, $\Phi(t; x, y, z)$; see Eq. (1.78). In complementary fashions, the continuity equation (1.77) is identically solved, by representing the velocity field as the rotation of a vector potential $\mathbf{A}$,

$$\mathbf{v} = \text{curl } \mathbf{A} \ . \tag{13.46}$$

In the case of irrotational motion, the vector potential is a solution of

$$\text{curl } \mathbf{v} = \text{curl curl } \mathbf{A} = \mathbf{0} \; . \tag{13.47}$$

The vector potential is extremely useful when dealing with two-dimensional flows in a plane (x, y) , or if the flow is axisymmetric (r, x) . In both cases, the vector potential depends on a single function ψ . In the former, plane flow $\mathbf{A} = \psi(x, y)\,\mathbf{e}_z$, and with Eq. (13.44),

$$\mathbf{v} = \text{curl } \mathbf{A} = \frac{\partial \psi}{\partial y}\,\mathbf{e}_x - \frac{\partial \psi}{\partial x}\,\mathbf{e}_y \;,\; u = \frac{\partial \psi}{\partial y} = \frac{\partial \Phi}{\partial x} \;,\; v = -\frac{\partial \psi}{\partial x} = \frac{\partial \Phi}{\partial y} \; . \tag{13.48a}$$

Substitution in Eq. (13.47) renders $\psi(x, y)$ as a proper solution of the two-dimensional *Laplace* equation

$$\text{curl curl } \mathbf{A} = -\left(\frac{\partial^2 \psi}{\partial x^2} + \frac{\partial^2 \psi}{\partial y^2}\right)\mathbf{e}_z = 0 \; . \tag{13.48b}$$

The radial component $v(r, x)$ of an axisymmetric velocity field and the axial one, $u(r, x)$, are related, when premultiplied by the radial coordinate r , to the derivatives of the function $\psi(r, x)$ by

$$u = \frac{1}{r}\frac{\partial \psi}{\partial r} = \frac{\partial \Phi}{\partial x} \;,\; v = -\frac{1}{r}\frac{\partial \psi}{\partial x} = \frac{\partial \Phi}{\partial r} \;\rightarrow\; \text{div } \mathbf{v} = \frac{\partial u}{\partial x} + \frac{\partial v}{\partial r} + \frac{v}{r} = 0 \; . \tag{13.48c}$$

The contour lines of the surface $z = \psi(x, y)$, namely, $z = const$, are the streamlines C of the incompressible flow: The total derivative vanishes along these lines

$$d\psi|_C = \frac{\partial \psi}{\partial x}\,dx + \frac{\partial \psi}{\partial y}\,dy = -v\,dx + u\,dy = 0 \;\rightarrow\; \frac{dy}{dx}\Big|_C = \frac{v}{u} \; . \tag{13.49}$$

That is, the tangent of the contour lines of the stream function has the same direction as the velocity $\mathbf{v}$; that property explains its name. In those cases where compressibility of the fluid must be considered, the concept of stream function is still valuable when the specific mass flow rate, eg ρu for a stationary flow is substituted for the velocity components, eg u , and Eq. (1.74) with $\partial \rho / \partial t = 0$ is identically satisfied.

Equation (13.48a) of a two-dimensional and incompressible flow already indicates that the potential function and stream function are related by the *Cauchy-Riemann* differential equations. Hence, both functions can be combined to form an analytic complex

potential $F(z)$ [where $z = x + iy$ is a point in the (x, y) flow plane
and $i = \sqrt{-1}$ is the imaginary unit]

$$F(z) = \Phi(x, y) + i\,\psi(x, y) \ , \ \Delta F = 0 \ , \ \Delta = 4\frac{\partial^2}{\partial z \partial \bar{z}} \ , \ \bar{z} = x - iy \ ,$$

(13.50)

since real and imaginary part are also harmonic functions.
The isotropic complex derivative of the potential

$$\frac{dF}{dz} = \frac{\partial \Phi}{\partial x} + i\frac{\partial \psi}{\partial x} = u - iv \ ,$$

(13.51)

gives the conjugate complex velocity vector, ie the real part of the
complex derivative is the x component and the imaginary part is the
negative y component of the velocity field. Considering some simple
complex analytic functions will well illustrate elements of
potential flow and their streamlines. The application of *Bernoulli's*
equation (8.40) gives the associated pressure field. By means of
superposition, quite complicated flow patterns can be conveniently
determined. Also, the vast field of conformal mappings becomes
applicable.

13.4.1. Illustrative Examples

(§)   Two-Dimensional Potential Flow Toward a Rigid Wall. Putting
the complex velocity potential as $F(z) = a\,z^2$, with a real coefficient
a , renders the real part, $\Phi = a\,(x^2 - y^2)$, and the imaginary part, the
stream function $\psi = 2a\,xy$. The contour lines $\psi = const$ are
orthogonal hyperbolas, filling all four quadrants in the (x, y) plane.
The derivative $dF/dz = 2a\,z$ gives $u = 2a\,x$ and $v = -2a\,y$, ie
linearly varying velocity components. The streamlines in the upper
half-plane, $y \geq 0$, of the contour map are shown in Fig. 13.11.
      Since none of the fluid enters the lower half-plane through the
abscissa x , which is an axis of symmetry and is part of the
bifurcated streamline through the stagnation point S as well, it can
be materialized as a plane rigid wall, without changing the inviscid
flow pattern. Similarly, the flow in a quarter space results when the
y axis is also materialized. The lines of constant pressure are
concentric circles with a maximal pressure at the stagnation point
S, Eq. (8.40).

(§)   Two-Dimensional Flow in a Corner Space and Round a Sharp
Edge. Selecting the function $F(z) = a\,z^v$, $(a, v$ real) and using polar
coordinates, where $z = r\,exp\,(i\varphi)$, give the stream function $\psi = [a\,r^v$
$sin\,(v\varphi)]$. By avoiding any ambiguity arising from the multivalued
functions by materializing some of the straight lines by rigid walls,

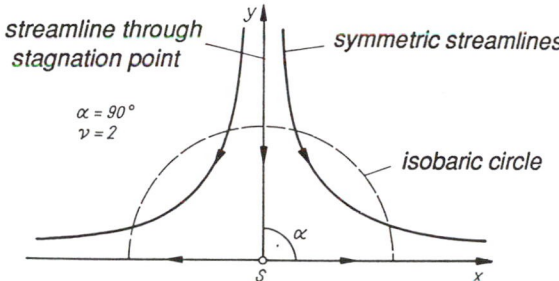

Fig. 13.11. Two-dimensional flow toward a rigid wall. Stagnation point S with bifurcation of the streamline

a value of $v = 3$ renders, eg the flow pattern of Fig. 13.12(a), where $\alpha = \pi/3$ (in general, $\alpha = \pi/v$, $v > 1$).

The flow of the ideal fluid round an edge is given for exponents in the range $1/2 < v < 1$; Fig. 13.12(b) illustrates the flow when $v = 3/5$. The ideal, singular flow round the edge of a thin and semiinfinite plate results by putting $v = 1/2$.

(§) Singular Potential Flows. The singular three-dimensional flow element that is the only one considered here is the isotropic point source of constant strength. The conservation of mass requires the rate of discharge through all control spheres of radius R surrounding the singular point to be equal and the same as that at which fluid is presumed to be created (source) or destroyed (sink) at that point. Such a flow must be radial. The two-dimensional counterpart considers the mass flow rate constant through a circular cylindrical control surface of radius r centered at the line source of constant strength per unit of length. The latter

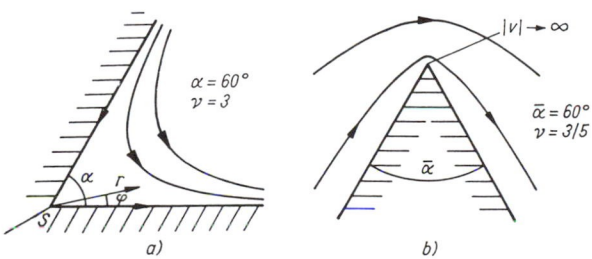

Fig. 13.12. Two-dimensional potential flow in a corner space (a) and round a sharp edge (b). Note the stagnation point S in (a) and the unrealistic infinite value of the speed at the edge in (b)

corresponds to a point source in a plane. Hence, the radial velocities decay quite differently under spatial or plane flow conditions, with the distance from the singular point,

$$\left\{\begin{matrix} \dot{m} = \rho\, 4\, \pi\, R^2\, v_R(R) = const \\ v_R(R) = \dfrac{\dot{m}}{\rho\, 4\, \pi}\, \dfrac{1}{R^2} \end{matrix}\right\} \quad,\quad \left\{\begin{matrix} \dot{m} = \rho\, 2\, \pi\, r\, v_r(r) = const \\ v_r(r) = \dfrac{\dot{m}}{\rho\, 2\, \pi}\, \dfrac{1}{r} \end{matrix}\right\}\ . \qquad (13.52)$$

Integration renders the velocity potentials of the singular radial source flows quite differently. For the two-dimensional flow also, the stream function and complex potential are given below

$$\Phi(R) = \int v_R(R)\, dR = -\frac{\dot{m}}{\rho\, 4\, \pi}\, \frac{1}{R}\ ,\quad \Phi(r) = \int v_r(r)\, dr = \frac{\dot{m}}{\rho\, 2\, \pi}\, \ln r\ ,\quad \psi = \frac{\dot{m}}{\rho\, 2\, \pi}\, \varphi\ ,$$

$$\rightarrow F(z) = \frac{\dot{m}}{\rho\, 2\, \pi}\, \ln z\ . \qquad (13.53a)$$

The velocity vector and complex conjugate one are determined by taking the derivatives

$$\mathbf{v}(R) = grad\, \Phi(R) \quad;\quad F'(z) = \frac{\dot{m}}{\rho\, 2\, \pi}\, \frac{\bar{z}}{r^2}\ ,\quad \bar{z} = r\, e^{-i\varphi}\ . \qquad (13.53b)$$

The flow toward a small orifice in the side wall or at the bottom of an extremely large tank can be visualized by taking one-half the flow pattern toward a sink. The flow resulting from the superposition of a source and a sink is found by adding their potentials.

Another important element of a singular irrotational two-dimensional flow results from a single vortex line: Considering a circular flow around an axis with circulation vanishing for any closed curve not enclosing the point of the axis requires that two circular arcs be selected for such a loop, bounded by any two radial lines, $\Gamma = v_2\, r_2\, \varphi - v_1\, r_1\, \varphi = 0$. Hence, the velocity $v_\varphi(r) = C/r$. The velocity vectors are $\mathbf{v} = v_\varphi(r)\, \mathbf{e}_\varphi$. Any closed curve surrounding the singular point gives constant circulation, eg integration over a concentric circle renders the constant, ie physically, the strength of the vortex line,

$$\Gamma = \int_0^{2\pi} v_\varphi\, r\, d\varphi = 2\pi\, C = const\ . \qquad (13.54a)$$

Since the ratio Γ/A becomes infinitely large as the area of the circle approaches zero, the rotation ω is infinite along the axis (the axis being a vortex line), but equal to zero at every other point in the flow. The potentials of that irrotational flow expressed in centered polar coordinates are easily derived as

$$\Phi = \frac{\Gamma}{2\pi} \varphi \quad , \quad \psi = -\frac{\Gamma}{2\pi} \ln r \quad \rightarrow \quad F(z) = \frac{\Gamma}{2\pi i} \ln z \quad . \tag{13.54b}$$

The potentials are complementary to those of the plane point source. Note the difference in the rigid-body rotation of the fluid body where $v_\varphi = r\omega$, which is not an irrotational motion. The two flows may be combined to remove the singularity at the vortex line; matching is done at a cylindrical interface $r = r_c$.

13.4.2. The Singularity Method

Combinations of sources and sinks, as well as of vortex lines, with the potential of a parallel and homogeneous main stream render important flow patterns. The simplest combinations in two-dimensional flow are considered below. A comprehensive review is given by *F. Keune and K. Burg: Singularity Methods of Fluid Mechanics. Braun, Karlsruhe, 1975, (in German)*.

(§) Superposition of a Line Source and a Parallel Main Stream. The differentiation of the linear complex potential $u_\infty z$ determines a parallel stream in the x direction with constant speed. Adding a plane point source gives

$$F(z) = u_\infty z + \frac{\dot{m}}{2\pi \rho} \ln z \quad . \tag{13.55}$$

The complex potential determines a flow with a single stagnation point. The streamline $y = 0$ bifurcates at that point, and the coordinates of the stagnation point S are the solution of

$$\frac{dF(z)}{dz} = u - i\,v = u_\infty + \frac{\dot{m}}{2\pi \rho}\,\frac{\bar{z}}{r^2} = 0 \quad , \quad r^2 = x^2 + y^2 \quad \rightarrow \quad r_S = |x_S| \;,\; \varphi_S = \pi \;,$$

$$x_S = -\frac{\dot{m}}{2\pi \rho\, u_\infty} \quad , \quad y_S = 0 \quad . \tag{13.56}$$

The stream function ψ takes on the value ψ_S at the stagnation point and the streamline that passes through that point is determined by the equation

$$\psi = u_\infty y + \frac{\dot{m}}{2\pi \rho} \varphi = \psi_S = \frac{\dot{m}}{2 \rho} \quad . \tag{13.57}$$

Its asymptotes are parallel to the abscissa and are determined by

$$\lim_{x \to \infty} y(x) = \frac{\dot{m}}{2\rho u_\infty} \ .$$

(13.58)

The lower half of Fig. 13.13(a) illustrates that the efflux of the point source stays within the streamline through the stagnation point. Hence, the latter can be materialized and the remaining flow pattern corresponds to the inviscid flow round a cylindrical nose of a semiinfinite body ($x > x_S$).

Putting a sink in some distance $x = a$, which has the same strength, renders a second stagnation point. The potential is easily derived by superposition. By noting the difference of two logarithms, taking the derivative, solving for the coordinates of the stagnation points, the equation of the bifurcating streamline results

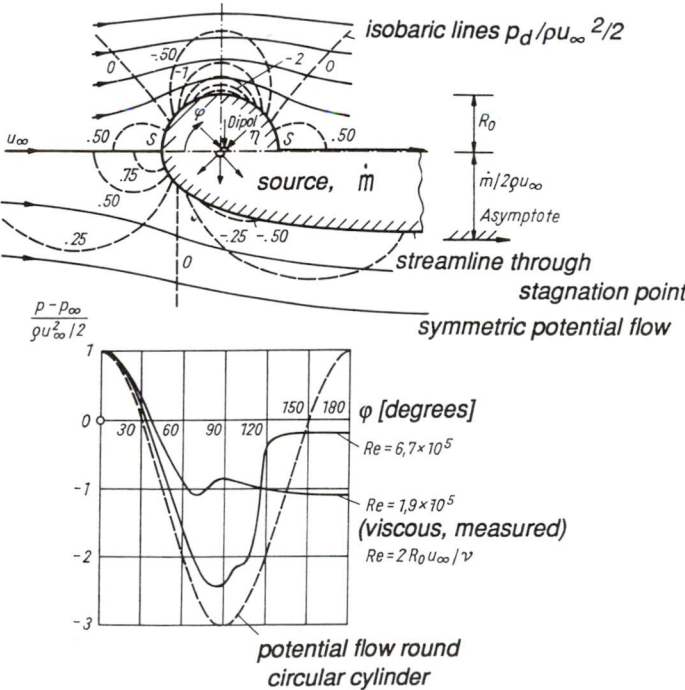

Fig. 13.13(a). Potential flow patterns: Parallel main stream with point source (semiinfinite body) and with dipole (circular cylinder). Diagram of the pressure distribution at the surface of a circular cylinder. Influence of viscosity

$$F(z) = u_\infty z + \frac{\dot{m}}{2\pi\,\rho}\,\ln\frac{z}{z-a}\ ,\ \ F'(z) = u_\infty + \frac{\dot{m}}{2\pi\,\rho}\left(\frac{1}{z} - \frac{1}{z-a}\right) = 0 \ \rightarrow$$

$$x_S = \frac{a}{2}\left(1 \pm \sqrt{1 + 2\,\frac{\dot{m}}{\pi\,a\,\rho\,u_\infty}}\right)\ ,\ \ y_S = 0\ ,$$

$$\psi = u_\infty\,y - \frac{\dot{m}}{2\pi\,\rho}\left[\tan^{-1}\frac{y}{r} - \tan^{-1}\frac{y}{\sqrt{(x-a)^2 + y^2}}\right] = \psi_S = 0\ .$$

$$(13.59a)$$

The outer flow is that round a finite cylinder, which is named the *Rankine* body.

With decreasing values of the distance *a* between the source and the sink, the shape of the body becomes more and more circular. In the limit $a \rightarrow 0$, but simultaneously increasing the strength of the sources beyond any bounds, a so-called *dipole* results with a finite dipole moment η , and the streamline passing through the stagnation points and thus separating the inner from the outer flow becomes circular The expansion of the logarithmic function in Eq. (13.59a) and taking the limit give

$$\lim_{\substack{a\rightarrow 0\\ \dot{m}\rightarrow\infty}}\left\{\frac{\dot{m}}{\rho}\,a = \eta\right\} \rightarrow F(z) = u_\infty\,z + \frac{\eta}{2\pi}\,\frac{1}{z}\ ,$$

$$\psi = u_\infty\,y - \frac{\eta}{2\pi}\,\frac{y}{r^2} = \psi_S = 0 \ \rightarrow\ \left\{\begin{array}{l} r = R_0 = \sqrt{\dfrac{\eta}{2\pi\,u_\infty}}\\[2mm] \eta = 2\pi\,R_0^2\,u_\infty \end{array}\right\}.$$

$$(13.59b)$$

The strength of the dipole is determined by the given radius of the circular cylinder R_0 . The upper half of Fig. 13.13(a) illustrates the resulting flow pattern of the outer flow. The streamlines as well as isobars of the potential flow round such blunt finite cylinders correspond to the actual (viscous) flow only in some neighborhood of the stagnation point upstream. The influence of the *Reynolds* number on the pressure distribution is also shown in Fig. 13.13(a).

(§)   Superposition of Potential Vortex Lines and a Parallel Main Stream. Two parallel vortex lines of strengths Γ_1 and Γ_2 are considered when located at points $z_1\,(t)$ and $z_2\,(t)$, respectively, in an infinite fluid body. The complex potential of the resulting irrotational flow is given by the superposition

$$F_{1,2}(z) = \frac{\Gamma_1}{2\pi\,i}\,\ln\,(z - z_1) + \frac{\Gamma_2}{2\pi\,i}\,\ln\,(z - z_2).$$

$$(13.60)$$

The vortices propel each other in a direction normal to the plane spanned by their axes. The induced velocities are then given from the derivatives of the potential taken at the sites

$$\dot{z}_1 = \frac{\Gamma_2}{2\pi i}\frac{1}{(z_1 - z_2)} \quad , \quad \dot{z}_2 = \frac{\Gamma_1}{2\pi i}\frac{1}{(z_2 - z_1)} \ .$$

$$(13.61)$$

There must be an axis parallel to the vortex lines and passing through the "vortex center" where the velocity is permanently zero. The coordinates are easily derived from the condition

$$\frac{dF_{1,2}}{dz} = 0 \ \rightarrow \ z_\Gamma = \frac{\Gamma_1 z_1 + \Gamma_2 z_2}{\Gamma_1 + \Gamma_2} \ , \ \dot{z}_\Gamma \equiv 0 \ .$$

$$(13.62)$$

Note the static moments of the circulation, and see Eq. (2.76) for the definition of centers. The rigid line connecting the singular vortex points rotates about this center with the angular velocity

$$\omega = \frac{|\dot{z}_1|}{|z_1 - z_\Gamma|} = \frac{\Gamma_1 + \Gamma_2}{2\pi a} = \text{const} \ , \ a = |z_1 - z_2| = \text{const} \ .$$

$$(13.63)$$

According to Eq. (13.62), the circulation-center moves to infinity for counterrotating vortices of equal strength, $\Gamma_1 = -\Gamma_2 = \Gamma$, and $\omega \rightarrow 0$. Hence, in that case, the vortex lines move along parallel straight paths with speed $\Gamma/2\pi a$. The axis of symmetry of the induced flow separates the flow pattern into two halves, and thus, it can be materialized by a rigid wall. That is, a single vortex line a distance $a/2$ from a rigid wall moves along like the speed induced by its image behind that wall.

Another important flow pattern results, if a parallel main stream is superposed, such that the pair of vortices of equal strengths becomes stationary. Superposition gives

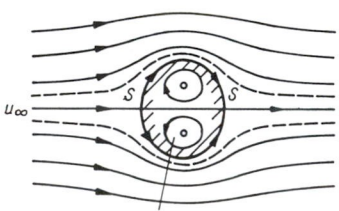

internal flow (vortices not moving)

Fig. 13.13(b). Streamlines produced by a stationary pair of potential vortices in a parallel main stream. Separation of the flow pattern by the streamline bifurcating at the stagnation points S

$$F(z) = \frac{i\Gamma}{2\pi}\left(\frac{z}{a} + \ln\frac{z + a/2}{z - a/2}\right) .$$

(13.64)

Two stagnation points of the resulting flow are located on the x axis. The contour of the streamline bifurcating at these points renders a cylinder, x is an axis of symmetry, and the outer flow gives the pattern of the potential flow round that blunt body. Figure 13.13(b) depicts the streamlines of a parallel main stream displaced in the near field of the cylinder. The superposition of sources and sinks (including dipoles) with vortex lines renders flow patterns with the production of hydrodynamic lift, eg of airfoils. The *Joukowsky* foil is a classic solution found by conformal mapping. References are given in the classical literature of fluid mechanics.

13.4.3. Hydrodynamic Forces in Two-Dimensional and Stationary Potential Flow, the Blasius Formula

The resultants of the pressure distribution acting on a rigid body that translates with constant speed through an ideal fluid are elegantly determined by a complex analysis. The force per unit of length on an arc element of a cylindrical body is simply given by the components (see Fig. 13.14), $dy/dx > 0$,

$$dF_x = -p\, dy \quad , \quad dF_y = p\, dx \quad .$$

(13.65)

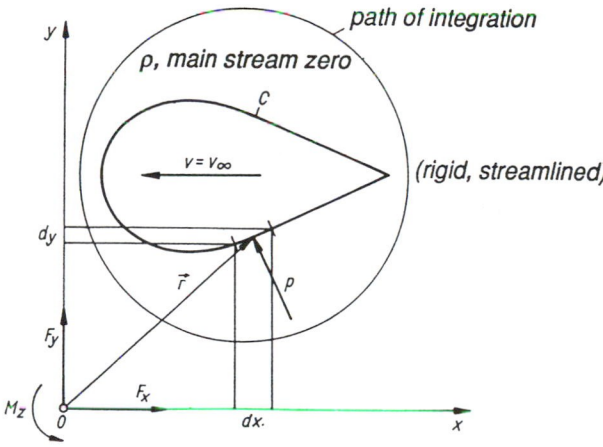

Fig. 13.14. Surface tractions (pressure) acting on a rigid body translating steadily through an ideal fluid

The axial moment with respect to the origin 0 is

$$dM_z = p\,(x\,dx + y\,dy) \quad . \tag{13.66}$$

For the sake of complex notation, an imaginary component of the axial moment is constructed by the "virial" of the force dF , which may be considered a generalization of the static moment to vector quantities,

$$dN = (\mathbf{r} \cdot \mathbf{dF}) = p\,(-x\,dy + y\,dx). \tag{13.67}$$

In irrotational motion, the pressure results from *Bernoulli´s* equation (8.42) with a universal constant

$$p = Const - \frac{\rho}{2}\,|\mathbf{v}|^2 \quad . \tag{13.68}$$

In terms of the complex velocity potential, the squared speed becomes

$$|\mathbf{v}|^2 = (u - i\,v)(u + i\,v) = \frac{dF}{dz}\,\frac{d\overline{F}}{d\overline{z}} \quad . \tag{13.69}$$

Upon substitution, the complex representation results, where *Bernoulli´s Constant* is omitted, since constant pressure acting on a closed contour is self-equilibrating

$$dF_x - i\,dF_y = -p\,(dy + i\,dx) = -i\,p\,d\overline{z} = \frac{i\rho}{2}\,\frac{dF}{dz}\,d\overline{F},$$

$$dM_z + i\,dN = p\,z\,d\overline{z} = -\frac{\rho}{2}\,z\,\frac{dF}{dz}\,d\overline{F}. \tag{13.70}$$

The body contour is part of a (bifurcating) streamline where $\psi = \psi_S$, and hence, the total derivative $d\psi = 0$ along that line, ie the differential of the complex potential dF , is real and thus equal to the differential of its conjugate. Loop integration of Eq. (13.70) over the surface C of the body renders *Blasius ´* formula

$$F_x - i\,F_y = \frac{i\rho}{2}\,\oint_C \left(\frac{dF}{dz}\right)^2 dz \quad , \quad M_z + i\,N = -\frac{\rho}{2}\,\oint_C z\left(\frac{dF}{dz}\right)^2 dz \quad . \tag{13.71}$$

According to the rules of complex integration, the results become independent of the special contour C, if $F(z)$ and its derivative dF/dz are analytic functions in the whole complex plane of definition of the outer flow. In such a case, no singularities are encountered in the outer fluid flow and the integration contour can be selected arbitrarily as indicated in the Fig. 13.14. In the cases of flow round bodies with isolated singularities (see, eg the *von Karman* trail of vortices in Sec. 13.4.4), the integration path is still to be selected arbitrarily under the condition of properly cutting away these singularities.

In the important case of a parallel main stream round a rigid body at rest (see Sec. 13.4.2) the derivative of the complex velocity potential can be expanded into the *Laurent* series (no singularities in the outer flow are assumed to exist)

$$\left(\frac{dF}{dz}\right)^2 = v_\infty^2 + 2\, v_\infty \frac{\dot{m}/\rho - i\,\Gamma}{2\pi\, z} + \frac{1}{2\pi\, z^2}\left[-2\eta\, v_\infty + \frac{(\dot{m}/\rho - i\,\Gamma)^2}{2\pi}\right] + O(z^3)\,,$$

(13.72)

where $\eta = \eta_x + i\,\eta_y$ is the resulting complex moment of all dipoles and Γ denotes the resulting circulation. After substitution in Eq. (13.71), integration over a circle of sufficiently large radius R (to enclose the body) can be performed. The origin of the coordinate system is located in the interior of the cylinder. Only first-order poles (in the interior) contribute to the integrals and the residue renders the formulas for the complex forces, which turn out to be exact, despite the expansion of Eq. (13.72),

$$F_x - i\, F_y = 2\pi\, i\,\frac{i\rho}{2}\, 2\, v_\infty \frac{\dot{m}/\rho - i\,\Gamma}{2\pi}\,,$$

$$M_z + i\, N = -2\pi\, i\,\frac{\rho}{2}\,\frac{1}{2\pi}\left[-2\,\eta\, v_\infty + \frac{(\dot{m}/\rho - i\,\Gamma)^2}{2\pi}\right].$$

(13.73a)

Comparing the real and imaginary parts yields the explicit formula for the hydrodynamic resultants in a parallel main stream

$\qquad F_x = -\dot{m}\, v_\infty\,$,   propelling force of a plane point source,

$\qquad F_y = \rho\, v_\infty\,\Gamma\,$,   lateral force (lift) of a potential vortex,

(13.73b)

$M_z = -\rho\left(\eta_y\, v_\infty + \frac{\dot{m}\,\Gamma}{2\pi\,\rho}\right)\,$,   hydrodynamic moment with respect to

the origin O, source, dipole and circulation contribute.

13.4.4. *von Karman Trail of Vortices, the Strouhal Number*

Periodic phenomena are observed in the wake behind cylinders at rest in a parallel main stream. The initiation of such vortices is typically a viscous effect. They are shed alternately from the top and bottom side of the cylinder in parallel rows. But the flow pattern can still be approximated by adding two rows of potential

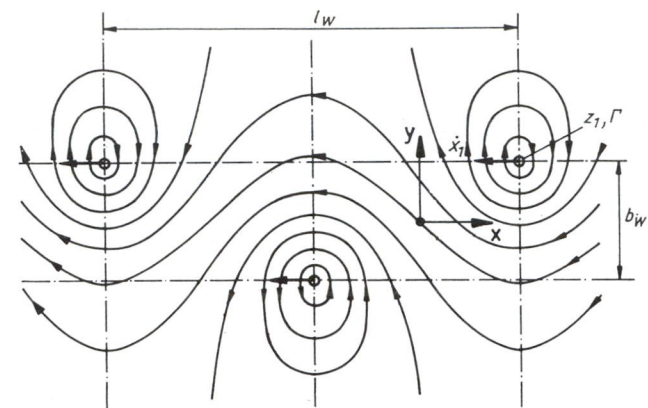

Fig. 13.15. Periodic street of potential vortices. Velocity induced at points of
vortex lines. Resulting streamlines

vortex lines with counterrotating circulation and equal strength to
the stationary inviscid parallel main flow. Within such an
approximation, the diffusion of the circulation of individual vortex
lines with distance, also a viscous effect, is neglected; see Exercise
A 13.4. The trail is sketched in Fig. 13.15 with width b_w and
equidistant spacing l_w. The coordinate system is arranged such that
the "first" vortex line is located at $z_1 = l_w /4 + i\, b_w /2$. The total
complex velocity potential results by summation of the individual
contributions and by noting the representation of the complex sine
function by an infinite product

$$\sin \pi z/l = \pi \frac{z}{l} \prod_{k=1}^{\infty} \left(1 - \frac{(z/l)^2}{k^2}\right) \;\to\; F_w(z) = -\frac{\Gamma}{2\pi i} \ln \frac{\sin (z - z_1)\, \pi/l_w}{\sin (z + z_1)\, \pi/l_w} \; . \quad (13.74)$$

The velocity induced by the infinite trail at the point of a vortex
line, say, of number 1, is given by the derivative, leaving out the
vortex line 1 in the complex potential,

$$\dot{z}_1 = \left.\frac{dF_w(z)}{dz}\right|_{z = z_1 \notin z_1} = -\frac{\Gamma}{2 l_w} \tanh \pi b_w/l_w = \dot{x}_1 \; , \; \dot{y}_1 = 0 \; . \quad (13.75)$$

The trail moves along with the constant speed given in Eq. (13.75)
from the right to the left. Such a flow pattern is seen to be related
to the wake behind a rigid blunt cylinder moving with some constant
but greater speed u_∞ through a viscous fluid. The application of
conservation of momentum to a control volume, properly attached to
the moving body, $w_x = -\,u_\infty$ [see Eqs. (7.13) and (7.14)], and
enclosing a sufficiently large number of vortices, renders a force

that can be interpreted as the drag. It is due to the periodic formation of new vortices that follow from the alternating separation of the boundary layer at the body's surface. The flow becomes highly rotational in the viscous fluid, but the approximation by potential vortices leaves the ideal flow irrotational. The force is quite easily found to be

$$F_D = \rho \, \Gamma \left(u_\infty - 2 \, \dot{x}_1 \right) b_w/l_w + \rho \, \Gamma^2/2 \, \pi \, l_w \; . \tag{13.76}$$

Theoretical investigations, which are well confirmed by measurements, render the least unstable configuration of the trail, with respect to small oscillations, when the ratio $b_w/l_w = 0.281$. In that case, the circulation can be eliminated, and a theoretical drag coefficient results in a second-order approximation

$$c_D = \frac{F_D}{\frac{\rho}{2} \, u_\infty^2 \, d} \approx \frac{l_w}{d} \left[1.59 \, \frac{|\dot{x}_1|}{u_\infty} - 0.63 \left(\frac{\dot{x}_1}{u_\infty} \right)^2 \right] \; , \; \dot{x}_1 = \Gamma/2 \, l_w \, \sqrt{2} \; . \tag{13.77}$$

The shedding frequency f (ie the frequency of vortex production), and hence, the geometry of the *von Karman* trail are strongly dependent on the lateral width d of the cylinder and its shape. From now on the body is assumed to be at rest. In a parallel main stream of assigned speed u_∞ (the vortex street in that case drifts to the right), a dimensionless coefficient, the *Strouhal* number determines similarity

$$St = \frac{f \, d}{u_\infty} \; , \; f = \frac{u_\infty - |\dot{x}_1|}{l_w} \; , \; [Hz] \; . \tag{13.78}$$

In the wide range of *Reynolds* numbers $500 < Re = u_\infty \, d/v < 5 \times 10^4$ of an irrotational parallel main stream, the *Strouhal* number is narrowly bounded, eg for circular cylinders between $0.18 \leq St \leq 0.20$, and for plates with angles of attack $\alpha > 30°$ between $0.15 \leq St \leq 0.18$. Sensitivity is observed, however, in any rotation present in the approaching main stream, eg a *Strouhal* number as low as $St = 0.1$ was observed in natural wind flow round steel columns of circular cross-section. It should be mentioned that surface roughness also influences the *Strouhal* number.

The periodic unbalanced force due to the alternating formation of the vortices causes a vibration that gives rise to such phenomena as the *Aeolian* harp, the singing of telephone wires, and, in general, the howling of winds. When the shedding frequency corresponds to the (basic) natural frequency of the structure, resonance occurs, which, after a sufficiently long time of exposure, can cause severe damage to the structure. Slender chimneys and long-span suspension bridges are subject to such excitations and they need proper measures of control. The combination of bending and torsional vibrations with additional forcing by alternating lift is extremely dangerous and may cause failure, as was documented in a movie on the disastrous collapse of the Tacoma Narrows Bridge in the state of Washington (see also Sec. 9.6).

The total drag coefficient of Fig. 13.8 is dominated by the portion of pressure drag. The friction drag decreases with the increase of the *Reynolds* number, while the former remains fairly constant. Formation of the *von Karman* trail of vortices is observed for *Re > 40*. The sharp drop in drag at the critical *Re = (2 ÷ 4) x10⁵*, from a value of about *1.2* to about *0.3*, indicates a sudden change in the flow pattern: The laminar boundary layer with a separation point at about $\varphi = 80°$ changes to a turbulent layer before separation occurs, which is then shifted to an angle of about $\varphi = 105°$. Despite the slight increase in friction drag associated with the turbulent boundary layer, the decrease in pressure drag dominates. The width of the *von Karman* street becomes much smaller and the dissipation of ordered kinetic energy is reduced.

13.4.5. The Hydrodynamic Pressure at the Face of a Moving Dam

In the case of a horizontal ground motion, eg that due to an earthquake, the wetted face of a dam or the plane wall of a container moves against the fluid that is assumed to have a free surface. Figure 13.16 depicts the two-dimensional configuration with the assumption of a rigid and long dam. The flow is assumed to be incompressible and inviscid and, hence, irrotational. The velocity potential is sought in the form of a separable solution of the *Laplace* equation

$$\Delta\Phi = 0 \quad , \quad \Delta = \frac{\partial^2}{\partial x^2} + \frac{\partial^2}{\partial y^2} \quad , \quad \Phi = g(x)\, h(y) \ .$$

$$(13.79)$$

Substitution and subsequently separation of variables render two ordinary differential equations,

$$\frac{g''}{g} = -\frac{h''}{h} = -\lambda^2 = \text{const} \ .$$

$$(13.80)$$

With the basic solutions $g = \cos \lambda x$ and $h = \exp(-\lambda y)$ (decaying properly with $y \to \infty$, $\lambda > 0$) by superposition, a series solution of the time-dependent potential is constructed, which satisfies Eq. (13.79) for any *f(t)*,

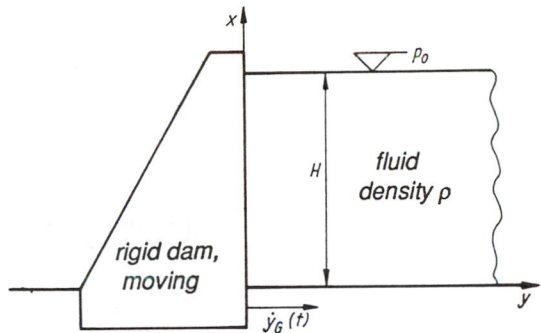

Fig. 13.16. Moving boundary. Incompressible potential flow

$$\Phi(x, y; t) = f(t) \sum_n C_n \, e^{-\lambda_n y} \cos(\lambda_n x).$$

(13.81)

The boundary conditions according to Fig. 13.16 and the radiation condition (the fluid strip is assumed to be semiinfinite) require

$$x = 0: \; v_x = \text{grad } \Phi \cdot \mathbf{e}_x = \frac{\partial \Phi}{\partial x} = 0 \; \rightarrow \; -f(t) \sum_n \lambda_n C_n \, e^{-\lambda_n y} \sin(\lambda_n x)\big|_{x=0} = 0 \, ,$$

(13.82)

$$y = 0: \; v_y = \text{grad } \Phi \cdot \mathbf{e}_y = \frac{\partial \Phi}{\partial y} = \dot{y}_G \; \rightarrow \; -f(t) \sum_n \lambda_n C_n \, e^{-\lambda_n y} \cos(\lambda_n x)\big|_{y=0} = \dot{y}_G \, ,$$

(13.83)

$$x = H: \; p = -\rho \frac{\partial \Phi}{\partial t} = 0 \; \rightarrow \dot{f}(t) \sum_n C_n \, e^{-\lambda_n y} \cos(\lambda_n x)\big|_{x=H} = 0 \, ,$$

(13.84)

$$y \rightarrow \infty: \; \lim_{y \to \infty} \Phi = \lim_{y \to \infty} \frac{\partial \Phi}{\partial y} = 0 \; \rightarrow \lambda_n > 0 \, .$$

(13.85)

Variation of the pressure due to surface waves is neglected in Eq. (13.84). Otherwise, a linearized (mixed) boundary condition at $x = H$ results when taking into account the surface wave in a first-order approximation. It results from the quasihydrostatic pressure increase under the wave crest $x_S(y, t)$, $p = -\rho \, \partial \Phi / \partial t = \rho g (x_S - H)$, by time differentiation, and after substituting the vertical velocity component: $g \, \partial \Phi / \partial x + \partial^2 \Phi / \partial t^2 \, |_{x=H} = 0$. For celerity, see Eq. (9.135). The simplified condition (13.84) indicates that the characteristic values are given by

$$\cos \lambda_n H = 0 \rightarrow \lambda_n H = \frac{\pi}{2}, \frac{3\pi}{2}, ..., \frac{(2n-1)\pi}{2}, \quad n = 1, 2, ... \quad (13.86)$$

Thus, Eq. (13.83) remains to be solved for the constants C_n

$$\sum_n \lambda_n C_n \cos \lambda_n x = -1 \rightarrow f(t) = \ddot{y}_G . \quad (13.87)$$

Such a problem is to be tackled, in general, by the periodic succession of the (constant) right-hand side, such that an expansion of the resulting periodic function into a *Fourier cosine* series becomes possible: The interval of definition of the constant in Fig. 13.16 is $0 < x < H$, and hence, the period must be $4H$. The *Fourier* series of such a box-type function can be taken from the literature; see, eg *I. N. Bronstein and K. A. Semendjajew: Taschenbuch der Mathematik. Deutsch, Frankfurt/M, 1980, p. 663.*

$$-1 = -\frac{4}{\pi} \left(\cos \frac{\pi x}{2H} - \frac{1}{3} \cos \frac{3\pi x}{2H} + \frac{1}{5} \cos \frac{5\pi x}{2H} ... \right) = \frac{4}{\pi} \sum_{n=1,2,..} \frac{(-1)^n}{2n-1} \cos \lambda_n x .$$
$$(13.88)$$

Equating Eqs. (13.87) and (13.88) and comparing coefficients yield

$$C_n = \frac{4}{\pi} \frac{(-1)^n}{2n-1} \frac{2H}{(2n-1)\pi} = \frac{(-1)^n 8H}{\pi^2 (2n-1)^2} , \quad n = 1, 2, ... \quad (13.89)$$

The periodic succession of Eq. (13.87) is the crucial step in solving such a fluid-solid interaction problem. Analogously, the more complicated boundary value problem of an elastic wall interacting with a (linearly) compressible fluid is to be tackled; see Exercise A 13.2. Finally, the series solution for the hydrodynamic pressure results. When taken at $y = 0$, it determines the dynamic loading of the dam that must be considered in addition to the hydrostatic pressure

$$p(x, y; t) = -\rho \frac{\partial \Phi}{\partial t} = 2 \rho H \ddot{y}_G \sum_{n=1,2,..} \frac{(-1)^n}{\mu_n^2} e^{-\mu_n y/H} \cos \frac{\mu_n x}{H} , \quad (13.90)$$

$$\mu_n = (2n-1) \pi/2 .$$

Note that the dynamic pressure is proportional to the ground acceleration. Equation (13.90) can be generalized to include linear compressibility of the fluid. The influence function in the kernel of the resulting integral is the *Bessel* function of order zero. Resonance at a rather low frequency may be observed.

13.4.6. Stationary Efflux of Gas from a Pressure Vessel

The flow through a small nozzle forms a free jet under atmospheric pressure p_0. The expansion from an interior pressure p_1 is assumed under isentropic conditions

$$\frac{\rho}{\rho_1} = \left(\frac{p}{p_1}\right)^{1/\kappa} \quad , \quad \kappa = c_p / c_v \quad , \tag{13.91}$$

where c_p and c_v is the specific heat of the ideal gas at constant pressure and constant density (specific volume), respectively; see also Eq. (2.90). Bernoulli's equation (8.31), with compressibility taken into account, renders along a streamline from a point in the interior of the large vessel, where the gas is at rest, to a point in the free jet of cross-section A_0 (no body forces are considered)

$$\frac{v_0^2}{2} - 0 = - \int_{p_1}^{p_0} \frac{dp}{\rho} = - \frac{p_1}{\rho_1} \int_1^{p_0/p_1} \left(\frac{p}{p_1}\right)^{-1/\kappa} d\left(\frac{p}{p_1}\right) = - \frac{p_1}{\rho_1} \frac{\kappa}{\kappa-1} \left[\left(\frac{p_0}{p_1}\right)^{(\kappa-1)/\kappa} - 1\right]. \tag{13.92}$$

The mass flow rate is given by the formula

$$\dot{m} = \rho_0 \, v_0 \, A_0 = A_0 \left(\frac{p_0}{p_1}\right)^{1/\kappa} \sqrt{\frac{2\,\kappa}{\kappa-1}\, p_1 \, \rho_1 \left[1 - \left(\frac{p_0}{p_1}\right)^{(\kappa-1)/\kappa}\right]} \quad , \tag{13.93}$$

and vanishes, as expected, if $p_1 = p_0$, but also quite unnaturally if the vessel is surrounded by a vacuum, ie for $p_0 = 0$. The latter case is unphysical, and thus, it must be concluded that Eq. (13.93) is valid only until the mass flow rate becomes maximal at a critical outer pressure $p_0 = p^*$. The critical pressure ratio (p_1 is assigned)

$$\left(\frac{p^*}{p_1}\right) = \left(\frac{2}{\kappa+1}\right)^{\kappa/(\kappa-1)} \quad , \quad \kappa = 1.4 \text{ (eg for dry air)} \rightarrow \left(\frac{p^*}{p_1}\right) = 0.528 \, , \tag{13.94}$$

gives the maximum of the mass flow rate

$$\max \dot{m} = A_0 \left(\frac{2}{\kappa+1}\right)^{1/(\kappa-1)} \sqrt{\frac{2\,\kappa}{\kappa+1}\, p_1 \, \rho_1} \, . \tag{13.95}$$

In that case, the speed of the jet equals the local sound speed in the gas at the critical pressure

$$c = \sqrt{\frac{dp}{d\rho}\bigg|_{p^*}} = v_0^* = \sqrt{\frac{\kappa\, p_1}{\rho_1}\left(\frac{p^*}{p_1}\right)^{(\kappa-1)/\kappa}} = c^* \, . \tag{13.96}$$

In terms of the *Mach* number $Ma = v/c$, the critical conditions are reached if the local *Mach* number $Ma^* = v_0/c = 1$ in the free jet. Equation (13.93) remains valid only for subsonic flows, ie if $p_0 \geq p^*$. Any reduction of the outer pressure below p^* cannot be felt upstream, since the speed of the flow has already reached the speed of sound. The mass flow rate through such a simple opening cannot be increased above the maximal value given in Eq. (13.95). For $p_0 < p^*$, it becomes independent of the counter pressure. The mass flow rate density $\rho^* v^*$ is maximal at the critical pressure p^* . If the pressure in the free jet $p^* > p_0$, lateral vibrations are observed, where supersonic speeds are reached in the regions of enlarged cross-sections. Commonly, the counter pressure p_0 is assigned, and for a gas with $\kappa = 1.4$, the pressure in the vessel, $p_1 = 1.89\, p_0$, renders the mass flow rate maximal. Any further increase of the interior pressure has no effect on the mass flow rate.

Also in the case of a supercritical pressure ratio, a controlled expansion of the free jet is desirable. The nozzle must be redesigned by adding a divergent pipe behind a narrowing section with minimal cross-section A_0 : This renders a *Laval* nozzle, where critical pressure p^* and local *Mach* number $Ma^* = 1$ are assigned to A_0 . Contrary to the commonly encountered subsonic flow, the pressure decreases in the supersonic flow through the divergent portion of the *Laval* nozzle. The supersonic flow is accelerated when it enters the portion of the pipe with increasing cross-section. After passing through such a *Laval* nozzle, the free jet flowing into a vacuum takes on maximal speed

$$v_{max} = \sqrt{\frac{2\kappa}{\kappa-1}\frac{p_1}{\rho_1}} = c_1 \sqrt{\frac{2}{\kappa-1}} .$$
$$(13.97)$$

The mass flow rate is still given by Eq. (13.95) since the critical pressure $p^* = p_0$ is effective in the narrowest cross-section A_0 . Two helpful books on gas dynamics are *K. G. Guderley: Theorie schallnaher Strömungen. Springer-Verlag, Berlin, 1957.*, and, *K. Oswatitsch: Gasdynamik. Springer-Verlag, Wien, 1976.*

13.5. Momentum Integral Method of Boundary-Layer Analysis

This method, originated by *von Karman* , is applicable to laminar as well as to turbulent boundary layers. In the latter case, an additional assumption must be made about the wall shear stress (a laminar sublayer should be taken into account). A common approximation is just to disregard the (small) shear stress at the edge of the boundary layer, $y = \delta(x)$; see, eg Eq. (13.39), where a smooth transition to the outer, inviscid flow is assumed. By considering a control volume of a two-dimensional stationary and incompressible flow in the boundary layer, bounded by an element dx of the wall and extending to the edge, where the velocity $U = U(x)$ is prescribed, Eq. (7.13) is applied in the x direction, ie parallel to the wall, and the mass flow rates are taken per unit of length. Expansion of all functions of $(x + dx)$ at the neighboring location x, dividing through dx, and taking the limit $dx \to 0$ yield (τ_0 is the wall shear stress; see Fig. 13.17)

$$\frac{d}{dx} \int_0^{\delta(x)} \rho\, u^2 \, dy + \dot{m}_y' \, U(x) = -\tau_0 - \delta(x) \frac{dp}{dx} \quad , \quad \dot{m}_y' \, dx = -\frac{d}{dx} \left(\int_0^{\delta(x)} \rho\, u \, dy \right) dx \, .$$

$$(13.98a)$$

Note that the mass flow rate through the edge must equal the net influx through the cross-sections $\delta(x)$ and $\delta(x + dx)$ of the boundary layer. No body forces are considered. The pressure gradient can be expressed in terms of velocity by means of the stationary *Euler* equation of the inviscid outer flow, the x component of Eq. (8.30),

$$\rho\, U \frac{dU}{dx} = -\frac{dp}{dx} \, .$$

$$(13.98b)$$

With an identical transformation with respect to the derivative of the momentum flux through the edge, the shear stress at the wall is expressed by

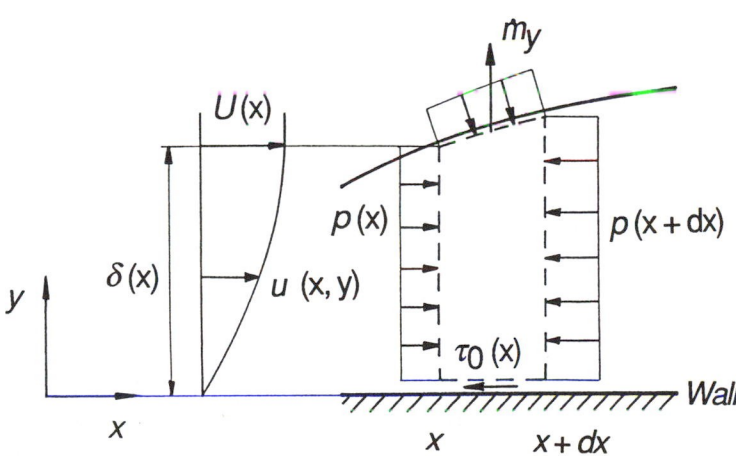

Fig. 13.17. Stationary incompressible flow through a control volume of the boundary layer of infinitesimal length dx. Velocity of inviscid flow with pressure gradient, $U(x)$

$$\tau_0 = \rho \frac{d}{dx} \left[U^2 \int_0^{\delta(x)} \frac{u}{U}\left(1 - \frac{u}{U}\right) dy \right] + \rho U \frac{dU}{dx} \int_0^{\delta(x)} \left(1 - \frac{u}{U}\right) dy \ .$$

(13.99)

The remaining integrals with the dimension of length quite naturally determine the boundary-layer thickness with respect to momentum flux, the momentum thickness $\theta(x)$, and with respect to mass flow rate, the displacement thickness $\delta^*(x)$, respectively,

$$\theta(x) = \int_0^\infty \frac{u}{U}\left(1 - \frac{u}{U}\right) dy \quad , \quad \delta^*(x) = \int_0^\infty \left(1 - \frac{u}{U}\right) dy \ .$$

(13.100)

Integration can be extended to infinity, since the integrand vanishes for $y > \delta(x)$. Equations (13.99) and (13.100) are useful if a velocity distribution is either known or its shape is assumed. For the flow over a flat plate, a case considered in Sec.13.3.2, ie with no pressure gradient, $U = const$, Eq. (13.99) reduces to

$$\tau_0 = \rho \, U^2 \frac{d\theta}{dx} \quad \rightarrow \quad c_f = \frac{\tau_0}{\rho U^2 / 2} = 2 \frac{d\theta}{dx} \ .$$

(13.101)

c_f is the local friction coefficient. The simplest approximation of the laminar velocity profile is the parabolic one

$$u/U = 2 \, y/\delta - (y/\delta)^2 \ .$$

(13.102)

It satisfies the boundary condition of no-slip, $u = 0$ at $y = 0$, and the transition condition $u = U$ and $du/dy = 0$ at the edge $y = \delta(x)$. See Sec. 13.3.2.

13.6. Exercises A 13.1 to A 13.4 and Solutions

A 13.1: A viscous fluid in a spinning cylindrical container with a vertical axis may rotate stationary like a frozen rigid body, $\Omega = const$. Determine the surfaces of constant pressure and, hence, the shape of the interface with air at an assigned atmospheric pressure $p = p_0$; see Fig. A 13.1.

Solution: The rigid-body rotation is not irrotational [see Eq. (13.54) for the irrotational flow of a potential vortex]. Along the straight vertical axis, $r = 0$, there is a hydrostatic pressure distribution assigned: $p(r = 0, h) = p_0 + \rho gh$, $h = -z$. The radial component of the

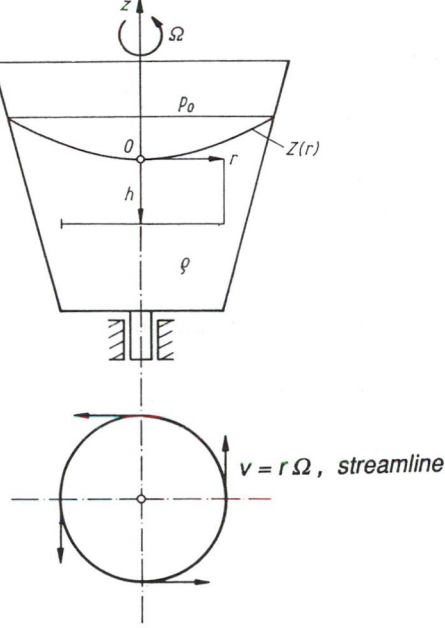

Fig. A 13.1.

Navier-Stokes equation, lateral to the circular streamlines, renders $(v_r = 0, v = r\,\Omega)$

$$\rho\,\frac{v^2}{r} = \frac{\partial p}{\partial r} = \rho\,r\,\Omega^2.$$

Integration from the free surface of assigned pressure downward yields

$$0 = p_0 + \rho g h - p_0(r, h) + \rho\,\frac{r^2\,\Omega^2}{2}.$$

Formally, the relation is verified by considering the *Bernoulli* equation (8.66) in the rotating coordinate system with no relative flow left over. The surfaces of constant pressure are paraboloids of revolution affined to the free surface

$$p(r, h = -Z) = p_0 \;\rightarrow\; Z(r) = \frac{r^2\,\Omega^2}{2g}.$$

A 13.2: The hydrodynamic pressure distribution of Eq. (13.90) should be extended to include the contribution of the flow due to the linear elastic deflections of the cantilever plate strip shown in Fig. A 13.2. For simplicity, bending stiffness $K = Eh^3/12(1 - v^2)$ is assumed to be constant. The midplane deforms into a shallow cylindrical surface. Its shape is approximated by a finite series according to *Ritz* . Proper functions $\varphi_j (x)$, $j = 1, 2, ..., n$ are selected from an orthogonal system of functions in *[0, H]* . When we substitute K for *(EJ)* in Eq. (7.147), *Galerkin's* rule (11.11) becomes applicable. Determine the hydrodynamically coupled linear equations of motion for the generalized coordinates $q_i (t)$.

Solution: The solution of the linearized incompressible flow is given by superposition of the velocity potential (13.81), where C_n is to be determined by Eq. (13.89), with a second potential Φ_D of an analogous series representation to account for the effect of dam (plate) deformations, *w(x, t)* ,

$$\Phi_D(x, y, t) = f(t) \sum_n D_n \cos \lambda_n x \ e^{-\lambda_n y} \ , \ \lambda_n \text{ of Eq. (13.86).}$$

The new constants D_n result from the remaining nonhomogeneous boundary condition at $y = 0$

$$\frac{\partial \Phi_D}{\partial y} = \dot{w}(x, t) = - f(t) \sum_n \lambda_n D_n \cos \lambda_n x = \sum_{i=1}^{m} \dot{q}_i(t) \ \varphi_i(x) .$$

Analogously to Eq. (13.88), the right-hand side must be expanded into a *Fourier cosine* series. The shape functions $\varphi_i (x)$, with period *4 H* symmetric at *x = 0* , have an associated set of *Fourier*

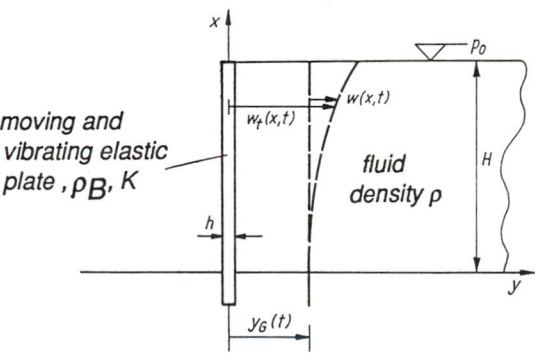

Fig. A 13.2.

coefficients, and when we compare coefficients, the result is

$$C_{ni} = \frac{2}{H} \int_0^H \varphi_i(x) \cos \lambda_n x \, dx \;\rightarrow\; -f(t) \, D_n = \lambda_n^{-1} \sum_{i=1}^m \dot{q}_i(t) \, C_{ni} \;.$$

The additional hydrodynamic pressure that is due to the deformation of the dam (plate) takes on the form of a double series

$$p_D(x, y, t) = -\rho \frac{\partial \Phi_D}{\partial t} = \sum_{i=1}^m \rho \, \ddot{q}_i(t) \sum_n \left(\lambda_n^{-1} \, C_{ni} \, e^{-\lambda_n y} \cos \lambda_n x \right) .$$

The dynamic loading at the face $y = 0$ by superposition becomes

$$\bar{p} = p(x, 0, t) + p_D(x, 0, t) \;, \quad p \text{ of Eq. (13.90)}.$$

Substitution into Eq. (7.147) yields the error

$$\sum_{k=1}^m \left\{ K \frac{\partial^4 \varphi_k}{\partial x^4} \, q_k(t) + h \, \rho_B \left[\ddot{y}_G(t) + \ddot{q}_k(t) \, \varphi_k(x) \right] \right\} + \bar{p}(x, 0, t) = p^* \;.$$

Galerkin's rule gives, if we take into account the orthogonality of the shape functions and the parameters,

$$\int_0^H \varphi_k(x) \, h \, \rho_B \, \varphi_j(x) \, dx = m_j^{(B)} \, \delta_{kj} \;, \quad \int_0^H \varphi_j(x) \, K \frac{\partial^4 \varphi_k}{\partial x^4} \, dx = k_j \, \delta_{kj} \;,$$

$$h \, \rho_B \int_0^H \varphi_j(x) \, dx = L_j \;,$$

the hydrodynamically coupled set of m equations of motion (material damping is neglected)

$$m_j^{(B)} \, \ddot{q}_j + k_j \, q_j = - L_j \, \ddot{y}_G(t) - m_j \, \ddot{y}_G(t) - \sum_{i=1}^m m_{ji} \, \ddot{q}_i \;, \quad j = 1, 2, \ldots, m,$$

$$m_j = \rho \sum_n \frac{(-1)^n}{\lambda_n^2} \, C_{nj} \;, \quad m_{ji} = \rho \frac{H}{2} \sum_n \frac{1}{\lambda_n} \, C_{ni} \, C_{nj} \;.$$

Note the off-diagonal, ie the coupling mass coefficients m_{ji}. The natural frequencies of the plate vibrating in a vacuum are approximately given by $\sqrt{[k_j / m_j^{(B)}]}$. The solution of the equations of motion is more involved due to hydrodynamic coupling. Dynamic stresses in the plate are calculated by static methods, keeping time constant and considering the dynamic loading

$$\hat{p}(x, t) = \bar{p}(x, t) - h \, \rho_B \left[\ddot{y}_G(t) + \ddot{w}^*(x, t) \right] \;.$$

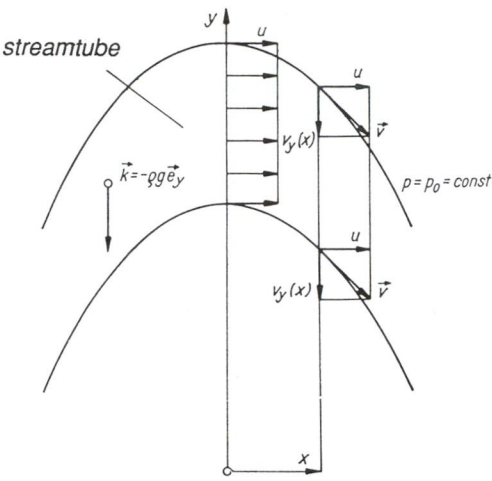

Fig. A 13.3. The hydraulic arch

A 13.3: Show that the two-dimensional inviscid and incompressible flow in a homogeneous and parallel gravity field, $v_x = u = const$, $v_y = -g\,x/u$ (y points vertically upward), is an exact solution of the *Euler* equation and solves the equation of continuity as well. Furthermore, any streamtube can be isolated to form a "hydraulic arch" with both free surfaces of constant pressure. See, *Chia-Shun-Yih: "The Hydraulic Arch." Quarterly Applied Mathematics. Vol. 31, (1973), pp. 377-378.*

Solution: The two components of the *Euler* equation (8.30)

$$-\frac{1}{\rho}\frac{\partial p}{\partial x} = v_x\frac{\partial v_x}{\partial x} + v_y\frac{\partial v_x}{\partial y} = 0 \quad , \quad -g - \frac{1}{\rho}\frac{\partial p}{\partial y} = v_x\frac{\partial v_y}{\partial x} + v_y\frac{\partial v_y}{\partial y} = -g\,,$$

yield $p = const$ in the whole (x, y) plane. The flow is not irrotational since $curl\ \mathbf{v} = (-g/u)\ \mathbf{e}_z \neq \mathbf{0}$. The streamfunction of that flow follows by the integration of Eq. (13.48a) as

$$\psi(x, y) = u\,y + \frac{g}{2}\frac{x^2}{u}\ .$$

The streamlines are congruent parabolas, $\psi = const$, and the pressure is constant along these lines. Hence, free surfaces do exist as shown in Fig. A 13.3.

A 13.4: The diffusion of a vortex in a viscous fluid of infinite extent is a special two-dimensional solution of the nonstationary *Navier-Stokes* equations. The initial condition is given by the velocity distribution of a potential vortex.

Solution: The vertical vortex line is z. If we put $v_\varphi = v$ in polar coordinates, a condition for the pressure gradient lateral to the circular streamlines results: $\rho v^2/r = \partial p/\partial r$, since $v_r = \partial v_r/\partial t = 0$. The second component of the *Navier-Stokes* vector equation gives $\partial v/\partial t = v(\Delta v - v/r^2)$. Vorticity has a single component $\omega_z = \omega = (1/2r)\,\partial(rv)/\partial r$. The velocity v can be eliminated to render the diffusion equation [see also Eq. (6.354) and specify for a nonstationary axisymmetric temperature]

$$\frac{\partial \omega}{\partial t} = \frac{v}{r}\frac{\partial}{\partial r}\left(r\,\frac{\partial \omega}{\partial r}\right) .$$

By means of the new variable $\zeta = r/2\sqrt{(vt)}$ and the dimensionless angular speed $\Omega = \omega t$, an ordinary differential equation results with variable coefficients

$$\Omega'' + \left(2\zeta + \frac{1}{\zeta}\right)\Omega' + 4\Omega = 0 , \quad \Omega' \equiv \frac{d\Omega}{d\zeta} .$$

Usually the equation presents itself in the form multiplied by ζ, the solutions of which were extensively discussed by *H. Goertler*. A special solution is given by the *Gauss* function $\Omega_1 = C_1\,\exp(-\zeta^2)$. Additional solutions may be found by differentiating with respect to time t. Hence, it suffices to consider

$$\omega_1 = \frac{C_1}{t}\,\exp\left[-\left(r^2/4vt\right)\right] , \quad v_1 = \frac{2C_1 v}{r}\left\{1 - \exp\left[-\left(r^2/4vt\right)\right]\right\} ,$$

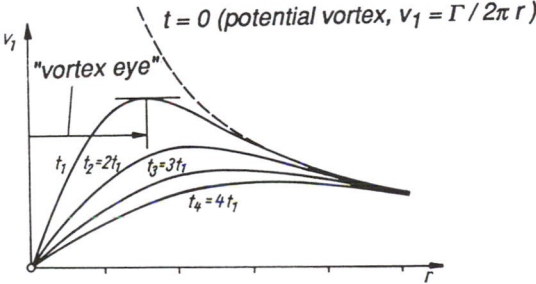

Fig. A 13.4. Diffusion of a vortex. Free flow without external energy supply

since the limit $t \to 0$ gives the initially prescribed potential vortex
[see Eq. (13.54) and Fig. A 13.4]

$$\lim_{t \to 0} \omega_1|_{r>0} = 0 \quad , \quad \lim_{t \to 0} v_1|_{r>0} = \frac{2C_1 v}{r} = \frac{\Gamma}{2\pi r} \; .$$

The free viscous flow decays with time and a monotonically
increasing inner portion of the fluid rotates like a rigid body.

Appendix

Table A.

Some Average Values of Mechanical Material Parameters

(Use for numerical calculations of illustrative examples in the text).

	Density, $\rho \times 10^{-3}$ kg/m³	Young's modulus, $E \times 10^{-3}$ N/mm²	Poisson's ratio, ν	Yield limit, σ_Y N/mm²	Lin. thermal exp. coeff., $\alpha \times 10^6$ m/(m × K)	Temperat. conductivity, $a = k/\rho c$ mm²/s
Highly alloyed steel	7.85	210	0.3	240-750 up to 1650	10-12	12.8
Aluminum alloy	2.8	71	0.35	30-270	25	60
Nickel	8.8	197	0.3	140-750	13.3	15
Wood[1]	0.3-0.7	‖ 11-15 ⊥ 0.4-1.1	0.3-0.45		‖ 5.4 ⊥ 34	0.1-0.4
Glass	2.5	50-90	0.2-0.28	viscous→ brittle	3-10	0.3-0.5
Concrete	2.0	21 (compr.)	0.16		10-12	0.3-0.5
Bedrock	2.5	35-90	0.15-0.30		8-12	0.8-1.5
Water	≈1				−28 (0°C) 69 (20°C)	0.135 0.143
Oil (20°C)	0.89-0.96				230	0.09
Air[2]	1.29 x 10⁻³				(1/3T) ≈1200	22

[1] Typical anisotropic material: ‖ in direction of fibers, ⊥ lateral to fibers .
[2] State at normal atmospheric conditions

Bibliography

Journals

Applied Mechanics Reviews (AMR), ASME, New York
Journal Applied Mechanics, ASME Quarterly Transactions, New York
ZAMM, Zeitschrift f. Angewandte Mathematik u. Mechanik, Berlin
Ingenieur-Archiv, Springer-Verlag, Berlin
Acta Mechanica, Springer-Verlag, Wien
Int. Journal of Solids and Structures, Pergamon Press, Oxford
Journal of Fluid Mechanics, University Press, Cambridge

Handbooks and Surveys

Advances in Applied Mechanics, Academic Press, New York
Mechanics Today, Pergamon Press, Oxford
Progress in Solid Mechanics, Noth-Holland, Amsterdam
Szabo, I., Geschichte der mechanischen Prinzipien. Birkhäuser, Basel, 1979
Todhunter, I. and Pearson, K., A History of the Theory of Elasticity and of the Strength of Materials. Reprint, Dover, New York, 1960
Timoshenko, S. P., History of Strength of Materials. Reprint, Dover, New York, 1983
Benvenuto, E., An Introduction to the History of Structural Mechanics. 2 vols., Springer-Verlag, New York, 1991
Morse, P. M. and Feshbach, H., Methods of Theoretical Physics. McGraw-Hill, New York, 1953
Flügge, S. (Ed.), Handbook of Physics. Vols. III/1, 3; VIa/1...4; VIII/1, 2; IX, Springer-Verlag, Berlin.
Flügge, W. (Ed.), Handbook of Engineering Mechanics. McGraw-Hill, New York, 1962
Streeter, V. L. (Ed.), Handbook of Fluid Dynamics. McGraw-Hill, New York, 1961
Harris, O. M. and Crede, C. E. (Eds.), Shock and Vibration Handbook. McGraw-Hill, New York, 3rd ed., 1988
Achenbach, J. (General Ed.), Mechanics and Mathematical Methods. A Series of Handbooks. North-Holland, Amsterdam, (since 1983)
Proceedings IUTAM-Congresses and Symposia
Proceedings US-National Congresses Applied Mechanics
CISM-Courses and Lectures. Springer-Verlag, Wien
Newmark, N. M. and Rosenblueth, E., Fundamentals of Earthquake Engineering. Prentice-Hall, Englewood Cliffs, N. J., 1971
Brebbia, C. A. (Ed.), Finite Element Systems. A Handbook. Springer-Verlag, New York, 2nd ed., 1982
Schielen, W. (Ed.), Multibody Systems Handbook. Springer-Verlag, Berlin, 1990
Blevins, R. D., Formulas for Natural Frequency and Mode Shape. Van Nostrad Reinhold, New York, 1979
Leissa, A. W., Vibration of Plates, SP-160 (1969), Vibration of Shells, SP-288 (1973), NASA, Washington

Young, W. C., ROARK´S Formulas for Stress and Strain. McGraw-Hill, New York, 6th ed., 1989
Hütte. Die Grundlagen der Ingenieurwissenschaften. *Czichos, H.* (Ed.), Springer-Verlag, Berlin, 29th ed., 1989
Dubbels Taschenbuch für den Maschinenbau. *Beitz, W. and Küttner, K. H.* (Eds.), Springer-Verlag, Berlin, 17th ed., 1990
Beton-Kalender. Ernst & Sohn, Berlin
Moon, P. and Spencer, D. E., Field Theory Handbook. Springer-Verlag, Berlin, 1971
Abramowitz, M. and Stegun, I. A. (Eds.), Handbook of Mathematical Functions. Dover, New York, 1965
Bronstein, I. N. and Semendjajew, K. A., Taschenbuch der Mathematik. 2 vols., Deutsch, Frankfurt/M, 1980
Doetsch, G., Anleitung zum praktischen Gebrauch der Laplace-Transformation. Oldenbourg, München, 1961

Suggested References

Etkin, B., Dynamics of Flight. Wiley, New York, 1959
Fraeijs de Veubeke, B. M., A Course in Elasticity. Springer-Verlag, New York, 1979
Fung, Y. C., Foundations of Solid Mechanics. Prentice-Hall, Englewood Cliffs, N. J., 1965
Guderley, K. G., Theorie schallnaher Strömungen. Springer-Verlag, Berlin, 1957
Hoff, N. J., The Analysis of Structures. Wiley, New York, 1956
Malvern, L. E., Introduction to the Mechanics of a Continuous Medium. Prentice-Hall, Englewood Cliffs, N. J., 1969
Moon, F. C., Chaotic Vibrations. Wiley, New York, 1987
Oswatitsch, K., Grundlagen der Gasdynamik. Springer-Verlag, Wien, 1976
Oswatitsch, K., Spezialgebiete der Gasdynamik. Springer-Verlag, Wien, 1977
Panton, R. L., Incompressible Flow. Wiley, New York, 1984
Rolfe, S. T. and Barson, J. M., Fracture and Fatigue Control in Structures. Prentice-Hall, Englewood Cliffs, N. J., 1977
Rouse, H., Fluid Mechanics of Hydraulic Engineers. Reprint, Dover, New York, 1961
Schlichting, H., Grenzschicht-Theorie. Braun, Karlsruhe, 1965
Van Dyke, M., Perturbation Methods in Fluid Mechanics. Academic Press, New York, 1964

Chapter 1

Beyer, R., Technische Raumkinematik. Springer-Verlag, Berlin, 1963
Rauh, K., Praktische Getriebelehre. 2 vols., Springer-Verlag, Berlin, 1951 and 1954
Wunderlich, W., Ebene Kinematik. Vol. 447/447a, Bibliographisches Institut, Mannheim, 1970
Eringen, A. C., Nonlinear Theory of Continuous Media. McGraw-Hill, New York, 1962
Prager, W., Introduction to Mechanics of Continua. Reprint, Dover, New York, 1973

Chapter 2

Hirschfeld, K., Baustatik. Theorie und Beispiele. Springer-Verlag, Berlin, 3. Aufl. 1969
Pflüger, A., Statik der Stabwerke. Springer-Verlag, Berlin, 1978
Sattler, K., Lehrbuch der Statik. Bd. I/A bis II/B, Springer-Verlag, Berlin, 1969
Stüssi, F., Vorlesungen über Baustatik. Vol. 1, Birkhäuser, Basel, 1962
Girkmann, K. and Königshofer, E., Die Hochspannungs-Freileitungen. Springer-Verlag, Wien, 1952
Czitary, E., Seilschwebebahnen. Springer-Verlag, Wien, 1951

Chapter 4

Bland, D. R., The Theory of Linear Viscoelasticity. Pergamon Press, Oxford, 1960
Desai, C. S., et al (Eds.), Constitutive Laws for Engineering Materials. Theory and Applications. Vol. 1 (1982), 2 vols. (1987), Elsevier, New York

Hill, R., The Mathematical Theory of Plasticity. Clarendon Press, Oxford, 1956
Hult, J., Creep in Engineering Structures. Blaisdell, Waltham, Mass., 1966
Lekhnitskii, S. G., Theory of Elasticity of an Anisotropic Body. Holden-Day, San Francisco, 1963
Lippmann, H., Mechanik des Plastischen Fließens. Springer-Verlag, Berlin, 1981
Odqvist, F. K. G. and Hult, J., Kriechfestigkeit metallischer Werkstoffe. Springer-Verlag, Berlin, 1962
Stüwe, H. P. (Ed.), Mechanische Anisotropie. Springer-Verlag, Wien, 1974
Zeman, J. L. and Ziegler, F. (Eds.), Topics in Applied Continuum Mechanics. Springer-Verlag, Wien, 1974

Chapter 5

Langhaar, H. L., Energy Methods in Applied Mechanics. Wiley, New York, 1962
Washizu, K., Variational Methods in Elasticity and Plasticity. Pergamon Press, Oxford, 1974

Chapter 6

Timoshenko, S. P. and Goodier, J. N., Theory of Elasticity. McGraw-Hill, New York, 1970
Sokolnikoff, I. S., Mathematical Theory of Elasticity. McGraw-Hill, New York, 1956
Carslaw, H. S. and Jaeger, J. C., Conduction of Heat in Solids. Clarendon Press, Oxford, 1959
Schuh, H., Heat Transfer in Structures. Pergamon Press, Oxford, 1965
Boley, B. A. and Weiner, J., Theory of Thermal Stresses. Wiley, New York, 1960
Hetnarski, R. B. (Ed.), Thermal Stresses. 3 vols., North-Holland, Amsterdam, 1989
Parkus, H., Thermoelasticity. Springer-Verlag, Wien, 1976
Szilard, R., Theory and Analysis of Plates. Prentice-Hall, Englewood Cliffs, N. J., 1974
Girkmann, K., Flächentragwerke. Springer-Verlag, Wien, 1963
Timoshenko, S. P. and Woinowsky-Krieger, S., Theory of Plates and Shells. McGraw-Hill, New York, 1959
Flügge, W., Stresses in Shells. Springer-Verlag, Berlin, 1962
Goldenveiser, A. L., Theory of Thin Elastic Shells. Pergamon Press, Oxford, 1961
Seide, P., Small Elastic Deformations of Thin Shells. Noordhoff, Leyden, 1975
Neuber, H., Kerbspannungslehre. Springer-Verlag, Berlin, 1958
Liebowitz, H. (Ed.), Fracture. An Advanced Treatise. Vol. I to VII. Academic Press, New York, 1968-1972
Nadai, A., Theory of Flow and Fracture of Solids. McGraw-Hill, New York, 1950
Peterson, R. E., Stress Concentration Factors. Wiley, New York, 1974

Chapter 7

Kane, T. R. and Levinson, D. A., Dynamics: Theory and Applications. McGraw-Hill, New York, 1985
Wittenburg, J., Dynamics of Systems of Rigid Bodies. Teubner, Stuttgart, 1977
Slibar, A. and Springer, H. (Eds.), The Dynamics of Vehicles. Proceedings 5th VSD-2nd IUTAM-Symposium, Swets and Zeitlinger, Amsterdam, 1978
Bianchi, G. and Schiehlen, W. (Eds.), Dynamics of Multibody Systems. Proceedings IUTAM-Symposium, Springer-Verlag, Berlin, 1986
Schiehlen, W. (Ed.), Nonlinear Dynamics in Engineering Systems. Proceedings IUTAM-Symposium, Springer-Verlag, Berlin, 1990
Rimrott, F. P. J., Introductory Orbit Dynamics. Vieweg, Braunschweig, 1989
Timoshenko, S. P., Young, D. H. and Weaver, W., Vibration Problems In Engineering. Wiley, New York, 4th ed., 1974
DenHartog, J. P., Mechanical Vibration. McGraw-Hill, New York, 4th ed., 1956
Meirovitch, L., Elements of Vibration Analysis. McGraw-Hill, New York, 1975
Clough, R. W. and Penzien, J., Dynamics of Structures. McGraw-Hill, New York, 1975
Hurty, W. C. and Rubinstein, M. F., Dynamics of Structures. Prentice Hall, Englewood Cliffs, N. J., 1964
Yang, C. Y., Random Vibrations of Structures. Wiley, New York, 1986

Pestel, E. C. and Leckie, F. L., Matrix Methods in Elastomechanics. McGraw-Hill, New York, 1963

Magrab, E. B., Vibrations of Elastic Structural Members. Sijthoff & Noordhoff, Alphen aan der Rijn, 1979

Schmidt, G., Parametererregte Schwingungen. VEB Verlag Technik, Berlin, 1975

Achenbach, J. D., Wave Propagation in Elastic Solids. North-Holland, Amsterdam, 1975

Ewing, W. M., Jardetsky, W. S. and Press, F., Elastic Waves in Layered Media. McGraw-Hill, New York, 1957

Pao, Y. H. and Mow, C. C., Diffraction of Elastic Waves and Dynamic Stress Concentrations. Crane, Russak, New York, 1973

Chapter 9

Ziegler, H., Principles of Structural Stability. Blaisdell, Waltham, Mass., 1968

Dym, C. L., Stability Theory and its Application to Structural Mechanics. Noordhoff, Leyden, 1974

Müller, P. C., Stabilität und Matrizen. Springer-Verlag, Berlin, 1977

Horne, M. R., Plastic Theory of Structures. Pergamon Press, Oxford, 2nd ed., 1979

Zyczkowski, M., Combined Loadings in the Theory of Plasticity. PWN-Polish Scientific Publisher, Cracow, 1981

Dowell, E. H. (Ed.), A Modern Course in Aeroelasticity. Kluwer Academic Press, Dordrecht, 2nd ed., !989

Chapter 10

Hamel, G., Theoretische Mechanik. Springer-Verlag, Berlin, 1949

Päsler, M., Prinzipe der Mechanik. De Gruyter, Berlin, 1968

Goldstein, H., Classical Mechanics. Addison-Wesley, Reading, Mass., 1981

Chapter 11

Biezeno, C. B. and Grammel, R., Technische Dynamik. Springer-Verlag, Berlin, 1953

Meirovitch, L., Computational Methods in Structural Dynamics. Sijthoff & Noordhoff, Groningen, 1980

Fryba, L., Vibrations of Solids and Structures under Moving Loads. Sijthoff & Noordhoff, Groningen, 1972

Kauderer, H., Nichtlineare Mechanik. Springer-Verlag, Berlin, 1958

Stoker, J. J., Nonlinear Vibrations in Mechanical and Electrical Systems. Wiley, New York, 1950

Chapter 12

Goldsmith, W., Impact. The Theory and Physical Behaviour of Colliding Solids. Arnold, London, 1960

Chapter 13

Lamb, H., Hydrodynamics. University Press, Cambridge, 6th ed., 1932

Landau, L. D. and Lifshitz, E. M., Fluid Mechanics. Pergamon Press, Oxford, 1959

Kotschin, N. J., Kibel, I. A. and Rose, N. W., Theoretische Hydromechanik. Akademie-Verlag, Berlin, Vol. I 1954, Vol. II 1955

Betz, A., Konforme Abbildung. Springer-Verlag, Berlin, 1964

Schneider, W., Mathematische Methoden der Strömungsmechanik. Vieweg, Braunschweig, 1978

Ashley, H. and Landahl, M., Aerodynamics of Wings and Bodies. Addison-Wesley, Reading, Mass., 1965

Launders, B. E. and Spalding, D. B., Mathematical Models of Turbulence. Academic Press, New York, 1972

Whitham, G. B., Linear and Nonlinear Waves. Wiley, New York, 1974

Bisplinghoff, R. L., Ashley, H. and Halfmann, R. L., Aeroelasticity. Addison-Wesley, Reading, Mass., 1965

Index

WIDENER UNIVERSITY
WOLFGRAM
LIBRARY
CHESTER PA